D0162967

BRADY

Fire Administration I

Randy R. Bruegman

Upper Saddle River, New Jersey

Library of Congress Control Number: 2008921119

Publisher: Julie Levin Alexander
Publisher's Assistant: Regina Bruno
Executive Editor: Marlene McHugh Pratt
Senior Acquisitions Editor: Stephen Smith
Associate Editor: Monica Moosang
Editorial Assistant: Patricia Linard
Director of Marketing: Karen Allman
Executive Marketing Manager: Katrin Beacom
Marketing Specialist: Michael Sirinides
Managing Production Editor: Patrick Walsh
Production Liaison: Julie Li
Production Editor: Lisa Garboski, bookworks
Media Product Manager: John Jordan
New Media Project Manager: Stephen J. Hartner
Manufacturing Manager: Ilene Sanford
Manufacturing Buyer: Pat Brown
Senior Design Coordinator: Christopher Weigand
Interior Designer: Lee Goldstein
Cover Designer: Blair Brown
Composition: Aptara
Printing and Binding: Edwards Brothers
Cover Printer: Phoenix Color Corporation

Pearson Education Ltd., London
Pearson Education Singapore Pte. Ltd.
Pearson Education Canada, Ltd.
Pearson Education—Japan
Pearson Educatin Australia Pty. Limited
Pearson Education North Asia Ltd.
Pearson Educación de Mexico, S.A. de C.V.
Pearson Education Malaysia Pte. Ltd.
Pearson Education, Inc., Upper Saddle River, New Jersey

10 9 8 7 6 5 4 3 2 1
ISBN 13: 978-0-13-172084-8
ISBN 10: 0-13-172084-8

A special thanks to my wife, Susan, for her patience and editing ability during the writing of this book, as I spent many of my evenings and weekends for almost two years working in my office, and to my executive assistant Maria Campos, who worked the same schedule and was instrumental in the preparation of the manuscript. Thanks to all of the professionals I have had the opportunity to learn from and who have been mentors to me throughout my career.

To each of you, be safe in the completion of the mission, and commit to making a positive difference in your life each and every day.

Chief Bruegman

Contents

Preface *xv*

Reviewers List *xvii*

About the Author *xix*

About the Experts *xxi*

Chapter 1 *Our Heritage and Our History 1*

Codes as a Means to Regulate Fire 4
The Colonial Fire Experience 5
Societal Impacts of the Fire Service 9
Culture and Diversity: The Changing Composition of the United States 9
Trends Impacting the Workplace 11
Employee Empowerment 11
Total Quality Management 11
Traditional Power Structures Are Crumbling 12
Demographic Shifts 12
The Twenty-First-Century Workforce 12
Differing Values 13
Changing Education Levels 13
Legislative Mandates 14
The Civil Rights Act of 1964 14
Equal Employment Opportunity 14
Sexual Harassment 15
Use Commonsense Supervision to Detect Discrimination 16
The Americans with Disabilities Act 17
Local and State Antidiscrimination Laws 17
The Age Discrimination Act 17
Impacts and Challenges to the Fire Service 18
Human Relations in the Fire Service 18
The Fire Service Mosaic 19
Pride, Service, Commitment 21
The Introduction of the Maltese Cross into the Order 23
Other Insignia of the Fire Service 26
On Leadership 30
The Trumpets Worn 31
Through the Eyes of a Nation 31

Advice from the Experts 36
Review Questions 40
References 40

Chapter 2 *Preparing for Your Future 42*

From a Blue-Collar Job to a Profession 42
Professional Status: The Future of Fire Service Training and Education 44
Part One—Introduction 44
Training and Education 47
A National System for Fire-Related Higher Education 51
The Future: Where Do We Go from Here 54
The Future 56
IAFC *Officer Development Handbook* 57
Volunteer Advocates for Professional Development 59
History 60
What Does Your Future Hold 61
Mentoring 61
Mapping 62
Measuring for Success 62
Chief Fire Officer (CFO) Designation 79
Professional Development Portfolio Setup 79
Chief Medical Officer (CMO) Designation 80
Executive Fire Officer Program 81
Degrees at a Distance Program 81
Frequently Asked Questions 81
Distance-Learning Programs 82
International Fire Service Accreditation Congress (IFSAC) 83
National Board on Fire Service Professional Qualifications (NBFSPQ) 84
Organization 84
Accreditation 84
Certification 84
Benefits 84
Developing Your Road Map to Success 85
Advice from the Experts 85
Review Questions 89
References 90

Chapter 3 *Principles of Leadership and Management 91*

Early Thinking About Management 92
Why Study Management Theory? 93
The Classical School of Management Thought 95
Bureaucracy 95
Scientific Management 97
Administrative Management 102
The Behavioral Movement 104
The Hawthorne Effect 106
Human Relations Management 108
The Needs Hierachy Model 110
Theory X and Theory Y 112

Theory Z 113
The Managerial/Leadership Grid 114
The Motivation-Hygiene Model 116
Human Resources Theory 119
Integrating the Management Theories 119
Systems Theory 119
Contingency View 120
Emerging Management Trends 121
W. Edwards Deming (1900–1993) 121
Peter Drucker (1909–2005) 127
Leadership Changes for the Twenty-First Century 128
The Changing Nature of Expectations 128
Leadership 129
Building the Bridge to the Future 131
In Search of Successful Transformations 132
Management Challenges for the Twenty-First Century 132
Seven Old Assumptions of Management 133
Eight New Management Assumptions 133
Advice from the Experts 135
Review Questions 139
References 139

Chapter 4 ***What Is Your Leadership Style? 141***

Leadership Styles 141
Situational Leadership: Changing Your Style to Reflect Different Needs 142
The Managerial Grid 142
Likert's Leadership Styles 144
Lewin's Leadership Styles 144
Six Emotional Leadership Styles 145
Leaders Create Resonance 146
Charismatic Leadership 146
Participative Leadership 148
Situational Leadership 148
Transactional Leadership 149
Ivan Pavlov 149
Transformational Leadership 150
The Quiet Leader 151
Servant Leadership 151
Leadership Starts with Knowing Yourself 152
Exploring the Johari Window 153
Schemata 154
How Our Personal Framework Shapes Our Leadership Style 156
How Leadership Style Can Impact Results 157
Your Personal Frame of Reference Is the Foundation 161
My Personal Frame of Reference 161
What Is Your Leadership Style? 165
The Jungian Type Inventory 167

The Importance of Style 174
Advice from the Experts 174
Review Questions 180
References 180

Chapter 5 *Leading and Managing in a Changing Environment 182*

Don't Play a New Game by the Old Rules 183
Leading Change 185
It Is All Up to You 185
The Boiled Frog Syndrome 187
Big Hat, No Cattle! 193
Focus on the Mission 193
Groups 195
The Dynamics of Groups 195
Good to Great 196
Stewardship and the Core Value of Service 198
The United States Navy 201
The United States Marine Corps 201
The United States Air Force 202
The United States Army 203
What We Can Learn from Others 204
Agenda for Change 207
Nature of Culture 208
Vision Is a Must 211
Building a Culture of Our Choice 216
The Psychology of Change 223
How Do We Get Them to Let Go? 224
The Neutral Zone 225
Case Study 225
Advice from the Experts 229
Review Questions 233
References 234

Chapter 6 *Leadership Ethics 235*

Values-Based Leadership: The Only Approach to Success 235
Ethics and Their Impact 236
Identifying Ethic Codes 240
2003 National Business Ethics Survey 242
Modeling Ethical Behavior 245
Make It Acceptable to Talk About Ethics 245
Create a Habit of Respecting Organizational Lore 246
Include Ethical Conduct as a Measure on Performance Evaluations 246
Make Ethics More Than an Annual Discussion 246
The Role of Leadership in Organizational Integrity and Ethical Leadership 246
Mandatory Ethics 248
Aspirational Ethics 249

Personal Orientation 249
Ethical Decision-Making Process 249
Ethics, Politics, and Leadership 256
Our Past Is Often a Window to Our Future 257
Commission on Professional Credentialing (CFO) Code of Conduct 258
The Political Dynamic 262
The Ten Commandments of Political Survival 263
Scoring Political Points 265
Leadership Strategies for Personal Success 265
Our Own Personal Code of Conduct 266
Guidelines and Resources 267
Government Leadership 269
Developing an Ethics Code 270
Sample Code of Ethics for Elected and Appointed Officials and Employees 271
What Is an Ethics Audit? 273
Key Ethics Law Principles for Public Servants 273
Making Ethics a Core Value 275
Advice from the Experts 277
Review Questions 280
References 281

Chapter 7 ***Personnel Management: Building Your Team 283***

Building Your Team 283
Whose Side Are They On? 284
A New Beginning 288
Building High-Performance Teams 290
Characteristics of a High-Performance Team 294
Task/Process Orientation: The Importance of Task and Process Characteristics 297
Measuring Your Leadership Style 298
Leadership Styles 302
Characteristics of High-Performance Teams 303
A Diversified Workforce 304
What Does This Mean to the Fire Service? 307
Bridging the Generations 309
Recruitment 310
Managing 310
Retention 310
Employee Morale Is Critical to Team Success 313
Improving Morale: We Must Ask the Right Question 314
Four Points to Guide Your Morale Building Efforts 315
Focus on What You Can Control, Not on What You Cannot 317
Empower Your Team 319
What Is Empowerment? 319
Access to Information 320
Inclusion and Participation 321
Accountability 321

Employee Empowerment: A Crucial Ingredient in Promoting Quality Service 322
Understanding the Paradox of Empowerment 324
Advice from the Experts 325
Review Questions 329
References 330

Chapter 8 ***Managing Emergency Services 331***

Local Models of Organization and Government 332
Council-Manager Government 332
Mayor-Council Government 334
City Commission Government 334
Town 335
Organizational Structure 337
Informal Organization 338
Organizational Design 338
Organizing Functions of Work 340
Authority 340
Origins and Historical Context of Budgeting 343
Budget Reforms and Innovations 344
Line-Item Budgeting 344
Performance Budgeting 346
Program Budgeting 346
Planning, Programming, and Budgeting System (PPBS) 346
Zero-Based Budgeting 348
Budget Reform in the 1990s: The "New" Performance Budgeting 349
Federal Reform Efforts in the 1990s 349
State Reform Efforts in the 1990s 350
Local Reform Efforts in the 1990s 350
Prospects for Budgeting in the Twenty-First Century 350
Fiscal Management 350
Employment Law 355
Basis for Employment Law 355
Equal Employment Opportunity 355
Employee and Labor Relations 360
Societal Trends 361
"Right-to-Work" States 362
Organized Labor Actions 362
Labor Relations in Volunteer and Nonunionized Departments 362
Collective Bargaining 363
Settling Disputes 364
Power and Mutual Gains Bargaining 365
The General Process of Negotiating 365
Performance Appraisal Methods 368
Appraisal Judgments 368
Objectives of Performance Appraisals 369
Agenda for the Performance Appraisal Conference 369

Conducting the Evaluation 369
Evaluation Methods and Instruments 370
Promotional Process 372
Disciplinary Problems 373
Progressive Discipline 374
Disciplinary Action 374
Discipline 375
Documentation 376
Disciplinary Interview 377
General Guidelines 378
Due Process 378
Appeal Process 379
Grievance Procedures 380
Discipline and Volunteers 381
Employee Assistance Programs 381
Advice from the Experts 382
Review Questions 385
References 386

Chapter 9 ***Analytical Approaches to Public Fire Protection 388***

The History of Resource Deployment 389
The Development of National Standards of Fire Cover 390
Community Risk Management 391
Life Safety 392
Responder Risk 392
Property Loss 395
Risk and Planning 395
The Physical Protection of Critical Infrastructures and Key Assets 396
A New Mission 397
Homeland Security and Infrastructure Protection: A Shared Responsibility 397
The Significance of Critical Infrastructures and Key Assets 398
National Resilience: Sustaining Protection for the Long Term 400
Probability and Consequences 401
Community Response to Risks 402
Community Risk Assessment—Fire Suppression 402
Standards of Response Coverage: Integrated Risk Management Planning 403
Risk Assessment Model 404
The Cascade of Events 405
Evaluating Fire Suppression Capabilities 406
The Stages of Fire Growth 407
Scene Operations 408
What Is an Effective Response Force? 409
Evaluating EMS Capabilities 410

Evaluation of Fire Departments as a Factor in Property Insurance Rating 412
Early History of Evaluating Fire Departments 413
A Permanent Fire Protection Grading System 414
Keeping the Schedule Up to Date 415
ISO Takes Responsibility for Municipal Grading 416
The Fire Suppression Rating Schedule Today 416
Determining the PPC for a Community 419
The Effect of PPC on Insurance Premiums 420
How Communities Rely on the PPC Program 420
Summary 421
The CFAI 421
Performance Indicators 423
Objective of the CFAI Program 424
Effects on Levels of Service 425
Assessing Benefit 426
Creating and Evaluating an Integrated Risk Management Plan: Standards of Response 427
Risk Assessment Model 427
Standards of Measurement 428
Developing a Standards of Response Coverage 429
Risk Assessment 430
CFAI Benchmarks and Baselines of Performance 430
Integration, Reporting, and Policy Decisions 433
NFPA 1710 and 1720 433
Origin and Development of NFPA 1710 433
Origin and Development of NFPA 1720 441
Advice from the Experts 445
Review Questions 449
References 449

Chapter 10 ***Quality of the Fire Service 450***
Quality Is Just Good Business! 451
Why Six Sigma? 451
The Impact of Quality 452
What Quality Is Not 453
Challenges to the Firefighting Community 455
Today's Fire Department 456
The Need for Quality Improvement 456
What Does This Mean to You? 458
The Elements of Quality Improvement 460
Introduction to Total Quality Leadership 463
Quality and Change 464
Benchmarking for Top Performance—Introduction 470
Cycle Time 473
Performance Measurements 477
Performance Measurement Dimensions 479
Lessons Learned 483
The Commission on Fire Accreditation International (CFAI) Benchmarking Survey 484

Measuring Customer Satisfaction 488
Quality Customer Service 488
Customer-Centered Continuous Improvement Process 489
Advice from the Experts 491
Review Questions 496
References 496

Chapter 11 ***Community Disaster Planning 497***

What Is a Disaster? 498
The Risk 499
The Plan 500
The Government's Role in Disaster Response 501
Preparedness: A Community Effort 502
FEMA History 503
Preparedness 506
Response 508
Media Communications If You Are the Spokesperson 509
Recovery 509
The Federal Response Plan (FRP) 509
Mitigation 511
StormReady 512
The Integrated Emergency Management System (IEMS) Concept 513
Hazard Assessment: Why Is It Important? 514
Hazard Identification 515
The National Response Plan 515
Introduction to the National Response Plan (NRP) 516
Incidents of National Significance 516
Roles and Responsibilities 516
Nongovernmental and Volunteer Organizations (NGO) 523
Citizen Involvement 525
Citizen Corps 525
Programs and Partners 525
Managing the National Response Plan 527
Organizational Structure 527
National Incident Management System (NIMS) 528
The Incident Command System 530
ICS Organization 530
The Operations Section 532
The Planning Section 532
The Logistics Section 533
The Finance/Administration Section 534
Variations in Preparedness at the Local Level 534
The National Infrastructure Protection Plan 535
The Goal 535
Authorities, Roles, and Responsibilities 536
The Value of the NIPP Plan 536
Sector Partnership Model 537
Risk Management Framework 538

Coordination 538
Communication 539
Funding 540
The Six Mission Areas of Homeland Security 541
Advice from the Experts 541
Review Questions 545
References 546

Chapter 12 ***Shaping the Future 547***

Fire Suppression 548
Emergency Medical Service 549
Specialized Response 551
Hazardous Materials 551
Technical Response 552
Wildland Firefighting 553
Aircraft Rescue Firefighting 554
Federal and Military Fire Protection 555
Industrial Fire Protection 555
Code Enforcement 556
Fire Prevention 556
Human Behaviors, Values, and Impacts on Fire and Life Safety 560
Integrated Risk Management 560
Technology 561
Firefighter Safety 563
National Mutual-Aid System 565
Fire Service Research 566
National Fire Service Doctrine 568
The Future Is Up to You 569
Advice from the Experts 570
References 574

Appendix A ***Measuring Customer Satisfaction 575***

Appendix B ***WMD Response Assets Guide 593***

Appendix C ***Assessing Your Local Capacity to Respond to a Disaster 624***

Glossary 652

Index 684

Preface

When I was asked by Prentice Hall to consider writing a textbook for the fire service in leading and managing today's organization, I was humbled by the request and soon became overwhelmed by the task at hand.

I began my career as a volunteer firefighter in a small community in western Nebraska in 1975, and the fire service as an industry was much different than it is today. The rules, regulations, and standards by which we operated; the technology we used; and the caliber of people who were engaged in the profession were all quite different then.

During the course of my career over the past 30 years, I have seen not an evolution but a revolution in the way we conduct our business, the scope of services we offer, and the complexity of the issues we have to deal with as leaders, administrators, and firefighters on a daily basis.

So as I sat down to outline this book, I asked myself, what is important from a leader's perspective in today's fire service? What will make a leader more successful in the future? What information will be needed to become an excellent company officer, a shift battalion commander, a deputy chief in charge of a division or bureau, or the chief executive officer? The text is built around the career development model of the National Fire Academy and the model curriculum that has been identified by the Fire Emergency Service Higher Education (FESHE) group, which is made up of the industry leaders in fire service education and training.

I concluded that there needs to be a balance of information to include a level of academics introduced in respect to the science, leadership, and management in the fire service. But to make it meaningful, there must be a focus on what happens in the real world to accomplish this. The text is divided into 12 chapters.

Chapter 1, *Our Heritage and Our History*, is focused on those historical events that have forged our industry today and can be seen in every one of our organizations.

Chapter 2, *Preparing for Your Future*, is designed to provide information by which you can develop your game plan for personal success.

Chapter 3, *Principles of Leadership and Management*, blends the academics of leadership and management research into what occurs in our organizations on a daily basis.

Chapter 4, *What Is Your Leadership Style?* will help you to understand how you lead and manage and, as important, why you do.

Chapter 5, *Leading and Managing in a Changing Environment*, provides an insightful look into how we handle change personally and organizationally.

Chapter 6, *Leadership Ethics*, is focused on the elements critical to ethical leadership and management practices.

Chapter 7, *Personnel Management: Building Your Team*, explores the elements of team building and provides a depth of understanding on how to blend various styles and personalities to get the most from your people.

Chapter 8, *Managing Emergency Services*, is targeted on the support elements so vital to every organization, budget, and personnel management.

Chapter 9, *Analytical Approaches to Public Fire Protection*, provides an in-depth look at the history of deployment practices in the United States and provides the basis to begin developing a standard of coverage model for your own community.

Chapter 10, *Quality of the Fire Service*, looks at methods of quality improvement and its application to improving the services delivered to citizens every day.

Chapter 11, *Community Disaster Planning*, provides an in-depth overview of the changes in disaster planning and response since 9-11, and includes several appendices to help with your community evaluation and preparedness process.

Chapter 12, *Shaping the Future*, explores the possibilities of what may occur in the fire service, and how you can play an important role in helping to shape the fire service of the future.

Examples from my own experiences have been incorporated throughout the text so that you may learn from mistakes that I have made along the way. In addition, at the end of each chapter, I have included advice from successful fire service leaders to gain their perspectives on the issues they have faced and the lessons they have learned. Our fire service experts come from a broad spectrum of our industry and include:

Michael Chiaramonte, Chief Fire Inspector, Retired, Lynbrook Fire Department (New York)
Bett Clark, Former Fire Chief, Bernalillo County Fire and Rescue (New Mexico)
Kelvin Cochran, Fire Chief, Atlanta Fire Department (Georgia)
Cliff Jones, Fire Chief, Tempe Fire Department (Arizona)
Bill Killen, Retired Fire Chief and IAFC President, 2005–2006
Kevin King, U.S. Marine Corps Fire Service
Mark Light, Executive Director, International Association of Fire Chiefs
Lori Moore-Merrell, Assistant to the General President, International Association of Fire Fighters
Jay Reardon, Fire Chief, Retired, Northbrook Fire Department (Illinois), currently President of the Mutual Aid Box Alarm System (MABAS)
Gary W. Smith, Retired Fire Chief, Aptos–La Salva Fire Protection District (California)
Steve Westermann, Fire Chief, Central Jackson County Fire Protection District (Missouri), and IAFC President 2007–2008
Thomas Wieczorek, Executive Director, Center for Public Safety Excellence, and Retired City Manager
Janet Wilmoth, Editorial Director, *Fire Chief* magazine
Fred Windisch, Fire Chief, Ponderosa Volunteer Fire Department (Texas)

Experience is often the best teacher; if we can learn from those who have gone before us and who have dealt with difficult issues in their careers, it will help all of us to be more effective leaders in the future.

My objective throughout this text is that you come to know yourself in your role as leader and manager, so that you can be successful in your organization. I have found throughout my own experience that those individuals who are the most successful leaders and managers also have the most insight into themselves as individuals. It matters not whether you are leading a career department, a combination system, or an all-volunteer organization. The principles of leadership and their application are the same. Knowing yourself allows you the opportunity to play to your strengths and overcome your weaknesses. I often find that leaders and managers who are struggling in the fire service have the academics down. However, they have failed to recognize those aspects about themselves that can often limit their successes or leverage their strengths to become great leaders and have the positive impact upon their organizations and the people they serve.

So let's get started.

Randy R. Bruegman

Reviewers List

John P. Alexander, Adjunct Instructor
Connecticut Fire Academy
Windsor Locks, Connecticut

Ronald R. Bowser
Adjunct Assistant Professor
Emergency Management, Fire Science, and Homeland Security
Department of Business and Professional Programs
School of Undergraduate Studies
University of Maryland University College
Adelphi, Maryland

Cliff Jones
Fire Chief
Tempe, Arizona

Matthew Marcarelli
Lieutenant, City of New Haven Fire Department
Instructor, Connecticut Fire Academy
Middlesex County Fire School
Rescue Team Manager
Connecticut Urban Search and Rescue Task Force
Northford, Connecticut

Mark Martin
Division Chief (Retired)
City of Stow Fire Department
Stow, Ohio

Joseph Mercieri, Chief
Littleton Fire Rescue
Littleton, New Hampshire

Gary W. Smith
Fire Chief (Retired)
Instructor, Cogswell College
Sunnyvale, California

Jason J. Zigmont, BS, FSI, NREMT-P
Executive Director
The Center for Public Safety Education
East Berlin, Connecticut

About the Author

Fire Chief Randy R. Bruegman began his career as a volunteer firefighter in Nebraska. He was hired as a firefighter in Ft. Collins, Colorado, where he served in a variety of positions including engineer, inspector, lieutenant, captain, and battalion chief. He has served as the fire chief for the city of Campbell, California; the Village of Hoffman Estates, Illinois; Clackamas County Fire District No. 1 (Oregon); and the city of Fresno, California, the last since September 2003.

He is a noted author and lecturer on such topics as leadership and managing change in the fire service and a contributing author of fire service literature, including Fire Attack: The Strategy and Tactics of Initial Company Response; Making a Difference: The Fire Officer Role; Surviving Haz-Mat, Haz-Mat for First Responders; The Volunteer Firefighter, A Breed Apart; Fire Survivor Today; Managing a Changing Role and Mission; and Chief Fire Officer's Desk Reference. He has authored two books: *Exceeding Customer Expectations* and *The Chief Officer: A Symbol Is a Promise*.

He served on the Accreditation Task Force for the International Association of Fire Chiefs for eight years, six as vice chairman. Chief Bruegman served as chairman of the Commission of Fire Accreditation International (CFAI) for three years, 1997–2000. He holds member status with the Institute of Fire Engineers, is a Chief Fire Officer Designate, and is a member of NFPA. He serves on the editorial advisory board for *Fire Chief* magazine and as a member of the NFPA 1710 Technical Committee. He was elected as president of the International Association of Fire Chiefs in August 2002 and has served as the president of the Board of Directors of the Center for Public Safety Excellence since 2003. He has an associate's degree in fire science, a bachelor's degree in business, and a master's degree in management.

About the Experts

MICHAEL D. CHIARAMONTE, CFO

Michael Chiaramonte has a Bachelor of Science degree from the University of Houston, a Master of Arts from Hofstra University, and 75 additional education credits from Brooklyn College. He is a retired secondary school English and Communications teacher (35 years) and he is a 39-year member of the fire service. He served as the chief of the Lynbrook Volunteer Fire Department and since 1975 has served as the Chief Fire Inspector of the Lynbrook Volunteer Fire Department. He is currently EMT-B Certified. He is past chairman and board member of the IAFC Volunteer and Combination Officers Section, President of IAFC Eastern Division, certified New York State Code Enforcement Official, National Accredited Chief Fire Officer Designee, member of *Fire Chief* magazine Advisory Board, and author of numerous fire service articles. He is also a member of the Advisory Board of the National Fallen Fighters Foundation, a National Fire Academy instructor, an instructor at the Nassau County EMS Academy, a speaker, and a presenter at many major conferences internationally.

BETT CLARK

Chief Clark has an extensive association with the fire service and EMS in Bernalillo County, New Mexico. She became acquainted with BCFRD as a young girl when her mother was a dispatcher and her father was a firefighter and chief at District No. 6. She served as a volunteer from 1980 to 1985, as well as a dispatcher, firefighter, paramedic, and lieutenant. She was a system status manager and paramedic with Albuquerque Ambulance for five years before she was hired by the fire department in 1985 as a firefighter/paramedic. She has worked up through the ranks of lieutenant, captain, fire marshal, interim deputy chief, and chief. She is a member of the International Association of Fire Chiefs and Women Chief Officers. She was installed as the president of the IAFC Southwestern Division in October 2005. She also sits on the Fire Rescue International Program Planning Committee. Chief Clark is very active on several boards and committees of the United Way and New Mexico Governor's Fire Service Task Force. As chief she leads a combination department, providing ALS level with 200 paid and 60 volunteer members. They cover 500 square miles, with 12 main districts, responding to more than 11,000 calls in the unincorporated area surrounding Albuquerque, New Mexico. The department also has a model Fire Corps and Explorer Post Program.

KELVIN COCHRAN

Chief Cochran was with the city of Shreveport Fire Department for 26 years and became fire chief there in 1999. Chief Cochran became the fire chief of the Atlanta

Fire Department, Georgia, in January 2008. His education includes a bachelor's degree in Organizational Management (Wiley College, Texas) and a master's degree in Industrial and Organizational Psychology (Louisiana Tech University, Louisiana). He served as the second vice president of the International Association of Fire Chiefs (IAFC), a member of the Board of Visitors for the U.S. Fire Administration/National Fire Academy; former president of the Metropolitan Fire Chiefs Association of the IAFC and National Fire Protection Association (NFPA); and former chairman of the IAFC Program Planning Committee. His fire service passion is in the areas of leadership, training, conference speaking, and strategic planning.

CLIFF JONES

Cliff Jones has been in the fire service for over 35 years, including 18 years as fire chief for the City of Tempe. He serves as an instructor in the Arizona State University Fire Services Program in the areas of fire service program management, leadership, and fire service administration. Currently, he serves on the Board of Trustees of the Commission on Fire Accreditation International (CFAI), which is headquartered in Chantilly, Virginia. His educational background includes an Associate of Arts Degree in Fire Science from Phoenix College, a Bachelor of Science in Political Science, and a master's degree in Public Administration from Arizona State University. In 1999, Chief Jones was selected for, and attended, Harvard University's Program for Senior Executives in State and Local Government. In 1997, Chief Jones was selected Career Fire Chief of the Year by *Fire Chief* magazine. Professional memberships include the International Association of Fire Chiefs (IAFC), the National Fire Protection Association (NFPA), the Institution of Fire Engineers–United States of America Branch, and the City of Tempe Professional Development Club.

BILL KILLEN

Bill Killen retired as fire chief, Holston Army Ammunition Plant, Wackenhut Services Inc., Kingsport, Tennessee, on March 31, 2006. He served as director of Navy Fire and Emergency Services from 1985 to September 3, 2004. He has more than 50 years' experience as a volunteer and career firefighter and chief officer in career, municipal, industrial, federal, and military fire departments. As an active member of the IAFC for 32 years, Chief Killen has served on several committees and was the founder and charter chairman of the IAFC Federal and Military Section. He served as president of the IAFC from 2005 to 2006. He earned a bachelor's degree in Fire Administration from the University College, University of Maryland, and holds the Chief Fire Officer designation.

KEVIN KING

Mr. King has served as the Headquarters Marine Corps Fire Protection Program Manager and Administrator of the Marine Corps Fire Service since September 1991. He is responsible for the management of 14 Marine Corps Fire and Emergency Services Departments and over 950 fire and emergency services personnel protecting Marine Corps installations located throughout the United States and Japan. He is also responsible for the Marine Corps Fire Protection Engineering Program covering over $27 billion worth of facilities. Mr. King received his bachelor of science degree in Fire Protection Engineering from the University of Maryland in 1981 and is a registered professional engineer in Maryland and Virginia. He is a graduate of the National

Fire Academy Executive Fire Officer Program and has designation as a Chief Fire Officer. Mr. King is a life member and former president of the Stonewall Jackson Volunteer Fire Department in Prince William County, Virginia. Mr. King is a member of the National Fire Protection Association, International Association of Fire Chiefs, and the National Society of Executive Fire Officers.

MARK LIGHT

Mark Light is currently the executive director of the International Association of Fire Chiefs (IAFC). Prior to coming to the IAFC, he was the fire chief in Henrico County, Virginia, where he served for over six years. Mr. Light is a peer assessor for both the Commission on Fire Accreditation International and the Commission on Chief Fire Officer Designation. Prior to being appointed as the fire chief in Henrico County, he served as the deputy chief of the Roanoke County Fire and Rescue Department. He has a Masters of Public Administration from Virginia Polytechnic and State University; a Bachelors of Arts in Business Administration with a minor in Political Science from Mary Baldwin College; and an associate's degree in Business Administration from Virginia Western Community College. He is a graduate of the Executive Fire Officer Program at the National Fire Academy and completed a fire service fellowship at Harvard University at the John F. Kennedy School of Government. He serves as a committee member of the International Association of Fire Chiefs Labor/Management Partnership program for fire chiefs and union presidents. He recently was awarded the Certified Association Executive (CAE) designation by the American Society of Association Executives.

LORI MOORE-MERRELL, DrPH, MPH, EMT-P

Lori Moore-Merrell is an assistant to the general president of the International Association of Fire Fighters (IAFF) in charge of Technical Assistance for Labor Issues and Collective Bargaining, Fire and EMS Operations, and IAFF Field Services. Lori's expertise is in emergency response system design; staffing; and deployment of mobile resources, system performance measurement, and evaluation. Lori's experience and educational background includes seven years of service as a firefighter/paramedic with the City of Memphis Fire Department and a member of IAFF Local 1784. She received her EMT-Paramedic license in 1984 and became an instructor-coordinator for the state of Tennessee in 1991. She joined IAFF Headquarters Operations in 1993 as an EMS specialist and was promoted to EMS director in 1995. She holds a Bachelor of Science degree in education and EMS from the University of Memphis and a Master of Public Health degree in Epidemiology from the George Washington University School of Public Health. Lori has also completed a Doctor of Public Health degree in Health Policy and Quality Performance Measurement at the George Washington University. She was awarded the James O. Page Achievement award by the International Association of Fire Chiefs and serves as a commissioner to the Commission for Fire Service Accreditation (CFAI). Dr. Moore-Merrell also serves as a gubernatorial appointee to the Commonwealth of Virginia Emergency Medical Services Advisory Board and has been appointed to the International Fire Service Training Association (IFSTA) Executive Board. Dr. Moore-Merrell has written numerous publications. Dr. Moore-Merrell has served on various federal EMS task forces for the National Highway Traffic Safety Administration (NHTSA/DOT) and participated as an expert peer reviewer of many federal publications, including Department of Transportation First Responder, EMT-Basic, and paramedic Curriculum Revision; Peer Review /Blue Ribbon Panel—*National Injury Prevention* white paper;

Peer Review/Blue Ribbon Panel—*EMS Agenda for the Future*; NHTSA Task Force—*EMS Education Agenda for the Future*; EMS-C/MCH Federal Grant Administrator—research firefighter first responders pediatric EMS; National EMS-C Resource Alliance Developed Pediatric Equipment Guidelines; HCFA negotiated rule-making for national ambulance fee structure; National EMS Information System Task Force; NHTSA/APHA roundtable; and has hosted EMS orientation sessions for international groups from the Netherlands, Russia, and the United Arab Emirates. At present, Dr. Moore-Merrell serves as the principal investigator on a USFA cooperative agreement to develop fire service risk management models for staffing and resource deployment. Dr. Moore-Merrell is a member of the National Association of EMS Physicians, the Society for Academic Emergency Medicine, and the American Public Health Association. She serves as an organizational liaison to the American College of Emergency Physicians, the National Association of State EMS Directors, the National Institutes of Health (NIH) Heart, Lung, and Blood Institute, and the National Association of EMS Educators.

JAMES P. (JAY) REARDON

Jay Reardon is a retired fire chief for the Village of Northbrook and is currently the President of the Mutual Aid Box Alarm System (MABAS). Chief Reardon's career includes various governmental positions in local and federal government over 30 years in the emergency service field, including in Evergreen Park, Illinois, where he served as a cadet, firefighter, and apparatus engineer; in Hoffman Estates, Illinois, where he served as a firefighter, received paramedic certification, and was promoted to the rank of tactical lieutenant and captain-shift commander; in Portage, Michigan, where he served as assistant fire chief of Emergency Operations and was promoted to fire chief; and in Collier County, Florida, where he served as administrator of the division of emergency services. Chief Reardon's career also involved services within the federal government. His responsibilities included liaison between the Department of Defense and local communities in the reuse, economic redevelopment, and civilian conversion of a military institution scheduled for closure. Chief Reardon was an active reservist with the U.S. Air Force previously assigned to F-16 Tactical Units and C-130 Airlift Wings. His career has been distinguished as a noncommissioned and commissioned officer, selected as Disaster Preparedness Officer of the Year for the Air Force Reserves. Currently, Chief Reardon serves as Officer in Charge of Readiness for the 22nd Air Force, Dobbins Air Force Base, Georgia. Chief Reardon's military career concluded in November 2004 when he retired after 28 years of service. Chief Reardon is an executive fire officer, has a Bachelor of Science degree in Fire Service Management from Southern Illinois University, and graduated, with honors, from Western Michigan University with a master's degree in Public Administration. He was recently awarded a postgraduate Harvard University Scholarship to the John F. Kennedy School of Government. In 2003 he was awarded Chief Fire Officer Designation by the Commission on Fire Accreditation International, Inc. Chief Reardon's background in disaster planning, incident management, and recovery operations is extremely broad based. He has instructed at various colleges and conferences, as well as authored numerous articles published in various trade journals. Chief Reardon currently serves as president of MABAS involving over 900 fire agencies; past president, Illinois Fire Chiefs Association; appointed by Illinois Governor Ryan to serve on the 2000 Blue Ribbon Panel for the fire service issues; and statewide Terrorism Preparedness Task Force. Chief Reardon previously served as Illinois Fire Chiefs' representative to the Great Lakes Division Board of the International Association of Fire Chiefs. During the 2002 Illinois Fire Chiefs Annual Symposium, he was named "Illinois Fire Chief of the Year" by his peers and colleagues. Shortly thereafter, he was named *Fire Chief* magazine's International Career Fire Chief of the Year 2002. In 2003, Chief Reardon was awarded the International Fire Chiefs prestigious Presidents Leadership Award for championing interstate mutual aid initiatives. Chief Reardon was also selected by the International Association of Fire Chiefs as one of three selected "experts" in the development and publication of its *NFPA 1710 Implementation Guide* for fire chiefs. Chief Reardon also serves on several national committees, including the International Association of Fire Chiefs.

GARY W. SMITH

Gary was born and raised in Fort Bragg, California. He started his fire service career as a firefighter for the city of Davis in 1970 while attending college. In 1979, he was promoted to fire marshal for the city of Manteca. Gary received a Bachelor of Science degree in Fire Administration from Cogswell College in 1984. That same year he was promoted to fire chief for the city of Watsonville and continued to serve in that position for over 15 years. In February 1999, he was asked to take over as fire chief for the Aptos–La Salva Fire Protection District. Gary was also very active with the International Association of Fire Chiefs, specifically by organizing the first Wildland Fire Policy Committee. In July 2003, Gary retired from the fire service or, in his words, "transitioned" to a life of personal involvement in key community and national issues. Today, he is very committed to working as executive director of Leadership Santa Cruz County and is the president of the Ammonia Safety Training Institute, making ammonia the safest managed hazardous material in the world. Gary enjoys being an active player with the Aptos Chamber, Salvation Army, and Dominican Hospital Foundation.

STEVEN WESTERMANN

Steven Westermann has served as chief of the department in the Central Jackson County Fire Protection District since 1988 and has been in the fire service since 1972. He was elected as the president of the International Association of Fire Chiefs from 2007 to 2008. Prior to serving as president, he served as the international director for the Missouri Valley Division, was president of the Missouri Valley Division, and was on the IAFC's NFPA 1710 Implementation Guide Task Force. He also currently serves as the chair of the National Policy Centers Task Force for the IAFC. He has served as president of both the Heart of America Kansas City Metro Fire Chiefs Council and the Missouri Association of Fire Chiefs. Westermann has a master's in Public Administration from the University of Missouri–Kansas City, is an EFO graduate, and is a graduate of the Senior Executives in State and Local Government program at Harvard University.

THOMAS WIECZOREK

Tom has held a number of positions both in and out of local government. He began his professional career in the field of journalism, working in Michigan and Colorado as a reporter, special section editor, and managing editor. He next held positions in law enforcement before taking over fire functions, which led to his tenure as fire chief, deputy director of public safety, and director of public safety in Ionia, Michigan. He was named city manager for Ionia in 1989 and served in that capacity until retiring in 2005. His education includes Kalamazoo Valley Community College, Grand

Rapids Junior College, Grand Valley State University, and Michigan State University. He has taught or assisted in teaching a number of programs at Grand Valley State University, the National Highway Traffic Safety Administration (NHTSA), American Public Works Michigan branch, Michigan Rural Water Association, and Grand Rapids Junior College. He has testified frequently for the Michigan Municipal League before the legislature and in several courts as an expert in the field of accident reconstruction. Tom is active in a number of associations and is a speaker at numerous state and national events. He is the president of the Michigan Local Government Manager's Association (MLGMA); serves as vice chairman of the State of Michigan Department of Transportation Asset Management Council; served as the vice chairperson of the Commission on Fire Officer Designation; served as president of the Ionia County Memorial Hospital Board of Directors; president of the Ionia-Montcalm Domestic Violence Program; was president of the West Michigan City Management Association; and is a member of the American Public Works Association and numerous other groups. He serves as a representative of ICMA on the NFPA 1710 career committee. He received the Mark E. Keane "Award for Excellence" in 2000 from the ICMA, which is the association's highest award. He has been honored with the John B. Swainson Award for Historic Preservation from the State of Michigan Bureau of History (1997) and was honored as City Manager of the Year (1999) and Person of the Year (2003) by the Rural Water Association of Michigan. He currently serves as the executive director of the Center for Public Safety Excellence (CPSE).

JANET WILMOTH

Janet Wilmoth grew up in a family of firefighters in a Chicago suburb. Wilmoth first worked for *Fire Chief* magazine in 1986 as an associate editor and later served as *Fire Chief's* international correspondent while living in Europe and Asia. Wilmoth returned to *Fire Chief* in 1998 and was instrumental in expanding the magazine franchise to 18 different media and products. Wilmoth currently serves on the Fire Emergency Manufacturers and Services Association Board of Directors, the National Fallen Firefighters Foundation's Corporate Advisory Board, as well as the International Association of Fire Chiefs Foundation Board of Directors. She is on the steering committee for the Fire Department Safety Officers Association's annual Apparatus Specification & Vehicle Maintenance Symposium. Returning to her hometown, Janet and her husband once again make their home in Lisle, Illinois, where she enjoys a variety of activities including riding on their Harley-Davidson.

FRED C. WINDISCH

Fred C. Windisch, EFO CFO, is the fire chief of the Ponderosa Volunter Fire Department (VFD) near Houston, Texas. Fred began his volunteer fire service career in 1972 in Illinois and joined Ponderosa in 1975 after Shell Oil Company transferred him and his family to Houston. Fred currently serves on the Board of Directors of the International Association of Fire Chiefs, representing the Volunteer and Combination Officers Section. He retired from Shell in 2000, was appointed fire marshal in Harris County, Texas (the third most populous county in the country) in 2000, and was appointed Ponderosa's first paid fire chief in 2004. Ponderosa VFD is a three-fire-station ISO3 combination department protecting a population of about 50,000 citizens and a huge business base in unincorporated north Harris County.

The following grid outlines the course requirements of the Fire Administration I course developed as part of the FESHE Model Curriculum. For your convenience, we have indicated specific chapters where these requirements are located in the text:

Course Requirements	***1***	***2***	***3***	***4***	***5***	***6***	***7***	***8***	***9***	***10***	***11***	***12***
Identify career development opportunities and strategies for success.	X	X	X	X	X	X	X	X	X	X	X	X
Explain the need for effective communication skills, both written and verbal.		X	X	X	X	X	X	X	X	X	X	
						Chapter Number						
	1	***2***	***3***	***4***	***5***	***6***	***7***	***8***	***9***	***10***	***11***	***12***
Articulate the concepts of span and control, effective delegation, and division of labor.		X	X	X		X	X	X	X	X		
Recognize appropriate appraising and disciplinary actions and the impact on employee behavior.	X							X				
Examine the history and development of management and supervision.	X		X	X					X			
Evaluate methods of managing available resources.			X		X		X	X	X	X	X	X
Identify roles and responsibilities of leaders in organizations.	X	X	X	X	X	X	X	X	X	X	X	X
Compare and contrast the traits of effective versus ineffective supervision and management styles.			X	X	X	X	X					
Identify and assess safety needs for both emergency and non-emergency situations.									X		X	X
Identify the importance of ethics as it applies to supervisors.	X		X	X		X	X	X				
Identify the role of a company officer in the Incident Command System (ICS).									X		X	
Describe the benefits of documentation.	X					X		X	X	X	X	
Identify and analyze the major causes involved in line of duty firefighter deaths related to health, wellness, fitness, and vehicle operations.												X

Our Heritage and Our History

Key Terms

act phase, p. 11
check phase, p. 11
demographic, p. 12
diversity, p. 10
do phase, p. 11
empowerment, p. 11
Equal Employment Opportunity Commission (EEOC), p. 14
Junto, p. 6
PDCA cycle, p. 11
plan phase, p. 11
values, p. 13

Objectives

After completing this chapter, you should be able to:

- Describe how history impacts the fire industry today.
- Describe the changing culture of the fire service industry.
- Explain the importance of cultural and ethnic diversity in the hiring of employees.
- Define the legal issues that have impacted the fire service.
- Describe the significance of fire service symbols and insignia.
- Describe how major fires and events have helped shape the fire industry.

Throughout the history of the human race, fire has served as an agent of progress, a protector, a means of survival, and also a devastating force of destruction. Although we have made great progress in accumulating the knowledge and technology to harvest fire and use it to our advantage, all too frequently we still see fire out of control as it assumes its destructive role.

A firefighter entering the fire service should have an understanding of the history of the fire service because of its profound influence on the fire service today. Since the earliest period of recorded history, as people began to assemble into specific geographic areas, cities and counties were formed, and the protection of life and property was of paramount importance. Men united together in small communities and villages with many walled fortresses with the chief aim being security from attack. The gates were generally closed from sunset to sunrise, and a watchman stood near the gates and announced approaching danger. Organizations formed to protect their communities during this time period would become the precursors to modern fire protection.

One of the first organized firefighting forces was established in Rome by Caesar Augustus. Augustus created a force of some 600 men belonging to the "*familia publica*, servants of commonwealth," stationed near the city gates, for the specific purpose of fighting fire. Unfortunately, these firefighters were in fact slaves and had little or no stake in the society that they were forced to protect. They were often accused of being too slow to react to fires and reluctant to endure the physical risks to save the lives and properties of their Roman masters. Eventually their work was given to companies of volunteers, who never had the support or favor of the government. After one particularly disastrous fire, the Emperor Augustus passed responsibility for firefighting duties to the Roman army and established the *Corps of Vigiles* to protect the city of Rome and its empire for the next five hundred years. Only citizens of Rome could join the Roman army, those free at birth. But because the first *Vigiles* were recruited from freed men, they were never considered true Roman soldiers. However, because recruitment was slow, the inducement of full citizenship was offered to freed men after six years' service. Consequently, service in the *Corps of Vigiles* was seen as an honorable means of obtaining full Roman citizenship. A normal period of service with the corps lasted 26 years. Quite interestingly, firefighting was not the only duty of the *Vigiles*. At night they would "police" the cities and towns and quite often be required to track down and recapture runaway slaves. During the day they would be stationed at public baths to watch over the clothes of the bathers to prevent pilfering by the public and bath attendants.

The *Vigiles* were distributed throughout the city. Each battalion was responsible for two of the 14 wards into which Augustus had divided the city. The cost of maintaining this force was paid by the public treasury, and each fire prompted an official inquiry, which in the case of fires judged to be caused by negligence resulted in the punishment of the careless citizen. The roots of our paramilitary organizations can be traced to this first organized firefighting force. The *Vigiles* carried quite sophisticated firefighting equipment, including water pumps and hoses, and even *ballista* catapults to which fire hooks were attached for climbing ropes or as a means to demolish burning buildings in order to prevent the spread of fire. It is believed they wore helmets for protection, but it is unlikely they wore any other kind of armor.

Rome, just a small village on the Tiber River about 900 B.C., found itself at the crossroads of culture and commerce. As Rome grew into one of the world's first metropolitan areas, the issue of fire and how to protect the citizenry from it became a growing concern. By the year 100 B.C. Rome was a burgeoning, bustling city that was periodically impacted by serious fire that destroyed property and lives. Moreover, the city had a well-developed bureaucracy and other governmental support systems. It does not come as a complete surprise that these conditions were at the genesis of the creation of community fire protection.[1]

Blackstone's *History of the British Fire Service* provides one of the best descriptions of how and when the Roman fire brigade came into existence. Modern firefighters often recognize the fact that their profession is based upon the structure created so many years ago but do not fully appreciate how colorful and complicated fire protection was in that era. For example, it is often not discussed that the creation of some of the first brigades was based upon the entrepreneurial, rather than the altruistic, spirit. Crassus, the Roman emperor who was referred to as one of the greediest of the Roman aristocracy, built his future upon the existence of fires. His force of firefighters would respond but would not engage in fighting the fire on request of the home or business owners until they sold Crassus the property. He would arrive on the scene and offer the hapless fire victim a ridiculously low price for the property. As the building burned down, the price would decrease. More than a few victims, realizing their life savings were going up in smoke, accepted the deal. Crassus became very wealthy by exploiting such tragic situations. But the emperor's greed was not the only driving force. By about 100 B.C. Rome was a literal firetrap with large four- or five-story apartment buildings constructed throughout the city. These wooden buildings were home to tens of thousands of families. The conditions found in the buildings were dismal, and when a fire broke out, the chances of total destruction of several buildings were quite high.

Government participation in the creation of fire brigades took many of its basic ideas from Crassus. The difference was the groups were organized along the line of the military order. The organization was known as the *Vigiles*, Latin for "vigilant." Organized as "companies," the firefighters were used to patrol the streets of Rome after dark. Always alert for fire, these same men also served as an early form of law enforcement. The *Corps of Vigiles* was commanded by a Roman military officer of "equestrian" rank, an officer on horseback, considered to be higher than the infantry officer. Three ranks have been identified within the *Corps of Vigiles*.

- **The Aquarius.** This firefighter's main duties were the supply of water to the siphons or pumps and the organization of "bucket chains."
- **The Siphonarius.** This firefighter was responsible for the supervision and operation of the water pumps.
- **The Uncinarius.** This firefighter was a "hook" man who carried a large fire hook for pulling off burning roofs.

All firefighters carried knives, axes, and spades to help them in their duties.

There is evidence the *Corps of Vigiles* was also organized in cities other than Rome. Barracks for the *Vigiles* were located in many cities where large concentrations of people and property were found. The credit for formalizing the *Corps of Vigiles* belongs to Augustus Caesar. Although few records exist of these early brigades, significant information is available related to the major fires that struck the capital of the empire. Most are familiar with the Great Fire of Rome that reportedly destroyed over 10,000 dwellings, killed hundreds of people, and reduced about a square mile of the city to rubble in less than 48 hours. At the height of the empire glory, approximately 10,000 *Vigiles* were in service. No real information exists that describes how well distributed the barracks facilities were for the corps, but there is evidence of multiple sites when the city was at its peak of commerce and political influence.

One Roman soldier exemplified the firefighting spirit and his influence is evident throughout the fire service today. His name was *Florian von Cetum*, the patron saint of firefighters who lived during the beginning of the first century (circa A.D. 30). He was a high-ranking officer in the Roman army, first holding the title of centurion, then advancing to *prefectus vigilum* (fire chief).[2]

The men Florian commanded were regular members of the Roman army assigned to the fire brigade. Only the most-skilled soldiers were permitted to take part in the brigade. They not only protected Rome and other locales but also were a mobile force. The brigade would protect the Roman army on its marches and conquests. They were especially useful when the soldiers erected tent cities or other temporary, flammable living quarters. Like other Roman officers of the time, Florian wore a crest atop his helmet so that his men could easily recognize and follow him into the battle against fire. His crest was bright red with purple just below. (Purple was the most expensive dye at the time, reserved for only the highest-ranking generals.[3]) He had a reputation of energy in the task of fighting fires in villages and towns where he was stationed while in the military. Florian was also a devoted Christian who unfortunately became a martyr for his beliefs. According to legend he was drowned in a river as punishment for his beliefs. The Catholic Church made him the patron saint of firefighters.

European firefighters often display a small statue of St. Florian, shown throwing a bucket of water on a burning house, as part of the station decor or in their personal memorabilia. Many firefighters throughout the world, irrespective of religious affiliation, wear a St. Florian medallion as a symbol of protection in their daily duties.

The Roman Empire prevailed for over 1,000 years; with the decline of the empire there was also a noted decline in the *Corps of Vigiles*. Brigades were abandoned as a result of a shift to local control of many governmental services. When the world entered the Dark Ages, fire protection entered a period of reduced viability. The only means of firefighting through this time period appeared to be neighbor helping neighbor. The need to fight fire did not diminish, but the degree of organization and the effort did for quite some time. The fire service would not emerge again as a viable organized service until the 1600s.[4]

CODES AS A MEANS TO REGULATE FIRE

One, if not the earliest, known code of law regulating building construction is that of *Hammurabi* (1792–1750 B.C.), founder of the Babylonian Empire that was eventually destroyed by the Asia Minor raids. Hammurabi may have begun building the tower of Babel, which can now be identified with the temple tower in Babylon known as Etemenanki. His code of laws is one of the greatest of ancient codes. It is carved on a diorite column, in 3,600 lines of cuneiform, and was found at Susa in 1902. The code, which addresses such issues as business and family relations, labor, private property, and personal injuries, is generally humanitarian. However, one severe feature is the retributive nature of the punishment, which follows "an eye for an eye" literally. Much of this code is drawn from earlier Sumerian and Semitic laws, which seem to provide the basis for its harsh nature. A translation of the table containing the code is in Table 1.1.[5]

TABLE 1.1 Table Containing Code

228: If a builder built a house for a man and completed it, that man shall pay him two shekels of silver per sar (approximately 12 square feet) of house as his wage.

229: If a builder has built a house for a man and his work is not strong, and the house he has built falls in and kills the householder, the builder shall be slain.

230: If the child of the householder is killed, the child of the builder shall be slain.

231: If the slave of the householder is killed, the builder shall give a slave for a slave to the householder.

232: If goods have been destroyed, he shall replace all that has been destroyed. Because the house was not made strong and has fallen in, he shall restore the fallen house from his own material.

233: If a builder has built a house for a man and his work is not done properly and a wall shifts, then that builder shall make that wall good with his own silver.

THE COLONIAL FIRE EXPERIENCE

As immigrants from England, France, Spain, the Netherlands, and Scandinavian lands colonized the eastern seaboard of the United States, they founded the first permanent settlement in Jamestown, Virginia, in 1607. Shortly thereafter on January 7, 1608, the young colony was destroyed by fire. As a result of this first American conflagration, most of the colonists' lodging and provisions were destroyed. Many died during that winter from exposure and hunger due to the lack of protection from the environment.[6] *Captain John Smith* made a concise assessment of the situation: "I begin to think that it is safer for me to dwell in the wild Indian country than in this stockade, where fools accidentally discharge their muskets and others burn down their homes at night."[7]

In 1620 the colony of Plymouth, Massachusetts, was settled. Three years later a fire spread out of control and destroyed seven buildings and almost all the settlement's provisions for the coming winter, nearly ending the existence of the colony. Boston had the distinction of having the most numerous and most destructive fires of the colonial towns, experiencing nine conflagrations. At the same time America's larger cities, Philadelphia and New York, had yet to have one. Although luck may have played a role, more likely it was the construction techniques utilized by the three cities. Philadelphia and New York had brick and stone available for construction purposes.[8] During this time Boston's building stone was inferior, resulting in more of the buildings being built of wood. Flammable construction undoubtedly played a major factor in Boston's poor fire record. Also contributing was the lack of enforcement of the fire prevention building laws. It was not until after numerous fires occurred that Massachusetts passed a law in 1638 prohibiting smoking outdoors. It was the first "no smoking" ban in America. A number of colonial settlements soon followed this lead.[9]

Governor Peter Stuyvesant of New Amsterdam (later to become New York City) was truly one of the pioneers of fire prevention. Recognizing the hazards of combustible chimneys, Governor Stuyvesant succeeded in having laws passed in 1648 prohibiting construction of wooden or plaster chimneys. These were the first of many such laws passed by American colonists in an effort to prevent fire disasters. His next step was to appoint volunteer fire wardens to enforce the laws and inspect the chimneys. These wardens were required to levy fines to owners of faulty chimneys. Understanding that the greatest threat of major fire spread was presented by fires occurring at night, Governor Stuyvesant established a curfew, as did other towns at the time. A ringing of the bell at 9 P.M. ordered the extinguishing or covering of all fires until 4:30 A.M. An additional step in 1658 by Peter Stuyvesant was his appointment of eight young men to roam the streets of New Amsterdam at night to watch for fires. These men, clad in long capes and carrying wooden rattles that twisted and sounded the alarm, are considered the first step in organized firefighting in America. The rattle watch soon grew from 8 to 50 members and began to arouse opposition among the townspeople. Many considered the rattle watch more as prowlers than protectors. This resentment of the authoritarian rule of Peter Stuyvesant and of the impositions of the fire wardens and the rattle watch helped to fuel a takeover by the British in 1664. The British renamed the colony New York and took over where Governor Stuyvesant had left off, with the adoption of ordinances regulating burning out of chimneys at regular intervals and mandating the use of chimney sweeps.[10]

FIGURE 1.1 ◆ Benjamin Franklin
(Courtesy of Randy Bruegman, The Chief Officer: A Symbol Is a Promise, *p. 8)*

Junto
In 1927 Benjamin Franklin organized a group of friends to provide a structured forum for discussion. The group, initially composed of 12 members, called itself Junto. The members of the Junto were drawn from diverse occupations and backgrounds, but they shared a spirit of inquiry and a desire to improve themselves, their community, and to help others. Among the original members were printers, surveyors, a cabinetmaker, a cobbler, a clerk, and a merchant.

Benjamin Franklin, one of the Founding Fathers of America (see Figure 1.1), was an early champion of fire prevention and most notable among the famous Americans who helped shape the country and the fire service. He was a writer, printer, philosopher, scientist, statesman of the American Revolution, as well as a firefighter. Franklin helped draft the Declaration of Independence, served as a diplomat, and invented items ranging from lightning rods to bifocal eyeglasses.

His writings in his *Pennsylvania Gazette* had much to do with increasing the public's awareness of safety and forming opinions of the importance of fire prevention. Franklin coined one of his most familiar epigrams, "An ounce of prevention is worth a pound of cure," in a letter warning Philadelphia citizens about the hazards of carrying burning firebrands or coals in a full shovel from one room to another, and recommended the use of a closed warming pan for this purpose. Franklin also campaigned for clean chimneys and for officially appointing chimney sweeps. In addition to his other contributions to fire prevention, Franklin founded the volunteer Union Fire Company in 1736.[11]

In 1727, Franklin organized a group of friends to provide a structured forum for discussion. The group, initially composed of 12 members, called itself the **Junto**. The members of the Junto were drawn from diverse occupations and backgrounds, but

FIGURE 1.2 ◆ Junto Meets

they all shared a spirit of inquiry and a desire to improve themselves and their community, and to help others. Among the original members were printers, surveyors, a cabinetmaker, a cobbler, a clerk, and a merchant. Although most of the members were older than Franklin, he was clearly their leader.[12] The results of the Junto are still evident today as an integral part of American society. The Junto gave us our first library, volunteer fire departments, the first public hospital, the police department, paved streets, and the University of Pennsylvania[13] (see Figure 1.2).

Although the first attempt at providing fire insurance was a failure after a devastating fire in Charlestown, Massachusetts, in 1736, the concept did not die. Ben Franklin is credited with founding the first successful American fire insurance company on April 13, 1752. Initially this company was named the Philadelphia Contributorship for the Insurance of Houses from loss of fire, but the name was later changed to Hand in Hand Insurance Company, which remains in business today.[14]

The company adopted fire marks to be affixed to the front of the insured property for easy identification in determining whether the property was owned by a paying customer. These marks often dictated which fire company would fight the fire. During this time period the ingenuity of the American fire service also began to emerge. Two important "tools" utilized by early American firefighters were the bed key and salvage bags. With firefighting apparatus able to supply only a small stream of water, a fire gaining headway was soon out of control. Arriving firefighters quite often opted for immediate salvage efforts in the burning building and surrounding exposures. The bed key was a small metal tool that allowed the men quickly to disassemble the wooden bed frame, often the most valuable item owned by a family, and remove it to safety. Other valuable household goods were placed in salvage bags and carried to safety.

As the fire service progressed into the nineteenth century, a very exciting time emerged for the United States as it expanded westward as thousands of immigrants swelled its population. Industry was expanding in cities to the west and with that came several significant fires, which helped shape the course of the American fire service into the twentieth and twenty-first centuries. On October 8, 1871, the *Chicago Tribune's* front page read, "Firefighters prepare for fall and winter fires." That night

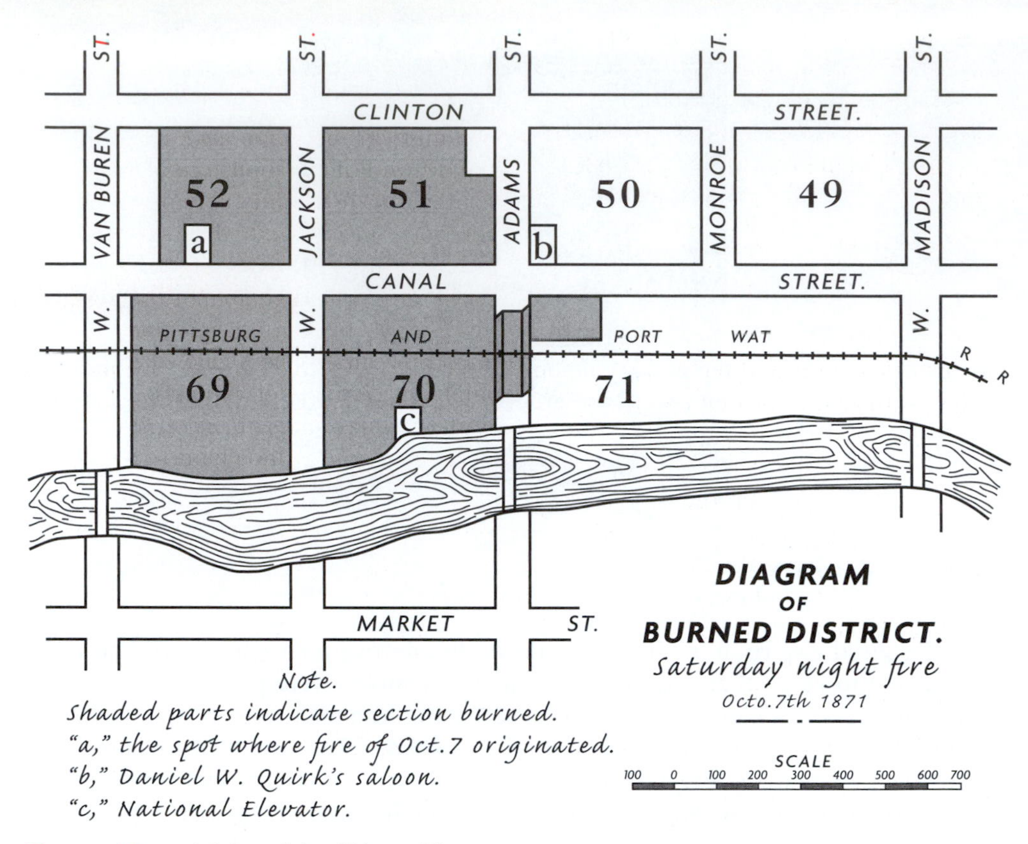

FIGURE 1.3 ◆ A Map of the Chicago Fire

fire broke out in the vicinity of a barn owned by Mr. and Mrs. Patrick O'Leary at 137 Dekoven Street (now the site of the Chicago Fire Training Academy) (see Figure 1.3). Tradition lays the blame for the start of the conflagration on Mrs. O'Leary's cow. Supposedly, the cow kicked over a kerosene lantern, which ignited the fire. Whether or not this is true, the extent of the losses is well known. The fire burned for 27 hours, destroyed 17,500 buildings, killed 250 to 300 people, and left approximately 100,000 homeless. Fire loss was estimated at approximately $200 million, of which only $88 million was insured and only $45 million was actually paid in claims. This resulted in multiple bankruptcies of American insurance companies.

On the same day that Chicago was experiencing its devastating event, a small lumbering community in Peshtigo, Wisconsin, was also suffering one of the most significant fires in terms of loss of life in the history of the United States. On October 8, 1871, a forest fire in the area surrounding Peshtigo developed into a firestorm that swept through the town, destroying every building except for one house that was under construction. The loss of life was nearly 800 people.

The Great Boston Fire of 1872 destroyed over 776 buildings in one area, the business district covering approximately one square mile. The fire claimed the lives of 13 victims, including two firefighters, and losses were estimated at $75 million. Boston's fire protection was poor at best at the time. The inadequate water supply was being provided through four-inch mains, while hydrants were fed by three-inch branch lines. Witnesses on the scene testified the hose streams did not go above the third floor. Many of the buildings burned that day were five- to six-story structures; however, the longest ladder reached only 40 feet. As a result of the great conflagrations of the 1800s, the *ISO Fire Insurance Grading Schedule* was developed, which is still in use today.[15]

History has taught us many lessons that have helped shape today's fire industry, from how buildings are constructed, how resources are deployed, the adoption of codes and standards, to the equipment firefighters use. In fact, many aspects of the current fire industry are a reflection of past events.

SOCIETAL IMPACTS OF THE FIRE SERVICE

As historic events and technology inventions have helped to shape the fire industry today, so have the societal impacts of an ever-changing nation.[16] With those changes, the socioeconomic environment of the fire service has also changed. Today, the social factors affecting the fire service seem to be changing more rapidly than ever; therefore, it is important that the fire service examine and understand these factors, which have helped to shape its identity. While looking back at the history of this country and the events that have affected its culture, the fire industry must also look at recent social development and examine why it needs to be alert for changes during the coming decade. In doing so, the fire service will begin to see the beneficial challenges it faces as it builds a multicultural workforce. The fire industry must also be aware of the prominent legal factors that have come into play that affect the responsibilities of fire officers and have changed the way business is conducted on a daily basis.

CULTURE AND DIVERSITY: THE CHANGING COMPOSITION OF THE UNITED STATES

The composition of the U.S. population is changing. Society is vastly different today from the society that existed when the country was founded and, in fact, has experienced dramatic shifts and changes in the last 30 years.

1770 to 1870 In the late 1700s, American society was made up of predominantly Anglo-Saxon settlers struggling to establish livelihoods, businesses, and families. The government consisted of Caucasian males. This was the way society functioned. Virtually all of the laws that were passed and the businesses that were founded were owned by Caucasian men. African Americans were present during the early years of the country as slaves. The new United States of America evolved slowly during its first century. The Civil War demonstrated that this organization of states would, indeed, remain united; it was the Civil War that once and for all helped to settle the question regarding Black Americans. They were no longer slaves to a White society but were "freemen."

1870 to 1970 The next one hundred years saw dramatic change. Between 1870 and 1970 the power structure of society began to shift. Former slaves gained full citizen status. The early civil rights movement began. African Americans began to demand their full civil rights and in 1909 the National Association for the Advancement of Colored People (NAACP) was founded. In the 1920s Marcus Garvey, founder of the Universal Negro Improvement Association, was responsible for the flowering of African American creative talent in literature, music, and the arts. This movement came to be known as the Harlem Renaissance and resulted in a rise in race consciousness among African Americans in the New York area. The Industrial Revolution ignited a fire of enthusiasm among Americans and those who wished to become Americans.

Immigrants primarily from northern Europe poured into the country. They saw in America an opportunity to excel and achieve. Immigrants by the thousands settled in the big cities, forming neighborhood enclaves and establishing common support for one another—churches, grocery stores, clothing outlets, and social clubs.

Shortly after the turn of the century, women secured the right to vote. Women also played a prominent role in the massive work effort undertaken during World War II. In that war, women moved out of the home and into the factory, taking on responsibilities that had previously belonged only to men.

The hundred years following the Civil War saw significant changes wrought with judicial and legislative initiatives. Ten amendments to the Constitution were passed during this period, many of which dealt with the provision of rights to women and minorities. During the later part of this period, the modern civil rights movement was born. The Civil Rights Act of 1957 (the first federal civil rights legislation to be passed since 1875) authorized the federal government to take legal measures to prevent a citizen from being denied voting rights. The Civil Rights Act of 1964 was passed, introducing the age of equal opportunity. In 1964 the Twenty-Fourth Amendment to the Constitution banned the poll tax, and in 1965 the Voting Rights Act eliminated all discriminatory qualifying tests for voter registrants. During the 1960s and early 1970s, guidelines for nondiscriminatory hiring were developed, and court orders and consent decrees combined to assure affirmative action.

1970 and Beyond Since the early 1970s, the nation has seen new power struggles in American society. Blacks, Hispanics, Native Americans, and women all have emerged to demand their rightful places as citizens and workers. Civil disturbances erupted in the large cities, and the strain of growth coupled with a new diversity presented challenges for many communities. Judicial cases have seen the courts issue major rulings in regard to racial, sexual, age, and religious discrimination. Special-interest groups have emerged as powerful voting blocs—the National Association for the Advancement of Colored People (NAACP) and the National Organization for Women (NOW), among them. Education has seen the demand for alternative language courses in schools and the design of single-culture study programs such as Afrocentric and Judaic studies. Figures released by the United States Census Bureau in 1992 indicated that Americans claimed dozens of different ancestries. More than 30 different ethnic designations were named by more than 100,000 persons, and numerous others were listed by a smaller number of respondents. Among the largest groups were German, 58 million; Irish, 38.8 million; English, 32.7 million; African American, 30.2 million; Hispanic, 22.4 million; Italian, 14.7 million; American, 13.0 million; French, 10.3 million; and Polish, 9.3 million.

diversity

The presence of a wide range of variation in the qualities or attributes under discussion. When used to describe people and population groups, diversity encompasses such factors as age, gender, race, ethnicity, ability, and religion, as well as education, professional background, and marital and parental status.

Diversity Defined The result of this historic social evolution is a country seeking to accept the differences among its people, who realize they are truly a kaleidoscope of diversity in the workplace and society. Such **diversity**, once found only in metropolitan areas, is now seen throughout the country in even the smallest communities. What does this mean? It means people believe they can maintain their cultural characteristics while cooperatively working, living, and sharing mutual goals. Diversity is defined as the characteristics that make people distinct—age, gender, race, ethnicity, ability, and religion.

Culture Defined *Culture* is defined as learned behavior patterns of people (what they think, say, do, make, believe, value, and feel). It is now expected that the workplace will reflect the society in which it operates. This is no different for the fire service.

In fact, the fire service has a larger obligation to reflect society because it functions as an element of a democratic society. Representation in the workforce is a critical component of future organizational design. Thus, the fire service must represent the diversity of the community it serves, and the fire service must adopt an open, supportive environment for diversity. Organizational policies should reflect this commitment, and fire officers should understand and enforce the laws that assure equality. Finally, the fire service must remain constantly aware of changing social influences and their potential impacts on the organization.

TRENDS IMPACTING THE WORKPLACE

The fire service is faced with some unique challenges. Just as the private sector has been drawn into a competitive world market during the past three decades, the fire service is now confronted with the need to compete. Publicly funded, the fire service has struggled to gain a diminishing share of citizen dollars—through taxation or donation. The volunteer fire service faces the same critical competitive demands as the career fire service. Drawn into the midst of these issues is the mid-level manager, who forms the linchpin between fire department leadership and the groups who deliver services directly to the community and to fellow employees.

EMPLOYEE EMPOWERMENT

Today, the fire service must empower its employees, an alien concept to management structures in many fire departments prior to the 1990s. Empowering employees means allowing employees at their respective levels of responsibility to make decisions without hierarchical permission, regarding service delivery initiatives. **Empowerment** means autonomy for the service-level employee (the firefighter, the inspector, the public educator, the paramedic, the apparatus mechanic, the driver/engineer). Beginning in the 1970s and 1980s, as the empowerment collaboration approached the public sector, many fire department officers were simply not ready to relinquish their power. Empowerment requires managers to establish boundaries within which subordinates can make decisions. Boundary setting is complex and time consuming to managers, but nothing is simple in this modern age. As the fire service has evolved, it has come to realize that if it is to remain competitive, it must move forward with empowered employees.

TOTAL QUALITY MANAGEMENT

Closely tied to empowerment is the concept of *total quality management* (TQM). Local government including the fire service has lagged behind the private sector in adopting this creative approach to leading organizations. With TQM, groups of employees work to overcome the daily problems confronting them in their jobs. The use of a basic process called the plan-do-check-act cycle solves problems, increases service quality, and saves money. This whole process is centered on employee empowerment. Employees **plan** a change or a test in the way they are currently performing their jobs. Then, they implement (**do**) the change on a small scale to see whether it will work. They **check** to see whether the change has worked or whether it needs modification. Finally, they **act** on what was learned, modify it, improve it, and try it again—**PDCA Cycle**.

Without empowerment, the process grinds to a halt. Getting permission through a hierarchical chain of command causes the process to fail. The process is built upon autonomy of the work group to make its own decisions. In the TQM process, mid-level

empowerment
The process of increasing the capacity of individuals or groups to make choices and to transform those choices into desired actions and outcomes. Central to this process are actions that both build individual and collective assets, and improve the efficiency and fairness of the organizational and institutional context that govern the use of these assets.

plan phase
The first phase of the plan-do-check-act (PDCA) cycle. A plan identifying what needs to be improved, how it is to be implemented, and how the results are to be evaluated is developed by the organization.

do phase
The second phase of the plan-do-check-act (PDCA) cycle. Changes that are expected to improve processes are tried or made during this phase.

check phase
The third phase of the plan-do-check-act cycle. The effects of

act phase
The fourth phase of the plan-do-check-act (PDCA) cycle. Decisions are made whether to adopt the changes that were tested, propose new changes, or run through the cycle once more.

PDCA cycle
See Plan-Do-Check-Act Cycle.

managers become facilitators and suppliers to the decision groups. Such a role is far different from being a supervisor, telling others what to do, how to do it, and when to do it. But the mid-level manager under TQM has a far more challenging job of facilitating decision groups comprised of different races, sexes, and values. The real challenges for mid-level managers under these new management concepts are those of managing diversity and instilling motivation. This topic will be covered in more detail in Chapter 10.

TRADITIONAL POWER STRUCTURES ARE CRUMBLING

Another change confronting the fire service culture is the change in the *traditional power structures* that many organizations have used. Today Caucasian males are no longer the primary group from which department leadership is drawn. The pool of fire department entrants now contains fewer Caucasian males, in part because of past legal mandates that have provided equity and fairness in the entry-level process, which opened positions to women and minorities.

In 1990, Caucasian males composed 61 percent of the workforce, and by the year 2000, they composed approximately 50 percent of the total workforce. Communities expect the fire service to be representative of the local labor market. In fact, most consent decrees and court orders to date base racial and sexual hiring goals on the distribution among those groups within the community—the relevant labor market.

DEMOGRAPHIC SHIFTS

demographic
Pertaining to the study of human population characteristics including size, growth rates, density, distribution, migration, birth rates, and mortality rates.

Changing Face of the Community A social factor affecting the modern fire department is the **demographic** shifts of people in the community. According to U.S. Census figures,[17] most individuals move to a different location every four to five years. For many years, urban areas have seen the abandonment of the inner city by middle-class families, who have sought the quiet refuge of the suburbs and chosen to commute to work (in some cases, a long commute). Because of this movement to the suburbs, neighborhoods have seen dramatic transformations that, in some cases, changed a neighborhood from one ethnicity to another. Sometimes these transformations have been stimulated by government-funded programs to revitalize the inner city. Today the inner city is making a comeback, as many are moving back to the metropolitan centers.

Changing Face of the Family Another demographic factor having an impact on modern society is the changing concept of family. Recent events have drawn attention to the societal impact of family structure, with single-parent and divorced-parent families increasing. According to Morris Massey,[18] the family is the single, most important factor in forming values and behaviors. Recent figures show that 24 percent of children in America now live with a single parent. Roughly 50 percent of U.S. children live in a nontraditional home arrangement in which one or both of their biological parents are missing. These societal problems add to the challenges the fire service faces. In managing a diverse workforce, many fire departments now have assumed the task of accommodating nontraditional parents in their daily work demands. The demands of the fire service are also impacted by the instability that some home lives often create.

THE TWENTY-FIRST CENTURY WORKFORCE

The changing labor force is a significant social factor affecting the fire service. In 1987, the Department of Labor sponsored a study of the labor issues confronting the United States as it moved into the twenty-first century. The report, *Workforce 2000*,[19] was created by the Hudson Institute of Indianapolis. Researchers found some interesting results that will have a direct impact on the fire service during the coming decades.

Among those findings was that during the 1990s the workforce population grew at its slowest rate since the 1930s. In the year 2000, the average age of the worker was 39 (the current age is 33); one-third of all workers were at least 65 years old; women composed 47 percent of the workforce (up from the current 43 percent). In the year 2000, 61 percent of all women in the United States were working. By the year 2000, 29 percent of all new entrants to the job market were nonwhite. The figures generated from the *Workforce 2000* report illustrate the need for the fire service to plan for and implement a multicultural approach to labor acquisition.

DIFFERING VALUES

An important social factor posing a challenge for fire officers is the issue of conflicting values. Depending on circumstances, age differences of 10 to 40 years present significantly different perspectives on how problems are viewed and solved. Value conflicts easily arise between the "now" generation and older generations. We have all experienced it. The modern fire officer must understand the elements that combine to cause the conflicts.

Morris Massey differentiates between those he calls baby boomers and synthesizers. Baby boomers (born from 1946 to 1965) typically reflect values of a strong work ethic, the maintenance of the family, a trust of government, and a career devotion to a single employer. Synthesizers (born during and after the 1970s), on the other hand, may value leisure, accept divorce more readily, tend to distrust government, and change jobs more frequently. This clash of values provides the forum for misunderstanding and lack of mutual appreciation between these two generations. The clashes are usually manifested as a conflict between extremely opposite perspectives. For example, baby boomers tend to value group commitments, respect authority through obedience, believe in a melting pot of cultures, and desire materialism with resulting wealth. Synthesizers tend to value individualism, yet expect participation in decisions. They view culture through diversity with appreciation of differences and seek new experiences and artistic freedom in their career choices. It is easy to see through these descriptions how different generations can view the same things from vastly different perspectives.

Values are imprinted in us by the age of seven. These values are influenced by families, religion, school, friends, and experiences. Once imprinted, these values are reinforced through the modeling behaviors of those around us. Finally, these values are completely socialized within us by the age of 20. The values do not change except under the most extreme and trying of circumstances, which are referred to as "significant emotional events." Even then, values are rarely altered totally. What does this mean to the fire officer? It means that, to understand the values of others from different cultures and different ages, fire officers have to walk in others' shoes to see where they have been.

values

Enduring beliefs and assumptions about specific modes of conduct or states of existence that are preferable to opposite or converse modes of conduct or states of existence. Values are the general guiding principles that govern our actions.

CHANGING EDUCATION LEVELS

The final social factor to consider is the changing educational background of those who will compose the fire service in the coming decades. The fire service, because of diversity, will require officers with higher levels of interpersonal skills. The role of today's officer is that of leader, manager, coach, mentor, and facilitator. The future fire officer will need to possess a higher level of education and breadth of experience to meet the demands of the job. The educational challenge for the fire service is two-fold. First, many entrants now have higher education levels. In most fire departments, 50 percent of the new employees or volunteers have some college education. Second, many entrants have fewer technical or hands-on skills. Fewer entrants possess military experience or trade backgrounds; this may change in the coming years due to the recent war effort in the Middle East. Thus, the fire service must meet the challenge of dealing with a more intellectually talented group of employees who, at the same time, may require more extensive training to gain the necessary hands-on firefighting technical skills. This topic will be discussed in more detail in Chapter 2.

LEGISLATIVE MANDATES

A significant part of recent history in this country having a tremendous impact on all aspects of life today is legislative mandates. One of the driving forces that has compelled not only the fire service but also the United States toward acceptance of diversity is the body of laws that has evolved over the past 30 years. These laws have provided equal opportunities to the groups that had been deprived of these freedoms in the past. The following chronicles the laws, with a special emphasis on sexual harassment and disabilities. Each has helped to shape today's workplace and marks a period in this country's history of dramatic change.

THE CIVIL RIGHTS ACT OF 1964

The Civil Rights Act of 1964 arose from the cries of citizens and politicians who recognized the inequalities still existing in American society one hundred years after the Civil War. The act made common the term *illegal discrimination* and opened the door to subsequent legislation seeking to guarantee the rights of the oppressed. The term *discrimination* is defined as the process of distinguishing, on the basis of a perceived feature or characteristic, one item or person from a group of items or persons. Discrimination, in and of itself, is not wrong; it is the primary means utilized to differentiate among things mentally. However, illegal discrimination is defined by the Civil Rights Act as conduct that unjustifiably distinguishes among similarly situated people on a basis prohibited by law, such as race, color, religion, national origin, gender, age, and handicap. Title VII of the Civil Rights Act formed the cornerstone of equal opportunity as it is known today. From Title VII a multitude of policies and rules has flowed that seeks to provide equal opportunity. Equal opportunity is defined as the right of all persons without regard to race, color, religion, national origin, gender, age, or handicap to apply for and receive benefits and services. Originally, equal opportunity was guaranteed only in programs involving federal funds. Since then, the expansion of equal opportunity to virtually all local and state programs has occurred.

Equal Employment Opportunity Commission (EEOC)

To administer its responsibilities, the EEOC accepts written charges filed against an employer alleging that it has engaged in unlawful employment practice in violation of Title VII or other federal civil rights laws and has the power to bring suits, subpoena witnesses, issue guidelines that have the force of law, render decisions, provide legal assistance to complainants, etc., in regard to fair employment.

EQUAL EMPLOYMENT OPPORTUNITY

Since 1964, state and local governments have passed laws and adopted legislation, ordinances, and policies ensuring equal opportunity for citizens and employees. In 1972, the **Equal Employment Opportunity Commission (EEOC)** was formed for the purpose of issuing rules and monitoring the intent of the Civil Rights Act. Volunteer firefighters may be affected by an unfortunate liability concerning their civil rights as volunteers. Courts have ruled that volunteers are not "employees"; therefore, they are not entitled to civil rights protection under Title VII. This is particularly troublesome for men and women in sexual harassment cases. Some states have also passed legislation that classifies volunteers as employees of the state when they are acting on behalf of the state or local government. It is advisable for volunteers to verify their legal position under these conditions.

SEXUAL HARASSMENT

In 1974 the EEOC issued rules that made sexual harassment unlawful. It defined sexual harassment as harassment on the basis of sex and a violation of the law. Unwelcome sexual advances, requests for sexual favors, and other verbal or physical conduct of a sexual nature constitute sexual harassment when:

- submission to such conduct is made . . . a condition of employment;
- submission/rejection of such conduct is the basis for employment decisions; and/or
- such conduct affects performance . . . or creates a hostile work environment.

Is sexual harassment a problem for the fire service? A study conducted in 1990 surveyed female firefighters across the United States Females compose approximately 1 percent of the nation's fire service. The study found 57.0 percent of the females reported having been sexually harassed. Of these, 41.0 percent had reported the incident to their supervisors; and 12.5 percent had received no reply to their reports, whereas 35.4 percent saw the perpetrators disciplined. Remarkably, 8.4 percent of the female complainants were disciplined; 10.4 percent of the complainants were told to ignore the incidents; and 37.5 percent said their incidents were handled in another manner. The results indicate sexual harassment is a problem in the fire service as well as in the workforce at large in the United States.

What are the underlying causes of sexual harassment? First, the work environment is changing; it now has more diversity, including women. Second, the role of women in the workplace is changing. Women now can ascend to supervisory and managerial levels, areas previously denied to them. Third, only now is the problem of gender violence and negative self-perception surfacing. Anger and violence between men and women likely were the emerging topics of the 1990s. Women are overcoming the compulsion to see themselves as below the worth of men. In many cases, this stigma came as a result of broken homes or inappropriate or absent role models. However, as they move ahead, women are achieving for themselves and their organizations in ways never before seen. The fourth area of sexual harassment is that of workplace control. Men who view their declining power roles in relation to women will use harassment as a means of controlling women, in the hope of maintaining their organizational positioning.

Types of Sexual Harassment There are two types of sexual harassment, *quid pro quo* and hostile environment. *Quid pro quo* is a Latin term meaning "the granting of a benefit in return for favor." It is the "you scratch my back and I'll scratch yours" principle. It was the *quid pro quo* charge of sexual harassment that Anita Hill brought against Supreme Court nominee Clarence Thomas. Hostile environment is defined as conditions that cause the employee's performance to suffer because of sexually related pressures (language, pictures, jokes, comments, etc.).

Supervisor's Responsibility The supervisor's responsibility regarding sexual harassment incidents is not that complex. It is fairly easy to detect sexual harassment. The supervisor can often rely on the "gut feeling" that the incident is wrong. In fact, the gut feeling is usually one of the best indicators that the supervisor should take some action to stop the behavior or correct it. Nonetheless, supervisors are expected to administer department policies with vigilance. When a complaint is made, the supervisor must act promptly and decisively. Once action is determined, the supervisor is expected to document the incident and the actions taken. Anything short of prompt and decisive action by a supervisor in sexual harassment complaints places the organization and the supervisor at legal risk.

Organizational Liability Organizational liability varies with the type of sexual harassment. *Quid pro quo* cases tend to subject the organization to strict liability. In such cases, the supervisor is viewed legally as the employer, even though he or she is only a supervisor. In hostile environment cases, the courts typically want to know whether the supervisor knew of the hostile conditions and acted to correct them. This "knew or should have known" rule is used frequently in court cases.

Court Cases on Sexual Harassment Some prominent court cases have formed the basis by which courts view and rule on sexual harassment. A 1986 case, *Meritor Savings Bank v. Vinson*, set a standard for determining a *quid pro quo* situation. Ms. Vinson worked for the Meritor Savings Bank. She was asked to dinner by her supervisor to discuss business. During dinner the supervisor suggested that Ms. Vinson accompany him to a motel for sexual intercourse. Ms. Vinson refused but, after additional prompting, she then agreed to go along. Subsequently, Ms. Vinson had sexual intercourse with her supervisor over 40 times in four years. Finally, Ms. Vinson was terminated for violation of sick leave policies. She filed suit against the bank and her supervisor and won. The court stated that Ms. Vinson's initial refusal to enter into a sexual relationship was indicative of her opposition to the relationship. The answer "no" meant no! Ms. Vinson felt compelled to agree to the relationship because of the continued pressures brought to bear by her supervisor.

In California another case involved a hostile work environment—the existence of nude pictures. The court determined the standard by which the environment would be judged was "what a reasonable woman would think about the conditions of the workplace." Thus, if a reasonable woman would have found the workplace objectionable, then a hostile environment would be said to exist. This "reasonable woman" rule led to the notion that you should not say or do anything at work that you would not say or do with your mother present (*Ellison v. Brady, 1991*). Another reasonable woman case was decided by a Florida District Court in 1991. Ms. Robinson worked for the Jacksonville Shipyards. In her job she was exposed to nude pictures of women as well as sexual comments by her predominantly male coworkers. She complained, but the problems were never corrected. She eventually filed suit and won (*Robinson v. Jacksonville Shipyards, 1991*).

What does all this mean for the modern fire officer? Supervisors have a strict responsibility to maintain a work environment free from discriminatory behavior. Should this behavior occur, the supervisor must act quickly and thoroughly to correct the problem.

USE COMMONSENSE SUPERVISION TO DETECT DISCRIMINATION

Detecting sexual harassment or any form of illegal discrimination in the workplace is not a difficult task. The supervisor can use some very commonsense approaches as a guide. First, listen to the gut feeling that tells you the situation is odd. Second, ask yourself, "Would I be comfortable with this situation if my mother were present?"

Third, ask, "Would I like these actions if they were directed at me?" Finally, ask, "Would I like it if this situation were printed on the front page of the newspaper?" The answers to these questions can serve well in determining whether illegal discrimination in any form is occurring.

THE AMERICANS WITH DISABILITIES ACT

The Americans with Disabilities Act (ADA) of 1990 provides civil rights protection to persons with disabilities and guarantees equal opportunity in employment, public accommodations, transportation, government services, and telecommunications. There are five sections of the act.

1. Employment regulations apply to employers of more than 15 employees. The act prohibits pre-hiring inquiries in the following areas: any disability (visual, speech, hearing), diseases (epilepsy, muscular dystrophy, multiple sclerosis, AIDS, cancer, heart disease, diabetes, mental retardation, high blood pressure, and others), glasses or contact lenses, color blindness, prescriptions or medications, and treatment for substance abuse or smoking. Job offers must be made prior to testing or inquiry about any disability.
2. Public sector regulations primarily related to transportation.
3. Public accommodations requiring accessibility for all people.
4. Telecommunication services to ensure accommodation for the hearing and speech impaired.
5. Miscellaneous including medication reliance and disabilities.

The issue of AIDS/HIV protection usually is contained within the ADA. Recent court cases have treated the AIDS issue as one of illness; therefore, a disability. For the fire service, the ADA means reasonable accommodation must be made to assist qualified employees with disabilities. It also means a person does not have to be hired or retained in employment if his or her condition presents a direct threat, risk, or harm to the public or other employees. However, such rejection must be based upon job-related criteria that can be clearly documented and sustained under legal scrutiny. Clearly, rejection of an employee cannot be related to the fact that the accommodation merely costs the department more money.

LOCAL AND STATE ANTIDISCRIMINATION LAWS

Local and state laws generally have replicated those of the federal government. The passage of these laws has assured virtually universal protection of civil rights to all individuals, thus overcoming the shortfalls of federal legislation, which could have limits. As mentioned previously, civil rights coverage to volunteer firefighters may not exist if the state does not classify them as employees. Volunteers should check their state laws to determine their rights. Sexual preference rights have not yet been extended by federal legislation. Gay and lesbian issues are certainly popular topics, but the only rights extended in this area are to be found at the state or local level. The possibility exists that the federal government could extend sexual preference rights under civil rights laws within the next decade.

THE AGE DISCRIMINATION ACT

The Age Discrimination Act (1975) provides that no person shall, on the basis of age, be excluded from participation in, be denied benefits of, or be subjected to discrimination under any program or activity. The act essentially extends civil rights benefits to age. The act and subsequent rules issued from it effectively eliminated age-mandatory retirement policies. Employers, particularly the fire service, still can require retirement at a specific age but only under proof generated through job-related studies or job analysis. The job studies must be completed for each position in the organization for which a mandatory retirement age is required. Such studies are expensive and open to challenge in the courts. Thus, many fire departments have totally removed the mandatory retirement age from their policies.

IMPACTS AND CHALLENGES TO THE FIRE SERVICE

A diverse workforce is a fact of the twenty-first century. The challenge for the fire service is to train employees to accept diversity and work together to realize the benefits each of us brings to the organization. Experience has shown diverse people can learn to work and live with one another. However, such arrangements do not come without some effort. One important way to make diversity work is by developing and enforcing organizational policies that support diversity. Coincidentally, it is the need to relate to one another that can result in the benefits derived from a multicultural workforce.

A small effort to accept the differences in culture can result in compassion and acceptance of people who are different. With acceptance comes trust in the workplace, enabling concentration on the development of life-safety objectives, which are responsive to the whole community served. All you have to do is look at the faces of the people on the street as fire trucks pass by in the performance of their duties. The young girls on the street stare with excitement when they see a female firefighter. Young Black children see hope in their ambitions when they see a Black firefighter. When that Hispanic firefighter rolls by on that huge piece of apparatus, young Hispanic boys and girls realize their future is within their control. When young Asian children see an Asian firefighter, it opens the door for them to realize the fire service is open to them as a profession.

Acceptance of diversity will do much to generate the trust and acceptance needed in the modern fire service. It is diversity that will enlighten the fire service to the needs and expectations of the public it serves. It is diversity that will contribute to creative ways of problem solving and the formulation of strategies to overcome community life-safety problems. Without it, we cannot form empowered teams capable of service delivery in a dynamic environment that demands total quality.

HUMAN RELATIONS IN THE FIRE SERVICE

The fire service of the twenty-first century must be built upon a strong foundation of human relations that can manage the benefits diversity brings to the department and forms the building blocks for the future. First, the fire service should embrace the benefits of building diversity. Officers should be educated to appreciate these benefits as well as trained to facilitate the contributions of all fire department employees. It is hoped fire officers will be able to promote the positive gains found through differing viewpoints, differing perspectives of culture, and the opportunities presented by diverse work teams. Interpersonal skills can be strengthened by the process of diversity. Different people and different points of views blend together to make an organization greater than the sum of its parts: a synergistic organization through diversity.

Second, fire officers should look to those fire departments that are making positive strides in incorporating diversity in their delivery of services. Typically, large metropolitan fire departments are composed of the most diverse workforces in the country. Using the International Association of Fire Chiefs (IAFC), the Center for Public Safety Excellence (CPSE), and the International Association of Fire Fighters (IAFF), computer networks are often a good source to find answers and comparative data about fire departments. Some benchmark programs can be found on the cutting edge of multicultural efforts to promote representative community services. The process of benchmarking diversity holds a great deal of promise for those departments wishing to explore it.

Third, fire departments should plan for diversity, review current policies, look at demographic trends, and project workforce representation for the next 10 to 20 years. Fire department leaders should ensure entry-level and promotional selection procedures are in compliance with the Federal Uniform Guidelines on Employee Selection issued by the Department of Labor. Compliance with the guidelines assures selection procedures are fair and nondiscriminatory.

Fourth, the fire service must adopt policies that specify the intent of the employer and the philosophy of the employer's commitment to a diverse workforce. Fire officers must receive thorough training on the new policies; these policies must be strictly enforced; and violators must be promptly and firmly disciplined.

Finally, diverse communities must be treated equally. How many times have firefighters been part of a response to low-income homes and witnessed the property of those who live there being treated as though it were not valuable? Emergency responders must remain conscious that all people served represent organizational constituency—the citizen customer. All property and people must be treated in the same manner as the firefighters would wish to be treated if they were in a similar situation.

THE FIRE SERVICE MOSAIC

The United States is no longer a melting pot in which people are expected to unilaterally adopt a single approach to living, to speaking, to thinking, or to acting. It is a collection of diverse cultures, each contributing a unique view of life, filled with values and experiences, to what becomes a new and exciting way of life. Diversity in the fire service contributes added value to the delivery of services to the public. True concern for citizens arises from the shared views of employees who represent a cross section of the community, a macrocosm.

Legal mandates stimulate the acceptance of diversity. Civil rights are protected and equal opportunity is ensured. It is incumbent upon officers to make certain the law is obeyed. When speaking of diversity, many automatically think of race or religion. For the fire service, there are other important aspects to diversity—experience and knowledge. Many departments today have very specific entry-level requirements. Applicants must be a firefighter I, EMT, and sometimes a paramedic, and the result is recruitment from a very small segment of society. The same colleges, same skill sets, and much the same job experience are found in the pool of applicants. As diversity is important in respect to race, religion, and culture, it is also in the diversity of employee knowledge and experience base. If recruitment is from only a small segment of society, then the opportunity to gain from the knowledge and experience of a diverse workforce will be lost because of this limited recruiting. In such cases the organization will never see the benefit of this diversity of knowledge and experience, which may have had a very positive impact on the long-term health of the organization.

The future success of the fire service is dependent upon the ability to accept the fact that the diverse workforce possesses the talent to improve the level of services collectively; it can meet the challenges of an ever-demanding public. Together, a diverse workforce can exemplify the true multicultural unity of Americans.

At the President's Conference on Fire Prevention in May 1947, President Harry S. Truman wrote,

> The serious losses in life and property resulting annually from fires cause me deep concern. I am sure that such unnecessary waste can be reduced. The substantial progress made in the science of fire prevention and fire protection in this country during the past forty years convinces me that the means are available for limiting this unnecessary destruction.

President Truman's perspective is as valid today as it was then. Much of that committee's report focused on the three E's of safety—*engineering*, *enforcement*, and *education*. The significance of the report is that it was the first time the federal government focused on and recommended a more coordinated approach to fire protection in the United States.

Twenty-six years later, the National Commission on Fire Prevention and Control released in 1973 the report *America Burning*. This truly landmark report has set the course of what many of us have experienced in the American fire service. The report focused on the nation's fire problems and noted the following needs: (1) for more of an emphasis on fire prevention; (2) for a better trained and educated fire service; (3) to educate the American public in fire safety; (4) to understand that the designs and materials in the environment in which Americans live and work present unnecessary hazards; (5) to improve fire protection of building features; and (6) to recognize the importance of research. To encourage solutions to these problems, over 90 recommendations and important tasks were outlined for the proposed National Fire Prevention and Control Administration, which in 1978 became the U.S. Fire Administration. Those recommendations included:

- Developing a comprehensive national fire data system to help establish priorities for research and action
- Monitoring this research in both the public and private sectors to assist in the interchange of information
- Providing grants to states, allowing local government to develop comprehensive fire protection plans
- Establishing a national fire academy for the advancement of fire service education
- Undertaking a major effort to educate Americans in fire safety

In June 2002, *America at Risk, America Burning Recommissioned* was released. The report states that to a great extent the fire problem in America remains as severe as it was 30 years ago. If progress is measured in terms of loss of life, then the progress in addressing the problem, which began with the first *America Burning* report in 1973, has come to a virtual standstill. The "indifference with which Americans confront the subject," which the 1973 Commission found so striking, continues today. Yet fire departments currently face expanded responsibilities and broader assignments than traditional structural fire response and suppression.

In that 2002 report the Commission reached two major conclusions: (1) The frequency and severity of fires in America do not result from a lack of knowledge of the causes, means of prevention, or methods of suppression. We have a fire "problem" because our nation has failed to apply and fund known loss reduction strategies adequately. Had past recommendations of *America Burning* and subsequent reports been implemented, there would have been no need for this Commission. Unless those recommendations and the ones that follow are funded and implemented, the Commission's efforts will have been an exercise in futility. (2) The responsibilities of today's fire departments extend well beyond the traditional fire hazard. The fire service is the primary responder to almost all local hazards, protecting a community's commercial as well as human assets, and firehouses are the closest connection government has to disaster-threatened neighborhoods. Firefighters, who frequently expose themselves to unnecessary risk, and the communities they serve would all benefit if there were the same dedication to the avoidance of loss from fires and other hazards existing in the conduct of fire suppression and rescue operations.

These two reports, while reflective of the time in which they were written, offer common themes that are insightful about our challenges for the future.

1. *Cultural orientation.* Understand that the most critical factors to be considered are those of public attitude, behavior, and values contributing to America's high fire loss. A high level of safety cannot be achieved unless the views and attitudes of the American public are changed.
2. *Political action infrastructure.* There must be an organized and coordinated capability that, at the national level, can identify overall problems, establish priorities, and make things happen. Unlike many other social and economic issues that have coordinated national representation (lobbying and public relations programs), fire safety often receives extensive public attention only following a major incident.
3. *Development of new and improved fire protection technology.* While knowledge is needed in the physics and chemistry of combustion, so is continued research on a new generation of affordable smart detection and fire suppression systems.
4. *Fire protection information.* The continuous and complete data collection analysis to identify fire protection problem solutions has been and still is a high priority. Unfortunately, the data collection analysis efforts being conducted within the fire community are in disarray and not achieving the intended objectives.
5. *Redefine the traditional public safety service delivery system.* Roles and responsibilities of the fire service will continue to evolve to meet the changing local government environment and an expanded mission.

While much has changed in the fire service over the past 50 years, our industry has repeatedly identified those issues that will impact the citizens we serve. The question we must ask ourselves is, "Will we be addressing these very same issues 50 years from now?"[20]

PRIDE, SERVICE, COMMITMENT[21]

The evolution of the fire service has been driven not only by events but also by the leaders of the fire service who have established a vision and a path for the fire service of the future. Many in the fire service today have little understanding of past experiences, traditions, and history but yet continue to push to evolve to a higher level of professionalism. The blending of traditions, history, and heritage (with the new level of professionalism and expertise) is required of those in positions of authority today.

> Without belittling the courage with which men have died, we should not forget those acts of courage with which men . . . have lived. The courage of life is often a less dramatic spectacle than the courage of a final moment, but it is

> no less a magnificent mixture of triumph and tragedy. A man does what he must—in spite of personal consequences, in spite of obstacles and dangers and pressures—and that is the basis of all human morality. . . .
>
> In whatever arena of life one may meet the challenge of courage, whatever may be the sacrifices he faces if he follows his conscience—the loss of his friends, his fortune, his contentment, even the esteem of his fellow men—each man must decide for himself the course he will follow.
>
> *John F. Kennedy*
> *JFK Library and Museum*
> *www.jfklibrary.org*

The North American fire service that originated in local communities is based on the tradition of neighbor protecting neighbor. With its roots in the colonial era, the image of the fire service recalls the bucket brigade passing pails of water from person to person toward the burning home. From these humble beginnings, the fire service has continued to evolve into a unique mix of highly trained men and women, both career and volunteer, who respond daily to a wide variety of hazards.

Today, people hear and read a lot about what *professionalism* is. In the fire service, a great deal of time is spent discussing what defines professionalism and how it can be achieved. Those who take the oath to become a firefighter take a pledge of service over self. For those who become fire officers, the badge they wear to lead and manage is a symbol that has become part of the fire service tradition. This tradition is linked to a wonderful heritage of bravery, loyalty, honor, ability, and devotion to duty resulting from a history of heroic deeds. The uniforms, the badges, and the insignia of the fire service symbolize this tradition. People have made use of badges and insignia as emblems and symbols through experience, extending from antiquity through the Middle Ages to the present day, to make known the ideals of a service and to instill loyalty to the organization and pride in its accomplishments.

Most are familiar to some degree with the uniforms worn by the men and women of the United States military forces: Army, Navy, Air Force, Coast Guard and Marine Corps. Those who have the privilege of wearing the uniform are often called upon to risk their lives for the principles of freedom. Upon these uniforms are worn the badges and insignia of the division or service, symbolizing within their compass the traditions of the particular unit or regiment. The emblems are reflective of their tradition and patriotism, and are symbolic of heroic acts of past members. It is within such symbols that can be found spirit and loyalty, the *esprit de corps*. Just as on the uniforms of the military, badges, insignias, and emblems that symbolize the historic traditions and background of the fire service are displayed on the uniforms of the fire service of local governments.

Many of the emblems most frequently used are not understood; therefore, it is important to trace their symbolism and significance. A *symbol* is a visible sign (badge, emblem, figure) representing a quality or an object. The flag of the United States not only is emblematic of this country but also symbolizes the principles upon which the country was founded. The wedding ring, an unbroken circle, is a token of love and fidelity. The scales or balances represent justice. The olive branch is symbolic of peace, the palm leaf of victory, and the laurel wreath of excellence. None of these qualities can be pictured or photographed, so the symbol or emblem representing each becomes a visual picture of the indicated characteristic.[21]

The badges and insignia of the fire service are examples of a rich and historic symbolism, deeply rooted in history. The cross displayed on many fire service symbols, known as the "Maltese Cross," was first worn by the Christian knights who shielded the weak. It is reflective of the modern fire service and has become a symbol of the protection of life itself. The badges that firefighters wear today are an outgrowth of the shields and coats of arms worn by the Christian knights to distinguish them as friend or foe in battle. The emblem of the fire service is taken from the cross of the Knights of Saint John of Jerusalem, founded as a charitable nonmilitary organization, that existed during the eleventh and twelfth centuries. A white and silver cross on a dark background was adopted by these knights, or "hospitallers," as they were also known because of their work in setting up hospices and hospitals for the sick and the poor.

THE INTRODUCTION OF THE MALTESE CROSS INTO THE ORDER

In the mid-1500s when the knights were at Malta, the familiar design now known as the Maltese Cross made its appearance. The first evidence of the modern Maltese Cross appeared on the 2 Tari and 4 Tari copper coins of 1557 to 1568.

It is not surprising that the symbols of the fire service can be linked back to the Knights of Saint John. Saint John of Jerusalem Order of the Hospital was a religious and military order of hospitallers founded in Jerusalem in the Middle Ages. This Order continues today in its humanitarian tasks in many parts of the world under several slightly different names. The origin of the hospitallers was in eleventh-century Jerusalem in the church of Saint John the Baptist, founded by Italian merchants from Amalfi to care for sick pilgrims. After the crusaders conquered Jerusalem in 1099, the hospital superior, a monk named Gerard, started his work in Jerusalem and founded hospitallers in a hospice called Provencal, an Italian city. The hospital soon began to evolve and took on entirely new responsibilities, transcending its original mission and character, and providing outreach to those who had been injured in battle. It was these great grateful crusader knights, after healing of their wounded in the hospital, who bestowed portions of their estates to the Order, while others remained in the Holy Land as members of the hospital. The Order of Saint John soon became a very wealthy and powerful body dedicated to combining the task of attending the sick and the poor while waging war on Islam in the Middle East.

As the Maltese Cross is used in much of the symbolism and insignia of the fire service today, there is evidence of the use of the cross and similar symbols in the oldest civilizations. The emblem can be seen in a picture represented by a knight of the Order of the Hospital (see Figure 1.4).

The knights, many of whom had a certain flair, often dressed in vivid colors to dramatize their presence and wore large capes in addition to their suits of armor. The capes were made of crimson-colored cloth on the inside, reflective of the blood of Christ, and black on the outside, representing the knights' sacrifice for humanity. These capes came to play a critical role on the battlefield, and just as the Knights of Saint John were engaged to fight many battles, so are the men and women in the fire service as they go forth to face one of the most relentless enemies humankind has ever known, fire. The fire service linkage to the Knights of Saint John is not by mistake nor by design. It may be due to the similarity of the shared ideals and the traditions of the fire service and of the Knights of Saint John, helping those in need.

The emblems embossed on the shields of the knights became crucial markings in battle, as the armor they wore rendered them otherwise unrecognizable. The Maltese Cross (see Figure 1.5) provided an excellent means of identification on the battlefield. Today, the symbols that adorn fire service uniforms are in part symbols of protection and badges of honor, whose essence has been defined over the past thousand years.

Hospitaller Knight

FIGURE 1.4 ◆ Hospitaller Knight

The Knights of Saint John (see Figure 1.6) encountered a new weapon, simple but horrible, unknown to European warriors at the time—fire. As they advanced on the walls of a city or fortress, glass bombs containing naphtha were tossed among them. Many of the crusaders were struck, becoming saturated with the volatile liquid. The breaking of the glass bombs signaled the hurling of a flaming tree or some other burning object into their midst. As a result, hundreds of knights were burned alive. At the same time, the knights' capes became saturated with this liquid, so many

FIGURE 1.5 ◆ Maltese Cross *(Courtesy of Randy Bruegman,* The Chief Officer: A Symbol Is a Promise, *p. 4)*

FIGURE 1.6 ◆ Knight of Saint John *(Courtesy of Randy Bruegman,* The Chief Officer: A Symbol Is a Promise, *p. 5)*

had to become firefighters just to survive. Others tore off their capes and threw them over their burning comrades. With the use of their capes, the Knights of Saint John risked their lives to save their brother combatants from dying a fiery death. In essence, these men became the first in a long list of courageous firefighters. Their heroic efforts were recognized and rewarded by their fellow crusaders; each was presented with a badge of honor—a cross similar to the one firefighters and chief fire officers wear today.

In recognition, the Maltese Cross has become a medal of honor and a symbol of courage, and forever linked to the fire service. The Maltese Cross has come to represent the ideal of the fire service: the commitment to saving lives and property. Today, this symbol represents all that firefighters do to protect their communities and their commitment to provide for the safety of the people who live and work within the jurisdictions they serve.

The brass trumpets that adorn a fire officer's collar brass or badge are symbolic as well. Their origin was first documented in 1752, when six speaking trumpets were purchased for the New York City volunteer fire department to enhance the ability of fire officers to communicate on the fire ground. The first trumpets were nothing more than megaphones constructed of tin, but soon other materials such as copper, steel, and brass were utilized. Each volunteer company enhanced its trumpet with insignia, silver plating, and markings to denote its identity. In the 1800s, chief fire officers utilized the trumpet to amplify their orders given on the fire ground. Today, the trumpet symbolizes the rank, authority, and responsibility of one who has been promoted in the fire service. Out of the necessity to communicate on the fire ground, a long tradition of fire service identification, recognition, and tradition began. The multiple crossed images of the speaking trumpet are forever linked to the ranks of chief fire officers. The trumpet (see Figure 1.7) not only is a symbolic remembrance of those who have gone before us but also is symbolic of the roles and responsibilities chief fire officers and those in leadership roles fulfill in performing their day-to-day responsibilities.

OTHER INSIGNIA OF THE FIRE SERVICE

According to the history of heraldry, a shield upon a coat of arms implies a "defender." In many cases, the fire service badges are in the form of a shield containing the emblem of the Maltese Cross. Just as the shield carried by the knights was protective armor, serving to guard the wearer from the swords and arrows of the enemy, the fire service is a protective armor for those in need of protection from fire and other hazards.

Border Among the decorations usually found upon a fire badge is a rope, or cord, forming a border. The rope or cord border is unbroken, emblematic of unbroken service and loyalty. The crossed arms, axes, ladders, and so on, of the crusaders in the

FIGURE 1.7 ◆ Trumpet
(*Source: Fresno Fire Department*)

form of the "Saltier" or "St. Andrew's Cross" are symbolic of resolution and a reward of such for those who have scaled the walls of cities—a decoration reserved for the "bravest." The crossed scaling ladders denote one who is fearless in attacking, meaning the "extremely brave." Hence, to the present day found upon uniforms of the military are crossed swords, crossed guns, and the like, and upon the badges of ranking officers are the speaking trumpets.

Phoenix The emblem of the phoenix (not to be confused with the eagle), found upon helmets or fire hats, badges, flags, and seals of fire departments, was used by the early Christians as a symbol of the resurrection of Christ. To them it symbolized the pledge of eternal life. It was also used as a sign of friendship, a secret sign of recognition by which each Christian knew the other. Legend has it that this bird, about the size of an eagle, lived at the time of Adam and Eve and did not eat of the forbidden fruit, as all other birds did. As a reward for its loyalty, the phoenix will live forever. When this bird senses old age approaching, it builds a nest of twigs, setting it afire with friction caused by the flapping of its wings. During the fire it is renewed in youth and strength, rising from the ashes more beautiful than ever before. The head of the phoenix is finely crested with beautiful golden plumage; its neck is adorned with golden-colored feathers; its tail is white; and its body is purple or crimson.

Eagle The eagle, majestic in flight and dauntless in courage, is admired by humankind. The eagle is usually displayed surmounting the shield, with wings extended, signifying that the wearer is a person of authority. The extended wings, like the extension of a cross, signify protection and charity; the beak and the exposed talons represent authority and protection.

Colors Colors, too, were used as symbols of certain qualities in heraldry. For example, the Knights of Saint John used a silver cross on a dark background; the Knight Templars, a red cross on a white background; and the Teutonic Knights, a black cross on a white background. Overall, blue signifies sincerity of purpose, loyalty, and truth; gold or yellow stands for generosity and elevation of mind; white or silver signifies peace and sincerity; red represents magnanimity and nobleness; and black signifies grief.

When these various devices and colors are assembled with regard to the fire service, they represent the following:

- blue of the uniform for sincerity of purpose, loyalty, and truth;
- Maltese Cross of white upon a dark background (blue), for protective service and comradeship, symbolizing the ideal of the service;
- border of the badge or crest of unbroken cord for the unending service and loyalty to the citizens served;
- phoenix emblematic of ever-renewed life and service;
- eagle emblem of authority and protection;
- badge, crest, and front piece of the helmet in the form of a shield with gold and white indicating a chief; white and black, an officer below the rank of chief; black with white lettering, a firefighter; and
- trumpets upon the badge denoting a particular rank: one silver trumpet, a lieutenant; two silver trumpets, a captain; two gold trumpets, a battalion chief, and so on, to the five crossed trumpets of the chief engineer.

Too frequently, firefighters being part of the picture lose sight of the true meaning of the fire service. To them, the unremitting labor, devotion to service, and daily acts of heroism, spectacular as they are to the general public, become matter of fact and ordinary. These men and women in every community from the largest metropolitan areas to the smallest villages, by their spirit, tradition, and commitment, protect their communities on a daily basis.

Many members of the fire service have been called upon in the past to make the supreme sacrifice in the public's interest. Loyal to the ideal of the service, they gave their lives in the fulfillment of their duty. They had learned there was something more to life than merely living, and that was "duty and service to their fellow men." The fraternity of firefighters is a comradeship of long traditions, cherished not only because of its skill but also because of its enduring spirit of loyal service. Men and women who have faced death and peril together have consummated a peculiar and sacred comradeship. No man or woman can have a finer fellowship than association with such men and women as they band together in the combat of fire. This is the tradition upon which has formed the organization of the fire service, unique and unshared by any other group as symbolized by the history and symbols of the fire service.[22]

The evolution of the U.S. fire service is reminiscent of its history: the colonial era, the Civil War, the Industrial Revolution, the post–World War I and II eras, and the technology revolution. The heritage and traditions of this service have been instrumental in shaping the fire service of today. Many of the Founding Fathers of the United States were volunteer firefighters, including George Washington, Thomas Jefferson, Benjamin Franklin, Samuel Adams, Paul Revere, Alexander Hamilton, John Hancock, John Jay, and Aaron Burr. As previously discussed, Benjamin Franklin formed one of the first volunteer fire companies, the Union Fire Company, in 1736 in Philadelphia, and made many notable contributions to fire protection during his lifetime (see Figure 1.8).

Today's public perception of the fire service is molded less on fire service history and its rich traditions or on the reality of what occurs from day to day, and more on what people see on television as the depiction of the fire industry. On television programs such as *Rescue 9-1-1*, after someone dials 9-1-1, everything seems to work like clockwork. People respond quickly and arrive just in time to produce a positive outcome in every situation. Movies such as *Backdraft* and *Ladder 49* capture some of the aspects of the fire service, helping to create a public awareness of what the fire service is about. *Third Watch* or *Rescue Me* are gritty shows of how it is on the streets for firefighters and paramedics in New York City. Those perceptions clashed with harsh reality on September 11, 2001 (see Figure 1.9).

For many reasons, the fire service will never be the same. The loss of so many firefighters in one incident, the courage they displayed, and their commitment to duty have changed the nation's perception of the fire service. The responsibilities and roles of the fire service leadership have changed as well. There have been in the past and will be in the future defining moments in the fire service that help to create and frame what the industry is today and what it will become. Since September 11, a new day has dawned for this profession, related not only to the challenges that the fire service faces but also to the responsibilities of those who will lead it. As never before, the fire service requires leaders who can articulate the needs of this industry at the local, state, federal, and international levels. Each day that passes is a day away from the tragedy that helped to redefine the fire service forever. Those who are leaders today, those considering leadership roles in the future, as well as the 30-year firefighters can never become complacent about what is necessary when they accept the badge of a firefighter and the responsibility that comes with it.

FIGURE 1.8 ◆ Bucket Brigade
(*Source: Randy Bruegman,* The Chief Officer: A Symbol Is a Promise, *p. 7*)

duty and service
While the way in which we respond to emergencies has changed over time, the commitment to service has not.

FIGURE 1.9 ◆ The World Trade Center, New York City *(Courtesy Randy Bruegman,* The Chief Officer: A Symbol Is a Promise, *p. 9)*

ON LEADERSHIP

Many chief officers and captains grow complacent and lose focus as they become mired in their day-to-day duties and operations.[23] Many company officers do little more than show up for work and hide out in the firehouse. Yet fire officers have the responsibility to anticipate the future needs of their organizations, to coach and mentor their personnel, and, above all, to be ready to do what the community and nation expect them to do! Equally important, fire officers must not mistake how business is conducted today for what will happen in the future.

> Don't mistake the edge of the rut for the horizon.
>
> *Anonymous*

This is an important statement. As firefighters focus on their daily challenges, the rut is often an easy and comfortable place to be. However, in the future the fire service will look very different than it does today, and those who have the ability to look beyond the horizon will be the ones who provide the architecture for the future.

The fire industry cannot continue to approach its issues and problems with the same reasoning used for the past two decades. By looking at the industry's mind-set, the frame of reference from which it operates, fire officers must begin to ask themselves whether they are limiting their ability to address the problems that most concern chief fire officers today. They must understand that by limiting their vision of what needs to happen, they also limit their ability to address critical problems and find the solutions to them. Leaders' frame of reference often dictates how they process information and thus determines what they believe is possible for the future. This belief determines how each fire officer will plan and move forward to implement the plan. Over the next several years, the fire service will need to create new frameworks and look at issues from fresh and varying perspectives. Those who are leaders must remove the blinders they often use and have the courage to move the fire industry into uncharted waters. What the fire service will become in the next 10 to 15 years depends on today's chief fire executives and the fire service leaders of today. What the fire service will become during the next 20 to 30 years depends on the young captains and firefighters in today's fire departments. Each group will be not only the architects of what is occurring today but also the bridge builders to what will be in the future.

Today, the opportunity exists to bridge the gap between the traditions that have made this profession what it is and what it will be in the future. The future will demand greater accountability and responsiveness, and a rethinking of how the fire service approaches its customers, the services it delivers, and how it delivers them. This will not be a simple task. It is often much easier to define the architecture than actually to construct it. Fire service leaders must address many significant issues yet are often overcome by the complexity of the task. As a result, instead of taking incremental steps to achieve small successes, leaders in the fire service have allowed the industry to avoid addressing the root causes of the problem. The significant questions for leaders in the fire service today are, "What is the vision of the future?" "How will the fire service get there?" "Who will provide the leadership to achieve it?" Many chief fire officers do not want to upset the status quo, even if they are locked into dysfunctional systems. Whether it is an organizational culture of bureaucracy or a culture that perpetuates poor behavior, the leaders of tomorrow may have to change their own attitudes and approach to produce a meaningful change in the fire service. Leaders often know the direction that their industry must take, but orchestrating a major change is a momentous job, whether on a national level or in an individual department. From experience, it is impossible to do it alone.

The fire service of the twenty-first century is ready to build on the impressive accomplishments of the past 30 years. Its abilities have grown exponentially in that time, as have the demands for its services. The fire service no longer is simply a local resource. Meeting the challenges of the future will take strong leadership and commitment from the fire officers of today and tomorrow. If leaders are to position the fire service to take full advantage of future opportunities, then they must be good stewards of the public resources given to them. This begins with a clear understanding of the history of the service. It is this rich history and tradition of the fire service, coupled with the constant changes in the industry, that will require leaders to meet future challenges with innovation, tempered with the realities of their own experiences.

THE TRUMPETS WORN

Trumpets that adorn the uniform to distinguish the rank are emblematic of the qualities it takes to be a good chief fire officer, a lieutenant captain, a good leader. As you contemplate taking this step in your own careers, or if you are already a chief fire officer, you may find yourself struggling with the day-to-day duties that you are called upon to carry out. On days such as these, it is helpful to consider the badge and the trumpets on your collar as reflecting the characteristics of successful fire officers—courage, integrity, work ethic, and empathy.

This book will explore many of the characteristics that must be developed to be a successful leader. If you are already a leader and wish to stay at the top of your game, what do you need to do to avoid becoming stagnant and ineffective? Today, the unfortunate reality is that many who wear the badge of a fire officer are not really leaders. There is also reluctance among young fire officers to take the next step and become a chief fire officer for a variety of reasons, and many who do not want to promote to be company officers are riding in the backseat.

So let's start by going back to the day of your graduation from the recruit academy, with your family and friends present to witness the pinning of your first badge. This is a proud moment not only for you as a graduate but also for your family. As you look to promote through the ranks and possibly to take on the role of a chief fire officer, this has to be done with an acceptance of the responsibility, an understanding of the long tradition of service, and reflection on the day you received your first badge. The *badge is a symbol* (see Figure 1.10) of several hundred years of heritage and obligation. Unfortunately, many people have lost sight of this or simply do not understand it. It is time to consider what the badge represents, understanding the rich history and the traditions that have made the fire service what it is today. Firefighters must understand why their commitment to serve is based on a history that extends back almost a thousand years and why, more than ever, the symbols worn are a promise to deliver.

THROUGH THE EYES OF A NATION

On the morning of April 19, 1995, an age of innocence was lost as a terrorist's bomb ripped through the Alfred P. Murrah Federal Building in Oklahoma City. As Americans focused on this horrible tragedy, the images they saw daily were of the lives lost

FIGURE 1.10 ◆ Firefighter Badge
(*Source: Fresno Fire Department*)

of family members who would be seen no more; grief; and the struggle to survive, to understand why, and to reflect. Out of this chaos emerged a sense of order, calmness, and resolve to get through it.

The Oklahoma City Fire Department was thrust onto the world stage, and its performance was superb. Within hours of the incident, factual, professionally delivered media reports were given that kept the public informed and made people feel part of the event. In the days and weeks that followed, urban search and rescue (USAR) teams and teams from other federal and state agencies arrived from all parts of the country. The high level of professionalism displayed by all was seen in the images broadcast of the rescues made and of those that were not (see Figure 1.11).

Technology has provided us with wondrous tools to look at places throughout the world and instantly become a part of life-changing events. With the flip of a television channel, the images of war, starvation, holocaust, tragedy, triumph, and victory are brought to us as they happen. With the ability to see events in real time, the public perception of the American fire service and fire services throughout the world has changed. Think back over the major incidents that have occurred within the past few years, when those in the fire service were critical of each other after watching events unfold on CNN and knew that things were not going according to plan. The question is, "Does the public share that same perception?" Often it does not. Fire companies have all been to a fire where they have created a parking lot by mistake. Firefighters have seen people die and thought if only they had done certain things, the outcome would have been different. Yet even when the building has burned to the ground, the owners or occupants still thank the firefighters for their heroic efforts, though they may not have done their best.

This is what impressed me about the Oklahoma City incident. It is where the world's perception met the reality of what first responders are called upon to do. As this event unfolded in the first hours after the explosion and in the following days as the public learned about the response of the Oklahoma City Fire Department and related agencies, it became apparent that the public's perception of their heroic efforts and the fire service's reality of a job well done were both well founded. In many situations we encounter within our own jurisdictions, this is not the case.

FIGURE 1.11 ◆ Oklahoma City Alfred P. Murrah Federal Building *(Courtesy of Randy Bruegman,* The Chief Officer: A Symbol Is a Promise, p. 14*)*

The images we are left with when a tragedy occurs can remain with us for a lifetime: the assassinations of John F. Kennedy and Martin Luther King, Jr., the explosion of the *Challenger*, and the loss of the *Columbia* upon reentry. Most of us know exactly where we were when these events took place. The same is true of the images that emerged from the Oklahoma City incident—the firefighter leading a frightened and injured occupant down an aerial ladder saying, "Look at me, lean on me, trust in me, I will get you down." These images repeated time and again throughout this tragedy will remain with the public for many years to come (see Figure 1.12).

This was never more evident than on September 11, 2001, a day of tragedy for the American people and a day of great loss for the American fire service. It is also a day that reflects the commitment firefighters must bring to the job each day when they put on their uniforms. The firefighters who responded in New York City, in Arlington, Virginia, and to the fields of Pennsylvania that day did so because they had taken an oath to serve.

The unfolding of the flag on the side of the Pentagon (see Figure 1.13) was that same display from a national perspective. The flag, which symbolizes shared values,

FIGURE 1.12 ◆ Oklahoma City Incident *(Courtesy of Randy Bruegman,* The Chief Officer: A Symbol Is a Promise, p. 15*)*

freedoms, and loyalty, was a promise to the nation that, while bent, it would not be broken. The same holds true for those in the fire service and is reflected in the symbols that adorn their uniforms.

When firefighters take their oath, they are given a badge symbolizing the fire service. This badge is also a pledge that when a crisis occurs, they will respond and do everything they can to bring order out of chaos. Every officer has another core mission: to protect his or her personnel. Every year in October at the National Fire Academy in Emittsburg, Maryland (see Figure 1.14), the fire service comes together to honor those firefighters who have died in the line of duty. Family, friends, and fire service personnel from throughout the United States gather to recognize, remember, and honor all of the firefighters, both career and volunteer, who were lost in the line of duty the previous year.

In 2001, due to the September 11 attacks, 442 were honored—343 from the World Trade Center attack and 99 from other tragic events that occurred that year. The numbers do not tell the story, but the families do. Those firefighters lost, their ages ranging from the teens to the sixties, from different backgrounds, different parts of the country, and different departments, shared the same sense of pride, commitment, and dedication that make the common theme of service over self apparent to all of us in the fire service. Different as they were, they were connected in life by this bond that links everyone in the fire service. In death, they were connected by their families, friends, and colleagues who gathered to grieve, to heal, to share their experiences, and to remember, laugh, cry, and move forward.

The gathering of America's fire service each year to pay tribute to the men and women who have made the ultimate sacrifice is a time of healing and support. The ceremony demonstrates to the families of those who died that we collectively share their pain and that they need not make this very difficult journey alone. It also refocuses our commitment to quality leadership and the understanding that for those who assume a fire officer's position, it is their time to lead. Part of that leadership involves making a commitment to improve the safety of firefighting personnel. Leaders of this industry must continue to take aggressive actions to stop the preventable accidents that lead to children without parents and families in pain. When we make a commitment to address this issue as leaders at the local and national levels, we provide a basis on which our profession will grow in the future.

FIGURE 1.13 ◆ Flag on Side of Pentagon *(Courtesy of Jocelyn Augustino/FEMA News Photo)*

FIGURE 1.14 ◆ National Fallen Firefighters Memorial *(Courtesy of Rick Kemenyas/FEMA News Photo)*

Something significant happened on September 11, 2001, in the fire service. Not only did the world see, perhaps for the first time, the fire service's commitment and dedication to service over self, the fire service once again had the opportunity to see itself, to understand just how deep the commitment to serve is and how important it is to our communities and to our nation. Since September 11, it has been interesting to observe the newfound respect for the fire service from elected offices and the general public, as well as how we in the fire service have come to perceive ourselves. For most of us it is not about a paycheck, the recognition, or the number of hours we volunteer; it is about service. In many respects, we have reconnected to that thousand years of history. We have refocused on our mission: the protection of life and property and the core value of placing service over self. This reflects back on our true beginnings—the Christian knights and the Knights of Saint John, who not only helped each other but also took up the call to serve others. This is what we do, and this is why it is so important we understand today, more than ever, that the symbols we wear are truly a commitment and a promise to deliver to those we serve[24] (see Figure 1.15).

ADVICE FROM THE EXPERTS

How important is tradition in the service?

MICHAEL CHIARAMONTE

Chief Fire Inspector, Retired, Lynbrook Fire Department (New York): Tradition has its place in the fire service, but it should not dominate our service. Although tradition bonds us together, it should not cloud change and progress.

FIGURE 1.15 ◆ Chief Officer Badge *(Courtesy of Randy R. Bruegman)*

BETT CLARK

Former Fire Chief, Bernalillo County Fire and Rescue (New Mexico): I think too much emphasis is placed on tradition. We are proud of our history and our forefathers and mothers. However, we have a daily opportunity to create the future of every fire service department and member. If we still followed tradition, we would have no women in the fire service; no EMS component; no progress in technology for firefighting tools and equipment; and most certainly no codes or industry safety standards. It good to know why we are what we are; it is even more rewarding to be who we can be in the future.

KELVIN COCHRAN

Fire Chief, Atlanta Fire Department (Georgia): Traditions are essential to the furtherance of the rich culture and values that have transcended generational and geographical boundaries in the fire service. As the fire service evolved from the developmental days of Benjamin Franklin, so did fire service traditions—some good and some bad. The good traditions range from red fire trucks to badges with bugles (or trumpets, according to Ron Coleman), or from clean fire trucks and firehouses to eating dinner together. Those traditions that are not so good today or in the future include majority male organizations, hazing rookies, autocratic leadership culture, and the mentality that we exist only to fight fires. The list could go on. It is vitally important for fire service organizations and leaders to strengthen the institution, cementing those traditions that have made us great.

CLIFF JONES

Fire Chief, Tempe Fire Department (Arizona): Tradition is very important in and to the fire service. The key is to identify what traditions support the fire service and what traditions hurt it. Pride in apparatus appearance and reliability is an example of a supportive tradition, and hazing is an example of a tradition that hurts the fire service. Change is now coming about at such a rapid pace that managing the benefits of tradition is more difficult than it was when there was simply less going on. New employees are exposed to such a wide array of topics when they go through recruit training that indoctrinating them on the positive traditions of the fire service becomes more difficult. To continue traditions, and for them to continue to have a positive impact on image and service, traditions have to be managed by those entrusted to care for them, which includes the chief officers and company officers of each fire department. The lesson here is that it is critical that positive traditions be identified and that labor and management work together to make these traditions part of the fire department/fire service culture.

BILL KILLEN

Retied Fire Chief and IAFC President 2005–2006: Tradition is the tie that binds the fire service and reinforces the values of community and/or public service. I am not a subscriber to the adage, "The fire service is 250 years of tradition unimpeded by progress." I believe the average fire service member views his or her role as a career and/or volunteer and derives satisfaction from the job and its many traditions. Tradition is very important in that it reveals where we came from.

KEVIN KING

U.S. Marine Corps Fire Service: I believe tradition is very important in terms of understanding why the fire service exists and the sacrifices those before us have made to make the fire service better for us. In that manner, it is somewhat like history. Our traditions define us, give us a platform to build on, and hopefully allow us to learn from the past. However, traditions can become problematic if they prevent us from moving forward and from being willing to change. (These are opinions of Mr. King and do not reflect official policy of the Department of Defense or the United States Marine Corps.)

MARK LIGHT

Executive Director, International Association of Fire Chiefs: Tradition is important in learning why we exist and who we are. In many cases, tradition gives us a foundation of our history, and how we fought to move a profession forward. Unfortunately, many in the fire service see tradition as sacred, and this is often used as a reason not to change or be changed. It is important for the fire service to know our traditions, respect the past, take forward the positive, and shed the negative. By showing respect to tradition, we can acknowledge our predecessors, maintain the positive, and put the negative traditions into the history books.

LORI MOORE-MERRELL

Assistant to the General President, International Association of Fire Fighters: Tradition is very important in the fire service. However, having said this, it is necessary to note that it is a two-edged sword with both positive and negative attributes. On the positive side, maintaining tradition should support many characteristics of firefighters, including pride in the job, loyalty to each other, integrity, honor, and courage. Unfortunately, in the recent past these attributes are not being carried forward to new recruits, and firefighting (including all emergency response) has become just a job rather than a way of life. We must hold on to traditional characteristics that define "who" a firefighter is and should always be regardless of what is unfolding around him or her. Ours is a noble profession, and those who are a part of it should exhibit appropriate characteristics or find another field. On the negative side, maintaining tradition in some practices and procedures of the job is stifling to growth and necessary change. It may also contribute to unsafe practices being perpetuated when technology and science offer a better and safer way. For example, several years ago when Nomex hoods first hit the industry, many firefighters refused to wear them because they wanted to "feel the heat on their ears" as a gauge of when a fire was too hot. Fire service leaders must take the initiative to determine when tradition is on the positive or negative edge of the sword and sustain or eliminate it as is necessary or prudent.

JAY REARDON

Fire Chief, Retired, Northbrook Fire Department, (Illinois); Currently President of the Mutual Aid Box Alarm System (MABAS): Tradition is the needed element for defining our profession's qualities, values, pride, and reason for being. Tradition tells us what should remain important to us as our heritage. Tradition loses its value (its historical guidance) when it ignores societal change and the public trust we are given as the fire service tradition. When society's expectations and perceptions are violated because tradition is used as a crutch to defend stupidity, we cheapen the richness past firefighters graciously have given us. Tradition explains the past and should help us to see our duty for the future.

GARY W. SMITH

Retired Fire Chief, Aptos-La Salva Fire Protection District (Aptos, CA): Traditions are experiences that for one reason or another were so strongly felt that future generations, specifically those who lead and manage, decide to carry that same experience forward. After a while that "repeated experience" we call tradition may not work so well, but for one reason or another the practice becomes a part of the culture of the organization/profession. At that point, even the leaders and managers have a difficult time changing the tradition. So the lesson is that we must be careful about which practices we carry forward as traditions. A tradition of customer-friendly service is obviously a good thing for most fire departments, whereas a tradition of suppression-focused services, with no interest in any other services, may not be so good to continue. This would be a tradition that new leadership will have to work hard to change because the culture of the organization will embrace and strongly hold on to the suppression-focused operation.

STEVE WESTERMANN

Fire Chief, Central Jackson County Fire Protection District (Blue Springs, Missouri), and IAFC President 2007–2008: Ceremonial tradition can be a useful tool to remind us of where we as the fire service have been. But as our world gets smaller, operational tradition can be a hindrance to providing a quality, efficient service to our constituents. With the increasing demands placed on today's and tomorrow's fire service (economy, training demands, legal liability, firefighter safety), the fire service must understand that we have to find new methods or practices to become more efficient in the delivery of our services. If you are going to be bound to a practice because "that's the way we've always done it," you will fail.

THOMAS WIECZOREK

Executive Director, Center for Public Safety Excellence, and Retired City Manager: Fortunately or unfortunately, tradition seems to be a significant part of the fire service. Fortunate in that institutional memory is often developed that can be handed from one generation to another; unfortunate in that it is often difficult to introduce new concepts or ideas into the profession. I believe that the claim to tradition has hindered the fire service when compared to other functions such as police. Police are viewed as progressive even when they revert to methodology that was successfully used one hundred years ago (community policing). Too often the fire service is viewed as using the same models and concepts that we did one hundred years ago, often not so successfully. Unfortunately, this may often be true.

JANET WILMOTH

Editorial Director, *Fire Chief* Magazine: I believe tradition is important when it involves swearing in or promoting new personnel or burying firefighters. It appears the upcoming generations are less attracted by the formality and "old" ways, versus the fact that this is a good job and there are new ways to do the job.

FRED WINDISCH

Fire Chief, Ponderosa Volunteer Fire Department (Houston, Texas): On a scale of one to ten (high), I believe tradition is about a five. I base this belief on the necessity of understanding where we come from with the hope of not repeating past errors. Tradition is OK to know, but it certainly is not necessary to remain in a rut. After all, a rut is an uncovered grave.

Review Questions

1. Describe the *Corps of Vigiles* and how it was part of history and the fire service.
2. Ben Franklin convinced 11 of his friends to join him in forming a club called the Junto (pronounced "who-n'-toe"). What significance did it have for the United States?
3. Describe the influences two of the major fires in America had on the fire service and the culture in general.
4. How has the culture of the fire service changed since you were hired?
5. Research the demographic shifts within your community. What impact have they had on your department?
6. List and describe two cultural changes that have had an impact on the fire service.
7. Describe the impact of legal issues on the fire service.
8. What legislative mandates have been introduced since you were hired that forced a change in business practice?
9. Discuss the four tips in order to avoid a sexual harassment lawsuit.
10. Describe the potential impacts on the fire service in the twenty-first century.
11. How has your department promoted the hiring of diversity within the organization?
12. Describe how diversity is affecting and will affect the fire service.
13. Is your department reflective of the community it serves?
14. Describe the significance behind the Maltese Cross. Select one of the symbols in the fire service and describe its significance.
15. What insignia used by your department can be traced to the early origins of the fire service?
16. What was the motivating factor(s) for the design of your patch or department emblem?
17. Provide a history of your own department.
18. What has changed in your department since September 11, 2001?

References

1. *Age* (Norman, OK: University of Oklahoma Press, 1962), p. 116; Rodolfo Lanciani, *Ancient Rome* (New York: Houghton Mifflin, 1888), pp. 116–222.
2. G. V. Blackstone, *A History of the British Fire Service*, Rutledge and Kegan Paul, London, 1957.

3. Chief Michael L. Kuk, Saint Florian *American Fire Journal*, April 1999.
4. Ron Coleman, *The Vigiles: Roman Fire Brigade*, nonpublished, 2005, pp. 1–5.
5. *Columbia Encyclopedia*, 6th ed. Copyright © 2003–2005 Columbia University Press, NY.
6. Robert E. O'Bannon, *Building Department Administration* (Whittier, CA: International Conference of Building Officials, 1973), p. 9.
7. Unless otherwise indicated, the text under this heading draws on the following book: Paul Robet Lyons, *Fire in America!* (Boston: National Fire Protection Association, 1976), pp. 1–9. This material is used with the permission of the publisher: the National Fire Protection Association, Boston.
8. Paul Hashagen, "Firefighting in Colonial America," *Firehouse Magazine*, September 1998.
9. Paul C. Ditzel, *Fire Engines, Firefighters* (New York: Crown, 1976), p. 16.
10. Ibid., p. 18.
11. Unless otherwise indicated, the material under this heading draws on the following book: Ditzel, *Fire Engines, Firefighters*, pp. 18–19.
12. www.pbs.org/benfranklin/13_citizen_networker.html
13. www.juntosociety.com
14. Ditzel, *Fire Engines, Firefighters*, p. 29.
15. Ibid.
16. Federal Emergency Management Agency/ United States Fire Administration/National Fire Academy, *Managing in a Changing Environment, Social Impacts of the Fire Service*, Student Manual (September 1995).
17. U.S. Census Bureau, www.census.gov
18. Morris Massey, *Flashpoint: When Values Collide*, Video Marketing Resources, Inc., 1993.
19. W. B. Johnson and A. H. Packer, *Workforce 2000: Work and Workers for the 21st Century*, Indianapolis: Hudson Institute, 1987.
20. Randy R. Bruegman "America Burning," *Fire Chief*, September 2006.
21. Randy R. Bruegman, *The Chief Officer: A Symbol Is a Promise*, Chapter 1, pp 3–18, Pearson Prentice Hall, NJ.
22. Coleman, *Introduction to the History of the Fire Service*, pp. 5–16.
23. Bruegman, *The Chief Officer: A Symbol Is a Promise,* Chapter 1, pp 3–18, Pearson Prentice Hall, NJ.
24. Ibid.

CHAPTER 2

Preparing for Your Future

Key Terms

accreditation, p. 63
ACE, p. 55
certification, p. 45
CFAI, p. 53
CPC, p. 63
EFOP, p. 80
FESHE, p. 49
IAFC, p. 53
NFA, p. 47
NVFC, p. 59
TRADE, p. 48
USFA, p. 47

Objectives

After completing this chapter, you should be able to:

- Define the role of training and education for the fire service of the future.
- Describe how the National Fire Academy has influenced professional development in the fire service.
- Describe the role FESHE has had in professional development in the fire service.
- Describe the influence the Wingspread conferences have had on the fire service.
- Define the difference between certification and accreditation.
- Develop a personal career development plan.

FROM A BLUE-COLLAR JOB TO A PROFESSION

One of the most interesting transitions seen in the past 30 years in the fire service is the demand placed on fire personnel in respect to the skills and education level they need to acquire to be successful. Only a generation ago this profession was perceived simply as a blue-collar job. Don't get me wrong; there is nothing wrong with that. Yet what has occurred during the past 25 to 30 years is the evolution of the skill and knowledge needed to be effective as a firefighter and an officer. Today, fire service personnel are paramedics, hazardous materials technicians, urban search and rescue specialists, confined space rescue specialists, emergency operations managers, interoperability coordinators, and trained responders to weapons of mass destruction. Chiefs are called upon not only to lead multimillion-dollar organizations but also to be adept at political strategy, budget applications, and community risk analysis.

The fire captain, who once focused on an area's fire problem, is now asked to be an expert in multiple disciplines, with the content of each far exceeding that of his or her predecessor. Today's firefighter needs to multitask a variety of specialties, and within each the level of expected performance is much higher than ever before.

Today, from the fire chief executive to the firefighter, each must embrace a much larger community of interest than the fire service has historically recognized. Those interests now include a more sophisticated system of response, legal mandates, governance, and higher expectation of the community regarding the services provided. As such, fire service personnel have been challenged to learn complex zoning and land use issues, private sector business interests, and localized neighborhood safety concerns, and to become competent in local and national politics. This new era includes a vast array of regulatory mandates that necessitates coordination among law, health, planning, and public works interests. The fire chief executive must now employ the same planning tools and business logic used by the private sector executive to develop contracts and agreements. The who, what, why, where, and how of management must be developed with an understanding of political and multidisciplinary executive insight. The fire leader of today must coordinate with other government organizations to develop successful and creative innovations, which lead to a safe and healthy community. For the firefighter, the company officer, and those who deliver service in the street every day, their role has changed just as dramatically. Today, as emergency medical specialists ranging from first responders to paramedics; to specialty skilled teams for hazardous materials, urban search and rescue, and confined space; to geographical informational specialists; to communications experts, the role of the line firefighter has seen a dramatic change in just the last three decades. The question is, "What will the future hold?" (See Table 2.1.)

If the past is any indication of what our future will be, the continuum of expertise will continue to evolve. This evolution will move the industry to one that is based upon professional standards, certifications, and recognized credentials.

So where does the fire service go from here? It has become evident in the last decade that the training and educational requirements for the fire service have increased in direct proportion to an expanded mission and increased community expectations. As we have experienced an increase in the skills required, we have also been part of a revolution of standard(s), a development that has guided our industry to new levels of professionalism and accountability. In a series of articles written by Dr. Denis Onieal (superintendent of the National Fire Academy) in 2004, he articulated the evolution of the fire service from a blue-collar perspective to a profession. Dr. Onieal has graciously allowed me to use portions of his articles in this text. His writing provides a vision of the future of fire service education and offers a perspective on how the fire service is positioning itself to get there (see Figure 2.1).

TABLE 2.1 ◆ Training and Education Continuum

Entry-Level Training	Professional Fire Service Education
	Specialty Skill Progression Training
	Academic Training
	Company Training

FIGURE 2.1 ◆ Dr. Onieal Speaking, National Fire Academy *(Courtesy of Dr. Denis Onieal)*

PROFESSIONAL STATUS: THE FUTURE OF FIRE SERVICE TRAINING AND EDUCATION

PART ONE—INTRODUCTION

Dr. Onieal explored three areas of critical importance regarding fire service education: (1) Training and Education, a Model for Training and Education; (2) Independent Assessment of Skills and Reciprocity; and (3) The Future—Where We Go from Here.[1]

Sir Eyre Massey-Shaw, fire chief of the London Fire Brigade in 1873, when speaking of the people in the fire service, said, ". . . that the business [fire], if properly studied and understood, is worth being regarded as a profession."[2]

Think about this. You are the fire chief in your community, and your son or daughter expresses a desire to become a physician. He or she asks whether you know what training and education are required. "Sure," you say, "four years of college, four years of medical school, internship, residency, and pass the medical boards." "How about an attorney?" "A little different," you say. "Four years of college, three years of law school, pass the bar exam." And then another of your children asks, "I want to be the fire chief, just like you. What do I need to do?"

This is not as easy to answer. It varies from place to place, depending upon the organization, the structure of the department, and the governing agency. The process is not the same wherever you go; frequently, it is a slow and uneven process, or one based solely on popularity. Too often, the process frustrates talented men and women, losing our best and brightest. These are the very people who epitomize the word *professional*: those who have the aptitude and drive to help the department face new challenges.

Professional status as a term has been bandied about in the fire and emergency services for years. What constitutes professional status is in the eye of the beholder. Are we to look at a "professional" independent of the fire service? To some, it means the performance of a series of skills in a manner that is far above average. To others, a professional is associated with performing skills "full time" for a living. Many believe the distinguishing characteristics of a profession are years of formal education, approval of an accrediting board, and continuing education requirements. More than likely, most would agree with the last statement.

Definitions aside, it is the walk down the main street in any city or town in America that demonstrates who in the community is a professional. The physicians and nurses, the architects and engineers, the attorneys and accountants are among the top professions in any community. What makes them so?

Each has a unique set of knowledge and skills independent of a particular organization or place; these "portable" skills are held in equal regard no matter where the person practices. In the process of becoming a professional, an accredited and independent testing process assures competency to the public. Professionals are associated with others in their profession through some formal organization; they typically put service to others as more important than profit; and they assume responsibility for their professional acts. Typically, their profession has some continuing education requirements, and the work is client centered. Interestingly enough, the fire and emergency services have most of these. In theory (although perhaps not in current practice), providing emergency services is a portable skill; many professionals move from department to department, from state to state. We have independent testing and assurance of competencies, for example, NFPA standards, **certifications,** and in some cases requirements for continuing education. The fire and emergency services have several professional organizations, and the services delivered are certainly client centered. Profit just is not in the lexicon; all we concern ourselves with is people.

certification
A process whereby an individual is tested and evaluated in order to determine his or her mastery of a specific body of knowledge or some portion of a body of knowledge.

Then why are we not given the professional status of physicians and nurses, architects and engineers, and attorneys and accountants? Well, those professions have some things that the fire and emergency services do not yet have; they involve a few more steps. Those six professions (and others like them) have other substantive tenets; principal among them is a universally recognized system to acquire the knowledge and skills to practice. Their systems of acquiring knowledge are reciprocal among all states. When physicians or lawyers or nurses move from state to state, they may have to present their credentials to the professional board in the new state. They may have to take an exam or perhaps take some refresher courses, but they do not have to go back to school to learn the basics all over again. You can learn surgery in Texas and perform surgeries in Minnesota. You can attend law school in Massachusetts and appear in court in Washington. You can learn electrical engineering in Montana and design computers in Silicon Valley in California. However, if fire officers with up-to-date professional training decide to "practice" their profession in another state, they may have to attend rookie school. That's right, rookie school—learning about classes of fire, types of extinguishers, coupling hose, and raising ladders.

Today, there is no universally recognized and reciprocal system to acquire the knowledge and skills required in the fire and emergency services. None. It is the largest hurdle associated with professional status that we have yet to overcome; although it is not the only thing, it is the most significant. It is interesting to see where current professions were one hundred years ago. Most people probably do not realize that medical education was haphazard in this country until 1910. In the late 1700s most physicians apprenticed and a few attended medical schools in Europe. In the 1800s many "for-profit" schools of medicine in the United States were of questionable quality. It was not until 1910 that Abraham Flexner, the American education reformer, wrote *Medical Education in the United States and Canada*. He exposed the inadequacies of most of these private medical schools. Subsequently, the American Medical Association and the Association of American Medical Colleges established standards for course content, qualifications of teachers, laboratory facilities, affiliation with teaching hospitals, and licensing of practitioners, which have survived to this day.[3]

Many people also do not realize that although Abraham Lincoln was a lawyer, he never went to law school; he apprenticed. Law schools began in this country about 1875. Less than one hundred years ago, babies were born at home, delivered by midwives; some dental care was provided by barbers (yes, barbers!) known as "sanitaries."

Professions have been specialized too. Fifty years ago, pediatricians removed tonsils in their office; today surgeons do this in hospitals. Forty years ago, most nurses were RNs with diplomas from three-year nursing schools. Today, higher-educated nurses called nurse practitioners can diagnose illness, order medical testing, and prescribe medications. Attorneys have specialized practices too—corporate, civil, criminal, personnel, and a host of others. This increased specialization is a natural outgrowth of the complexity and increased requirements of practice.

Are the fire and emergency services becoming specialized? You bet. The principal responsibilities of the fire and emergency professions are the reduction of community risk, public education, fire prevention, code enforcement, and health and accident risk reduction.

When those prevention activities fail, what once was the fire department is now the emergency response of first and last resort. Citizens know if they call, they will hear sirens in a few short minutes. You are first on the scene—from a heart attack to a car accident, from a hazardous materials release to a trench rescue. For fire, earthquake, flood, hurricane, emergency birth, airplane crash, train derailment, or terrorist event, all anyone needs to do is dial 9-1-1. The public is expecting you.

Twenty or 30 years ago, the foundations for professional status for the fire and emergency services were laid, and performance standards were established. Colleges and universities recognized the need for formal education and began degree programs. Fire departments began to require certifications, and many started to require degrees or advanced degrees for hiring or promoting. Uncommon 30 years ago, but quite common today, was the hiring of people with professional training and education from outside the organization (instead of through the ranks) to run it. This is evidence we are ready to make the next move up the ladder of professions.

One of the principal challenges is that aspiring fire service professionals are staggered by the number of independent systems of training and education. There is no "one way" for the student to determine which is the most appropriate training and/or education. Neither is there one way to become the chief. The problem is exacerbated by the reality that there is little chance that one system will recognize the student's performance in another system. Moving from fire department to fire department (or, even more difficult, from a fire department in one state to a fire department in another), training or education already received may not be recognized.

The fire and emergency services today are assuredly further along the path to professional status than those in medicine and law were one hundred years ago. We have a body of knowledge, standards, and processes to assure competency (available through the International Fire Service Accreditation Congress, IFSAC, and the National Board on Fire Service Professional Qualifications, NBFSPQ or Pro Board). We have places to acquire professional knowledge, but right now, they are locally based; they are not part of a system that everyone recognizes. The missing link is a nationally recognized, reciprocal *system of training and education*. The good news is we have all the parts; nothing has to be invented or established, but they just need to be integrated:

- Training systems (available through local, state, and the National Fire Academy)
- Education systems (available through two-year, four-year, graduate programs, and the National Fire Academy)
- Independent assessment of skills (IFSAC and Pro Board)
- Reciprocity among systems of training and education

TRAINING AND EDUCATION

How do the training and education systems available to the fire service today at the local, state, and national programs complement and supplement each other?

The current roles of local, state, and national emergency services training generally establish the boundaries for each to prevent costly duplication. Locally, larger departments are capable of training their own people to certain levels of competency. Smaller departments will seek training from a larger organization, work with other small departments to combine training resources, or seek training from another government agency—either the county or the state training system. Depending on the size of the organization and its needs, local training tends more toward recruit, refresher, and hands-on training.

State training organizations generally attempt to provide training not available locally—ranging from basic recruit training to courses for chief fire officers, from hazardous materials awareness to firefighting strategies at petroleum facilities, and from farm rescue to wildland firefighting. State training organizations vary in their size and capacity, from a few people to a complex, university-based system.

At the national level, each state fire training system works with the United States Fire Administration's National Fire Academy (**USFA/NFA**) to deliver USFA/NFA curriculum. The USFA/NFA develops and delivers training not available at the local or state level. Community Risk Reduction, Public Education, Codes and Standards, Detection and Suppression Systems, Executive Development, Terrorism, Command and Control of Incidents, Strategic Planning, Information Systems, and Budgeting are among the USFA/NFA's curriculum areas (see Figure 2.2).

USFA
United States Fire Administration.

NFA
National Fire Academy.

FIGURE 2.2 ◆ The National Fire Academy in Emmitsburg, Maryland

This system is not something that is planned for the future; it exists today. At the national level most would probably be surprised to learn that the USFA/NFA does the least amount of training at its Emmitsburg campus—about 8,000 students per year. Most of the USFA/NFA training occurs off-campus through the cooperative efforts of state and metropolitan-size fire training organizations. The USFA/NFA trains over 87,000 fire and emergency services personnel in off-campus course deliveries, self-study courses, CD-based simulation training, and other alternative deliveries through its virtual campus (see www.training.fema.gov).

TRADE
Training Resources and Data Exchange.

Through the organization of state training systems and metropolitan-size fire departments (called the Training Resources and Data Exchange network, **TRADE**), the USFA/NFA also provides the instructors, course materials, site support, assistance, and a small student stipend for the delivery of three 6-day courses in each of the ten federal regions.

Working with over 100 two- and four-year colleges, the FESHE National Fire Science Curriculum Committee has developed a model core and noncore curriculum, courses, syllabi, and content for associate's and bachelor's degree programs. It is a *model*, not a requirement; but the work has developed a direction for college programs that establishes a base for the transferability of credits (you do not have to start all over again) and ease of understanding the meaning of a fire degree. Many current college programs have committed to following the model as revisions in their programs are made.

From an education point of view, model course descriptions, courses, syllabi, and content increase the understanding of what a fire degree means for students, schools, and employers. The model creates an atmosphere in which schools will be more accepting of transfer credits from other degree programs and encourages the writing of new textbooks specifically for college courses. As more Fire Science associate's degree programs adopt these model courses, future leaders of the fire service will have had the same courses and content as part of their professional development and credentialing, just like doctors, lawyers, nurses, and other professions.

To further encourage this effort, representatives of state fire training systems have committed to "crosswalk" the 13 DDP courses to the Pro Board Qualification standards. Both the colleges and the state fire training systems now have a basis to exchange the credit.

The Problem: A Fragmented System of Professional Development Most firefighters and officers have earned college credits and training certificates since their first day in the fire service. However, this professional development is usually uncoordinated and fragmented, resulting in duplications of effort and inefficiencies for students. Lack of coordination among fire-related training, higher education, and certification contributes to this problem.

Collaboration and coordination are needed among all service providers responsible for fire and emergency services professional development. Each has a major role to play. This concept has evolved from the past annual Fire and Emergency Services Higher Education (FESHE) conferences. Combined, these products and outcomes represent a new strategic approach to professional development, which will help move the fire and emergency services from a technical occupation to a full-fledged profession—a profession similar to physicians, nurses, lawyers, and architects, who, unlike fire service personnel, have common course requirements within their respective degree programs (see Figure 2.3).

There are several major tenets on which a "profession" is built, including reciprocity for practicing in different states (with an exam), universally accepted standards of

FIGURE 2.3 ◆ The FESHE Program Mark

practice, and a professional development model, among others. The work accomplished during the FESHE conferences helps to address each.

The Role of FESHE Conferences The U.S. Fire Administration (USFA) hosts the annual **FESHE** conference on its campus in Emmitsburg, Maryland. These conferences are a combination of presentations, problem solving, and consensus-building sessions, which results in higher education–related products or recommendations for national adoption, such as:

FESHE
Fire Emergency Services Higher Education.

- degree programs that teach critical thinking skills by requiring significant numbers of general education, rather than mostly fire science, courses;
- appropriate recognition of certification for academic credit, and vice versa;
- associate's degree programs that are transferable to baccalaureate programs;
- a model fire science curriculum at the associate level that universally standardizes what students learn and facilitates the application of these courses toward certification goals; and
- collaboration between fire service certification and training agencies and academic fire programs.

Fire and Emergency Services Professional Development Model Suggested roles and responsibilities are shown in Table 2.2. The model curriculum for the associate's degree and baccalaureate degree programs follow.

Associate's Degree Programs FESHE has identified six core associate-level courses in the model curriculum, including:

- *Building Construction for Fire Protection (BC)*
- *Fire Behavior and Combustion (FB&C)*
- *Fire Prevention (FP)*
- *Fire Protection Hydraulics and Water Supply (FH&WS)*
- *Fire Protection Systems (FPS)*
- *Principles of Emergency Services (PES)*

It was recommended that all fire science associate's degree programs require these courses as the theoretical core on which their major is based. The course outlines the need for a uniformity of curriculum and content among the fire science courses in

TABLE 2.2 ◆ Training and Higher Education

	Training	*Higher Education*
Learning Outcomes	Provide students with practical applications that give them the "ability to do the work," using skills- or competency-based approaches.	Provide graduates with cognitive skills that give them the "ability to manage."
Firefighter I and II; Special Certification, and Fire Officer I–IV	Deliver courses that directly support Firefighter I and II, special certifications, and Fire Officer I–IV standards, as appropriate.	Provide "officer development" and deliver courses that address Fire Officer I–IV certification. At the executive officer's level, a master's degree in public administration (or related disciplines) and applied training in strategic policy making are desirable professional preparations.
Risk Management Oriented	Direct all relevant and applicable curricula toward "risk management" because the fire service's response and mitigation missions have expanded greatly over the years to include all disasters, natural and humanmade. Address "all hazards" rather than solely fire-related incidents. This coordination of training and higher education provides a professional development path for transforming chief fire officers into "all-risk managers."	
Standards "Crosswalks"	Certification agencies identify the standards addressed by the fire science courses offered within their states, particularly those in the model curricula. Fire science publishers for the model associate courses identify standards addressed in their textbooks. NFA standards "crosswalks" for its resident, field, and baccalaureate courses are available on the USFA web page www.usfa.fema.gov/fire-service/nfa-abt7.cfm. Fire-related training, higher education, and certification service providers collaborate to promote students' eligibility to apply academic credits toward appropriate standards, and vice versa.	

two-year programs within the United States. Many schools already offer these courses in their programs, while others are in the process of adopting them. Once adopted, these model courses address the need for problem-free student transfers between schools. Likewise, they promote crosswalks for those who apply their academic course work toward satisfaction of the national qualification standards necessary for fire-fighter certifications and degrees.

FESHE also developed similar outlines for other courses that are commonly offered in fire science programs. If a school offers any of these noncore courses, it is suggested these outlines be adopted, as well. The noncore courses are:

- *Fire Administration I (FA-I)*
- *Occupational Health and Safety (OHS)*

- *Legal Aspects (LA)*
- *Hazardous Materials Chemistry (HMC)*
- *Strategy and Tactics (S&T)*
- *Fire Investigation I (FI-I)*
- *Fire Investigation II (FI-II)*

Baccalaureate Degree Programs NFA has released its 13-course, upper-level degrees at a Degrees at a Distance Program (DDP) curriculum to accredited baccalaureate degree programs, which have signed agreements with their state's fire service training agency. DDP will remain as NFA's delivery system for the 13 courses; however, their release to other schools enables the formation of model curriculum at this level. The courses are:

- *Advanced Fire Administration (AFA)*
- *Analytical Approaches to Public Fire Protection (AAPFP)*
- *Applications of Fire Research (AFR)*
- *Community and the Fire Threat (CFT)*
- *Disaster and Fire Defense Planning (DFDP)*
- *Fire Dynamics (FD)*
- *Fire Prevention Organization and Management (FPOM)*
- *Fire Protection Structures and Systems Design (FPSSD)*
- *Fire-Related Human Behavior (FRHB)*
- *Incendiary Fire Analysis and Investigation (IFAI)*
- *Managerial Issues in Hazardous Materials (MIHM)*
- *Personnel Management for the Fire Service (PMFS)*
- *Political and Legal Foundations of Fire Protection (PLFFP)*

A NATIONAL SYSTEM FOR FIRE-RELATED HIGHER EDUCATION

With the development of model lower-level (associate) curriculum outlines and the establishment of upper-level (baccalaureate) courses available, the major components are in place to move toward a national system for fire-related higher education.

Most core and noncore courses line up with baccalaureate courses of similar content, thus preparing associate's degree graduates for their bachelor's degree studies (see Table 2.3).

This national system for fire-related higher education is important because, as with other professions, a theoretical core of academic courses should be a prerequisite for entering these fields. As more schools adopt these curricula, the fire and emergency services move toward becoming a full-fledged profession (see Figure 2.4).

A Call for Collaboration There are no easy paths to uniting the fire and emergency services professional development system. The relationships among the providers of training, certification, and higher education are varied across the country. In most states, levels of cooperation among the three range from excellent to nonexistent.

Some exceptional state models of cooperation do exist, including California and Oregon. The models' similarities demonstrate that partnerships can solve training, education, and turf battles by bringing together stakeholders in some formal or informal organization or consortium; and through cooperation, a professional development delivery system that works for the state can be created and maintained.

TABLE 2.3 ◆ Lower-Level Courses and Corresponding Upper-Level Courses

Lower-Level Course	*Corresponding Upper-Level Course*
Fire Behavior and Combustion (FB&C)	*Fire Dynamics (FD)*
Fire Prevention (FP)	*Fire Prevention Organization and Management (FPOM)*
Fire Protection Hydraulics and Water Supply/Fire Protection Systems (FH&WS/FPS)	*Fire Protection Structures and Systems Design (FPSSD)*
Hazardous Materials Chemistry (HMC)	*Managerial Issues in Hazardous Materials (MIHM)*
Strategy and Tactics (S&T)	*Disaster and Fire Defense Planning (DFDP)*
Fire Administration I (FA-I)	*Advanced Fire Administration (AFA)*
Legal Aspects (LA)	*Political and Legal Foundations of Fire Protection (PLFFP)*
Fire Investigation I & II (FI-I & II)	*Incendiary Fire Analysis and Investigation (IFAI)*

Combining the Systems: Independent Assessment of Skills/Reciprocity Currently, the principal parts of a system of professional development exist. Each local, state, higher education, and federal training organization, in its own way, has been working toward the same goal—the training and education of the men and women in the fire and emergency services toward professional status.

The most practical approach to accomplishing the next step—assembling the training and education into one professional and reciprocal system. While this is a

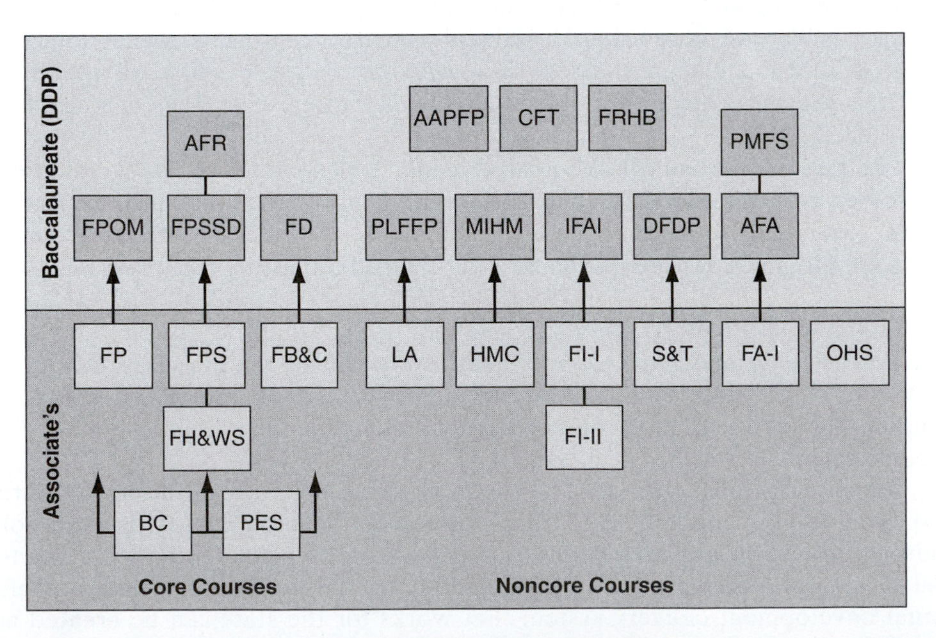

FIGURE 2.4 ◆ National Model for Fire-Related Higher Education

voluntary step, it should lead to a cooperative effort that will provide benefits to both the training and education systems and the students. It is what other professions have done in the past. It is what the fire and emergency services need to do.

Independent Assessment of Skills Part of this system must include some process by which individuals are certified as competent to practice. Assessment of knowledge, skills, and abilities is completed after a particular course of study. In the medical profession, the state medical association may be the agency, not the medical school. In most states, attorneys must "pass the bar" to qualify to practice law, and the bar exam is independent of the law school. Nursing, engineering, architecture, and accounting are other examples of professions with certifying agencies that are independent of the professional school. The assessment of individual knowledge, skills, and abilities in every profession is independent of the school.

Fortunately, the fire service already has certifying agencies that award certifications to individuals who have demonstrated competency, and the certifying agency assures that competency to the public. The more familiar certifications to those in the fire and emergency services are Firefighter I, II, III; Fire Officer I, II, III, IV; and Fire Instructor I, II. Other professional associations offer certification in related fields; fire investigation, fire inspector, and emergency medical technician are a few.

Another important part of the equation is the Professional Development Committee of the International Association of Fire Chiefs (**IAFC**). Committee members are representatives of the professional practitioners. For years, the IAFC has worked on a number of initiatives to assist and promote the development of systems to credential chief fire officers (CFO) through the Commission on Professional Credentialing and fire departments through the Commission on Fire Accreditation International (**CFAI**) to recognize personal and organizational achievement and assure some level of competency to the public.

IAFC
International Association of Fire Chiefs.

CFAI
Commission on Fire Accreditation International.

Reciprocity With the descriptions of the systems that the fire service enjoys—training, education, and independent assessment—the next logical step is to combine these parts into one system of professional development for the fire service that is universally recognized and reciprocal.

1. Strengthen reciprocity by having all of the state fire training systems participate as full partners in the USFA/NFA system of training and education.
2. Strengthening reciprocity is to recognize that state fire training systems have developed courses that meet very high quality standards and at the same time meet local need. Throughout the nation, certain needs for professional training are not national in scope. For example, the New York City Fire Department (FDNY) may need a course on subway fires, whereas the state of Kansas may have a need for a farm rescue course. It is obvious the FDNY will probably never need a farm rescue course, and it will be a while before the state of Kansas has its own subway system. There has to be a way to meet these needs.

Working with the state fire training systems, the USFA/NFA not only has addressed a way that individual regions can meet their needs but also has done it in a manner that strengthens reciprocity. The USFA/NFA has given state training systems a way to include their top-level courses into the national curriculum. To accomplish this, a partnership was formed with the state training directors, who agreed upon the criteria and standards for selection and approval of these courses. If the course meets the criteria, then it becomes a part of the national curriculum. These state-developed courses, which are peer reviewed and approved, are *endorsed* courses. Students who pass an endorsed course may receive USFA/NFA certificates and be registered in the NFA database. One key benefit of an endorsed course is that it is taught by local instructors.

3. Strengthening reciprocity and increasing the number of courses delivered will give states the opportunity to deliver train-the-trainer courses for any of the 37 two-day Direct Delivery USFA/NFA courses. These local trainers, working through the state system, may issue NFA certificates and register the students in the NFA database.

Benefits of a Common System

1. *Professional status.* Establishment of the professional status of the men and women in the fire and emergency services is a desired benefit. There is already a recognized body of professional knowledge, and the USFA/NFA is beginning a universal system that allows everyone equal access to this knowledge.
2. *Participation in USFA/NFA courses.* Everyone cannot attend classes in Emmitsburg; therefore, enfranchisement permits states to deliver NFA courses locally at local training sites, using NFA instructors, with full college credit recommendation.
3. *Reduction in course development costs.* Currently, fire departments 50 miles away from each other are spending time, effort, and money to develop the very same course. They have no idea that another department close by is working just as hard, spending just as much money, and facing the same development obstacles they are. With an endorsement system available, departments can contact the state for information on what courses are already available before they decide to begin developing a course. No training system in this country has all of the people and money needed to develop courses; this solves a lot of those development problems.
4. *Reciprocity.* Each state is now a part of a national system, empowered to issue USFA/NFA certificates for training and education. Logically, states would *accept* certificates as evidence of training received in some other jurisdiction. These basics are already built into the system; it simply saves training time and money. No one has to repeat a course because of having moved; no department has to retrain a person in courses already taken. It is similar to the status enjoyed by physicians, nurses, attorneys, engineers, architects, accountants, and others.
5. *Training.* Increase the number of training courses available to state and local training systems.
6. *Education.* Promote colleges and universities to be a part of a system. Build an environment in which colleges can award credit for certification received and state fire training systems may accept some college credit toward certification requirements. Following the model curriculum, students should be able to transfer college credit between systems, and employers would have a firm understanding of the knowledge, skills, and abilities of those who hold degrees in the fire field.
7. *Training and education.* Allow fire service training and education to follow a logical sequence, endorsed by the International Association of Fire Chiefs Professional Development Committee's *Officer Development Handbook*.

WHERE DO WE GO FROM HERE

Any trip, from a leisurely drive to a cross-country excursion, begins with "where you are." You cannot go anywhere, or find any place, unless you know where you are. Rather than try to describe the future, it might be helpful to cover how the system of training and education helps individuals at particular times in their fire and emergency services career determine where they are.

Understand that each of us has our own goals and ambitions. This written material may have described your situation or that of someone you know. There is no "one best way"; we each must decide for ourselves the paths to take. These examples are meant to expose some of the potential opportunities for someone in the fire and emergency services, but may not suit a specific individual's needs. Right now, when someone begins a career in the fire and emergency services, he or she is faced with professional development choices that have consequences that may not be fully appreciated. Some particular aspects of the job influence the individual; for others, a colleague, close friend, or officer may influence them. In any event, the individual has several paths to choose from, but in many cases, he or she may see only one or two.

One path leads toward training and certification in particular disciplines. Of course, the ones familiar to all are Firefighter I, II, and III. Other disciplines in the training and certification include inspection, training, fire officer, and a host of others. Depending upon the level of certification desired, these certifications can take a long time to complete. Depending upon the department and the personnel selection system, these certifications may lead to promotion to a higher rank.

A second path leads toward a degree—associate's, bachelor's, master's, or graduate. This path involves years of college course work, research, and writing. Formal academic pursuits typically occur outside the fire department on the student's own time and often at his or her own cost. Like certification, education may or may not increase the likelihood of promotion, depending upon the department's personnel practices.

A third path deals solely with the department's promotion practices. In most departments, this involves some form of competitive examination, ranging from a multiple-choice test and interview to an assessment center. Any of these promotion processes may include an interview with the chief, city manager, or mayor.

Up until now, these paths were viewed as mutually exclusive; an individual chose one over others in order to achieve his or her goal. With a system of training and education, this is not the case. With an agreement between a state fire training system and a college in the state, as a firefighter moves through the certification processes, there is an opportunity to receive some college credit toward certification (and the training behind that certification). As an example, if someone takes a Management 101 course in college, he or she may receive some credit toward certification in Fire Officer I.

Because state fire training systems may issue certificates for the NFA courses they deliver, they must also accept them. This is one step toward reciprocity. Hopefully, before long, all states will accept certification awarded by another state, as many do now.

At the national level, and in cooperation with the American Council on Education (**ACE**), most of the courses receive college credit recommendation. It is up to the local college as to whether it will accept this recommendation, but most do. On the certification side, a panel of state and local fire academy representatives convened in Emmitsburg to crosswalk the NFA courses with the applicable standards. The crosswalks may be reviewed on the NFA web site, www.usfa.fema.gov/dhtml/fire-service/nfa-abt7.cfm. Look at a course and you will find the standards it meets. Type in a standard and it will identify which NFA courses include the standard. It is up to the local jurisdiction as to whether it chooses to accept this; however, it is a fully peer-reviewed process accepted by many.

ACE
American Council on Education.

With model degree programs and syllabi developed by the Fire and Emergency Services Higher Education (FESHE) initiative, students will have more opportunity to transfer model course credits between colleges. Once established, employers will have a better understanding of the education underlying the degree. With concurrence by the International Association of Fire Chiefs (IAFC) Professional Development

Committee, this same training and education path is the one both have chosen to be used for chief fire officer development. Simply stated, there is a common system for certification, a common system for education, and the ability to have them work together toward one integrated system that leads to professional status. It is a path one can identify and choose to follow. No one must choose between one path or another. They are complementary.

THE FUTURE

This analysis began with being able to answer the question, "How do you become a physician, a nurse, an attorney, an engineer, or an accountant?" For perhaps the first time, there is one answer to the question, "How do you become a fire chief?" As fire service personnel continue on this path well traveled by other professions, other elements of professionalism will emerge. One is a research journal, refereed by peer scholars. Another will probably be some level of continuing education requirements. But the final step, the end of the road, the time at which the fire service will become a profession like all the others, will be when that professional (career or volunteer) firefighter or officer can have his or her professional status rescinded independent of the employer. Such accountability will be required if the fire service is to be deemed as professional. The path is not easy, and it will not be quick enough for some. To achieve the professional status enjoyed by others, those in the fire service must do more than demand it. The path taken by the other professions is the model firefighters or officers must follow. It works, the time is right, and the agreements have been made. The elements and systems are aligned; it is up to you to control and advance the fire service profession.

Dr. Onieal has helped to frame the issue of fire service training from a national perspective. It is exciting that these stakeholders have integrated their efforts to promote a more cohesive national approach to the education and development of future fire officers for the first time in the history of the fire service. A likely result will be a recognized system of the professionalism of the fire service. Now this is not to say the firefighters will not continue to get their hands dirty when they perform their duties. It goes without saying there still will be fires to fight, medical aid to provide, and hazardous materials to contain, none of which can be done by sitting at a desk. For the new firefighter or seasoned veteran who is creating the road map for the next stage of career development, it is important to understand how the pieces of professional development fit together today and what the picture may look like in the future.

However, one thing is evident. We can raise the level of professionalism as viewed by other professions, by local leaders, and by the residents of our communities. During the last 20 years, the International Association of Fire Chiefs' working committee on professional development has focused on educational activities and the preparation of firefighters to become chief officers in the future. In conjunction with the work being done at the National Fire Academy and throughout the university system in the United States, the IAFC created an *Officer Development Handbook*, which provides a comprehensive road map for individuals in the fire service that will help them to prepare themselves academically for the roles they wish to pursue (see Figure 2.5).

Although this is not a new concept for firefighters entering the service today, their path to promotions, officer development, and leadership will be much more structured and articulated than it has been in the past. As firefighters move to a more

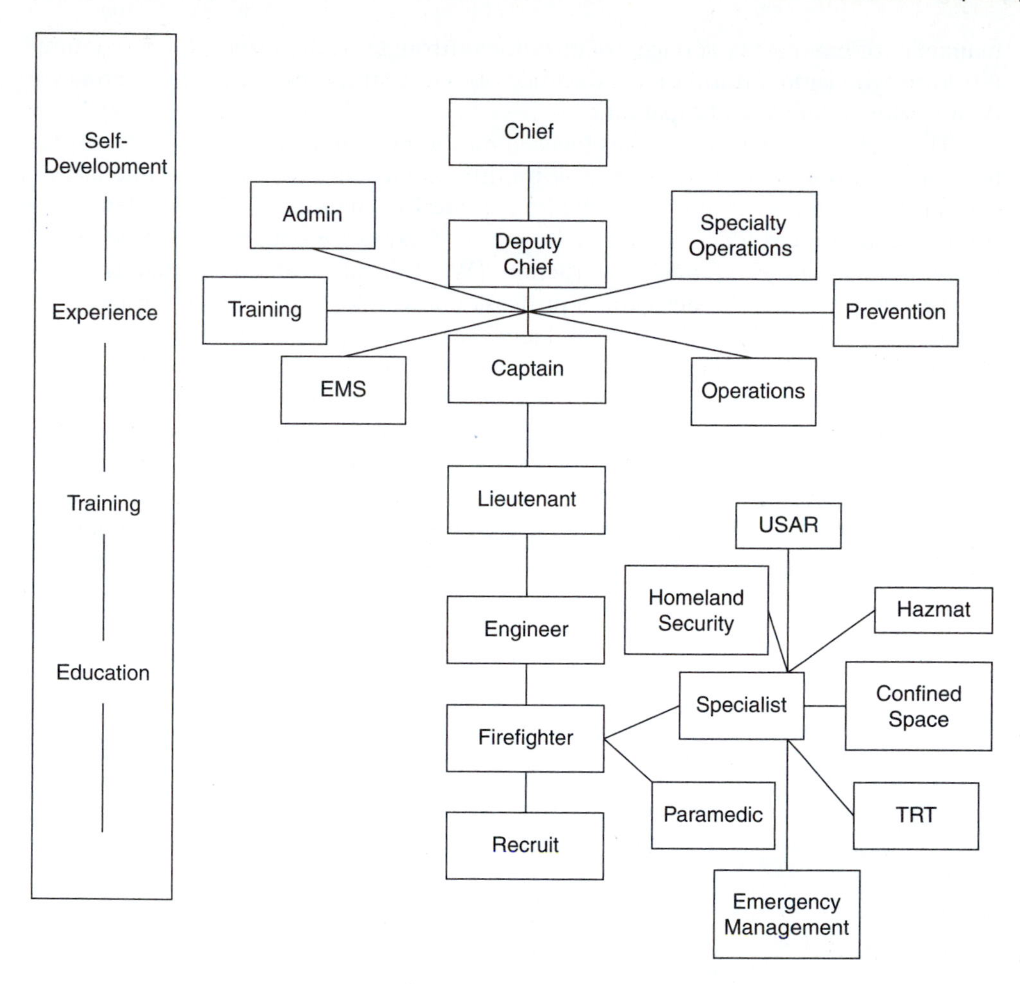

FIGURE 2.5 ◆ The Comprehensive Road Map

professional structure within the fire industry, there must be an established set of basic criteria that is universally applied throughout this profession. After over 225 years of fire service evolution, firefighters find this industry at a turning point in the development of standards, education, and professional development. This alignment of thought and vision will serve to promote the rapid expansion of professional development and will raise the level of professionalism, both real and perceived, in the communities being served.

IAFC OFFICER DEVELOPMENT HANDBOOK

Established in 1873, the IAFC is a powerful network of more than 12,000 chief fire and emergency officers.[4] It is also one of the oldest professional associations in the United States. The members of the IAFC are the world's leading experts in firefighting, emergency medical services, terrorism response, hazardous materials spills, natural disasters, search and rescue, and public safety legislation. The mission of the IAFC is to provide leadership to career and volunteer chiefs, chief fire officers, and managers of emergency service organizations throughout the international community through vision, information, education, services, and representation to enhance their professionalism and capabilities.

The IAFC (see Figure 2.6) has focused on the need to evaluate the level of professionalism, education, and training of the fire service by hosting one of the largest fire service conferences in the world: Fire Rescue International (FRI). By delivering training conferences focused on specific areas of expertise—such as EMS, wildland firefighting, weapons of mass destruction (WMD), and hazardous materials response—encouraging the design of a professional development model, the IAFC, in partnership with the NFA and state and local academies, produced the *Officer Development Handbook* (see Figure 2.7).

> This *Officer Development Handbook* was the result of a three-year work effort by the IAFC's Professional Development Committee. Beginning with a single-minded vision to provide a clear roadmap for success as a fire service officer, the committee has worked in an energetic and determined manner to bring this vision to reality. The committee members melded diverse points

Figure 2.6 ◆ IAFC Headquarters *(Courtesy www.iafc.org)*

Figure 2.7 ◆ IAFC Logo *(Courtesy International Association of Fire Chiefs, from www.iafc.org)*

of view into a professional development planning tool that will serve both incumbent and aspiring officers. As a group, they are committed to moving the fire service profession toward professional status, and this handbook is a significant step in that direction. I recommend this handbook with great confidence. Please take the opportunity to recognize the consummate professionals listed below who came together to achieve this goal.

Jim Broman, Chair
IAFC Professional Development Committee

VOLUNTEER ADVOCATES FOR PROFESSIONAL DEVELOPMENT

Two organizations that support officers and members who represent the volunteer and combination organizations are the National Volunteer Fire Council (**NVFC**) and the National Volunteer and Combination Officers Section (VCOS). For both, professional development and education are key components of their efforts.

NVFC
National Volunteer Fire Council.

NVFC The NVFC is a nonprofit membership association representing the interests of the volunteer fire, EMS, and rescue services. The NVFC serves as the information source regarding legislation, standards, and regulatory issues and is made up of 49 state firefighters associations (state organizations), which appoint a member from their organization to serve as the director from their respective state.

Beyond the state associations, NVFC offers membership to the individual firefighter (personal members), the department (department membership), and the corporate world (sustaining membership). Its mission is to provide a unified voice for volunteer fire/EMS organizations by:

representing the interests of the volunteer fire/EMS organizations at the U.S. Congress and federal agencies;
promoting the interests of the state and local organizations at the national level;
promoting and providing education and training for the volunteer fire/EMS organizations;
providing representation on national standards setting committees and projects; and
gathering information from and disseminating information to the volunteer fire/EMS organizations.

VOCS The Volunteer and Combination Officers Section of the IAFC officially began in 1994 when the section was formed at Fire-Rescue International held in St. Louis. The mission of the section was primarily focused on educating and representing volunteer chief officers, which morphed into leadership courses and efforts on Capitol Hill. During the growth years, the focus was modified to include "combination" leadership because it is the fastest growing segment of the volunteer fire service. The VCOS is currently the largest section within the IAFC, boasting about 2,700 members.

The VCOS created many success stories that have benefited the volunteer fire service. The most important was establishing a partner arrangement with Pierce Manufacturing when it agreed to provide the VCOS with $100 of profit from apparatus sold to volunteer and combination departments. This is referred to as the Pierce/VCOS Education Initiative. The revenue has resulted in providing a cadre of experienced volunteer chiefs presenting the Beyond Hoses and Helmet education series. Over 2,000 chief officers have participated in the series.

The initiative has also funded the Blue, Red, and White Ribbon Reports (www.vcos.org). These reports address governmental suggestions to produce quality organizations, the right way to lead combination systems, and the best way to be a high-performance leader in your organization. The next report (Orange) will be about EMS in the volunteer and combination fire service and its partnership with the EMS section of the IAFC.

The Annual Symposium in the Sun, now in its eleventh year, has turned into the only conference totally focused on leading volunteer and combination departments. Its high-value speakers share their expertise to strengthen individuals and produce networking that keeps attendees at the top of their game.

Various VCOS leaders present at many conferences and specialty events in the United States, Canada, and the Dominican Republic. Utilizing lessons learned from the vast knowledge base of experienced leaders has resulted in influencing thousands more volunteer and combination leaders. The formula is simple; networking, sharing, and presenting produce better-led organizations.

The VCOS web site provides information and various resources that assist volunteer and combination leadership.

The VCOS has collaborated with other fire service partners to assist the fire service:

- VFIS—Safe Driving Award
- *National Fire and Rescue* Magazine—the VCOS publishing partner
- Provident Insurance—the John M. Buckman III Leadership Award
- Paratech—funds the quarterly newsletter
- Target Safety—provided a $1.5 million in-kind grant for free online training and education

HISTORY

The need for professional development, especially for fire service officers, is not a new issue. As early as 1966, this issue drew international attention as a key component of the report from the first Wingspread Conference—*Statements of National Significance to the Fire Problem in the United States*. This conference convened top fire service leaders on five occasions at ten-year intervals. Each conference continued to emphasize the need for the development of effective leadership.

In the initial report, the committee notes that all too often "success is largely dependent upon the caliber of leadership of the individual fire chiefs, and there is no assurance that this progress will continue . . . when there is a change of leadership."[5] "The career of the fire executive must be systematic and deliberate."[6]

Wingspread II 1976, Statement 6: A means of deliberate and systematic development of all fire service personnel through the executive level is still needed. There is an educational void near the top.[7]

Wingspread III 1986, Statement 3: Professional development in the fire service has made significant strides, but improvement is still needed.[8]

Wingspread IV 1996, Statements 7 and 9: "Leadership: To move successfully into the future, the fire service needs leaders capable of developing and managing their organizations in dramatically changed environments." "Training and Education: Fire service managers must increase their professional standing in order to remain credible to community policymakers and the public." This professionalism should be grounded firmly in an integrated system of nationally recognized and/or certified education and training.[9]

Wingspread V 2006, Statements 12 and 16: Firefighter Credentials—A standardized and simple system for providing credentials for qualified firefighters and fire officers is needed to ensure qualified people are enlisted to support major emergency operations and regional emergencies. Professional Development—Professional development in the fire service has made significant strides, but improvement is still needed. The fire service needs to continue to evolve as a profession as have other governmental entities that operate in the environments where we work as well as other governmental organizations and the private sector. These skills are as important in the volunteer and combination fire services as they are in the career fire service.[10]

Each report points out the ineffective fire service practice of promoting personnel into higher ranks and then attempting to train or educate them. This practice of on-the-job training, rather than systematic skills building and preparation, is in direct contrast to the methodologies employed by virtually any other profession.

While the fire service has made progress, much remains to be accomplished. With a well-crafted professional development plan and personal commitment, the incumbent fire officer or the new recruit firefighter can make a career difference now and for the future with the use of the IAFC development model.

Each of us must carefully consider what our motivations are in respect to promoting and/or seeking expanded responsibilities.

- Do you want to have a greater influence on your work environment?
- Do you have an interest in the challenges of leadership?
- Do you value status within the organization?
- Are you interested in higher levels of compensation?
- Do you have a personal commitment to public service?

Whatever the motivation, all these possibilities will be more achiavable through professional development. However, they will not occur without a significant investment of your talent, time, energy, and in some cases financial support. Each progression in rank comes with added challenges and complexity. Knowing what you want from your career is of vital importance in preparing your future game plan for success. Remember the quote from Don Quixote, "Make it thy business to know thyself, which is the most difficult lesson in the world." This will be discussed in greater detail in Chapter 4, which explores leadership styles.

MENTORING

Mentoring is an important aspect of career development. A successful mentor helps to guide and coach you through the development experience and growth. A good mentor does not tell you what to do but rather gives you options, challenges you to see the big picture, encourages, identifies areas for improvement, and helps you refine your skills.

It is quite likely that, over the course of time, you have had several people who have provided mentoring at different times and in different ways. Mentors are people who have "been there . . . done that," are willing to share their experience(s), and are a resource you can draw upon to shape your own skills and style as a leader. Remember, those who will follow you are in need of this same support. Be ready to invest when the opportunity presents itself to mentor others. The mentor can benefit as much from the relationship as the protégé.

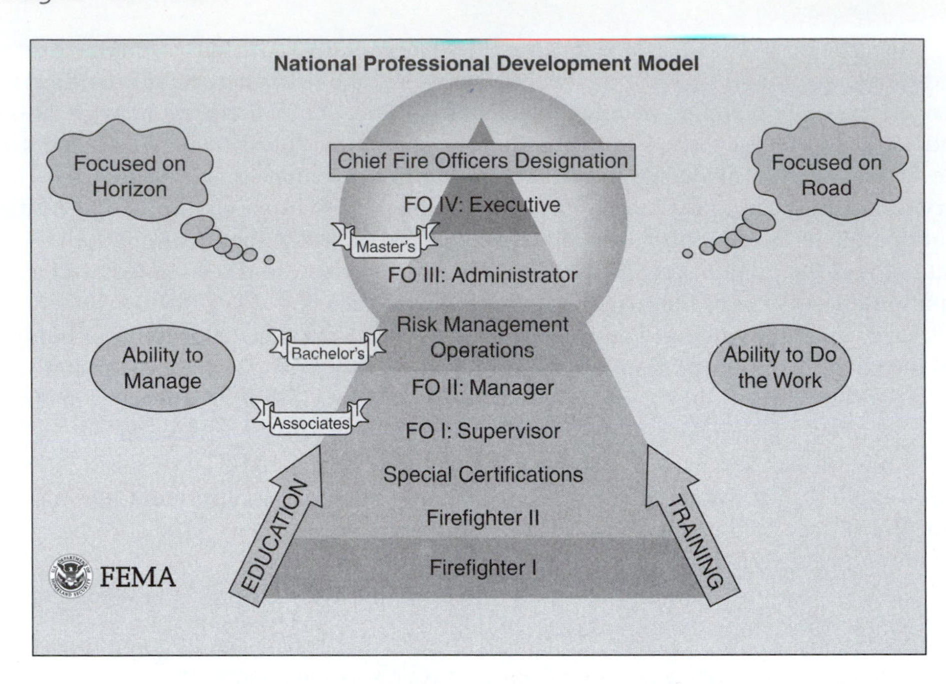

FIGURE 2.8 ◆ Fire and Emergency Services Professional Development Model

MAPPING

The IAFC's definition of mapping is, "Professional development is the planned, progressive, life-long process of education, training, self-development and experience." Mapping contains four distinct elements, on which the IAFC *Officer Development Handbook* is based: education, training, experience, and self-development.

The first two elements are especially critical and merit special attention. They form the basis of a nationally recognized model for fire service professional development (see Figure 2.8). This model clearly illustrates the importance of both education and training. It also reflects the fact that emergency response training activities are more prevalent during the first part of your career, whereas the focus on organizational skills and education is emphasized as you are promoted and take on more departmental responsibility.

MEASURING FOR SUCCESS

Dr. Onieal spoke to the journey of professional development. Much of a journey's success is measured by the progress made along the way. These measurements are the means whereby you gauge your success and appreciate what has been accomplished and what remains ahead of you. The journey of professional development will be no different.

In this context, progress is measured through credentials. These credentials may come in the form of certifications, designations, academic degrees, diplomas, licenses, certificates, transcripts, and continuing education units (CEUs). The IAFC handbook focuses primarily upon fire service certification(s), designation as a chief fire officer, and academic transcripts. You should clearly understand that your objective is not to achieve the credential but rather to possess the knowledge, skill, or ability to which the credential attests. The credential documents your achievements; therefore, it is an essential component.

Certifications typically are granted through national/international certifying bodies such as the International Fire Service Accreditation Congress (IFSAC) or the National Board on Fire Service Professional Qualifications (NBFSPQ). Subsequent to recognized training, you will be tested in the areas of knowledge and/or skills by agents of the certifying body. If successful, you will be granted certification for the appropriate level of professional achievement. These certifications begin with the entry-level firefighter and progress to the fire officer levels.

As you progress beyond the managing fire officer level, the nature of the work and the career preparation become more subjective. In response to the unique nature of this work, designation as a chief fire officer has emerged as an effective credential for those at the administrative fire officer and executive fire officer levels.

The Commission on Professional Credentialing (**CPC**) and its parent organization, the Center for Public Safety Excellence, guide this process. This process employs a portfolio approach, whereby the aspiring fire officer can plan, track, and present those professional development accomplishments for peer assessment. Academic progress is measured through the use of transcripts, formal records of a student's performance that are maintained by the learning institutions. Upon successful completion of recommended or required course work, the college or university grants credit for the accomplishment and enters it on the individual's transcript. As the student completes a prescribed block of education, the granting of academic degrees further recognizes that accomplishment, for example, baccalaureate degree, master's degree, and doctoral degree.

CPC
Commission on Professional Credentialing.

As the options for pursuing education have expanded dramatically over the past decade, it is even more important to explore the qualifications of the college or university where you consider earning a degree. One of the most important considerations is accreditation by a recognized accrediting organization.

In order to assure a basic level of quality in education, the practice of **accreditation** arose as a means of conducting nongovernmental, peer evaluation of educational institutions and programs. There are two basic types of educational accreditation. One, identified as "institutional" accreditation, normally applies to an entire college or university. The other, identified as "programmatic" accreditation, applies only to programs, departments, or schools that are a part of the institution. Accreditation does not provide automatic acceptance by one institution of credit earned at another institution, nor does it guarantee acceptance by an employer. However, it is often the first question others will ask when reviewing your educational qualifications. Although the government does not accredit educational institutions and/or programs, the secretary of education is required by law to publish a list of nationally recognized accrediting agencies that the secretary determines to be reliable authorities. There are also accrediting organizations that the secretary does not consider as reliable because they do not assure the quality of education the accreditation process is intended to provide. For further information about accreditation of educational institutions, and for a list of nationally recognized accrediting agencies, visit the Department of Education web site at www.ed.gov.

accreditation
A process by which an association or agency evaluates and recognizes a program of study or an institution as meeting certain predetermined standards or qualifications. It applies only to institutions and their programs of study or their services.

A stated objective in the IAFC *Officer Development Handbook* is *"To develop officers to have the knowledge and skills necessary to be successful in supervisory, management, administrative, and executive positions."* This key statement is given special emphasis so users will understand that professional development is not solely about certifications and degrees. These benchmarks are useful in documenting achievements in education and training and as such may be predictors of the recipient possessing the requisite knowledge and skills.

For example, the first element, *education*, is crucial. A given college degree may be from a fully accredited higher education institution, but the course work may lack one or more subject areas essential to success as an officer/leader in your agency. The model focuses on key elements and targeted learning outcomes arranged and organized consistent with typical certification and degree programs.

The second element, *training*, is also vital. Fire service technical certifications are based primarily upon NFPA Professional Qualification Standards (1000 series) with the balance based upon other key national standards. Chief Fire Officer Designation, by comparison, is based upon a blend of technical competencies, college education, leadership experience, and job-related activities.

The educational requirements are consistent with those published through the National Fire Academy by the Fire and Emergency Services Higher Education Conference in its Model Fire Science Curriculum. The completion of all stipulated higher education course work should enable the student to qualify for the commensurate academic degree(s).

The third element, *experience*, should be self-evident. It is tied to those work experiences that are important to fostering the mastery of basic skills, instilling self-confidence in the officer's ability to assess situations, and improving upon them.

The final element, *self-development*, is more subjective. It deals with awareness, personal attributes, and attitudes, which are individually developed and refined. It results from how one has grown, matured, and evolved over time. The expectations listed are based upon key indicators, activities, and experiences in a person's self-development during his or her preparation to assume the challenges of supervision and leadership. This self-development includes seminars, self-study, being mentored, and similar experiences. This continuous process enables the individual to maintain his or her knowledge, skills, and abilities now and into the future.

Currently, fire and emergency services professionals utilize professional associations, conferences, seminars, workshops, and similar offerings to meet their individual needs. A consensus plan or structure has yet to be defined to accomplish this. Regardless of your current career progress, the IAFC *Officer Development Handbook* encourages everyone in the fire service to incorporate systematic and deliberate professional development in the fire and emergency services career. Whether your involvement is through full-time employment or volunteer service, the challenges of today's fire service demand your commitment to education, training, experience, and self-development. An important reminder: *Your professional development is a journey, not a destination.*

As outlined in the IAFC handbook, the development process is broken down for Supervising Fire Officer, Managing Fire Officer, Administrative Fire Officer, and Executive Fire Officer. Each is subdivided into the related NFPA standards, elements of training, education, experience, and self-development (see Figure 2.9). The process provides an excellent road map for young firefighters and young officers to evaluate where they currently are in their career development plan and to begin to develop a long-term strategy that will lead them to future promotional opportunities.

The NFPA promulgates minimum fire officer professional qualification standards for use in certification through an independent examination process. The applicable standards from NFPA 1021, for each of the four officer development levels, are included for reference. Although it is recommended that firefighters include them in their professional development planning process, it is important to stress that they compose only a portion of the total development process (see Table 2.4 through Table 2.23).

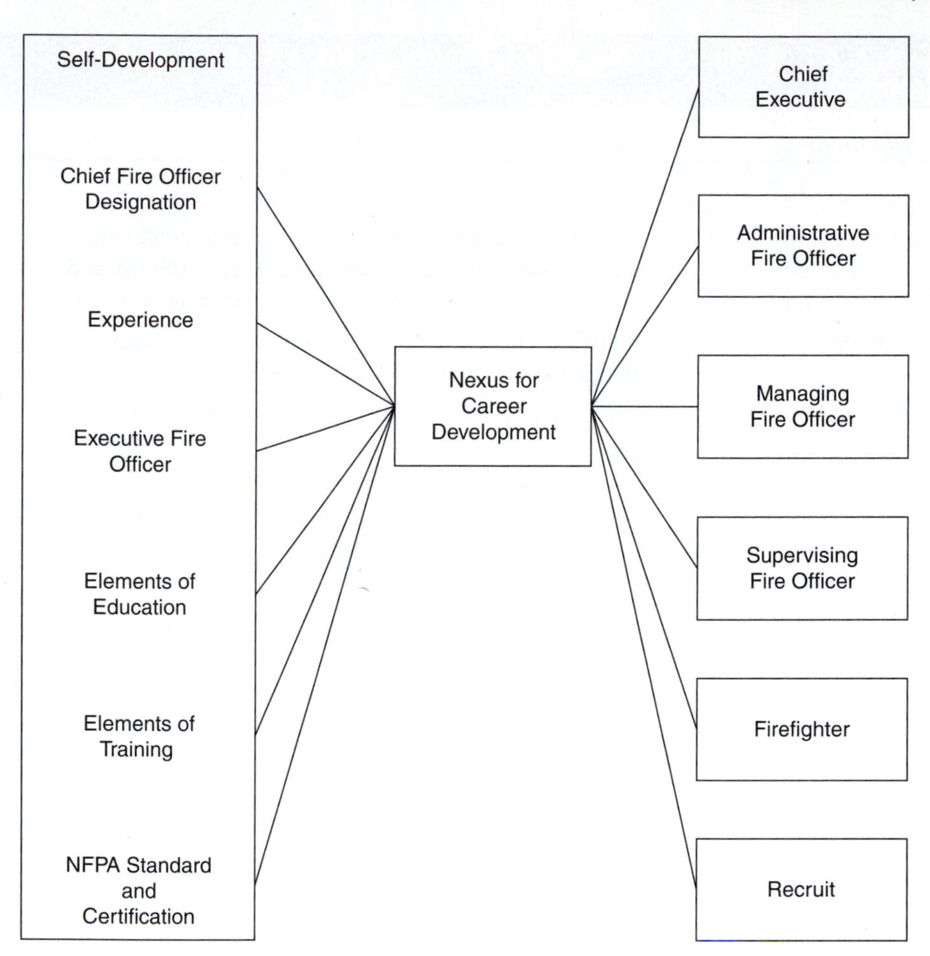

FIGURE 2.9 ◆ The Self-Development Road Map

TABLE 2.4 ◆ Supervising Fire Officer: NFPA Fire Officer I Standards

Component	*Content*
General	Firefighter II.
General Knowledge	Organizational structure; procedures; operations; budget; records; codes and ordinances; IMS; social, political, and cultural factors; supervisory methods; labor agreements.
General Skills	Verbal and written communication; report writing; incident management system.
Human Resource Management	Use human resources to accomplish assignments safely during emergency, nonemergency, and training work periods; recommend action for member problems; apply policies and procedures; coordinate the completion of tasks and projects.
Community and Government Relations	Deal with public inquiries and concerns according to policy and procedure.
Administration	Implement departmental policy and procedure at the unit level; complete assigned reports, logs, and files.
Inspection and Investigation	Determine preliminary fire cause; secure a scene; preserve evidence.
Emergency Service Delivery	Conduct pre-incident planning; develop incident action plans; implement resource deployment; implement emergency incident scene supervision.
Health and Safety	Integrate health and safety plans, policies, and procedures into daily unit work activities; conduct initial accident investigations.

TABLE 2.5 ◆ Supervising Fire Officer: Training

Element	*Note*
Firefighter I	NFPA 1001; Firefighter I
Firefighter II	NFPA 1001; Firefighter II
Fire Officer I	NFPA 1021; Fire Officer I
Incident Safety Officer	NFPA 1521
IMS	NIIMS
Instructor I	NFPA 1041 or equivalent
Inspector I	NFPA 1031 or equivalent
Emergency Medical Services	Per state/local requirements
Valid Driver's License + related endorsements	Per state/local requirements
HazMat; Operations Level	NFPA 472

TABLE 2.6 ◆ Supervising Fire Officer: Education

SFO-01 Outcome	***DISCIPLINE:***	***Communications***
Ability to write detailed prose.	LEVEL: SUGGESTED:	100 English Composition
SFO-02 Outcome	***DISCIPLINE:***	***Communications***
Understanding and using basic interpersonal, group, and public communication skills.	LEVEL: SUGGESTED:	100 Public Speaking
SFO-03 Outcome	***DISCIPLINE:***	***Communications***
Ability to write accurate and clear letters, memos, technical reports, and business communications.	LEVEL: SUGGESTED:	100 Business Communications
SFO-04 Outcome	***DISCIPLINE:***	***Science***
Understanding about ecosystem construction and destruction, energy production, and use and waste generation and disposal.	LEVEL: SUGGESTED:	100 Biology
SFO-05 Outcome	***DISCIPLINE:***	***Science***
Understanding basic principles of general chemistry including the metric system theory and structure.	LEVEL: SUGGESTED:	100 Chemistry
SFO-06 Outcome	***DISCIPLINE:***	***Science***
Understanding basic principles of areas of psychology: physiology, cognition, motivation, learning, intelligence, personality, and mental heath.	LEVEL: SUGGESTED:	100 Psychology
SFO-07 Outcome	***DISCIPLINE:***	***Science***
Understanding basic principles of social groups, forces, structures, processes, institutions, and events.	LEVEL: SUGGESTED:	100 Sociology
SFO-08 Outcome	***DISCIPLINE:***	***Science***
Understanding and using the basics of mathematical models; elementary concepts of probability and simulation; emphasis on business applications.	LEVEL: SUGGESTED:	100/200 Intro to Finite Math; Algebra
SFO-09 Outcome	***DISCIPLINE:***	***Science***
Understanding basic principles of information technology and business computer systems for effective daily use.	LEVEL: SUGGESTED:	100 Business Computer Systems
SFO-10 Outcome	***DISCIPLINE:***	***Science***
Understanding and implementing basic principles of health, fitness, and wellness.	LEVEL: SUGGESTED:	100 Health/Wellness
SFO-11 Outcome	***DISCIPLINE:***	***Science***
Understanding basic concepts of government at the federal, state, and local levels.	LEVEL: SUGGESTED:	100 American Government

(Continued)

TABLE 2.6 ◆ Supervising Fire Officer: Education (*Continued*)

SFO-12 Outcome	***DISCIPLINE:***	***Science***
Understanding functional areas of human resource management and laws; job analysis, testing; performing interviewing, selection, training, and performance evaluation.	LEVEL: SUGGESTED:	100 Human Resource Management
SFO-13 Outcome	***DISCIPLINE:***	***Science***
Understanding basic theories and fundamentals of how and why fires start, spread, and are controlled.	LEVEL: SUGGESTED:	100 Fire Behavior and Combustion
SFO-14 Outcome	***DISCIPLINE:***	***Science***
Understanding the components of building construction related to fire and life safety, including inspections, pre-incident planning, and emergency operations.	LEVEL: SUGGESTED:	100 Building Construction
SFO-15 Outcome	***DISCIPLINE:***	***Science***
Understanding and performing basic responsibilities of company officers including supervision, delegation, problem solving, decision making, communications, and leadership.	LEVEL: SUGGESTED:	200 Fire Administration I

TABLE 2.7 ◆ Supervising Fire Officer: Experience

Element	***Application***
Agency Operations	Qualified Responder: 3–5 years.
Coaching	Peer coaching: e.g., recruits and other organizational workgroups; small group leadership; sports teams, youth clubs, etc.
Directing Resources	Acting officer: 200 hours, include emergency response and nonemergency activities.
Incident Management	Function as the supervisor of a single resource unit.
Planning	Participate in a planning process.
Instruction	Develop and deliver training classes.
Human Resource Management	Develop teamwork skills.
Financial Resource Management	Participate in or contribute to a station, project, or small program budget.
Project Management	Participate in an organizational work project.
Interagency	N/A
Emergency Management	Participate in mass casualty training, exercises, and incidents.
Community Involvement	Interact with homeowners associations, service clubs, etc.
Professional Associations	Network with others in the service; involvement in local, state, and/or regional professional association(s); e.g., instructors, EMS, inspectors, investigators, safety officers.

TABLE 2.8 ◆ Supervising Fire Officer: Self-Development

Element	*Application*
Health/Fitness	Ongoing health and wellness program.
Physical Ability	Maintain according to job requirements.
Career Mapping	Personal and professional inventory; identify personal traits, strengths, and areas for development.
Communication	Written and oral communication; listening; giving/receiving constructive feedback.
Interpersonal Dynamics/Skills	Customer service skills, teamwork, conflict resolution.
Diversity	Understanding the value/importance of organizational and community diversity.
Ethics	Understand, demonstrate, and promote ethical behavior for the individual.
Legal Issues	Understanding the value/importance of law in its application to the organizational work unit.
Technology	Awareness of the importance and value of technology in the work unit; develop/maintain skills to use technology in the work unit.
Local and/or Contemporary Hazards/Issues	Develop a current awareness and understanding of unique local hazards and emerging issues.

TABLE 2.9 ◆ Managing Fire Officer: NFPA Fire Officer II Standards

Component	*Content*
General	Fire Officer I and Instructor I.
General Knowledge	Organization of local government; legislative processes; functions of related divisions, bureaus, agencies, and organizations.
General Skills	Intergovernmental and interagency cooperation.
Human Resources Management	Evaluate member performance; maximize performance and/or correct unacceptable performance; complete formal performance appraisal process.
Community and Government Relations	Deliver public fire and life-safety educational programs.
Administration	Prepare budget requests; news releases; recommended policy changes; basic analytical reports.
Inspection and Investigation	Conduct hazard inspections; documentation of violations; fire investigations to determine origin and preliminary cause.
Emergency Service Delivery	Supervise multicompany emergency incident operations; hazardous materials response.
Health and Safety	Review injury, accident, and exposure reports; identify unsafe work environments or behaviors; initiate action to correct the problem.

TABLE 2.10 ◆ Managing Fire Officer: Training

Element	*Note*
Fire Officer II	NFPA 1021
Multicompany Incident Management	MCTO and MCI
Public Information Officer	Media Relations
Fire Investigator I	NFPA 1033 or equivalent
Public Educator I	NFPA 1035 or equivalent
Leadership Development Series	National Fire Academy

TABLE 2.11 ◆ Managing Fire Officer: Education

MFO-01 Outcome	***DISCIPLINE:***	***Quantitative (Math)***
Understanding and using statistical data for basic descriptive measures, statistical inference, and forecasting.	LEVEL: SUGGESTED:	100 Intro to Studies
MFO-02 Outcome	***DISCIPLINE:***	***Communications***
Understanding and practicing basic interpersonal communication skills, including perception, listening, and conflict resolution.	LEVEL: SUGGESTED:	200 Interpersonal Communication
MFO-03 Outcome	***DISCIPLINE:***	***Humanities***
Understanding American political philosophy, social justice, and systems of American politics.	LEVEL: SUGGESTED:	100 Philosophy
MFO-04 Outcome	***DISCIPLINE:***	***Humanities***
Understanding and using basic methods for critical analysis of arguments, including inductive and statistical inference, scientific reasoning and argument structure.	LEVEL: SUGGESTED:	100 Critical Reasoning
MFO-05 Outcome	***DISCIPLINE:***	***Humanities***
Understanding ethical issues, including whistle-blowing, discrimination, social responsibility, honesty in the workplace, and setting appropriate workplace standards.	LEVEL: SUGGESTED:	200 Professional Ethics
MFO-06 Outcome	***DISCIPLINE:***	***Communications***
Understanding and demonstrating analysis, research, problem solving, organization, and expression of ideas in typical staff reports.	LEVEL: SUGGESTED:	200 Professional Report Writing
MFO-07 Outcome	***DISCIPLINE:***	***Business***
Understanding accounting information as part of the control, planning, and decision-making processes.	LEVEL: SUGGESTED:	200 Accounting Analysis
MFO-08 Outcome	***DISCIPLINE:***	***Management***
Understanding basic principles of organization and management as applied to fire service agencies; apply theories to management problems.	LEVEL: SUGGESTED:	200 Fire Service Management

(Continued)

MFO-09 Outcome	***DISCIPLINE:***	***Law***
Understanding the basic legal system structures and content as they affect local government and employers.	LEVEL: SUGGESTED:	100 Intro to Law

MFO-10 Outcome	***DISCIPLINE:***	***Management***
Understanding and using the principles and techniques for effective project planning.	LEVEL: SUGGESTED:	200 Intro to Planning

MFO-11 Outcome	***DISCIPLINE:***	***Fire Science***
Understanding the basic philosophy, organization, and operation of fire and injury prevention programs.	LEVEL: SUGGESTED:	100 Prevention and Education

MFO-12 Outcome	***DISCIPLINE:***	***Fire Science***
Understanding the basic design and operation of fire detection, alarm, and suppression systems.	LEVEL: SUGGESTED:	100 Fire Protection Systems

MFO-13 Outcome	***DISCIPLINE:***	***Fire Science***
Understanding the theory and principles for the use of water in fire suppression activities; includes hydraulic principles.	LEVEL: SUGGESTED:	100 Fire Protection Hydraulics

TABLE 2.12 ◆ Managing Fire Officer: Experience

Element	***Application***
Agency Operations	Qualified SFO 2–4 years.
Coaching/Counseling	Provide coaching/counseling to new members; involvement in Critical Incident Stress Management.
Directing Resources	Acting officer for multicompany operations, include emergency response and nonemergency activities.
Incident Management	Function as the supervisor or an aide to the incident commander of a multicompany operation.
Planning	Develop, implement, or manage a planning process.
Instruction	Develop/implement company training plan.
Human Resource Management	Participate in human resource functions involving individuals, e.g., performance appraisal, accountability, and discipline, as well as group dynamics, e.g., facilitation, conflict resolution, diversity, and staffing.
Financial Resource Management	Manage a station, project, or small program budget.
Program/Project Management	Responsible for the planning, budgeting, implementation, management, and/or reporting on a significant project or program.
Interagency	Participate in an interagency committee, team, or work effort.
Emergency Management	Participate in the development and/or updating of local emergency management plans.
Community Involvement	Participate in non–fire service groups, e.g., charitable organizations, youth clubs, service clubs, sports teams, etc.
Professional Associations	Involvement in local and state professional association(s).

TABLE 2.13 ◆ Managing Fire Officer: Self-Development

Element	*Application*
Health/Fitness	Ongoing health and wellness program.
Physical Ability	Maintain according to job requirements.
Career Mapping	Explore career areas of special interest; seek a mentor.
Communication	Speaking before small groups.
Interpersonal Dynamics/Skills	Group facilitation; coaching/counseling.
Diversity	Embrace organizational and community diversity.
Ethics	Understand, demonstrate, and promote ethical behavior for the team.
Legal Issues	Understanding the value/importance of law in its application to organizational programs.
Technology	Develop/maintain skills to manage the use of technology in the work unit; develop/maintain skills to use technology appropriate to work responsibilities.
Local and/or Contemporary Hazards/Issues	Develop and communicate a current awareness and understanding of unique local hazards and emerging issues.

TABLE 2.14 ◆ Administrative Fire Officer: NFPA Fire Officer III Standards

Component	*Content*
General	Fire Officer II and Instructor II.
General Knowledge	National and international trends related to fire service organization, management, and administrative principles; public and private organizations that support the fire service.
General Skills	Evaluative methods; analytical methods; verbal and written communication; influence members.
Human Resources Management	Establish procedures for hiring, training, assigning, and promoting members; promote professional development of members.
Community and Government Relations	Develop programs to improve and expand services; build partnerships with the public to provide increased safety and quality of life.
Administration	Prepare and manage a budget; acquire resources through a proper competitive bidding process; direct the operation of an agency records management system; analyze and interpret records and data; develop a resource deployment plan.
Inspection and Investigation	Evaluate inspection programs and code requirements as to their effectiveness in ensuring the protection of life and property; evaluate pre-incident plans.
Emergency Service Delivery	Manage multi-agency planning, response, deployment, and operations.
Health and Safety	Develop, manage, and evaluate a departmental health and safety program; develop a measurable accident and injury prevention program.

TABLE 2.15 ◆ Administrative Fire Officer: Training

Element	*Note*
Fire Officer III	NFPA 1021
Interjurisdictional Incident Management	
IT Applications; Database Management	
Leading Change	National Fire Academy
Negotiation; Mediation; Facilitation	"Getting to Yes"
Research and Technical Reporting	
Strategic Planning; Deployment Planning	

TABLE 2.16 ◆ Administrative Fire Officer: Education

AFO-01 Outcome	***DISCIPLINE:***	***Business***
Understanding basic concepts of economic thinking; basic understanding of the complex economic problems in modern society.	LEVEL: SUGGESTED:	100/200 Intro to Economics
AFO-02 Outcome	***DISCIPLINE:***	***Management***
Understanding the field of management including planning, motivation, group dynamics, decision making, organizing, and group organizational change.	LEVEL: SUGGESTED:	300 Principles of Management
AFO-03 Outcome	***DISCIPLINE:***	***Management***
Basic concepts of management and decision making in a political environment; how these concepts relate to practical problems faced by public administrators.	LEVEL: SUGGESTED:	300 Management in the Public Sector
AFO-04 Outcome	***DISCIPLINE:***	***Humanities***
Understand the historical examples of leadership throughout history from medieval times to present day.	LEVEL: SUGGESTED:	300 Leadership
AFO-05 Outcome	***DISCIPLINE:***	***Management***
Understand the theory and practice of personnel administration and human resource management, including recruiting, selection, compensation, performance appraisal, training, and labor relations.	LEVEL: SUGGESTED:	300 Human Resource Management

(Continued)

TABLE 2.16 ◆ Administrative Fire Officer: Education (*Continued*)

AFO-06 Outcome	***DISCIPLINE:***	***Management***
Understand the factors that shape risk and the strategies for fire and injury prevention, including risk reduction, education, enforcement, investigation, research, and planning.	LEVEL: SUGGESTED:	300 Risk Management
AFO-07 Outcome	***DISCIPLINE:***	***Administration***
Understand and implement an organization and its management in the fire service; organizational structures; resources; finance; planning.	LEVEL: SUGGESTED:	300 Advanced Fire Administration
AFO-08 Outcome	***DISCIPLINE:***	***Administration***
Understand the tools and techniques of rational decision making in fire departments, including data, statistics, probability, decision analysis, modeling, cost-benefit analysis, and linear programming.	LEVEL: SUGGESTED:	300 Analytical Approaches to Public Fire Protection
AFO-09 Outcome	***DISCIPLINE:***	***Law***
Understand and function effectively in the legal, political, and social aspects of government's role in public safety, including the legal system, department operations, personnel issues, and legislation.	LEVEL: SUGGESTED:	300 Political and Legal Foundations of Fire Protection
AFO-10 Outcome	***DISCIPLINE:***	***Quantitative (Math)***
Understand the principles of budgeting, financial reporting, and management in governmental organizations; emphasis on the use of financial data in planning, control, and decision making.	LEVEL: SUGGESTED:	400 Management Budgeting and Accounting
AFO-11 Outcome	***DISCIPLINE:***	***Communications***
Understand the psychological and social factors affecting human work behavior and performance, including communication, motivation, leadership, social influence, and group dynamics.	LEVEL: SUGGESTED:	400 Organizational Behavior
AFO-12 Outcome	***DISCIPLINE:***	***Humanities***
Develop skills for moral decision making in professional life; explore styles of moral reasoning based on the differing premises of duty and ethics.	LEVEL: SUGGESTED:	400 Professional Ethics

TABLE 2.17 ◆ Administrative Fire Officer: Experience

Element	*Application*
Agency Operations	Qualified MFO 3–5 years.
Coaching/Counseling	Provide coaching/counseling to new members and subordinate officers; provide member development.
Directing Resources/Influencing	Participate in multiple function program management; participate in events, presentations, and other interactions with elected officials, business community, media, and special-interest groups.
Incident Management	Serve as an incident commander at a significant incident managed under ICS.
Planning	Inter/intra agency project or committee leadership.
Instruction	Develop/implement organizational training effort.
Human Resource Management	Responsible for human resource functions including staffing, diversity, performance appraisal, accountability; also, the investigation, documentation, and reporting on personnel issues including matters of discipline.
Financial Resource Management	Plan, implement, manage, and report budget functions at a program or divisional level.
Program/Project Management	Responsible for managing significant organizational project(s).
Interagency	Guide/direct an interagency committee or team effort; serve as organizational liaison with other agencies.
Emergency Management	Participate in emergency management planning and activities for mitigation and recovery.
Community Involvement	Participate in the planning and implementation of community events.
Professional Associations	Membership in local, state, regional, or national fire service association(s); serve on committees.
Professional Contribution	Prepare or assist with the preparation of instructional/informational material for publication/presentations; make presentations.

TABLE 2.18 ◆ Administrative Fire Officer: Self-Development

Element	*Application*
Health/Fitness	Ongoing health and wellness program.
Physical Ability	Maintain according to job requirements.
Career Mapping	Begin CFOD process; learn mentorship.
Communication	Large group/public presentations.
Interpersonal Dynamics/Skills	Time management; building teams; becoming a mentor.
Diversity	Promote and reinforce organizational and community diversity.
Ethics	Understand, demonstrate, and promote ethical behavior for the organization.
Legal Issues	Understanding the value/importance of law in its application to the organization.
Technology	Develop/maintain skills to integrate and coordinate the use of technology throughout the agency; develop/maintain skills to use technology appropriate to work responsibilities.
Local and/or Contemporary Hazards/Issues	Assess and analyze unique community risks and emerging issues.

TABLE 2.19 ◆ Executive Fire Officer: NFPA Fire Officer IV Standards

Component	*Content*
General	Fire Officer III.
General Knowledge	Advanced administrative, financial, communications, political, legal, managerial, analytical, and information management.
General Skills	Effectively apply prerequisite knowledge.
Human Resources Management	Administer job performance; evaluate and improve department performance; appraise and direct a grievance program, training and education program, a member assistance program, and inventive program(s).
Community and Government Relations	Project a positive image of the department; assume a leadership role in community events; effectively interact with community leaders.
Administration	Coordinate long-range planning, fiscal projections; evaluate training system requirements and establish goals.
Inspection and Investigation	No additional duties.
Emergency Service Delivery	Establish an ongoing program of comprehensive preparedness for natural or human-caused disaster incidents.
Health and Safety	Establish a comprehensive risk management program.

TABLE 2.20 ◆ Executive Fire Officer: Training

Element	*Note*
Influencing and Presentation Skills	
Meeting Facilitation	
Risk Assessment/Management	"Cause and Effect" Analysis
Disaster Incident Management	Emergency Management Institute
Emergency Operations Center Management	Emergency Management Institute

TABLE 2.21 ◆ Executive Fire Officer: Education

EFO-01 Outcome	*DISCIPLINE:*	
Understands organizational life and key challenges/opportunities of managing public organizations; organizational mission, values, communication, culture, policy process, legislative–executive relations, and media relations.	LEVEL: SUGGESTED:	Graduate Public Management I
EFO-02 Outcome	***DISCIPLINE:***	
Understands organizational design, personnel, and management in mission-driven organizations; includes organizational design, networks, service delivery, managing for performance, and ethical leadership.	LEVEL: SUGGESTED:	Graduate Public Management I

(Continued)

EFO-03 Outcome	***DISCIPLINE:***	
Understands decision making from normative, prescriptive, and descriptive perspectives; individual decision-making and organizational decision practice; decision analysis.	LEVEL: SUGGESTED:	Graduate Management of Public Process
EFO-04 Outcome	***DISCIPLINE:***	
Understands managerial uses of accounting and financial management in the public sector; includes fund accounting, cost accounting, asset accounting, internal controls, auditing, financial analysis, and reporting.	LEVEL: SUGGESTED:	Graduate Financial Management in the Public Sector
EFO-05 Outcome	***DISCIPLINE:***	
Understands the issues involved in the implementation of public policy and programs, and the institutional and political constraints on policy making, and the skills needed to address them.	LEVEL: SUGGESTED:	Graduate Management of Policy Process
EFO-06 Outcome	***DISCIPLINE:***	
Understands the nature of public sector executive life; the function of leadership in implementing and changing policy; leadership styles, and the relation of leadership to its constituencies.	LEVEL: SUGGESTED:	Graduate Executive Leadership
EFO-07 Outcome	***DISCIPLINE:***	
Understands the legal framework of administrative action; constitutional requirements; operation of the administrative process; and judicial review of administrative activity.	LEVEL: SUGGESTED:	Graduate Public Administrative Law
EFO-08 Outcome	***DISCIPLINE:***	
Understands moral issues in public life; integration of moral concerns into public discussion resulting in good policy without polarization.	LEVEL: SUGGESTED:	Graduate Ethics and Public Policy
EFO-09 Outcome	***DISCIPLINE:***	
Understands possibilities offered by mediation and negotiation techniques to resolve disputes and disagreements over public-policy issues.	LEVEL: SUGGESTED:	Graduate Mediation and Negotiation
EFO-10 Outcome	***DISCIPLINE:***	
Understands theories and models of behavioral science in organizational diagnosis and development (OD); review of the OD approach; diagnosis, problem confrontation, and team building.	LEVEL: SUGGESTED:	Graduate Organizational Development in Public Agencies
EFO-11 Outcome	***DISCIPLINE:***	
Understands theory, practice, and politics of program evaluation, from simple feedback mechanisms to evaluation of large-scale programs.	LEVEL: SUGGESTED:	Graduate Program Evaluation
EFO-12 Outcome	***DISCIPLINE:***	
Understands theory, practice, and politics of developing an organizational strategic plan; incorporating multiple, diverse stakeholders.	LEVEL: SUGGESTED:	Graduate Strategic Planning
EFO-13 Outcome	***DISCIPLINE:***	
Understands theory, practice, and politics of developing and carrying out an implementation for an organizational strategic plan.	LEVEL: SUGGESTED:	Graduate Strategic Plan Implementation
EFO-14 Outcome	***DISCIPLINE:***	
Understands how to formulate research questions, conduct research, and assess statistical tools or research methods to answer different types of policy or management questions.	LEVEL: SUGGESTED:	Graduate Quantitative Analysis

TABLE 2.22 ◆ Executive Fire Officer: Experience

Element	*Application*
Agency Operations	Qualified AFO 4 years
Coaching/Counseling	Participate in interagency coaching/counseling efforts; direct member development programs.
Directing Resources/ Influencing	Manage one or more functional areas of the organization; manage organizational change efforts.
Incident Management	Multiple experiences as an incident commander at significant incidents managed under ICS and/or function as a section chief of an ICS overhead team.
Planning/Research	Assist/lead a strategic level of planning for a program or division; participate in the analysis, interpretation, and reporting of empirical data.
Instruction	Assess/evaluate organizational professional development needs; establish and communicate strategic direction.
Human Resource Management	Participate in the development of human resource strategies for the agency.
Financial Resource Management	Participate in the development of strategic financial planning, e.g., revenue projections, capital budgeting, fiscal controls, audits.
Program/Project Management	Direct/manage the development and implementation of a significant policy change or addition.
Interagency	Develop interagency agreements, contracts, MOUs, etc.; develop regional protocols/procedures.
Emergency Management	Lead the planning, training, and the exercise of emergency management preparation and response activities; work in an EOC; serve on multi-agency projects and teams.
Community Involvement	Represent the agency with community groups or agencies.
Professional Associations	Active membership and involvement in local, state, regional, or national association(s).
Professional Contribution	Serve on state and/or national boards, committees, task forces, and related policy work groups.

TABLE 2.23 ◆ Executive Fire Officer: Self-Development

Element	*Application*
Health/Fitness	Ongoing health and wellness program.
Physical Ability	Maintain according to job requirements.
Career Mapping	Complete CFOD process; be a mentor.
Communication	Interagency relations; interest-based negotiations.
Interpersonal Dynamics/Skills	Professional development executive programs, e.g., Harvard Program.
Diversity	Celebrate organizational and community diversity.
Ethics	Understand, demonstrate, and promote ethical behavior for the profession.
Legal Issues	Understand the value/importance of law in its application to the community; influence/participate in the development of law.
Technology	Provide strategic direction on the use of technology within the organization; develop/maintain skills to use technology appropriate to work responsibilities.
Local and/or Contemporary Hazards/Issues	Predict emerging local issues and trends.

IAFC Officer Development Handbook, November 2003, pp. 12–38.[11]

FIGURE 2.10 ◆ CFO Logo

FIGURE 2.11 ◆ CPC Logo

CHIEF FIRE OFFICER

The Chief Fire Officer (CFO) Program is intended for aspiring fire officers and existing chief officers. The program was designed with the assistance of a task force and the IAFC Professional Development Committee (PDC) to help incumbent and future officers with a tool for measuring their success as chief fire officers. The CFO program is now administered by the Commission on Professional Credentialing.

The mission of the Commission on Professional Credentialing (see Figure 2.11) is to promote professional development of fire and emergency services personnel by providing guidance for career planning through participation in the professional designation program. It should also be noted that the CFO program will continue to interface with the Professional Development Committee of the IAFC and the NFA to ensure career development processes are up to date and utilized to promote professional development (see Figure 2.12).

An officer must have a minimum of 150 points in education and experience to complete the competency portion of the CFO application and submit a portfolio to the Commission on Chief Fire Officer Designation. As long as an officer has followed the IAFC Professional Development model, the officer should not have any difficulty meeting the minimum requirements.

PROFESSIONAL DEVELOPMENT PORTFOLIO SETUP

As new issues and skill sets emerge, the fire service will continue to evolve as a profession, and so will the chief fire officer program (see Figure 2.13).

FIGURE 2.12 ◆ Steven Strawderman Receives CFO Designation (*Source: CPSE*)

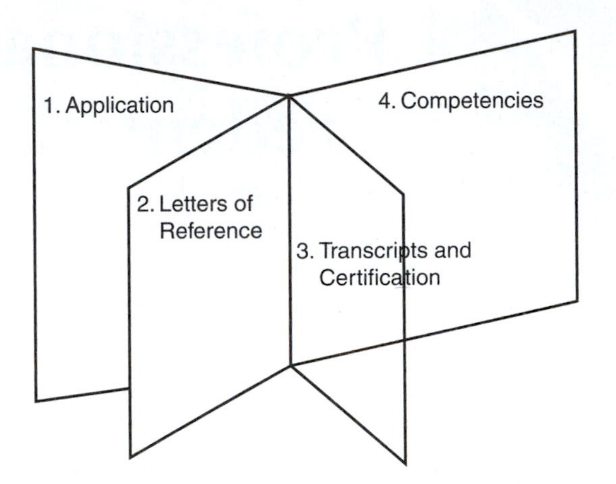

FIGURE 2.13 ◆ Chief Fire Officer Designation Portfolio

CHIEF MEDICAL OFFICER (CMO) DESIGNATION

The Chief Medical Officer (CMO) Designation closely mirrors the Chief Fire Officer (CFO) Designation program with adjustments to cover the essential mission parameters of a high-functioning emergency medical services (EMS) system. This designation is not a training program but a verification and recognition of past accomplishments and a starting point for future achievements.

The CMO candidate will submit a portfolio demonstrating excellence in seven areas: education, experience, professional development, professional contributions, professional membership/affiliations, community involvement, and technical competencies. Those candidates exhibiting extensive experience and educational background may meet the eligibility requirements to exempt them from having to complete the technical competency portion of the application. The portfolios are evaluated by a team of peer reviewers appointed by the Commission on Professional Credentialing (CPC). The CPC is the conferring body for the designation, which is valid for three years. The CMO designation will be open to EMS leaders from fire, private, hospital, and third-service-based providers.

EXECUTIVE FIRE OFFICER PROGRAM

The Executive Fire Officer Program (**EFOP**) is an initiative of the U.S. Fire Administration/National Fire Academy designed to provide senior officers and others in key leadership roles with enhanced executive-level knowledge, skills, and abilities necessary to lead these transformations, conduct research, and engage in lifelong learning. The program also provides an understanding of the following:

EFOP
Executive Fire Officer Program.

- The need to transform fire and emergency service organizations from being reactive to being proactive, with an emphasis on leadership development, prevention, and risk reduction
- The transformation of fire and emergency service organizations to reflect the diversity of America's communities
- The value of research and its application to the profession
- The value of lifelong learning

The officers enhance their professional development through a unique series of four graduate and upper-division baccalaureate equivalent courses. The EFOP spans a four-year period with four core courses. Each course is two weeks long. EFOP participants must complete an applied research project (ARP) that relates to the individual organization within six months after the completion of each of the four courses. A certificate of completion for the entire EFOP is awarded only after the successful completion of the final research project. *Note: Completion of the ARP is a prerequisite for attending the next course in the sequence of the program.* Additional information is available online at www.usfa.fema.gov/fire-service/nfa/courses/oncampus/ nfa-on2.shtm.

DEGREES AT A DISTANCE PROGRAM

Since 1986, the USFA/NFA has been administering the Degrees at a Distance Program (DDP) in cooperation with seven universities around the country. These DDP schools use distance-learning technologies to permit students to earn their bachelor's degrees. The program enrolls about 1,000 students from all over the country, with approximately 100 students graduating each year. These schools use standardized courses (including course title), syllabi, and content provided by the USFA/NFA. The USFA/NFA has sought to expand these kinds of educational opportunities for degree-seeking students, while at the same time seeking to strengthen the meaning and understanding of the value of the degree through its Fire and Emergency Services Higher Education (**FESHE**) initiative.

FESHE
Fire and Emergency Services Higher Education.

DDP provides an alternative means for fire service personnel to earn a bachelor's degree or to pursue college-level learning in a fire-related course concentration without the requirement of having to attend on-campus classes. While independent study and distance learning have appealed to working adults nationally in growing numbers in the past few years, DDP is particularly attractive to fire service personnel whose fire department work shifts normally make classroom attendance difficult. DDP institutions emphasize faculty–student interaction through written and telephone contact. Students receive detailed guidance and feedback on the required assignments and take proctored final exams at hometown locations.

FREQUENTLY ASKED QUESTIONS

What Exactly Are the Degrees at a Distance Program? The Degrees at a Distance Program (DDP) is a way to take college courses that can be used toward a bachelor's degree with concentration in the area of fire administration or fire prevention technology. Managed by the NFA, the program is offered through a national network of four-year colleges and universities. A regional college provides the opportunity to get a college education through independent study.

Without Being in a Classroom Environment, What Kind of Academic Interaction Can I Expect? Each DDP college or university emphasizes faculty–student contact. Students maintain contact with their instructors by mail, telephone, and/or computer communication. Students receive detailed guidance and analytical comments on each required assignment and may take proctored exams at convenient locations.

May I Take a Course Even if I Don't Want to Pursue a Degree at This Time? Yes. Those students who simply wish to upgrade their professional skills may take individual courses for credit. NFA certificates are awarded for the successful completion of six courses.

What if I Have an Associate's Degree? This program is perfect for you. DDP courses are junior-senior level. Your regional college will give you guidelines and will map your bachelor's degree route.

How Is This Program Different from Going on Campus and Taking a Regular College Course? In this program you get full upper-level college credit, but the emphasis is on independent study with no classroom attendance required. When you register for a DDP course, you obtain an instructional package that includes the course guide and required texts. This complete learning package makes it possible for you to learn without classroom attendance as you maintain your work schedule.

What Impact Could This Program Have on My Career? Education is often a key to advancement. By increasing your academic qualifications and your professional experience, your opportunities are likely to increase and your sense of personal accomplishment can be heightened.

Additional information is available online at www.usfa.fema.gov/fire-service/nfa/higher-ed/nfa-high.shtm.

DISTANCE LEARNING PROGRAMS

Alternative distance learning programs are available and have become an adjunct to the traditional classroom experience. One such program is Charter Oak State College, which is part of the state system in Connecticut and was founded to serve adult students through nontraditional assessment and delivery methods. It is accredited by New England Association of Schools and Colleges. Charter Oak offers four general studies degrees—an associate in arts, associate in science, bachelor of arts, and bachelor of science. Within the bachelor's degrees, students can earn content-specific concentrations in over 40 areas. Charter Oak is reflective of the many distance learning programs developed in the last two decades.

Students earn credits by taking online or video courses through a process of portfolio review, contract learning, and special assessment by taking standardized examinations or by taking courses recommended for credit by the American Council on Education or National Program on Noncollegiate Sponsored Instruction, including military credit. Many such schools have a very liberal transfer policy, and in numerous cases the entire degree can be earned elsewhere.

Most courses are either 8 or 15 weeks in length, with a few that are 5 weeks in length. This often allows working adults the flexibility to take two to three courses per semester. Such courses allow the students to take courses within a structure that provides flexibility to learn from others around the world. Numerous schools are utilizing this approach, so check with the institutions in your area or search online to determine whether distance learning is right for you.

INTERNATIONAL FIRE SERVICE ACCREDITATION CONGRESS (IFSAC)

This peer-driven, self-governing system accredits both fire service certification programs and higher education fire-related degree programs. IFSAC is a nonprofit project authorized by the Board of Regents of Oklahoma State University as a part of the fire service programs mission of the College of Engineering, Architecture, and Technology. The IFSAC administrative offices are located on the Oklahoma State University campus in Stillwater, Oklahoma. The administrative staff consists of the IFSAC manager, a unit assistant, and student staff technicians.

What Is the Difference Between Certification and Accreditation? To accredit is to give official authorization to or approval of; to provide with credentials; to recognize or vouch for as conforming to a standard; to recognize (an educational institution) as maintaining standards that qualify its graduates for admission to higher or more specialized institutions or for professional practice. *Accredit* is often confused with *certify*. Accreditation involves the program or institution itself, whereas certification is a function of the program or institution and applies to individuals.

Is My Fire Department Accredited? IFSAC does not accredit fire departments. Actually, this is offered through the Commission on Fire Accreditation International (CFAI). The commission was developed through a joint effort of the International Association of Fire Chiefs (IAFC) and the International City/County Management Association (ICMA). Today, the CFAI is one commission that is part of the Center for Public Safety Excellence.

What Type of Training Does IFSAC Offer? IFSAC does not provide training of any sort. It is the responsibility of IFSAC to accredit certificate programs.

How Do I Obtain My Transcripts and Verify Any College Credit That I Have Received Through the Courses I Have Taken? Unfortunately, IFSAC administration cannot provide transcripts and/or earned credits for these courses. IFSAC is able only to confirm you have been certified for the courses you have taken and you are located within its registry. Students should contact the entity from which they received their certificate regarding this question. This entity should be able to provide the transcripts or training records. Most entities will call them training records because some colleges get confused with the wording.

Also, bear in mind that most certificate courses, such as Firefighter I, II, and so forth, are not considered as college credit earning courses. They are usually classified as continuing education courses. In this case, you will not have earned any college credit. Either way, it is suggested you contact the entity from which you earned the certificate(s). It will be able to tell you whether or not you have earned college credit and how to obtain a copy of your training records. Further information is available online at www.ifsac.org.

NATIONAL BOARD ON FIRE SERVICE PROFESSIONAL QUALIFICATIONS (NBFSPQ)

The purpose of the National Board on Fire Service Professional Qualifications (Pro Board) is to establish an internationally recognized means of acknowledging professional achievement in the fire service and related fields. The primary goal is the accreditation of organizations that certify uniform members of public fire departments, both career and volunteer. However, other organizations with fire protection interests also may be considered for participation. Accreditation is generally provided at the state or provincial level to the certifying authority of that jurisdiction.

ORGANIZATION

The Pro Board is sponsored by five prominent fire service organizations. Each organization seats one member on the board of directors whose task it is to set policy and oversee operations of the system. The Committee on Accreditation (COA) is composed of representatives appointed by each of the sponsoring organizations and representatives elected by the accredited agencies. They are charged with the task of accreditation through review of applications, site visits, and implementation of policy set by the board of directors. The advisory committee is composed of delegates from all of the accredited agencies. It serves as a conduit for policy questions and suggestions to be addressed by the board of directors and/or the COA. Further information is available online at www.theproboard.org.

ACCREDITATION

The Pro Board accredits fire service training agencies that use the National Fire Protection Association (NFPA) professional qualification standards. The accreditation process begins with the submission of an application, including a detailed self-study document, by the organization seeking accreditation. The application package is then reviewed by the members of the Committee on Accreditation (COA) for completeness and compliance with the bylaws of the Pro Board. The next step is a site visit by a team of COA members, usually two, who perform an extensive on-site review of the organization's testing and certification processes. The site visit team prepares a report and presents it to the COA, which decides whether accreditation is to be granted.

CERTIFICATION

An agency accredited by the Pro Board makes national certification available to its members. Members are then eligible to be placed on the Pro Board's national register and receive a certificate of national certification. In addition, the Pro Board encourages reciprocity among certifying agencies. This helps assure that Pro Board certification will be recognized by the department as the members seek advancement and by other departments should personnel seek to transfer within the fire service.

BENEFITS

Professionalism has long been a goal sought by the fire service. Within the past 25 years a system has evolved to produce national professional qualifications standards that an agency can use to establish performance measures for training programs. Agencies that achieve Pro Board accreditation are recognized as having met the rigors of review by an independent organization. This independent review is the best way to assure candidates and governance bodies that the training agency's program meets the national standards.

Certification from a nationally accredited agency is a statement of success, an indisputable mark of performance belonging to individual fire service professionals. Each successful candidate for certification from an accredited agency knows that he or she has been measured against peers and meets rigorous national standards. National certification affords the individual a uniformity and portability of qualifications. In addition, the credibility of an organization is enhanced by having members certified to national consensus standards. A high percentage of certified members within a department should certainly help managers in their pursuit of adequate funding at budget time.

DEVELOPING YOUR ROAD MAP TO SUCCESS

The last 30 years in the fire service have seen dramatic changes in the skills needed and the educational levels required to be an effective firefighter, company officer, and chief officer. The next 30 years are anticipated to be no different. In fact, as the fire service's level of professionalism continues to evolve, an even more defined path of skill sets and educational levels for the various daily jobs and duties in the fire service will be seen. In mentoring young firefighters and fire officers, one of the things to stress with each one is the need to develop a personal game plan for success. Often a young firefighter decides to compete for a promotion three to four weeks before the test actually occurs. In today's world, this just does not cut it. Professional development begins the day you enter the recruit academy. By the time you leave the academy, you should already be thinking about what skills will be necessary and the educational requirements that must be met to promote successfully within your organization.

Throughout the United States the requirements to be a firefighter and to promote can be vastly different. However, we are beginning to see a collective vision emerge on the design of a common system. One day, in the not too distant future, a common set of standards and requirements at the state and local levels may become a reality. From New York to California, to Florida, to Washington State, to Illinois, the professional requirements to be in the fire service may be the same. When this occurs, we will have arrived at what Dr. Onieal referred to as the establishment of a professional basis: the skill and education that can be recognized universally in every state as well as national acceptance in respect to the minimum universal requirements for each position in our industry. Today, fire service organizations are not quite there yet. For each firefighter, fire officer, and chief officer who is aspiring to continue to prepare for the next step in the profession, it is essential to develop a personal game plan for success. However, that should not deter us from an industry objective to continue to raise the bar of our level of professionalism. After all, if the fire service industry is to move from a blue-collar job to a profession and wishes to be viewed as such, nothing less will do.

ADVICE FROM THE EXPERTS

What advice would you give a young firefighter or fire officer in regard to career development?

MICHAEL CHIARAMONTE

Chief Fire Inspector, Retired, Lynbrook Fire Department (New York): Learn all you can and have the goal of being chief of the department some day.

BETT CLARK

Former Fire Chief, Bernalillo County Fire and Rescue (New Mexico): I tell the new recruits and fire explorers, your career and development start the day you make the commitment to become a member of the fire service. When you are hired as a firefighter, start learning the requirements for promotion. In a career as rewarding as the fire service, the time passes very quickly; tomorrow will be here before you know it. Be prepared. Even if you do not desire to promote, your preparation will ensure that you will be a better firefighter and leader in your community. Go to school. Even if you take only one course a semester, you can complete a degree program. Many opportunities are now available to give formal education credit based on life experiences, and firefighters certainly have some great and yet challenging experiences. You can earn college credit just by being you. As we build the future of the fire service, degrees will become the norm in promoted positions. Knowledge is power and no one can ever take away your knowledge. I have a very successful career as fire service member and leader, but I have one regret that I often think about . . . no college degree. It would add to my credibility now and self-marketing for a future after my fire service tenure ends.

KELVIN COCHRAN

Fire Chief, Atlanta Fire Department (Georgia): First things first. Be absolutely sure that being in the fire service is a calling on your life, not just another job. Motives play a major role in career development and advancement. Once a firefighter or fire officer is convinced of his or her calling, the passion for the profession will drive that individual to higher levels of development and achievement. In addition to pursuing advances in education and technical training, fervently pursue developing character traits that lead to advancement and sustained success. Have a vision for advancement, but do your best in that part of the organization of which you are presently assigned. Volunteer for everything. Do whatever you can to contribute to the success of your superiors and peers. On your way up the ladder of success, build credibility on every rung along the way.

CLIFF JONES

Fire Chief, Tempe Fire Department (Arizona): Career development is much more complex now than it once was in the past. My advice is to sit down and make a conscious decision on what you want to do with your fire service career. Talk with people who are in positions that you aspire to be in, and ask them what it is going to take for you to get to their level in the future. Then carefully choose a mentor or mentors within the department or in the general area. Select a mentor who exhibits the qualities of a positive and successful leader. Select a person who takes pride in himself or herself and in his or her organization. Select a person who has a continuous positive impact on the department and the fire service. Choose an educational goal and stick with it. With more and more fire service people taking advantage of higher education, it is important to make every class count and attain good grades in the process. See educational requirements for promotion as minimum requirements and work to exceed those requirements in areas that are relevant to your goals and your personal improvement.

BILL KILLEN

Retired Fire Chief and IAFC President 2005–2006: Conduct a personal self-assessment and inventory of your skills, knowledge, abilities. List the personal goals you would like to accomplish in the fire service. Analyze your assets and develop your own personal plan to prepare you to achieve your goals.

KEVIN KING

U. S. Marine Corps Fire Service: First and foremost, continue your learning process. This should include formal education, fire service technical training, and certification training. In addition, consider non-fire-service-related training, such as computer skills, financial management, law, and public relations, as these and other skills are important to the future fire service. Also, I would encourage up and coming personnel to try something new and different within the organization. Broaden your horizons and learn all you can about your organization. Finally, be a team player. You can certainly debate issues and programs, but support the organization once a decision is made. (These are opinions of Mr. King and do not reflect official policy of the Department of Defense or the United States Marine Corps.)

MARK LIGHT

Executive Director, International Association of Fire Chiefs: Every new firefighter should take every opportunity to learn. This learning can be through various station assignments, assignments to different divisions of the department, formal education, training, and just overall experience. I often see new firefighters who want to wait five years before they start to develop themselves. Every recruit firefighter should look at his or her entire career as a journey, one that starts the day of hire and ends the day of retirement. It is difficult to get recruits to look at 5, 10, or even 20 years downstream, yet those who are successful, start now to prepare for the future. Some people start their careers with the goal of being the fire chief, but in many cases, that burning desire surfaces only through the years, when they have developed themselves and realize they have a lot of talent, skills, and abilities to share with others.

LORI MOORE-MERRELL

Assistant to the General President, International Association of Fire Fighters: NEVER STOP LEARNING! You must keep up with the profession. Do not become stagnant and rest on your laurels, believing they will sustain you in all situations. Seek continuous improvement of yourself and those around you. Also get in and stay in the union and be active. Be a part of the continuous improvement of the job through good labor–management relations. Finally, remember that you are to be both a leader and a servant in the community. Be respectful to everyone.

JAY REARDON

Fire Chief, Retired, Northbrook Fire Department (Illinois); Currently President of the Mutual Aid Box Alarm System (MABAS): If you do not invest in yourself, no one will ever invest in you. Every day is an assessment center; you can expect to be tested every day. Your performance is based on your preparation. Perception becomes reality unless interrupted by fact. Your behavior creates perceptions; your efforts and achievements create fact. You do not need to master technology, but you will need to understand and accept it.

GARY W. SMITH

Retired Fire Chief, Aptos-La Salva Fire Protection District (Aptos, CA): Take the time to get in touch with your passions and strengths and play to them. Every person is different . . . it is in our DNA. We have developed talents and skills based on our natural skills, so why work in areas that do not give us opportunity to use those skills? Some people are very social and do well selling a point of view. They have passion for doing good for society and the team; obviously leadership is in their genes. Others have analytical skills; they enjoy taking things apart and working on the details of how things work. They may love mechanical and engineering work-related tasks. Obviously there are other career tracks and work responsibilities that play well for these individuals. So figure it out . . . what makes you tick? What puts "fire in the belly" for your passions? List those things; put your family and personal desires at the top of the list, and your professional desires to follow. Then look at the list, work with the list, and lay out a vision for yourself, one or two well-thought-out sentences that tell YOU what direction you want to go, knowing that each day that vision may change in some way. It is only a starting point. After getting in touch with your personal vision, work on three other lists laid out on a single piece of paper; the paper would have *the vision, knowledge, network*, and *doing* listed across the top of the page. Working from the vision, begin defining the knowledge, network of supporters who could help you attain the vision, and learn more about finding about what you do not know and then list some "doing" experiences. Do not be afraid to venture forward, even if it means failure. This can be the best way to learn! Always evaluate your doing experiences and determine what you need to do to improve for the next experience. Maybe it means that the vision will change; it is up to you. Use your network and knowledge bases to help make sure your doing experiences are more successful. Someday you will look back and say, "Dang, I did it!"

STEVE WESTERMANN

Fire Chief, Central Jackson County Fire Protection District (Blue Springs, Missouri), and IAFC President 2007–2008: (1) Get as much education as you can. (2) Read everything, even if it is a topic that is not in your normal interests. (3) Keep up with technology (which usually is not too hard for new firefighters) because the technology changes in the future will be exponentially greater and come at an even faster pace. (4) Pick a service organization and a professional organization. Do not just become a member; get involved!

THOMAS WIECZOREK

Executive Director, Center for Public Safety Excellence, and Retired City Manager: During the past year I have had the opportunity to talk to a number of groups ranging from managers to mayors. The comment that struck me most was "we don't want to hire someone who can fight fires anymore." When I pressed for an explanation, the reasoning that went behind the comment was most agencies have older or more experienced personnel who can "read the fire" and conduct the operations at a fire scene. However, what mayors, managers, and leaders are looking for are individuals who can look beyond outputs and begin moving departments toward successful outcomes based on data. They want analysts who can assemble reports and statistics to support the conclusions and requests made to councils. I would encourage young firefighters to become exposed to and become expert in geographic information systems. The fire service is the perfect place to integrate GIS, much more so than the police. Fire officials have credibility in communities because they are viewed as "friends" and not intruders. As such they can access information and locations that would not be available to police. They also are able to see how the data can be used on a more day-to-day approach. Having the ability to capture and analyze the data puts a fire officer in a unique position of selecting the outcome and justifying it by a huge data set including maps, charts, graphs, and other details. The fire official becomes an integral part of the planning process rather than a casual observer, recommending capital projects for water systems, hydrants, and other similar investments. The fire official becomes the leader in an emergency situation, having access to the records and data upon which plans can be implemented, amended, or revised. I would guide the young fire officer toward the ability to communicate, both written and verbal. Too often, I have seen résumés and other work products come across my desk that are horribly written. Do not rely on spell-check, as it will catch only the obvious. In my years as city manager I saw that many professionals, whom I viewed as highly intelligent and excellent individuals, could not carry on a formal presentation in public without putting people to sleep. I would also tell the young fire official to be exposed to planning; know what "Smart Growth" is, the "new urbanism" concepts, asset management, and integrated risk management planning. Of course, this official will need to know about fire and firefighting as well as medical service, but these other areas will put an individual desiring a career as a fire professional above the rest.

JANET WILMOTH

Editorial Director, *Fire Chief* Magazine: Get a copy of the IAFC's *Officer Development Handbook*, but do not limit yourself to those guidelines. Read *Fortune* (or *Forbes*), read business books, and become proficient at public speaking.

FRED WINDISCH

Fire Chief, Ponderosa Volunteer Fire Department (Houston, Texas): A person's personal development is a long-term goal that can be accomplished, to a point, prior to entering the fire service. I have witnessed young firefighters who have made the decision to stop a degree program because the "job" became a reality. In these cases, the person should clearly continue the education track as soon as possible. When the decision is a possibility of getting the job, the individual must consider the long-term effects of NOT achieving formal education. For too many years, the fire service has focused on "squirting water" training and has forgotten that learning capabilities are developed in formal education. This has to change.

Review Questions

1. Summarize Dr. Denis Onieal's writings regarding professional status of the fire service. This should be composed of three parts:
    - **a.** Training and Education, a Model for Training and Education;
    - **b.** Independent Assessment of Skills and Reciprocity; and
    - **c.** The Future—Where We Go from Here.
2. What is the difference between certification and accreditation?

3. What are the certifications, educational requirements, or time/grade requirements for your next promotional opportunity?
4. Define the various training tracks that are offered by your state training system. What certifications are required to attain each level?
5. Describe the relevance, importance, and purpose of the Wingspread Conferences.
6. What is the history of the Wingspread Conferences, and what recommendations were made in respect to professional development?
7. What is the professional development policy or guide used by your department?
8. How has the National Fire Academy influenced the professional development of the fire service?
9. The IAFC *Officer Development Handbook* consists of several levels of responsibility and authority. What are they and how are they different?
10. Describe the Chief Fire Officer Designation and the value of obtaining the designation.
11. Describe the Executive Fire Officer Program.
12. Describe motivation, mentorship, measuring, and mapping as they pertain to the fire service.
13. Develop a five- and ten-year professional development plan.

References

1. Dr. Denis Onieal, *Professional Status: The Future of Fire Service Training and Education*. Information about the USFA/NFA programs can be found in its catalog or web page: www.usfa.fema.gov/dhtml/fire-service/nfa.cfm
2. Ron Coleman, *Masters Degree Thesis*, 1994.
3. Abraham Flexner, *Medical Education in the United States and Canada* (1910).
4. International Association of Fire Chiefs, *Officer Development Handbook*, First Edition, November 2003, Printed in the USA, pp 2–9.
5. Wingspread Conference, *Statements of National Significance to the Fire Problem in the United States*, The Johnson Foundation, 1996, p 5.
6. Ibid, Statement 9, p. 2.
7. Wingspread II Conference, *Statements of National Significance to the Fire Problem in the United States*, The Johnson Foundation, 1976, p. 3.
8. Wingspread III Conference, *Statements of National Significance to the Fire Problem in the United States*, The Johnson Foundation, 1986.
9. Wingspread IV Conference, *Statements of National Significance to the Fire Problem in the United States*, The Johnson Foundation, 1996, p. 3.
10. Wingspread V Conference, *Statements of National Significance to the Fire Service and to Those Served*, 2006, pp. 2–3 of Final Draft.
11. International Association of Fire Chiefs, *Officer Development Handbook*, First Edition, November 2003, Printed in the USA, pp. 46–55.

Principles of Leadership and Management

Key Terms

British Standards (BS) Institute, p. 102
Gantt Chart, p. 101
Hawthorne experiments, p. 105
management by objectives (MBO), p. 127
managerial grid model, p. 114
motivation, p. 109
power, p. 117
scientific management, p. 97
synergy, p. 120
systems approach, p. 119
Theory X, p. 112
Theory Y, p. 113
Theory Z, p. 113
total quality leadership (TQL), p. 123

Objectives

After completing this chapter, you should be able to:

- Describe the history of management theories.
- Explain the differences found between leading and managing.
- Explain concepts found in the managerial grid and how they impact organizations and the people in them.
- Explain modern management theories such as Theory X, Theory Y, Theory Z, Maslow's hierarchy, motivation-hygiene model, and others.
- Explain the influence that quality improvement and management by objectives have had on modern organizations.

The study of the history of leadership and management concepts is a useful tool in understanding the theories that help to characterize many of the texts written on leadership and management. The basis for this discussion is to define leadership and management. Often the terms are used interchangeably; but, in reality, as they play out in an organization, they are quite different. The critical difference between leadership and management is the idea that employees willingly follow leaders because they want to, not because they have to. In the fire service, leaders are often seen who do not possess the formal power by rank or some other designation to reward or sanction performance; however, the employees will give these leaders power by complying with what they request. These informal leaders are found throughout every organization. On the other hand, managers who do not possess good leadership skills often rely on more formal authority to attain organizational objectives and to motivate employees.

In Chapter 1 of *A Force for Change*, John Kotter captures the essence of the discussion. Management and leadership, so defined, are clearly in some ways similar.[1] They both involve deciding what needs to be done, creating networks of people and relationships that can accomplish an agenda, and then trying to ensure these people actually get the job done. They are both, in this sense, complete action systems; neither is simply one aspect of the other. People who think of management as being only the implementation part of leadership ignore the fact that leadership has its own implementation processes: aligning people to new directions and then inspiring them to make this happen. Similarly, people who think of leadership as only part of the implementation aspect of management (the motivational part) ignore the direction-setting aspect of leadership. Despite some similarities, certain differences make management and leadership very distinct. The planning and budgeting processes of management tend to focus on time frames ranging from months to a few years, on details, on eliminating risks, and on incremental rationality. By contrast, this part of the leadership process establishes a direction that focuses on longer time frames, the big picture, strategies that take calculated risks, and people's values. In a similar way, organizing and staffing tend to focus on specialization, getting the right person into or trained for the right job, and compliance, whereas aligning people tends to focus on integration, getting the whole group lined up in the right direction, and commitment. Controlling and problem solving usually focus on containment, control, and predictability, whereas motivating and inspiring focus on empowerment, expansion, and creating an occasional surprise that energizes people. But even more fundamentally, leadership and management differ in terms of their primary functions. The first can produce useful change; the second can create orderly results, which keep something working efficiently. This does not mean management is never associated with change; in tandem with effective leadership, it can help produce a more orderly change process. Nor does this mean that leadership is never associated with order; to the contrary, in tandem with effective management, an effective leadership process can help produce the changes necessary to bring a chaotic situation under control. Leadership by itself never keeps an operation on time and on budget year after year, and management by itself never creates significant useful change.

Taken together, all of these differences in function and form create the potential for conflict. Strong leadership, for example, can disrupt an orderly planning system and undermine the management hierarchy, whereas strong management can discourage the risk taking and enthusiasm needed for leadership. Examples of such conflict have been reported many times over the years, usually between individuals who personify only one of the two sets of processes: "pure managers" fighting it out with "pure leaders." Despite this potential for conflict, the only logical conclusion one can draw is that both are needed if organizations are to prosper.[2]

EARLY THINKING ABOUT MANAGEMENT

People have been shaping and reshaping organizations for many centuries. Throughout world history, we can trace the stories of people working together in formal organizations such as the Greek and Roman armies, the Roman Catholic Church, the East India

Company, and the Hudson Bay Company. People have been writing about how to make organizations efficient and effective since long before terms such as *leadership* and *management* became common usage. Certainly, a large number of people have been working together throughout history to accomplish great things. It is seen today in historic structures such as pyramids and other huge monuments and buildings, and in recent history the laying of the tracks for railroads, the pouring of thousands of miles of concrete for interstate highways across the United States, and the formation of armies and governments. Each has taken a strong organization and people in charge who possess both good management and leadership skills. Yet the true discipline and academic research into the dynamics of leading and managing have truly taken shape during the last two centuries.

The Industrial Revolution (1820–1870) started with the development of large factories in the eighteenth century, and the amassing of a workforce into one location created a new dynamic in labor needs and understanding. While this revolution of process and technology led to significant changes in the mass production of textiles and many other products, it also opened the door to the need for more in-depth study and understanding of leadership and management. As these factories continued to evolve, become larger, and increase in complexity, tremendous challenges were created in how these new organizations were to be structured and how they were to manage such a large amount of material, people, and information. This was often to be accomplished over much longer distances through the use of multiple facilities, which created the need to deal with a whole new set of leadership and management issues. The Industrial Revolution helped to create the basic framework for the research that has taken place, the managerial processes used today, and the structures in which many of today's organizations operate.

In today's world of public administration in the fire service, we have witnessed a strong need for those who will ascend to key roles of responsibility in the future to possess the skills of both management and leadership if they are to be successful. This is especially true if they are the chief fire executive. In recent history it was sufficient to be an expert field tactician, recognized as such, and promoted to fulfill the needs of the role of the chief fire executive. Frankly, those days are gone. Today, as we ascend through the ranks, the need to acquire a diverse set of skills is a key component to the success of those in positions of authority.

WHY STUDY MANAGEMENT THEORY?

Theories are perspectives by which people make sense of their world experiences. In the formal sense, a theory is a coherent group of assumptions put forth to explain the relationship between two or more observable facts. Yet personal perspectives are invisible powers that emphasize the unseen ways in which we individually approach the world in which we live. This chapter will cover several of the behavioral and research models that are critical links to how we lead and manage in the fire service today. Theories provide a stable focus for understanding what we experience and provide criteria for determining what is relevant. Theories also enable us to communicate efficiently and thus move into more and more complex relationships with other people. Imagine the frustration you would encounter if, in dealing with other people, you always had to define even the most basic assumptions you make about the world in which you live and work. Theories also challenge us to keep learning about the world in which we live and our assumptions in respect to leadership and management. By definition, theories have boundaries; there is only so much that can be covered by any one theory. Once aware of this, we are better able to ask ourselves whether there are alternative ways of looking at the world (especially when our theories no longer seem to fit our experience) and to consider the consequences of adopting alternative beliefs and processes.

The *evolution of modern management* thought began in the nineteenth century and flourished during the twentieth century as witnessed by a revolution in management theory ranging from classical theory to the Japanese management approach. Today's management theory is the result of the interdisciplinary efforts of many people; significant academic research; and a rapidly changing cultural, technological, and economical dynamic during the last century. The beginning of the modern organization occurred primarily during the middle of the nineteenth century with the rise of the factory system, principally in the textile industry, in which automation and mass production became the cornerstone of productivity. The need existed to define what management was in the first instance, as well as to operationalize it in meaningful terms for an organization. During this period two principal management theorists accepted this challenge and emerged as the so-called pre-classicists of management thought. Those two were, Robert Owen and Charles Babbage who began to address the development of management theory seriously. Robert Owen (1771–1858) was an entrepreneur and social reformer, whereas Babbage was a noted mathematician with a strong managerial interest.

Robert Owen's ideas stemmed from his ownership of a cotton mill in New Lanark, Scotland, where he developed a strong interest in the welfare of his 400 to 500 child employees. Owen spearheaded a legislative movement to limit child employment to those over the age of 10 while reducing the workday to 10½ hours. In 1813 Owen published a pamphlet, *A New View of Society*, describing his vision of society. He also became active in upgrading employees' living conditions through the implementation of improvements in housing, sanitation, and public works and by establishing schools for the children. Owen strongly believed that character is a product of circumstances, and environment and early education are critical in forming good character. Although his ideas were extremely controversial during his lifetime, Owen is credited with being the forerunner of the modern human relations school of management.[3]

Charles Babbage (1791–1871), a noted English mathematician, is credited as being the "father of the modern computer" for performing the fundamental research for the development of the first practical mechanical calculator. In addition, Babbage conducted basic research on the development of an "analytical engine," acknowledged to be the forerunner of today's modern computer. His interest in management stemmed largely from his concerns with work specialization or the degree to which work is divided into its parts. This is now recognized as being the forerunner of contemporary operations research. Babbage's other major management contribution came from the development of a modern profit-sharing plan including an employee bonus for useful suggestions as well as a share of the company's profits.

Although both Owen and Babbage were important management innovators, their efforts lacked the central tenets of a theory of management. Owen was primarily credited with recognizing the need for the theory of management and making specific suggestions regarding management techniques in the areas of human relations. Babbage is credited with developing the concepts of specialization of labor and profit sharing.[4]

THE CLASSICAL SCHOOL OF MANAGEMENT THOUGHT

The twentieth century witnessed a period of tremendous management theory research and activity. The classical school of management was primarily concerned with developing such a theory to improve management effectiveness in organizations. However, the classical school theorists went a step further. Not only did they seek to develop a comprehensive theory of management, but they also wanted to provide the tools a manager required for dealing with organizational challenges. Within the classical school there are the bureaucratic management, scientific management, and administrative management branches.[5] The classical school of thought began around 1900 and continued into the 1920s. Traditional or classical management focuses on efficiency, including bureaucratic, scientific, and administrative management. Bureaucratic management relies on a rational set of structuring guidelines, such as rules and procedures, hierarchy, and a clear division of labor. Scientific management focuses on the "one best way" to do a job. Administrative management emphasizes the flow of information in the operation of the organization.

BUREAUCRACY

Max Weber (1864–1920), known as the father of modern sociology, analyzed bureaucracy as the most logical and rational structure for large organizations. The source and formation of the definition of bureaucracy can be found in the French *bureaucratie* (see Table 3.1).

Bureaucratic Management Max Weber is associated with the bureaucratic management branch of the classical school. Weber, the son of a prominent Bismarckian era German politician, was raised in Berlin and studied law at the University of Berlin. Weber's interest in organizations evolves from his view of the institutionalization of power and and authority in the modern Western world. He constructed a "rational-legal

TABLE 3.1 ◆ Bureaucracy

bureau, office; see bureau + =cratie, rule (from Old French; see -cracy)

Bureaucracy defined:

- Administration of a government chiefly through bureaus or departments staffed with nonelected officials.
- The departments and their officials as a group: promised to reorganize the federal bureaucracy.
- Management or administration marked by hierarchical authority among numerous offices and by fixed procedures.
- The administrative structure of a large or complex organization: a mid-level manager in a corporate bureaucracy.
- An administrative system in which the need or inclination to follow rigid or complex procedures impedes effective action: innovative ideas that get bogged down in red tape and bureaucracy.

Source: Education.yahoo.com/reference/dictionary/entr?id+b0557400

authority" model of an ideal-type bureaucracy. This ideal-type bureaucracy rested on a belief of the "legality" of patterns, normative rules, and the right of those elevated to authority to issue commands (legal authority). Weber believed all bureaucracies have certain characteristics:

1. *A well-defined hierarchy.* All positions within a bureaucracy are structured in a way permitting the higher positions to supervise and control the lower positions. This provides a clear chain of command, facilitating control and order throughout the organization.
2. *Division of labor and specialization.* All responsibilities in an organization are rationalized to the point where each employee will have the necessary expertise to master a particular task. This necessitates granting each employee the requisite authority to complete all such tasks.
3. *Rules and regulations.* All organization activities should be rationalized to the point where standard operating procedures are developed to provide certainty and facilitate coordination.
4. *Impersonal relationships between managers and employees.* It is necessary for managers to maintain an impersonal relationship with the employees because of the need to have a rational decision-making process rather than one influenced by favoritism and personal prejudice. This organizational atmosphere would also facilitate rational evaluation of employee outcomes in which personal prejudice would not be a dominant consideration.
5. *Competence.* Competence should be the basis for all decisions made in hiring, job assignments, and promotions. This would eliminate personal bias and the significance of "knowing someone" in central personnel decisions. This fosters ability and merit as the primary characteristics of a bureaucratic organization.
6. *Records.* It is absolutely essential for a bureaucracy to maintain complete files regarding all its activities. This advances an accurate organizational "memory" whereby accurate and complete documents will be available concerning all bureaucratic actions and determinations.

Weber's bureaucratic principles have been widely adopted throughout the world. Yet there also are many critics of his philosophies. Weber is well known for his rational way in which formal social organizations apply the ideal characteristics of a bureaucracy. Many aspects of modern public administration are associated with him. While Max Weber is best known and recognized today as one of the leading scholars and founders of modern sociology, he also accomplished much in the field of economics; however, he was not distinguished in this particular area during his lifetime.[6]

Dysfunctional Aspects of Bureaucracy The primary criticism of Weber's theory of bureaucracy is the overwhelming acceptance of authority as its central tenet. This inevitably fosters an unrelenting need to develop additional authority, causing the bureaucracy to be unresponsive and lack effectiveness. The emphasis on impersonality can lead to personal frustration for its employees while generating red tape to reinforce previously authorized decisions. The bureaucracy is increasingly viewed by the employees and the public as a passionless instrument for responding to human needs. The need to divide and specialize labor can engender feelings of employee frustration, alienation, and estrangement from the organization. As the demands of society become ever more complex, the need increases for interpersonal communication and sharing among employees within the organization. Unwittingly, Weber helped to foster an extremely negative attitude toward the concept of bureaucracy, conjuring up images of a highly inflexible and inhumane organization often working at cross purposes with the needs of those it is supposed to serve.

The American *Robert K. Merton* (1910–2003) was among the first sociologists to emphasize systematically the now-familiar side of the bureaucratic picture: its red tape and inefficiency. According to Merton, if, as Weber thought, the predominance of rational rules and their close control of all actions favor the reliability and predictability of the bureaucrat's behavior, they also account for the lack of flexibility and tendency to turn means into ends. Indeed, the emphasis on conformity and strict observance of the rules induce the individual to internalize them. Instead of simple means, procedural rules become ends in themselves. As a result, a kind of "goal displacement" occurs. The instrumental and formalistic aspect of the bureaucratic role becomes more important than the substantive one, the achievement of the main organizational goals. According to Merton, when one leaves the sphere of the ideal and studies a real organization, one can see how a certain bureaucratic characteristic (such as strict control by rules) can both promote and hinder organizational efficiency and can result in both functional effects (predictability, precision) and dysfunctional effects (rigidity).[7]

SCIENTIFIC MANAGEMENT

Another branch of the classical school of management is the scientific management approach. The **scientific management** approach emphasizes empirical research for developing a comprehensive management solution. Scientific management principles are applied by managers in a very specific fashion. A fundamental implication of scientific management is the manager is primarily responsible for increasing an organization's productivity. This has major implications for the American economy in the face of a consistent lack of competitive productivity and GNP growth. The major representatives of this school of thought were Frederick Winslow Taylor and Frank and Lillian Gilbreth. Scientific management focuses on worker and machine relationships. Organizational productivity can be increased by increasing the efficiency of production processes. The efficiency perspective is concerned with creating jobs that economize on time, human energy, and other productive resources. Jobs are designed so each worker has a specified, well-controlled task that can be performed as instructed. Specific procedures and methods for each job must be followed with no exceptions.

scientific management
A management approach, formulated by Frederick Taylor and others between 1890 and 1930, that sought to determine scientifically the best methods for performing any task and for selecting, training, and motivating workers.

Frederick Taylor (1856–1916) One of the most important scholars who began to create the principles of scientific management was Frederick Winslow Taylor. Taylor was one of the first to analyze human behavior at work systemically. His model was the machine, seeing the workforce as interchangeable parts, performing together to achieve a specific function. Taylor attempted to analyze complex organizations, as engineers had done to machines. As the parts of the machine were easily interchangeable, so too should the human parts be within the machine model of an organization. Many of Frederick Taylor's definitive studies were performed at the Bethlehem Steel Company in Pittsburgh. To improve productivity, Taylor would examine the time and motion details of a job, develop a better method for performing the job, and train the worker. Furthermore, Taylor offered a piece rate (pay) that increased as workers produced more.

The Soldiering Analysis Working in the steel industry, Taylor had observed the phenomenon of workers purposely operating well below their capacity, which he referred to as soldiering. He attributed soldiering to three causes:

1. The almost universally held belief among workers was if they became more productive, fewer of them would be needed and jobs would be eliminated.
2. Non-incentive wage systems encouraged low productivity if the employee received the same pay regardless of how much was produced, assuming the employee could convince

the employer the slow pace really was a good (or right) pace for the job. Employees took great care not to work at an increased pace for fear the faster pace would become the new standard. If employees were paid by the quantity they produced, they feared management would decrease their per-unit pay if the quantity increases were an underlying theme.

3. Workers wasted much of their effort by relying on rule-of-thumb methods rather than on optimal work methods that could be determined by the scientific study of the task.

In his analysis of soldiering and to improve efficiency, Taylor began to conduct experiments to determine the best level of performance for certain jobs and what was necessary to achieve this performance. During this period at the steel mill, he performed exhaustive experiments on worker productivity and tested what he called the "task system," later developing into the Taylor System and eventually progressing into scientific management. His experiments involved determining the best way of performing each work operation, the time it required, the materials needed, and the work sequence. He sought to establish a clear division of labor between management and employees. Taylor's task management methodology rested on a fundamental belief that management, the entrepreneurs in Taylor's day, was not only superior intellectually to the average employees but also had a duty to supervise them and organize their work activities. This would eliminate what Taylor called "the natural tendency of workers to soldier" on the job. In 1911 a paper Taylor originally prepared for presentation to the American Society of Mechanical Engineers (ASME) was published as *The Principles of Scientific Management.* Taylor positioned scientific management as the best management approach for achieving productivity increases. It rested on the manager's superior ability and responsibility to apply systematic knowledge to the organizational work setting.

The Pig Iron Experiment Taylor applied his research to application in his famous Pig Iron Experiment. He surmised if workers were moving 12½ tons of pig iron per day, they could be incentivized to try to move 47½ tons per day. Left on their own to determine how to achieve this goal, they probably would become exhausted, after only a few hours would become discouraged, and would fail to reach their goal. However, by first conducting experiments to determine the amount of resting that was necessary, the worker's manager could determine the optimal timing of lifting and resting so the worker could move the 47½ tons per day without tiring. Not all the workers were physically capable of moving 47½ tons per day; perhaps only ⅛ of the pig iron handlers were capable of doing so. While this ⅛ of the handlers were not extraordinary people, their physical capabilities were well suited to moving pig iron. This example suggests workers should be selected according to how well they are suited for a particular job. Taylor broke the job down into its smallest constituent movements, timing each one with a stopwatch. The job was then redesigned to reduce the number of motions as well as effort and risk of error. Rest periods of specific interval and duration were implemented, and a differential pay scale was used to improve the output. With scientific management, Taylor increased the worker's output from 12 to 47 tons per day! The Taylor model gave rise to dramatic productivity increases.

The Science of Shoveling In another study of the "science of shoveling," Taylor conducted time studies and determined the optimal weight a worker should lift with a shovel was 21 pounds. Since there is a wide range of densities of materials, the shovel should be sized so that it would hold 21 pounds (research had indicated this to be the optimal size) of the substance being shoveled. The company provided the workers with optimal shovel size. The result was a three- to fourfold increase in productivity, which rewarded workers with pay increases. Prior to scientific management, workers used their own shovels of varying sizes and design and rarely had the optimal shovel for the job.

Basic Framework of Scientific Management

- Describe and break down the task to its smallest unit; test for each element of work.
- Restrict behavioral alternatives facing the worker and remove worker discretion in planning, organizing, controlling.
- Use time and motion studies to find one best way to do the job.
- Provide incentives to perform the job one best way and tie pay to performance.
- Use experts (industrial engineers) to establish various conditions of work.

Key Results of the Scientific Management Movement

- New departments emerged—industrial engineering, personnel, quality control
- Growth in middle management; separation of planning from operations
- Rational rules and procedures; increase in efficiency
- Formalized management; mass production
- Human problems—dehumanization of work, sabotage, group resistance, hatred
- Attempted to make organizations adjunct to machines
- Look at interaction of human characteristics, social environment, task, physical environment, capacity, speed, durability, and cost
- Reduce human variability

Taylor's Four Principles of Scientific Management After years of various experiments to determine optimal work methods, Taylor proposed the following four principles of scientific management.

1. Replace rule-of-thumb work methods with methods based on a scientific study of the tasks.
2. Scientifically select, train, and develop each worker rather than passively leaving them to train themselves.
3. Cooperate with the workers to ensure that the scientifically developed methods are being followed.
4. Divide work equally between managers and workers, so the managers apply scientific management principles to planning the work and the workers actually perform the tasks.

These principles were implemented in many factories, often increasing productivity by a factor of three or more. Henry Ford applied Taylor's principles in his automobile factories, and families even began to perform their household tasks based on the results of time and motion studies.

Drawbacks of Scientific Management While scientific management principles improved productivity and had a substantial impact on industry, they also increased the monotony of work. The core job dimensions of skill variety, task identity, task significance, autonomy, and feedback were all missing from the picture of scientific management. In many cases the new ways of working were accepted by the workers; in some cases they were not. The use of stopwatches often was a protested issue and led to a strike at one factory where "Taylorism" was being tested. Complaints that Taylorism was dehumanizing led to an investigation by the United States Congress. Despite its controversy, scientific management changed the way work was done, and many aspects of the basic research continue to be used today.[8]

Frank (1868–1924) and Lillian (1878–1972) Gilbreth Frank and Lillian Gilbreth focused on identifying the elemental motions in work, the way these motions were combined to form methods of operation, and the basic time each motion took. Frank Gilbreth, born on July 7, 1868, in Fairfield, Maine, was a bricklayer, a building contractor, and a management engineer. He was a member of the ASME and the Taylor Society, and a lecturer at Purdue University. Lillian Moller, born on May 24, 1878, in Oakland, California, graduated from the University of California with a B.A. and M.A. and earned a Ph.D. from Brown University. She earned membership in the ASME and, like her husband, lectured at Purdue University. They believed it was possible to design work methods by which times could be estimated in advance, rather than relying upon observation-based time studies. Frank Gilbreth, known as the father of time and motion studies, filmed individual physical labor movements. This enabled the manager to break down a job into its component parts and to redesign processes to streamline the effort. His wife, Lillian Gilbreth, was a psychologist and author of *The Psychology of Work*. In 1911 Frank Gilbreth wrote *Motion Study*, and in 1919 the couple wrote *Applied Motion Study*. Frank and Lillian had 12 children; two of their children, Frank B. Gilbreth, Jr., and Ernestine Gilbreth Careyone, wrote their story, *Cheaper by the Dozen*.

One of Frank Gilbreth's first studies concerned bricklaying. Having worked as an apprentice bricklayer, he had an appreciation not only for this task but also for the labor involved. After extensive studies of bricklayers, he was able to reduce the motions in bricklaying from 18½ to 4, producing a 120 percent increase in productivity. He also designed and patented special scaffolding to reduce bending and reaching, which increased output from 1,600 to 2,700 bricks per day. However, unions resisted his improvements, and most workers persisted in using the old, fatiguing methods. From their various studies the Gilbreths developed the laws of human motion from which evolved the principles of motion economy. They coined the term *motion study* to cover their field of research and as a way of distinguishing it from those involved in time studies. It was a technique they believed should always precede method study and still holds true today. The use of cameras in motion study stems from this time period, and the Gilbreths used micromotion study. They called a breakdown of work into fundamental elements *Therbligs* (derived from *Gilbreth* spelled backwards). These elements were studied by means of a motion picture camera and a timing device, which indicated time intervals on the film as it was exposed. This was done in order to record and examine detailed opportunity improvements as well as to embed this information into graphs to observe rhythm and movement. As a result they invented cylographs and chronocycle graphs to observe rhythm and movement.[9] The Gilbreths believed there was "one best way" to perform an operation. However, this one best way could be replaced when a better way was discovered. The Gilbreths defined motion study as dividing work into the most fundamental elements possible. Studying those elements separately and in relation to one another built methods of least waste from the studied and timed elements. They defined time study as searching scientific analysis of methods and equipment used or planned in doing a task, the development in practical detail of the best way of doing it, and the determination of the time required to complete the task. The Gilbreths drew symbols on operator charts to represent various elements of a task such as search, select, grasp, transport, hold, delay, and others.

After Frank Gilbreth's death, Dr. Lillian Gilbreth continued the work and extended it into the home in an effort to find the one best way to perform household tasks. She also worked in the area of assistance to those with disabilities, for instance, her design of an ideal kitchen layout for the person afflicted with heart disease. Widely recognized as one of the world's great industrial and management engineers, she traveled and worked in many countries of the world.

Henry Gantt (1861–1919) Another well-known pioneer in the early days of scientific management was Henry Gantt. Gantt was a mechanical engineer and management consultant who is most recognized for developing the **Gantt Chart** in the 1910s. The Gantt Chart, employed on such major projects such as the Hoover Dam and the interstate highway system, is a visual display used for scheduling multiple overlapping tasks over a time period. Gantt worked for Frederick Winslow Taylor in the United States and is to be remembered for his humanizing influence on management, emphasizing the conditions that have favorable psychological effect on the workforce. He focused on motivational schemes, emphasizing the greater effectiveness of rewards for good work rather than penalties for poor work. He developed a pay incentive system with a guaranteed minimum wage and bonus system for people on fixed wages. Gantt also focused on the importance of the qualities of leadership and management skills in building effective industrial organizations. Scientific management led to rapid developments in machinery and technology and, with the improvement of materials, came the moving assembly line.

Gantt Chart
Graph presenting the different steps of a project in a structured and chronological way.

The Production Assembly Line Toward the end of the nineteenth century, the internal combustion engine was invented, leading to the development of the motor car. There was a move toward streamlining production, spawning the first assembly line method of manufacture. This production method was attributed to the mail-order business of Sears and Roebuck of America. At one time even a house could be ordered through its catalog. More famous was, of course, Henry Ford. His automobile factory in the United States is the best example of the change to modern assembly-line techniques. Before the "line" was set up, each car chassis was assembled by one man, taking approximately 728 minutes, or 12½ hours. Eight months later, with standardization and division of labor, the total labor time had been reduced to just 93 minutes, or 1 hour 33 minutes per car. Using a constantly moving assembly line, subdivision of labor, and careful coordination of operations, Ford realized huge gains in productivity. Ford's mass-production techniques would eventually allow for the manufacture of a Model T every 24 seconds.

His innovations made him an international celebrity. It is interesting to note the idea of an assembly line came to Henry Ford when he was watching a moving conveyor of carcasses in a Chicago slaughterhouse.

Charles E. Bedaux (1897–1944) Another pioneering contributor to the field of scientific management was Charles Bedaux. Although not embarking on his career until after Taylor's death, he was to have widespread influence, first in the United States and later in Europe. Charles Bedaux is best known for his pioneering work on work measurement, mainly by people engaged in this area of study. His is an incredible story of a man whose brilliance and meteoric rise shot him from being a bottle washer to a millionaire. In his time he was recognized by world leaders and statesmen, often rubbing shoulders with British royalty. Bedaux's achievements are often overlooked as he was arrested and charged with treason against the United States. He committed suicide before he was brought to trial.

Charles Bedaux, born in Charenton, Paris, in 1887, immigrated to America in 1906 and worked as a sand hog digging foundations for skyscrapers in Manhattan. He "graduated" to dishwasher and later was a petty salesman selling wares such as toothpaste, life insurance, and cattle. It was in Grand Rapids, Michigan, in 1917 while in furniture manufacturing that he saw the potential in exploiting the inefficient methods employed in the working situation. In one year he had established the Charles E. Bedaux Company and by 1925 had become a millionaire with 19 offices over the world with 600 clients, encompassing most of the big corporate clients and well-known household names of the time, including Campbell's Soup, Kodak, General Electric, Lyons, and many others. He was loved by businessmen and hated by the workforce due to his success in applying his methods improvements, work measurement, and incentive schemes. Particularly, he developed time study (or "speed-up systems," as the unions referred to it) from the earlier studies of pioneers such as Frederick W. Taylor.

While Taylor used "selected times," Bedaux introduced the concept of the standard minute value based on the "speed and effort" rating system.[10] *Rating* is a term used in work measurement to assess the speed and effort put into a job by the worker. The British Standards Institute definition of the verb *to rate* is to assess the worker's rate of working relative to the observer's concept of the rate corresponding to standard rating. The observer may take into account, separately or in combination, one or more factors necessary to perform the task, for example, speed of movement, effort, dexterity, and consistency. The basic concept is to determine the time necessary to perform a task or job. It is not sufficient just to assess this by timing with a chronometer, a worker performing the task, or even estimating it, but to assess the time actually taken by a qualified worker who knows the job and is properly trained to do it and then adjust this actual time to what it would have been had that worker been working at the standard rate.

Rating eliminates the need to search for the mythical standard worker and takes out of the equation the need for the worker to adjust his or her pace to the standard rate of working, something which is difficult to do.

One of the original rating scales was *Bedaux's 60/80 scale*. Bedaux considered workers paid on a fixed-day work system without any financial incentive would normally do 60 minutes' worth of work in an hour, whereas one on a financial bonus scheme would get the work done on average one-third faster, doing 80 minutes' worth of work in an hour (incentive rate). The rest of this 60/80 scale was prorated. So, for example, a worker working twice as fast as this perceived "normal" 60 rating would be assessed as working at the 120 rating. The Bedaux scale was later converted to decimal form accepted by **British Standards (BS) Institute**, which allocated a rating of 75 BS, in place of Bedaux's 60, and 100 BS rating, replacing Bedaux's 80 rating. The complete BS scale supersedes the corresponding Bedaux scale pro rata.[11] Clearly such ratings can be highly subjective. To aid raters in conforming to the universally accepted concept of rating, there are sets of films/videos and CDs that demonstrate various jobs with their rates and have tests for training purposes.

British Standards (BS) Institute

British Standards is the new name of the British Standards Institute and is part of BSI Group, which also includes a testing organization. British Standards has a royal charter to act as the standards organization for the United Kingdom.

ADMINISTRATIVE MANAGEMENT

Henri Fayol (1841–1925) Known as the father of modern management, Henri Fayol was a French industrialist who developed a framework for studying management. His theories on organization of labor were very influential in the beginning of the twentieth century. He belonged to the administrative management branch of the classical school. His entire working career was spent with a mining company, Commentary-Fourchambault Company, in which he rose from an apprentice to general manager in 1888, remaining there until his retirement in 1918. He is credited with turning the company around from a threatened bankruptcy into a strong financial position by the time of his retirement at age 77. As a result of his management experience, Fayol strongly believed management theories could be developed and taught to others.

Fayol saw the job of the manager as planning, organizing, commanding, coordinating activities, and controlling performance. Henri Fayol wrote down his concepts of administration, based largely on his own management experience. Fayol's 14 principles of management were discussed in detail in his book published in 1916, *Administration Industrielle et General.* The book became prominent in the United States after a second English translation appeared in 1949 under the title *General and Industrial Management.*[12] Fayol's 14 principles include the following:

1. *Division of labor.* The more people specialize, the more efficiently they can perform their work. This principle is epitomized by the modern assembly line.
2. *Authority.* Managers must give orders so that they can get things done. Although their formal authority gives them the right to command, managers will not always compel obedience unless they have personal authority (such as relevant expertise) as well.
3. *Discipline.* Members in an organization need to respect the rules and agreements that govern the organization. To Fayol, discipline results from good leadership at all levels of the organization, fair agreements (such as provisions for rewarding superior performance), and judiciously enforced penalties for infractions.
4. *Unity of command.* Each employee must receive instructions from only one person. Fayol believed that when an employee reported to more than one manager, conflicts in instructions and confusion of authority would result.
5. *Unity of direction.* Those operations within the organization that have the same objective should be directed by only one manager using one plan. For example, the personnel department in a company should not have two directors, each with a different hiring policy.
6. *Subordination of individual interest to the common good.* In any undertaking, the interests of employees should not take precedence over the interests of the organization as a whole.
7. *Remuneration.* Compensation for work done should be fair to both employees and employers.
8. *Centralization.* Decreasing the role of subordinates in decision making is centralization; increasing their role is decentralization. Fayol believed that managers should retain final responsibility, but should at the same time give their subordinates enough authority to do their jobs properly. The problem is to find the proper degree of centralization in each case.
9. *The hierarchy.* The line of authority in an organization, often represented today by the neat boxes and lines of the organization chart, runs in order of rank from top management to the lowest level of enterprise.
10. *Order.* Materials and people should be in the right place at the right time. People, in particular, should be in the jobs or positions they are most suited to.
11. *Equity.* Managers should be both friendly and fair to subordinates.
12. *Stability of staff.* A high employee turnover rate undermines the efficient functioning of an organization.
13. *Initiative.* Subordinates should be given the freedom to conceive and carry out their plans, even though some mistakes may result.
14. *Esprit de corps.* Promoting team spirit will give the organization a sense of unity. To Fayol, even small factors should help to develop the spirit. He suggested, for example, the use of verbal communications instead of formal, written communication whenever possible.[13]

Chester Barnard (1886–1961) When Chester Barnard retired as the CEO of New Jersey Bell Telephone, he recorded his insights about management in his book, *The*

Functions of the Executive.[14] It outlined the legitimacy of the supervisor's directives and the extent of the subordinates' acceptance. He developed the concepts of strategic planning and the acceptance theory of authority. Strategic planning is the formulation of major plans or strategies, which guide the organization in pursuit of major objectives. Barnard taught the three top functions of the executive were to (1) establish and maintain an effective community system, (2) hire and retain effective personnel, and (3) motivate those employees. His acceptance theory of authority states managers have only as much authority as employees allow them to have. The acceptance theory of authority suggests authority flows downward but depends on acceptance by the subordinates. The acceptance of authority depends on four conditions:

1. Employees must understand what the manager wants them to do.
2. Employees must be able to comply with the directive.
3. Employees must think the directive is in keeping with organizational objectives.
4. Employees must think the directive is not contrary to their personal goals.

Barnard felt organizations were communication systems. He wrote, "it is particularly important for managers to develop a sense of common purpose where a willingness to cooperate is strongly encouraged." He is credited with developing the acceptance theory of management, emphasizing the willingness of people to accept those having authority to act. He feels the manager's ability to exercise authority is strongly determined by the employee's "zone of indifference," in which orders are accepted without undue question. Barnard believed each person had a zone of indifference or a range within each individual in which he or she would willingly accept orders without consciously questioning authority. It was up to the organization to provide sufficient inducements to broaden each employee's zone of indifference so that the manager's orders would be obeyed. Barnard also believed informal organizations within formal organizations perform necessary and vital communication functions for the overall organization. This is consistent with his belief that the executive's main organizational function is acting as a channel of communication and maintaining the organization in operation. Barnard's sympathy for and understanding of employee needs in the dynamics of the organizational communication process positions him as a bridge to the behavioral school of management, many of whose early members were his contemporaries.[15]

THE BEHAVIORAL MOVEMENT

As management research continued to evolve in the twentieth century, questions were increasingly raised regarding the interactions and motivations of the individual in organizations. Management principles developed during the classical period were simply not useful in dealing with many management situations and could not explain the behavior of individual employees. The principles of classical management theory were helpful in placing management objectives in the perspective of an organization; however, they failed to fulfill one of their earliest goals—providing management tools for dealing with organizational personnel challenges. In short, classical theory ignored employee motivation and behavior.

The behavioral school emerged partly because the classical approach did not achieve sufficient production efficiency and workplace harmony. To managers' frustration, people did not always follow predicted or expected patterns of behavior. Thus there was increased interest in helping managers deal more effectively with the "people side" of their organizations. Several theorists tried to strengthen classical organization theory with the insights of sociology and psychology. Curiously, it was a set of experiments that shed light on several underlying behaviors that impact how we lead and manage. The **Hawthorne experiments** (see page 106) rigorously applied classical management theory only to reveal its shortcomings. The behavioral school was a natural outgrowth of this revolutionary, yet controversial, management experiment. Its theorists include Mary Parker Follett and Herbert Simon as well as numerous psychologists who turned from studying individual behavior to organizational behavior.

Hawthorne experiments
The Hawthorne experiments, conducted at Western Electric's Hawthorne plant outside Chicago, starting in 1927 and running through 1932, were intended to bring about a greater understanding of the effects of working conditions on worker productivity. The results of the experiments were contrary to the management theory of the time (see *scientific management*) and were key in bringing about an understanding of motivation factors in employment.

Mary Parker Follett (1868–1933) Follett was a key historical figure in the field of organizational communications who spent her life working in poor Boston neighborhoods. Follett served on minimum wage boards and aided suffrage organizations. Being a teacher, author, scholar, feminist, and public servant certainly influenced her work relating to organizational relationships. Follett has been called a social anthropologist, recognized as a scholar, far ahead of her time, and deemed a pioneer in integrative negotiation. Physiologist Charles A. Elwood, one of Follett's contemporaries, wrote, "Follett was easily the most recognizable woman thinker among social and political lines and perhaps one of the most philosophical thinkers in the field of social theory of all time. This should be highly gratifying to all advocates of the emancipation and education of women, as well as to all who seek to further the progress of the social sciences."[16] Mary Parker Follett's concepts included the universal goal, the universal principle, and the Law of the Situation. The universal goal of organizations is an integration of individual effort into a synergistic whole. The universal principle is a circular or reciprocal response emphasizing feedback to the sender (the concept of two-way communication). The Law of the Situation emphasizes there is no one best way to do anything, because the best way is dependent on the situation at hand.

Follett's work as a management and political scholar introduced such phrases as *conflict resolution, authority and power*, and the *task of leadership*. She was definitely breaking new ground in respect to the understanding of interpersonal dynamics in the workplace and opening doors for women during this era.

> Unity, not uniformity, must be our aim. We attain unity only through variety. Differences must be integrated, not annihilated, not absorbed.
>
> *Mary Parker Follett*

Mary Parker Follett made very important contributions to the field of human resource management, suggesting organizations function on the principle of people power, not the power over people; the principle of integration; and the sharing of power. She strongly believed in the inherent problem-solving ability of people working in groups. Rather than assuming classical management's strongly hierarchical position of power in organizations, Follett asserts power should be cooperatively shared for the purpose of resolving conflict. She is best known for her integration method of conflict resolution as opposed to the three choices she cites of domination, compromise, or voluntary submission by one side to another. Follett stated the best way to handle this situation is to resolve the issue jointly through *creative conflict resolution*. Creative conflict resolution involves cooperatively working with others to devise inventive, new ideas often providing strong interpersonal benefits. We deal with daily issues within all our organizations as we try to integrate a more cohesive dynamic to empower people through facilitating, good coaching, mentoring relationships, and inclusion. Much of Mary Parker Follett's work is the basis for what we continue to do today in respect to personal aspects of being good leaders and managers.[17]

Behavioral or human relations management emerged in the 1920s and dealt with the human aspects of organizations. It has been referred to as the neoclassical school because it was initially a reaction to the shortcomings of the classical approaches to management. The human relations movement is often associated with the Hawthorne Studies, which were conducted from 1927 to 1932 at the Hawthorne plant of the Western Electric Company in Cicero, Illinois.

THE HAWTHORNE EFFECT

During the early 1900s, gas and electric companies fought for control of the lighting business in industry and homes. Industries eventually turned to the less expensive tungsten filament electric lightbulb.[18] The electric companies predicted less energy usage due to the efficiency of the bulbs, and they began to advertise that better (more) illumination increased productivity, as had been stated in several studies funded by those same industries. To overcome skepticism, the electric companies formed a committee, with Thomas Edison as honorary chair, and convinced the National Academy of Sciences at the Massachusetts Institute of Technology to conduct additional studies. Harvard Business School researchers T. N. Whitehead, Elton Mayo, and George Homans were led by Fritz Roethlisberger. Elton Mayo, known as the father of the Hawthorne Studies, identified the Hawthorne effect as the bias that occurs when people know they are being studied. The Hawthorne Studies are significant because they demonstrated the important influence of human factors on worker productivity.

In all, seven studies were conducted at the Hawthorne plant between 1927 and 1932. A well-known early study involved a highly repetitive task of assembling telephone relays by six women. Accurate records were kept of such factors as light intensity, room temperature, humidity, and production. Production was measured by a continuously running output recorder, and information about the rate of production was available to individual workers. As the research continued, the results became more puzzling; production rates kept rising no matter what was done. The researchers decided illumination was not an important variable and began examining other factors, including weather conditions, wage payments, rest periods, and the workers' physical and emotional conditions. Apparently no single factor, least of all the level of lighting, could account for productivity levels.[19]

There were four major phases to the Hawthorne Studies: the illumination experiments, the relay assembly group experiments, the interviewing program, and the bank wiring group studies. The intent of all these studies was to determine the effect of working conditions on productivity. The illumination experiments tried to determine whether better lighting would lead to increased productivity. Both the controlled group and the experimental group of female employees produced more whether the lights were turned up or down. It was discovered this increased productivity was a result of the attention received by the group. In the relay assembly group experiments, six female employees who worked in a special separate area were given breaks; had the freedom to talk; and were continuously observed by a researcher who served as the supervisor. The supervisor consulted the employees prior to any change. The interviewing stage examined over 21,000 employees in respect to how they felt about work and the company.

The bank wiring group studies were analyzed thoroughly and were included in a now classic book, *The Human Group*.[20] The reported study was done in the bank wiring observation room, where no changes were made in physical working conditions, methods of payment, or anything else. The workers were simply observed. There were only 14 men in this department, divided into three semiautonomous but interdependent work groups. The men did piecework, and their output was easily measured. The study showed the group had a clear-cut standard for production levels—*a standard established by the group itself, not by management.* Another finding, given little or no emphasis at the time, was that the workers did not increase productivity because of the observer's "interest." For years, several important findings were that workers' individual characteristics and the social characteristics of both the work and its environment were highly important. "Solitary" people do not behave the same as people who belong to a group. The group itself is important. A group's production is better predicted by the establishment of a standard of productivity and when there are identified punitive measures should the group fail in its mission. One of the most widely known "facts" about the studies is the *Hawthorne effect*, which is when workers' behavior changes and productivity increases because the workers are aware that persons important in their lives are taking an interest in them.

Relay Assembly Test Room Experiments

- Examined relation of light intensity and worker efficiency
- Failed to find simple relationship
- Believed behavior is not merely physiological but also psychological
- Decided to learn more about workers, for example, worker attitude

Relay Assembly Test II, 1927

- Selected six workers from large shop floor; average worker completed five relays in six minutes
- Kept record of output for five years—quality, weather conditions, worker health, sleep
- Had no supervision; workers told of experiment, could suggest changes
- Work conditions varied, for example, rest periods, length of workday
- Looked at effect of changes
- Results—output rose slowly and steadily even with shorter workday
- Workers said experiment was "fun"; liked absence of supervision; group developed socially, informal leadership, common purpose

Interviewing Stage, 1928

- Examined how 21,000 employees felt about work and company
- Learned how to improve supervisory training
- Found supervision improved as supervisors began to look at employees differently
- Found managers knew little about good supervision
- Concluded that employees could not be viewed as individuals, but rather as part of organized social groups, families, neighborhoods, working groups
- Workers band together for protection; purposely restrict output to norm; resent group piecework; punish rate busters; enjoyed fooling management
- Informal leaders kept group together

Bank Wiring Observation Room, 1931–1932

- Chose nine workers, three soldermen, two inspectors to assemble terminal banks
- Group piecework used—guaranteed base rate; pay reflected both group and individual effort
- What happened—employees had notion of proper day's work; most work done in morning; when they felt they had done what they considered enough, they slacked off so output was constant
- Wage incentive did not work; informal social organization evolved
- Workers often traded jobs and helped one another; supervisor often looked the other way
- Why did workers restrict output—did not want management to know they could do more
- Complex social system evolved—common sentiments, relationships
- What is critical is not what is, but what is perceived
- Because workers could not affect management, group gave meaning and significance to work
- Workers resisted formal changes in management to break up loyalties and routine

Herbert Simon (1916–2001) The death knell of classical management theory was pronounced by Herbert Simon in his book *Administrative Behavior: A Study of Decision-Making Processes in Administration Organization*, published in 1947.[21] Simon is particularly critical of the principles of administration, including span of control and unity of command, while saying all of the principles collectively were "no more than proverbs." Simon found the principles of classical administration to be contradictory and vague. Simon's greatest management contribution is in decision-making theory for which he received a Nobel Prize in 1978. Simon states decision makers perform in an arena of *bounded rationality* and that the approach to decision making must be one of *satisficing*, whereby satisfactory rather than optimum decisions are often reached. *Satisficing* successfully adapts to and is a realistic solution for the limited time and resources a manager has when considering alternatives in the decision-making process.

HUMAN RELATIONS MANAGEMENT

The work of Mary Parker Follett, the Hawthorne experiments, and the criticism of the classical school by Herbert Simon led to a deeper consideration of the needs of the employees and the role of management as a provider for these needs. *Human relations* is usually used as a general term to describe the ways in which managers interact with their employees. When "employee management" stimulates more and better work, the organization has effective human relations; when morale and efficiency deteriorate, its human relations are said to be ineffective. The human relations movement arose from early systematic attempts to discover the social and psychological factors that would create effective human relations.

The application of these human relations theories can be seen in today's work environment. For example, with the restructuring of today's competitive global economy, many companies have made the decision to "downsize," or reduce the numbers of managers and workers. However, some companies, well aware of the dynamics pointed out by the Hawthorne Studies, have approached employee cutbacks or organizational reengineering with great care. Behavioral scientists brought new dimensions to the study of management. *The two major organizational theorists in the human relations movement are Abraham H. Maslow* and *Douglas McGregor.* Abraham Maslow and Douglas McGregor, among others, wrote about "self-actualizing" people. Their work spawned a new way of thinking about how relationships can be beneficially arranged in organizations. They also determined that people wanted more than "instantaneous" pleasure or rewards. If people were this complex in the way they led their lives, then their organizational relationships needed to support that complexity.

Second, behavioral scientists applied the methods of scientific investigation to the study of how people behaved in organizations as whole entities. The work of James March and Herbert Simon in the late 1950s developed hundreds of propositions for scientific investigation about patterns of behavior, particularly with regard to communication, in organizations. Their influence in the development of subsequent management theory has been significant and ongoing.

According to Maslow, the needs people are motivated to satisfy fall into a hierarchy. Physical and safety needs are at the bottom of the hierarchy, and at the top are ego needs (the need for respect, for example) and self-actualizing needs (such as the need for meaning and personal growth). In general, Maslow said lower-level needs must be satisfied before higher-level needs can be met. Because many lower-level needs are routinely satisfied in contemporary society, most people are motivated more by the higher-level ego and self-actualizing needs. Later behavioral scientists feel even this model cannot explain all the factors that may motivate people in the workplace. They argue not everyone goes predictably from one level of need to the next. For some people, work is only a means for meeting lower-level needs. Others are satisfied with nothing less than the fulfillment of their highest-level needs; they may even choose to work in jobs that threaten their safety, if by doing so they can attain personal goals. (Sound familiar?!) The more realistic model of human motivation, these behavioral scientists argue, is "complex person." Using this model, the effective manager is aware no two people are exactly alike and tailors motivational approaches according to individual needs.

As American corporations increasingly do business with other cultures, it is important to remember theories can be culturally bounded and limited in their applications. For example, Maslow's hierarchy of needs is not a description of a universal motivational process. In other nations the order of the hierarchy might be quite different depending on the values and economics of the specific country. In Sweden, quality of life is ranked most important, while in Japan and Germany, security is ranked the highest; and in India it is based on survival needs for many.

McGregor provided another angle on this complex person idea. He distinguished two alternative basic assumptions about people, their approach to work, their motivation while at work, how they interact, and how they manage people. These two assumptions, which he called Theory X and Theory Y, take opposite views of people's commitment to work in organizations. Theory X managers, McGregor proposed, assume people must be constantly coaxed into putting forth effort in their jobs. Theory Y managers, on the other hand, assume people relish work and eagerly approach it as an opportunity to develop their creative capacities. Theory Y was an example of a complex person perspective. Theory Y management, McGregor claimed, was stymied by the prevalence of Theory X practices in organizations of the 1950s. The roots of Theory X can be traced to the days of scientific management and the managing of factories based on these principles.

motivation
The factors that cause, channel, and sustain an individual's behavior.

The word **motivation** comes from the Latin word *movere*, meaning "to move." In common usage, a motivator is anything causing a person to change behavior, or

TABLE 3.2 ◆ The Motivational Framework

Need (Deficiency)	→	Search for ways to satisfy need	→	Choice of behavior to satisfy need ↓
↖		Determination of future needs and search/choice for satisfaction	←	Evaluation of need satisfaction

"move," and a motive is an incentive to act. Obviously, what motivates one person may not motivate another, and the same motivator may not always have the same impact on the same person (see Table 3.2).

Leaders need to identify legitimate and satisfactory ways to convince workers to improve their behavior and productivity on the job. Getting people to change basic values is difficult. Managers often seek to change the job or the environment surrounding the job. Motivation "models" and theories show how to achieve this most efficiently. One of the best-known models on motivation is the pyramid of needs.

THE NEEDS HIERARCHY MODEL

Abraham Harold Maslow (1908–1970), was the first of several children born in Brooklyn, New York, to parents who were uneducated Jewish immigrants from Russia. While attending the University of Wisconsin, he became interested in psychology and spent time working with Harry Harlow, who was famous for his experiments with baby Rhesus monkeys and detachment behavior. Maslow became interested in the *concept of self-actualization*, started his own theoretical work, and began his crusade for humanistic psychology.

The needs hierarchy model developed by Maslow[22] shows people have five levels of needs. The concept is based upon the theory that the lower-level needs must be satisfied, or at least partially satisfied, before the higher-level needs emerge and can be addressed. The five levels, from lowest to highest, are physiological (food, clothing, shelter), safety, social, ego, and self-actualization (developing full potential) (see Figure 3.1).

Maslow's pyramid of needs models the basis of motivation, which he developed over years of study. The first and broadest level of needs is physiological needs. This level provides the basics of life, such as food, shelter, and physical comforts. The second level involves safety (physical and job security), a high priority for most people. The top three levels of motivation (belonging, esteem, and self-actualization) present the biggest leadership challenges. It is difficult to develop an environment that allows team players to find opportunities for self-actualization and to be respected and em-

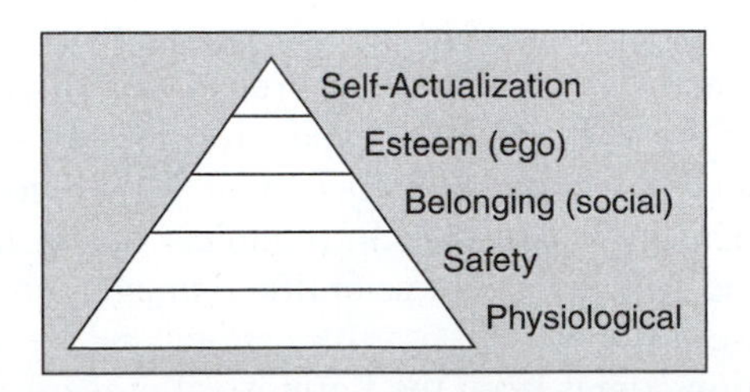

FIGURE 3.1 ◆ Maslow's Pyramid of Needs

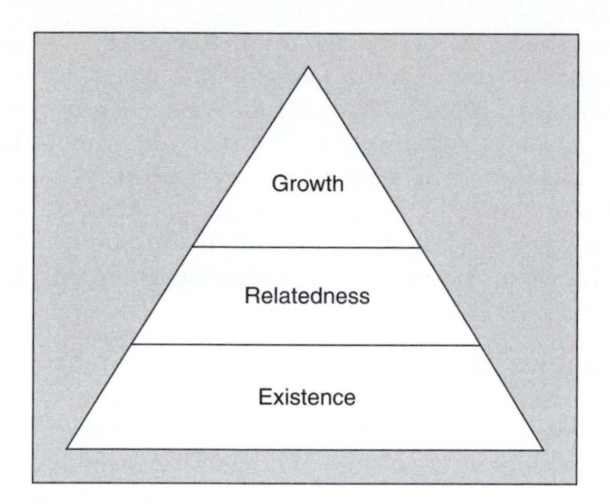

FIGURE 3.2 ◆ Three Levels of Need

powered by fellow team players to continue positive and productive individual efforts. Employees move up and down the pyramid during their employment experience. The highest and most influential level of motivation (self-actualization) is often short lived. Employees can attain a temporary "high" from personal achievement and then fall back to another level on the pyramid very quickly. Work experiences offer continued challenges while employees are pursuing self-actualization. Good leaders work hard to provide the necessary resources, planning, guidance, and support to achieve desired results (on time and under budget). A positive work environment—which supports employee achievement, recognizes success, and adjusts for failures—is necessary when motivating employees.

You may see Maslow's five levels reduced to three: *existence* (the lowest level), *relatedness* (the level encompassing social needs), and *growth* (including the ego and self-actualization levels), as some of today's researchers have adopted his basic concept (see Figure 3.2). These make up what is sometimes called the *ERG* model. People often behave on two or more levels simultaneously. For example, an employee may require social acceptance even though his or her self (ego) is fulfilled after a personal accomplishment. Also, people may shift quickly from one need level to another, as when facing physical danger or when employment is lost.

In applying the motivational theories to fire departments, there are distinct differences among organizations. For example, in a department that is respected and secure within the community and whose services are well used, existence is not threatened. A leadership style that promotes a safe work environment, performs ongoing training, establishes safe work standards, and upgrades safety equipment to the "state-of-the-art" upholds the level of security necessary under the existence theories. A leadership style that uses the tenets of team play by diffusing authority through the ranks and using communities, quality circles, brainstorming, and other group skills to determine departmental direction builds a solid base for social relatedness needs. When team players realize they have a vested interest in achieving desired results (as validated by team goal setting and buy-in to the mission), they exercise the opportunity to grow within the organization and achieve self-actualization. If the leaders support and recognize personal growth and provide opportunity for the team players to achieve desired results, the morale of the organization is enhanced.

Behavior of personnel can be affected by extrinsic and intrinsic rewards. Extrinsic rewards, such as working conditions, promotions, and commendations, can be given by the department. Intrinsic rewards, such as a sense of worth and accomplishment, can be given only by oneself. However, department leadership can design and organize the work environment and procedures to enable individuals to reward themselves intrinsically. An example occurs when departments hold critiques of fires that focus on what was done well in addition to what could stand improvement. This allows for positive communication about performance as well as constructive critical review.[23]

THEORY X AND THEORY Y

In his 1960 management book, *The Human Side of Enterprise*,[24] Douglas McGregor (1906–1964) made his mark on the history of organizational management when he proposed two motivational theories by which managers perceived employees' motivation.

McGregor's work was based on Maslow's hierarchy of needs. He grouped Maslow's hierarchy into "lower-order" (Theory X) needs and "higher-order" (Theory Y) needs. He suggested management could use either set of needs to motivate employees, but better results could be obtained by meeting the Theory Y needs. These opposing motivation theories are referred to as Theory X and Theory Y. Each theory assumes management's role is to organize resources, including people, to the best benefit of the company. Beyond this common benefit, they are quite dissimilar. McGregor's Theory X and Theory Y are still featured in textbooks on management and organizational theories. However, contemporary management scholars have largely rejected McGregor's arguments, preferring contingency theories emerging from more empirical studies and analysis. Written in 1960, McGregor's book is undoubtedly reflective of the times, yet captures a fundamental core issue of how people treat one another in the workplace. Although much has changed since 1960, many of the issues brought about in Theory X and Theory Y are still alive and well in many organizations.

Theory X

Developed by Douglas McGregor in the 1960s. The assumptions that the average employee dislikes work, is lazy, has little ambition, and must be directed, coerced, or threatened with punishment to perform adequately.

Theory X assumes the average human being has an inherited dislike of work and will avoid it if he or she can. Essentially, Theory X assumes the primary source of most employee motivation is monetary, with security a strong second. In this theory, management assumes employees are inherently lazy because the workers need to be closely supervised and a comprehensive systems of controls developed. A hierarchical structure is needed with a narrow span of control at each level. According to this theory, employees will show little ambition without an enticing incentive program and will avoid responsibility whenever they can.

According to McGregor, most managers (in the 1960s) tend to subscribe to Theory X, taking a rather pessimistic view of their employees. A Theory X manager believes his or her employees do not really want to work, they would rather avoid responsibility, and it is the manager's job to structure the work and energize the employee. The result of this line of thought is Theory X managers naturally adopt a more authoritarian style based on the threat of punishment:

- Because of their dislike of work, most people must be controlled and threatened before they will work harder.
- The average person prefers to be directed, has little ambition, dislikes responsibilities, and desires security above everything else.
- Employees must be coerced, controlled, directed, and threatened with punishment to get them to put forth adequate effort toward the achievement of organizational objectives.

In **Theory Y** management assumes employees are ambitious, self-motivated, and anxious to accept greater responsibility, and exercise self-control and self-direction. It is believed that employees enjoy their mental and physical work activities. It is also believed employees have the desire to be imaginative and creative in their jobs if they are given a chance. There is an opportunity for greater productivity by giving employees the freedom to be their best. Theory Y managers believe, given the right conditions, most people will want to do well at work and there is a pool of unused creativity in the workforce. They believe the satisfaction of doing a good job is a strong motivation in and of itself. Theory Y managers will try to remove the barriers that prevent workers from fully actualizing their potential. In contrast to Theory X managers, Theory Y managers tend to believe:

Theory Y
Developed by Douglas McGregor in the 1960s. Assumes employees are ambitious, self-motivated, and anxious to accept greater responsibility and exercise self-control and self-direction. It is believed that employees enjoy their mental and physical work activities, and have the desire to be imaginative and creative in their jobs if they are given a chance.

- Given the right conditions for employees, their application, and physical and mental health, work is as natural as rest or play. Work is play that offers satisfaction and meaning.
- If people feel committed, they will exercise self-direction and self-control in support of the organization's objective(s).
- The accomplishment of these objectives will support the organization's mission and provide intrinsic rewards associated with their achievements. Theory Y managers recognize the influence of learning. They believe if the right conditions are created, the average person learns to accept and seek responsibility.
- The capacity to exercise imagination, ingenuity, and creativity in the solution of organizational problems is distributed throughout the workforce.
- In modern organizations the intellectual potential of the average person is only partially utilized. People are capable of handling much more complex problems, given the opportunity.

These assumptions are based on social science research, which observed the potential that is present in people that the organization must recognize in order for it to become more effective. McGregor saw these two theories as quite separate attitudes. Theory Y would be difficult to put into practice on the shop floor in a large mass production operation. It could be effectively used initially managing managers and professionals. In *The Human Side of Enterprise*, McGregor showed how Theory Y affects the management of promotions and salaries and the development of effective managers. McGregor also saw Theory Y as conducive to participative problem solving. There are times when a manager must exercise authority, and in certain cases this is the only method of achieving those desired results, for example, when subordinates do not agree with a proposed change, yet it is a new departmental policy that must be implemented.

THEORY Z

Theory Z is not a McGregor idea and, as such, is not McGregor's extension of his XY theory. **Theory Z** was developed by William Ouchi, in his 1981 book, *Theory Z: How American Management Can Meet the Japanese Challenge*.[25] William Ouchi is professor of management at UCLA, Los Angeles, and a board member of several large U.S. organizations.

Theory Z
According to William Ouchi, the management belief that the key to productivity and quality is the development and participation of all employees.

Theory Z is often referred to as the Japanese management style. It is interesting Ouchi chose to name his model Theory Z, which, apart from anything else, tends to give the impression it is a McGregor idea. Nevertheless, Theory Z advocates a combination of all that is best about Theory Y and modern Japanese management, which places a large amount of freedom and trust with workers because it assumes workers

TABLE 3.3 ◆ Theory X, Theory Y, and Theory Z Models

Assumptions About:	*Theory X*	*Theory Y*	*Theory Z*
Workers' motivation	The Theory X manager assumes that the only motivation that works for employees is money.	The Theory Y manager assumes that employees are motivated by their needs to fulfill their social, esteem, self-actualization, and security needs.	The Theory Z manager assumes that employees are motivated by a strong sense of commitment to be a part of something worthwhile—the self-actualization need.
Workers' attitude toward work	The Theory X manager assumes that the employees dislike work, avoid responsibilities, and seek only security from work (the paycheck).	The Theory Y manager believes that employees see work as a natural activity and will seek out opportunities to have increased responsibility and understanding of their tasks.	The Theory Z manager believes that employees will not only seek out opportunities for responsibility but also crave opportunities to advance and learn more about the company.
What will work with employees	The Theory X manager believes that workers will respond only to coercion, control, direction (telling them exactly what to do), or threatening punishment or firing.	The Theory Y manager believes that workers will respond best to favorable working conditions that do not pose threats or strong control.	The Theory Z manager believes that employees should learn the business through the various departments, coming up through the ranks slowly, and that the company will get the best benefits from that employee by making it possible for him or her to have "lifetime employment." The result will be strong bonds of loyalty developed by long-term employment and shared responsibility for decisions.

have a strong loyalty and interest in teamwork and the organization. Theory Z also places more reliance on the attitude and responsibilities of the workers, whereas McGregor's XY theories are focused mainly on management and motivation from the manager's and organization's perspectives. There is no doubt that Ouchi's Theory Z model offers excellent ideas (see Table 3.3).

managerial grid model
A behavioral leadership model developed by Robert Blake and Jane Mouton. This model identifies five different leadership styles based on the *concern for people* and the *concern for production*.

THE MANAGERIAL/LEADERSHIP GRID

Another classic in management research, dealing with motivation of personnel, was reported in 1964 by the team of Robert R. Blake and Jane S. Mouton.[26] Their **managerial grid model** illustrates the concern for people and the concern for production found in successful managers. There is an implication that the most successful managers emphasize both concerns strongly. This means workers want leadership to provide good working conditions, job security, recognition, respect, and praise. It

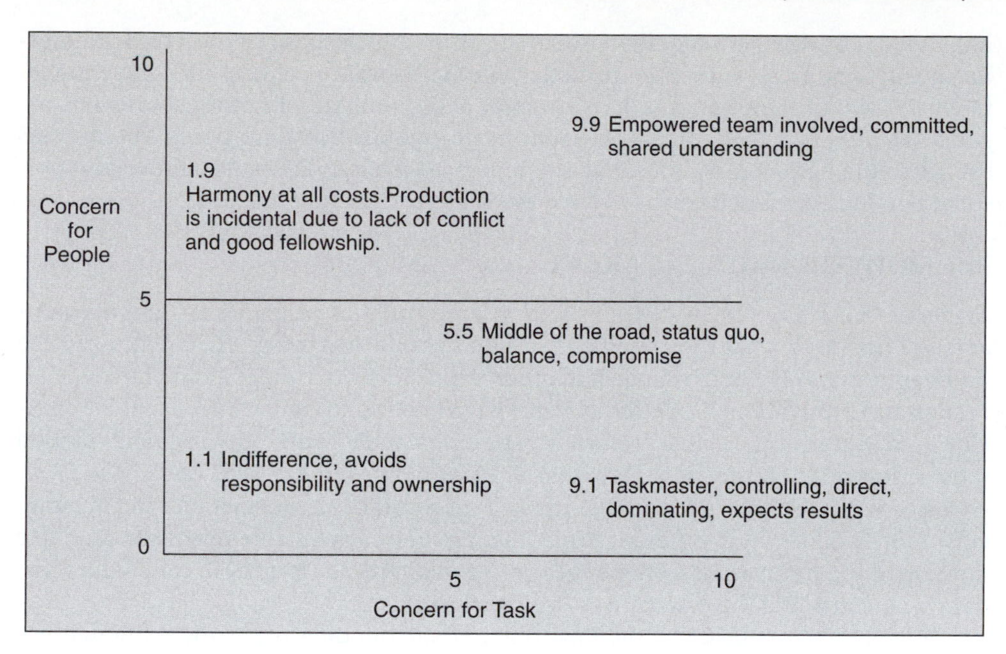

FIGURE 3.3 ◆ Managerial/Leadership Grid

means leaders should use management skills to push for production goals (however they may be defined) as well as team harmony (see Figure 3.3).

The managerial grid provides a behavioral science schematic for comparing nine theories of intersection between production and human relationships. The horizontal axis represents concern for production. The vertical axis represents concern for relationships among those engaged in production. Each is expressed as a nine-point scale with 1 representing minimum interest/concern and 9 presenting maximum interest/concern. The managerial grid recognizes the possibility some managers may lean more toward one of the two concerns than the other, or the leader's concerns may shift depending on the particular situation. An experienced leader can read the status of an organization and understand the current motivational experiences. Teams framed and inculturated under a certain type of leadership orientation may not react well when exposed to another type of leadership (though the change of focus may be necessary).

Leaders should use the managerial grid more as a scale than as a thermostat that can be adjusted one way or the other. We cannot simply turn on the X or the Y to move the orientation to the most desirable position. The experiences that create a team feeling, yielding an X or Y orientation, are complicated and take years to develop. It is not uncommon for teams to take years to adjust to a different leadership style.[27]

E. Wright Bakke[28] (1932–1971) proposed a concept that helps explain what may underlie work behavior, and thus productivity. Bakke's concept deals with the conflicts resulting from the mixture of organizational and individual demands. Organizational leadership impresses a pattern upon the individual employee and attempts to make the person an agent who will work toward the department's goals. Conversely, the employee seeks to impress an orientation pattern upon the organization for personal purposes. The simultaneous operation of these two processes reconstructs both the individual and the organization in what Bakke terms a fusing of the two. This "fusion process" implies that the goals and aspirations of both the department and its individual personnel are modified over time, probably because of their own counter-balancing pressures. Neither gets exactly what it sets out to achieve. Of course, sometimes the will of the organization is stronger and dominates. In other cases workers' goals are obtained that may be different from organizational purposes. The process speaks to the basic motivations that are important from both organizational and personal perspectives. Each helps to shape the culture of the organization.

THE MOTIVATION-HYGIENE MODEL

Frederick Herzberg (1923–2000)[29] asked a large number of engineers and accountants in Pittsburgh what they liked and disliked about their work. He categorized the findings into two factors: hygiene and motivation. To better understand employee attitudes and motivation, Frederick Herzberg performed studies, which included interviews, to determine which factors in an employee's work environment caused satisfaction or dissatisfaction. He published his findings in the 1959 book, *The Motivation to Work*. Herzberg found that the factors causing job satisfaction (and presumably motivation) were different from the factors causing job dissatisfaction. He developed the motivation-hygiene theory to explain these results. He called the satisfiers *motivators* and the dissatisfiers *hygiene factors*.

Hygiene factors describe the environment surrounding the work. Although not personally motivational, they are critical to the well-being of the organization. Like preventive inoculations, they are not cures, but they can prevent sickness (hence the term *hygiene factors*). Hygiene factors include company policy and administration, the development of promotional procedures, the kind of supervision people are exposed to, training standards, security, salary, working conditions, and relationships with others. Although these factors do not lead to higher levels of motivation, without them there would be dissatisfaction.

Motivation factors include the nature of the work itself and the opportunity for job-enriching experiences such as personal achievement of a goal or challenge. Motivation factors increase job satisfaction, raise motivation, and help improve productivity. Herzberg decided motivation stems from a challenging job encouraging personal growth, development, recognition, enjoyment of the work, and job-enriching experiences. Poor hygiene and leadership practices that fail to recognize the importance of team effort and proper discipline are discouraging. Table 3.4 presents the top six factors causing dissatisfaction and the top six factors causing satisfaction, listed in the order of higher to lower importance.

Herzberg reasoned that because the factors causing satisfaction are different from those causing dissatisfaction, the two feelings cannot simply be treated as opposites of

TABLE 3.4 ◆ Factors Affecting Job Attitudes

Leading to Dissatisfaction	*Leading to Satisfaction*
Company policy	Achievement
Supervision	Recognition
Relationship with boss	Work itself
Work conditions	Responsibility
Salary	Advancement
Relationship with peers	Growth

one another. The opposite of satisfaction is not dissatisfaction, but rather, no satisfaction. Similarly, the opposite of dissatisfaction is no dissatisfaction. Although at first glance this distinction between the two opposites may sound like a play on words, Herzberg argued that there are no distinct human needs portrayed. First, there are physiological needs that can be fulfilled by money, for example, to purchase food and shelter. Second, there is the psychological need to achieve and grow, and this need is fulfilled by activities that cause one to grow. In some cases the factors that determine whether there is dissatisfaction or no dissatisfaction are not part of the work itself but, rather, are external factors. Herzberg often referred to these hygiene factors as *KITA factors*. *KITA* is an acronym for Kick in the A . . . , the process of providing incentives or a threat of punishment to cause someone to do something. Herzberg argues these KITA factors provide only short-run success because the motivator factors that determine whether there is satisfaction or no satisfaction are intrinsic to the job itself, and do not result from carrot and stick incentives.

Implications for Management If the motivation-hygiene theory holds, management not only must provide hygiene factors to avoid employee dissatisfaction but also must provide factors intrinsic to the work itself in order for employees to be satisfied with their jobs. Herzberg argued job enrichment is required for intrinsic motivation, and it is a continuous management process. According to Herzberg, the job should have sufficient challenge to utilize the full ability of the employee; employees who demonstrate increasing levels of ability should be given increasing levels of responsibility; and if a job cannot be designed to use an employee's full abilities, then the organization should consider automating the task or replacing the employee with one who has a lower level of skill. If a person cannot be fully utilized, then there will be a motivation problem.

Critics of Herzberg's theory argue the two-factor result is observed because it is natural for people to take credit for satisfaction and to blame dissatisfaction on external factors. Furthermore, job satisfaction does not necessarily imply a high level of motivation or productivity. Herzberg's theory has been broadly read, and its enduring value is its recognition that true motivation comes from within a person and not from KITA factors. The motivation-hygiene model provides leadership with a better understanding of motivation and job satisfaction. The salary system is a form of hygiene. The payroll system provides satisfaction if developed appropriately and dissatisfaction if it is "out of sync" with employee expectations. A work project for which overtime is paid, not the money itself, can be a motivational force. If the work assignment provides the employee with the opportunity to achieve a higher level of self-esteem and a feeling of personal accomplishment, motivation will occur.[30]

David C. McClelland (1917–1998) David C. McClelland of Harvard University performed research on motivation patterns. McClelland spent much of his career studying motivation and how it affects leadership behavior. He identified *achievement*—meeting or exceeding a standard of excellence of improving personal performance—as one of three internal drivers (he called them *social motives*) that explain how we behave. The other two are *affiliation*, maintaining close personal relationships, and **power**, which involves being strong and influencing or having an impact on others. He said the power motive comes in two forms: personalized—the leader draws strength from controlling others and making them feel weak; and socialized—the leader's strength comes from empowering people. Studies show that great charismatic leaders are highly motivated by socialized power; personalized power is often associated with the exploitation of subordinates (see Table 3.5).

power
The ability to exert influence; that is, the ability to change the attitudes or behavior of individuals or groups.

TABLE 3.5 ◆ What's Your Motivation?

A small set of motives, present to some extent in all people, helps explain how leaders behave. The motives generate needs, which lead to aspirations, which in turn drive behavior.

	Achievement	*Affiliation*	*Power*	
			Personalized Power	*Socialized Power*
When this motive is aroused in the, leaders experience a need to:	Improve their personal performance and meet or exceed standards of excellence	Maintain close, friendly relationships	Be strong and influence others, making them feel weak	Help people feel stronger and more capable
As a result, they wish to:	Meet or surpass a self-imposed standard	Establish, restore, or maintain warm relationships	Perform powerful actions	Empower people
	Accomplish something new	Be liked and accepted	Control, influence, or persuade people	Persuade people
	Plan the long-term advancement of their careers	Participate in group activities, primarily for social reasons	Impress people inside or outside the company	Impress people inside or outside the company
			Generate strong positive or negative emotions in others	Generate strong positive emotions in others
			Maintain their reputations, positions, or strengths	Maintain their reputations, positions, or strengths
				Give help, advice, or support
These aspirations lead them to:	Micromanage	Avoid confrontation	Be coercive and ruthless	Coach and teach
	Try to do things or set the pace themselves	Worry more about people the performance	Control or manipulate others	Be democratic and involve others
	Express impatience with poor performers	Look for ways to create harmony	Manage up—that is, focus more on making a good impression than on managing their subordinates	Be highly supportive
	Give little positive feedback	Avoid giving negative feedback	Look out for their own interests and reputations	Focus on the team or group rather than themselves
	Cut corners			
	Focus on goals and outcomes rather than people			

Source: Scott W. Spreier, Mary H. Fontaine, and Ruth L. Malloy, "Leadership Run Amok: The Destructive Potential of Overachievers," *Harvard Business Review*, June 1, 2006.

McClelland's research showed that all three motives are present to some extent in everyone. Although we are not usually conscious of them, they give rise in us to needs and concerns that lead to certain behaviors. Meeting those needs gives us a sense of satisfaction and energizes us, so we keep repeating the behaviors, whether or not they result in the outcomes we desire.

McClelland initially believed that, of the three motives, achievement was the most critical to organizational, even national, success. In *The Achieving Society*, his seminal study on the subject, first published in 1961, he reported that a high concern with achievement within a country was followed by rapid national growth, whereas a drop led to a decline in economic welfare. In another study, he reported a direct correlation between the number of patents generated in a country and the level of achievement as a motivation.

But McClelland also recognized the downside of achievement: the tendencies to cheat and cut corners and to leave people out of the loop. Some high achievers "are so fixated on finding a shortcut to the goal," he noted, "that they may not be too particular about the means they use to reach it." In later work, he argues that the most effective leaders were primarily motivated by socialized power: They channeled their efforts into helping others be successful[31] (see Table 3.5).

HUMAN RESOURCES THEORY

Beginning in the early 1950s, the human resources theory represented a substantial progression from the human relations approach. The behavioral approach did not always increase productivity. Thus, motivation and leadership techniques became a topic of great interest. The human resources theory began to explore and understand that employees are very creative and competent, and that much of their talent is largely untapped by their employers. Employees want meaningful work; they want to contribute; they want to participate in decision making and leadership functions.

INTEGRATING THE MANAGEMENT THEORIES

An overview of systems and contingency theories can help integrate the theories of management and promote an understanding when appropriate managerial techniques can be applied as required by environmental conditions. A broad perspective is valuable to managers when overseeing one unit or the total integration of many departments or divisions.

SYSTEMS THEORY

During the 1940s and World War II, systems analysis emerged and used a system concept and quantitative approach from mathematics, statistics, engineering, and related fields to solve problems. The point of the **systems approach** is that managers cannot function wholly within the confines of the traditional organization chart. They must mesh their department with the whole enterprise. To do this, they have to communicate not only with other employees and departments but also frequently with representatives of other organizations as well. Many of the concepts of

systems approach
View of the organization as a unified, directed system of interrelated parts.

general systems theory have found their way into the language of today's management processes.

Subsystems. The parts that make up the whole of a system are called subsystems. Each system in turn may be a subsystem of a still larger whole.

Synergy. **Synergy** means that the whole is greater than the sum of its parts. In organizational terms, synergy means that as separate departments within an organization cooperate and interact, they become more productive than if each were to act in isolation.

Open and closed systems. A system is considered an open system if it interacts with its environment; it is considered a closed system if it does not.

System boundary. Each system has a boundary that separates it from its environment. In a closed system, the system boundary is rigid; in an open system, the boundary is more flexible.

Flow. A system has flows of information, materials, and energy (including human energy). These enter the system from the environment as inputs (raw materials, for example), undergo transformation processes within the system (operations that alter them), and exit the system as outputs (goods and services).

Feedback. Feedback is the key to system controls. As operations of the system proceed, information is fed back to the appropriate people, and perhaps to a computer, so that the work can be assessed and, if necessary, corrected.

synergy
The Greek root of this word means "working together." The concept of synergy is that every individual working on a team achieves more than he or she would as an individual. The situation in which the whole is greater than its parts. In organizational terms, the fact that departments that interact cooperatively can be more productive than if they operate in isolation.

System theory calls attention to the dynamic and interrelated nature of organizations and the management task. Thus, it provides a framework within which we can plan actions and anticipate both immediate and far-reaching consequences while allowing us to understand unanticipated consequences as they develop. With a systems perspective, managers can more easily maintain a balance between the needs of the various parts of the enterprise and the needs and goals of the organization.[32]

CONTINGENCY VIEW

In the mid-1960s, the contingency view of management, or situational approach, emerged. This view emphasizes the fit between organization processes and the characteristics of the situation. It calls for fitting the structure of the organization to various possible or chance events. It questions the use of universal management practices and advocates using traditional, behavioral, and systems viewpoints independently or in combination to deal with various circumstances. The contingency approach assumes that managerial behavior is dependent on a wide variety of elements. As such it provides a framework for integrating the knowledge of management thought.

The contingency approach (sometimes called the situational approach) was developed by managers, consultants, and researchers who tried to apply the concepts of the major schools of management study to real-life situations. When methods highly effective in one situation failed to work in other situations, they sought an explanation. Why, for example, did an organizational development program work brilliantly in one situation and fail miserably in another? Advocates of the contingency approach had a logical answer to all such questions: results differ because situations differ; a technique that works in one case will not necessarily work in all cases. According to the contingency approach, the manager's task is to identify what technique will, in a particular situation, under particular circumstances, and at a particular time, best contribute to the attainment of management goals. Where workers need to be encouraged to increase productivity, for example, the classical theorist may prescribe a new work-simplification scheme. The behavioral scientist may instead seek to create a psychologically motivating climate and recommend an approach such as job enrichment—the combination of tasks that are different in scope and responsibility and allow the worker greater autonomy in making decisions. But the manager trained in the contingency approach will ask, "What method will work better here?" If the workers are unskilled and training opportunities and resources are limited, work simplification would be the best solution. However, with skilled workers driven by pride in their abilities, a job-enrichment program might be more effective. The contingency approach represents an important turn in modern management theory, because it portrays each set of organizational relationships in its unique circumstances.

EMERGING MANAGEMENT TRENDS

In contemporary management thought, management theorists recognize that many management theories have been developed since the beginning of the twentieth century. These include classical management, scientific management, the behavioral movement, the human relations approach, behavioral research models, management by objectives (MBO), total quality management (TQM), and the list continues to evolve. Although the findings of one management school of thought often overlap those of others, at times the findings may be contradictory.

Several management paradigms that have emerged subsequent to the behavioral school have merged or meshed the various management schools of thought. This includes the two major management theorists, W. E. Deming and Peter Drucker.

W. EDWARDS DEMING (1900–1993)

Deming (see Figure 3.4) received his doctorate in physics from Yale and was invited by the Union of Japanese Scientists and Engineers to visit Japan in 1950. In addition

FIGURE 3.4 ◆ W. Edwards Deming
(Courtesy of Randy R. Bruegman, Exceeding Customer Expectations: Quality Concepts for the Fire Service, *p. 11)*

to urging the Japanese to use sampling methods to test for quality control, he also taught them the best way to lowered production costs was quality improvement. Dr. Deming was concerned with increasing organizational productivity by applying statistical quality controls as well as improving organizational communication.

Although well known in Japan, Deming was ignored for years in the United States; however, this began to change after he was featured on an American TV show dealing with the reasons why the Japanese competition was threatening American business. Dr. Deming is known as the father of the Japanese postwar industrial revival and was regarded by many as the leading quality guru in the United States. Trained as a statistician, his expertise was used during World War II to assist the United States in its effort to improve the quality of war materials. Dr. Deming was quoted as saying, "We have learned to live in a world of mistakes and defective products as if they were necessary to life. It is time to adopt a new philosophy in America." Invited by Japanese industrial leaders and engineers to Japan at the end of World War II, Dr. Deming was asked how long it would take to shift the perception of the world from the existing paradigm that Japan produced cheap, shoddy imitations to one of producing innovative quality products. Dr. Deming told the group members if they would follow his directions, they could achieve the desired outcome in five years. Few of the leaders believed him, but they were ashamed to say so and would have been embarrassed if they failed to follow his suggestions. As Dr. Deming told it, "They surprised me and did it in four years." He was invited back to Japan time after time, where he became a revered counselor. For his efforts he was awarded the Second Order of the Sacred Treasure by former Emperor Hirohito.

Japanese scientists and engineers named the famed Deming Prize after him. It is bestowed on organizations that apply and achieve stringent quality-performance criteria.

Total Quality Leadership Today, much of the quality leadership and management philosophies taught are based on the research and teachings of Dr. Deming's system of profound knowledge (see Figure 3.5).

The basic aim of any quality program is to better meet the needs of the organization's customers by using people and qualitative methods to continuously improve processes at all levels of the organization. It is a philosophy, not only about how the

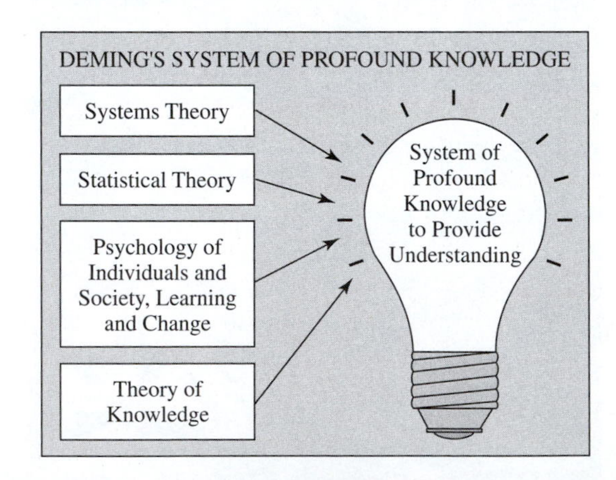

FIGURE 3.5 ◆ Dr. Deming's System of Profound Knowledge
(Courtesy of Randy R. Bruegman, Exceeding Customer Expectations: Quality Concepts for the Fire Service, *p. 11)*

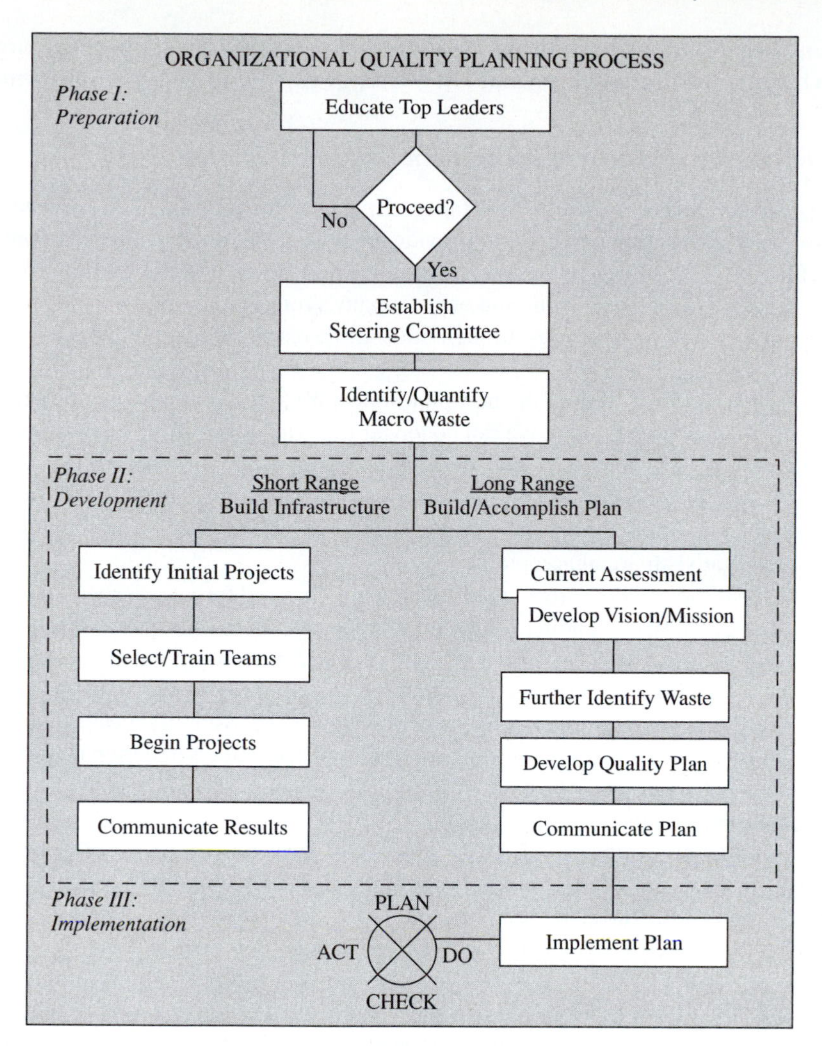

FIGURE 3.6 ◆ Organizational Quality Planning Process
(Courtesy of Randy R. Bruegman, Exceeding Customer Expectations: Quality Concepts for the Fire Service, *p. 13)*

system works but also about how the strategic leaders enhance systems and processes, and work to provide better service to their customers. Cooperation and coordination at all levels within every division of a department will be necessary to implement an effective **total quality leadership (TQL)** system (see Figure 3.6).

total quality leadership (TQL)
The application of quantitative methods and people to assess and improve materials and services supplied to the organization and all significant processes within the organization and to meet the needs of the end user now and in the future.

Dr. Deming's approach to process improvement includes:

Phase 1: Take a "picture of your process."
Phase 2: Analyze the "picture."
Phase 3: Make improvements and monitor the results.

Dr. Deming's business philosophy is summarized in his famous *14 Points*, listed next. These points have inspired significant changes among a number of leading U.S. companies striving to compete in the world's increasingly competitive environment. But the 14 Points pose a challenge for many organizations to figure out how to apply them in a meaningful way that will result in continual improvement. To accomplish this, many organizations have developed internal processes for their entire organizations to begin to institutionalize what Dr. Deming referred to as "the transformation."

Dr. Deming's 14 Points of Management

1. *Constancy of purpose.* Create constancy of purpose for improvement of products and service to society, allocating resources to provide for long-range needs rather than only short-term profitability, with a plan to become competitive, to stay in business, and to provide jobs. Deming suggests a radical new definition of a company's role. Rather than to make money, it is to stay in business and provide jobs through innovation, research, constant improvement, and maintenance.[33]

2. *The new philosophy.* Adopt the new philosophy. We are in a new economic age, created in Japan but certainly global today. We can no longer live with commonly accepted levels of delays, mistakes, defective materials, and defective workmanship. Transformation of Western management style is necessary to halt the continued decline of business and industry. Americans are too tolerant of poor workmanship and sullen service. We need a new religion in which mistakes and negativism are unacceptable.

3. *Cease dependence on mass inspection.* Eliminate the need for mass inspection as the way of life to achieve quality by building quality into the product in the first place. Require statistical evidence of built-in quality in both manufacturing and purchasing functions. American organizations typically inspect a product as it comes off the assembly line or at major stages along the way. Defective products are either thrown out or reworked. Both practices are unnecessarily expensive. In effect, a company is paying workers to make defects and then to correct them. Quality comes not from inspection but from improvement in the process. With instruction, workers can be enlisted in this improvement.

4. *End lowest tender contracts.* End the practice of awarding business solely on the basis of price tag. Instead, require meaningful measures of quality along with price. Reduce the number of suppliers for the same item by eliminating those that do not qualify with statistical and other evidence of quality. The aim is to minimize total cost, not merely initial cost, by minimizing variation. This may be achieved by moving toward a single supplier for any one item, on a long-term relationship of loyalty and trust. Purchasing departments customarily operate on orders to seek the lowest price vendor. Frequently, this leads to supplies of low quality.

5. *Improve every process.* Improve constantly and forever every process for planning, production, and service. Search continually for problems in order to improve quality and productivity, to constantly decrease costs, to institute innovation, and to constantly improve product, service, and process. It is management's job to work continually on the system (design, incoming materials, maintenance, improvement of machines, supervision, and training). Improvement is not a one-time effort. Management is obligated to continually reduce waste and improve quality.

6. *Institute training on the job.* Institute modern methods of training on the job for all, including management, to make better use of every employee. New skills are required to keep up with changes in materials, methods, production, and service design, machinery, techniques, and service. Too often, workers have learned their job from another worker who was never trained properly. They end up following unintelligible instructions and can't do their jobs well because of improper training.

7. *Institute leadership.* Adopt and institute leadership aimed at helping people do a better job. The responsibility of managers and supervisors must be changed from sheer numbers to quality. Improvement of quality will automatically improve productivity. Management must ensure that immediate action is taken on reports of inherited defects, maintenance requirements, poor tools, fuzzy operational definitions, and all conditions detrimental to quality. The job of a supervisor is not to tell people what to do or to punish them. A supervisor must lead. Leading consists of helping people do a better job by removing barriers or addressing issues that are keeping an individual from being successful.

8. *Drive out fear.* Encourage effective two-way communication and other means to drive out fear throughout the organization so everybody may work effectively and more productively for the company. Many employees are afraid to ask questions or to take a position, even when they do not understand what their job is or what is right and wrong. They continue to do things the wrong way or not to do them at all. The economic losses from fear are appalling. To ensure better quality and productivity, it is necessary that people feel secure.

9. *Break down barriers.* Break down barriers between departments and staff areas. People in different areas, such as leasing, maintenance, administration, must work in teams to tackle problems that may be encountered with products or service. Often a company's departments or units compete with one another. They do not work as a team, and so they cannot solve or foresee problems. Worse, one department's goals may conflict with another's.

10. *Eliminate exhortations.* Eliminate the use of slogans, posters, and exhortations for the workforce, demanding zero defects and new levels of productivity, without providing methods. Such exhortations only create adversarial relationships; the bulk of the causes of low quality and low productivity belong to the system and thus lie beyond the power of the workforce. These never helped anybody do a good job. Let workers formulate their own ideals.

11. *Eliminate arbitrary numerical targets.* Eliminate work standards that prescribe quotas for the workforce and numerical goals for people in management. Substitute aids and helpful leadership in order to achieve continual improvement of quality and productivity. Quotas take into account only numbers, not quality or methods. They are usually a guarantee of inefficiency and high cost. To hold a job, some people will meet a quota at any cost without regard to damage to the company.

12. *Permit pride of workmanship.* Remove barriers that rob hourly workers and people in management of their right to pride of workmanship. This implies, among other things, abolition of the annual merit rating (appraisal of performance) and of management by objectives. Again, the responsibility of managers, supervisors, and foremen must be changed from sheer numbers to quality. People are eager to do a good job and are distressed when they cannot do so. Too often, misguided supervisors, faulty equipment, and defective materials stand in the way of good performance.

13. *Encourage education.* Institute a vigorous program of education, and encourage self-improvement for everyone. What an organization needs is not just good people; it needs people who are improving with education. Advances in competitive position will have their roots in knowledge. Both management and the workforce must be educated in the new methods, including teamwork and statistical techniques.

14. *Top management commitment and action.* Clearly define top management's permanent commitment to ever-improving quality and productivity, and its obligation to implement all of these principles. Indeed it is not enough that top managers commit themselves for life to quality and productivity. They must know what it is they are committed to, what they must do. Create a structure in top management that will push every day on the preceding 13 Points, and take action in order to accomplish the transformation. Support is not enough: action is required! Involve everyone in the organization in accomplishing the transformation. A special top management team with a plan of action will be needed to carry out the quality mission. Neither workers nor managers can do it on their own. A critical mass of people in the company must understand the 14 Points (see Figure 3.7).

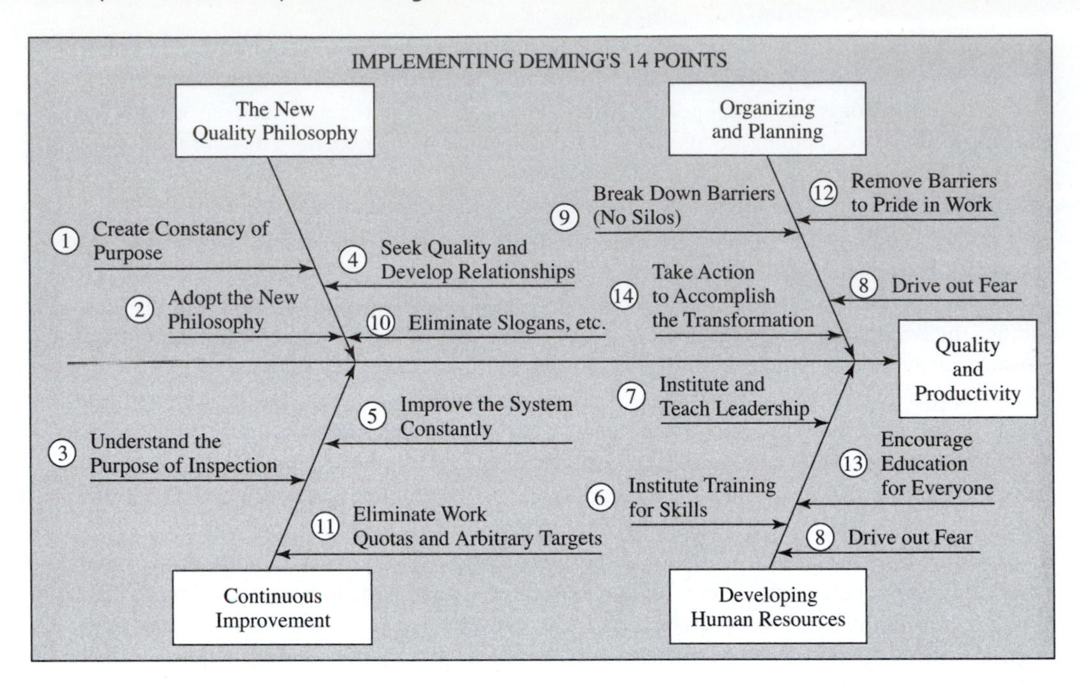

FIGURE 3.7 ◆ Implementing Dr. Deming's 14 Points
(Courtesy of Randy R. Bruegman, Exceeding Customer Expectations: Quality Concepts for the Fire Service, *p. 12)*

What Are the Seven Deadly Diseases? Although the intention of Deming's theory of management (as outlined in his 14 Points) is to achieve positive change and improve quality, it does not mean there are no obstacles to overcome in the process. Deming identified seven of those obstacles he referred to as the seven deadly diseases. These existing aspects of an organization can have a catastrophic, negative impact on an organization (see Figure 3.8).

FIGURE 3.8 ◆ Seven Deadly Diseases
(Courtesy of Randy R. Bruegman, Exceeding Customer Expectations: Quality Concepts for the Fire Service, *p. 13)*

The seven deadly diseases include the following:[34]

1. A lack of constancy of purpose to plan products and services that will have a market, keep the company in business, and provide jobs. (Think about this deadly disease and its application to the fire service and the trends in expanding programs and services!)
2. An emphasis on short-term profits. (In the fire service, this can equate to a one-time political or budget victory.)
3. Evaluations of performance, merit ratings, or annual reviews. Is your system having the desired results or just pushing paper?
4. Mobility of management and job hopping. Are you succession planning within your organization?
5. Management by using only visible figures (counting the money), with little or no consideration of the human aspects of the organization.
6. Excessive medical costs. What has your agency experienced in the past five years?
7. Excessive costs of liability and lawyers' fees.

PETER DRUCKER (1909–2005)

Peter Ferdinand Drucker was a management theorist who created many of the leadership and management principles in use today. Drucker, born in a suburb of Vienna, Austria, moved to the United States in 1937. He wrote influential works about management starting in the 1940s. He wrote over 30 books and was an editorial columnist for *The Wall Street Journal* from 1975 to 1995. He was also a frequent contributor to *Harvard Business Review*. Drucker continued to consult for businesses and nonprofit organizations well into his nineties until his death.

Drucker was awarded the Presidential Medal of Freedom by U.S. President George W. Bush on July 9, 2002. His most controversial work was on compensation schemes, in which he said senior management should not be compensated more than 20 times the lowest-paid employees. This made him an enemy of some of the same people who had previously praised him.

Management by Objectives Peter Drucker popularized the administrative style called **management by objectives (MBO)**, a system whereby supervisors and subordinates agree to certain goals. The employee's subsequent performance is rated according to how well he or she accomplishes those goals.

management by objectives (MBO)
A formal set of procedures that establishes and reviews progress toward common goals for managers and subordinates. Management by objectives specifies that superiors and subordinates will jointly set goals for a specified period of time and then meet again to evaluate the subordinate's performance in terms of the previously established goals.

Management by objectives (MBO) is a process of agreeing upon objectives within an organization so management and employees buy in to the objectives and understand what they are. It is all too easy for managers to fail to outline and agree with their employees about what it is everyone is trying to achieve. MBO substitutes for good intentions a process that requires a rather precise written description of objectives (for the period ahead) and timelines for their monitoring and achievement. The process requires that the manager and the employee agree to what the employee will attempt to achieve in the period ahead, and (very important) that the employee accept and buy in to the objectives (otherwise commitment will be lacking). For example, let us suppose they feel that it will be sensible to introduce a key performance indicator to show the development of a new fee revenue in fire prevention. Then the manager and the employee need to discuss what is being planned, what the time schedule is, and what the performance indicators should be. Thereafter, the two meet regularly to ensure the objective is being attended to and will be delivered on time.

Organizations have scarce resources, and so it is incumbent on the managers to consider both the level of resourcing and whether the objectives that are jointly agreed upon within the organization are the right ones and represent the best allocation of effort. Also, reliable management information systems are needed to establish relevant objectives and monitor their performance toward the achievement of those objectives in a meaningful way. The concept of MBO is readily evident in most organizations today, although you may not hear it being referred to as such. The basic concepts of aligning the mission core values and vision of the organization to that of the employees are the basis for most organizational employee relationships. In part, the management by objectives concept was a foundational piece to a significant shift in leading and managing. During the last 25 years, the relationship between labor and management has, to a great degree, shifted to a more collaborative approach. Our economic/budget processes are much more defined in respect to short-term goals and long-term strategic objectives and our collaboration or the creation of synergistic partnerships are in part a result of Drucker's introduction of MBO.

LEADERSHIP CHANGES FOR THE TWENTY-FIRST CENTURY

In 1989 in his book, *Why Leaders Can't Lead*, Warren Bennis wrote about where all the leaders had gone and why those who remained simply could not lead.[35] He argued there was an unconscious conspiracy afoot to prevent organizational leaders from innovating or, as he said, "to immerse us in the routine."

The perspective that many have is that you are either a leader or a manager. The reality is that to be successful we need strength in both areas. What Bennis was observing was the beginnings of a significant transformation in how we lead and manage in the United States today. Over the last two decades, we have witnessed this fundamental change in how we lead and manage within our departments. This change has virtually redefined the relationship between leaders and followers. This transformation has paralleled a move away from an authoritarian toward a more collaborative leadership style. We have discussed the history of much of the research that has led us to this point in time. The transformation taking place today will have implications for our organizations for many years to come.

THE CHANGING NATURE OF EXPECTATIONS

One clearly evident change involves the expectations people have of individuals in leadership roles. The change in expectations has ignited discussions over what should represent the core of leadership developmental potential. What most organizational scholars would agree on, however, is that their organizations must move toward creating an environment in which employees' intellectual capital is nurtured, developed, and more directly rewarded. In the early stages of change, in which Bennis's comments were originally embedded, there were clear signs emerging of what was about to occur as organizations entered into the 1990s. People such as Deming in the area of quality and Drucker in the area of management development had, for some time, made very clear statements that things were radically changing in the way organizations functioned and would be functioning into the foreseeable future. Now, as we have moved into the twenty-first century, another realization has become evident: to be a "sustainable" organization, significant change is required in the strategies used for developing human resources throughout all levels of an organization. To move organizations into the future will require a dramatic change in the way our schools, businesses, and institutions are currently led. The alignment of degree requirements and reciprocity of certifications for the fire service are an indication that such a shift is taking place within our industry.

LEADERSHIP

It is clear the challenges organizations are faced with today (staying abreast with a rapidly changing marketplace, public safety revenues, meeting the expectations of a diverse workforce, leadership, and management) have never been more relevant to the future of organizations. Yet, consistent with calls for rethinking leadership models and practices, the process of leadership itself has undergone significant, if not radical, change, as organizations and communities have transformed to take advantage of opportunities and challenges that each is faced with today. Yet it is important to realize we are in the midst of transforming to a new system of leadership in organizations, and in the course of such dramatic change, there are likely to be periods of instability, chaos, and doubt. The most significant change may be that leadership now appears more frequently at all levels of the organization in varying degrees and complexity.

Perhaps when Bennis speaks to unconscious conspiracy, what we are actually experiencing is a change about leaders and what leadership is within our organizations. To coincide with the changes in the dynamics of our environment, the values and attitudes of our workforce, and our management techniques, one might then ask, "In what way have the expectations changed?" Much has been written regarding the need to reconfigure the control systems in organizations and to begin to *re-engineer* and *continuously improve*, in order to let leadership practices migrate to levels where they could do the most good for the organization and its members. Deming realized long ago one could not achieve total quality through fear, nor through command and control systems.

Evidence of the transformational leadership in many organizations has resulted in the development of self-directed teams, with shared decision-making responsibility. This will continue to be an essential component of leadership in the future. The current workforce, for the first time in history, has four different generations working at the same time, each with its own unique set of values and job expectations. We are witnessing the advent of more dispersed leadership as a result. To be successful, leadership has migrated to be more inclusive. It has taken a considerable amount of time for this aspect of leadership to become embedded in the organization's way of thinking. Yet, in many organizations today, it is still not evident. The delay in moving toward a new way of thinking about leadership can be attributed to the inertia in many bureaucratic systems. Part of it is due to politics, and the concern over losing control, and part of it is due to a fear and resistance to changing to something people do not understand. Yet the process of migration and change is continuing in varying degrees across all organizations, with some organizations well into the process of change and others still denying "it's going to occur at all."

As organizations have downsized, there appears to be an ongoing migration of leadership to all levels of organizations and society. This migration is fueled in part by the information revolution, in which leadership has become energized by the new information systems and technology, which allow near-instant access to a plethora of information. This has made it possible for people at the lowest levels of organizations and throughout our communities to influence key decisions more directly. More people can now react, manage, and lead intelligently as they have the information to do so.

Leadership has also emerged, in the recent wave of change that has occurred, with the establishment of self-directed teams and a greater emphasis on labor–management collaboration. Leadership processes have also moved outside the doors of organizations, in an effort to develop more empowered and intelligent customers. Taking an optimistic view of the future, we may be witnessing the full development of an "empowered workforce," "empowered electorate," and "empowered citizenry," who each are now more likely to question, challenge, and offer their own solutions to problems. The follower as leader has become a reality in many institutions. Yet some leaders still treat followers as the low-class subordinates, offering them little opportunity to provide input and to participate in critical decisions. However, the change no longer appears to be as "unconscious" as Bennis first observed. In fact, many organizations have deliberately questioned their own leadership in an attempt to "reinvent" themselves and their workforce. Has leadership as a process fundamentally changed? Today, on a one-to-one basis, it probably has not changed, except some styles are much more evident than others. Managers have become more developmentally oriented, not that some weren't 50 years ago. But if we observe at an organizational or community level, then the change has been "almost" revolutionary in terms of the number of managers and supervisors who consider developing others a core job requirement. Today, mentoring and succession planning are crucial elements of leadership.

Have the expectations changed of leaders themselves? The answer is YES! What constitutes "exemplary" leadership in today's fire service? Leadership must be visionary, developmental, service oriented, ethical, stimulating, facilitative, and clear in establishing expectations. The dimensions of leadership that dominated discussions for nearly 40 years are only a small portion of what is discussed in terms of the "full range" of leadership behavior and styles today.

Many corporate leaders began significant re-engineering and radical transformations of their organizations in the late 1980s and on into the 1990s. Some of our current leaders in the fire service are still in a "Rip van Winkle" state, failing to wake up to the realities of significant changes that have occurred and are occurring around them. But I believe the number is much less than it was just a decade ago. The rapid transformations occurring in our organizations parallel the transformations occurring with leadership. Yet one of the greatest difficulties in changing and advancing the human versus the technical systems is that human systems are embedded in an old system of behaving, long after the change has "supposedly" taken place. CULTURE!

Have we taught individuals to lead in certain ways, either formally through academics or informally within the organization through patterned behavior? The question is how do we retrain them to learn new ways of leading to satisfy the expectations of such a diverse workforce? At the same time, the employees have to be willing to work with their leaders to achieve a successful transformation in how each influences the other. Their traditional base of thinking can keep each of them from changing. In reality, nothing should be further from the truth, as witnessed by the extensive discussions regarding the importance of understanding the "new forms" of leadership: team initiatives, total quality efforts, re-engineering interventions, attention to customer focus, and working with a new and more diverse workforce. Servant leadership is leadership that has reconceptualized the relationship between a leader and follower.

Robert Greenleaf in his book, *Servant Leadership*, writes "one great effect of *the revolution of expectation among young people* has been to force companies to try to make their work more significant for their employees. In the end, internal pressure may break down the concept of chief executive officer and force a council of equals. In order to attract and hold enough very able young people to deal with the mounting complexities, the existing chief executive officer may be forced to raise the immediate administrative group more to an equal status. And if one is to preside over a successful business, one's major talent will need to evolve from being *the chief* into being the *builder of the team*. Furthermore, if the idealism of this generation of young people persists, as I think it will, and as these young people move into top positions, *they* will insist on more determined efforts to provide significant and meaningful work to more people. This in the end will result in a new business ethic."[36]

Unfortunately, in many organizations, leading and managing are still discussed in terms of "them versus us." As organizations transform into intelligent network systems, the positive influence of leadership will reside more in the network of partnerships rather than in one single individual. What is happening today is a fundamental change in the way leadership is exercised in organizations and institutions. It is more shared within and between levels, and may have far greater potential and substantially less risk, as a consequence of it being embedded in networks versus any one individual's office. Today, we have flatter hierarchies, virtual information, enhanced autonomy, a more rapid pace of change, and a much greater need for interdependencies. The question is whether the leadership systems have changed to accommodate these dramatic transformations. Transformations that are evolving in many organizations today pose a tremendous threat for leadership researchers and practitioners, unless they reevaluate their thinking about leadership processes and their applications. Old paradigms of leadership, not leadership itself, are becoming more irrelevant to explaining the full range of leadership processes observed in organizations and institutions today, and those likely to emerge into the foreseeable future.

Technical and social trends that are transforming organizations today demand a shift in attention toward understanding the "collective forces" of leadership, which are emerging in institutions at all levels. Yet we cannot neglect the individual, who remains a potent force in an expanded leadership system and/or framework. What we must do is rethink current models of leadership to fully integrate the individual and collective forces shaping our institutions and society. With this orientation in mind, a whole new frontier for dialogue on leadership systems and practices has emerged—one that is both multidimensional and multilevel. Harnessing the collective leadership forces in our organizational systems has, perhaps, become one of the most critical and current challenges confronting organizations today. Yet this challenge will not be realized if we do not change the installed base of thinking in those who study leadership, as well as in those who are supposed to demonstrate it.

BUILDING THE BRIDGE TO THE FUTURE

At the present time, many organizations are at a fragile place in their evolution—the time between the past and the future. For some leaders, the present looks very much like the past, particularly if their recent past is embedded in success. In the midst of change, and without any clear direction, such individuals view significant shifts in control as potential threats to the stability and harmony of their system—a system that in many cases they helped to create. These individuals can delay, if not prevent, the "natural" evolution of their organizational system by taking "control" and stifling the redistribution of leadership. They lock themselves into ways of interacting that ultimately lock them into obsolete practices. If the change does finally occur, it may need to be radical and revolutionary to overcome the forces that have kept the organization and/or institution from changing and developing. Why? Because in the midst of significant change, these individuals are often poorly equipped to transfer control to a larger collective.

Typically this is risky because, even with the best intentions, the larger employee group may be poorly equipped to take on greater control and responsibility. Thus to fundamentally change leadership practices, we must also change the followers' ways of thinking, acting, and reacting.

IN SEARCH OF SUCCESSFUL TRANSFORMATIONS

Those organizations that have been most successful in achieving dramatic transformations in their leadership systems have had several distinguishing characteristics.

- They have articulated in terms of new systems and processes what they are hoping and expecting to evolve.
- They have realized the need to replace old systems and have spent time retiring them, while also creating new ones.
- They have considered the needs of individuals at all levels, which are used to operating in a different way, and have taken the time to explain, justify, and ultimately reward operating in new and substantially different ways. In this regard, they have involved those being affected in the process of change, by including their ideas, needs, concerns, and aspirations as part of the process of change.
- They have made it worthwhile to change, motivating individuals to operate in line with the new system's requirements.
- They have provided the necessary education to change rather than simply assuming people know how to change.
- They have demonstrated the courage to stay on course, regardless of the resistance to change, and have been patient, allowing for mistakes to occur along the way.

Interestingly enough, each of the distinguishing characteristics mentioned represents a form of leadership, often described and observed at a systemic and/or organizational level. Indeed, it requires leadership to transform the leadership systems in our organizations successfully.

So leadership has not disappeared; rather it has been migrating to many different levels in organizations and society, where it is shared among a much broader range of individuals. The cultures that have become rooted in our organizations can be a tremendous impediment to introducing new leadership methods and models. It can also stifle the course of migration. Perhaps this is one of the reasons why a significant number of the re-engineering and TQM efforts have failed. Today, we have a brighter and more challenging workforce. There is very little doubt that transformational leadership will have a very positive impact on the levels of trust, performance, and innovativeness of future teams and organizations.

MANAGEMENT CHALLENGES FOR THE TWENTY-FIRST CENTURY

Peter F. Drucker, in his book, *Management Challenges for the 21st Century*, provides insightful and timely information for individuals and organizations as they work toward common goals in the next one hundred years. Drucker reviews the seven major assumptions that have been held by experts in the field of management for most of the twentieth century and shows why they are now obsolete. He goes on to give eight new assumptions for the twenty-first century, ones that are essential for viewing the roles of individuals and management in both profit and not-for-profit organizations.

Neither individuals nor organizations can be successful if they stick with the old assumptions, according to Drucker, just as the horse and carriage can no longer compete with the automobile.

SEVEN OLD ASSUMPTIONS OF MANAGEMENT

There is a critical difference between a natural science and a social discipline, according to Drucker. The physical universe displays natural laws that describe objective reality. Natural laws are constrained by what can be observed, and these laws tend to be stable or change slowly and incrementally over time. A natural science deals with the behavior of objects. A social discipline such as management deals with the behavior of people and human institutions. The social universe has no natural laws of this kind. It is thus subject to continuous change; this means assumptions that were valid yesterday can become invalid and, indeed, totally misleading in no time at all. Drucker identifies the following old assumptions for the social discipline of management.

Three Old Assumptions for the Discipline of Management

1. Management is business management.
2. There is, or there must be, one right organization structure.
3. There is, or there must be, one right way to manage people.

Four Old Assumptions for the Practice of Management

4. Technologies, markets, and end users are given.
5. Management's scope is legally defined.
6. Management is internally focused.
7. The economy as defined by national boundaries is the "ecology" of enterprise and management.

According to Drucker, six out of the seven assumptions (2, 3, 4, 5, 6, and 7) were close enough to reality to be useful until the early 1980s. However, all are now hopelessly outdated. "They are now so far removed from actual reality that they are becoming obstacles to the theory and even more serious obstacles to the practice of management. Indeed reality is fast becoming the very opposite of what these assumptions claim it to be."

EIGHT NEW MANAGEMENT ASSUMPTIONS

1. Management is not only for profit-making businesses. Management is the specific and distinguishing organ of any and all organizations.
2. There is not only one right organization. The right organization is the organization that fits the task.
3. There is NOT one right way to manage people. One does not "manage" people. The task is to lead people. And the goal is to make productive the specific strengths and knowledge of each individual.
4. Technologies and end users are not fixed and given. Increasingly, neither technology nor end use is a foundation of management policy. They are limitations. The foundations have to be customer values and customer decisions on the distribution of their disposable income. It is with those that management policy and management strategy increasingly will have to start.
5. Management's scope is not only legally defined. The new assumption on which management, both as a discipline and as a practice, will increasingly have to base itself is that the scope of management is not legal. It has to be operational. It has to embrace the entire process. It has to be focused on results and performance across the entire economic chain.

6. Management's scope is not only politically defined. National boundaries are important primarily as restraints. The practice of management, and by no means for business only, will increasingly have to be defined operationally rather than politically.
7. The inside is not the only management domain. The results of any institution exist only on the outside. Management exists for the sake of the institution's results. It has to start with the intended results and organize the resources of the institution to attain these results. It is the organ that renders the institution, whether business, church, university, hospital, or a battered woman's shelter, capable of producing results outside itself.
8. Management's concern and management's responsibility are everything that affects the performance of the institution and its results, whether inside or outside, whether under the institution's control or totally beyond it.

Drucker's new set of assumptions recognizes the complexity of leading and managing in today's organizational environment. "One does not 'manage' people," Drucker says. "The task is to lead people. And the goal is to make productive the specific strengths and knowledge of each individual." Drucker reviews the history of manual-worker productivity in manufacturing during the twentieth century (which saw a fiftyfold increase) and speaks to the need for new methods that will make the improvements in knowledge-worker productivity that will be required in the twenty-first century.

Frederick Winslow Taylor's pioneering study of manual labor in manufacturing processes is credited with starting the revolution in manufacturing efficiency that took place at the time. According to Drucker, scientific management, industrial engineering, and even total quality management by W. Edwards Deming are rooted in the basic strategy that Taylor articulated. Taylor's principles for manual-worker productivity emphasized effective and efficient motion of the object to ensure the most successful outcome. The object, its necessary and sufficient motion in time and space, and the manual worker's movements are integrated to achieve control over manufacturing variables and meet requirements for a quality product. The manual-worker in manufacturing conforms to the needs of the job.

The knowledge-worker has a different job description from the manual-worker on a production line. Drucker identifies these six major factors for knowledge-worker productivity in the future.

1. The knowledge-worker's question is, "What is the task?"
2. Knowledge-workers have to manage themselves and have autonomy.
3. Continuing innovation has to be part of the work, the task, and the responsibility of knowledge workers.
4. Knowledge work requires continuous learning and continuous teaching by the knowledge worker.
5. Productivity of the knowledge worker is not primarily a matter of quantity of output. Quality is at least as important.
6. Knowledge workers must be treated as "assets" rather than "costs." They must prefer to work for the organization, over all other opportunities.

The "care and feeding" of autonomous individual knowledge-workers will become more and more important as new management assumptions replace old ones. These new assumptions will require new process models of human consciousness. Manual-worker productivity was made possible by Taylor's four dimensions of object-motion-time-task study. We must assume three- and four-dimensional knowledge-task modeling will be needed for similar breakthroughs in knowledge-worker productivity in the twenty-first century.[37]

In the fire service we have been in the midst of our own evolution of how we lead and manage our departments in the twenty-first century. We have been forced to re-think our own industry assumptions, re-engineer many of our processes, adapt to emerging technology, create a new dynamic in employee/management relations, and create bold visions for the future. To do so will take cutting edge leadership and state of the art management.

ADVICE FROM THE EXPERTS

How have leading and managing changed in the span of your career?

MICHAEL CHIARAMONTE

Chief Fire Inspector, Retired, Lynbrook Fire Department (New York): It went from a rigid military approach with instant obedience and no thought about an order to having to give a reasonable rationale for each directive in order to get commitment. A leader must change with the times.

BETT CLARK

Former Fire Chief, Bernalillo County Fire and Rescue (New Mexico): The needs and expectations of firefighters change as society changes those expectations. As chief officers we must be problem solvers, financial counselors, facilitators, teachers, parents. The workforce is younger with less taught traditional values and job ethics coming from high school graduates. The new recruits have grown up in an instant gratification atmosphere . . . fast food, instant messaging, more technology, and less labor oriented. The fire service is still a hands-on, feeling, and compassionate career field that does not always fulfill the need for instant gratification. Leading and managing have also expanded to include prevention and education that are proactive rather than the reactive "put the wet stuff on the red stuff." A successful chief officer will need to remember to think, act, and lead in a proactive manner for our future. As call volume decreases with prevention, education, and safer technology, firefighters are bored. A good leader can embrace proactive programs and activities that benefit the community and give the firefighters an opportunity to fulfill another self-achievement need.

KELVIN COCHRAN

Fire Chief, Atlanta Fire Department (Georgia): In the early 1980s when I entered the fire service, World War II and Korean War veterans were officers; Vietnam veterans were senior driver operators. The leadership philosophy was simple and understood: "When I say jump, you jump. No questions asked!" By and large, it worked. The culture placed more emphasis on managing than leading. In the minds of many officers, leading and managing were one and the same. In today's fire service organizations, when officers say jump, firefighters want to know how high and why. Today, fire service leaders must recognize the distinction between managing and leading and be responsive to the needs of subordinates to be included in decision making, understand how their role is going to make a difference, and have specifics on what part they are to contribute to task objectives.

CLIFF JONES

Fire Chief, Tempe Fire Department (Arizona): Both are now more important than ever. Leadership has changed in that, in the span of my career, it is a major focus of personal and professional development. Literally hundreds of books are available on the subject of leadership, and it is seen by many as the key to success. Managing, on the other hand, now gets less notoriety and is not discussed and taught as much as leadership. Many of the scandal-type problems seen in the fire service today are allowed to cause damage to people and departments because someone failed to manage people. In the same vein, many of these same scandals could have been preempted if someone had exercised management and leadership. In my experience, both are critical every day. They are not mutually exclusive traits but complementary traits for success. If leadership is the art and science of influence, then management is the art and science of progress. Both are important to fire officers in the day-to-day operations of their respective departments.

BILL KILLEN

Retired Fire Chief and IAFC President 2005–2006: Leading and managing have changed considerably since I entered the fire service in the mid-1950s. I believe the biggest change in leading and managing can be directly attributed to the community college system and the many opportunities provided to the fire service in obtaining two-year degrees, followed by the Open Learning Fire Service Programs leading to four-year degrees in fire administration and management. Higher education provided the forum to study management styles, concepts, and the leadership styles of successful and, in some instances, not so successful managers and leaders. Advances in technology in industry and the military complex created new standards and systems that were applied to other disciplines with great success. Societal changes broadened the ethnic and cultural makeup of the fire service and added considerable impact to how we lead and manage people of diverse backgrounds.

KEVIN KING

U.S. Marine Corps Fire Service: Probably the biggest change has been the focus on leading and managing our people. If they are truly our most important assets, then it makes sense for leaders and managers to spend the most time working with our personnel. If we focus on improving the leadership and management of the people in our organizations, the products and services portion of the fire service will generally succeed. (These are opinions of Mr. King and do not reflect official policy of the Department of Defense or the United States Marine Corps.)

MARK LIGHT

Executive Director, International Association of Fire Chiefs: In times past, the fire officer/fire chief could pretty much demand things to be done based on rank. In those times, being the chief was about position power and, in many cases, not about expert power. Over the years, this has changed. Leaders must now be able to use a variety of skills to manage and lead their fire departments. We can't just issue decrees and expect them to be followed, as today's workforce is more educated and willing to challenge authority. In the past, the fire chief was considered the "guy" who ran the fire department and was often considered the "best" firefighter. Today, the chief may be a woman or a man and is a critical part of the city/county/town's management team. The fire chief brings a myriad of skills to the municipal management team and can offer perspectives that others simply don't have. In short, leading and managing a fire department today takes a tremendous amount of talent, and one must be willing to continually learn and develop as a leader based on today's culture, not rely on the past tradition.

LORI MOORE-MERRELL

Assistant to the General President, International Association of Fire Fighters: Leadership and management are very different roles. Though one individual can exhibit attributes of both, it is rare that you naturally find a good manager and a good leader in the same individual. Leaders may learn to be good managers, and managers may learn to be good leaders, but they do so by first recognizing their weaknesses. In the fire service, good leaders should be on the fire ground (or scene of other emergencies), and good managers behind a desk. Unfortunately, fire officers today are expected to be both. They must move from the role of incident commander to paper pusher in the span of a few hours. This is not a natural transition and many find it difficult. As stated previously, rarely is one individual good at both. Firefighters (or paramedics) preparing to promote through the ranks as well as active fire officers should understand in which areas their strengths are found and work to shore up weaknesses in the others.

JAY REARDON

Fire Chief, Retired, Northbrook Fire Department (Illinois); Currently President of the Mutual Aid Box Alarm System (MABAS): Management and leadership are two completely different things. You might have one, but it does not necessarily suggest you have the other. Management is mechanical in nature with steps in a process. Some might argue management is of scientific design. The trick with mastering management is understanding the concepts of process and then successfully merging with human dynamics (people). Leadership is where no instruction manual or process exists. It is the thin ice, going often where others have never gone. Leadership is recognized by people; they allow you to lead, they expect you to lead, and they trust you to lead. Leadership is driven by trust and an extremely high success rate. Today's society (often influenced by the media and today's in-your-face technological capabilities) is by nature skeptical. Today's new fire service employees are Generation X ("prove it to me first" types, different from when I began my career in the late 1960s). Chapter 4 talks about style and leadership; perhaps the changes we have experienced are better answered there.

GARY W. SMITH

Retired Fire Chief, Aptos-La Salva Fire Protection District (Aptos, CA): I have learned to lead by doing it, venturing forward, failing and succeeding, learning the dynamics of change, and understanding how teams work. A big part of the challenge is to figure myself out. Looking back I wish I would have spent more time figuring out my passions and need for new knowledge, and taken advantage of my network of friends and compatriots with a better focus on how to build to success. As a manager I would have spent more time figuring out the strengths and weaknesses of each of my team players, especially those who manage and lead within the organization. I would have played much more to each of their strengths and not have wasted time trying to make square pegs fit into round holes.

STEVE WESTERMANN

Fire Chief, Central Jackson County Fire Protection District (Blue Springs, Missouri), and IAFC President 2007–2008: My first chief was an autocratic manager. He was from the old school of "I'm the boss; just follow orders." You could not (or would not dare) ask any questions, and you certainly did not have consensus decisions or any leadership partnerships at that time. If you were in a department with an employee group, it was definitely a we versus them attitude/atmosphere on both sides. To be successful today a chief officer must be willing to share his or her power in order to form better collaborative partnerships internally and external to the organization. For example, even on the fire ground where one might think that autocracy would be accepted, there is now the practice of crew resource management, where, to ensure everyone's safety and to prevent groupthink, everyone on the team has the ability for input. To be successful you must have an innate desire and willingness to be part of a team and to seek out all possible relationships.

THOMAS WIECZOREK

Executive Director, Center for Public Safety Excellence, and Retired City Manager: When I first started, the concept was "I am the boss; do as I say, not as I do. Don't ask questions and refer to the first part of the statement." Just like the generation categories that have been assigned (Generation X, etc.), I believe managing today is much more about team. One thing that I believe has led the fire service to the successes it enjoys is that the public, elected officials, and appointed officials value the camaraderie of the fire service. This can be a detriment that leads to "pack governance" and so it is a fine line between the two. The National Incident Management System is derived from the Incident Command System, successfully used in the fire service for many years. Having been exposed to both, the police side of the operation is not as versatile in this approach or method simply because police are not trained to be. Police are trained to be "the Army of One." Fire are trained to act in teams. The generation that we are seeing about to enter the workforce has been structured around "teams" since they were born. They have been carted off to soccer, to football, to play areas where everything is done in the team format. Any leader who does not or cannot acknowledge this will find it difficult to lead in the future.

JANET WILMOTH

Editorial Director, *Fire Chief* Magazine: In the past 20 years, I have seen chiefs' leadership styles move from a range of dictatorial styles to micromanagers to management by objective. The future is in empowering and delegating (we hope!).

FRED WINDISCH

Fire Chief, Ponderosa Volunteer Fire Department (Houston, Texas): Leading is learned; managing change is a process. I had many opportunities via my former employer (Petro Chemical) to learn about leading and managing changes because that business was in a constant state of flux and was driven by the necessity of making a profit. My EFO experience was second to none in the education areas because it linked values and capabilities into real-life scenarios, adapting leadership methods to change for improvement.

Review Questions

1. Give a brief overview of the history of management theories.
2. Weber believed all bureaucracies have certain characteristics. Explain the benefits and disadvantages of this management theory.
3. What was the classical school of management thought?
4. Describe the dysfunctional aspects of bureaucracy.
5. Explain the difference between management and leadership.
6. Describe Taylor's four principles of scientific management. Explain the benefits and disadvantages of this management theory.
7. Discuss Fayol's 14 principles of management. Explain the benefits and disadvantages of these principles.
8. Explain the behavioral movement. Describe the positives and negatives of this movement. Relate this to how it would effect fire service management.
9. Describe the Hawthorne effect. What were some of the main effects it had on society in terms of management and leadership?
10. Explain the difference between Theory X and Theory Y management. Relate this to Theory Z management.
11. Describe Maslow's hierarchy of needs and the effect this has on individuals in terms of managing personnel.
12. Explain the managerial grid model. Describe how this can be used.
13. Explain the motivation-hygiene model.
14. What impact did W. Edwards Deming have on modern management trends?
15. Describe Deming's 14 Points of management.
16. Explain the seven deadly diseases.
17. What is MBO and how has it influenced today's organizations?
18. What are Drucker's eight new management assumptions and how would they impact your organization?
19. What will leading and managing in the fire service evolve into during the next 25 years?

References

1. John P. Kotter, *A Force for Change: How Leadership Differs from Management* (New York: Free Press, 1990).
2. Ibid.
3. From the Robert Owen Memorial Museum (The Cross, Broad Street, Newtown, Powys, SY2162BB, UK), Copyright 1997.
4. Charles Babbage, *Science and Reform: Selected Works of Charles Babbage* (New York: Cambridge University Press, 1989).
5. *History of Management Thought*, "TYPE=PICT:ALT=Openthisresultin newwindow
6. Max Weber, *Max Weber: The Theory of Social and Economic Organization*, trans. A. M. Henderson, ed. with introduction by Talcott Parsons (New York: Oxford University Press, 1947).
7. Robert K. Merton, "The Unanticipated Consequences of Purposive Social Action," *America Sociological Review*, 1, no. 6 (December 1936), 894–904.
8. Robert Kanigel, *The One Best Way: Frederick Winslow Taylor and the Enigma of Efficiency* (New York: Viking, 1997), p. 675.

9. Frank Gilbreth, *Motion Study* (Hive Management History Series, No. 14) (Easton, PA: Hive Publishing, 1972).
10. Jim Christy, *Price of Power, A Biography of Charles Eugene Bedaux* (New York: Doubleday, 1985).
11. British Standards Institute, *Glossary of Terms Used in Management Services* No. 3138, 1992; D. A. Whitmore, *Work Measurement* (Staffordshire, UK: Institute of Management Services, 1975).
12. Henri Fayol, *Industrial and General Administration*, trans. J. A. Caubrough (Paris France: General International Management Institute, 1930).
13. Ibid.
14. Chester I. Barnard, *The Functions of the Executive*, 30th ed., introduction by Kenneth R. Andrews (Cambridge, MA: Harvard University Press, 1968).
15. Ibid.
16. Mary Parker Follett Foundation, www.follettfoundation.org
17. *Management Theory of Mary Parker Follett*, Information at Business.com
18. Summarized from E. F. Huse, *The Modern Manager* (St. Paul, MN: West, 1979), pp. 48–50. The original research is reported in F. Roethlisberger and W. Dickson, *Management and the Worker—An Account of a Research Program Conducted by the Western Electric Company, Hawthorne Works, Chicago* (Cambridge, MA: Harvard University Press, 1939). Numerous criticisms of the study and its findings appear in the literature. See, for example, H. Parson, "What Happened at Hawthorne?" *Science*, 183 (March 1974), 922–932; and A. Carey, "The Hawthorne Studies: Radical Criticism," *American Sociological Review*, 32 (June 1967), 403–416.
19. Ibid., pp. 49–50.
20. George Homans, *The Human Group* (New York: Harcourt College, 1950).
21. Herbert Simon, *Administrative Behavior: A Study of Decision-Making Processes in Administration Organization* (New York: Free Press, 1947).
22. Abraham H. Maslow, with Deborah C. Stephens and Gary Heil, *Maslow on Management* (New York: John Wiley, 1998).
23. Thomas S. Bateman (University of Virginia) and Michael J. Crant (University of Notre Dame), "Revisiting Intrinsic and Extrinsic Motivation," *The Academy of Management Annual Meeting*, Seattle, WA.
24. D. McGregor, *The Human Side of Enterprise* (New York: McGraw-Hill, 1960). British Standards Institute, *Glossary of Terms*; Whitmore, *Work Measurement*.
25. William G. Ouchi, *Theory Z: How American Business Can Meet the Japanese Challenge* (Reading MA: Addison, 1981).
26. R. R. Blake and J. S. Mouton, *The Managerial Grid* (Houston, TX: Gulf Publishing, 1964).
27. R. R. Blake, *Leadership Dilemmas–Grid Solutions* (Blake/Mouton Grid Management and Organization Development Series) (Oxford, England: Butterworth-Heinemann, 1991).
28. E. W. Bakke, *The Fusion Process* (New Haven, CT: Yale University, 1953). See also C. Argyris, *Personality and Organization* (New York: Harper and Row, 1957).
29. F. Herzberg, B. Mausner, and B. Snyderman, *The Motivation to Work* (New York: John Wiley, 1959).
30. Ibid.
31. Scott W. Spreier, Mary H. Fontaine, and Ruth L. Malloy, "Leadership Run Amok: The Destructive Potential of Overachievers," *Harvard Business Review*, June 1, 2006.
32. *20th Century Management Theories*, Business.com Directory
33. W. Edwards Deming, *Out of the Crisis* (Massachusetts Institute of Technology, Center for Advanced Engineering Study) (Cambridge, MA: MIT Press, 1989).
34. Randy R. Bruegman, *Exceeding Customer Expectations, Quality Concepts for the Fire Service* (Upper Saddle River, NJ: Prentice Hall, 2003).
35. Warren Bennis, *Why Leaders Can't Lead: The Unconscious Conspiracy Continues* (Los Angeles: University of Southern California Graduate School of Business, 1990).
36. Robert Greenleaf, *Servant Leadership* (Mahwah, NJ: Paulist Press, 1977).
37. Peter F. Drucker, *Management Challenges for the 21st Century* (New York: HarperCollins, 1999).

What Is Your Leadership Style?

4 CHAPTER

Key Terms

Johari Window, p. 152
Jungian Type Inventory, p. 167
Keirsey Temperament Sorter, p. 168
schema, p. 154

Objectives

After completing this chapter, you should be able to:

- Define the various leadership styles in use today.
- Describe when each leadership style is appropriate.
- Define the importance of the Johari Window in enhancing leadership ability.
- Describe the four preference styles and compare and contrast each type.
- Define schemata and describe how they can impact leadership and managerial abilities.
- Articulate your own personal framework that shapes your leadership style.

Over the past three decades, the fire service has experienced numerous significant changes, as our industry has strived to meet new objectives, an expanded mission, more extensive education requirements, and enhanced safety for the public and firefighters. This new environment has forced leaders to rethink many of their established and managerial processes. At times, leaders have been reluctant to make needed changes in their organizations. This is often a direct result of their failure to understand how to perceive, understand, and interpret their surrounding environment, industry, department, and, in some cases, themselves.

LEADERSHIP STYLES

There are a number of ways to explore leadership style.[1] If those in leadership positions can successfully understand their natural style of leadership and explore how their style might be adapted to fit different circumstances and different people, then their leadership will be much more effective. Knowing one's own strengths and abilities and creating synergy with the leadership ability of others is another vital element of being a great leader. For example, during World War II, British Prime Minister Winston Churchill knew he had a vital task in building the morale of the British people. He chose to step aside from several important cabinet discussions in order to spend time delivering speeches to his country. He recognized the importance his words would have on his country and the world and had faith in his leadership team to allow it the freedom to continue without his presence.

CHANGING YOUR STYLE TO REFLECT DIFFERENT NEEDS

Good leaders are able to adopt differing leadership styles with different people or with the same people at different times depending on the situation.[2] The key factors likely to influence the style a leader adopts at a particular time include the nature of the work to be done, the skill level of the person being asked to do the work, and the ongoing needs of the leader's relationship with that person. This practice is quite common in the fire service. Fire leaders who command at a fire will not lead the same in the firehouse or when interacting with other departments.

This way of thinking is based on the work by Blake and Mouton, who developed a simple model around two axes, concern with people and task. This model was covered in Chapter 3 but is worth reviewing. The 9.1 leadership style is focused solely on getting the job done, which in the medium or long term is likely to prove highly unpopular, especially in a voluntary organization. Conversely, a 1.9 style leads to a warm and cozy country club where there is loving concern for people and their needs, but nothing happens. A 1.1 leadership style suggests the worst of both worlds: little or nothing is happening. Clearly, the safest place to be over time, where the needs of the task and the people are held in balance, is 5.5. This is a halfway house, where compromise is the order of the day. However, 9.9 indicates the leadership objective, where people and task are integrated, providing maximum results. Although this 9.9 leadership style may be the desired objective, different situations and different people will require leaders to use a range of styles at different times. For example, when a fire breaks out, concern for the task of evacuating a building in a safe and orderly fashion is of prime importance. Asking your firefighters to gather into small groups and reflect on how the building should be evacuated is clearly not helpful. At other times, greater concern for people over task would be appropriate (see Figure 4.1).

THE MANAGERIAL GRID

Leaders may be concerned for their people, and they must also have some concern for the work to be done.[3] The question is, how much attention do they pay to one or the other? (See Table 4.1.)

Impoverished management. Managers have a low concern for both people and production.

Authority-compliance. Strong focus on task but with little concern for people. Focus on efficiency, including the elimination of people wherever possible.

Country club management. Care and concern for people, with a comfortable and friendly environment and collegial style, but a low focus on task may give questionable results.

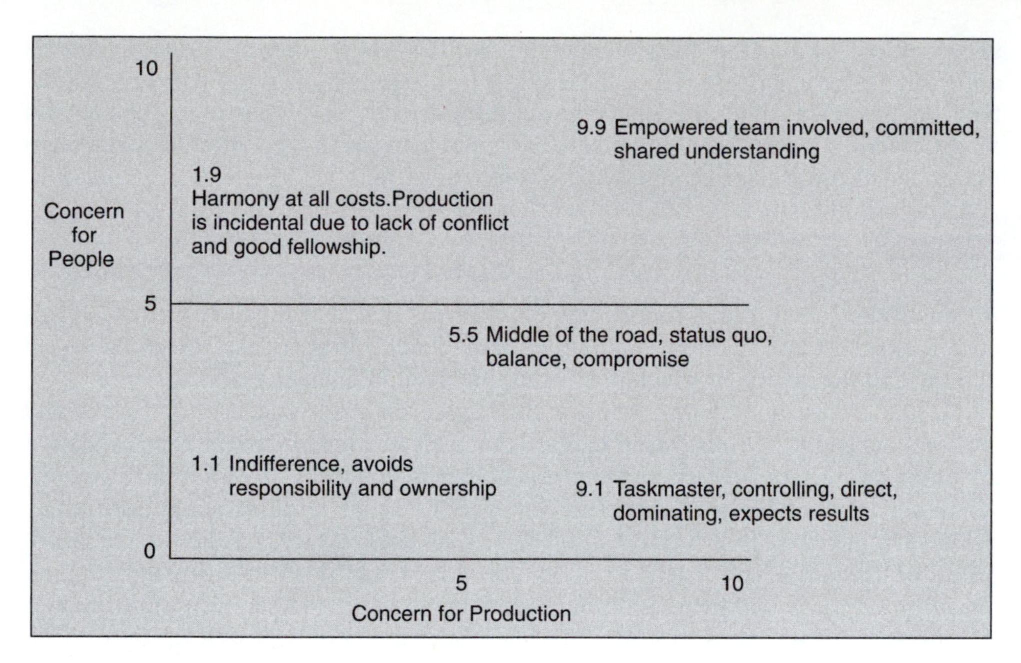

FIGURE 4.1 ◆ Managerial/Leadership Grid
(Courtesy of Randy R. Bruegman)

Middle of the road management. A weak balance of focus on both people and the work. Doing enough to get things done but not pushing the boundaries of what may be possible from both a productivity and personnel growth standpoint.

Team management. Firing on all cylinders. People are committed to task, and the leader is committed to people (as well as task).

Relate this to your own organization—where on this chart are your department, your division, your assigned company, and you? This well-known grid, using the *task versus person preference*, appears in many other studies, such as the Michigan Leader-

TABLE 4.1 ◆ Concern for People/Production

10	Country Club Management 1–Task 9–People		Team Management 9–People 9–Task
Concern for People		Middle of the Road Management 5–People 5–Task	
5			
	Impoverished Management 1–People 1–Task		Authority- Compliance 9–Task 1–People
		Concern for Production (Task)	
0		5	10

ship Studies and the Ohio State Leadership Studies. Many other task–people models and variants have appeared since then. While task and person are both clearly important dimensions, as other models point out, they are not all there is to leadership and management. It should be noted that the original name of the model was the managerial grid, which was later changed to the leadership grid.

LIKERT'S LEADERSHIP STYLES

Rensis Likert (1903–1981), founder of the University of Michigan Institute for Science Research, identified four main styles of leadership, in particular around decision making and the degree to which people are involved in the process.[4]

In the *exploitive authoritative style*, the leader has a low concern for people and uses threats and other fear-based methods to achieve conformance. Communication is almost entirely downward and the psychological concerns of people are ignored.

When the leader adds concern for people to an authoritative position, a *benevolent dictatorship* is formed. The leader now uses rewards to encourage appropriate performance and listens more to concerns lower down the organization, although what the leader hears is often filtered, being limited to what the subordinates think that the boss wants to hear. Although there may be some delegation of decisions, almost all major decisions are still made centrally.

The upward flow of information in the *consultative style* is cautious and tempered, although the leader is making genuine efforts to listen carefully to ideas. Nevertheless, major decisions are still largely centralized.

At the *participative level*, the leader makes maximum use of participative methods, engaging people lower down the organization in decision making. People across the organization are psychologically closer and work well together at all levels and across divisions.

LEWIN'S LEADERSHIP STYLES

Kurt Lewin (1890–1947), a German-born psychologist, and his colleagues did leadership decision experiments in 1939 and identified three different styles of leadership, in particular in regard to decision making.[5]

In the *autocratic style*, the leader makes decisions without consulting with others. In Lewin's experiments, he found this caused the highest level of discontent. An autocratic style works when there is no need for input on the decision, when the decision would not change as a result of input, and when the motivation of people to perform subsequent actions would not be affected whether they were or were not involved in the decision making.

In the *democratic style*, the leader involves the people in the decision making. The process for the final decision may vary from the leader having the final say, to the leader facilitating consensus in the group. Democratic decision making is usually appreciated by the people, especially if they have been accustomed to autocratic decisions with which they disagree. It can be problematic when there is a wide range of opinions and no clear way of reaching an equitable final decision.

The *laissez-faire style* minimizes the leader's involvement in decision making, hence allowing people to make their own decisions, although the leader may still be responsible for the outcome. Laissez-faire works best when people are capable and motivated in making their own decisions and when there is no requirement for a central coordination (for example, in sharing resources across a range of different people and groups).

In Lewin's experiments, the most effective style was discovered to be democratic. Excessive autocratic styles led to organizational revolution and disconnect, and were not conducive to developing a positive labor–management relationship. While under a laissez-faire approach, people were not coherent in their work and did not put in the energy they did when being actively led. These experiments, actually done with groups of children, were early in the modern era of management research and were consequently highly influential.

SIX EMOTIONAL LEADERSHIP STYLES

Daniel Goleman, Richard Boyatzis, and Annie McKee, in *Primal Leadership*, described six styles of leading that have different effects on the emotions of the target followers. These are styles, not types.[6] Any leader can use any style, and a good mix that is customized to the situation is generally the most effective approach.

The Visionary Leader The visionary leaders move people toward a shared vision, telling them where to go but not how to get there, motivating them to focus on the vision. They openly share information, hence giving knowledge power to others. They can fail while trying to motivate more experienced experts or peers. This style is best when a new direction is needed. Overall, it has a very strong impact on the work environment.

The Coaching Leader The coaching leaders connect to organizational goals, holding long conversations that reach beyond the workplace, helping people find strengths and weaknesses, and tying these to career aspirations and actions. They are good at delegating challenging assignments and demonstrating faith that demands justification, which leads to high levels of loyalty. Done badly, this style is micromanaging. It is best used when individuals need to build long-term capabilities. It has a highly positive impact on the work environment.

The Affiliative Leader The affiliative leader creates people harmony within the organization. This very collaborative style focuses on emotional needs over work needs. When done badly, it avoids emotionally distressing situations such as negative feedback. Done well, it is often used alongside visionary leadership. It is best used for healing rifts and getting through stressful situations. It can have a positive impact on the work environment.

The Democratic Leader The democratic leader values input and commitment via participation, listening to both bad news and good news. When done badly, there is a lot of listening but very little effective action. It is best used to gain buy-in or when simple inputs are needed (when you are uncertain). It has a positive impact on the work environment.

The Pacesetting Leader The pacesetting leaders build challenging and exciting goals for people, expecting excellence and often exemplifying it themselves. They identify poor performers and demand more of them. If necessary, they will roll up their sleeves and rescue the situation themselves. These leaders tend to be low on guidance, expecting people to know what to do. They get short-term results, but over the long term, this style can lead to exhaustion and decline. Done badly, it lacks emotional intelligence, especially self-management. It is best used for results from a motivated and competent team. Because it is often poorly done, it has a very negative effect on the work environment.

The Commanding Leader The commanding leader soothes fears and gives clear directions by his or her powerful stance, commanding and expecting full compliance (agreement is not needed). These leaders need emotional self-control for success and can seem cold and distant. This approach is best in times of crisis when unquestioned rapid action is needed and with problem employees who do not respond to other methods.

LEADERS CREATE RESONANCE

Resonance comes from the Latin word *resonare*[7] meaning "again creating sound." Effective leaders are attuned to other people's feelings and move them in a positive emotional direction. They speak authentically about their own values, direction, and priorities and resonate with the emotions of surrounding people. Under the guidance of an effective leader, people feel a mutual comfort level. Resonance comes naturally to people with a high degree of emotional intelligence (self-awareness, self-management, social awareness, and relationship management) but also involves intellectual aspects.

Creation of resonance can be done in six ways, leading to six leadership styles (see Table 4.2). Typically, the most effective leaders can act according to and even skillfully switch between the various styles, depending on the situation.

CHARISMATIC LEADERSHIP

The charismatic leader gathers followers through personality and charm, rather than any form of external power or authority.[8] *Charismatic leaders* work the room as they move from person to person. They pay attention to the person they are talking to at the moment, making the person feel as if he or she is, for that moment, the most important person in the world. Charismatic leaders pay a great deal of attention to scanning and reading their environment and are astute in picking up the moods and concerns of both individuals and larger audiences. They then will hone their actions and words to suit the situation. Charismatic leaders use a wide range of methods to manage their image and, if they are not naturally charismatic, may practice assiduously at developing their skills. They may engender trust through visible self-sacrifice and taking personal risks in the name of their beliefs. They will show great confidence in their followers. They are very persuasive and make very effective use of body language as well as verbal language. Many politicians use a charismatic style, as they need to gather a large number of followers. If you want to increase your charisma, studying videos of their speeches and the way they interact with others is a great source of learning. Religious leaders, too, may well use charisma, as do cult leaders.

Charismatic leaders who are building a group will often focus strongly on making the group very clear and distinct, separating it from other groups. They will then build the image of the group, in particular in the minds of their followers, as being far superior to all others. The charismatic leader will typically attach him- or herself firmly to the identity of the group, such as to join the group is to become one with the leader. In doing so, these leaders create an unchallengeable position for themselves.

TABLE 4.2 ◆ Six Leadership Styles

	Visionary Leadership	*Coaching Leadership*	*Affiliative Leadership*	*Democratic Leadership*	*Pacesetting Leadership*	*Commanding Leadership*
Leader Characteristics	Inspires. Believes in own vision. Empathetic. Explains how and why people's efforts contribute to the "dream."	Listens. Helps people identify their own strengths and weaknesses	Promotes harmony. Friendly. Empathetic. Boosts morale. Solves conflicts.	Superb listener. Team worker. Collaborator. Influencer.	Strong urge to achieve. High personal standards. Initiative. Low on empathy and collaboration. Impatience. Micromanaging. Numbers-driver.	Commanding. "Do it because I say so." Threatening. Tight control. Monitoring studiously. Creating dissonance. Contaminates everyone's moods. Drives away talent.
How Style Builds Resonance	Moves people toward shared dreams.	Connects what a person wants with the organization's goals.	Creates harmony by connecting people to each other.	Appreciates people's input and gets commitment through participation.	Realizes challenging and exciting goals.	Decreases fear by giving clear direction in an emergency.
The Impact of the Style on the (Business) Climate	+++	++	+	+	Often −~ when used too exclusively or poorly.	Often −~
When Style Is Appropriate	When changes require a new vision. Or when a clear direction is needed. Radical change.	To help competent, motivated employees to improve performance by building long-term capabilities.	To heal rifts in a team. To motivate during stressful times. Or to strengthen connections.	To build support or consensus. Or to get valuable input from employees.	To get high-quality results from a motivated and competent team. Sales.	In a grave crisis. Or with problem employees. To start an urgent organizational turnaround. Traditional military.

TABLE 4.3 ◆ Participative Leadership

<Not Participative			*Highly Participative>*	
Autocratic decision by leader	Leader proposes decision, listens to feedback, then decides	Team proposes decision; leader makes final decision	Joint decision with team as equals	Full delegation of decision to team

PARTICIPATIVE LEADERSHIP

A participative leader, rather than making autocratic decisions, seeks to involve other people in the process, possibly including subordinates, peers, superiors, and other stakeholders.[9] Often, however, as it is within the manager's whim to give or deny control to his or her subordinates, most participative activity is within the immediate team. The question of how much influence others are given may vary according to the manager's preferences and beliefs, and a whole spectrum of participation is possible (see Table 4.3), including stages at which the leader sells the idea to the team. Another variant is for the leader to describe the "what" of objectives or goals and let the team or individuals decide the "how" of the process by which the "how" was achieved in the past. This was often called management by objectives. The level of participation may also depend on the type of decisions being made. Decisions on how to implement goals may be highly participative, whereas decisions during subordinate performance evaluations are more likely to be made by the manager.

SITUATIONAL LEADERSHIP

When a decision is needed, an effective leader does not just fall into a single preferred style.[10] Things are not that simple. Factors that affect situational decisions include motivation and capability of followers: factors within the particular circumstances being dealt with and the time frame available to make the decision. The relationship between followers and the leader may be another factor that affects leader behavior as much as it does follower behavior.

The leader's perception of the followers and the situation will affect what is done rather than the truth of the situation. This and other factors such as stress and mood will also modify the leader's behavior.

Gary Yukl, at the University at Albany School of Business, has combined other approaches and identified six variables.

- *Subordinate effort:* the motivation and actual effort expended
- *Subordinate ability and role clarity:* followers knowing what to do and how to do it
- *Organization of the work:* the structure of the work and utilization of resources
- *Cooperation and cohesiveness:* of the group in working together
- *Resources and support:* the availability of tools, materials, people, and the like
- *External coordination:* the need to collaborate with other groups

Situational leaders work on such factors as external relationships, acquisition of resources, managing demands on the group, and managing the structures and culture of the group. Tannenbaum and Schmidt identified three forces that led to the leader's action: the forces in the situation, the forces in the follower, and the forces in the leader. This recognizes the leader's style is highly variable, and even such distant events as a family argument can lead to the displacement activity of a more aggressive stance in a work disagreement that becomes more hostile than usual. Maier noted that leaders consider not only the likelihood of a follower accepting a suggestion but also the overall importance of getting things done. In critical situations, a leader is more likely to be directive in style simply because of the implications of failure.

TRANSACTIONAL LEADERSHIP

The transactional leaders work through creating clear structures whereby they manifest what is required of their subordinates and the rewards they get for following orders.[11] Punishments are not always mentioned, but they are also well understood and formal systems of discipline are usually in place. The early stage of transactional leadership is in negotiating the contract whereby the subordinate is given a salary and other benefits, and the company (and by implication the subordinate's manager) gets authority over the subordinate. When the transactional leader allocates work to a subordinate, the subordinate is considered to be fully responsible for it, whether or not he or she has the resources or capability to carry it out. When things go wrong, the subordinate is considered to be personally at fault and is punished for the failure (just as he or she is rewarded for succeeding). Whereas transformational leadership has more of a "selling" style, transactional leadership, once the contract is in place, takes more of a "telling" style or approach. Despite much research that highlights its limitations, transactional leadership is still a popular approach with many managers. Indeed, in the leadership versus management spectrum, this style is very much toward the management end of the scale.

The main limitation is the assumption of a person largely motivated by money and simple reward, hence whose behavior is predictable. The underlying psychology is behaviorism, including the classical conditioning experiments of Ivan Pavlov. These theories are largely based on controlled laboratory experiments (often with animals) and ignore complex emotional factors and social values.

IVAN PAVLOV

Ivan Petrovich Pavlov (1849–1936), a Russian physiologist, first described the phenomenon now known as conditioning in experiments with dogs. He was awarded the Nobel Prize in Physiology or Medicine in 1904. Pavlov was investigating the gastric function of dogs by externalizing a salivary gland so he could collect, measure, and analyze the saliva produced in response to food under different conditions. He noticed that the dogs tended to salivate before food was actually delivered to their mouths and set out to investigate this "psychic secretion," as he called it. He decided that this was more interesting than the chemistry of saliva and changed the focus of his research, conducting a long series of experiments in which he manipulated the stimuli occurring before the presentation of food. Thereby, he established the basic laws for the establishment and extinction of what he called "conditional reflexes"—that is, reflex responses, such as salivation—that occurred only conditionally upon specific previous experiences of the animal. These experiments, performed in the 1890s and 1900s, were known to Western scientists through translations of individual accounts but first became fully available in English in a book published in 1927.

Perhaps unfortunately, Pavlov's phrase "conditional reflex" was mistranslated from the Russian as "conditioned reflex," and other scientists reading his work concluded that because such reflexes were conditioned, they must be produced by a process called conditioning. As Pavlov's work became known in the West, particularly through the writings of John B. Watson, the idea of conditioning as an automatic form of learning became a key concept in the developing specialty of comparative psychology and the general approach to psychology to understand such behaviorism.[12] In the concept of the fire service or any other organization, have you seen examples of such conditional behavior? In each department there is certain patterning creating predictable response. The natural tensions between chief officers and rank and file—the "us versus them," police versus fire, engine versus truck company—mind-sets are broad-based examples of conditioned firehouse behavior.

TRANSFORMATIONAL LEADERSHIP

Working for a *transformational leader* can be a wonderful and uplifting experience.[13] These leaders put passion and energy into everything. They care about their employees and want them to succeed. Transformational leadership starts with the development of a vision, a view of the future that will excite and convert potential followers. This vision may be developed by the leader or by the senior team, or it may emerge from a broad series of discussions. The important factor is the leader buys into it—hook, line, and sinker. The transformational leader constantly sells the vision, which takes energy and commitment, as few people will immediately buy into a radical vision and some will join the show much more slowly than others. The transformational leader takes every opportunity and will use many techniques to convince others to climb on board the bandwagon.

In order to recruit followers, transformational leaders have to be very careful in creating trust, and their personal integrity is a critical part of the package they are selling. In effect, they are selling themselves as well as the vision. Parallel with the selling activity is seeking the way forward. Some transformational leaders know the way and simply want others to follow them. Others do not have a ready strategy but will happily lead the exploration of possible routes to the promised land. The route forward may not be obvious and may not be plotted in detail, but with a clear vision, the direction will always be known by all; finding the way forward can be an ongoing process of course correction. Transformational leaders will accept that there will be failures, roadblocks, and barriers. As long as they believe progress is being made, they will be happy.

Transformational leaders are always visible and will stand up to be counted rather than hide behind the troops. They show by their attitudes and actions how everyone else should behave. They also make continued efforts to motivate and rally their followers, constantly doing the rounds, listening, soothing, and enthusing. It is their unswerving commitment, as much as anything else, which keeps people going, particularly through the difficult times when some may question whether the vision can ever be achieved. If people do not believe they can succeed, then their efforts will fail. Transformational leaders seek to infect and reinfect their followers with a high level of commitment to the vision. Overall, they balance their attention among action that creates progress, organizational wins, and the mental state of their followers. Perhaps, more than leaders of other approaches, they are people oriented and believe that success comes first and last through a deep and sustained commitment to the people they work with.

THE QUIET LEADER

In his book, *Good to Great*, Jim Collins identified five levels of effectiveness people can take in organizations.[14] At level four is the merely effective leader, whereas at level five is the leader who combines professional will with personal humility. The "professional will" indicates how far these leaders are from being timid wilting flowers and will march against any advice if they believe it is the right thing to do. In "personal humility" they put the well-being of others before their own personal needs, for example, giving others credit after successes but taking personal responsibility for failures. The quiet leader is not a modern concept. Lao Tzu, in the classic Taoist text *Tao Te Ching*, was discussing the same characteristic around 500 B.C.

> The very highest is barely known by men,
>
> Then comes that which they know and love,
>
> Then that which is feared,
>
> Then that which is despised.
>
> He who does not trust enough will not be trusted.
>
> When actions are performed
>
> Without unnecessary speech,
>
> People say "We did it!"

Here again, the highest level of leadership is virtually invisible.

The emphasis on the quiet leader is a reaction against the lauding of charismatic leaders in the media. In particular, during the heyday of the dot-com boom of the 1990s, some very verbal leaders received much coverage, while the quiet leaders were getting the job done. Of course, being quiet is not the secret of the universe, and leaders still need to see the way forward. Their job can be harder when they are faced with people who are accustomed to an extroverted charismatic style. A quiet style of leadership can be very confusing to some work groups, and they may downplay that leader, which is usually a mistake. Successful quiet leaders often play the values card to persuade others, showing selfishness and lack of emotional control as being unworthy characteristics. Again, there is a trap in this; leadership teams can fall into patterns of behavior whereby peace and harmony are prized over any form of challenge and conflict.

SERVANT LEADERSHIP

The servant leader serves others, rather than others serving the leader. Serving others thus comes by helping them to achieve and improve.[15] There are two criteria of servant leadership:

- The people served grow as individuals, becoming "healthier, wiser, more autonomous, and more likely themselves to become servants."
- The extent to which the leadership benefits those who are least advantaged in society (or at least does not disadvantage them).

An excellent example of a servant leader is Ernest Shackleton, the early twentieth-century explorer who, after his ship became frozen in the Antarctic, brought every one of his 27 crew home alive. This consisted of an 800-mile journey in open boats across the winter Antarctic seas that took two years. However, Shackleton's sense of responsibility toward his men never wavered.

Robert Greenleaf states true leadership "emerges from those whose primary motivation is a deep desire to help others." Servant leadership is a very moral position, putting the well-being of the followers before other goals. It may be easy to dismiss servant leadership as soft and easy, though this is not necessarily so, as individual followers may be expected to make sacrifices for the good of the whole.

The focus on the less privileged in society shows the servant leader as serving not only his or her followers but also the whole of society. As such, servant leadership is a natural model for working in the public sector. It requires more careful interpretation in the private sector due to the needs of the shareholders and customers, and the rigors of market competition are lost. Servant leadership also aligns closely with religious morals and has been adopted by several Christian organizations.

A challenge to servant leadership is in the assumption of the leader that the followers want to change. There is also the question of what "better" is and who decides this.

LEADERSHIP STARTS WITH KNOWING YOURSELF

A good friend of mine, who lives in southern California, is author and motivational speaker Tom Bay. Tom became interested in the fire service several years ago and can often be found riding with the Orange County Fire Authority. In several of his books, Tom has written that each of us has a "life window" to write on, which helps to provide a frame of reference on how we process and accept information.[16] Therefore, our life window dictates the directions we as individuals take and will subsequently help determine how we lead and manage our organizations. Think about yourself, the experiences you have had, and any life-changing events that have had an impact on you. What would be written on your window? What life events have formed your values, work ethic, likes, and dislikes? These all have a direct impact on how you process information; how you look at the world around you; and how you feel about an issue, people, family, politics, and religion. Get the picture? All of these experiences help to make up your own internal processor of information. Within that processor are filters that provide the framework by which we make decisions, treat people, and view our external environment. Within that framework are our own biases, prejudices, and blind spots that influence how we lead and manage on a daily basis.

Johari Window
A metaphorical tool created by Joseph Luft and Harry Ingham—thus the Joe-Harry/Johari window—in 1955 in the United States; used to help people better understand their interpersonal communication and relationships.

The **Johari Window** model[17] is based on the disclosure/feedback model of awareness, named after Joseph Luft and Harry Ingham. It was first used in an informational session at the Western Training Laboratory in Group Development in 1955. Luft and Ingham called their Johari Window model "Johari" after combining their first names, Joe and Harry. The Johari Window soon became a widely used model for understanding and training self-awareness, personal development, improving communications, interpersonal relationships, group dynamics, team development, and intergroup relationships. The Johari Window model is also referred to as a "disclosure/feedback model of self awareness" and by some as an "information processing tool." The Johari Window actually represents information (feelings, experience, views, attitudes, skills, intentions, motivation, etc.) within or about a person in relation to his or her group, from four perspectives. The Johari Window model can also be used to represent the same information for a group in relation to other groups or for a person in relation to

TABLE 4.4 ◆ Johari Window

		Feedback	
		Known to Self	**Unknown to Self**
Disclosure	Known to Others	OPEN AREA The Arena	BLIND AREA The Blind Spot
	Unknown to Others	HIDDEN AREA The Façade	THE UNKNOWN

another person. Johari Window terminology refers to "self" and "others." "Self" means oneself, that is, the person subject to the Johari Window analysis. "Others" means other people in the person's group or team. When the Johari Window model is used to assess and develop groups in relation to other groups, the self would be the group, and others would be other groups. The four Johari Window perspectives are called "regions," "areas," or "quadrants." Each of these regions contains and represents the information (feelings, motivation, etc.) known about the person, in terms of whether the information is known or unknown by the person and by others in the group (see Table 4.4).

EXPLAINING THE JOHARI WINDOW

The Johari Window model consists of a foursquare grid (think of taking a piece of paper and dividing it into four parts by drawing one line down the middle of the paper from top to bottom, and another line through the middle of the paper from side to side).

Using the Johari model, each person is represented by his or her own four-quadrant, or four-pane, window. Each of these contains and represents personal information—feelings, motivation—about the person and shows whether the information is known or not known by that person or other people. The four quadrants follow:

Quadrant 1: open area. What is known by the person about him- or herself and is also known by others.

Quadrant 2: blind area or "blind spot." What is unknown by the person about him- or herself but which others know. This can be simple information or can involve deep issues (for example, feelings of inadequacy, incompetence, unworthiness, rejection) that are difficult for individuals to face directly and yet others can see.

Quadrant 3: hidden or avoided area. What the person knows about him- or herself that others do not.

Quadrant 4: unknown area. What is unknown by the person about him- or herself and is also unknown by others.

The process of enlarging the open quadrant vertically is called self-disclosure, a give-and-take process between the person and the people with whom he or she interacts. As information is shared, the boundary with the hidden quadrant moves downward. And as other people reciprocate, trust tends to build between them.

As leaders, the smaller our blind spot, the more receptive we often are to the feedback and input on our leadership style. The result is, as we are open to more information, we can use it to become more effective leaders.

SCHEMATA

schema
In psychology and cognitive science, a mental structure that represents some aspect of the world. People use schemata to organize current knowledge and provide a framework for future understanding. Examples of schemata include stereotypes, social roles, scripts, worldviews, and archetypes.

A **schema** (plural *schemata*) is a mental structure used to organize and simplify knowledge of the world around us.[18] We have schemata about ourselves, other people, mechanical devices, food, and in fact almost everything. Schemata affect what we notice, how we interpret things, and how we make decisions and act. They serve as filters, accentuating and downplaying various elements. We use them to classify things, such as when we "pigeonhole" people based upon a prejudice or bias. They also help us forecast, predicting what will happen. We even remember and recall things via schemata, using them to "encode" memories. We tend to have favorite schemata that we use often. When interpreting the world, we try to use these first, going on to others if they do not sufficiently fit. Schemata are also self-sustaining and will persist even in the face of disconfirming evidence. This is because, if something does not match the schema, such as evidence against it, it is ignored. Some schemata are easier to change than others, and some people are more open about changing any of their schemata than are other people.

Other types of schemata include the following:

- Social schemata are about general social knowledge.
- Person schemata are about individual people or groups of people.
- Idealized person schemata are called prototypes.
- Self-schemata are about oneself.
- Role schemata are about proper behaviors in given situations.
- Event schemata (or scripts) are about what happens in specific situations.

A schema is a hypothetical mental structure for representing generic concepts stored in memory, providing a sort of framework or plan or script. Schemata are thought to have these features: be composed of generic or abstract knowledge used to guide encoding, organization, and retrieval of information; reflect properties of experiences encountered by an individual, integrated over many experiences; and may be formed and used without the individual's conscious awareness.

Although schemata are assumed to reflect an individual's experience, they are also assumed to be shared across individuals, perhaps in a culture. Once formed, schemata are thought to be relatively stable over time. We know more about how schemata are used than we do about how they are acquired.

Like a play, a schema has a basic script, but each time it is performed, the details will differ.

Like a theory, a schema enables us to make predictions from incomplete information by filling in the missing details with "default values." (Of course, this can be a problem when it causes us to remember things we never actually saw.) Our schemata are our own personal computer programs that enable us to actively evaluate and purge incoming information.

How Schemata Are Created and Modified Schemata are created through experience with people, objects, and events in the world. When we encounter something repeatedly, such as a restaurant, we begin to generalize across our restaurant experiences to develop an abstracted, generic set of expectations about what we will encounter in a restaurant. This is useful because if we are told a story about eating in a restaurant, the person does not have to provide all of the details about being seated, giving the order to the server, leaving a tip, and so on. Our schema for the restaurant experience can fill in these missing details. Not all of the information we have about restaurants necessarily gets added to our schema. Three processes are proposed to account for the modification of schemata: accretion, tuning, and restructuring.

- *Accretion.* New information is remembered in the context of an existing schema, without altering that schema. For example, suppose you go to a Starbucks and everything you experience there is consistent with your expectations of a Starbucks "experience." This is because you have been there on many occasions in different cities, and it has been the same each time. You can remember the details of the visit, but because they match an existing schema, they don't really alter that schema in any significant way.
- *Tuning.* New information or experience cannot be fully accommodated under an existing schema, so the schema evolves to become more consistent with experience. For example, when you first encountered a Starbucks, you began to develop a schema to accommodate this experience.
- *Restructuring.* When new information cannot be accommodated merely by tuning in an existing schema, it results in the creation of a new schema. For example, let's say that Starbucks changes everything, from its logo to the drinks it serves, to the color schemes of its stores; experience with the new Starbucks may be so different from your previous experience with the conventional ones that you would be forced to create a new schema.

For example, read the following sentence and tell how many F's are in it: "FINISHED FILES ARE THE RESULT OF YEARS OF SCIENTIFIC STUDY COMBINED WITH THE EXPERIENCE OF MANY YEARS."

Did you get the right answer?* If not, did you have to read it several times to get the right answer? How can you look at something and not see it? Beliefs and past conditioning affect your perception, opportunity, and resources.

Any book that is published is read possibly hundreds of times, including by professional proofreaders. Yet grammatical and other errors still get into print. Why? Because the mind is very kind and corrects the errors that our eyes see.

During the past 30 years, innovation and improved science have resulted in the fire industry's experiencing a revolutionary change, from the equipment used on a daily basis, the technology accessed, and the construction of the buildings to be protected. The fire service has also experienced many changes in how its leaders lead and manage people. The shifting generation values and expectations and the move away from a strict command and control organization to participative management philosophies have forced all of us to take a hard look at our traditions and our old ways of thinking. In many cases, they have challenged us to explore our individual frameworks and have forced many organizations to focus on culture change. Change can never occur without a conscious effort on our part as well as effective leadership that challenges us to examine our frameworks and test them against the realities of the day. To do so, we must be willing to listen to others and be receptive to their feedback and perceptions. At times, our schemata can get in the way of this.

A great story demonstrates this on a very basic level yet is so insightful into what often happens within our own departments. It begins with two battleships assigned to a training squadron that were at sea on maneuvers for several days in heavy weather. Sailors serving on the lead battleship were assigned to watch (lookout) duty on the bridge as night fell. Visibility was poor, with patchy fog and driving rain. Due to the bad weather, the captain had remained on the bridge, keeping an eye on all activities. Shortly after dark, the lookout on the boom of the bridge reported a light bearing off the starboard bow. "Is it steady or moving extremely?" the captain called out. The

*(*Answer: six*)

lookout replied, "Steady, captain." This meant the ship was on a dangerous collision course with the other vessel. The captain then called to the signalman, "Signal that ship that we're on a collision course and advise it to change course by 20 degrees." Back came the reply, "Advisable that you change course immediately 20 degrees, Seaman Second Class Jones." By this time the captain was quite angry. "Would you tell Seaman Second Class Jones that his ship had better change course 20 degrees." He shouted to the signalman also to send the message, "I'm Captain Smith and we are a battleship. Change your course 20 degrees." Back came the flashing light telling the captain, "I am a lighthouse, Seaman Jones." The battleship changed course. Sound familiar? How many times have you witnessed similar interactions in your own organization? Whether it was because of rank, personal biases, personality conflicts, or blind spots, the leadership style displayed was less than positive to the outcome.

Never let your rank, your ego, your biases, or your blind spots take you where you should not have gone. Each of us is the leader of our own ship and, as leaders in our industry, we are accountable for the welfare of many others. This alone requires us to be aware of control and take responsibility for our actions as we help to mentor and guide the actions of others. These basic frameworks are part of the larger picture and are essential to understand if we hope to provide the leadership to guide and motivate others to lead and manage our department, if we hope to make a difference in our career.

HOW OUR PERSONAL FRAMEWORK SHAPES OUR LEADERSHIP STYLE

In the world today, those in leadership and management positions are often challenged to keep their skill sets sharp enough to be effective. The challenges we face today are different from and, in many respects, more complex than those in the past. These challenges include a new economic landscape of a global economy that often drives the way we do business in significantly different ways, even in just the last two decades. We are challenged by the rapidness of informational exchange via the Internet and other media today, which have caused us to be in touch 24/7, 365 days of the year. We are challenged by the ever-changing workforce. For the first time in history, four generations are working together in the same organizations due to increased longevity and a healthier older workforce. As many studies have indicated, each generation brings with it a different value system, workplace expectations, work ethic, and expectation of their leadership.

All of us are challenged by the complexity of our organizations today in the services and the products that must be delivered more efficiently and effectively than ever before. Another challenge is in the adaptation of our own leadership and management skills to meet the emerging trends of fire industry and the changing landscape of local government. All of these factors are integral pieces of the mosaic of today's leadership challenges. The objective of leaders and managers is to provide a workplace environment that is stimulating, inviting, and meeting the increasing demands of their customers, while maintaining the bottom line, profit (staying within budget), and quality service. As a fire chief in a large metropolitan city who has worked in the fire service for over 25 years, my experience and my tenure on the job have allowed me to understand better what has shaped my leadership style. This experience has given me opportunities to start and run my own very successful business; to work in five different fire departments, each with a different organizational design; and to work with a diversity of professionals providing the spectrum of leadership and management skills. It is fascinating to watch the focus on the latest fad or trend on the best-seller list that drives many organizations to re-engineer themselves, yet never addresses the real issues of the organization. So many leaders today try to emulate a new way of thinking or create a new mind-set within their organizations, yet never understand the core values of leadership and management, which truly drive organizational change. The fact of the matter is that there is no magic bullet in today's world of leadership. No one thing makes organizations run better, makes better leaders, or makes organizations provide better service. Often it comes back to a general philosophy used in the fire service when operating during an emergency—the command toolbox.

Up until the moment the alarm goes off in the firehouse, you go about your day-to-day routine. In an instant, with the sounding of the alarm, you are moved from the normalcy of your daily activities to emergency response. When you are notified of the emergency event—whether a medical response, structural fire, hazardous materials incident, or some other type of incident—good fire ground officers switch to their checklist of those important items needing to be addressed, depending upon the incident. In the fire service, this is referred to as the command toolbox. When commanding and controlling an emergency, many tools are in the toolbox in the form of incident command structures, various pieces of equipment, specialty trained teams, and different types of resource deployment strategies. However, not all of them are used on every incident.

What is occurring in many organizations today is that leaders have become dependent upon the latest fad or management craze to attempt to lead and manage their organizations. It seems whenever a new buzzword, terminology, or concept gains some positive response for a given period of time, all of a sudden it appears to determine new leadership styles and becomes the focus in many organizations. For many who find themselves "in charge," it has become the nexus to their next leadership and management style for their organization. As leaders, something catches their fancy and says, Here use this, it will make you a better leader/manager, will increase your current production, will enhance your revenue, and will make your organization better. In most cases, these fads and management fantasies distributed to senior executives are often just a placebo that will work for only a short period of time. In fact, in many instances, the organization ultimately reverts to its old way of conducting business. Many in leadership and management positions fail to recognize the key element to becoming a good leader is to understand one's own leadership style(s). (When I became a fire chief, I quickly realized that I had a great deal to learn regarding the process of management. It took me several years to realize that I needed to learn what drove my leadership styles and my leadership tendencies, if I ever wanted to be a good leader.) Several very good testing instruments are available to help you define your leadership tendencies, but it's the understanding of the basis of those tendencies that enables you to become a powerful leader. When you understand your own framework and are cognizant of your potential blind spots, your strengths and weaknesses will become clear. Only then can you play to your strengths and develop strategies to overcome your weaknesses.

HOW LEADERSHIP STYLE CAN IMPACT RESULTS

At a leadership class I attended with employees from all areas of a business with over 4,000 employees, the general feeling of the group was "good class, but it will never happen here." For five days the comments reinforced "why it won't happen here." It boiled down to the simple dynamic: With every change in management came the introduction of the new management or leadership craze. Even though the training was excellent, the employees felt in three to four years there would be a new manager at the helm, and they would be off on a new adventure. In this case, after the training was completed, management began to focus immediately on organizational performance and forgot the human side of change. In all business, public or private, at the end of the day it all comes down to quality service, efficiency, effectiveness, profit, and, for those of us in government, staying within budget. The focus and efforts of many in leadership and management positions often zero in on counting widgets and evaluating daily performance. Although these are important issues, in the big picture they may not make much of a long-term difference if they do not deal with the heart of the organization; therefore, counting widgets probably will not matter.

Growing up in Nebraska, one of the sayings I heard a lot was "Son, you are talking to the wrong end of the horse." I did not really understand this until I had the opportunity to manage and lead people. The saying speaks of two very important issues often overlooked by leaders, and by doing so they end up in a place they did not choose. So what does this mean? Think about it. If you are trying to ride a horse, especially one that may be a little skittish, the wrong place to approach the horse from is the backside, unless you wish to find yourself in the hospital. If you have ever watched a horse trainer, or a good horse person, he or she is always in eye contact with the horse to establish a bond, then moves in slowly and begins to establish a relationship. The trainer takes time in allowing the horse to become accustomed to his or her presence and touch, until there is a level of trust. The trainer does this well before attempting to ride the horse.

In many cases, when we focus on our organizational leadership and managerial development, our efforts are misguided. Frankly, we are talking to the wrong end of the horse. Simply, the focus is on the immediacy of the outcomes and how to get there. We forget that without building a relationship with components of trust and mutual respect, no change will likely be institutionalized. In many organizations the focus is on measurement with the assumption "if you can measure it, you can change it." In most cases this is true, but if we wish to institutionalize the change, we have to focus on the right end of the horse or the outcome will not change.

What does institutionalization mean? It is the legacy. Ask yourself, if you left your department and revisited in five years, would there by any visible evidence you had ever been there? You do not have to be the chief executive or a chief officer to leave a legacy. Good company officers or engineers can leave a legacy in the firehouse in which they worked, the projects for which they were responsible, their work ethic, or their ethical leadership. You probably have worked or are working for such individuals. To institutionalize and leave a legacy of impact, we have to engage the heart and the mind of the organization and its people; otherwise, we are doing nothing more than going through a bureaucratic process. We become so concentrated in our effort to find the magic bullet and the next management craze we can hand off to our organizations to make them better. We look for the next leadership craze to bring to our organizations, prophesying on how it will improve our organizations—the quick fix to organizational wellness. But, in reality, quick fixes are short lived.

Having written a couple of books about quality improvement, I am a big believer in the quality initiatives undertaken in this country in the past 30 years. People with whom I have worked will tell you I am about measuring to manage: managing performance and measuring to facilitate change. However, we cannot just measure to lead. To be effective leaders, we have to institutionalize, and this is talking to the right end of the horse. When we focus on people first, we engage both their heart and their intellect. When we provide clarity of expectations, treat our people with fairness, and show them we care, their performance will follow. If we focus solely on performance, output, or the latest fad, our people will not bring their best to the table. This is what

I witnessed occurring in the class mentioned previously. In fact, several weeks after the class, the leadership of the corporation instituted a detailed reporting process to measure what else—employee time and activity. What do you think the impact was on all the good that came from the leadership training? You are right; there was an immediate high degree of skepticism regarding the intent, which nullified much of the positives created by the training effort. The immediate reaction was one of paranoia: "See, I told you this was all just smoke and mirrors; the real goal is to watch us." "I knew they really didn't care about us, and this proves it." These types of comments were verbalized throughout all levels of the organization when the new plan was announced! Now this company will be able to measure a plethora of information, and it will be able to make decisions based upon that information. Do you think, by taking the tactics this company did, the leadership will promote a culture that will be able to institutionalize these changes and continue with this new leadership program? Probably not. So one has to ask, why did they do it? It is all about style, leadership style. The fact is that all organizations are reflective to a great degree of the leadership within the organization, within the company, within the department, within the division. In most cases, the success of a department, a division, or a unit is directly linked to either the leader of the organization or the specific areas of responsibility.

Leaders and managers are reflective of their experiences and levels of comfort in how they lead on a daily basis. Where do you think the comfort zone was for the leader of the company just mentioned? Was it in the analytical or the interpersonal? Was it in counting widgets? Or was it in empowering the workforce? This is not to say one way or the other is wrong, but remember the tendency of our leadership styles is born in the events of our past experiences. Take a look at many of the successful corporations today and one can palpate the style of leadership existing within them. It is seen in their corporate philosophy, in their actions, and in how they portray their corporations to the general public. Leaders often fail to recognize what drives them to make decisions, what motivates them to establish the visions they have for their company, and, as important, how they process information to arrive at critical decisions. In the scenario presented, do you think a CEO more aware of his or her leadership style would have approached the strategy of implementing this new training program differently? Failure to understand one's leadership style can often result in incongruence between the leadership and the organization. In this scenario, if the leader were cognizant of or more in tune with his or her leadership style and tendencies, in respect to the organization culture, a much better outcome would have been the result. I have worked for and with many who think they can continue to hand off all these magic bullets for organizational development and then in six months to a year wonder why the company or the division has not changed. The same is true for private sector corporations and public sector general government. CEOs are brought in who have been very successful in other organizations and are simply ineffective within 12 months on the new job. Why does this happen? While there can be many reasons, one of the relevant underlying themes is the mismatch of the organizational culture to the leadership tendencies of the new CEO. Let me give you an example.

You are hired as a CEO of a well-established firm with one hundred years of experience, a firm that is vested in the traditional aspects of its service, has a strong labor organization, and has a reputation as a very pragmatic company within the industry. What type of leadership style do you think would best fit this organization? Your experience is in quickly taking several start-up firms to the top. You are the entrepreneur's entrepreneur. You like to empower your people to move fast and get things done on the fly. You do not focus on the process but on the outcome. So do you think there may be any potential future conflict here? The answer is maybe. While this may be a marriage made in heaven, the potential for significant conflict exists due to the divergent backgrounds of the organization and the CEO. The only way this marriage will be successful is if you as the CEO recognize this divergence and can meld your style into the organization and also blend the organization's style with yours in a way acceptable to both.

Today, many CEOs and public leaders fail to recognize that just being in charge does not mean they do not have to adapt themselves to the present situation. In many cases adaptation does not happen, and the CEO and the company suffer. This does not happen only at the senior management position but also in the firehouse and in the field. In the promotion of some officers, in certain cases due to their leadership style, they were never as successful as their potential through the testing process indicated. Good leaders have the ability to adapt their style to fit the situation present while not abdicating what has made them successful. They become the most accomplished, and the companies, public entities, fire companies, and battalions they lead are recognized as the best. This is because they know their own core values, recognize their own framework of leadership and management, have the unique ability to utilize their strengths based upon the situation presented, and have learned from this experience.

Whether you are already in a leadership role, aspire to be a leader, or realize the need to improve your own innate abilities to lead and manage, the ability to play to your strengths and blend them to the organization's needs is an art. To be a good leader you must approach the job with an understanding of who you are, to understand what your frame of reference is, and to comprehend what motivates you to make decisions from an organizational context. Too often, we become so focused on what we are doing with our companies and our organizations on a day-to-day basis that we lose sight of the big picture. When this happens we often fail to recognize one of the most important areas of understanding and observable behavior, which is ourselves. As a leader, senior manager, or chief executive of a company or unit, if you do not understand what drives you to make decisions, you cannot grow as a leader. The reality of the situation is that ultimately your company, your department, or your unit will suffer. So I hope through the course of this chapter, you begin to understand a little about what makes you tick, what drives you to do the things you do, and how you can maximize your performance and minimize your shortcomings. When you achieve this, you will move to a new level of understanding and achievement as a leader and manager.

In fact, the company previously mentioned made a change at the top—possibly in part because the leadership style of the CEO was not congruent with that of the company, in part because the implementation of this new corporate philosophy was poorly executed, not by intent, but from a failure to understand the dynamics of the organization. This is an example of the importance of building key relationships prior to the focus on performance and measurement. In retrospect, much of this was driven by the CEO's leadership framework. Our leadership style is a result of our life experiences, and we carry them with us in our daily work endeavors.

We each have a Johari Window and within the blind spot are our unknown strengths and weaknesses. We each have our own schemata—our own wiring schematic that filters, receives, categorizes, bundles the information we are exposed to daily—and our own set of rules or scripts that we use to interpret the world. Information that does not fit our schema may not register to us or we may not comprehend the information correctly. We tend to perceive things according to our beliefs and what we believe to be true and react accordingly. This selective perception is how a placebo works. If we think it is the real medicine, then we feel better. To understand our own personal frame of reference is to open our potential to become a better leader and manager. It is our own personal baggage that we carry with us every day that helps or hinders each of us to filter information, from personal and professional preferences, and impacts how we make our decisions. This happens many times each and every day!

YOUR PERSONAL FRAME OF REFERENCE IS THE FOUNDATION

Many people sometimes find it difficult to convey in words what has helped to establish their frame of reference and points of contact on the compass. You need to begin to think in terms of your own frame of reference—what life events, work, and personal experiences have become part of the fabric that influences your decisions?

We are not born with a slate that says chief executive, fire chief, or captain. In fact, the life window for the most part is blank. Although several genetic factors are at play that will factor into the making of our personality and our likes and dislikes, our experiences as we mature help define who we ultimately become. If you have children, look at how different they can be, even though they were raised in the same household. We share certain personality traits with our siblings, yet our approaches to problems and issues can be vastly different. Part of this is, of course, the personality traits that are uniquely individual, but a significant factor is the experiences that leave a lasting impression on who we are and who we become. At the beginning of life we begin mentally to write down or imprint things important to us. In our life journey, many experiences will have an impact on us, which will shape our values and beliefs and define the way we approach and deal with people and problems in our lives and our organizations. This is a powerful force of which great leaders are cognizant. They understand what makes them tick as well as how these experiential impacts can influence their decision-making process, their management practices, and their ability to lead.

This is a crucial element to becoming a good leader, to gain a better understanding of how you make decisions. When you understand your framework, your perceptions, and your own realities, you can often see what limits your ability to be an effective leader in your organization. Our framework works to guide our management decisions and leadership style. Our experiences also act as a filter in respect to how we process, see, accept, and ultimately use the information presented. As an example of what many of us have experienced, you meet someone for the first time who reminds you of someone you have previously known. You immediately connect your feelings to your past experience with the other person, whether positive or negative. Often your perception of this person is immediately linked to a past experience that has no substantive relevance to the current situation. The feeling is there and may impact your relationship with this individual. The same is true for organizational situations. Have you ever found yourself in similar dialog on the same issue in different organizations and your immediate reaction is to make the same decision as you made before, even though the players and circumstances may be quite different? By developing the ability to recognize this, you open the door to your true leadership potential.

Great leaders understand they have limited vision in some areas but are able to adapt and overcome those limitations so they can lead a country, a company, a battalion, or a fire company. More important, they can do so effectively.

MY PERSONAL FRAME OF REFERENCE

I thought it may be helpful to share my frame of reference to give you an idea of the concept of framing. I was born in 1955, toward the end of the baby boomer era, in Albuquerque, New Mexico, the last of three children. My father was a plumber who owned a small, successful plumbing business. My mother was a stay-at-home housewife, and my two sisters were 10 and 14 years older than I, so I was pretty much raised as an only child.

At the age of nine my father passed away. Losing a parent at a young age can have a significant impact on your entire life, as it did mine. As I look back on the whole experience, I can now see that event had a cascade effect from that moment forward and has helped to shape who I am today. *Were there significant events in your adolescence that have had a lasting impact or left a lasting impression on you?*

Understanding that a single mother raising a child in a large city such as Albuquerque was going to be difficult at best, my mother relocated us to Nebraska. This event had a significant impact on who and what I am today. Anytime you move there can be a fair amount of disruption in your life, which was the case with me. However, the relocation to a midwestern state, which had a very highly structured value system, and to a small community from a large city, began to lay the foundation for me to develop and understand the value of a good work ethic and the value of honesty and integrity. In my leadership positions, these values are very, very important. Watching my mother work two jobs to support the family and living over a garage in a two-room unit, which was built more for storage than anything else, make me appreciate the house I live in today. Growing up in a small farming community and having to work mowing lawns at a very early age taught me the value of a good day's work for a good day's pay. It also taught me to appreciate the value everyone brings to a job. *What were your formative years like? Home life? Economic situation? Did you grow up in a small town or large city? What impact did that experience have upon you? Is it in play today?*

In high school, like many of you, I was active in high school sports, but probably more important, I worked summers helping carpenters and bricklayers and worked in a local lumberyard. These jobs gave me an appreciation for the folks who work hard every day for the money they earn. When I finished high school, I could have gone straight to college but instead went straight to work. It wasn't until three years later when I became a volunteer firefighter in a small community in western Nebraska that I realized I had a career aspiration to become a firefighter, which ultimately I fulfilled. I was selected to be a firefighter for the city of Fort Collins, Colorado.

What was the motivation for your wanting to enter the fire service? During my time in Fort Collins, which was a little over 13 years, I had the opportunity to be a firefighter, an emergency medical technician, a fire engineer, a lieutenant, a captain, and finally a battalion commander.

As a hazardous materials specialist, probably one of the most harrowing experiences I have had in the field was removing chemicals from the local university chemistry building. Many of the chemicals had crystallized, making them volatile and explosive. There are certain times when you are put in harm's way and see people die. These situations leave a lasting and formative impact on how we approach our lives and our problems today. For all the bad situations encountered on calls, the ones that did not go well, the ones that were emotionally taxing, there are so many others where lives and property are saved and we made a difference in someone's life. These experiences are just as formative. *Have you been on calls that have left a lasting impression on you? Have they impacted the way you approach life today?*

My first taste of leadership was in Fort Collins. I started in 1979 and found myself working for an organization experiencing rapid growth and rapid change. In fact, the community and surrounding area were developing and growing so quickly that change became the organizational norm. *What were your first five years on the job like? Do you think they impacted your leadership and management style?*

I have visited Fort Collins several times since I left, and there is still a level of frustration when things are not moving as rapidly as people think they should, because that is the culture that has developed over the course of the last 20 years. The experience left a lasting imprint on the importance of organizational culture and the ability for one to lead and empower an organization. During this time, I also started a business with two other firefighters conducting hazardous materials audits, training, and video production. To raise money to produce the video, we sold three limited partnerships for over $500,000 to our friends and family. No pressure to succeed! But we did. In fact, the rate of return was over 100 percent, and the firm grew to over 30 people in less than two years. Yet there were times when everything we owned was on the line. The experience taught me business sense and true appreciation for those small business owners who do it every day. *Have you ever taken a career or business risk in pursuit of a vision you had?*

Also during my time in Fort Collins, I went through a divorce and subsequently remarried, blending a family. Challenging, fun, frustrating at times, but it taught me patience. From Fort Collins I became the fire chief in Campbell, California, a job that lasted 12 months due to consolidation of departments, which eliminated my job. For those of you in the private sector, this is known as outsourcing and relocation. Forced to look for a job within six months of my arrival—although stressful—was a great experience because it was my first stab at being a fire chief. I look back today and realize how much I did not know, and how accepting the workforce was to this young fire chief coming in to try and provide good leadership.

I relocated to the village of Hoffman Estates, a community on the northwest side of Chicago. This beautiful bedroom community, developed in the mid-1950s, was emerging in the 1990s as a corporate center, as many companies in and around Chicago were relocating from the inner city to suburbia. It was a great experience because, for the first time, I had to deal with some very significant labor relations issues. We know and have read the stories about the tough union mentality existing in and around Chicago, and I found this to be a very harsh reality. Yes, the old labor mentality was alive and well in Hoffman Estates at that time. A lot of backroom deals, a lot of clandestine politics, and for a young fire chief who did not have a lot of experience, it was truly a test by fire. The political dynamics and the confrontations with the union leadership were very intense, to say the least. Yet the firefighters and my senior staff were a very talented group. This experience taught me the value of organizational culture, and how it can become so institutionalized that it has an impact on almost every decision made.

I worked for a city manager named Pete Burchard who had a great influence on me in respect to watching him maneuver through a very difficult, much politicized process. How he was able to keep focused on the target and not get distracted by all the nonsense was remarkable. He managed not to become too discouraged and too engaged in the rhetoric and kept the organization moving forward in a very positive way. It was a great learning experience on how to blend leadership and management and make a difference for the community and the people with whom you work. Always keep focused on the target(s) even when there are multiple distractions. This great learning opportunity has become part of my frame of reference in how I lead today. *Have you worked for or with someone who has influenced and continues to influence you today?*

I went from Hoffman Estates to Clackamas County, Oregon, a fire district just south of Portland. *Growing*, *dynamic*, *innovative*, and *progressive* are adjectives that best sum up the organization. I was the first outside chief, so as you might expect there was a high degree of reluctance by many. Yet there was a willingness to try new ideas, to push the envelope, and we did. This department became the first accredited fire department in Oregon and has some of the highest caliber people assembled in one agency in the fire service. During this time, I was also involved in the development of the fire service accreditation project and the Chief Fire Officer Designation process. In addition, I was elected as a presidential officer for the International Association of Fire Chiefs (IAFC), serving as the president in 2002–2003. Also during this time, my personal frame of reference to the fire service changed quite dramatically. As a presidential officer in the IAFC, my role was to advocate for the international leadership of the fire service, to promote the direction for the fire service, and to establish the vision for those critical issues during my term.

September 11, 2001, changed the vision and the role the fire service was to play in respect to domestic response and preparedness. *Has your perspective of being in the fire service changed since September 11? How?* As a result, I found myself, along with several of my colleagues, spending a great deal of time on Capitol Hill, interacting with some of the highest ranking leaders of the federal government, being indoctrinated into a whole new perspective of the federal system, how it works and does not work, the aspects of terrorism, and our part in the protection of the homeland. My frame of reference expanded, as did my appreciation and respect for the men and women who work for me in the fire service. This is not to say I did not respect them before, because I did. However, my leadership and managerial approach changed from those experiences. I am no longer content to downplay the needs of the fire service, be it locally or nationally. I am much more forthright in the discussion of needs with elected officials, doing so respectfully, of course, and never trying to sugarcoat the issues. I also found the importance of a collective vision.

Another experience occurred during this time that caused me to reflect on why I am in this industry in the first place. Toward the end of September 2001 my secretary informed me there was somebody in the lobby to see me. I walked out and there stood an eight- to nine-year-old boy holding a two- to three-gallon pickle jar full of money. I noticed right away that he had the "We will never forget" ribbon pinned to his shirt. As I approached him, I bent down, and, without introducing himself, he handed me the pickle jar full of money and said, "Can you make sure the firefighters' families in New York City get this money?" Of course, I said I would and made sure it did. His mother was with him, and I asked her how he raised this money. She said he had taken his allowance to a local grocery store and bought candy bars and, for the past three weeks, had gone door to door selling them for a dollar each and collected enough money to fill the jar. It totaled $187 and some change. As I went back to my office, I knew that young man would grow up to be a firefighter, police officer, or paramedic. He will engage in some profession that provides service to others because he has a servant's heart. This experience caused me to reflect on why I am in the fire service. It is not about the adrenaline rush, the lights, the siren, and the challenges; it is about the commitment to service and leadership.

At that point in time, our country was in a state of change, trying to realign the government to address the issue of terrorism, and the fire service leadership was redefining its role. As the incoming president of the IAFC in August 2002, the world, our country, and the fire service found themselves in a place like no other in our history. The attacks of September 11 not only had a profound impact on the world and our country but also hit the core of the fire service. The loss of 343 firefighters in a matter of 90 minutes had challenged some of our very foundations; in an instant our mission as local firefighters changed to a realization that we were America's first responders, and our role was a critical link in providing for the common defense of our country. As events and experiences help to frame our leadership and managerial styles, they can also shape the visions we impart to others. I knew my role in this period of history was to paint a collective vision, which the fire service could rally around and focus its energy on.

After several months of soul searching before taking the office of IAFC president, I chose the theme, "A Time to Lead." I felt that it captured what we in the fire service were faced with in overcoming our emotional grief, in facing a new set of challenges, and in doing our part in the protection of our country. (I will share it with you in Chapter 10 when change and the importance of vision are discussed.) It captures the importance of establishing a common vision and helps us gain a perspective on how our own experiences help to shape our leadership and managerial styles on a daily basis. *Have you had a similar experience that helped to bring clarity to your personal or professional life?*

From the Clackamas County Fire District, I was hired as the fire chief of Fresno, California, a city approaching a half million people. In 1980, the Fresno Fire Department was a Class 1 fire department, with almost 340 personnel for a city of 185,000, but over time the department had been underinvested in. When I arrived it had 40 fewer personnel than it had had in 1980, the city had more than doubled in size to over 470,000, call volume was high, fire loss was high, most stations were in despicable condition, and companies were being closed because enough equipment could not be kept running. Coming into a city bureaucracy that was not totally functional, there existed a work ethic in the department that was amazing. I guess when you have done without for so long, you adapt to overcome, and they certainly had done that. While the personnel were great, the existing infrastructure, from the rolling stock to computer systems, was a mess. I also found the ability of the city support system to react to the needed changes to be cumbersome, with many organization silos and territories to work through. This is not to say they did not want them fixed; however, the ability to react quickly to the needed changes meant jumping through many hoops. This taxed my leadership skills as well as my patience, but once again I learned the value of a collective vision. With so many things to fix, it would have been very easy to become mired in the enormity of the issues. I found the need to paint a picture that everyone could rally around, to focus on the needed changes, and to provide the resources necessary to get the job done.

In the fire service, what is one of the most important things from a customer's perspective that provides for positive outcomes in the type of calls you respond to? Response times! So we built our community vision around that, and called it "4 Minutes to Excellence." We incorporated that concept into a Public Safety Commission Plan and into our budget messages and related goals; we went out into the community to local service clubs and talked departmental needs and the importance for the fire department to arrive at the scene of an emergency within four minutes of being dispatched. We put it in our brochures and on our coffee mugs and focused our culture and our efforts on attaining this goal. Not only was this an external message, but it forced every aspect of the organization to focus its efforts on the four-minute goal. In fact, we found that we were not walking our own talk in our actions and daily activities, which impacted our response times. No, it did not change overnight, but it did provide for a common community vision of the target, and we provided the plan for how to get there. The interesting aspect of this, after about a year, council members, local media, and other department directors in the city would ask when proposing a new program or making a budget request, "How does this help your '4 Minutes to Excellence' goal?" *Vision can be a powerful force!* (See Figure 4.2.)

Although other events and experiences have helped shape my leadership style, these significant events have helped to frame it. Now let's frame yours (see Table 4.5).

WHAT IS YOUR LEADERSHIP STYLE?

Let us begin to build your leadership framework. As leadership is the process of influencing others toward the achievement of organizational goals, then successful organizations must have dynamic and effective leaders. Leaders must be dynamic,

FIGURE 4.2 ◆ Fresno *4 Minutes to Excellence* Brochure
(Source: Fresno Fire Department)

TABLE 4.5 ◆ Significant Events That Frame Your Leadership Style

Window of Significant Events That Frame Your Leadership and Management Style			
Firefighter/EMT Lieutenant Captain/Training	B/C Shift Chief, Chief Hoffman Estates, Illinois Chief, Clackamas County, Oregon	Chief, Fresno Fire Department, California	CPSE Commission Member and Chairperson
Marriage Divorce Marriage Kids	Business Owner IAFC President, 2002–2003	CPSE Board President Lecturer	Authored books on quality, hazmat, and leadership

Note: Diagram should be a window or picture frame.

responsive to the needs of their organization, and effective in accomplishing the organizational goals through a group of people who are willing to follow them. To do so, you must know yourself.

THE JUNGIAN TYPE INVENTORY

Jungian Type Inventory Model based on the types and preferences of Carl Gustav Jung, who wrote *Psychological Types* in 1921.

The **Jungian Type Inventory** is based on the types and preferences of Carl Gustav Jung, who wrote *Psychological Types* in 1921. The model is based on Jung's four psychological functions of thinking, feeling, sensing, and intuition. Thinking and feeling are the judging functions, and sensing and intuition are the perceiving functions. The judging functions are those we use for making decisions, the perceiving functions for gathering information. According to Jung, the personality of each person is characterized by the dominance of one of these functions in either an extroverted or introverted way. Someone using his or her dominant function in an extroverted way would rely on it when dealing with the external environment, such as in public settings. Someone using his or her dominant function in an introverted way would rely on it when alone or with intimates.[19]

Katherine Briggs and Isobel Briggs Myers, a mother-and-daughter team, built the modern system based upon psychological difference according to the theories of Carl Jung that is probably the most popular type of system in the world today. In particular, they devised a written test (the *Myers-Briggs Type Inventory*, or *MBTI*®) to identify the person's type. Other variants have evolved that are also based on the Jung typology. The most well known of these is David Keirsey's Temperament Sorter.

Preferences The Jungian inventory measures on four preference scales, giving a variable score to show the strength of each one. In Table 4.6, the standard terms are shown first, with alternatives in parentheses.

The Myers-Briggs Type Indicator, based on the work of Carl Jung, provides an understanding of individual differences based on preferences on four scales. The four types follow:

1. *Introversion/extroversion.* This scale indicates where people get their energy and focus their attention. Introverts focus on the inner world of ideas and concepts. Extroverts focus on the outer world of people and things.

TABLE 4.6 ◆ Preferences

Preference	*From . . .*	*To . . .*
Energizing (Motivation)	E = Extroversion (Expressive, External)	I = Introversion (Reserved, Internal)
Attending (Acquiring Information, Inferring Meaning)	S = Sensing (Observant, Facts)	N = Intuiting (Introspective, Ideas)
Deciding (Formulating Intent)	T = Thinking (Tough-minded, Logic)	F = Feeling (Friendly, Emotion)
Living	J = Judging (Scheduling, Structured)	P = Perceiving (Probing, Flexible, Open)

TABLE 4-7 ◆ 16 Characteristic Types

ISTJ (12%) Doing what should be done	**ISFJ** (8%) A high sense of duty	**INFJ** (4%) An inspiration to others	**INTJ** (6%) Everything has room for improvement
ISTP (4%) Ready to try anything once	**ISFP** (4%) Sees much but shares little	**INFP** (4%) Performing noble service to help society	**INTP** (4%) A love of problem solving
ESTP (3%) The ultimate realists	**ESFP** (5%) You go around only once in life	**ENFP** (8%) Giving life an extra squeeze	**ENTP** (5%) One exciting challenge after another
ESTJ (12%) Life's administrators	**ESFJ** (8%) Hosts and hostesses of the world	**ENFJ** (5%) Smooth-talking persuaders	**ENTJ** (6%) Life's natural leaders

2. *Sensing/intuition.* This scale describes opposite ways of obtaining information from the environment. Sensing types trust and rely most on their five senses. Intuitive types trust and use their intuition, extrapolating meaning, and develop relationships based upon their previous experiences.
3. *Thinking/feeling.* This scale refers to how people reach conclusions, make decisions, and form opinions. Thinking types use logical, objective analysis. Feeling types weigh alternatives based on values, how important something is, and how much people care about it.
4. *Judging/perceptive.* This scale refers to the type of lifestyle individuals prefer and refers back to the previous two scales. Judging types tend to live more in the decision-making mode and prefer planned, orderly lives. Perceptive types spend more time taking in information and tend to like flexibility and spontaneity.

Table 4.7 outlines the 16 characteristic types, the percentage of the population that falls within each characteristic type, and a description of each major characteristic.

Keirsey Temperament Sorter
A personality test that attempts to identify which of four temperaments, and which of 16 types, a person prefers. Hippocrates, a Greek philosopher who lived circa 460–377 B.C., proposed four temperaments, which are related to the four humors. These were sanguine, choleric, phlegmatic, and melancholic.

Individual Styles The **Keirsey Temperament Sorter** accounts for the variations in cognitive styles that arise because people have different assumptions about the nature of truth, human nature, and relationships. The Keirsey Temperament Sorter[20] provides a means to identify the 16 characteristic styles and the preferences typically exhibited by people who fall into the different styles.

The Leadership Dimension of Communication and Problem Solving

Extroversion versus Introversion Extroversion and introversion are one of the preferences used in the Jungian Type Inventory. The naming is unfortunately a bit misleading, as extroversion is not about being loud and introversion is not about being shy. It is about where people get their energy and motivation from—other people or within themselves.

Extroversion (E) The energy of extroverts is outward toward people and things. They get their motivation from other people. They often want to change the world (rather than think about it). Extroverts like variety, action, and achievement; do well at school but may find college more difficult; have an often relaxed and confident attitude;

are understandable and accessible; and tend to act first and think later. At work, they seek variety and action and like working with other people. They prefer work that has breadth rather than depth. Introverts may see them as shallow and pushy.

Introversion (I) The energy of introverts is inward toward concepts and ideas. They need little external stimulation and, in fact, they can easily be overstimulated. It is possible that they focus more on their inner worlds because they suffer from sensory overload if they spend too much time outside and focusing on other people. Thus, they bottle up their own emotions, which can explode if pushed too far.

Rather than trying to change the world, they just want to understand it. Introverts think deeply about things and often do better at college than they did at school; have a reserved and questioning attitude, making them seem subtle and impenetrable; and tend to think before they act. At work, they like to work alone and often seek quiet for concentration. They prefer work that has depth rather than breadth. Extroverts may see them as egocentric and passive.

With Extroverts (E)

- Show energy and enthusiasm.
- Respond quickly without long pauses to think.
- Allow talking out loud without definite conclusions.
- Communicate openly—do not censure.
- Focus on the external world, the people, and things.
- Allow time for bouncing around ideas.
- Take words at face value.
- Do not assume they have made a commitment or decision.

With Introverts (I)

- Include introduction time to get to know you and trust you.
- Encourage responses with questions such as, "What do you think?"
- Use polling techniques for input and decision making.
- Allow time for thinking before responding and decision making.
- Make use of written responses where practical.
- Concentrate on one-on-one activities.
- Do not assume lack of interest on their part.

E's tend to think and problem solve out loud; you know that they are problem solving because you can see their lips moving and almost hear them thinking. E's often say, "Just let me talk long enough and I will figure out the answer." I's typically problem solve differently. They require a short period of quiet time to think about and reflect on the problem before they can discuss it. I's typically say, "If I can just get some quiet time to hear myself think, I can figure out the answer." A significant challenge occurs when you mix E's and I's together in problem solving. The E's immediately start talking out loud and the I's cannot hear themselves think. As a result the I's may not interact in the problem-solving process at all. Two E's (Is anyone listening?) or two I's (Is anyone talking?) also have challenges when communicating and problem solving (see Table 4.8).

The Leadership Dimension of Planning

Sensing versus Intuiting Sensing is more than touch and vision, and intuiting is not about gut feeling and fluffiness. It is about how we attend and create meaning from immediate data or after deeper thought.

TABLE 4.8 ◆ Characteristics of Extrovert and Introvert

Extrovert (E)	*Introvert (I)*
An E's essential stimulation is from the environment—people and things	An I's essential stimulation is from within—thoughts and reflections
Energized by other people, external experiences	Energized by inner resources, internal experiences
Does best work externally in action; interests have breadth	Does best work internally in reflection; interests have depth
Usually communicates freely—expressive	Usually reserved in communication until he or she knows and trusts a person
Acts, perhaps reflects, acts	Reflects, perhaps acts, reflects
Thinks best when talking with people	Thinks best when alone; shares with others when clear about what he or she believes
Usually takes the initiative in making contact with other people	Usually lets other people initiate contact
Has broad friendships with many people—gregarious	Has a few deep people contacts
Prefers to talk and listen	Prefers to read and write

Sensing (S) Sensors pay attention to immediate data both from their five senses and from their own direct experiences. They create meaning from conscious thought, rather than trusting their subconscious, limiting their attention to facts and solid data. As necessary, they will happily dig into the fine details of the situation; focus on what is immediate, practical, and real; live life as it is rather than trying to change the world; and like logic and tend to pursue things in a clear sequence. At work, they will have a clear schedule and prefer to use their proven skills in tactical situations. They may be seen as frivolous or shortsighted by intuitors.

Intuiting (N) Intuitors process data more deeply than sensors and are happy to trust their subconscious and "sixth sense," gut feeling, intuition, or whatever you want to call it. They are good at spotting patterns and taking a high-level view, as opposed to digging into the detail; like ideas and inspiration; and tend to focus on the future, where they will plan to change the world rather than continue to live in the imperfect present. At work, they like to acquire new skills and work at the strategic level. They may be seen as impractical, theoretical, and lacking determination by sensors.

With Sensors (S)

- Show evidence (e.g., facts, details, examples, etc.).
- Be practical, realistic, and grounded.
- Have a well-thought-out plan with details worked out in advance.
- Be direct.

- Show logical sequence of steps.
- Use concepts and strategies sparingly; concentrate more on the day-to-day consequences of a plan.

With Intuitors (N)

- Present ideas and global concept first, then draw out the details.
- Don't give details unless asked.
- When provided an idea or hypothesis or summary, don't ask for details; accept the intuitive conclusion at face value as working hypothesis.
- Be patient; work may come in spurts or bursts of energy.
- Let them dream; encourage imagination.

In the Keirsey Temperament Sorter, the sensing and intuiting functions are ways that people prefer to perceive and take in information. The sensing function takes in information by way of the five senses and likes to look at specific pieces of that information, deal with known facts, and live in the present, enjoying what exists. The intuiting function also takes in information via the five senses but then adds a sixth sense—a hunch or intuition. Most N's will state, "I make my worst decisions when I go against my intuition." N's are very conceptual and prefer to look at overall patterns and relationships. They like to deal with broad concepts or possibilities, and they plan for the future. They enjoy anticipating what might be. As a result of broad preferences, S's and N's tend to approach planning differently. N's prefer the broad, overall conceptual view, like to work with future possibilities, and are comfortable with envisioning processes. N's like to define where the organization is going and the possible attributes, conditions, and outcomes that it may seek to obtain. S's prefer step-by-step pragmatic planning based upon what can be accomplished today. They are most comfortable developing strategies, steps, and action plans to achieve certain goals. They prefer to define how the organization is going to achieve its goals. If N's conceive of putting a man on the moon, S's devise the systems and hardware to make it happen. Obviously, a good plan requires both perspectives—long-range conceptual goals (where) and pragmatic strategies and action plans (how). It is important to understand which strengths and preferences each partner brings to the planning process.

Partners with the same preferences may not focus on one of the two critical planning components—either the where or the how (see Table 4.9).

Thinking versus Feeling

Thinking and feeling are two of the preferences used in the Jungian Type Inventory. The naming may be a bit misleading, as thinking is more than thought, and feeling is not about being overemotional or fluffy. They are about how we decide through logic or through considering people.

Thinking (T)

Thinkers decide based primarily on logic. When they do, they consider a decision to be made. They tend to see the world in black and white and dislike fuzziness. Perhaps because people are so variable, they focus on tangible things, seeking truth and use of clear rules. At work, they are task oriented and seek to create clear value. Interacting with them tends to be brief and business-like. They may be seen as cold and heartless by feelers.

Feeling (F)

Feelers decide based primarily through social considerations, listening to their heart, and considering the feelings of others. They see life as a human existence and material things as being subservient. They value harmony and use tact in their interactions with others. At work, they are sociable and people oriented and make many decisions based on values (more than value). They may be seen as unreliable and emotional by thinkers.

TABLE 4.9 ◆ Characteristics of Sensing and Intuiting

Sensing (S)	*Intuiting (N)*
The S function takes in information via the five senses	The N function processes information via both the five senses and a sixth sense—hunch
Looks at specific parts and pieces	Looks at patterns and relationships
Deals with facts	Deals with possibilities
Lives in the present, enjoying what is there	Lives for the future, anticipating what might be
Trusts experience	Trusts theory more than experience
Tends to be seen as realistic	Tends to be seen as imaginative
Likes to apply reliable, proven solutions to problems	Likes problems that require new solutions
Likes the concrete	Likes the abstract
Learns sequentially, step by step	Learns by seeing connections—jumps in anywhere, leaps over steps
Tends to be good at precise work	Tends to be good at creating designs

With Thinkers (T)

- Be brief and concise.
- Be logical; don't ramble without any apparent purpose.
- Be intellectually critical and objective.
- Be calm and reasonable.
- Don't assume that feelings are unimportant; they may have a different value.
- Present feelings and emotions as additional facts to be weighted in a decision.

With Feelers (F)

- Introduce yourself and get to know the person; full acceptance may take a considerable amount of time.
- Be personable and friendly.
- Demonstrate empathy by showing areas of agreement first.
- Show how the idea will affect people and what people's reaction would be.
- Be aware that how you communicate is as important as what you are communicating.
- Let them talk about personal impact; accept decisions that may not be based on facts. See Table 4.10.

Judging versus Perceiving Judging is more than evaluation, and perceiving is not about looking at something. They are about how we approach life: in a structured way or an open, flexible way.

TABLE 4.10 ◆ Characteristics of Thinking and Feeling

Thinking	*Feeling*
Likes to decide things logically	Likes to decide things with personal feelings and values even if they are not logical
Wants to be treated with justice and fair play	Likes praise and pleasing people, even in unimportant things
May neglect and hurt others' feelings without knowing	Is aware of others' feelings
Gives more attention to ideas or things than relationships	Can predict how others will feel
Doesn't need harmony	Gets upset with conflict, values harmony

Judging (J) Judgers approach life in a structured way, creating plans and organizing their world to achieve their goals and desired results in a predictable way. They get their sense of control by taking charge of their environment and making choices early. Judgers are self-disciplined and decisive and seek closure in decisions. When they ask for things, they are specific and expect others to do as they say. They enjoy being experts. At work, they decide quickly and clearly and work to get the job done. Perceivers may see them as rigid and opinionated.

Perceiving (P) Perceivers view structure as being more limiting than enabling. They prefer to keep their choices open so they can cope with many problems that they know life will put in their way. They get their sense of control by keeping their options open and making choices only when they are necessary. Perceivers are generally curious and like to expand their knowledge, which they will freely acknowledge as being incomplete. They are tolerant of other people's differences and will adapt to fit into whatever the situation requires. At work, they tend to avoid or put off decisions and like most the exploration of problems and situations. Judgers may see them as aimless drifters.

With Judgers (J)

- Present a timetable and stick to it (or provide maximum warning if not).
- Allow them time to prepare.
- Show your achievements and results.
- Allow closure on consensus items; document those areas that require more work or discussion.
- Itemize achievements and decisions reached so far.

With Perceivers (P)

- Be personable and friendly.
- Demonstrate empathy by showing areas of agreement first.
- Show how the idea will affect people and what people's reaction would be.
- Be aware that how you communicate is as important as what you're communicating.
- Let them talk about personal impact; accept decisions that may not be based on facts. See Tables 4.11 and 4.12.

TABLE 4.11 ◆ Characteristics of Judging and Perceiving

Judging	*Perceiving*
Likes to have a plan that is settled and decided in advance	Likes to stay flexible and avoid fixed plans
Tries to make things come out the way they "ought to be"	Deals easily with unplanned and unexpected happenings
Likes to finish one project before starting another	Starts many projects but may have trouble finishing them all
Usually has mind made up	Usually looking for new information
May decide things too quickly	May decide things too slowly
Wants to be right	Wants to miss nothing
Lives by standards and schedules that are not easily changed	Lives by making changes to deal with problems as they come along

Utilizing the Keirsey Temperament Sorter from a team-building perspective will provide a personality portrait of team members and is an enjoyable exercise to promote positive group dynamics. It is a very easy instrument to use that takes less than 30 minutes. It has an uncanny ability to identify the individual traits that your leadership team will bring to the table. It is an excellent way to understand the psychological and cognitive styles of your senior leadership team in a very short time, and it will give you a good perspective on their cognitive style and how they process information and make decisions. Such a level of understanding will be very beneficial to you and the team.

THE IMPORTANCE OF STYLE

As discussed at the end of Chapter 3, organizations and the leadership styles within them are undergoing rapid transformations. Those transformations are changing our systems of behaving, our cultures, and, in many respects, our personal leadership styles. Many who hold leadership positions today, especially those who wear a uniform and a badge every day, see their power base in direct proportion to the number of bugles or trumpets they wear on their collar. But, in fact, we know with the changing dynamic of our workforce, our external environment, and the move toward a more inclusive system of leading and managing that this position is only one aspect of leadership. In fact, there are really two positions of power, one provided by the organization and one developed by the leader (see Table 4.13).

Leaders influence subordinates and peer groups to do the right thing. Their influence lies in their ability to alter people's beliefs and behaviors, and their power lies in their ability to influence. Table 4.13 is reflective of the current changing dynamic occurring in leadership in every organization. In the past and still in many organizations, leadership influence is based upon the position and the authority granted by the organization in which we work. You can see that the power base lies in authority and

TABLE 4.12 ◆ Characteristics with Each Type

Sensing Type		*Intuitive Type*	
ISTJ Serious, quiet, earn success by concentration and thoroughness. Practical, orderly, matter-of-fact, logical, realistic, and dependable. See to it that everything is well organized. Take responsibility. Make up their own minds as to what should be accomplished and work toward it steadily, regardless of protests or distractions.	ISFJ Quiet, friendly, responsible, and conscientious. Work devotedly to meet their obligations. Lend stability to any project or group. Thorough, painstaking, accurate. Their interests are usually not technical. Can be patient with necessary details. Loyal, considerate, perceptive, concerned with how other people feel.	INFJ Succeed by perseverance, originality, and desire to do whatever is needed or wanted. Put their best efforts into their work. Quietly forceful, conscientious, concerned for others. Respected for their firm principles. Likely to be honored and followed for their clear convictions as to how best to serve the common good.	INTJ Usually have original minds and great drive for their own ideas and purposes. In fields that appeal to them, they have a fine power to organize a job and carry it through with or without help. Skeptical, critical, independent, determined, sometimes stubborn. Must learn to yield to less important points in order to win the most important.
ISTP Cool with onlookers—quiet, reserved, observing and analyzing life with detached curiosity and unexpected flashes of original humor. Usually interested in cause and effect, how and why mechanical things work, and in organizing facts using logical principles.	ISFP Retiring, quietly friendly, sensitive, kind, modest about their abilities. Shun disagreements. Do not force their opinions or values on others. Usually do not care to lead but are often loyal followers. Often relaxed about getting things done, because they enjoy the present moment and do not want to spoil it by undue haste or exertion.	INFP Full of enthusiasms and loyalties but seldom talk of these until they know you well. Care about learning, ideas, language, and independent projects of their own. Tend to undertake too much, then somehow get it done. Friendly, but often too absorbed in what they are doing to be sociable. Little concern with possessions or physical surroundings.	INTP Quiet and reserved. Especially enjoy theoretical or scientific pursuits. Like solving problems with logic and analysis. Usually interested mainly in ideas, with little liking for parties or small talk. Tend to have sharply defined interests. Need careers where some strong interest can be used and useful.
ESTP Good at on-the-spot problem solving. Do not worry. Enjoy whatever comes along. Tend to like mechanical things and sports, with friends on the side. Adaptable, tolerant, generally conservative in values. Are best with real things that can be worked, handled, taken apart, or put together.	ESFP Outgoing, easygoing, accepting, friendly. Enjoy everything and make things more fun for others by their enjoyment. Like sports and making things happen. Know what's going on and join in eagerly. Find remembering facts easier than mastering theories. Are best in situations that need sound common sense and practical ability with people as well as things.	ENFP Warmly enthusiastic, high-spirited, ingenious, imaginative. Able to do anything that interests them. Quick with a solution for any difficulty and ready to help anyone with a problem. Often rely on their ability to improvise instead of preparing in advance. Can usually find compelling reasons for whatever they want.	ENTP Quick, ingenious, good at many things. Stimulating company, alert, and outspoken. May argue for fun on either side of question. Resourceful in solving new and challenging problems, but may neglect routine assignments. Apt to turn to one new interest after another. Skillful in finding logical reasons for what they want.
ESTJ Practical, realistic, matter of fact, with a natural head for business or mechanics. Not interested in subjects they see no use for, but can apply themselves when necessary. Like to organize and run activities. May make good administrators, especially if they remember to consider others' feelings and points of view.	ESFJ Warmhearted, talkative, popular, conscientious, born cooperators, active committee members. Need harmony and may be good at creating it. Always doing something nice for someone. Work best with encouragement and praise. Main interest is in things that directly and visibly affect people's lives.	ENFJ Responsive and responsible. Generally feel real concern for what others think or want, and try to handle things with due regard for the other person's feelings. Can present a proposal or lead a group discussion with ease and tact. Sociable, popular, sympathetic. Responsive to praise and criticism.	ENTJ Hearty, frank, decisive leaders in activities. Usually good in anything that requires reasoning intelligent talk, such as public speaking. Are usually well informed and enjoy adding to their fund of knowledge. May sometimes appear more positive and confident than their experience in an area warrants.

TABLE 4.13 ◆ Position Power Base

Personal Power Base (Developed by the Leader)		*Position Power Base (Provided by the Organization)*	
Power Bases	***Influence Effects***	***Power Bases***	***Influence Effects***
Information	Commitment	Authority	Compliance
Expertise	Commitment	Reward	Compliance
Goodwill	Commitment	Discipline	Resistance

Source: Adapted from the work of John R. P. French, Jr., and Bertram H. Raven. Five general power bases were described in their paper, "The Bases of Social Power," which appeared in *Studies in Social Power*, ed. Dorwin Cartwright (Ann Arbor, MI: University of Michigan, 1959).

our ability to reward and discipline our subordinates. Unfortunately, although we gain compliance, the influencing effects can also create resistance as we look into what our organizations will be in the future. It is an important fact that the leadership concepts used in the past simply will not work in the future.

As previously discussed in the changing dynamic of the workforce, the diversity, the values, the attitudes, and the generational aspects of today's employment will force all of us to be more inclusive in our leadership style. Successful leaders in the future will form a personal power base that is based on three critical elements: information, expertise, and goodwill. *Information* is based upon the facts or the reasoning a leader possesses and is able to share convincingly with members. This power base is at work when team members are influenced because they can see for themselves that some course of unified action is best. The power base of information is useful in the growth and development of the organizational team. It is important when team members need to understand the rationale to carry out the vision and the mission of the organization. Why is this important? It encourages a two-way information exchange and creates a more informed problem-solving technique. Perhaps more important, when you are working with professionals and an educated workforce, they expect to be treated in this way.

The second base is *expertise*, when a leader's judgment and knowledge in a specific area are well recognized. This base is at work when team members are influenced because they believe the leader knows what he or she is doing. They become confident in the merits of the decision and the rationale that was used to arrive at the conclusion and can be used to build confidence within your work group on the course of action that you have undertaken as a leader.

Last, but certainly not least, is the power base of *goodwill*, which is the feelings of support and respect that a leader has built with the people that he or she works with. This power base is at work when your team is influenced because they want to be supportive toward you as a leader. Built upon rapport and cooperation, good working relationships, and mutual respect, goodwill allows you to lead by example, providing a positive model for others to follow. Effective leaders strive to build a cooperative and cohesive team, provide and maintain harmony among team members, and gain cooperation on projects that are highly dependent upon the team members' abilities to get the job done. In the fire service, this happens many times a day because our whole process is developed around teamwork. Compare this to the positional power base provided by the organization built around authority, reward, and discipline, and the influence effects of compliance and resistance.

Future leadership trends will be built upon a commitment to the three factors of information, expertise, and goodwill. As we examine our own leadership styles, it is clearly evident in the academic research, in the changing dynamic within our workforce, and in the demands upon our organizations that without commitment from our employees, the success of our organizations will be limited. This is why it is so important that you understand your personal leadership style. It is not enough just to be an expert today. In the past in the fire service, it was acceptable, as we were promoted through the ranks, to be good fire ground tacticians. By doing so, we were almost assured the ability to take the next step up the promotional ladder. This is no longer an acceptable level of leadership. The complexities of leading and managing a fire service organization have never been greater. You need a clear understanding of what your personal frame of reference is. You have to appreciate that a personal frame of reference is engaged every day and plays a vital role in how you process information in your leadership or managerial role. You have to understand your strengths or weaknesses and your own schemata. Because when you truly assess your own strengths and weaknesses and your own blind spots, then and only then can you become an effective leader for your organization.

Remember, *leadership is the ability to take others where they would not have taken themselves.* It is the ability to influence those whom you work with to do the right thing, and to influence and alter people's beliefs and behaviors so that they become better at what they do and they help the organization to success. The base of power in leadership is very simple: the ability to positively influence individuals whom you work with and positively affect the ability for your organization to provide the services it needs to provide. To be a good leader you have to know yourself.

> When effective leadership accomplishes results, people say, "We did this ourselves!"

ADVICE FROM THE EXPERTS

How important is leadership style in the fire service today?

MICHAEL CHIARAMONTE

Chief Fire Inspector, Retired, Lynbrook Fire Department (New York): It is more important today than at any other time in the history of fire service.

BETT CLARK

Former Fire Chief, Bernalillo County Fire and Rescue (New Mexico): It is important to know your own leadership style, and what works well with the folks you lead. Only then can you be successful. I think this is the important message. There are many different defined leadership styles; however, you may not fit into any one definition. I am certain it is not important to fit into one; to be successful means fulfilling your needs while fulfilling the needs of others. I think personal values of caring, compassion, integrity, commitment, and dedication are the values that make any leadership style important in the fire service.

KELVIN J. COCHRAN

Fire Chief, Atlanta Fire Department (Georgia): The vast diversity of people and situations in fire departments today has created a demand for adaptive and cunning leaders. In today's fire service there is clearly no one best way to lead. Being autocratic has its place. Checking with the troops before making a decision has value. Letting the crew decide what is best has benefits. Using any one of these leadership styles at all times and under any circumstances is detrimental to the efficient and effective management of a fire department unit or division of labor. Leaders must possess the ability to diagnose situations and to discern the abilities of the crew or team, the communications skills to motivate them to work, and the flexibility to use the right leadership style and behavior for successful team performance.

CLIFF JONES

Fire Chief, Tempe Fire Department (Arizona): Leadership style can be very important. How leaders carry themselves in terms of leadership style can have direct bearing on their success with followers. Followers are looking for a person who cares about people and the organization. They are looking for a person who can exercise leadership in spite of the dangers of doing so. The leader's approach to people, approach to issue definition, and decision-making skills all help to make up leadership style. The key to style is to understand different styles and utilize the appropriate style for the situation.

BILL KILLEN

Retired Fire Chief and IAFC President 2005–2006: Leadership style is critical in today's fire service and will be even more so in the future as the ethnic culture of our nation evolves from the historical European background of English, German, Irish, Italian, and Polish to embrace a greater presence of Asian, Hispanic, and Middle East immigrants. Leadership style will demand greater ability to manage and lead diverse groups of people, dealing with the threats of world terrorism and constantly changing technology.

KEVIN KING

U.S. Marine Corps Fire Service: I am not sure the "style" is so important; rather I believe the focus on leadership is the crucial element. Given the dynamics in the fire service today, the multitudes of services expected by the public, and the fact that fire service remains a "people delivered" business, I believe the fire service continues to need strong leaders who can get their employees to provide high-quality products and services. This can be accomplished with many different styles of leadership, but I think it will be nearly impossible without quality leaders, both formal and informal. (These are opinions of Mr. King and do not reflect official policy of the Department of Defense or the United States Marine Corps.)

MARK LIGHT

Executive Director, International Association of Fire Chiefs: A fire service leader will be remembered more for his or her leadership style than almost any other part of his or her legacy. We all have officers we have worked for who imparted life and leadership lessons that have shaped who we are and how we respond today. Our styles should hopefully change with our organization and to meet the needs we encounter.

We often see chief officers who stay in the "command mode" of leadership, barking orders, not seeking input, not building coalitions. This style is very effective on the fire ground, but at city hall, this leadership style will be seen as a rogue player who doesn't play well with others. It is imperative that today's chiefs have a multitude of styles and, more important, know when to exercise the appropriate style to get their firefighters to perform at their best.

LORI MOORE-MERRELL

Assistant to the General President, International Association of Fire Fighters: Recognizing leadership style and matching it to situations and followers is important in any arena. Recognized leaders in one situation may not be the best person to have in the lead in another situation. For example, a hard-core militant, very organized incident commander may be exactly what is needed on a raging multi-alarm fire; however, he or she may not be the best person to have testify before the city council at budget time. Fire service leaders should know their styles and consequently who is likely to follow them and in what situations they lead best. Department chiefs should work to recognize leadership styles in their subordinate officers. By doing so, they can better match officers with roles where they will excel.

JAY REARDON

Fire Chief, Retired, Northbrook Fire Department (Illinois); Currently President of the Mutual Aid Box Alarm System (MABAS): Leadership style is the art of packaging a thought or a message in such a manner and at a rate that the recipients are able to accept the notion and digest it without emotion and rejection. Leadership and styles of leadership need to be adjustable for the mixed crowd—varied values and experiences in life and maturity. Accomplished leaders have an easier time at persuading listeners—the art of leadership is knowing how to deliver the message. Persuasive abilities—facts help tremendously—opinions alone will fail, but opinions often dictate the way we must deliver the message. Society is very complex, as is the fire service. Leadership and its styles are also more complex. Leadership is more difficult and more dangerous today than ever, and it will not get easier. One more point, successful leaders and their styles are directly proportionate to their perceptive abilities and authority to act once concurrence is reached with those being persuaded.

GARY W. SMITH

Retired Fire Chief, Aptos-La Salva Fire Protection District (Aptos, CA): I think many different leadership styles work. The key is to be consistent, trustworthy, and caring. The leader thinks about the good of those for whom he or she is working, policy makers, constituents, customers, as well as fellow team players. A person who leads needs to have great communication skills to set a clear and understandable vision that those being led can consider following. The vision needs a following, but not everyone is going to buy it, hook, line, and sinker. The world just does not work this way. The leader has patience and persistence to link the vision between policy makers and the team. Sometimes the vision is slow to catch on; only a line of successes will start to pull the doubters into the supporters column; but once the vision catches, it can snowball with a frenzy of success. The leader then knows it is time to start looking for the next hill to climb and the next set of challenges to take on.

STEVE WESTERMANN

Fire Chief, Central Jackson County Fire Protection District (Blue Springs, Missouri), and IAFC President 2007–2008: Extremely important and will continue to be important. You must be flexible in your leadership style for all types of situations that you may find yourself in and be comfortable in those various roles. While you must be flexible for the situation, you still must maintain a basic philosophy that you ascribe to. I like the servant leadership belief that once the goals are set, make sure the people you work with have all the resources they need to get the job done, and then let them do it. Continue to serve them by assuring they meet their goals.

THOMAS WIECZOREK

Executive Director, Center for Public Safety Excellence, and Retired City Manager: Leadership style is what separates the professional from the others. Being able to appear before public forums and communicate a cohesive vision based on data with visual aids puts a person in the "professional" category. Working with employee groups rather than dictating down to them, being inclusive yet being firm and fair—all of these are styles. The individual with the right style becomes an asset to the community's management team. Having the style that instead leads to constant conflict, difficulty, public perception issues, and questioning by elected officials does quite the opposite.

JANET WILMOTH

Editorial Director, *Fire Chief* Magazine: Critical. There must be substance to support leadership today. Firefighters and EMS personnel are more intelligent and demand to know reasons for decisions and directions.

FRED WINDISCH

Fire Chief, Ponderosa Volunteer Fire Department (Houston, Texas): Leadership is absolutely the most important capability that a chief officer must have in today's world. People desire strong leaders because leadership fosters success, and people want to be successful.

Review Questions

1. Describe the various leadership styles. Select one style and explain why you think it is a good one to emulate.
2. Explain the six emotional leadership styles. Describe when each style would be appropriate.
3. Explain the purpose behind the Johari Window. Describe how to use this effectively in honing your leadership style.
4. Describe schemata. Explain the three processes proposed to account for the modification of schemata.
5. Explain how your personal framework shapes your leadership style.
6. Explain how to determine your leadership style. Describe the four different types or styles. Compare and contrast each type.
7. Explain the importance of style.

References

1. *Leadership Styles*, www.teal.org.uk/situatio.htm
2. *Situational Leadership: Changing Your Style to Reflect Different Needs*, www.teal.org.uk/situatio.htm
3. *The Managerial Grid*, www.changingminds.org/disciplines/leadership/styles/managerial_grid.htm
4. *Likert's Leadership Styles*, www.changingminds.org/disciplines/leadership/styles/likert_style.htm

5. *Lewin's Leadership Styles*, www.changingminds.org/disciplines/leadership/styles/lewin_style.htm
6. *Six Emotional Leadership Styles*, www.changingminds.org/disciplines/leadership/styles/six_emotional_styles.htm;
7. *Leadership Styles*, www.12manage.com/methods_goleman_leadership_styles.html; Daniel Goleman, Richard Boyatzis, Annie McKee, *Primal Leadership* (Boston, MA 2002).
8. *Charismatic Leadership*, www.changingminds.org/disciplines/leadership/styles/charismatic_leadership.htm
9. *Participative Leadership*, www.changingminds.org/disciplines/leadership/styles/participative_leadership.htm
10. *Situational Leadership,* www.changingminds.org/disciplines/leadership/transational_leadership.htm; Gary Yukl, *Organizational Dynamics* (Albany School of Business, NY 2005); Robert Tannebaum and Warren H. Schmidt, *How to Chose a Leadership Pattern* (Harvard Business Review, 1958).
11. Transactional Leadership, www.changingminds.org/disciplines/leadership/styles/transactional_leadership.htm
12. *Ivan Pavlov*, http://en.wikipedia.org/wiki/Ivan_Pavlov
13. *Transactional Leadership*, www.changingminds.org/disciplines/leadership/styles/transformational_leadership.htm
14. *The Quiet Leader,* www.changingminds.org/disciplines/leadership/styles/servant_leadership.htm; Jim Collins, *Good to Great* (Harper Collins, NY 2001); William Bixby, *The Impossible Journey of Sir Ernest Shackelton* (Little Brown & Co., NY 1960).
15. *Servant Leadership*, www.changingminds.org/disciplines/leadership/styles/servant_leadership.htm
16. Tom Bay, *Look Within or Do Without* (Career Press: Franklin Lakes, NJ 2000).
17. *Johari Window Model*, www.businessballs.com/johariwindownmodel.htm
18. B. Armbruster, "Schema Theory and the Design of Content-Area Textbooks," *Educational Psychologist*, 21 (1996), 253–276.
19. Jolande Jacobi, *The Psychology of C. G. Jung* (Yale University Press 1973)
20. David Keirsey, Marilyn and Bates, *Please Understand Me* (Del Mar, CA: Prometheus Nemisis Book Co., 1984).

CHAPTER 5

Leading and Managing in a Changing Environment

Key Terms

core values, p. 199 **integrity, p. 202** **paradigm, p. 185**

Objectives

After completing this chapter, you should be able to:

- Describe what stewardship is.
- Describe why core values are important to an organization.
- Explain above-the-line accountability and why it is important.
- Define how organizational culture can impact change in a department.
- Define the key elements that can have a positive impact on organizational change.
- Describe the importance of leadership vision and its impact on the organization.

DON'T PLAY A NEW GAME BY THE OLD RULES

Albert Einstein once observed that the problems we face cannot be solved with the same level of thinking that created them. We must realize the significance of the change the fire service has experienced in just the past 20 years. Many aspects of our business today are much different than when many began their careers. With re-engineering, consolidations, political shifts, lack of resources, increased service-level demands, and outsourcing of services, the men and women who work in the fire service have found themselves in a perpetual state of change. It is important to remember, "In times of rapid change, experience can be your worst enemy." The rules of the game of public safety have changed. Whatever your position in the organization, we

FIGURE 5.1 ◆ Firefighters at Ground Zero (*Courtesy of Michael Rieger/FEMA News photo*)

all need to manage our own motivations to ensure we can make a difference in the teams we belong to, our organizations, and the community we serve. The steady waters of the past have given way to the turbulence of an open sea. This sea of change is driven by performance and quality initiatives, higher customer expectations, and limited resources, in which course corrections are the norm and must occur very quickly, if we are to be successful in the future. All of us have been challenged to help produce organizations that work better, cost less, and meet the safety and service needs of our local communities. After September 11, 2001, our mission expanded, and as first responders we became an integral part of the frontline of America's homeland security and defense (see Figure 5.1).

LEADING CHANGE

There seems to be an explosion of literature on the subject of organizational change. As a result, a number of consultants and professional companies have developed their services to assist in their general area of expertise. Change today means something more significant than when many of our generation started our careers 20 to 30 years ago. Changes within the fire service then were measured in terms of buying a new piece of equipment, opening a new fire station, instituting a new program, or hiring new firefighters. At present, the concept of organizational change is often measured in the expansion of our mission and services or in the restructuring of our operations. *Innovation* and *technologies, mergers* and *consolidations, intergovernmental collaborations, rightsizing, total quality management, re-engineering,* and *reinventing* are terms familiar to most of us today. Many organizations are not simply going through a change process but an organizational transformation. The differentiation between change and transformation is not a simple measurement. However, a transformation is often reflective of a fundamental and radical reorientation in the way the organization operates. In most cases, local government and the fire service have not gone through the degree of change experienced in the private sector. However, many of the companies and their employees who live in our communities have, and their expectations of your organization and the service you provide may reflect that experience. Ultimately, this can have an environmental impact on local government and subsequently on the local fire provider because status quo is no longer good enough for many of the residents served by your department.

Change should not be undertaken just for the sake of change but should be approached more strategically to accomplish an overall objective. Often organizational change is provoked by a significant external force that drives the change, such as funding reductions, city-wide expansion, or a dramatic need to increase productivity and service delivery systems. As organizations undertake this change process, they often find themselves moving away from an organization that is simply reactive to the surrounding environmental forces, to an entrepreneurial organization looking for opportunity to leverage its strengths, from a more stable and planned environment to one that is more free-wheeling and not as structured. A common force of change occurs with a change in leadership. The appointment of a new chief executive often promotes organization-wide change in accordance with the new leader's unique personality and leadership skill set. Whatever the driving force, there is typically a significant amount of resistance to change because people are afraid of the unknown, often do not understand the need for change, and may be cynical about the need to change (anything taking them out of their comfort zone). In any change process, conflicting goals in the organization are often created, highlighting the importance of leadership in this process. As we develop and begin to implement organization-wide changes, we often find ourselves in conflict with the very values important to the members in the organization. This is why successful organizational change is often developed around a vision that clearly articulates the desired state, yet is built upon the culture found within the organization and the values and beliefs of employees. Understanding this is a key component of any successful change process.

Leadership in the fire service must realize the future will not resemble the past. Leaders must focus on creating new opportunities rather than just solving problems. They must continually aim for something that will make a difference rather than something safe and simple. After all, isn't that what leadership is about? For many who find themselves in positions of authority, it is difficult to make the tough decisions, to set expectations, or to demand accountability. If you have ever worked for someone like this, while the demands upon you were low (I would wager), the organizational performance was less than desired. One of the problems the fire service has faced is that its heritage and traditions can often cloud the vision of what the future needs to be. In a number of organizations and with many captains to chief officers, it is often difficult to step outside the established frameworks of the organizational culture and provide the leadership needed for the health of the department.

In your own organization, have you witnessed inappropriate behavior that was overlooked, tolerated, and in some cases encouraged? In a firehouse located away from administrative oversight, the acceptable behavior level is dictated by the station captain or battalion commander. If the culture of the organization is accepting of this type of behavior, it will become pervasive. Professionally, you reach only the level of performance and behavior that you envision and are willing to pursue. When the fire service's culture is like a fraternity versus a paramilitary organization, the results are predictable. In a culture that is accepting of hazing and harassment, it is anticipated that the organization is not accepting of women, minorities, or anyone who is "not like them." If a lack of *esprit de corps* is accepted in the day-to-day activities, it will eventually be reflected on the emergency scene.

As an example, I recently visited a firehouse while traveling out of state. I had made prior arrangements to stop by, as I knew many of the firefighters in the department. While the hospitality was great, I was immediately struck by the lack of professional dress and the lackadaisical attitude of the crews. Their equipment was spread out on the ramp for my midafternoon visit. Every crew member was attired differently, from uniforms to T-shirts, from day pants to sweat pants, to ball caps turned backward. One crew member even had slippers on! What does this picture tell you? A lot! A walk through the station was equally interesting. The entire facility looked like a college dorm. So were these crews ready to respond and perform at their highest levels? I think not. You fight how you train, and I would surmise, based upon the attitude and culture displayed, that the crews approached training and the emergency scene in much the same way.

This may be one of the most significant challenges for leadership in the fire service today—stepping outside a framework that has developed over time, which integrates values, ethics, and culture, and has a direct impact daily on our professional ability. In many organizations, this framework has become a driving force, dictating how information is perceived and acted upon. Organizational norms become the filtration system, determining how information is shared, processed, and reacted to, and organizational culture is an unforeseen force that, if not addressed, will stifle the best ideas, the most detailed plans, and the most progressive leaders.

Dealing with organizational culture can be the most challenging and frustrating aspect of leadership yet can pay the most dividends. The ability to break down the cultural barriers existing within an organization, to provide new levels of thinking and openness, is a mark of true leadership. Joel Barker, a futurist, has written several books on how **paradigms** (frameworks) affect both the ability to lead change and the ability of people to accept it.[1] Never before in its history has the fire service faced so many challenges as it does today, new service-level demands, technological choices, and new workforce challenges. Never before have there been the opportunities to make revolutionary shifts in the fire industry. Exciting, isn't it? Never before has the fire service been armed with such a talented pool of firefighters and such progressive leadership. Yet if fire officers cannot recognize and understand the changing environment and help to lead their departments to a culture that expects excellence, then their true potential as a department will never be realized.

paradigm
A set of rules based on an explicit set of assumptions that explains how things work or ought to work. A paradigm is a framework of thought for understanding and explaining our perceived realities.

IT IS ALL UP TO YOU

How does one lead others to where they would not have gone themselves? We are only as good as our last performance. An interesting comparison to the Super Bowl sums up the world in which fire chiefs, chief officers, captains, firefighters, and paramedics operate today. Within the fire service, these competitive challenges can be seen in the shrinking tax dollar and in the significantly increased demand for service, not only in relation to some of the basic tenets of EMS, hazmat mitigation, and fire prevention, but also with the current mandate and responsibility to provide the first response to terrorist events. The competitive challenge is seen in the governing bodies, which continue to expand the scope of services fire departments provide, with limited or no increase in the resources available at the local, state, and national levels. Think about our competition as if it is the Super Bowl. We want to play hard for the season and win the big game, then sit around during the off-season and gloat about how great we are. We know our competitors in the private and the public sectors will not wait until next year for the rematch. They want to play again, next week and every week, until they finally win.

One of the leadership challenges in both the public and private sectors is that the rules of the game are continually changing, driven by shifting attitudes and values, technology, economics, community/consumer expectations, just to name a few. These also drive our competitive challenges.

> Through all of these changes, each of us has the opportunity to be evaluated and judged on how we deal with each issue or event we respond to. We can have a 20–1 record, but our leadership will often be judged by our latest performance.

As a fire chief, approach each day with an attitude of fielding the Super Bowl team every day. Anything less will not provide the needed level of service and may increase the risk to your firefighters. If your house is on fire or you are having a medical emergency, you do not want the team that is 4–16; you want the champs! There are fire service leaders who have unified organizations that were once in disarray. There are also leaders who take a good department and make it great. Unfortunately, many have witnessed departments that were at the top of their game and, due to ineffective leadership and experience, suffered a slow methodic decline, until one day they found themselves not in the Super Bowl but in the toilet bowl! To be successful, good leaders and managers must realize organizations need to be refueled, revitalized, and refocused periodically. Leaders understand the need periodically to create new visions as a vital link to developing high-performance, high-quality, service-oriented teams, leading their departments to what is often referred to as the "cutting edge." So when you report to work, whatever your position, what attitude do you bring to the game?

Fire service organizations also enter periods of stagnation. With very little creativity and innovation, as time passes, they are lethargic in their approach and actions on many levels. The common denominator in each case is a fire chief with a long career tenure whose leadership ability or interest peaked several years earlier; therefore, the department was placed on automatic pilot. Organizations not being led will settle into a comfort zone of mediocre performance. Where there is no vision and no motivation from the top to push the agenda, the organization will find a midline of performance. This topic is well covered in the fine book, *It's Your Ship,* by Captain D. Michael Abrashoff,[2] retired commander of the *USS Benfold*. It provides an insightful view of some basic leadership, motivation, and productivity concepts through the story of how Captain Abrashoff transformed the *Benfold,* a guided missile destroyer, in June 1997 from a ship that was mediocre to the best ship in the U.S. Navy as graded by its peers. There are parallels between what Captain Abrashoff faced when taking over the *Benfold* and what leaders confront when they take over a new organization, a new company, or a new assignment. He emphasizes the need to do the following:

- Lead by example.
- Listen aggressively.
- Communicate purpose and meaning.
- Create a climate of trust.
- Look for results.
- Don't look for salutes.

- Take calculated risks.
- Go beyond standard procedures.
- Build up your people's confidence.
- Generate unity.
- Improve your personnel's quality of life as much as possible.[3]

These parallels from a military standpoint are applicable to the bureaucracies leaders confront in their attempt to make change happen. The politics and barriers within each organization can create obstacles to performance and success. Abrashoff offers an enlightening, original, and commonsense point of view on leadership, motivation, and productivity. He also provides a fresh outlook on how to build an effective team. This is also an insightful story for those in leadership roles who find themselves in the position of just "minding the store." The leadership role will be as important a decade from now as it was the first day you joined the organization. This is why it is vital to realize that if your performance drops, so will your team's. It is up to the leader to keep on the leading edge through continued education, professional development, teaching, mentoring, or whatever keeps you at the top of your game.

THE BOILED FROG SYNDROME

Considerable attention has been focused on individual teams within the organization, but what about the organizational team as a whole? As noted previously, in the absence of leadership, organizations will arrive at a midline level of performance, driven not by the pursuit of excellence but by the preservation of the status quo. The entire team can become complacent in its attitudes and actions. Mediocrity becomes the benchmark, and the whole of the organization, from a performance perspective, is not what it should be. This gradual decline in leadership is a direct result of losing touch with the organization, resisting innovation and change, and failing to take advantage of technological breakthroughs. Look at a fire department as a system; many forces and interrelationships impact it. Each has a degree of influence in its ability to meet its mission and provide quality services (see Figure 5.2).

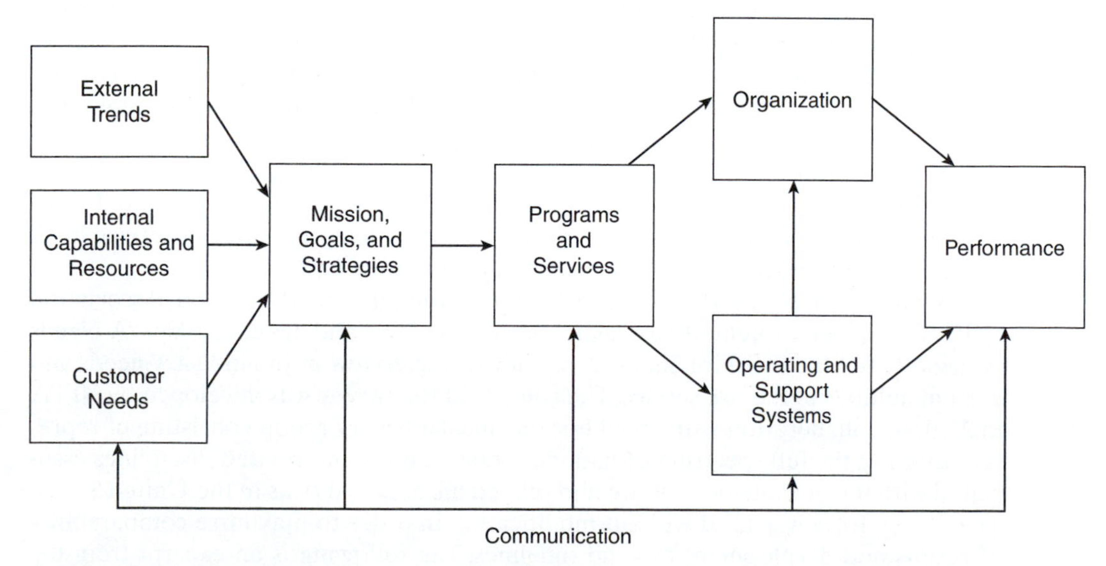

FIGURE 5.2 ◆ Communication Chart
(*Courtesy of Randy R. Bruegman,* The Chief Officer: A Symbol Is a Promise, *p. 103*)

In many departments change is often triggered by external events. It could be a significant fire, lawsuit, political election, or new city management. Many fire chiefs have lost their positions because they failed to recognize the need to refocus their organizations. Complacency left them vulnerable and created organizational apathy. When organizations are out of touch with the changing context of the industry and the community they serve, they find themselves in an organizational comfort zone. Experienced chiefs must avoid this especially vulnerable place, where nothing much happens, because the need for bureaucracy, fellowship, or socialization is greater than the need for innovation. The need for progress and new ideas in planning is not recognized. This is a dangerous place for the fire service profession and its firefighters, as well as for individual fire companies or divisions within an organization. In your own organization, have you witnessed or been part of either? These are the employees or volunteers who come to work or respond to a call banking on nothing of significance happening. We all know the rest of the story! When they are called upon, their performance is less than stellar, the outcome of the situation is not what it could have been, and the risk to those involved is increased. Unfortunately for those in this arena, it is easy to rationalize the lack of preparedness, performance, and outcome. Complacency is an organizational cancer, and if left untreated, it will suck the life out of the organization and its people.

Such complacency often creates an organizational culture that has minimal environmental perspectives and often cannot react when it needs to. This analogy has been referred to as the boiled frog syndrome, a story that is descriptive of what organizations experience when they become complacent about what is occurring within and around them. If a frog is placed in a pan of hot water, it will immediately jump out. More often than not, it will survive the experience. If you put the same frog in a pan of room-temperature water and then heat the water very slowly, the frog will stay in the pan until it boils to death. The frog could have jumped out, but the change happened so slowly that it did not realize the danger it was in.

At times, organizations can react in the same manner, becoming complacent to the point of self-destruction. Those who do not respond to internal or external forces in time to avoid significant damage often find themselves in the same situation as the frog. These organizations are "playing fire department," and then something catastrophic happens as a result of their complacent attitude. In today's fire service, there is no room for such organizations. With approximately 30,000 fire agencies across the United States, the range of ability can vary greatly. Unfortunately, the expectation of the community (whether a small rural community in the middle of Indiana or a major city such as Chicago) is that the organization is professionally staffed, fully equipped, well trained, and ready to handle any emergency. Some in the fire service have fooled themselves into thinking they can actually provide an adequate level of service in all of the communities being served. In reality, this is not the case.

A cooperative report published by the National Fire Protection Association (NFPA) and Department of Homeland Security in 2003 and again in 2006, "A Needs Assessment of the U.S. Fire Service," provides an overview of the industry needs and current abilities of the fire service. The content of the survey was developed by NFPA in 2001, in collaboration with an ad hoc technical advisory group consisting of representatives of the full spectrum of national organizations and related disciplines associated with the management of fire and related hazards and risks in the United States. The survey form was used without modification in order to maximize comparability of results and development of valid timelines. The following is an excerpt from the report, highlighting specific areas and the responses received.[4]

The US Fire Service—Revenues and Budgets

- Most of the revenues for all- or mostly-volunteer fire departments come from taxes, either a special fire district tax or some other tax, including an average of 64–68 percent of revenues covered for communities of less than 5,000 population.
- Other governmental payments—including reimbursements on a per-call basis, other local government payments, and state government payments—contributed an average of 11–13 percent of revenues for communities under 5,000 population.
- Fund-raising contributed an average of 19 percent of revenues for communities of less than 2,500 population.
- Used vehicles accounted for an average of 40 percent of apparatus purchased by or donated to departments protecting communities with less than 2,500 population.
- Converted vehicles accounted for an average of 14 percent of apparatus used by departments protecting communities with less than 2,500 population.

Personnel and Their Capabilities

- There are roughly 1.1 million active firefighters in the US, of which just under three-fourths (73 percent) are volunteer firefighters. Nearly half the volunteers serve in communities with less than 2,500 population. [See Table 5.1.]
- The number of volunteers has been declining and the number of career firefighters has been increasing for several years. Part of the reason is there has been a slight shift from all- or mostly-volunteer departments toward all- or mostly-career departments.
- In communities with less than 2,500 population, 21 percent of fire departments, nearly all of them all- or mostly-volunteer departments, deliver an average of four or fewer volunteer firefighters to a mid-day house fire. Because these departments average only one career firefighter per department, it is likely most of these departments often fail to deliver the minimum of four firefighters recognized by national standards as the necessary minimum for interior fire attack.

TABLE 5.1 ◆ Number of Career, Volunteer, and Total Firefighters by Size of Community

Population Protected	*Career Firefighters*	*Volunteer Firefighters*	*Total Firefighters*
1,000,000 or more	30,700	800	31,500
500,000 to 999,999	31,700	4,150	35,850
250,000 to 499,999	21,200	5,450	26,650
100,000 to 249,999	45,800	4,500	50,300
50,000 to 99,999	43,450	7,150	50,600
25,000 to 49,999	44,850	28,000	72,850
10,000 to 24,999	48,150	83,900	132,050
5,000 to 9,999	14,400	119,100	133,500
2,500 to 4,999	6,100	155,750	161,850
Under 2,500	7,750	398,350	406,100
Total	294,100	807,150	1,101,250

- An estimated 79,000 firefighters serve in fire departments that protect communities of at least 50,000 population and have fewer than four career firefighters assigned to first-due engine companies. It is likely, for many of these departments, the first arriving complement of firefighters often falls short of the minimum of four firefighters needed to initiate an interior attack on a structure fire, thereby requiring the first-arriving firefighters to wait until the rest of the first-alarm responders arrive.
- An estimated 214,000 firefighters, most of them volunteers serving in communities with less than 2,500 population, serve in departments that are involved in structural firefighting but have not formally trained all involved firefighters in those duties.
- An estimated 128,000 firefighters, most of them volunteers serving in communities with less than 2,500 population, serve in departments that are involved in structural firefighting but have not certified any firefighters to Firefighter Level I or II.
- An estimated 36 percent of fire departments are involved in delivering emergency medical services (EMS) but have not provided formal training in those duties to all involved personnel.
- The majority of fire departments do not have all their personnel involved in emergency medical services (EMS) certified to the level of Basic Life Support and almost no departments have all those personnel certified to the level of Advanced Life Support.
- An estimated 36 percent of fire departments involved in hazardous material response have not provided formal training in those duties to all involved personnel.
- More than four out of five fire departments do not have all their personnel involved in hazardous material response certified to the Operational level and almost no departments have all those personnel certified to the Technician level.
- An estimated 63 percent of fire departments involved in wildland firefighting have not provided formal training in those duties to all involved personnel.
- An estimated 50 percent of fire departments involved in technical rescue service have not provided formal training in those duties to all involved personnel.
- An estimated 737,000 firefighters serve in fire departments with no program to maintain basic firefighter fitness and health, most of them volunteers serving communities with less than 5,000 population.

Fire Prevention and Code Enforcement

- An estimated 67.0 million people (23 percent of the US resident population in 2005) are protected by fire departments that do not provide plans review.
- An estimated 118.9 million (40 percent) are protected by departments that do not provide permit approval.
- An estimated 128.9 million (44 percent) are protected by departments that do not provide routine testing of active systems (e.g., fire sprinklers).
- Each of the above services may be provided by another agency or organization in these communities.
- An estimated 103.6 million people (35 percent) are protected by fire departments that do not have a program for free distribution of home smoke alarms.
- An estimated 120.8 million people (41 percent) are protected by fire departments that do not have a juvenile firesetter program.
- An estimated 83.6 million people (28 percent) are protected by fire departments that do not have a school fire safety education program based on a national model curriculum. Moreover, independent data on the breadth of implementation of such curricula indicate most fire departments reporting programs provide only annual or occasional presentations based on material from such a curriculum.

- An estimated 20.3 million people (7 percent) live in communities where no one conducts fire-code inspections. Two-fifths of this population live in rural communities, with less than 2,500 population.

Facilities, Apparatus and Equipment

- Roughly 17,300 fire stations (36 percent of the estimated 48,400 total fire stations) are estimated to be at least 40 years old.
- Roughly 26,000 fire stations (54 percent) have no backup power.
- Roughly 35,000 fire stations (72 percent) are not equipped for exhaust emission control.
- Using maximum response distance guidelines from the Insurance Services Office and simple models of response distance as a function of community area and number of fire stations, developed by the Rand Corporation, it is estimated three-fifths to three-fourths of fire departments have too few fire stations to meet the guidelines.
- Roughly 14,000 fire engines (pumpers) (17 percent of all engines) are 15 to 19 years old, another 15,700 (19 percent) are 20 to 29 years old, and 10,900 (13 percent) are at least 30 years old. Therefore, roughly half (49 percent) of all engines are at least 15 years old.
- Among fire departments protecting communities with less than 2,500 population, at least 14 percent of departments are estimated to have no ladder/aerial apparatus but to have at least one building four stories high or higher in the community.
- An estimated 65 percent of fire departments do not have enough portable radios to equip all emergency responders on a shift. The percentage of departments that cannot provide radios to all emergency responders on a shift is highest for communities under 2,500 population.
- An estimated seven-tenths to three-fourths of fire departments have at least some portable radios that are not water-resistant. An estimated three-fourths to four-fifths of fire departments have at least some portable radios that lack intrinsic safety in an explosive atmosphere. The percentages are higher for small, rural communities.
- An estimated 60 percent of fire departments do not have enough self-contained breathing apparatus (SCBA) to equip all firefighters on a shift.
- Three-fifths (59 percent) of fire departments have at least some SCBA units that are at least 10 years old.
- An estimated half (48 percent) of fire departments do not have enough personal alert system (PASS) devices to equip all emergency responders on a shift.
- An estimated 8 percent of fire departments do not have enough personal protective clothing to equip all firefighters, most of them departments protecting communities with less than 2,500 population.
- An estimated two-thirds (66 percent) of departments have at least some personal protective clothing that is at least 10 years old.

Communications and Communications Equipment

- Three-fifths to four-fifths of fire departments (64–77 percent, by size of community protected) say they can communicate at incident scenes with their Federal, state, and local partners. Of these, though, only one-third say they can communicate with all their partners. This means only about one-fourth of departments overall can communicate with all partners.
- Roughly half of all fire departments have no map coordinate system. Most departments with a map coordinate system have only a local system. Interoperability of spatial-based plans, information systems, equipment, and procedures probably will not be possible under these circumstances, for multiple jurisdiction/agency catastrophic disaster response.

The U.S. National Grid (USNG-NAD83) standard was adopted by the Federal Geographic Data Committee (12/2001) as the system best suited for eventual national standardization (http://www.fgdc.gov/usng/index.html).

- One-fourth (28 percent) of departments (37 percent of rural fire departments) have 911-Basic for telephone communication. Two-thirds to three-fourths (71 percent) have 911-Enhanced, and only 1 percent have no special 3-digit number. Overall, one community in 16 (7 percent) has primary responsibility for dispatch operations lodged with the fire department, but that fraction rises to four-fifths for communities of at least 1 million population.
- One-third (30 percent) of communities have primary dispatch responsibility lodged with the police department, and another two-fifths (39 percent) with a combined public safety department.
- Two-fifths of departments (39 percent) lack a backup dispatch facility, including nearly half (46 percent) of departments protecting communities with less than 2,500 population.
- One-fourth (24 percent) of departments lack Internet access.

Ability to Handle Unusually Challenging Incidents

- 11 percent of fire departments can handle a technical rescue with EMS at a structural collapse of a building with 50 occupants with local trained personnel.
  - One-third (34 percent) of all departments consider such an incident outside their responsibility. This 2001 survey reported 44 percent of departments considered such an incident outside their responsibility.
  - 11 percent can handle the incident with local specialized equipment.
  - 26 percent have a written agreement to direct use of non-local resources.
  - All needs are greater for smaller communities.

- 12 percent of fire departments can handle a hazmat and EMS incident involving chemical/biological agents and 10 injuries with local trained personnel.
  - One-third (32 percent) of all departments consider such an incident outside their responsibility. The 2001 survey reported 42 percent of departments considered such an incident outside their responsibility.
  - 10 percent can handle the incident with local specialized equipment.
  - 30 percent have a written agreement to direct use of non-local resources.
  - All needs are greater for smaller communities.

- 24 percent of fire departments can handle a wildland/urban interface fire affecting 500 acres with local trained personnel.
  - One-fourth (27 percent) of all departments consider such an incident outside their responsibility.
  - 21 percent can handle the incident with local specialized equipment.
  - Roughly half the departments consider such an incident within their responsibility, and 40 percent overall have a written agreement to direct use of non-local resources.
  - All needs for local resources are less for the largest and smallest communities, and the need for written agreements is greater for smaller communities.

- 11 percent of fire departments can handle mitigation of a developing major flood with local trained personnel.
  - The majority of departments (52 percent) consider such an incident outside their responsibility.
  - 9 percent can handle the incident with local specialized equipment.
  - 18 percent have a written agreement to direct use of non-local resources.
  - All needs are greater for smaller communities.

New and Emerging Technology

- The majority of fire departments (55 percent) now own thermal imaging cameras, and the majority of those that do not have them now have plans to acquire them. The 2001 survey reported only 24 percent of departments had such cameras and most that did not have them professed no plans to acquire them.
- Only one department in 17 has mobile data terminals, only one in 31 has advanced personnel location equipment, and only one in 18 has equipment to collect chemical or biological samples for remote analysis. Most departments have no plans to acquire any of this equipment.

The 2006 report, as did the 2003 version, indicates numerous areas in which some in the fire service are neither equipped nor trained to respond to the type and complexity of events they are called to.

Many times, a significant fire or other event occurs because of lack of training, lack of fire ground discipline, and lack of needed equipment, and the community suffers a significant fire loss, a number of civilian fire deaths, or a firefighter fatality. In all of these situations, it was the chief fire officer who had to defend his or her actions, the actions of the community, and the actions of the firefighters. In many cases, it was due to the officer's complacency as a leader—not challenging the status quo, not challenging the cultural aspects of his or her organization, and not challenging city and county leaders to provide sufficient funding to ensure the job was done correctly. This complacency or denial (in the true ability of the organization) may be one of the biggest risks and challenges the American fire service faces today.

BIG HAT, NO CATTLE!

People in Texas have a saying for those who sport western wear and put a great deal of effort into dressing up to look like a cowboy: "Those are what we call 'Big Hat, No Cattle.'" These are people who dress the part but actually don't live it. Many fire service departments today would fall under the "big hat, no cattle" definition. On the surface these departments look to be well equipped, be well staffed, and have marketed their services well. Yet every time the bell rings, their performance in the field is lacking and the result is higher fire loss and life loss. These departments are focused more on their marketing efforts than their efforts to ensure the competency of those in the field of play. Unfortunately, in the fire service there is the tendency to be able to get by with this because most people, whom fire departments protect, are not tacticians of fire ground activity. Therefore, as long as fire departments respond with nice, shiny equipment and put significant amount of personnel on the scene, their performance is not really evaluated. This, too, is another tier of complacency. At an internal meeting in my department about culture and our future, one of my captains, Bob Van Tassel, made a very simple yet profound statement, "We have to be able to do what we say we can do." Simple, to the point, and dead-on accurate. Continued organizational revitalization is essential, but it is everyone's job no matter the rank or assignment. To strive for top-flight performance, every fire department is a large group working toward a common mission.

FOCUS ON THE MISSION

A very simple but effective method to focus on the mission can be found in the *Flawless Execution Model*^{sm}, the process used by fighter pilots to execute missions when failure is not an option. The system is built around four steps (*plan, brief, execute, debrief*) that help to bring focus on the critical objectives whether during an emergency event or when developing a game plan to accomplish an organization goal.[5]

Plan Every mission begins with a detailed plan. The plan starts with an objective that is clear, measurable, and achievable and that supports your department's overall vision and aligns with your organizational values.

Brief Communicate the plan. Those in the military call this a "briefing." It is essential that people on the frontlines know exactly what is expected of them. So often we ignore this and wonder why our objectives—whether on the fire ground or in one of the support services—are not achieved.

Execute Execute the mission. Generally, one thing gets in the way of flawlessly executing the mission; it is called "task saturation," which, simply put, is task overload. You and your team have so many competing priorities that you lock up and become dysfunctional. Many of us have had days when the pace of the day's business seems to be out of control. As managers, we must manage people, do our own work, and deal with hundreds of items that keep us from achieving our primary business objective. Task saturation causes an overwhelming sinking feeling, and when overwhelmed many cannot function effectively. The result is such fragmentation that it keeps you from accomplishing the goals of the organization.

Debrief Debrief after each mission, without fear of reprimand. Find what worked and what did not. Taking the lessons learned from a debriefing and applying them to the next plan increase the learning curve and allow fine-tuning of the processes that ensure organizational success.

These concepts are applicable not only to emergency incidents but also to effectively enhancing our daily operations. Although emergency postincident analysis is common, how many times have we utilized this concept as we have prepared for a council or board presentation, and then, when completed, debriefed on the lessons learned and the takeaways on how we can do it better the next time? This quality improvement model helps to focus on the mission and to ensure we do not fall into the ineffective behavior previously discussed.

Many studies have observed there are five phases in group dynamics—forming, norming, storming, performing, and mourning. Looking at fire departments from a team perspective, the fire officers must have the ability to persuade employees to take the risks and make the sacrifices necessary to achieve the goals of the department. Sun Tzu's book, *The Art of War*, written on the strategic elements of battle, offers a rather practical guide on how to plan and use personnel effectively. Tzu writes, "the way means inducing your people to have the same aim as their leadership so they will share death and share life without fear of danger."[6] In short, followers must have the same convictions as their leaders to be effective and get the most from the team. They must understand the dynamics that their entire organization and specific work groups experience, sometimes on a daily basis. As change is instituted within the organization, different groups will react differently. Understanding this as a chief fire officer, captain, or project manager is very helpful not only in reducing the tensions that sometimes accompany the change process but also in keeping the group focused on achieving a higher level of performance.

Groups such as organizations are dynamic. Each group will continually evolve and assume different behaviors and positions, depending upon what is occurring within the group, the personalities involved, and the issues at hand.

THE DYNAMICS OF GROUPS

A noted author since the 1980s on organizational leadership and change is Tom Peters. His book, *In Search of Excellence,* coauthored with Robert H. Waterman in 1982,[7] highlighted numerous companies that were at the pinnacle of success at the time. If you reread this book, it is interesting that many of the corporate leaders of that time noted for their vision have gone out of business, significantly restructured, or suffered significant losses. Many of the organizations that were experiencing so

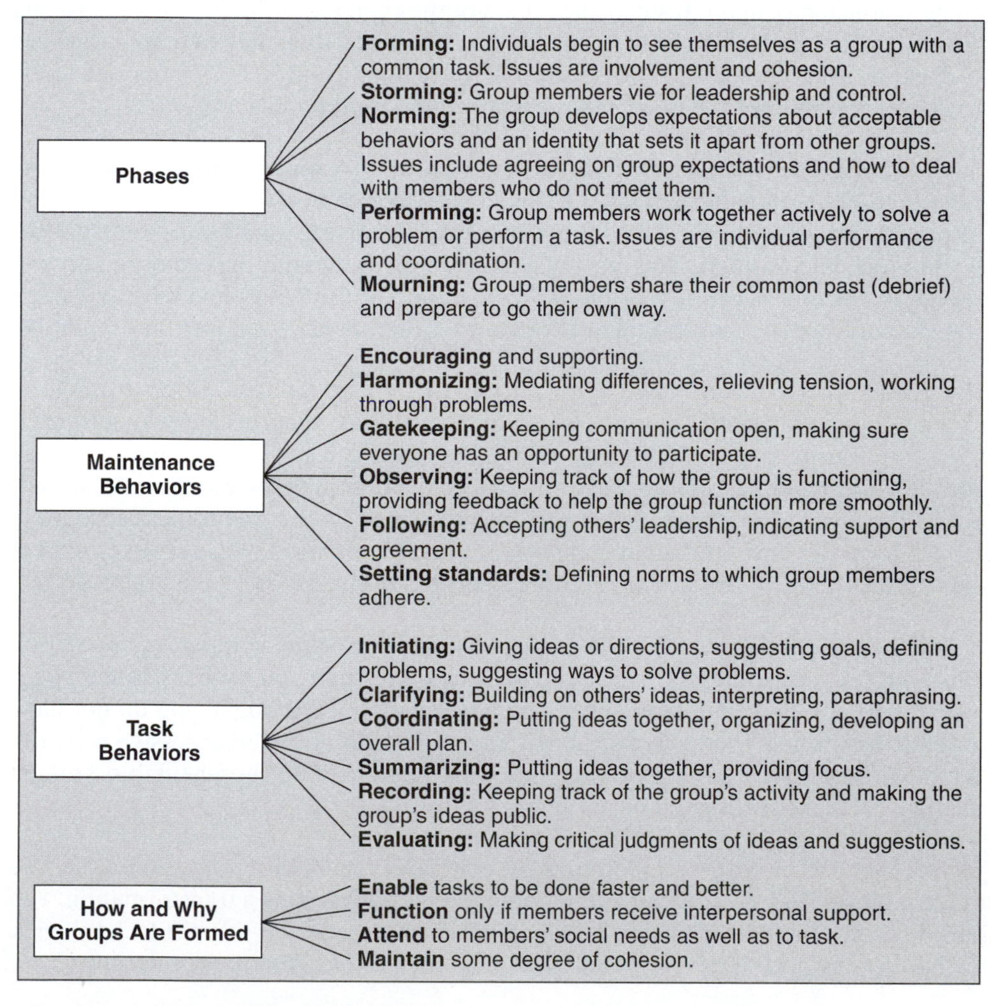

FIGURE 5.3 ◆ The Dynamics of Groups
(A Symbol Is a Promise, *p. 106)*

much success had been forced to relinquish their business operations in the 1990s. The message is worth noting: *"To remain successful, leaders have to understand that change is a continual process."* Transformational leaders often find themselves leading the charge in an attempt to get people and organizations to expand their frameworks and change old methods of doing business before it is too late.

GOOD TO GREAT

In Jim Collins's book, *Good to Great*,[8] he speaks to the Level 5 leadership needed to turn a good company into a great company. Compared to high-profile leaders with big personalities who make headlines and become celebrities, the good-to-great leaders seem to have come from Mars, self-effacing, quiet, reserved, even shy. These leaders are a paradoxical blend of personal humility and professional will. They are more like Lincoln and Socrates than Patton and Caesar.

First Who...Then What We might expect good-to-great leaders to begin by setting a new vision and strategy. Instead, they first got the right people on the bus, the wrong people off the bus, and the right people in the right seats; then they figured out where to drive it. The old adage, "People are our most important asset," turns out to be wrong. People are not the most important asset; the right people are.

Confront the Brutal Facts (yet Never Lose Faith) A former prisoner of war has more to teach us about what it takes to find a path to greatness than do most books on corporate strategy. Every good-to-great company embraced what has become known as the Stockdale Paradox: You must maintain unwavering faith that you can and will prevail in the end, regardless of the difficulties, and at the same time have the discipline to confront the most brutal facts of your current reality, whatever they might be.

The Hedgehog Concept (Simplicity Within the Three Circles) To go from good to great requires transcending the curse of competence. Just because something is your core business—just because you have been doing it for years or perhaps even decades—does not necessarily mean you can be the best in the world. And if you cannot be the best in the world at your core business, then your core business absolutely cannot form the basis of a great company. It must be replaced with a simple concept that reflects deep understanding of three intersecting circles.

A Culture of Discipline All companies have culture, some companies have discipline, but few companies have a *culture of discipline*. When you have disciplined people, you do not have hierarchy. When you have disciplined thought, you do not have bureaucracy. When you have disciplined action, you do not need excessive controls. When you combine a culture of discipline with an ethic of entrepreneurship, you get the magical alchemy of great performance.

Technology Accelerators Good-to-great companies think differently about the role of technology. They never use it as the primary means of igniting a transformation. Yet, paradoxically, they are pioneers in the application of *carefully selected* technologies. Technology by itself is never a primary root cause of either greatness or decline.[9]

The research conducted by Jim Collins and his staff, although focused on the private sector, has a direct correlation to what the public sector and, more specifically, the fire service is experiencing. The organizational shifts of corporate America noted in much of the current research are also occurring in many fire service organizations. Many organizations are in a state of complacency or disruption due to a lack of leadership and/or inability to accept and deal with change and innovation. Leaders who recognize the need to revitalize their organizations must create a new vision. In many cases, this means expanding the frameworks of the personnel to allow new information, thoughts, and processes to work. A fire chief has to develop a vision that is acceptable, is workable, and ignites a desire for change in his or her organization.

Of equal importance, this vision must be communicated in a manner that matches a chief officer's philosophy and leadership style. The same is true if you are a battalion chief or captain; although the vision shared may be more specific to your area of responsibility, it is equally important.

Fire officers have to recognize, despite the significant changes of the fire profession of the past 30 years, "We ain't seen nothing yet." In the next 25 years many fire service agencies may not be called the "fire department" but the "department of emergency services." In fact, the name could probably be changed today. Each year the percentage of fire suppression activities related to the other functions occurring within the fire service is much less than it was a decade ago and will be even less in a decade. In 2020, the issues facing local fire chiefs and the leadership at the national and international levels will be drastically different.

Of equal importance is how the fire service will continue to professionalize, not only at the chief fire officer level but also throughout every aspect of the organization. We must continue to raise the bar if we are to lead the changes in fire protection and emergency services in the future. We must not become good at what is worse for our organizations: the failure to look ahead and begin to revitalize and revamp our organizations to meet the challenges and opportunities of the future. Leaders need to perceive the role of leadership as involving much more than dealing with the day-to-day tasks of running a department. What is needed now and in the future are transformational leaders who can define what the issues will be. Fire officers must be aware of that responsibility if they are to take on those issues. As they accept the need to move the organizations to a higher level of performance, it will require personal risk. Leadership is never risk free; on the other hand, neither is complacency. As leaders prepare their organizations and people for the future, they have to understand, as they try to expand their organizational vision, they will often find themselves the only persons on the playing field—at least initially.

It is one thing to have a vision but quite another to motivate people to buy into it, believe in it, promote it, and ultimately implement it. Leaders must recognize the status quo is never enough. In doing their job, they must push the limits of the comfort zone of the organization and the people within. Be aware as change occurs, people in the organization will try to reestablish their comfort zones, often through unique and bizarre behavior—grievances, rumors/gossip, and passive-aggressive behavior toward their leader, individual programs, and the organization. In developing a new framework and a vision for organizations, it will be critical to maintain a common thread: the value of the organization, the ethics people operate by, the mission, and the long-term objectives must be clearly articulated. Warren Bennis, in *Why Leaders Can't Lead* (1990) and *On Becoming a Leader* (2003),[10] outlines the differences between a leader and a manager. There are several crucial differences in group dynamics and leading change. As you read the following list, ask yourself whether you are a leader, a manager, or both. If your title is chief fire officer, battalion chief, captain, or lieutenant, do you find yourself being a manager rather than a leader? If so, it is time for you to step back and retune your own method of operation. As Bennis puts it, *leaders conquer, while managers surrender*. Look at the differences:

- The manager administers; the leader innovates.
- The manager is a copy; the leader is an original.
- The manager maintains; the leader develops.
- The manager focuses on system and structure; the leader focuses on people.
- The manager relies on control; the leader inspires trust.
- The manager has a short-range view; the leader has a long-range perspective.
- The manager asks how and when; the leader asks what and why.
- The manager has his or her eye on the bottom line; the leader has his or her eye on the horizon.
- The manager accepts the status quo; the leader challenges it.
- The manager is a classic good soldier; the leader is his or her own person.
- The manager does things right; the leader does the right thing.

Leaders have a clear idea of what they want to do both personally and professionally, and have the strength and tenacity to be persistent even in the face of setbacks and failure. Managers understand the system, can maneuver the process of government to get things done, and are good followers. *To be a good fire service leader today leaders also have to be good managers, and managers must be good leaders*.

STEWARDSHIP AND THE CORE VALUE OF SERVICE

Leadership is more than command and control, and in the fire service, it means having command presence. Leaders can have those attributes, but they may not be effective, appreciated, or followed. Leaders who ensure service becomes a core value of the organization are the most successful. These leaders lead by example and will strive to institutionalize the basic concept of service over self. Organizations in the fire service and in the private sector that embody this core value of service in their day-to-day operations are truly the best organizations. Think about it. Would you rather buy a product from a company whose core value is to make you happy through "best in class" customer service or one whose core value is to make a profit, no matter what? Public service organizations must have a broad perspective on what is going on around them in their community. They cannot be self-absorbed; they must understand community dynamics and politics, and they must have the flexibility and ability to change rapidly. Core values separate good organizations from those that are world class.

In the fire service, it is important to ensure that basic skills and core competencies are relevant. These core competencies focus on what has to be done from a technical standpoint, but the core value of service goes beyond the ability to respond to an emergency once the alarm is sounded in the firehouse to embody who we are and the expectations we have for one another. The acceptance that stewardship is a core value underlying the services provided is the basis of the ability as an industry to grow.

Keshavan Nair, in counseling *Fortune* 500 companies on leadership and decision making, claims the core values are not only policy but also something deeper and more profound. He suggests every individual in the organization values being of service. *When service to others is valued, trust, loyalty, and truth flourish*. These create efficiencies of trust, trust creates positive energy, and positive energy fuels the organization.

Service, leadership, and stewardship are intertwined. In watching successful fire service leaders, those who have the core value of stewardship are the ones who place service above self and truly believe organizational service is a core value—a self-sustaining value, one that has tremendous power. Keshavan Nair also writes that the acceptance of a core value implies a commitment.[11] **Core values** within an organization and core values in service are long-term commitments. When leaders have both belief and commitment, they and their organizations can achieve great things. Core values often become the foundations of future traditions. Once leaders have articulated that service is a core value, the challenge is to determine how to begin to influence others within the organization to act in a way that supports core value. This starts with the leader leading by example to ensure he or she is doing more than he or she is paid to do. The model of good leadership is the most effective way to inspire others' commitment. Albert Schweitzer said, "Example is not the main thing in influencing others; it's the only thing." This is certainly true in this case.

core values
Attitudes and beliefs thought to pattern a culture uniquely.

In promoting stewardship and service as core values, there has to be a commitment to and from everyone in the organization. Leaders need to recognize those persons who provide an example of good stewardship and of service over self. I have used a document titled the "Hot Sheet" (a weekly two- to three-page rundown of what is occurring within the organization). E-mailing to all employees and persons in government within a region is an excellent way to showcase people who are role models by their actions showing the commitment to service and helps to instill the organizational core value of service. Here are a few other examples of real-life events that help do the same.

An engine company responded to a call from a man who had fallen off a ladder while fixing his gutter. As a result of the fall, he broke a couple of ribs and damaged his deck. After the crew arrived and took care of the man, loading him into the ambulance, crew members stayed a little longer to fix the deck. When the man came home from the hospital, he found that the deck had been repaired. Another story involves an 80-year-old man who was raking leaves and experienced chest pains. When the crew arrived, after taking care of his medical situation and placing him in the ambulance, some of them stayed behind and continued to rake the leaves. When he came home, he found all the yard work had been done. In another instance, an elderly woman was inappropriately burning debris in her backyard, and the local fire crew was called. They notified her of the fire code violation and extinguished the fire. The next morning when the captain of the crew went off duty, he stopped by with his personal truck, cleaned up the debris, and removed it. You may have similar stories in your own department. Are they shared? Are they celebrated? Creating this type of stewardship is an essential step in developing a core value of service, which is essential to making change work in the future. It transcends the day-to-day issues and helps to maintain the foundation on which the fire service has been built for the past 200 years.

During a promotional interview with a captain candidate whose name had repeatedly come up for providing good customer service and creating a stewardship process in his company, I asked how he had created the idea of service over self. He indicated it was a form of competition among his crew members to see what more they could do for the people in every emergency call to which they responded. Sometimes it was as simple as putting up a free smoke detector; at other times, it meant cleaning up the kitchen before they left if the occupant was transported to the hospital, ensuring the house was locked, and notifying the neighbors of the situation. In each case it was different, but it was clear the concept of stewardship had become a core value of this fire company. It was reflected not only in the comments of the general public but also in the recognition of the ability of other crews to go above and beyond the call of duty.

Vince Lombardi, the late coach of the Green Bay Packers and football coaching legend, said, "Excellence isn't a sometimes thing. You have to earn it and re-earn it every single day." Promoting excellence every day has to be the leader's goal to give customers what they expect and, more important, what they deserve. Leaders have to focus relentlessly on the importance of achieving excellence, making sure everyone has it as part of his or her mission and is committed to attaining it.

Patricia Aburdene, coauthor of the best-selling *Megatrends 2000,* in her new book, *Megatrends 2010*, investigates corporate social responsibility; finds significant numbers of companies are placing social, spiritual, and environmental values ahead of the bottom line; and reports data showing socially responsible practices actually help boost profits. Her book identifies seven new trends that redefine how we work, live, shop, and invest. This speaks directly to stewardship, value added, and the importance trust and integrity play in developing healthy organizations. Aburdene's list of megatrends, each of which serves as a chapter in her book, follows:

1. *The power of spirituality*. In turbulent times, we look within; 78 percent seek *more* spirit. Meditation and yoga soar. Divine presence spills into business. "Spiritual" CEOs as well as senior executives from Redken and Hewlett-Packard (HP) transform their companies.
2. *The dawn of conscious capitalism.* Top companies and leading CEOs are reinventing free enterprise to honor stakeholders and shareholders. Will it make the world a better place? Yes. Will it earn more money? This is the surprising part: Study after study shows the corporate good guys rack up great profits.
3. *Leading from the middle.* The charismatic, overpaid CEO is fading fast. Experts now say "ordinary" managers forge lasting change. How do they do it? Values, influence, moral authority.
4. *Spirituality in business.* This is springing up all over. Half speak of faith at work. Medtronic wins "Spirit at Work" awards. Ford, Intel, and other firms sponsor employee-based religious networks. Each month San Francisco's Chamber of Commerce sponsors a "spiritual" brown bag lunch.
5. *The values-driven consumer*. Conscious consumers, who have fled the mass market, are a multibillion-dollar "niche." Whether buying hybrid cars, green building supplies, or organic food, they vote with their values. So brands embodying positive values will attract them.
6. *The wave of conscious solutions*. Coming to a firm near you: Vision Quest, meditation, forgiveness training, HeartMath. They sound touchy-feely, but conscious business pioneers are tracking results that will blow your socks off.
7. *The socially responsible investment boom.* Today's stock portfolios are green in more ways than one, and the "social" investment trend is alive and well.

In this book's conclusion, "The Spiritual Transformation of Capitalism," Aburdene explores the underlying values of capitalism. She attempts to dispel the notion that free enterprise is rooted in greed. Conscious capitalism is not altruism, either; it relies instead on the wisdom of enlightened self-interest.[12] If Aburdene is correct in her forecast of future trends, what impacts may they have upon local government and the fire service? *Will stewardship and value play an even more important role in the future?*

Every organization will transition through a variety of leadership and management styles. A good set of core values is the guidepost that helps frame those things important to the organization. Organizations can never achieve excellence if their core values are not well understood and institutionalized throughout the organization. The best examples of core values can be found in the branches of the United States military service, where commitment to values is not only a phrase but also a way of life. Organizations that have a strong core value system help to frame and resolve ethical issues that arise. It also helps to frame the behavior of not only the organization but also the individuals within. Without a good core value system, an organization can never achieve the level of excellence it desires.

THE UNITED STATES NAVY

Throughout its history, the U.S. Navy has successfully met all its challenges.[13] America's naval service began during the American Revolution. On October 13, 1775, the Continental Congress authorized a few small ships, creating the Continental Navy. Esek Hopkins was appointed commander in chief and 22 officers were commissioned, including John Paul Jones. From those early days of naval service, certain bedrock principles or core values have carried on to today. They consist of three basic principles.

Honor: "I Will Bear True Faith and Allegiance. . . ." Accordingly, we will conduct ourselves in the highest ethical manner in all relationships with peers, superiors, and subordinates; be honest and truthful in our dealings with each other and with those outside the navy; be willing to make honest recommendations and accept those of junior personnel; encourage new ideas and deliver the bad news, even when it is unpopular; abide by an uncompromising code of integrity, taking responsibility for our actions and keeping our word; fulfill or exceed our legal and ethical responsibilities in our public and personal lives 24 hours a day. Illegal or improper behavior or even the appearance of such behavior will not be tolerated. We are accountable for our professional and personal behaviors. We will be mindful of the privilege to serve our follow Americans.

Courage: "I Will Support and Defend. . . ." Accordingly, we will have courage to meet the demands of our profession and the mission when it is hazardous, demanding, or otherwise difficult; make decisions in the best interest of the navy and the nation, without regard to personal consequences; meet these challenges while adhering to a higher standard of personal conduct and decency; be loyal to our nation, ensuring the resources entrusted to us are used in an honest, careful, and efficient way. Courage is the value that gives us the moral and mental strength to do what is right, even in the face of personal or professional adversity.

Commitment: "I Will Obey the Orders. . . ." Accordingly, we will demand respect up and down the chain of command; care for the safety, professional, personal, and spiritual well-being of our people; show respect toward all people without regard to race, religion, or gender; treat each individual with human dignity; be committed to positive change and constant improvement; exhibit the highest degree of moral character, technical excellence, quality, and competence in what we have been trained to do. The day-to-day duty of every navy man and woman is to work together as a team to improve the quality of our work, our people, and ourselves.

THE UNITED STATES MARINE CORPS

Generation after generation of American men and women have given special meaning to the title United States Marine.[14] These same men and women live by a set of enduring core values, which form the bedrock of their character. The core values give Marines strength and regulate their behavior; they bond the Marine Corps into a total force that can meet any challenge.

Honor Honor guides Marines to exemplify the ultimate in ethical and moral behavior; to never lie, cheat, or steal; to abide by an uncompromising code of integrity; to respect human dignity; and to respect others. The qualities of maturity, dedication, trust, and dependability commit Marines to act responsibly, to be accountable for their actions, to fulfill their obligations, and to hold others accountable for their actions.

Courage Courage is the mental, moral, and physical strength engrained in Marines. It carries them through the challenges of combat and helps them overcome fear. It is the inner strength that enables a Marine to do what is right, to adhere to a higher standard of personal conduct, and to make tough decisions under stress and pressure.

Commitment Commitment is the spirit of determination and dedication found in Marines. It leads to the highest order of discipline for individuals and units. It is the ingredient that enables 24-hour-a-day dedication to Corps and country. It inspires the unrelenting determination to achieve a standard of excellence in every endeavor.

THE UNITED STATES AIR FORCE

The core values are much more than minimum standards.[15] They remind all United States Air Force (USAF) members what it takes to get the mission done. They inspire its members to do the very best at all times. They are the common bond among all comrades in arms, and they are the glue that unifies the force and ties them together to the great warriors and public servants of the past.

The Air Force core values are *integrity first, service before self,* and *excellence in all we do.* Study them, understand them, follow them, and encourage others to do the same.

integrity
The personal inner sense of "wholeness" deriving from honesty and consistent uprightness of character. The etymology of the word relates it to the Latin adjective *integer* ("whole," "complete"). Evaluators, of course, usually assess integrity from some point of view, such as that of a given ethical tradition or in the context of an ethical relationship. Moral soundness; "he expects to find in us the common honesty and integrity of men of business"; "they admired his scrupulous professional integrity."

Integrity First **Integrity** is a character trait. It is the willingness to do what is right even when no one is looking. It is the "moral compass"; the inner voice; the voice of self-control; the basis for the trust imperative in today's military. Integrity is the ability to hold together and properly regulate all of the elements of a personality. A person of integrity, for example, is capable of acting on conviction and can control impulses and appetites.

Service Before Self Service before self tells us professional duties take precedence over personal desires.

- *Rule following.* To serve is to do one's duty, and our duties are most commonly expressed through rules.
- *Respect for others.* Service before self tells us also a good leader places the troops ahead of his or her personal comfort.
- *Discipline and self-control.* Professionals cannot indulge themselves in self-pity, discouragement, anger, frustration, or defeatism.
- *Faith in the system.* To lose faith in the system is to adopt the view you know better than those above you in the chain of command what should or should not be done.

Excellence in All We Do Excellence in all we do directs us to develop a sustained passion for the continuous improvement and innovation that will propel the Air Force into a long-term, upward spiral of accomplishment and performance.

- *Product/service excellence.* We must focus on providing services and generating products that fully respond to customer wants and anticipate customer needs, and we must do so within the boundaries established by the taxpaying public.

- *Personal excellence.* Military professionals must seek out and complete professional military education, stay in physical and mental shape, and continue to refresh their general educational backgrounds.
- *Community excellence.* Community excellence is achieved when the members of an organization can work together to successfully reach a common goal in an atmosphere free of fear that preserves individual self-worth. Some of the factors influencing interpersonal excellence are:
  - *Mutual respect.* Genuine respect involves viewing another person as an *individual* of fundamental worth.
  - *Benefit of the doubt.* Working hand in glove with mutual respect is the attitude that says all coworkers are "innocent until proven guilty."
- *Resources excellence.* Excellence in all we do also demands we aggressively implement policies to ensure the best possible cradle-to-grave management of resources.
  - *Material resources excellence.* Military professionals have an obligation to ensure all of the equipment and property they ask for is mission essential.
  - *Human resources excellence.* Human resources excellence means we recruit, train, promote, and retain those who can do the best job for us.
- *Operations excellence.* There are two kinds of operations excellence—internal and external.
  - *Excellence of internal operations.* This form of excellence pertains to the way we do business internal to the Air Force, from the unit level to Headquarters Air Force. It involves respect on the unit level and a total commitment to maximizing the Air Force team effort.
  - *Excellence of external operations.* This form of excellence pertains to the way in which we treat the world around us as we conduct our operations. In peacetime, for example, we must be sensitive to the rules governing environmental pollution, and in wartime we are required to obey the laws of war.

THE UNITED STATES ARMY

Loyalty. Bear true faith and allegiance to the Constitution, the Army, your unit, and other soldiers.[16]

Duty. Take responsibility and do what is right, no matter how tough it is, even when no one is watching.

Respect. Treat people as they should be treated.

Selfless service. Put welfare of Nation and mission ahead of personal desires. What is best for our Nation, Army, and organization must always come first.

Honor. Live up to all the Army values.

Integrity. Do what is right legally and morally. Integrity obliges one to act when duty calls.

Personal courage. Endure your fear and do the right thing. Persevere in what you know to be right and do not tolerate wrong behavior in others.

In going through a refocus within the Fresno Fire Department, we found the issue of core values had become very central to our discussions; the need for core values was exacerbated by the large influx of new personnel due to a significant number of retirements and adding new people to the department. Many organizations are faced with this same reality. With the loss of so many senior employees, the organizational loss of such a significant amount of history, experience, and institutional work ethic can have an adverse impact. Conversely, adding so many new employees brings an entire new generation into the workforce. With them, they bring a different generational experience, expectation of the job, and set of values.

Organizationally, it was becoming increasingly apparent the establishment of a set of departmental core values was needed to provide a common denominator to the workforce.

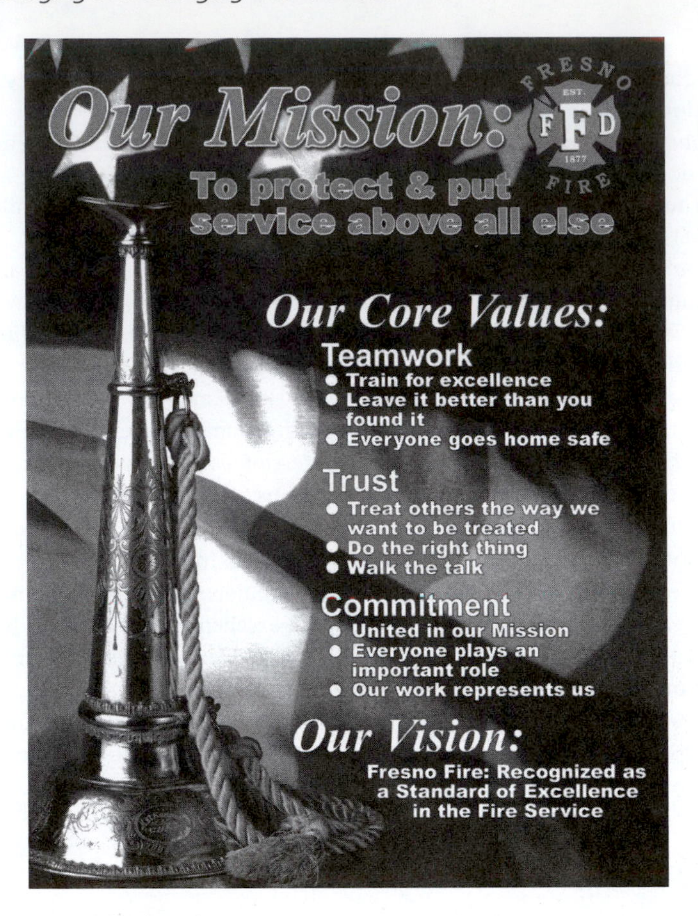

FIGURE 5.4 ◆ Fresno Mission
(Source: Fresno Fire Department)

To accomplish this, we brought a group of employees together who stratified ranks with the organization to begin to focus on what type of department they wanted to be in 10 years. While our mission was in sync and it was relatively easy to agree upon a vision, much discussion over the course of several meetings revolved around the core values of the organization. Through this process those involved came to realize, if we were to move toward our vision and if we were going to be successful in our mission, then we had to have a sound set of core values that the employees could believe in and would fuel our movements toward becoming an organization of excellence. Now while it will take time to institutionalize these core values, they would establish the framework of behavior and expectations to operate from within the organization. As we have seen with the military and other successful private and public corporations, core values, when institutionalized, provide the road map to sustain your mission and to reach your future vision (see Figure 5.4).

WHAT WE CAN LEARN FROM OTHERS

In his book, *Making the Corps*,[17] Thomas E. Ricks provides insight into the successful transition of men and women who entered the Marine Corps. In just 11 short weeks, these young people are molded into focused individuals and in many cases instilled with the characteristics of effective leaders. The Marine Corps also instills a set of core values and ownership into these young men and women that they carry with them for a lifetime. You have probably witnessed this yourself in conferences or other meetings when someone gives the Marine "hurrah," quickly to be joined by a chorus throughout the gathering. This from people who have not been in the Corps for 30 years! What can we learn from them? Ricks found what he termed "lessons from Parris Island": six key characteristics that have a great deal to do with leadership and offer powerful insights for anyone in a leadership role. They are reflective of the Marines' core values: honor, courage, commitment. Of all the things that can motivate people, the pursuit of excellence is one of the most effective and one of the least used in our society. Yet the Marines offer a powerful alternative to what is often seen today. The lessons from Parris Island are straightforward:

- Tell the truth.
- Do your best, no matter how trivial the task.
- Choose the difficult right, not the easy wrong.
- Look out for the group before you look out for yourself.
- Do not whine or make excuses.
- Judge others by their actions, not by their race or gender.

These lessons provide a very simple bridge to leadership: honesty, work ethic, integrity, and service. Take care of your people and go home. These seem to be basic steps, yet they are powerful in running an organization. They demonstrate the organization's leadership, its core value of service, and its commitment to pursue excellence in all the things it does.[18]

Paralleling *Making the Corps* is a book written in the early 1600s titled *A Book of Five Rings*. The author, Miyamoto Musashi, was a teacher of the samurai way of life, a philosophy and approach for those who desired to construct a full, dynamic, and successful life. It speaks of what is important in a leader in the same way as *Making the Corps* does. The way of the samurai consists of the following principles.

- *Act.* There is no perfect moment when the action must be done; concentrate on the action.
- *Don't complicate the situation.* Emphasize simplicity, naturalness, and down-to-earthiness.
- *Focus on your purpose.* What do you want to accomplish and what do you want to be? Find out what is important to you personally. Life is short, so confront it immediately. Whatever you do, think of what you are trying to accomplish (see Figure 5.5).
- *Always advance on life, never retreat from it.* If you feel shy, overcome the shyness so you can advance. Remember, it is better to die from a sword in the chest rather than an arrow in the back.
- *Leap into action.* If something is to be done, do it. Risk defeat. When you avoid action, you spend energy dealing with the unknown rather than on the job at hand.
- *Be absorbed in the action.* If you are absorbed in yourself, you tend not to act. You back off from the action, whether it involves making a decision, selling an idea, applying for a job, or making a career move.
- *Concentrate more on the action than on yourself.* It can be more effective. If you back off and think about what could go wrong or whether you made a bad decision, too much time will pass and the opportunity will be gone.[19]

Making the Corps and *A Book of Five Rings* offer a historic parallel regarding leadership, value, and service. They each demonstrate what leaders in today's fire service need to focus on if they are to be successful in promoting organizational change by establishing a core value of service and stewardship. Both books point out the ability of

FIGURE 5.5 ◆ Samurai

leaders to create a dynamic within their organizations. It can best be summed up by the saying "We must leave it better than it was when we found it." What we are trying to create in all of our organizations, whether public or private, is "above-the-line accountability." It expresses the idea, "If you see it, you own it. If you own it, you can help to solve it. If you can solve it, you should just do it." There is a classic line in the first *Star Wars* movie when Yoda is instructing Luke the Jedi on overcoming adversity: "Don't just try, do!" So often we become mired in the attempt, we fail to just do the task (see Figure 5.6).

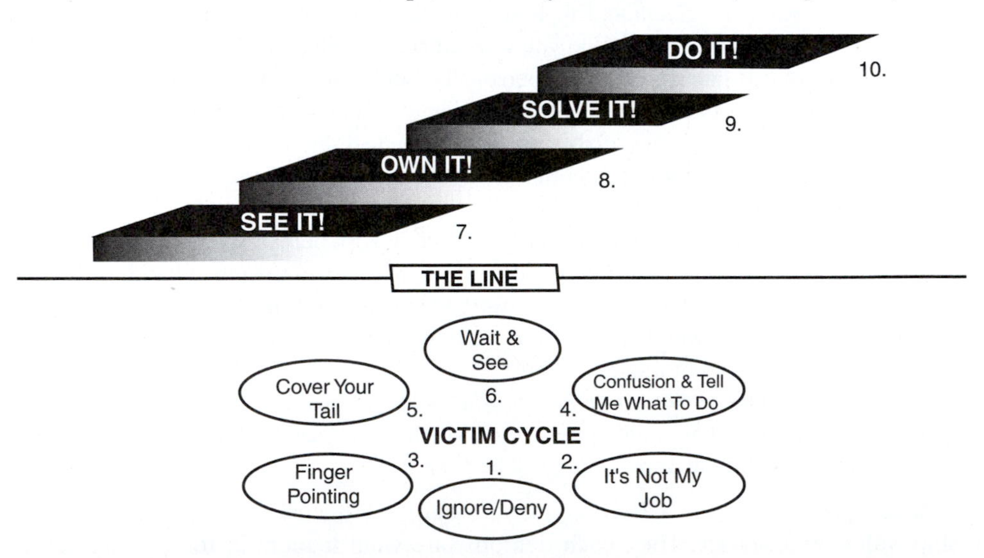

FIGURE 5.6 ◆ Above the Line
(Courtesy of Randy R. Bruegman, The Chief Officer: A Symbol Is a Promise, *p. 112)*

Another aspect to organizational culture that can be very debilitating is the opposite of above-the-line accountability, referred to as the victim mentality. Those who find themselves in this place often ignore what is happening around them, saying, "It's not my job," finger-point to protect themselves, refuse to get involved, wait to see what happens, and create confusion through rumors and passive-aggressive behavior. Can you think of a situation in your own department where you have seen this play out? The result would be to expend a greater degree of organizational energy to solve the problem because those involved were part of the victim cycle and did not deal with it in the first place. Many of the discussions about leadership and group dynamics can be summed up by the concept of above-the-line accountability within our organizations. Although there are many aspects and methods involved in creating an above-the-line culture of accountability, unless its importance is realized, organizations will never achieve it. Although a significant amount of time is spent training personnel on the technical areas of the job, leaders often fail to focus their efforts on those areas that can make an organization reach a level of individual accountability, which in turn can build an exceptional organization. This is where excellence will not be a sometime thing but an everyday thing.

Effective leaders try to impart this daily. Leaders must never underestimate the power of their own actions. Even the smallest gesture can change a person's life, for better or for worse. Their goal must be to create conditions that produce commitment and creative action by the people within their organizations. When they can create ownership using above-the-line accountability, they will be most effective. Mark Twain may have stated it best, "Always do what is right. It will gratify most of the people and astonish the rest."

AGENDA FOR CHANGE

> The past is gone: The present is full of confusion: and the future scares the hell out of me.
>
> *David L. Stein*

Going through any major change will challenge the way leaders view themselves and their organization. Significant organizational changes can be as stressful, on the whole, as life events such as death, birth, and divorce are for an individual. Living through this process is often like undertaking major remodeling in your home. To achieve the result you want, you must first rip out the old, leaving the basic structure intact, and begin to rebuild with new materials. Once you have added the final touches, you can move back in, begin to feel comfortable, and once again become productive. Change is a lot like this because it always takes a little longer and costs a little more than you originally thought. Structural and cultural changes require people to let go of the old way of doing things. In doing so, they will live through a period of doubt and uncertainty. Managing this process takes sensitivity and empathy, because change can be very frightening and unsettling to employees.

We all know change is inevitable. Whether you have been on the job for one year or 25, you have already been witness to change. Because this is the case, the good question is, "Will we always be riding this wave of transition?" The answer is "yes." For without change, people and organizations would become stale and unresponsive. The challenge is learning to move through the transition as easily and creatively as possible. What helps people navigate through this unknown territory is a map of what they can expect and information on how they can respond most effectively to challenges as they occur.

All in the fire service today face a unique challenge. The old adage, "two hundred years of tradition, unhampered by change," is simply not true anymore, and it may never have been. Yet one of the challenges the fire service must face, from an organizational standpoint, is to get personnel to look past the four walls of the fire station and understand that the world around them is changing. As these modifications occur, they place significantly new demands on everyone within the organization.

How do we navigate an ever-changing environment? First, the leadership of the department must make a commitment to share as much information as possible with everyone within the department. Second, there must be an expectation set that everyone will read the information. Those in leadership roles have to understand the resistance to change; it is part of our culture, and for most people it is their natural tendency. Change is easier said than done, and whether fire service tries to alter its environment, the workplace, or the services offered, the change destabilizes organizations. In some cases, it creates fear. A sense of permanence and tradition has been the backbone of the fire service and has become a significant influence in the development of the culture of the industry. Permanence and tradition have provided stability for fire service organizations, but in today's environment they have also produced many departments that are inflexible, unwilling to change, and bureaucratic in their approach.

It is safe to say the pace of change will only increase. As organizations move into the future, they must have their engines running at full throttle and a firm hand upon the wheel. Faith Popcorn, an author on change and trends, has predicted we will change as much in the next 10 years as we have in the past 50. Consider the ramifications. She states, "From 1940 to 1990 we experienced 50 years of change. From 1990 to 2000 we experienced 50 more. From 2000 to 2005 we experienced 50 more." What happens after that? If you plan to be part of the fire service for another 15 to 20 years, you may live through exponential change unlike anything ever seen or experienced before.[20] To gain a perspective on this, look back over the last five years at the technology change that has occurred. Look at a map of the world or of your own community and compare that map to one five years old. What has changed?

The need for change is evident; just look around. Almost everything in our environment is altering; therefore, we must change with it. Remember the old saying, "If you always do what you have always done, you will always get what you have always gotten," and in some cases, maybe less. This holds true for all of us in the fire service and for most organizations today. To successfully navigate the changes that will be needed under your leadership, you must understand the importance of culture, vision, and the need to possess a global view. The ideals you must promote to be successful—the importance of leadership, the empowerment of others, the ability to measure performance, and understanding how change affects people and organizations—are critical elements whether you are a fire chief, a battalion commander, a captain, or a young firefighter.

NATURE OF CULTURE

During times of great change, leaders have to ask their employees to be flexible; if they do not, leaders cannot position the department to be competitive in the future. Yet for 99 percent of the fire service, there is a monopoly on who is going to provide the service. *So why the need to be competitive?* We are in competition every year at budget time for money and resources to do our job efficiently, effectively, and safely. Therefore, the ability to maneuver and take advantage of opportunities when they arise is critical to the industry's long-term health. The forces of change experienced in the fire service will come from both inside and outside the organization. With more demands from constituents, the constant push for cost containment, the need for quality, shifts in the political arena, and an expanded mission to include homeland security issues and traditional culture, leaders will definitely be challenged. As leaders move through such changes, it is crucial they understand their organizations, their culture, and their effect on the change process. Whether change is initiated from inside or outside an organization, culture plays a large part in determining how it will be processed, accepted, and ultimately dealt with. Cultures and the strategies for change are often in conflict. When conflict occurs internally and the organizational culture does not embrace the changes that have been initiated, the change efforts often involve a struggle and in many cases will not be successful.

In contrast, when the change is externally driven, both the organization and the fire chief can find themselves in extremely vulnerable positions. In many cases, this form of change is often the most dangerous. Whether it is political, economic, or service driven, people find themselves in a crossfire in which there are often no wins for either side. Often, change caused by consolidation or a need to save money, or to expand service is met by organizational resistance. In each case the change effort did not fail, but the organization and its culture were permanently altered. As an example, for the past 30 years, the fire service has been expanding its role in emergency medicine—from first responder to basic life support (BLS), to advanced life support (ALS), to performing vaccinations and other community health care services. Today, EMS accounts for the majority of calls. Yet many metropolitan-sized organizations have resisted the incorporation of EMS into their fire service deployment package; or if they have, it is not embraced in the firehouse as an important service. In many cases, the resistance was due to the culture of the organization, which was so engrained that they were there to fight fires and provide rescue-related services, even though the number of fires and other such calls has continued to decline. The culture had created a level of arrogance that did not allow these people to see the changing context of the industry and their own environment. Then the terrorist attacks of September 11, 2001, occurred and after 24 months of economic downturn, many cities and counties were facing significant budget shortfalls. Many of these fire agencies began to experience reduced staffing and related budget cuts that included firefighter layoffs to balance the budget deficits. Here is the interesting fact: those metropolitan departments that were targeted for cuts had a common denominator; they did not provide ALS service.

So it all came down to a value-added question for the local governing body, and in several cases firefighters were laid off or the department was forced to run short of previous staffing levels. This is a good example of how the culture can move an organization or keep it in a place where it should not be. In larger, older, and more traditional departments, it is often difficult to make a shift in services, policy, or procedure due to the culture of the system. Throughout our industry, it is evident EMS is now and will be a critical element of our service delivery package in the future. Why did these organizations not see this? The answer is they saw it but could not move the organization in the new direction. Why? Culture! Guess what? They are moving in that direction now! In many of these situations, it will take years to overcome the emotional upheaval that change creates within the organization.

The future fire service must promote a culture that is much more responsive, adaptive, and accepting of change because that is the desired future, which will keep us competitive. The *desired future* will be one that not only maintains our heritage and traditions but also provides for the ability to adapt rapidly to the changing environment; provides for innovation and creativity, overcoming the mind-set "because we

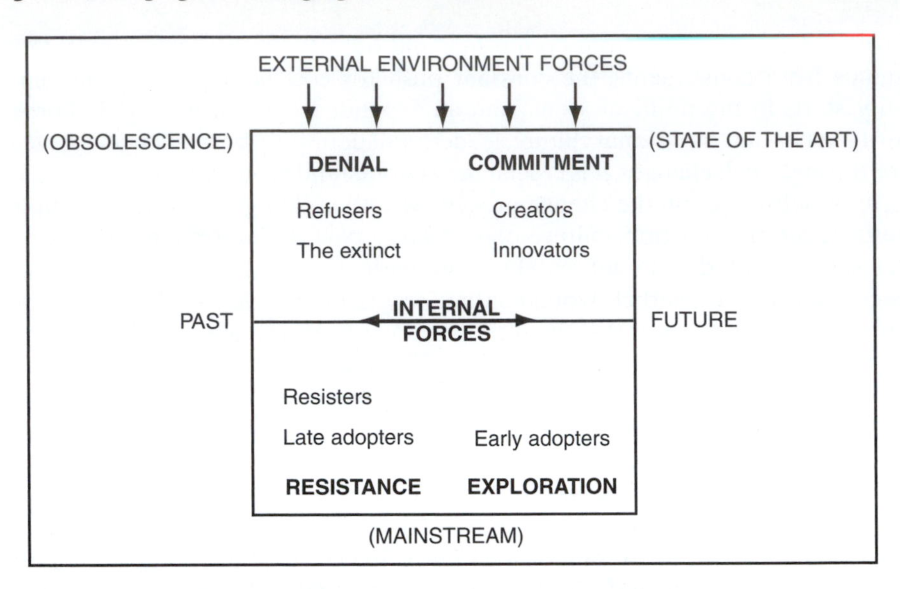

FIGURE 5.7 ◆ External Environmental Forces
(Courtesy of Randy R. Bruegman, The Chief Officer: A Symbol Is a Promise, *p. 116)*

have always done it this way"; and finds a proactive industry that focuses time and resources on preventing rather than reacting. It is important to understand the responses to change. Organizations, culture, and people respond to the pressure to change in vastly different ways, ranging from enthusiasm to indifference, to fear, to anger (see Figure 5.7).

People who have a firm commitment to the organization and are motivated to stay on top professionally are usually on a continual quest to be at the cutting edge in everything they do. These are the "innovators," often the most creative people in the organization, the first to embrace new ideas. Their commitment to achieve a competitive edge through knowledge is very high. These people are referred to as "lead ducks"; they are often shot at a lot because they will be the first out of the water on an issue, to the point where they can make their peers uncomfortable with their exuberance. They are usually the first to arrive to the appointed destination; then they are off to the next adventure.

In the next category are those referred to as the "explorers" or "early adopters." These are the people in our organizations who have a high degree of curiosity and a strong sense of mission. They are also on the cutting edge of change and do not mind taking a few risks along the way. They tend not to be lead ducks but are definitely near the front of the formation. They will research the project or issue before they commit to it, but once they do they will be a strong part of the team.

The next category consists of the "late adopters," who are somewhat resistant to change, slightly timid, and desire never to be wrong. They wait for ideas to be well proven before they jump on the bandwagon. They like to be in the safe zone, and their objective is to reduce their vulnerability and risk. They can be very strong performers but join only after some degree of validation has occurred for a concept or new idea.

The next group is the "resisters," who are uninformed and want to stay that way. They work hard to maintain the status quo based on their apathy, negativity, or simple ignorance. Change can overcome the resisters and, in most cases, they will adapt over time. However, it often takes an enormous amount of organizational energy to do so.

The last category, in which we often find too many people and, in some cases, organizations, are the "refusers." They have chosen not to progress. They refuse to accept any change within the organization or in the world around them. This is why we often refer to them as those about to become extinct. They choose to follow the road of the past that leads to a community called obsolescence. They live in a time warp, and the only way to deal with their denial of the change occurring is to wait for or motivate them to retire.

Whether an agenda for change is motivated by internal or external forces, there are some basic requirements you, as a leader, will need to address in order to succeed.

1. *Top management* must be involved. It needs to set the example and be active in the change process so others in the organization recognize its commitment.
2. *Measurement systems* must be used to track the progress of the change both at the upper level of the organization and in day-to-day operations.
3. As the leader, you need to *set the bar high* and push your organization, your division, or your company.
4. You must understand the need to provide education on how and why the change has to occur and the route you plan to take.
5. If you have implemented a change and it has been successful in your organization, *spread the story* within your organization and to the entire fire service. Having been hired by four different organizations either to lead a change process or to take the organization to the next level, my experience has taught me that leading a cultural or organizational change involves four key issues.

   a. *Information.* What is the change?
   b. *Inspiration.* Why is it needed?
   c. *Implementation.* How will it be done, both individually and organizationally?
   d. *Institutionalization.* How will we know when we have succeeded?

   If you deal with these four aspects of cultural change, you will be able to start leading your organization, battalion, or company in the direction you want to go. If you do not, you will spend most of your time putting out fires throughout your organization. Then one day you will look around and find that little has changed except your stress level.

VISION IS A MUST

> Give to us a clear vision that we may know where to stand and what to stand for. Let us not be content to wait and see what will happen but give us the determination to make the right things happen.
>
> *Peter Marshall*

When we talk about organizational change, one of the best places to start is with vision. This has to begin with you. You cannot impose your vision on others, but your clear view of where the organization needs to go can be a road map and inspiration for many. Start by formulating a vision statement—a declaration of the organization's most desirable future. In the fire service, although most of our mission statements are very similar, our visions for our organizations in five to ten years can be vastly different. You as a leader must make the vision part of your game plan. Strategies developed from a vision of how the transformation will occur are extremely important. A vision statement must also be emotionally charged. As a change agent, you must figure out a way to motivate your personnel to deal with the long-term goal as a necessity for the organization.

As president of the IAFC in 2002–2003, I had the opportunity to transform my vision into an action plan on a national basis. I chose the theme "A Time to Lead," as I believed it captured a renewed sense of purpose, pride, motivation, and challenge that the fire service faced after September 11 (see Figure 5.8a through Figure 5.8d).

The vision A Time to Lead expressed what the leadership of the fire service needed to do at this particular moment in time. The challenges we were facing with the expanded mission of homeland security, our ability to articulate our needs clearly on Capitol Hill, and the economic downturn of the country required leadership. A Time to Lead became the theme of the IAFC with a focus on a common vision. What was fascinating to observe over the course of that year was the number of other organizations that began to reiterate and make this vision their own. From IAFC divisions to committees, to state fire chiefs' associations, the theme was repeated. The result was more involvement of individual fire chiefs in the political process, a greater focus on common issues, and further recognition on Capitol Hill of the IAFC, the fire

FIGURE 5.8A ◆ A Time to Lead
(Courtesy of Randy R. Bruegman)

THE FIRE SERVICE: FROM HUMBLE BEGINNINGS

North America's fire service originated in local communities in the tradition of neighbor protecting neighbor. With its roots in the colonial era, the historical image of the fire service recalls a bucket brigade passing pails of water, person-to-person, toward a burning structure. From these humble beginnings, the fire service has grown into a cadre of highly trained volunteer and career men and women who respond to a wide variety of hazards.

From emergency medical response to hazardous materials mitigation, today's fire service protects our homes, our citizens and our nation from a variety of threats. Motivated by the same sense of duty and pride that has inspired generations of fire fighters, today's fire service continues to stand ready to protect the safety and security of all.

Fire Fighters: Protecting our Homes and Communities
The structure and organization of local 9-1-1 emergency response systems varies widely across North America. The only constant among these different systems is the critical role of the fire service in responding when someone calls for help. In almost every emergency situation—car wreck, hazardous materials spill, medical emergency, structural collapse and residential fire—local fire fighters are first on scene and the last to leave. Trained and experienced in handling these wide-ranging and highly technical emergencies, the fire service has the ability, equipment and personnel to protect our homes and communities in times of peril.

Emergency Medical Services (EMS): More than 108 million Americans enter the emergency room each year for medical or trauma emergencies. Approximately 16 million of these patients arrive after calling 9-1-1. In many of these cases, time is critical *(source: Centers for Disease Control and Prevention, National Center for Health Statistics)*. Minutes can mean the difference between life and death. A rapid response of trained emergency medical providers is a patient's best hope. Fire service EMTs and paramedics provide that rapid response in more than 90 percent of America's communities *(source: survey of IAFC membership)*. Based in firehouses strategically located throughout local communities, fire fighters are often the only trained personnel able to respond in minutes to a medical emergency.

In addition to emergency medical first response, the fire service is the largest provider of emergency ambulance transport in the country *(source: Statistics from the Centers for Medicare and Medicaid Services)*. Combining emergency medical first response and ambulance transport ensures the highest quality patient care by providing continuity of care throughout the pre-hospital environment. This minimizes the chances for error that arises as the patient is handed off from provider to provider.

Fire Prevention and Suppression*: In 2000, approximately 1.7 million fires occurred in the United States, resulting in 4,045 American fatalities. In addition, fire caused more than $11 billion in damage to homes and property in the United States. Although these statistics are staggering, there has been a significant and consistent decline in the number of fires, fire-related deaths and property damage over the last several decades.

Since 1978, the annual number of fires has dropped by nearly half from a high of almost three million. Fires in the home have seen a similar decline—52 percent from 1980-2000. During this time, civilian fire deaths in the home have plunged by more than 43 percent. There are several reasons for these positive trends—stricter building codes, increased fire prevention awareness and improved fire service delivery. These improvements all have been championed by fire service personnel. Whether testifying before a local governing board in favor of residential fire sprinklers, smoke alarms or carbon monoxide detectors, improved fire codes or instructing school children in fire safety, the men and women of the fire service have established an exceptional record in the area of fire safety, a record that has saved billions of dollars and countless lives.
**Statistics from the National Fire Protection Association (NFPA).*

The Fire Service—Protecting our Economy and Critical Infrastructure
Hazardous Materials: Every day, fire fighters respond to incidents involving hazardous materials. These incidents range from containing diesel spills on highways to answering "white powder" calls. For example, in July 2001, a train carrying hazardous materials derailed in a downtown Baltimore tunnel. The Baltimore Fire Department responded and brought the fire under control rapidly and contained the hazardous materials spill. The tunnel was able to reopen just a few days later with no loss of life or significant injuries to rescuers or local residents. However, this incident resulted in transportation and telecommunications shutdowns that affected the entire Atlantic seaboard. Rail traffic was diverted throughout the eastern United States, and Internet traffic slowed or stopped completely as cables that carry telecommunications traffic traveled through that tunnel were damaged. This is just one example of how the actions of the local first responder can have a dramatic effect on the nation's economy and critical infrastructure.

As the world's economies grow more integrated, events that once would have affected only a small area now have the

This vision document has been developed for discussion at the IAFC's Strategic Planning Conference that will be held in October 2002 in Colorado Springs
Invitations have been sent to the leadership of the 50 state fire chiefs' organizations
as well as the committees, task forces, sections, divisions and board of directors of the IAFC.
If you have comments about this document please e-mail them to: Bill Kang, IAFC Special Projects Coordinator, at bkang@iafc.org

FIGURE 5.8B ◆ A Time to Lead
(Courtesy of Randy R. Bruegman)

service, and the issues of importance to the fire service. The fire service of the future must remain competitive with the private sector, maintaining the adopted level of service within all jurisdictions and meeting the goals, demands, and expanded missions of our customers. A concise vision of where the organization needs to go is essential to making that happen.

Making the Vision a Reality As the chief fire officer, a leader, or a manager in your department, it is never easy to turn your vision into action. As we implement change, success depends on everyone's understanding of both the need for it and the process itself. It is important to match human issues with organizational issues. How you lead change is often dictated by the culture of the organization. If an organization readily accepts radical change, multiple changes can easily happen all at once.

potential to ripple outward and affect a far greater area. Critical transportation, communications and energy infrastructures bind our world together as never before. The shutdown of a single airport can cause delays and canceled flights across the continent. In turn, those delays cost businesses and consumers millions of dollars. The fire service plays a critical role in protecting our nation's economic infrastructure.

Situations similar to the Baltimore derailment, though not as large in scale, occur multiple times every day across North America. Thanks to the rapid response and professionalism of local fire departments, these events are contained quickly and commerce is free to continue in short order.

The Fire Service—Protecting our Nation

Terrorism Response: For years, much of what the fire service does and means to our nation had been taken for granted. This changed the morning of September 11, 2001. The world watched as the Fire Department of New York responded to the terrorist attacks that struck the World Trade Center. Hundreds of fire fighters entered the burning, structurally compromised twin towers to evacuate thousands of people trapped inside. When the twin towers collapsed less than two hours after being struck by hijacked planes, 343 New York City fire fighters perished. However, the selfless actions of these fire fighters are one of the key actions that saved an estimated 25,000 people on that fateful day.

That day, America's fire service was visibly on the front line of defense as the nation was attacked in New York, Washington, DC and in the skies over rural Pennsylvania. Terrorism is recognized now as an event that will be responded to by the fire service. The fire service will be on every scene within minutes.

Responding to terrorist incidents is now considered an expanded and essential new responsibility for America's fire service, and the nation's domestic defenders need the support of the federal government in order to fulfill this responsibility. Only direct federal funds and programs will ensure that local fire departments have the resources to train, equip and protect America's first responders—a need that the federal government must not fail to provide.

Wildfire: A Growing Threat

Wildland Fires: Wildland fire is the fastest-growing fire problem in North America with significant events occurring throughout the world. During the 1990s there were, on average, more than 106,000 wildland fires that burned more than four million acres per year (*source: National Interagency Fire Center*). Virtually every fire department faces some sort of wildland-urban interface threat, either directly within the community or through area, regional or statewide mutual aid. Nearly any department could be forced to operate in the interface danger zone. Effective fire management requires close coordination between federal, state and local communities. Because such a large percentage of the acreage lost to wildland fire occur on federal land, the federal government plays a significant role in wildland fire prevention and suppression. In recent years federal fire efforts have greatly expanded, and new strategies such as fuels management programs have been developed to minimize the occurrences of catastrophic fires.

Tomorrow's Fire Service

Building upon the progress of the last several decades, the fire service is poised to take greater strides in coming years. Future progress will be tied to improvements in technology, such as has been realized with the use of thermal imaging cameras and new radio communication systems. Improved technology will make many current fire-fighting practices safer and more effective.

Increased and improved fire prevention programs are key to a fire-safe future. Rapid mitigation through early detection and fixed fire extinguishing systems has reduced fire loss. The fire service will continue to educate families and businesses about effective fire prevention measures. In addition, we will aggressively advocate for an increased use of fixed fire protection including the increased use of smoke alarms, carbon monoxide detectors and residential sprinkler systems. As technology has improved, these systems have become more effective and affordable. Widespread implementation of these technologies can significantly reduce lives and properties lost to fire and enhance fire fighter safety.

Information technology will allow incident commanders to deploy resources more effectively and allow fire department leaders to analyze service delivery needs better. In the last decade, American businesses have used information technology to gather better customer and sales data to enhance their products and customer service and to improve productivity and profitably. In a similar manner, the fire service will use improved data collection and analysis to decrease response times, reduce danger to fire fighters, and, ultimately, save lives and property.

A Vision for the Future

The fire service stands at the dawn of the 21st century ready to build on the impressive accomplishments of the last 30 years. The abilities of the fire service have grown exponentially in that time. Fire departments are no longer simply a local resource. In order to meet the challenges of the future, fire departments must strengthen mutual aid agreements and forge new working partnerships with state and national agencies. As new threats emerge, fire fighters will train and respond when the call goes out for help. The threats will vary, the technology will improve, but the fire service will rely upon highly trained, skilled individuals dedicated to protecting their communities and their homeland.

The Role of the International Association of Fire Chiefs

Established in 1873, the IAFC will continue to be the most effective network of the world's fire and emergency service leadership; experts in the field of fire fighting, emergency medical services, terrorism events, hazardous material spills, natural disasters and public safety legislation.

FIGURE 5.8C ◆ A Time to Lead
(Courtesy of Randy R. Bruegman)

Conversely, in an organization that has no experience with change or is resistant, changes must be instituted slowly and gradually, with strong efforts to build a consensus. A key aspect of the change process is understanding that some of the most difficult challenges involve the refusers. It is very important we continue to keep an open mind and treat them fairly, but you as a leader must be mindful of your limited time and energy. Your efforts will be better spent trying to win over the resisters and those who have already made a commitment or who are exploring various options.

This is a classic example of the *Pareto's 80/20 rule* as applied to leadership and management. When leading significant change within your organization, you may find yourself spending 80 percent of your time on the 20 percent (the refusers), trying to get them to adapt to and/or accept the change. A decision from a leadership perspective that one has to make is this: Will the time and energy spent with a particular group of people in respect to the change process be worthwhile? Oftentimes, as the

A TIME TO LEAD
A MESSAGE FROM THE IAFC PRESIDENT

The IAFC will continue its leadership role in the promotion and support of the federal Assistance to Firefighters grant program (also known as the FIRE Act). We will actively call for the full authorized funding of $900 million for the Assistance to Firefighters grant program for fiscal year 2002-2003. We will work closely with the current administration and Congress to ensure that America's first line of defense, the first responder, is adequately funded to prepare to respond to acts of terrorism. We will continue to play an integral part with respect to the national strategy for homeland security. We must continue to work in partnership with allied organizations and private business to address the radio interoperability problem and to ensure that public safety has sufficient and dedicated radio frequency spectrum in the future.

Fire/Life Safety
The IAFC must take a more aggressive role in promoting fire prevention and life safety by ensuring that the chief fire officers have a greater degree of involvement with codes and standards. The formation of a fire marshal section within the IAFC will bring a greater voice to those issues that we know will save lives and property in the future. We must work aggressively with allied groups on the promotion of residential sprinkler systems. We must work with the building industry to gain their acceptance.

Wildfire & Wildland-Urban Interface
We must continue to be part of the development of a national strategy to address the wildland-urban interface fire problem. Given the increase in the sizes and numbers of wildland-urban interface fires—and the deaths of wildland fire fighters—the strategies of the past are not effective today.

Health and Safety
We must take a more substantial role in fire fighter health and safety issues. During this next year we will move forward with the creation of a fire fighter health and safety program within the IAFC to focus on those issues. In conjunction we will continue to further develop a near miss reporting system by which we can learn and share information throughout the fire service. We can, through the sharing of information, reduce the risk to our fire fighting personnel. In addition, the IAFC must continue to work closely with the U.S. Fire Administration and the National Institute of Standards and Technology to develop research projects on issues related to the health and safety of our fire fighting personnel.

The Leadership Institute
As a leadership organization, we must create the vehicle by which the vision of the future can be developed and articulated. In recognition of the pioneering work that has been performed on behalf of the nation's fire and emergency services in the U.S. Congress by Representative Curt Weldon, I am proposing that the IAFC seek the structure and funding to create a leadership institute unlike any other: ***The Curt Weldon Institute for Fire & Emergency Services Policy & Leadership.***

The Weldon Institute will be an *"institute without walls"* and will provide a vehicle for the current and future leaders of the fire and emergency services to discuss and envision what the future direction of the fire and emergency services will be, and to develop policy and strategic initiatives and partnerships to establish nothing less than a new model of excellence and impact.

Creating Top of the Mind Awareness
Over the past decade, the IAFC has become very well known within government circles as the association that represents the leaders of the fire and emergency service. Now we must take the necessary steps to increase our visibility, through our membership, to the general public. By aligning all of our divisions and sections with our corporate seal—a symbol that is the essence of our history, reflective of the leadership, expertise and integrity of our members—coupled with a corporate branding plan, the IAFC will become even more widely recognized as the leadership organization of the fire service.

Facing the Challenges of the Future Together
In the shadow of September 11, we begin a new century knowing that the challenges of the future must be met with a mixture of innovation tempered by the realities of our experience. As we meet this challenge, we must work more closely in partnership with local, state and national officials, carefully coordinating our efforts to improve the safety and security for all who we are charged to protect. While this century has brought new perils, in the fire service we stand ready to meet those challenges to protect our communities. As leaders of this industry I call upon you to join with me in a common voice as we move forward this next year, promoting the cause of protecting our homes, our economies, our nations, and to forge the future of the fire service.

Randy R. Bruegman
Chief Randy R. Bruegman
President IAFC 2002-2003

FIGURE 5.8D ◆ A Time to Lead
(Courtesy of Randy R. Bruegman)

Pareto principle teaches us, the time and energy may be better spent on those other groups that will adapt more quickly and help you to lead the change in the organization (see Table 5.2).

When we speak of change, we need leadership supported by effective managerial process. Change requires top-down commitment, not only from you but also from

TABLE 5.2 ◆ Organizational Time and Energy

Organizational Time and Energy 80%	→	20% Refusers
20%	→	Resisters 80% Late Adapter Early Adopter Creator Innovator

your senior staff, or the process simply will not work. Buy-in from all levels of the organization is critical to gain a consensus on the vision of the organization. You can accomplish a great deal when everyone is geared toward achieving the vision defined within the organization. Developing a common focus everyone can understand requires strong leadership that continues to articulate the vision. The ability to bridge the gaps between human and organizational issues is essential if we are to be successful in leading change. Change requires leading by example, flexibility toward the organizational structure, and an emphasis on team leadership. It also provides an opportunity to guide your organization, although it is not something you can totally control. However, you can anticipate, adapt, react, and act accordingly so your chances for success are enhanced. Leaders in the fire service today have to begin planting the seeds of change for the future, providing a means to move their organizations toward the change process and successfully implementing those changes for the organization and its personnel.

BUILDING A CULTURE OF OUR CHOICE

There is an interesting analogy to how cultural norms can impact and play out in our organizations. It is the story of the five apes (see Figure 5.9).

> The story begins with a cage containing five apes. In the cage hangs a banana on a string with stairs under it. Before long, an ape will go to the stairs and start to climb toward the banana. As soon as he touches the stairs, all the apes are sprayed with cold water. After a while, another ape will make an attempt with the same results—all the apes are sprayed with cold water. Turn off the cold water. If, later, another ape tries to climb the stairs, the other apes will try to prevent it, even though no water sprays them. Now, one ape is removed from the cage and is replaced with a new one. The new ape sees the banana and begins to climb the stairs. To his horror, all of the other apes attack him. After another attempt and another attack, he knows that if he tries to climb the stairs, he will be assaulted. Next, another of the original five apes is removed and is replaced with a new one. The newcomer goes to the stairs and

FIGURE 5.9 ◆ Five Apes

> is attacked. The previous newcomer takes part in the punishment with enthusiasm. Now, the third original ape is replaced with a new one. The new one makes it to the stairs and is attacked as well. Two of the four apes that beat him have no idea why they are not permitted to climb the stairs or why they are participating in the beating of the newest ape. After the fourth and fifth original apes have been replaced, all the apes that were originally sprayed with cold water are now gone. Nevertheless, no ape ever again approaches the stairs. Why not? "Because that's the way it's always been around here." "That is how organizational behavior is indoctrinated into social/corporate policy and a culture and becomes entrenched in our daily activity."
>
> *Author Unknown*

Have you ever experienced similar behavior in any organization you have worked in or volunteered for? Having worked in five different fire service organizations, I have experienced the five-ape phenomenon in each. It can play itself out in many different ways. You can find it in your own policy books when someone made a mistake over 25 years ago; therefore, a policy was written to correct the behavior of the entire organization. It has been institutionalized over a period of time within the organization. Today only a few, if any, remember the original motivation for the policy and/or change. You can see it when you sit down to negotiate a new labor agreement; either side brings forward concerns and issues that happened over 10 years ago and are not even relevant today. You can see it in the relationship between labor and management. I worked in organizations that believed if you wore a white shirt you were the enemy. Conversely, certain chief officers in the organizations I have led had little or no respect for the people who were riding on the fire trucks. This culture was institutionalized over time partly because there was no common set of core values.

My experience as a new fire chief with a new labor president was very indicative of the lack of organizational core values and how conditioning can lead you down a path you wish not to be on. Just like in the story of the five apes, this labor president had repeatedly been taught, much by observation, the only way to achieve the outcome desired by the labor group was to do battle on every issue, and so he did. The organization went through several mediations to resolve contracts and the filing of numerous unfair labor practices, and each time the labor group lost. The result perpetuated more of the same behavior.

Then one day the labor president told me that, even though he personally did not disagree with the direction the organization was going and my style of leadership, it was his job, as he put it, to "fight the man" and that was what he was going to do. It was not until then that I recognized the behavior being exhibited was not so much focused on me or the organization but was a result of three important facets of any organization: (1) the lack of core values, (2) culture, and (3) conditioned behavior. Because of the lack of organizational core values, behavior was all over the board. Second, the culture of the organization was that labor and management should not be collaborative; therefore, the conditioning of the "us versus them" mentality over the years perpetuated this attitude and culture. You can see when the behavior is repeated within an organization, it does not take long to become an organizational norm. As it illustrated in the story of the five apes, if such organizational norms are not conducive to supporting organizational core values, the vision, and the mission, the organization often becomes reactive or subversive to change. When this occurs, such behavior needs to be addressed. To do so, you must get to the root of the issue, which often requires a history lesson of your department.

Many fire service organizations today embody behaviors and norms that go back for generations. These cultural norms, in many cases, are the most limiting factor in the ability to address current issues. Creating the right culture, articulating your vision, focusing on the mission, and planting the seeds of the future are critical elements of successful change. In *The Adaptive Corporation,* Alvin Toffler stated,

> The adaptive corporation needs a new kind of leadership. These managers of adaptation are equipped with a whole new set of nonlinear skills. Above all, the adaptive manager, today, must be willing to think beyond the thinkable, to preconceptualize products, procedures, programs and purposes before crisis makes drastic changes inescapable. Warned of impending upheaval, most managers still pursue business as usual, yet business as usual is dangerous in an environment that has become, for all practical purposes, in constant change.[21]

Today in the fire service, many have placed themselves in a very dangerous position through their own complacency about the future. The need to reposition our organizations for future growth, and the competition our organizations face for revenue in conjunction with the ever-expanding list of services we deliver, have created a new reality for our industry. "Business as usual" is equivalent to a slow and certain death for many of our organizations and for you, whatever your rank. As we cope with growing expectations and with the increasing demand for services, our visions will become a critical factor in the future health of our organizations, providing leverage and influence with both the individuals and the organizations we lead. We know the pace of organizational change will continue to increase for many of the reasons previously discussed (technology, diversity, economics, expanded mission).

Learning new strategies on how to lead change will be critical for the successful fire officer of the future. Fire organizations, like most individuals, will not change until they must. Historically, the fire service has had an unimpressive track record in responding to change. For people to change, they must adapt in three ways—physically, intellectually, and emotionally. It all comes down to the universal question, "What is in it for me?" As a change strategist and as a leader in your organization, you must understand the reluctance you will face, the fear that will be created, the rumors that will be started, and the misunderstandings that will result when you begin to move people outside their comfort zone. There is no one solution to managing change; every initiative has its own challenges, and each organization is different. Social and cultural changes have forced fire service leaders to reexamine not only how they lead their organizations but also how they manage them on a day-to-day basis.

I have had the opportunity to lead several consolidation efforts. As you can imagine, when you discuss bringing two different organizations together in a merger, consolidation, or contract for service, literally hundreds of issues must be addressed. From pay to station selections, to what uniforms will be worn, to the patch on the uniform, to how the equipment will be marked, to how to standardize the SOPs, and the list goes on and on. In every situation I have been involved in, the rumors regarding what the organization was doing were, in many cases, so far-fetched they were almost laughable, but in each case, many people in the firehouse believed those rumors.

This points to a very important factor when going through a change process, simply that enough information can never be given. One of the things firefighters are excellent at, especially when they are sitting around the coffee table in the firehouse, is if they are not provided with sufficient and timely information, they will take the information they have on hand, or were told by a third party, and come to their own conclusion. In many cases, the conclusion or the picture they have established for themselves is totally incorrect.

There is another critical aspect of change in the fire service we must come to terms with. Certain people within the fire service start rumors to see how far they will perpetuate through the organization. Frankly, it is a game some in our profession like to play. As a leader, whether you are the chief of the department, a deputy chief, a battalion chief, a captain, or a formal or informal leader in some other aspect of the organization, you have to understand this. The best way to combat the rumor mill and facilitate a positive change process is to communicate as effectively and openly as possible to all personnel within the organization. To do anything less will find you spending an enormous amount of time and energy putting out brush fires on issues that have no credible relevance. Although there may not be a specific template for change, several common factors must be addressed as you and your organization move through the process. A strategy that is not revolutionary or unique includes components that have worked for many organizations, both private and public, and it emphasizes what is necessary for successful change: Establish your need, build relationships, understand the problem, research the past and look to the future, commit to a solution, and evaluate your progress.

Establish Your Need Successfully revitalized organizations start by communicating their vision on change using meaningful language. Classic examples are Motorola's six-sigma philosophy, General Electric's Boundary-less Quality Banner, Ford's "Quality Is Job #1," and Chevrolet's "Like a Rock." Each statement transmitted a message. Each message about change was a public declaration that the organization needed to rethink its old ways of doing business.

The problem for every organization is how to communicate to the elected officials and the community at large a vision they can grasp. As previously stated, in our business, time is a critical factor. We know that four minutes not only is important, as outlined in national standards, but also is based on scientific research. In medical emergencies, science tells us the sooner we intervene with proper medical intervention, the more likely a patient is to survive and subsequently enjoy a better quality of life. The same holds true from a fire perspective. If we arrive in four minutes and can set up our equipment and deploy into a burning building, we will more than likely intercede before flashover occurs. We know when flashover occurs, there is an increase in the loss of life, in property damage, and in risk to firefighters.

When we developed the "4 Minutes to Excellence" theme in Fresno, as previously discussed, it provided a way to frame a very complex issue in a manner that the elected officials and the public at large could understand. It not only became the focal point of the organization but also began to be played back from the elected officials, from other department heads, and actually from the media. We proposed new initiatives and new deployment strategies and the question always was, "How does this help your '4 Minutes to Excellence' goal?" Descriptive language helps to paint a picture and form a target on the wall so people can visualize it, begin to understand it, support it, and move toward it.

Inspirational leadership can often energize an organization to begin to accept this change by making public declarations that establish a vision and provide a target for the workforce. The task is for you to turn the vision into organizational needs and manage those needs to improve everyday performance. Establishing a need greatly depends upon how you sell the change. Is the change a result of a problem or opportunity? No component is as important to the outcome of any change process, or any change initiative within the organization, as your own attitude. It is also critical to start with a clear understanding of the organization's culture as well as the organizational paradigms that exist, the framework used to process information related to the change, how you will attempt to build a new culture, and the reaction to the culture.

Build Relationships The foundation for any change process is people. Finish this sentence: "My department could accomplish X if it were not for ____ or ____." You could probably select one or more names and/or organizational roadblocks. Most organizational problems are people problems. To be successful as you introduce change into the organization, you must build a support system and find organizational champions who can carry the banner in your absence. In times of change, people skills are definitely more important than the technical skills we bring to the job.

As in the story of the five apes, certain organizational norms have been developed as a result of traditional behaviors, autocratic leadership, union leadership that promotes mediocrity, and/or old-style politics that will attack anyone who differs from the norm. This disease afflicts many organizations today, and the only cure lies in the workforce itself. The fact is the workforce (usually one person at a time) determines when the old way of doing business is no longer productive. I saw this occur in an organization where, due to repressive labor leadership, the employees were reacting much like those apes in the cage. If they tried to do something new, they were chastised for trying to be a shining star—a classic example of promoting mediocrity so no one looks good or bad. Organizationally, this is like playing in a basketball game with one hand tied behind your back. With this type of culture, you can only hope to maintain your existing level of performance; it will be almost impossible to move the organization to the next level. The cycle is broken when one of two things occurs: (1) when enough members of the team are willing to change or to step outside the group norm; or (2) when a sufficient number of retirements, hirings, firings, or some other type of attrition occurs to allow a sufficient number of new attitudes to be infused into the group. The same can be true of the leadership of the organization whose culture consists of being exclusive, centralizing power, and treating employees without respect. The result is an unmotivated workforce that says nothing, never steps out of line, and just tries to survive the workplace each day. In either scenario, the resulting culture is not conducive to change or to addressing the difficult issues facing our industry today.

You need to know how much support you have and who your organizational champions are or will be. If you wait to build your relationships until after you have developed your game plan, you will probably fail. Remember, building relationships starts with listening. Take the time to listen to what you hear, to really observe what you see, to take into account what you and others in the organization are feeling, and to trust your intuition. As you gain more experience as a fire officer, you will become more effective at trusting your instinct about the right direction in which to lead or manage.

Understand the Problem In trying to understand what the issues are, be aware that what might seem to be a problem may often be only a symptom of the real problem, and try to understand the cause and scope of the problem. It is like looking at an iceberg; there are often many hidden dangers, out of sight under the water. Going beyond the surface issues and exploring your findings, keeping in mind solutions as you continue your analysis yet guarding against imposing your own favorite solution, are important. Do not fall into the trap of identifying or reacting to only those issues that seem immediate, are visible, look important, or are the most comfortable ones to address. Your purpose is not to develop a final solution but to gain an understanding of

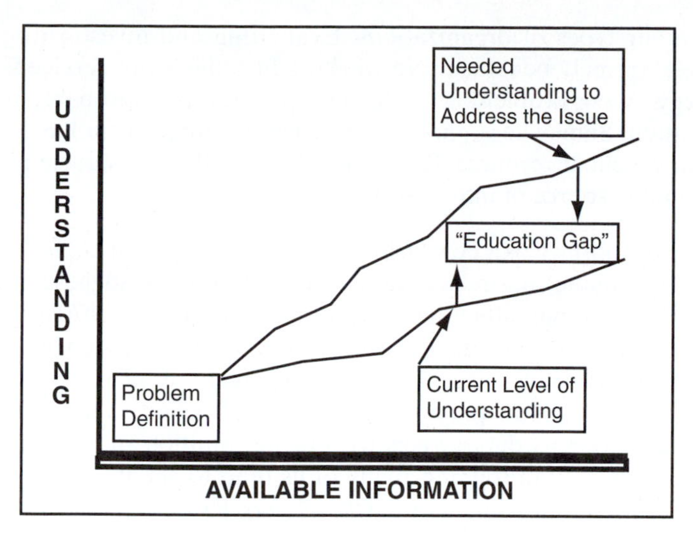

FIGURE 5.10 ◆ Available Information
(Courtesy of Randy R. Bruegman, The Chief Officer: A Symbol Is a Promise, *p. 124)*

the problem from a *360° perspective*. Do not become distracted from the purpose of your change strategy. Quick thinking and reacting can result in the implementation of quick fixes, but in many cases these do not solve the real problem. They turn out to be only a temporary solution, and you often find yourself, in a short period of time, back to square one dealing with the original issue, only it is bigger now than before. In your early exploration of the issue, try to determine the root cause of the problem (see Figure 5.10).

There is always a gap between what we know and what we should know before we set a course of action for our organization. To help address some of the cultural barriers and build the relationships that will be the foundation for your change process, articulating the problem and describing how to solve it are critical. In every major organizational shift, an amazing number of perspectives exist on this one topic. This underlines the importance of understanding the problem. Once you understand the problem, you can close the educational gap within the organization between what you currently know and what you need to know to address the problem adequately.

Research the Past and Look to the Future None of us have all the information we need on a day-to-day basis to make a 100 percent quantitative decision(s) in an environment that is becoming faster-paced each day. It is becoming increasingly important to access information. The problem for many of us is that research itself is a science. Use both qualitative research, which draws on statistic surveys, questionnaires, and computer modeling, and quantitative research, which draws upon data from your direct observation. Do not dismiss your personal experience as unimportant. Look back first. Try to understand how your organization got to the point where it is today. Part of the reason may be tradition, so you have to know your organization's history. Consider the past solutions that succeeded and then review those that failed. Oliver Wendell Holmes once said, "To understand what is happening today or what will happen in the future, I look back." Today, most leaders who take control of the future do so by examining the past as well as by taking full advantage of present opportunities. Look outside the fire service. Many problems having similar characteristics exist within other types of organizations. Evaluating and investigating other disciplines can be extremely beneficial. Networking in today's fire service is another excellent strategy. A big problem or a change opportunity is probably not unique to your organization. Somewhere another fire chief or officer who has faced a similar situation is an excellent resource. Reach out through the fire service network to tap into this invaluable source of information.

Commit to a Solution Colin Powell, in his book *My American Dream,* refers to a decision-making technique that relies heavily on intuition. He suggests the key to not making quick decisions but rather timely decisions is based on a 40/70 formula. If he has less than 40 percent of the information needed to make the decision, he chooses not to act. Conversely, he does not wait until he has 100 percent of the information either, because in many cases that is simply not possible. Powell believes that when he has 60 to 70 percent of the necessary data, he acts, trusting his intuition and experience.[22]

Due to our fire ground training, many of us tend to act at the 40 percent benchmark and than find our decision and outcome were less than what was desired. Very few pieces of literature exist today for fire chiefs on strategies for implementing change and building a new culture. The basis for what we do in the future lies in the answer to the questions, "Is it right for the organization?" "Is it right for the community?" and "Is it founded on providing good-quality service?" At times, our personnel can forget this and focus only on the issues that are central to them. Chief officers often do the same thing, forgetting what the demands are in the firehouse.

As you put your research and findings into your own words and share them with the people you have identified as champions within your organization, stay focused on the task at hand. Every change initiative has its own challenges, and sometimes you have to write your own book as you go. As a fire officer, you need to be truly committed to the change initiative. If you are not, you will likely fail. You have to lead by example much more than most people realize. Changing the organizational culture starts as a top-down activity, but make no mistake, it will be driven from below. If you observe companies in the private sector, those that succeeded in refocusing and re-engineering made it because their chief executives were willing to invest their time and energy in the organization's change. Changing the organizational culture requires great courage and commitment from the chief executive but requires the horsepower of the men and women who do the work every day.

Evaluate Your Progress Evaluating change requires critical thinking and specific performance measures to be established. Without them, it is often difficult to evaluate your change initiative objectively. Organizational environments are in a constant state of change, and yesterday's solutions clearly will not meet tomorrow's challenges. Performance measures and the ability to quantify improvements made within the organization are critical, not only to measuring the progress of the organization but also to articulating that you are on the right track. In his book, *Taking Charge of Change,* Doug Smith outlines 10 leadership principles and provides some insight into the change process. It is food for thought as you begin to move your organization into the future.

1. *Keep performance results* the primary objective of the change.
2. *Continually increase* the number of personnel taking responsibility for their own change (above-the-line accountability).
3. *Ensure* people always know why their changed performance matters to the organization as a whole.

4. Put people in a *position to learn* by doing, and provide the information and support needed just in time to perform.
5. *Embrace improvisation* as the best path of both performance and change. Encourage off-the-cuff creativity and imagination, not rigid adherence to management plans.
6. Use *team performance* to drive change whenever required.
7. *Focus the organizational structure* on the work people do, not on their decision-making authority.
8. *Create focused energy* and meaningful language because these are the scarcest resources during periods of change.
9. *Stimulate and sustain behavior-driven change* by harmonizing initiatives throughout the organization.
10. *Practice leadership* based on the creed to live the change you wish to bring about.

Initiating a change process will take courage, vision, and a clear understanding of the culture of your organization. Sometimes it does not matter where you start, but *as a leader your responsibility is to go first.*[23]

THE PSYCHOLOGY OF CHANGE

Another crucial component of all change processes is the psychology of change. In his book, *Managing Transitions, Making the Most of Change,* William Bridges provides a unique perspective on what people go through when coming to terms with change that is largely internal.[24]

The transition starts by letting go of your old reality. Think back to some of the initiatives that have been implemented in your organization, and remember the difficulties that some of the personnel, including senior staff, had grasping or supporting the new programs. The change can be a shift change of personnel, a new chief on board, or the addition of a new service. Some people are never able to make a successful transition. They simply cannot accept the change because they have never let go of the old reality. When Bridges talks about the psychology of change, he describes three distinct phases. No matter what level you work at in your organization, you can see them. The next time a change in your department is introduced, instead of moving to full engagement on the issue, step back and watch the actions and reactions of others. It will be a great learning experience and will help you become a better leader (see Figure 5.11).

The first is that a period of transition begins by letting go of the old reality. The second involves finding the neutral zone—the space between the old reality and the new one. This is the core of the transition process. It is the time when the old reality is

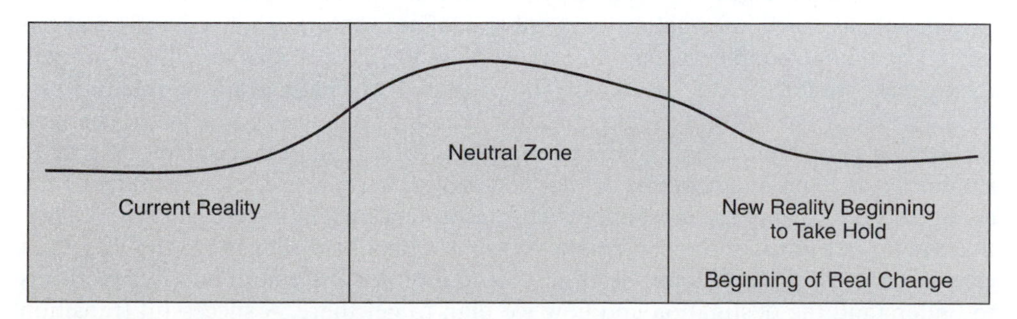

FIGURE 5.11 ◆ Change Continuum

gone, but the new one does not feel comfortable yet. People who have made this transition are often in a state of confusion in which the highest degree of mental anguish exists over having to let go of the old way of doing things. This is also when people feel most vulnerable. They are trying new things in a new way, and they feel this is the most dangerous time because things may not go as smoothly as they did in the past, at least for a while. This leads to the last phase, which is actually the beginning of real change. It is the time in a person's or organization's life that involves developing new understandings and attitudes and often creating new values. At this point you can truly begin to build a culture of your own choosing. New identities, new situations, new organizational values, and future leaders often emerge at this time.

HOW DO WE GET THEM TO LET GO?

Making a change within the organization involves some type of transition. In certain cases, people do not resist the change; they resist the loss of something they have done for a long time, which removes them from their own comfort zone. Understanding the psychology of change is important so we can identify early what will be perceived to be lost. It is vital to understand and describe in detail what the change will mean for all involved. Understand the chain reaction that will result when the change is made and who in the organization must adapt for the change to be successful. Two things are very important to comprehend: the current reality and the importance of the real and perceived loss to the personnel.

In our leadership role, we tend to look at things from an objective and often less emotional perspective. In many cases, our personnel may see them from a subjective and often emotional standpoint. So do not be surprised if, even on a minor issue, people seem to overreact because they are feeling a sense of loss. People can overreact from an accumulation of change. Bridges refers to this as a "transition deficit," a time when the old loss has not been dealt with. All of a sudden, you see an explosion of emotion; it is the baggage of the past. The individual can carry only so much of it for so long until giving out and letting go; the result is an emotional outburst. Looking at the psychology of change, we must acknowledge these losses openly and empathize with those who will ultimately have to make the change at all levels of the organization.

If you are making a large-scale change, you can expect to see classic signs of grieving within the organization. It is important to understand the components of this cycle, which are denial; anger; bargaining or unrealistic attempts to avoid the change situation; anxiety; sadness; depression; disorientation, confusion, and often forgetfulness; and acceptance and/or surrender.

One of the best ways to overcome this natural process of grieving and loss is to give people as much information as possible. Define what has changed and what has not; if you do not, people will continue to use both the old and new methods. They will then decide what to discard and keep. The result will be chaos. Many people just toss out everything and wait for the organization to come back with a new set of changes. So let your people take some of the old methods with them. It is crucial to be able to show what the end result will be so that you can ensure a degree of continuity as the change occurs. One of the most important reasons organizational change often fails is that we do not think of the end result. We do not plan how to manage the effects of change on our people. It is just as important to look for the emotional pitfalls as it is to understand the destination and how we plan to get there. A successful transition depends on our ability as leaders to convince people that it is safe to leave home.

THE NEUTRAL ZONE

The neutral zone is a place between what has existed and what will be. There are some definite organizational dangers there. Anxiety within the organization will be high, and the motivational level of people will drop. This will be a time when some in the department disengage and do not want any part of what is going on. There will also be some resentment toward you, as the leader of the change. People will become somewhat more protective of their territory, and their energy level will not be as high as usual, but this period will be short. It is also a time when people try to revisit problems that occurred as many as 10 to 20 years ago as change often reopens old wounds. Some people never forget, especially if they are focused on reliving the past or maintaining the status quo. People in the neutral zone may feel they are overloaded and getting mixed signals. The systems are in a constant state of flux. Great confusion exists because they are having difficulty letting go of the past and coming to terms with the new reality. This often causes polarization between those who want to undertake the change and those who resist it. Polarization often places leadership in the role of counselor, referee, or, in some cases, autocrat, as you must make the difficult decisions and set people on the right course. When you lead change you will find yourself in all three roles, sometimes daily.

It is also a time when the organization is trying to redefine and reorient itself. Some things can facilitate this process.

- Make sure your mission is clear.
- Establish a network consisting not only of senior management personnel but also of personnel throughout the organization to act as messengers of change.
- Make sure you are accessible for people who want to talk.
- Do not forget to go to them.
- Do not skirt the difficult issues; meet them head on.
- Be honest and up front.
- Remain focused on what you are trying to accomplish because it is very easy to become distracted.
- Establish a positive outlook by your actions. This is the time to consider problems and propose creative solutions.
- Encourage openness within the organization. This is an excellent way to start building your team.
- Empower people to experiment with new ideas.
- Turn losses, setbacks, and obstacles that have been overcome into a positive message.

Richard Nixon once said, "Only if you have been to the deepest valley can you ever know how magnificent it is to be on the highest mountain." When you are leading change, I can assure you that you will visit both places.

CASE STUDY

One of the best case studies in the last 25 years has been the creation of the Department of Homeland Security (DHS). When President Bush signed the national strategy for homeland security and the Homeland Security Act of 2002, it effectively mobilized and organized over 22 different agencies with 180,000 men and women into the federal department to work daily on the important task of homeland security. Think about the enormity of the challenge DHS has undertaken, and put it in context of what happens locally when we have even a moderate change within our local governing structure, a merger, a consolidation, or a change of the city manager. Needless to say, we have all read and followed what has occurred since its creation, and DHS has definitely had its challenges—consolidation, reorganization, streamlining of agencies, and performance issues.

Put this task in the context of the fire service. You have been given the task to consolidate every fire department in your state into one agency. What kind of issues do you think you would face if you had to do this? Say that you have 500 fire departments and 500 different fire chiefs whom you have to consolidate in your state. This means there will be reassignments; people who will be exited from the organization; standardization of the computer systems, hiring systems, and SOPs; creation of a hierarchy that will allow for effective communications both from the field and from the top; and development of a strategic plan with a common vision and a set of mission statements all personnel can support and buy in to. You need to accomplish this in a short period of time because you are a critical service provider.

This may be what the Department of Homeland Security faced with its reorganization plan. While many of us watched from the outside, it has been challenged on many levels as it has brought together different organizations. DHS has faced many of the issues we have covered about leading and managing in a changing environment. Both the emotional issues and the organizational issues must be addressed (see Figure 5.12).

It may take years for this new entity to totally resolve all those issues and probably a decade before all those issues are resolved in the final reorganization of this new federal agency, but it provides a good case study from a leadership-level perspective on the difficulty of large-scale organizational change. As David Osborne and Peter Hutchinson wrote in their book, *The Price of Government,* "Consider the federal

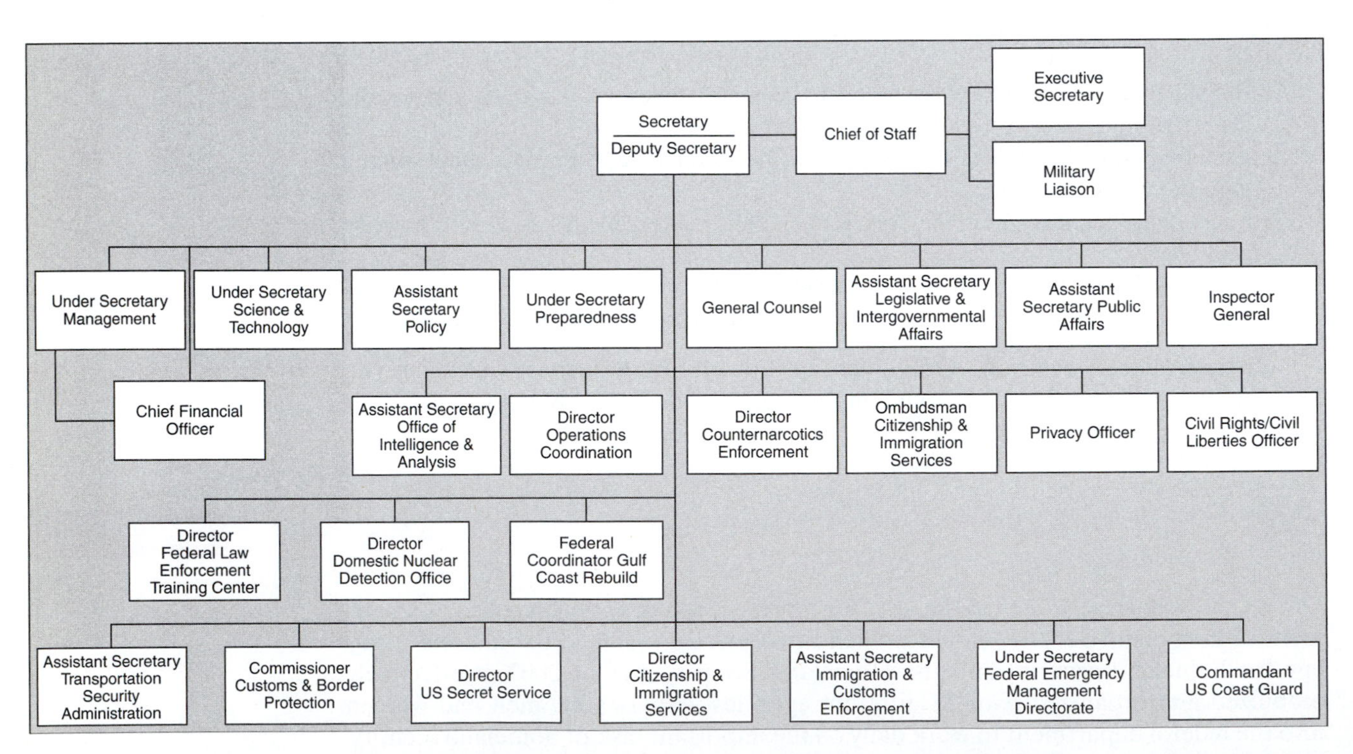

FIGURE 5.12 ◆ Department of Homeland Security Organization Chart *(Courtesy of www.dhs.gov)*

government's new Department of Homeland Security, born of the need to 'do something' after the September 11, 2001, terrorist attacks. It merges at least 170,000 federal employees from 22 agencies, working in areas ranging from agricultural research to port security to disaster assistance. It brings together the Transportation Security Administration (TSA), Customs Service, Immigration and Naturalization Service (INS), Secret Service, Coast Guard, and Federal Emergency Management Agency (FEMA).

Within Homeland Security, the *Information Analysis and Infrastructure Protection Directorate* absorbs the FBI's National Infrastructure Protection Center, the Defense Department's National Communications System, the Commerce Department's Critical Infrastructure Assurance Office, the Energy Department's National Infrastructure Simulation and Analysis Center, and the General Services Administration's Federal Computer Incident Response Center.

The *Border and Transportation Security Directorate* oversees the Coast Guard, TSA, much of the Customs Service, the Border Patrol, the Federal Protective Service and Federal Law Enforcement Training Center, the INS, Animal and Plant Health Inspection Service (formerly in the Agriculture Department), the Office of Domestic Preparedness (formerly at Justice and FEMA's former Office of National Preparedness).

The *Science and Technology Directorate,* focused on developing countermeasures to chemical, biological, radiological, and nuclear weapons, includes what was formerly Agriculture's Plum Island Animal Disaster Center, Energy's Lawrence Livermore National Laboratory, and Defense's National Bioweapons Defense Analysis Center.

The *Emergency Preparedness Division* hosts most of FEMA; the FBI's National Domestic Preparedness Office; the Health and Human Services Department's National Disaster Medical System, Metropolitan Medical Response System, and Office of Emergency Preparedness; the Justice Department's Domestic Emergency Support Teams; and the National Oceanic and Atmospheric Administration's Integrated Hazard Information System.

Few management experts expect the new department to succeed anytime soon. Homeland Security is a classic example of the knee-jerk impulse to consolidate. This impulse promises greater efficiency through elimination of duplication and overlap, but too often it delivers huge bureaucracies with so many layers that authority is fragmented, communication is difficult, and decisions take forever. Those who sponsor such mergers remind us of troubled young couples who decide getting married is the cure. No wedding ever solved the underlying problems in a relationship, and after the honeymoon the headaches remain.[25]

Slamming multiple organizations together often results in organizations with so many different missions that top leaders have no expertise in most of what the organization does. They have trouble agreeing on what is most important and cannot decide what to focus on. Large organizations with multiple missions are notoriously poor performers. Lower down, employees spend their time worrying about their futures and defending their turf, rather than doing their jobs. "Employees often have existing loyalties to the old chain of command, and it's hard to get people to switch over," says Princeton University sociologist Frank Dobbin. Creating a truly integrated corporate culture, he adds, "can take 10–15 years—enough time for a large portion of the workforce to turn over." Sometimes the cultural differences are inherent in the work itself.[26]

Before and during the period of transition, consolidations demand vast amounts of energy, most of which is directed at the bureaucratic structure and the people inside. They deliver energy *from the work* to the workplace. "It's going to take three to five years when all is said and done to get everything organized in the way that will be its final form," the Coast Guard's Rear Admiral Harvey Johnson said of the Homeland Security consolidation. Jesse Rasmussen, former director of human services in both Iowa and Nebraska, helped drive a massive consolidation of human service agencies in Nebraska. "I was arguing for this, because I thought it was the only way we'd get a different system," she says. But she learned a sad truth:

> I learned . . . that if you do that massive a restructuring, the restructuring itself is so distracting that people lose sight of why it's a good idea. You ought to just change business practices and business relationships, and then change structure when it's in the way. The same thing might have happened if the governor had said to the director of Human Services and director of Public Works, "I want your departments to do one plan for each child in your systems. Change the business practices and relationships, as opposed to rearranging the rooms in your house." It got down to people saying, "I paid for that wall, and I'm taking it with me." It was bizarre. And emotionally, people couldn't do the work.[27]

Five centuries ago, Machiavelli articulated a difficult truth about political reform:

> There is no more delicate matter to take in hand, nor more dangerous to conduct, nor more doubtful in its success, than to set up as a leader in the introduction of changes. For he who innovates will have for enemies all those who are well off under the existing order of things, and only lukewarm supporters in those who might be better off under the new.[26]

Today DHS continues to evolve into the organization it will become in the future. It began with the Homeland Security Act of 2002. The reason for the establishment of DHS was to provide the unifying core for the vast national network of organizations and institutions involved in efforts to secure our nation. In order to do this better and to provide guidance to the 180,000 DHS employees, the department developed its own high-level strategic plan.[28] The vision and mission statements, strategic goals, and objectives provide the framework guiding the actions that make up the daily operations of the department.

Vision: Preserving our freedoms, protecting America . . . we secure our homeland.

Mission: We will lead the unified national effort to secure America. We will prevent and deter terrorist attacks and protect against and respond to threats and hazards to the nation. We will ensure safe and secure borders, welcome lawful immigrants and visitors, and promote the free-flow of commerce.

Strategic Goals:

- *Awareness.* Identify and understand threats, assess vulnerabilities, determine potential impacts, and disseminate timely information to our homeland security partners and the American public.
- *Prevention.* Detect, deter, and mitigate threats to our homeland.
- *Protection.* Safeguard our people and their freedoms, critical infrastructure, property and the economy of our nation from acts of terrorism, natural disasters, or other emergencies.
- *Response.* Lead, manage, and coordinate the national response to acts of terrorism, natural disasters, or other emergencies.
- *Recovery.* Lead national, state, local, and private sector effort to restore services and rebuild communities after acts of terrorism, natural disasters, or other emergencies.

- *Service.* Serve the public effectively by facilitating lawful trade, travel, and immigration.
- *Organizational excellence.* Value our most important resource, our people. Create a culture that promotes a common identity, innovation, mutual respect, accountability, and teamwork to achieve efficiencies, effectiveness, and operational synergies.

These insights are helpful on the issues that emerge when undertaking significant organizational and environmental change such as the Department of Homeland Security has done. Only time will tell what has and has not been successful. Although they are on a much grander scale, the concepts and issues discussed in this chapter are relevant, whether you are bringing together 22 federal agencies or two local fire departments, or making a significant internal change within your department.

ADVICE FROM THE EXPERTS

The fire service has undergone a significant amount of change during the past 20 years. What advice would you give about leading and managing changes in the fire service?

MICHAEL CHIARAMONTE

Chief Fire Inspector, Retired, Lynbrook Fire Department (New York): Use baby steps and do not adopt change for the sake of change. Quality changes are effective and last.

BETT CLARK

Former Fire Chief, Bernalillo County Fire and Rescue (New Mexico): I think too much emphasis is placed on tradition. We are proud of our history and our forefathers and mothers. However, we have a daily opportunity to create the future of every fire service department and member. If we still followed tradition, we would not have women in the fire service; no EMS component; no progress in technology for firefighting tools and equipment; and most certainly no codes or industry safety standards. It is good to know why we are what we are; it is even more rewarding to be who we can be in the future. The needs and expectations of firefighters change as society changes those expectations. As chief officers we must be problem solvers, financial counselors, facilitators, teachers, and parents. The workforce is younger with fewer taught traditional values and job ethics coming from high school graduates. The new recruits have grown up in an instant gratification atmosphere . . . fast food, instant messaging, more technology, and less labor oriented. The fire service is still a hands-on, feeling, and compassionate career field that does not always fulfill the need for instant gratification. Leading and managing have also expanded to include prevention and education, which are proactive rather than the reactive "put the wet stuff on the red stuff." A successful chief officer needs to remember to think, act, and lead in a proactive manner for our future. As call volume decreases with prevention, education, and safer technology, firefighters are bored. A good leader can embrace proactive programs and activities that benefit the community and give the firefighters an opportunity to fulfill another need—self-achievement.

KELVIN COCHRAN

Fire Chief, Shreveport Fire Department (Louisiana): To manage and lead change in the fire service effectively, fire chiefs must stay abreast of social, political, technical, and economic factors necessitating change through continuous environmental scanning locally and globally. Firefighters respond better to change when they are well informed and are participating in the planning processes that will produce a change in their work environment. Fire chiefs are proactive in leading and managing change when attending and participating in meaningful conferences, researching, reading books and periodicals, and then adapting the fire department according to organizational needs and trends. Engaging other department members in these same activities creates a team of change agents to develop, market, and implement strategies for change within the organization.

CLIFF JONES

Fire Chief, Tempe Fire Department (Arizona): Leading change is one of the main leadership and management responsibilities of fire officers. My advice is to exercise leadership, management, and interpersonal skills in an effort to lead positive change. One of the key things is to analyze prospective change in terms of its positive and negative impacts as the organization goes through it. Next, find partners who can assist you in driving a particular change issue. Almost all change involves loss for people, and this loss very often causes people to resist change. Helping people through the loss and helping them to clearly understand the benefits of the change will be of considerable assistance in managing and leading the change process.

BILL KILLEN

Retired Fire Chief and IAFC President 2005–2006: In order to lead and manage changes in the fire service effectively, we have to stop talking about "thinking out of the box" and actually get out of the box; that is, walk the talk. It is interesting to note many of the problems identified after September 11, 2001, were the same ones identified after Hurricanes Katrina and Rita along the Gulf Coast. Often it is said history repeats itself, yet it is not history but we as individuals who continue to repeat the same mistakes.

KEVIN KING

U.S. Marine Corps Fire Service: My best advice is patience, patience, and more patience. Change is hard, especially significant change in an organization such as the fire service. In most cases it will not happen overnight, and many will resist the change. However, if it is a worthwhile change, then stick to your guns and your principles, and determination will eventually win people over. Also, do not forget to celebrate successful implementation of a significant change. Your people have come through a lot, and they need to know you appreciate their effort and determination. (These are opinions of Mr. King and do not reflect official policy of the Department of Defense or the United States Marine Corps.)

MARK LIGHT

Executive Director, International Association of Fire Chiefs: Chief officers must be open to question what they have done in the past and look at how changing may improve and enhance what they are doing. We all get comfortable operating a certain way and don't want to change, even when new ways may be light years ahead of where we are. To manage change, we must be willing to look at new things, and make decisions based on what is best for the organization, not on what is comfortable.

It is very easy to take the approach of "if it is not broken, don't fix it." Yet I would put forth that at times we need to take an approach of "if it is not broken, break it, so we can fix it." Obviously, this is not appropriate in many cases, but a leader must be willing to set the course for change based on his or her vision of the future and to put the resources and staff in place, with the proper training, to execute the change. Change may not be easy, and at times it may be painful, but as chief officers, we must learn to welcome change and use it to better our organizations.

LORI MOORE-MERRELL

Assistant to the General President, International Association of Fire Fighters: Be flexible, be open-minded, be positive, and be assertive where necessary, remembering the only constant is change.

JAY REARDON

Fire Chief, Retired, Northbrook Fire Department (Illinois), currently President of the Mutual Aid Box Alarm System (MABAS): Leadership goes in all directions: hierarchical, subordinate, collegial, public, and process. You first need to understand where you are when you believe you need to exercise leadership, what is your position, and rule. Leadership sometimes results in saying things and taking positions people are not expecting to hear. Leadership is usually difficult because you are attempting to persuade or direct people to do something they are not prepared to deliver. Leadership can shake one's confidence within oneself.

GARY W. SMITH

Retired Fire Chief, Aptos–La Salva Fire Protection District (Aptos, California): Change is the fodder for growth . . . it is the means to the end. Embrace change and dictate the future by being strong enough to set a vision and succeed by using great management skill to build the systems and prepare the team for making the venture. It is like personal success planning; the team needs to be educated, to network with those who have relationship to the cause, and to venture forward and "do." Great teams take care of themselves, understand the value and priority of safety, and work forward with trust in the system and fellow players because they are equipped and trained to succeed. This feeling of confidence empowers teams and builds great future managers and leaders who will continue the efforts of the team. This type of organization will thrive on change. Those who resist change generally are "flat" with their experiences and want to hold on to the past, tradition, hierarchy, and a strong sense of control on the day-to-day details, which are critical values for teams that avoid dealing with changing conditions. They venture forward very cautiously. It is okay if the team and the organization exist in an environment that wants to stay just as it is for the years to follow. The only "change" that they must consider is how to stay afloat when the environment and economy change around them, because as hard as we may try, the world is NOT going to stand still. People, society, and environment WILL CHANGE!

STEVE WESTERMANN

Fire Chief, Central Jackson County Fire Protection District (Blue Springs, Missouri), and IAFC President 2007–2008: Stay abreast of the times, whether economic times, professional trends, world and local news, or social changes that are occurring (for example, Generations X and Y). You have to stay up with the changes so you can understand what is happening and then shape your reactions or form your plans within your organization.

THOMAS WIECZOREK

Executive Director, Center for Public Safety Excellence, and Retired City Manager: Many things have been thrown on the fire service without a lot of foresight. Take EMS, for example. The underlying reasoning for many organizations taking on EMS was "that there wasn't enough to do in the fire service." Yes, it is true that fires are down across the country. But how is it that we communicate? Most chiefs produce a year-end report that gives data and numbers. What do those numbers usually report? Numbers killed, numbers injured, loss of property value. Now let's look at it from the police side. They report the numbers of crimes, but they show how many homicides were "solved" through arrest. They report how many larcenies and burglaries occurred but then report how many were "solved" or "closed" by prosecution. In other words, they take the same losses that the fire service does but show them in a positive light. As a result, they receive funds to add officers to play basketball, walk foot patrols (that have little to do with arrests but more to do with feelings of security), and will never have budget cuts because all those "successes" would evaporate. The fire service must show that aggressive prevention and mitigation, while lowering the risks associated with incidents, do not justify elimination of people but require paradigm shifts. In the area of asset management, a process used heavily in Europe, Australia, and now the U.S. Public Works (highways) is that for every $1 spent on prevention, $4 are saved in reconstruction. The argument made is that you cannot cut the overall budgets. Instead, you are spending the existing allocations wiser, smarter, and with better results. One area that I believe the fire service is best suited to lead in is the area of GIS. The fire station, located in every neighborhood across the country, should become the "information headquarters" for the city. What other group is still viewed as the hero in government? A recent article in *American City and County* magazine indicated that the public's disdain and distrust of all levels of government are at one of the highest points ever. Hurricane Katrina, ethic busts of federal legislators, and other events have tarnished the public perceptions of government. By and large, the fire service has not been affected by this and could be uniquely positioned to begin to develop the information technology centers that support other facets of government. By having the data, the fire service can then begin driving the decision-making processes. Do we need a staffed and equipped fire station of the same size every three miles in all cities? Or do we need to look at staffing some stations heavier with EMS? Do we need five, six, or eight staffed engine companies to conduct the initial deployment necessary to control situations in other parts of town and not just the standard three or four? Two of the reasons for the four-person engine companies are two-in, two-out and the need of two engines for water delivery. But with pumps available in larger sizes and actually needing five people if two-in, two-out will truly be utilized (one as pump operator, which is not yet automatic), should larger deployments be planned that would increase the chances of traffic accidents and other off-scene incidents that result in the majority of injuries and death? And could those six or eight be deployed on prevention in teams but then respond? Similarly, for dealing with EMS, do we need to look at "rapid response teams" in order to reach the patient's side in high-rise areas or congested zones? A ladder truck is difficult to maneuver in good conditions; putting it in bumper-to-bumper gridlock is asking for problems. Yet how many still send a six-person aerial truck because "otherwise they wouldn't do anything"?

JANET WILMOTH

Editorial Director, *Fire Chief* magazine: Critical. There must be substance to support leadership today. Firefighters and EMS personnel are more intelligent and demand to know reasons for decisions and directions. The big picture is getting bigger, and it's important for emergency service responders to understand that the scope of their responsibilities is not only broad but also overlaps with other agencies.

FRED WINDISCH

Fire Chief, Ponderosa Volunteer Fire Department (Houston, Texas): Be flexible, change for the better not just for change, and utilize your people for idea generation and implementation. A chief officer must be able to use active listening skills and filter out the chaff. With that in mind, the chief officer must check his or her ego at the door and always focus on the mission of providing high-quality services to the citizens.

Review Questions

1. What type of change has occurred in the last five years within the jurisdiction you serve and in your department?
2. Captain Abrashoff's book made parallels between what he faced when taking over the *Benfold* and what we face in taking over a new organization, a new company, or a new assignment. List the points he emphasized in his book.
3. Describe the boiled frog syndrome and how it can have a detrimental effect on an organization.
4. Discuss the "big hat, no cattle" philosophy and how it impacts the fire service.
5. What does stewardship mean to you?
6. Why are core values important for an organization?
7. Discuss the various group dynamics. Use forming, storming, norming, performing, and mourning to describe the various group dynamics.
8. Describe some of the highlights of the book *Good to Great* discussed in this chapter, and relate them to how they can influence you as a fire service leader.
9. Describe above-the-line accountability. How can you promote this with those you work with?
10. Compare and contrast the books *Making the Corps* and *A Book of Five Rings* as they relate to core values.
11. How does your organization react to change? Why do you think this is the case?
12. How does organizational culture impact the change process?
13. Have you witnessed or experienced behavior in your organization similar to that in the story of the five apes?
14. When undertaking an organizational change, what critical elements can have a positive impact on the process?
15. Describe the importance of vision and how a leader's vision sets the direction for the organization.
16. What three things would you change in your department and how would you implement/process each?

References

1. Joel Barker, *Discovering the Future: The Business of Paradigms* (Lake Elmo, MN: ILI Press, 1988).
2. Michael D. Abrashoff, *It's Your Ship* (New York: Warner Books, 2002).
3. Ibid.
4. U.S. Fire Administration, *Four Years Later—A Second Needs Assessment of the U.S. Fire Service,* a Cooperative Study Authorized by U.S. Public Law 108-767, Title XXXVI, FA-303/October 2006, Homeland Security.
5. www.afterburnerseminars.com/flawless_execution.asp?id=us
6. Sun Tzu, *The Art of War.* (Philadelphia, PA: Running Press Books, 2003).
7. Tom Peters and Robert H. Waterman, *In Search of Excellence* (New York: Warner Books, 1982).
8. Jim Collins, *Good to Great* (New York: HaperCollins, 2001), pp 12–13.
9. Ibid.
10. Warren Bennis, *Why Leaders Can't Lead: The Unconscious Conspiracy Continues* (Los Angeles: University of Southern California Graduate School of Business, 1990); *On Becoming a Leader* (New York: Perseus Books, 2003).
11. Keshavan Nair, *Beyond Winning: The Hand Book for the Leadership Revolution* (Hamilton, Ontario, Canada: Paradox Press, 1990).
12. Patricia Aburdene, *Megatrends 2010: The Rise of Capitalism* (VA: Hampton Roads, 2005).
13. *The United States Navy*, www.chinfo.navy.mil/navpalib/traditions/html/corvalu.html
14. *Marine Corps Core Values*, www.geocities.com/devildoggmarine03/corevalues.html?200516
15. *The United States Air Force Core Values,* www.usfa.af.mil/corevalue/essays.html
16. *The United States Army*, www.geocities.com/army_vent/armyvalues
17. Thomas E. Ricks, *Making the Corps* (New York: Touchstone Books, 1998).
18. Ibid.
19. Miyamoto Musashi, *A Book of Five Rings: The Classic Guide to Strategy* (Woodstock, NY: Overlook Press, 1974).
20. Faith Popcorn, *17 Trends That Drive Your Business—Your Life* (New York: HarperCollins, 1987).
21. Alvin Toffler, *The Adaptive Corporation* (New York: McGraw-Hill, 1984).
22. Colin Powell, *My American Dream* (New York: Random House, 1997).
23. Doug Smith, *Taking Charge of Change: 10 Principles for Managing People and Performance* (New York: Perseus Books, 1997).
24. William Bridges, *Managing Transitions, Making the Most of Change* (New York: Perseus Books, 1993).
25. David Osborne and Peter Hutchinson, *The Price of Government* (Cambridge, MA: Basic Books, 2004), p. 115.
26. Ibid., pp. 116–115.
27. Ibid., p. 326.
28. *Strategic Plan—Securing Our Homeland,* www.dhs.gov/xabout/strategicplan/index.shtm

Leadership Ethics

Key Term

ethics (ethos), p. 236

Objectives

After completing this chapter, you should be able to:

- Define the term *ethics* and describe how it is used in your department.
- Evaluate the ethical environment in your department.
- Describe how politics influences local governance decisions.
- Describe past events that have forced ethical legislation to be written.
- Articulate the role of ethics in the decision-making process.
- Describe how organizational culture can impact organizational ethics.

VALUES-BASED LEADERSHIP: THE ONLY APPROACH TO SUCCESS

In recent years, the business world has experienced a disturbing decline in respect to the positive values displayed by some of its leaders, which in turn has led to significant difficulties and even collapse of several major companies (e.g., Adelphia, Enron, HealthSouth, Tyco, WorldCom). These business failures were primarily caused by a lack of self-discipline by company leaders holding prominent positions of authority and the absence of suitable guidelines and government regulations. Similar indiscretions have also been witnessed in the public sector, such as the falsifying of personal résumés to obtain a position, the hiring of unqualified friends in prominent positions, and the inappropriate use of public funds for personal pleasure or gain. Although not often feature stories on CNN, these incidents make regional headlines that not only tarnish the reputation of the local officials involved but also raise suspicion of all who work in the public sector.

ETHICS AND THEIR IMPACT

ethics (ethos)
Analysis of the principles of human conduct in order to determine between right and wrong; philosophical principle used to determine correct and proper behavior by members of a society; sometimes called moral philosophy.

One of the greatest desires of most fire officers is to provide service in an honorable way and to follow a long history of leaders who have impacted the lives of a great number of people. What is the meaning of ethics? Dictionaries define **ethics (ethos)** as a given system of conduct; principles of honor or morality; guidelines for human actions; rules or standards for individuals or professions; and the character of a group based on its agreements about what is proper or expected behavior. The word *ethics* has its roots in ancient history, religion, law, social customs, and our own codc of conduct.

Ancient History

- The Greek culture adopted rules for its citizens, such as "Do the greatest good for the greatest number."

Religion

1. All of the great religions of the world have the equivalent of the Ten Commandments to guide the thinking and actions of their members.
2. These guidelines specifically state how people should act, especially in relationships with other human beings.
3. The golden rule, "Do unto others as you would have them do unto you," is universal, as are universal values of honesty, courage, justice, tolerance, and full use of talents.

Systems of Law

1. Laws summarize decisions of technically trained judges as well as juries of peers who decide whether a wrong has been committed.
2. New laws also arise as the values and attitudes of the community change.

Social Customs

1. Define what is acceptable and unacceptable in a certain community.
2. Members of any community either condone (support) or condemn (resist) specific actions.
3. There is a vast difference between which standards are upheld and which are relaxed.

Companies typically begin to struggle when corporate values are absent or differ from the values held by employees. The result is some employees will begin to bend the rules to fit their framework of what is ethical, while others who observe the organization's dynamic lose enthusiasm for their work and disengage from daily business and work-related social activities. Unfortunately, all too often, employees feel they can express their values at home but not at work. How many times have you seen this in the firehouse? A crew who leaves the station a mess, yet their own homes are immaculate; the company officer who expects quality personal service when off duty, yet has one of the most condescending, gruff personalities when dealing with the public on calls or inspections.

Values are those intrinsically desirable qualities that influence how people in organizations interact with each other and how leaders operate their organizations. All culminate in the reputation a department has in its local community as well as the perception held by others in the service from a regional and national perspective. Too often the discussion of values at work is the result of a person's or organization's ethical misadventure, which prompts the organization to conduct classes and training on the subject. This is important because each person has his or her own set of values, which are responsible for individual behavior. The ethical boundaries set by the organization must be clearly communicated repeatedly if they are to become the norm.[1] A good example of the opposite type of behavior is the true story of the teacher/coach in the Chicago public school system who not only encouraged his high school students to cheat on the city-wide academic decathlon contest but also gave them the answers. According to the team's 18-year-old student captain, "The coach gave us the answer key. . . . He told us everybody cheats, that's the way the world works, and we were fools to just play by the rules." Unfortunately, just as workers often mirror the standards set by their bosses, these students followed the guidance of their teacher.[2] What message did they take from this experience?

In his *Nicomachean Ethics*, Aristotle suggested morality cannot be learned simply by reading a treatise on virtue. The spirit of morality, said Aristotle, is awakened in the individual only through the witness and conduct of a moral person. The principle of the "witness of another," or what is now referred to as "patterning," "role modeling," or "mentoring," is predicated on the following: (1) as communal creatures, we learn to conduct ourselves primarily through the actions of significant others; (2) when the behavior of others is repeated often enough and proves to be peer-group positive, we emulate these actions; (3) if and when our actions are in turn reinforced by others, they become acquired characteristics or behavioral habits.

This concept is alive and well in the firehouse. Fire chiefs hire hundreds of new firefighters with positive attitudes and high expectations. Within a year or two after the characteristics of negative patterning, role modeling, and conditioning in the firehouse (depending upon their assigned crew), these same employees turn into disgruntled, complaining naysayers. Unfortunately, once many of these employees are inculcated into this mind-set, it is extremely difficult to return them to being positive and productive employees.

Values-based leadership communicates the ethics of institutions and establishes the desired standards and expectations leaders want and demand from their fellow workers and followers. Although it would be naïve to assert employees simply and unreflectively absorb the manners and mores of the workplace, it would be equally naïve to suggest they are unaffected by the modeling and standards of their respective places of employment. Work is where a great portion of our lives are spent, and the lessons learned there, good or bad, play a part in the development of moral perspectives and the manner in which ethical choices are formulated and adjudicated.[3] As a chief officer, battalion chief, or captain, if your firefighters see you doing one thing and then you expect a different behavior from them, it probably will not happen.

The late John Rawls, the James Bryant Conant University Professor Emeritus, wrote that, given the presence of others and their need to survive and to thrive, ethics is elementally the pursuit of justice, fair play, and equity. For Rawls, building on the cliché "ethics is how we decide to behave when we decide we belong together," the study of ethics has to do with developing standards for judging the conduct of one party whose behavior affects another. Minimally, "good behavior" intends no harm and respects the rights of all affected, and "bad behavior" willfully or negligently tramples on the rights and interests of others. Ethics, then, tries to find a way to protect one person's individual rights and needs against and alongside the rights and needs of others.[4] Of course, the paradox and central tension of ethics lie in the fact that while we are by nature communal and in need of others, at the same time we are by disposition more or less egocentric and self-serving.

If ethics are part of life, so too are work, labor, and business. Work is not something detached from the rest of human life; rather, "man is born to labor, as a bird to fly."[5] The fire service is more than a job; it is a way of life. The power and influence of the traditional cultures of the fire service are entwined in every aspect of the industry. The art of values-based leadership is to blend the positive aspects of traditions with the ethics required in today's society. For most firefighters this way of life is 24/7, 365 days a year. Whether on or off duty, the fire service is engrained into much of what firefighters do as people within their family structures and community relationships.

Henry Ford, Sr., once said, "For a long time people believed the only purpose of industry is to make a profit. They are wrong. Its purpose is to serve the general welfare."[6] What business ethics advocates is that people apply in the workplace the commonsense rules and standards learned at home, from the lectern, and from the pulpit. The moral issues facing each of us today are not new but essentially the same ones that we face publicly, only with a much larger script.[7] According to R. Edward Freeman, who heads the Darden Olsson Center for Applied Ethics in Virginia, ethics is "how we treat each other, every day, person to person. If you want to know a company's ethics, look at how it treats people—customers, suppliers, employees. Business is about people, and business ethics is about how customers and employees are treated."[8] The same is true in government. The day has long past when the conduct of day-to-day business was a private matter (if it ever was). In today's society, every act of business in both the public and private sectors has social consequences and may arouse public interest. Every time we hire, build, sell, or buy, we are acting on behalf of the people we serve. For those in the public sector, actions and relationships are under constant scrutiny, often being evaluated and judged in the local media.

Ethics is about the assessment and evaluation of values, because all of life is value laden. As Samuel Blumenfeld emphatically pointed out, "You have to be dead to be value-neutral." Look around the firehouse and the people you work with in your department; probably none are value neutral. Value-neutral people are neither the type of people who tend to pursue the fire service nor the type we tend to hire. Values are the ideas and beliefs that influence and direct our choices and actions. Whether they are right or wrong, good or bad, values, both consciously and unconsciously, guide how decisions are made. Eleanor Roosevelt once said, "If you want to know what people value, check their checkbooks!"[9] It would seem in today's society, with so many in the corporate world facing legal retribution, that statement is still valid.

Tom Peters and Bob Waterman were correct when they asserted, "The real role of leadership is to manage the values of an organization."[10] All leadership is value laden and is ideologically driven or motivated by a certain philosophical perspective, which upon analysis and judgment may or may not prove to be morally acceptable in the colloquial sense. All leaders have an agenda—a series of beliefs, proposals, values, ideas, and issues that they wish to "put on the table." It is a product of the personal framework explored in Chapter 4. In fact, leadership only asserts itself, and followers only become evident, when something is at stake—ideas to be clarified, issues to be determined, values to be adjudicated. Franklin D. Roosevelt, in a discussion on presidential leadership, stated, "The Presidency is . . . preeminently a place of moral leadership. All our great Presidents were leaders of thought at times when certain historic ideas in the life of the nation had to be clarified. . . ."[11] Although we would prefer to study the moral leadership of Abraham Lincoln, Winston Churchill, Mahatma Gandhi, and Mother Teresa, like it or not, we must also evaluate Adolph Hitler, Joseph Stalin, Sadam Hussein, and David Koresh within a moral context.

All ethical judgments are in some sense a "values versus values" or "rights versus rights" confrontation. Unfortunately, the question of "what we ought to do" in relation to the values and rights of others cannot be reduced to the analog of a simple litmus test. In fact, most ethics are based on what William James called the "will to believe." That is, we choose to believe, despite the ideas, arguments, and reasoning to the contrary, that individuals possess certain basic rights that cannot and should not be willfully disregarded or overridden by others. In "choosing to believe," said James, "we establish this belief as a factual baseline of our thought process for all considerations in regard to others." Without this "reasoned choice," he says, the ethical enterprise loses its "vitality" in human interactions.[12]

This is evident in the fire service. No matter how much factual information is provided or how many discussions held, if people choose to believe something else, its factual basis has very little relevance. We see this in the field on many occasions in what we call the rumor mill. Someone in the firehouse starts a rumor or takes a piece of information and spins it, in some cases just to cause organizational dissent. The interesting aspect of this phenomenon is how fast others believe it, take the story, and perpetuate it by telling it to others. The story often evolves and continues as more spin is added, and many never question whether the information was correct in the first place. Although some in the fire service profession think of this as almost a sport, the lack of ethical and professional behaviors in such scenarios is quite clear.

If ethical behavior intends no harm and respects the rights of all affected, and unethical behavior willfully or negligently tramples on the rights and interests of others, a leader, whatever position held in the organization, cannot deny or disregard the rights of others. The leader's worldview cannot be totally myopic nor his or her agenda purely self-serving. Leaders should not see followers as potential adversaries to be tested but rather as fellow travelers with similar aspirations and rights to be reckoned with. Even when they choose not to believe!

How should the ethics of a leader be judged? Clearly, perfection in every decision and action of a leader cannot be expected. As John Gardner has pointed out, particular consequences are never a reliable assessment of leadership.[13] Anyone in the fire service could write a book on the number of mistakes made based on lack of information, lack of experience, or too much ego. The quality and worth of leadership can be measured only in terms of what a leader intends, values, believes in, or stands for—in other words, character. In *Character: America's Search for Leadership*, Gail Sheehy argues, as did Aristotle before her, that character is the most crucial and most elusive element of leadership. The root of the word *character* comes from the Greek word for "engraving." As applied to human beings, it refers to the enduring marks, or etched-in factors, in our personality, which include our inborn talents as well as the learned and acquired traits imposed upon us by life and experience. These engravings define us, set us apart, and motivate our behaviors. On character, Abraham Lincoln one stated, "Nearly all men can stand adversity, but if you want to test a man's 'CHARACTER,' give him power."

Watergate of the early 1970s serves as a perfect example of the links between character and leadership. As Richard Nixon demonstrated so well, says Sheehy, "The Presidency is not the place to work out one's personal pathology. . . ."[14] Leaders rule us, run things, wield power. Therefore, says Sheehy, we must be careful because whom we choose to lead is what we shall be. If, as Heraclitus wrote, "character is fate," the fate our leaders reap will also be our own. Watergate has come to symbolize the failures of people in high places. Watergate now serves as a watershed, a turning point, in our nation's concern for integrity, honesty, and fair play from all leaders. It is not a mere coincidence that the birth of business ethics as an independent, academic discipline can be dated from Watergate and the legal trials resulting from it. No matter what our failures as individuals, Watergate sensitized us to the importance of ethical standards and conduct from those who direct the course of our political and public lives. What society is now demanding and business ethics is advocating is that our public servants and business leaders should be held accountable to an even higher standard of behavior than we might demand and expect of ourselves.

There is, unfortunately, a dark side to the theory of leadership ethics. Howard S. Schwartz, in his managerial article, "Narcissistic Process and Corporate Decay,"[15] argues that corporations are not bastions of benign, other-directed ethical reasoning; nor can corporations, because of the demands and requirements of business, be models and exemplars of moral behavior. The rule of business, says Schwartz, remains the "law of the jungle," "the survival of the fittest"; and the goal of survival engenders a combative "us-against-them mentality," which condones the moral imperative of getting ahead by any means necessary. Schwartz calls this phenomenon "organizational totalitarianism." Organizations and the people who manage them create for themselves a self-contained, self-serving worldview, which rationalizes anything done on their behalf and does not require justification on any grounds outside themselves.

The psychodynamics of this narcissistic perspective, says Schwartz, impose Draconian requirements on all participants in organizational life—do your work; achieve organizational goals; exhibit loyalty to your superiors; disregard personal values and beliefs; obey the law when necessary, and obfuscate it whenever possible; and deny internal or external discrepant information at odds with the stated organizational worldview. Within such a "totalitarian logic," neither leaders nor followers, rank nor file, operate as independent agents. To "maintain their place," to "get ahead," all must conform. The agenda of organizational totalitarianism is always the preservation of the status quo. Within such a logic, change is rarely possible. Except for extreme situations in which "systemic ineffectiveness" begins to breed "organization decay," transformation is never an option.

In those types of organizations, what is right in the corporation is not what is right in a person's home, life, or beliefs. What is right in the corporation is what the person above you wants from you.[16] This culture is reminiscent of the old command and control theory of running an organization. For the fire service, a paramilitary structure is a must to accomplish its mission, yet the balance of power versus inclusion built upon sound ethical standards and conduct is what sets organizations apart. Individuals who embrace the totalitarian approach always seem to be unbalanced with extreme behaviors. One minute they are right with you and the world around them; next they are pounding the table, combative, us against them, exhibiting a winning at all cost mentality. For these individuals there is often no ethical compass!

IDENTIFYING ETHICS CODES

Ethics codes and guidelines protect professionals from themselves, as well as from those who they perceive abuse the power of their profession.[17] Nonetheless, the inherent power of a code of ethics rises no higher than the collective moral character of those who subscribe to the code. Theoretically, a code of ethics sets guidelines for ideal behavior; in reality it represents minimum standards of behavior that

TABLE 6.1 ◆ Business Ethics Timeline

Decade	*Ethical Climate*	*Major Ethical Dilemmas*	*Business Ethics Developments*
1960s	Social unrest. Antiwar sentiment. Employees have an adversarial relationship with management. Values shift away from loyalty to an employer to loyalty to ideals. Old values are cast aside.	◆ Environmental issues ◆ Increased employee–employer tension ◆ Civil rights issues dominate ◆ Honesty ◆ The work ethic changes ◆ Drug use escalates	◆ Companies begin establishing codes of conduct and values statements ◆ Birth of social responsibility movement ◆ Corporations address ethics issues through legal or personnel departments
1970s	Defense contractors and other major industries riddled by scandal. The economy suffers through recession. Unemployment escalates. There are heightened environmental concerns. The public pushes to make businesses accountable for ethical shortcomings.	◆ Employee militancy (employee versus management mentality) ◆ Human rights issues surface (forced labor, substandard wages, unsafe practices) ◆ Some firms choose to cover rather than correct dilemmas	◆ ERC founded (1977) ◆ Compliance with laws highlighted ◆ Federal Corrupt Practices Act passed in 1977 ◆ Values movement begins to move ethics from compliance orientation to being "values centered"
1980s	The social contract between employers and employees is redefined. Defense contractors are required to conform to stringent rules. Corporations downsize and employees' attitudes about loyalty to the employer are eroded. Health care ethics emphasized.	◆ Bribes and illegal contracting practices ◆ Influence peddling ◆ Deceptive advertising ◆ Financial fraud (savings and loan scandal) ◆ Transparency issues arise	◆ ERC develops the U.S. Code of Ethics for Government Service (1980) ◆ ERC forms first business ethics office at General Dynamics (1985) ◆ Defense Industry Initiative established (1986) ◆ Some companies create ombudsman positions in addition to ethics officer roles ◆ False Claims Act (government contracting)
1990s	Global expansion brings new ethical challenges. There are major concerns about child labor, facilitation payments (bribes), and environmental issues. The emergence of the Internet challenges cultural borders. What was forbidden becomes common.	◆ Unsafe work practices in third world countries ◆ Increased corporate liability for personal damage (cigarette companies, Dow Chemical, etc.) ◆ Financial mismanagement and fraud	◆ Federal Sentencing Guidelines (1991) ◆ Class action lawsuits ◆ Global Sullivan Principles (1999) ◆ *In re Caremark* (Delaware Chancery Court ruling re Board responsibility for ethics) ◆ IGs requiring voluntary disclosure ◆ ERC establishes international business ethics centers ◆ Royal Dutch Shell International begins issuing annual reports on their ethical performance
2000s	Unprecedented economic growth is followed by financial failures. Ethics issues destroy some high profile firms. Personal data are collected and sold openly. Hackers and data thieves plague businesses and government agencies. Acts of terror and aggression occur internationally.	◆ Cyber crime ◆ Privacy issues (data mining) ◆ Financial mismanagement ◆ International corruption ◆ Loss of privacy—employees versus employers ◆ Intellectual property theft	◆ Business regulations mandate stronger ethical safeguards (Federal Sentencing Guidelines for Organizations; Sarbanes-Oxley Act of 2002) ◆ Anticorruption efforts grow ◆ Shift to emphasis on Corporate Social Responsibility and Integrity Management ◆ Formation of international ethics centers to serve the needs of global businesses ◆ OECD Convention on Bribery (1997–2000)

Note: The ethical climate and response to ethical dilemmas in the 1960s and 1970s blurs and loses some distinction across the decade boundaries due to the war in Vietnam, social upheaval, and resulting stress on businesses.

Source: © Ethics Resource Center, 1747 Pennsylvania Ave, Suite 400, Washington, DC 20006

often become the goals rather than a "trip wire" to signal unacceptable behavior. Typically, after achieving minimum standards, motivation to achieve higher moral and ethical standards becomes less ardent. Ethics codes encompass a wide range of issues but cannot include every possible scenario. Necessarily vague guidelines provide flexibility for individual interpretations and for unique circumstances. Nonspecific issues confound the ethical decision-making process, because individuals must rely on objective standards as well as subjective values when seeking solutions.

Much of the ethical law and norms we live with today have been driven by the societal climate and the events of the day that often promote new laws, standards, and expectations (see Table 6.1). We have witnessed substantial development of business and governmental ethics reform since the 1960s.

2003 NATIONAL BUSINESS ETHICS SURVEY

A 2003 National Business Ethics Survey (NBES) asked employees in the 48 contiguous states to share their views on ethics and compliance within their organizations.[18] The broad concept of business ethics may defy easy summary, but the survey questions in this third NBES report are specific and focused. They combine to yield answers on matters relating to how employees distinguish right from wrong behavior in their work, the availability of resources to aid in making appropriate decisions, and the general practice of values such as honesty and respect in the workplace. More specifically, the survey of 1,500 participants focused on a number of key areas including:

- Ethics practices of executives, supervisors, and coworkers
- Prevalence of formal ethics programs
- Pressures to compromise ethics standards
- Misconduct at work and the influences on reporting it
- Frequency with which certain ethical values are practiced
- Accountability for ethics violations

Context is critical in any survey. To make better sense of survey findings and trends in business ethics, look first to the larger business environment. Between the 2000 and 2003 surveys, we witnessed the end of the dot-com boom, a sizable downturn in the economy, and, beginning with Enron, a series of well-publicized scandals among major corporations and nonprofit organizations. New and emerging legislation, SEC regulations, and New York Stock Exchange rules have placed additional emphasis on executive conduct and organizational standards and practices. Of course, these related events form a backdrop for the findings of the surveys. Given these developments, one might expect employees in 2003 to be more aware of business ethics than in previous survey years. One might also expect business leaders to focus more attention on issues and concerns relating to ethics. But to what effect? Are negative events in business simply mirrored in how employees view the conduct of their leaders and assess their organizations? Or can corporate scandals in particular have the opposite effect, providing both the impetus for change and making employees' own organizations look better by comparison?

The answers from the survey are clear and compelling. Recent negative events have not resulted in employees' viewing business ethics in their own organizations more negatively. Though major vulnerabilities and challenges remain, findings from the 2003 NBES are surprisingly hopeful. Perceptions of ethics are generally positive.

Employees view ethics in their organizations more positively in 2003 as compared with 2000. The NBES survey data indicate the following:

- Employee perception that top management talked about the importance of ethics, kept promises, and modeled ethical behavior have all increased since 2000. For example, 82 percent of employees in 2003 said that top management in their organizations kept promises and commitments as compared with 77 percent in 2000. In addition, the increases between 2000 and 2003 tend to be more substantial in larger (over 500 employees) versus smaller organizations.
- Two key indicators of ethics-related problems in the workplace—(1) observed misconduct and (2) pressures to compromise ethics standards—have declined since the 2000 survey. Observed misconduct dropped from 31 percent in 2000 to 22 percent in 2003, while pressure to compromise fell from 13 percent to 10 percent during this time period.
- Declines in observed misconduct and pressure to compromise occurred primarily among nonmanagement employees.
- Reporting of misconduct by employees increased steadily in the surveys conducted in 1994 (48 percent), 2000 (57 percent), and 2003 (65 percent).
- Employees indicated values such as honesty and respect were practiced more frequently in their organizations in 2003.

Vulnerabilities and challenges remained. Some employee groups and organizations may have been more at risk for ethics-related problems. Potential risks and challenges were indicated by these findings:

- Nearly a third of respondents said their coworkers condone questionable ethics practices by showing respect for those who achieved success using them.
- The types of misconduct most frequently observed in 2003 included abusive or intimidating behavior (21 percent); misreporting of hours worked (20 percent); lying (19 percent); and withholding needed information (18 percent).
- Employees in transitioning organizations (undergoing mergers, acquisitions, or restructurings) observed misconduct and felt pressure at rates that were nearly double those in more stable organizations.
- Compared with other employees, younger managers (under age 30) with low tenure in their organizations (less than three years) were twice as likely to feel pressured to compromise ethics standards (21 percent versus 10 percent).
- Despite an overall increase in reporting misconduct, nearly half of all nonmanagement employees (44 percent) still did not report the misconduct they observed. The top two reasons given were (1) a belief that no corrective action would be taken and (2) fear that the report would not be kept confidential.
- Younger employees with low tenure were among the least likely to report misconduct (43 percent as compared with 69 percent for all other employees). They were also among the most likely to feel that management and coworkers would view them negatively if they reported.
- Fewer than three in five employees (58 percent) who reported misconduct were satisfied with the response of their organizations.
- In many areas views of ethics remained "rosier at the top." For example, senior and middle managers had less fear of reporting misconduct and were more satisfied with the response of their organizations. They also felt that honesty and respect were practiced more frequently than did lower-level employees.

Ethics programs made a difference. The NBES asked respondents about four elements of formal ethics programs: (1) written standards of conduct, (2) ethics training,

(3) ethics advice lines/offices, and (4) systems for anonymous reporting of misconduct. It was found that selected outcomes were related to the presence of these four program elements in organizations. Specifically, survey findings indicated the following:

- The presence of ethics program elements was associated with increased reporting of misconduct by employees. Specifically, employees were most likely to report in organizations with all four program elements in place (78 percent). Employee reporting declined steadily in organizations with fewer program elements such as written standards plus (67 percent), written standards only (52 percent), or none (39 percent).
- Ethics programs were associated with higher perceptions that employees were held accountable for ethics violations.
- In larger organizations (over 500 employees), ethics programs were associated with lower pressures on employees to compromise company standards of business conduct.

Actions count. Misconduct, pressure, reporting, and other outcomes were strongly related to the actions of top management, supervisors, and coworkers. Survey findings indicated the following:

- Employees who felt that top management acted ethically in four areas (talked about the importance of ethics, informed employees, kept promises, and modeled ethical behavior) observed far less misconduct (15 percent) than those who believed top management only talked about ethics or exhibited none of these actions (56 percent). Findings for pressure to compromise, accountability, and satisfaction were similar.
- When employees felt their supervisors and coworkers acted ethically, the relationships to observe misconduct, pressure, and related outcomes were similar to that for top management.

Size matters. There are important relationships between an organization's size and how its employees view ethics. Survey findings indicated the following:

- Smaller organizations (fewer than 500 employees) were less likely to have key elements of ethics programs in place than larger ones. For example, 41 percent of employees in smaller organizations said ethics training is provided as compared with 67 percent in larger organizations. Similarly, 77 percent of employees in larger organizations said that mechanisms to report misconduct anonymously were available versus 47 percent in smaller organizations.
- Employee perceptions of ethics in smaller and larger organizations converged in 2003. This finding contrasted with the 2000 survey findings, which showed employees in smaller organizations generally held more positive views of ethics. The convergence was due primarily to more positive ethics trends in larger organizations between 2000 and 2003.

Executives are increasingly called upon to "certify" the integrity of their organizations. Thus, ethical leaders should have a greater concern that the actions of one or more employees could eventually compromise their companies. Certainly, there was evidence in the 2003 NBES that concerns were warranted: views of ethics were still "rosier at the top," while pressures to compromise standards and fears of reporting misconduct were greater among particular groups of employees. Important challenges remain in a variety of related areas for leaders and their organizations and employees. But as a snapshot of business ethics in 2003, the NBES findings and trends were also positive and hopeful. Where leaders, supervisors, and coworkers were talking about ethics and setting the right example, employees had taken notice. Where systems were in place to help make ethics a priority, employees were responding. In both cases, organizations appeared to benefit. Where some corporations had recently failed due to ethical violations, their negative examples may have helped others to make better choices. As difficult as it may be, this is often how people learn and organizations improve.

> Ideals are like stars—we never reach them.
> But, like mariners at sea, we chart our course by them.
>
> *Author Unknown*

MODELING ETHICAL BEHAVIOR

Modeling is a powerful leadership strategy whereby leaders show their employees, through their own behavior, how they want them to behave with others. It has been used successfully to demonstrate flexibility, politeness, decisiveness, compassion, sharing, and numerous other desirable traits. It also works extremely well as an ethics teaching tool. In fact, one of the principal findings of the Ethics Resource Center's National Business Ethics Survey was that the modeling of ethical behavior by organizational leaders, managers, supervisors, and coworkers sets a good example for desired business behavior. When employees perceive formal and informal leaders are ethical, they feel less pressure to compromise ethical standards, observe less misconduct on the jobs, are more satisfied with their organization overall, and feel more valued as employees.

In other words, organizational ethics becomes real for employees when they see good ethics being applied. The problem with modeling is that all too often the audience present when ethical dilemmas arise is small, and the modeled behavior is missed by the majority of employees who are engaged with activities elsewhere. So, how can a leader multiply the effectiveness of modeling? Set the stage in advance, and then make sure you have a mechanism in place to discuss and share the story with others in the organization—the observed behavior that models good ethical conduct.

MAKE IT ACCEPTABLE TO TALK ABOUT ETHICS

As a topic, ethics, in many aspects of society, has taken a negative personification. The multitude of scandals and ethical shortcomings reported in the media paint a depressing picture of individuals who skirt their responsibilities or ignore conduct guidelines to the detriment of their organizations. Sadly, countless examples of modeled behavior of good ethical conduct are never reported, as good news does not sell newspapers or help ratings for local television or radio station(s).

- Take time during a staff meeting or other group event to commend someone on demonstrated ethical conduct.
- Talk about ethical conduct routinely and encourage your employees to seek guidance whenever they question whether or not an act is ethical.
- Keep the lines of communication open. When someone wants to talk about an ethical issue, the time to talk about it is then . . . not at some future date. Postponing the discussion diminishes the importance placed on ethics.

CREATE A HABIT OF REPEATING ORGANIZATIONAL SUCCESS

In every organization there are stories of individuals who have gone the extra mile to accomplish a seemingly impossible task. Stories also exist of how someone chose to do something right even though it may have been painful to do so.

- Capture organizational stories of proper conduct and use them to illustrate desired behavior.
- Write down the stories so they can be passed on to future employees.
- Use organizational newsletters and other publications to spread the word about ethical successes.

INCLUDE ETHICAL CONDUCT AS A MEASURE ON PERFORMANCE EVALUATIONS

Most people react more favorably to what is rewarded or measured. If the emphasis is on proactive, ethical behavior, more of it will be seen. If it is a measurable objective, employees will think more about it.

- Make your expectations for ethical conduct well known.
- Capture live examples of ethical conduct to use when delivering a performance evaluation.
- Provide some award or incentive for ethical performance. (One organization awarded "Caught you . . . doing something good" certificates, which were delivered along with a gourmet cookie. Another presented small department commitment plaques for ethical conduct.)

MAKE ETHICS MORE THAN AN ANNUAL DISCUSSION

Talk about ethics often and at all levels. Employees must be asked what they like and dislike about their jobs. Inquire whether there is anything they have to do at work that makes them uncomfortable or that seems to be a compromise of their own personal ethics. Listen to what they have to say. When an organization reaches the point of discussing ethics routinely and its team knows where the organization stands on the need for good conduct and behavior, it is what many refer to as the "straight face test." It is the ethical baseline of performance (the test). Would this action pass muster if it were the lead story in the paper? When personnel throughout the organization begin to frame their decisions and discussions in this manner, the institutionalization of ethics in the department begins.

THE ROLE OF LEADERSHIP IN ORGANIZATIONAL INTEGRITY AND ETHICAL LEADERSHIP

Components of Ethical Leadership Ethical leadership begins when leaders perceive and conceptualize the world around them. Ethical leadership, organizational ethics, and social responsibility are inseparable concepts. How ethical leaders relate to and come to understand the world around them involves judgment and action. The leader's role is to guide the human potential of the organization's stakeholders and to achieve organizational aspirations in ways that liberate rather than constrain their imaginations and judgment. In respect to ethics, everyone must take a leadership role. It is not enough to be ethical in one's individual actions to be an ethical leader. To be effective, four components of ethical leadership must be understood and developed: purpose, knowledge, authority, and trust. The relationship among these four components can be visualized as interrelated components. Attention to any one component alone is incomplete and misleading.

- *Purpose.* The ethical leader reasons and acts with organizational purposes firmly in mind. This provides focus and consistency.
- *Knowledge.* The ethical leader has the knowledge to judge and act prudently. This knowledge is found throughout the organization and its environment but must be shared by those who hold it.
- *Authority.* The ethical leader has the power to make decisions and act, but also recognizes all those involved and affected must have the authority to contribute what they have toward shared purposes.
- *Trust.* The ethical leader inspires—and is the beneficiary of—trust throughout the organization and its environment. Without trust and knowledge, people are afraid to exercise their authority.

Modes of Ethical Leadership It is often thought that ethical leadership must be "soft" leadership. Nothing could be further from the truth. Being an ethical leader means applying the right amount of authority in each situation. Sometimes the situation requires leadership that is anything but gentle. However, gratuitously tough leadership cannot be maintained for long without developing resentment and cynicism. It is helpful to think of the ethical leader as exercising authority within five levels of intervention into the judgments and actions of followers.

- *Inspiration.* Setting the example so that other committed members will contribute their fullest capabilities to achieve organizational purposes (the lowest degree of intervention)
- *Facilitation.* Supporting other committed members and guiding them where necessary so they are able to contribute their capabilities as fully as possible
- *Persuasion.* Appealing to reason to convince other members to contribute toward achieving organizational purposes
- *Incentives.* Offering incentives other than the intrinsic value of contributing to the achievement of organizational purposes where commitment is lacking
- *Force.* Forcing other members to contribute some degree of their capability where they have little or no commitment to do so on their own (the highest degree of intervention)

The leader must employ the authority granted him or her by the organization to achieve the purposes of the organization, while recognizing the knowledge needed to exercise this authority resides throughout the organization and its environment. The leader must ensure the purposes of the organization are known and shared, the capacity is in place to support its members' exercising their capabilities, and communication among managers and other employees is open and honest. The type of intervention selected will often depend upon the health of the organization and the pressures in its environment and culture; and, of course, the ideal is to inspire others as a steward of the vision, values, and excellence of the organization so that it reflects in its culture. However, often persuasion and facilitation are required of otherwise capable and committed members who may be unsure of their own capability. Sometimes where the organization is not healthy, the pressures to perform are intense or the culture of the organization is such that a more firm approach is necessitated.

The type of ethical leadership intervention depends, in large part, on the organizational culture. If the culture allows the organization to learn and grow within its environment, leadership may be largely inspirational. If the culture does not support organizational learning and growth within that environment, a more forceful leadership may be necessary. In any event, leaders must make their roles as integrity champions larger than life; otherwise, they and their examples may be lost in the pressures of organizational life. They must speak in terms of vision, values, and integrity. When the leader is not involved in a part of the organization's business, he or she must know who speaks for values and integrity. Moreover, the style of ethical leadership varies with the degree to which it reflects the organizational culture and the urgency of its situation in the environment. Ethical leadership is the stewardship that preserves the aspirations and the culture of the organization, while communicating the organization's core values. Ethical leadership balances achieving the organizational aspirations that are realistically attainable at the time and developing an ethical organizational culture. Different styles of leadership are necessary to maintain or implement change in the organizational culture that is optimal for it to survive and thrive within the organization's context. The specific culture required and the challenges it must face will be suggested by the nature of its essential social responsibility and the dynamics of its larger community.[19]

Ethical leadership practices are a necessary prerequisite for organization effectiveness. Leaders ultimately fail if they settle for less than the highest ethics standards. To increase the consistency of ethical practices:

1. *Publicize your standards.* Every leader/manager in the organization should be fully familiar with the ethical standards in place.
2. *Train leaders/managers on ethics requirements.* Leaders/managers need formal training on how they are expected to support the organization's ethics standards.
3. *Provide support systems.* Ethics systems are needed to support the ethics-related actions of leaders. These systems should include ethics policies, rewards, measurements, selection, and promotion systems that recognize the value of ethical leadership and managerial practices, as well as structures that allow employees to raise ethics concerns without fear of reprisal.
4. *Position senior management as the ethics models.* Senior managers set the standards for the organization through their decisions and actions. Their behavior must conform to the organization's strictest ethical standards.
5. *React ethically to critical events.* Ethical standards are most vulnerable in times of crisis. During a crisis leaders must conform to the highest ethical standards. It is also the time when there is the greatest temptation to cut corners and when the impact of their decisions and actions on those they lead is the greatest.
6. *Demonstrate concern.* People care about what their leaders care about. If leaders demonstrate concern for ethics, so will those they lead.[20]

MANDATORY ETHICS

The foundation of ethics codes rests either on the rule of law or administrative policies. Federal, state, and local governing bodies enact legislation to ensure a minimum standard of legal conformity. Ethics codes based on the rule of law carry legal sanctions. Administrative policies, often based on the rule of law, impact employment status or violate the values of the group that agreed to the set of self-imposed ethical standards. In either case, violating mandatory ethics can trigger legal or administrative sanctions, a change in job status, the permanent loss of employment, or any combination thereof. Most states and many local jurisdictions have statutes and ordinances that outline specific ethical behaviors and actions. Such ethical standards apply equally whether you are a career employee, volunteer, or elected official.

ASPIRATIONAL ETHICS

Aspirational ethics represents the optimum standard of behavior. Unlike mandatory ethics, aspirational ethics differs among individuals depending on their personal values, cultural influences, and sense of right and wrong. Aspirational ethics serves as an internal standard against which an individual judges personal behavior. For example, no law obligates a person strolling on a beach to save a child from drowning 50 feet from shore. Conversely, a person may feel a moral obligation to assist the drowning child because aspirational ethics compels a person to strive for optimal moral and ethical outcomes.

PERSONAL ORIENTATION

Personal orientation takes into account individual values, cultures, religious beliefs, personal biases, and other idiosyncrasies. The degree to which outward behavior differs from internal behavior expectations contributes to the amount of intrapersonal conflict experienced as a result of making an ethical decision. Conflicting feelings regarding a perceived duty and the need for peer acceptance also contribute to intrapersonal stress.

ETHICAL DECISION-MAKING PROCESS

The ethical decision-making process consists of three questions: "What should I do?" "What will I do?" "How does the decision I make match with my personal orientation?" Ethical decisions often engender fear—a fear of change in the status quo. People strive to maintain equilibrium in their lives and seldom act in a manner disrupting this balance. When confronted with an ethical decision, a person's ability to make objective decisions often becomes warped by this inherent tendency to maintain equilibrium. When making an ethical decision, a person conducts a personal risk-benefit analysis. Many ethical dilemmas present both short- and long-term solutions. An inverse relationship often exists between a short-term and a long-term ethical solution. In many cases a short-term solution often benefits the individual and can harm the organization, whereas a long-term decision may hurt the individual and provide benefit for the community or the department.

An ethical decision consists of a series of choices, not simply one decision. Making a bad primary ethical decision not only increases the number of future choices that will need to be made but often increases the impact of those choices. More important, a bad primary ethical decision spring-loads the ethical trap, resulting in an increased potential for legal or administrative action or unresolved intrapersonal conflict. Ethical dilemmas challenge the intellect because of the conflicting answers to the questions, "What should I do?" "What will I do?" If a person must choose between two options that do not oppose one another, selecting an option becomes a matter of choice and not a decision between right and wrong.

Let me share a real-life example. I had an engineer who caused damage to a ladder truck by running into a fence. His decision was to cover up the incident. By coercion of a younger firefighter to lie, a story was fabricated that the damage was done by a passing truck. When the owner of the fence called and wanted to know when we were going to fix it, the truth emerged. The result was termination of the engineer and a 30-day suspension for the firefighter, a short-term decision that had a long-term consequence.

In some cases, choosing right over wrong takes courage because people who make ethical choices often subject themselves to social and professional ridicule. Ethical decisions build personal character but not without pain. The vast majority of public servants are selfless, well-intended, and community-minded individuals. However, occasionally an agency will find itself dealing with individuals who are either unfamiliar with or disregard the norms and laws governing public service. What is a conscientious public servant to do if he or she suspects wrongdoing? Although it is difficult to give advice that addresses every situation, the following provides an analytical framework. There is a delicate balance between not turning a blind eye to potential wrongdoing and avoiding unjustly accusing someone.

First, contact your immediate supervisor, unless he or she is involved in the situation. Following the chain of command is important and provides the guidance needed to deal with the issue at hand. In many situations the employee may be reluctant and/or fearful to pursue the issue for a variety of reasons. Here are eight steps that are often helpful to walk through, when you or with an employee when a perceived or potential wrongdoing has occurred.

Step 1. Stop. Examine Your Motivations Ethics is about promoting fidelity to universal values (for example: trustworthiness, respect, responsibility, and fairness). In public service it is also about fostering the public's confidence in its governing institutions, employees, and public servants. A key goal is assuring the public that governmental decisions are made based on the public's interests, not on narrow private or self-serving ones. When considering what to do about someone else's perceived ethical or legal lapse, reflect on your motivations. Ask yourself whether your goal is truly to promote more ethical conduct in public service and increase public confidence in government. If the issue is a violation of the law, is it a technical "gotcha" violation, or does the violation truly represent a betrayal of the public's trust? Even if you are confident about the technical aspects, because the laws that address these issues are very complicated, consult your agency's counsel before moving forward.

There can be at least four motivations for calling a perceived transgression to light:

- **a.** *Organizational loyalty.* Individuals in this category are truly loyal to the organization and report concerns in order to remedy problems that could ultimately harm the organization. In addition, many people are strongly committed simply to doing the right thing.
- **b.** *Disillusionment.* Some people may be motivated to speak against perceived transgressions because their expectations exceed organizational realities. An important question to ask is whether these expectations reflect a full analysis of ethical considerations, including the fact that some ethical dilemmas reveal a conflict between competing legitimate ethical values. An example of such conflict is the tension between absolute fidelity to being honest (related to the value of trustworthiness) and avoiding unnecessarily hurting someone's feelings (related to the value of compassion).
- **c.** *Defensiveness.* Some people believe "the best defense is a good offense." They could be employees who anticipate disciplinary proceedings for poor performance or perhaps elected officials who fear that a transgression of their own is about to be revealed. The goal is not to vindicate ethical or

legal principles, but to lay a foundation for claiming retaliation when fault is found with their own conduct.

d. *Desire to harm.* Some individuals reveal or claim wrongdoing either to hurt or embarrass rivals or an organization, or as a form of retribution for perceived mistreatment.

If motivations fall into the latter two categories, carefully consider whether those involved are making an unethical use of ethics. The practice of co-opting ethics for personal or political advantage is known as "vigilante ethics." This destructive dynamic ultimately damages the public trust by impugning the motivations of public servants or would-be public servants for personal reasons that have nothing to do with creating a more ethical environment. Vigilante ethics is inherently a short-term strategy; people reap what they sow. Ultimately, an environment characterized by ethical charges and countercharges ends up reflecting badly on everyone, and if such behavior is an organizational norm, it acts like a disease that will ultimately destroy the heart of the organization.

Step 2. Figure Out What the "Wrong" Might Be. Ethics Versus the Law As you contemplate the nature of the "wrong" you observed, it can be helpful to keep in mind the distinction between the law and ethics. Following the law is what people *must* do; there are penalties and other consequences associated with violating the law. It is important whether you are a career or volunteer firefighter or an elected official to understand the laws that govern ethics in your local jurisdiction and state. Often local officials find themselves at odds with the laws governing public service, because they have no knowledge of what the laws state. Ethics tends to be what people *ought* to do based on commonly held values: trustworthiness, respect, fairness, compassion, loyalty, and responsibility (including public servants' responsibility to act in the best interests of the community as a whole). Although many laws reflect these values (for example, laws making it illegal to lie in government documents), conduct can technically be within the law but nevertheless be unethical. If you believe a colleague's conduct may be unethical—even if it is not unlawful—you need to consider carefully *why* you think it is unethical.

The Nature of an Ethical Dilemma. Some actions are clearly unethical, whereas others involve a more critical analysis. There are two kinds of ethical dilemmas: one involves conflicts between two "right" sets of values, and the other involves situations in which doing the right thing comes at a personal cost. Have you ever found yourself in either of these situations? Think about the ethical dilemma you have faced or witnessed in your organization and the kind of competing considerations that had to be weighed, in respect to both the individuals involved and how the organization dealt with it.

Does Your Agency Have a Code of Ethics? Referring to an agency code of ethics can be helpful. A code of ethics highlights the values important for those who serve the agency and how those values apply in the public service context. Furthermore, because there can be room for disagreement about what kind of conduct violates the letter or spirit of the code, it is useful if the code provides practical examples of the conduct that is consistent with its provisions (and, by negative implication, what kind of conduct is not).

To What Extent Is Ethics an Important Part of the Organizational Culture? Even if the agency does not have a formal ethics code, the community or the organization's leadership may have communicated in other ways its expectations about the importance of ethics and values in decision making and behavior. In an agency with a strong culture of ethics, it is easier to identify actions that are out of step with the organization's norms. In such a culture, leaders "walk the ethics talk," and there is a history of making difficult choices based on ethical considerations and doing what is right.

Step 3. Determine the Potential Consequences of Letting the Situation Go Unaddressed Keep in mind the consequences listed here are only the potential legal consequences. Just being *accused* of violating the law can have unpleasant results, including embarrassment (to the extent that some officials move out of their community); losing a good reputation and the community's respect; financial costs (hiring an attorney and the potential loss of one's job or professional license); and, for elected officials, being recalled or losing the next election before the legal process has concluded. If the conduct in question involves a violation of an agency ethics code, the code may provide accountability mechanisms. The availability and application of such mechanisms vary based on whether the individual in question is an employee or an elected official. For employees, feedback on ethical behaviors (or lapses in ethical behaviors) can be incorporated in the usual processes for providing input to employees and the employee review process.

Step 4. Speak with Others to See Whether They Share Your Concerns As an employee, talk with your supervisor or the next person in the chain of command. The agency's human resources department may also be able to serve as a sounding board. Frequently, such consultation will offer a more complete picture of what is going on and whether indeed the situation involves truly inappropriate conduct. This minimizes the likelihood of your misperceiving the situation based on speculation, conjecture, inaccuracies, or incomplete information. Such a consultation also widens the range of thinking about the best way to proceed in terms of achieving an overall positive result for the agency.

Communications Tips. Ethicist Michael Josephson offers the following tips on how to bring concerns about a particular situation to others' attention.

- **a.** Be prepared. Be sure you have your facts correct and are speaking with the right person.
- **b.** Be respectful. Watch your tone. Be earnest but not self-righteous or accusatory. Do not raise your voice or make threats. Be willing to listen as well as talk.
- **c.** Be fair. Do not assume bad motives. Be open to new facts and explanations. Do not equate someone's not agreeing with you with not listening, not caring, or being stupid.
- **d.** Be honest. Do not exaggerate or omit important facts.
- **e.** Stick to the point; stay focused.[21]

Step 5. Discuss the Issue with the Individual (or Have a Trusted Confidant Do So) Once it has been determined that someone is on a path to violate the law or the public's trust (or already has), the first goal is to get that person to stop. But how?

Figure Out the Motivation. One strategy is to try to determine what the individual's motivations are. For many, it is about outcomes—for example, personal financial gain or political advantage. This is why it can sometimes be a challenge to motivate people to "do the right thing" in the abstract, because the right thing can involve foregoing a benefit such as financial gain or perceived political advantage. For other individuals, the motivation can be a sense of self-importance. Author T. S. Eliot observed that half the wrongs in this world are caused by people who want to feel important.

Identify the Gaps in Analysis. For some public officials who step over the line, the thought process also can involve rationalizations. A common one is that somehow the individual "deserves" what might be considered an improper benefit because of his or her selfless commitment to public service. Another is that the law does not make sense. Yet another is "the end justifies the means"; in other words, a worthy goal justifies taking legal and/or ethical shortcuts. Neither the public nor the courts tend to buy these rationalizations. Another thought process can involve an incomplete assessment of the overall costs and benefits of a particular course of action. If an individual is on a path to violate either the law or the public's trust (or both), this cost-benefit assessment needs to include not only the anticipated benefits of this path but also the political, legal, financial, and emotional costs. These include loss of one's respected standing in the community (again, pride and feelings of importance can be powerful motivators), loss of office, and even the loss of one's freedom if the offense potentially involves incarceration. Be aware there is a strong human tendency to underestimate these likely costs ("The law isn't really clear." "I'll never get caught." "They won't be able to prove it." "After all I've done for the community, the public or the judge will surely go easy on me.") The objective reality is that prosecutors can offer strong incentives for people to testify against one another, and no one is particularly sympathetic to politicians and public employees who violate the public's trust.

Both the rationalization and underestimation thought processes tend to be fundamentally self-deluding. Because many people in this situation are motivated by outcomes, the task becomes one of diplomatically demonstrating the flawed nature of such reasoning and appealing to a person's sense of enlightened self-interest by helping him or her to appreciate the full range of potential consequences. The goal is to get the individual in question to stop the problematic behavior voluntarily and take whatever remedial steps are appropriate. If the individual denies the conduct or contends that there is no problem with it and the agency still has concerns, the next step may be either an internal investigation or the referral of the matter to an external enforcement agency.

Step 6. Determine Whether an Internal Investigation Is Appropriate An internal investigation may help an agency resolve the controversy over whether the conduct in question occurred and if so, whether it was improper. This enables an agency to respond proactively to allegations of misconduct. The scope of the investigation will likely be on whether the issue involves violations of internal procedures, standards, or the law. If it is the latter, consultation with the agency counsel will determine whether, and to what extent, an internal investigation is a constructive and helpful approach.

The Internal Investigations. The fundamental objective of an internal investigation is to determine the truth and demonstrate the agency's commitment to adhering to both the law and its internal procedures. Using a fair investigation process is paramount not only to achieving the goal of determining the truth but also in helping to establish the ethical culture of the organization.

Having an established investigation protocol can be helpful in this regard. This protocol enables the agency to explain the investigative process and its purpose (generally to discover the facts necessary to make a decision about a particular alleged behavior or action). It also communicates what to expect from the process and emphasizes its fairness and objectiveness. Issues that a protocol can cover include steps for initiating the process; the process to be used in the investigation (for example, fact-finding, interviewing witnesses, and assembling documentation); the final decision process; and how (including how widely) the results will be communicated.

Step 7. Determine Whether External Enforcement Authorities Should Be Contacted There are numerous types of external enforcement entities and mechanisms, including the following:

District attorney. District attorneys prosecute violations of state criminal laws.

U.S. attorney. These attorneys prosecute violations of federal law.

Grand jury. Grand juries have the authority to investigate public official misconduct.

State Commissions on Ethics. In many states such agencies are formed to investigate alleged violations of the Political Reform Act and other related state laws, impose penalties when appropriate, and assist state and local agencies in developing and enforcing conflict-of-interest codes. These agencies often interpret the laws relating to:

- campaign financing and spending;
- financial conflicts of interest;
- lobbyist registration and reporting at the state level;
- postgovernmental employment;
- mass mailings at public expense; and
- gifts and honoraria (speaking and writing fees) given to public officials and candidates.

Attorney general. State laws often require the attorney general to provide a whistleblower hotline to respond to concerns about potentially unlawful conduct. The attorney general also weighs in on the issue of whether an individual is unlawfully holding public office. Typically the attorney general's role is one of granting or denying permission to private individuals to bring such actions (known as quo warranto actions) on behalf of the public, although the attorney general has authority to bring these actions him- or herself. The attorney general may also issue opinions on general questions of law upon request by a state officer, legislator, county counsel, district attorney, sheriff, or city attorney.

Private right of action. Under certain circumstances an individual can bring a lawsuit to challenge unlawful behavior.

What About Going to the Media? Generally, the media should *not* be an early contact when a person believes an agency or someone within an agency has engaged in misconduct. Although many skilled investigative journalists exist, there are a number of reasons for this view:

- Approaching the media in the first instance calls the accuser's motivations into question.
- The media are unlikely to be able to conduct as thorough an investigation as a well-intended agency.
- Media attention may hinder an internal or external investigation.
- Allowing the accused and then the agency to take corrective action is more conducive to promoting public confidence.

In evaluating whether to contact the media, once again you have to examine the motivations of those involved. Is the motivation for taking action organizational loyalty? Disillusionment? Defensiveness? Desire to harm?

If the motivation is organizational loyalty, a key objective of any actions taken will be to determine whether a bona fide transgression has occurred and, if so, to pursue appropriate redress. The most likely way to achieve this objective is to first go through an organization's internal mechanisms and then, if those prove unavailing, to external law enforcement authorities. This is because the media typically will not have the expertise (specifically legal expertise) to evaluate whether a transgression has indeed occurred. The media may interview attorneys, but those interviewed are not likely to have the kind of in-depth information necessary to responsibly opine definitively whether a transgression has occurred. For these reasons, going to the media should generally be a last resort, perhaps only after law enforcement agencies have refused to act and there is a sound basis for believing a transgression damaging to the organization—and the public's trust in the organization—has occurred.

Reporting to the media initially suggests one's motivations are less ethical: for example, out of a *desire to harm* or to *secure strategic advantage*, should one's own conduct be subject to question (*defensiveness*). This is often seen in organizations with poor labor–management relations or a history of negative politics. Responsible members of the media adhere to their own code of ethics, whose central tenets include being fair and honest in reporting information. This code also encourages journalists to avoid allowing themselves to be manipulated by testing the accuracy of information from all sources and questioning sources' motives before promising anonymity. This is not to say media coverage of transgressions, perhaps as the result of investigative reporting, is in any way unethical. The media have an important role to play in fairly and factually reporting on transgressions. This notion is also part of journalists' ethics code that public enlightenment is the forerunner of justice and the foundation of democracy.

What Best Promotes Public Trust? One of the questions you should always ask yourself when confronted with an ethical dilemma is: "What course of action will most promote public confidence in my agency?" The reality is that the media are not likely to report on an ethical transgression in a way that promotes public confidence. Remember, the media are there to sell newspapers or gain market shares, so many like to sensationalize the event to capture the audience's attention. The only exception might be if the agency had been given an opportunity and took advantage of it to deal decisively with a claimed transgression. Even in this latter situation, there is no guarantee that the media would report favorably on the agency's actions.

Step 8. Consider Steps to Prevent the Situation from Recurring Education is often the best preventive measure. In most states the league of cities or city managers state association offers a number of resources to help local agencies address and educate their staff about ethics issues. They include identifying best practices for complying with ethics laws, fostering an ethical culture, promoting ethics through codes of ethics, promoting campaign ethics. and providing training and conducting workshops.

Suggested steps to take when you suspect a colleague has crossed over the ethical line:

a. Stop. Examine your motivations.
b. Figure out what the "wrong" might be.

c. Determine what the consequences might be of letting the situation go unaddressed.
d. Speak with others to see whether they share your concerns.
e. Discuss the issue with the individual (or have a trusted confidant do so).
f. Determine whether an internal investigation is appropriate.
g. Determine whether external enforcement authorities should be contacted.
h. Consider steps to prevent the situation from recurring.[22]

ETHICS, POLITICS, AND LEADERSHIP

The issue of politics and ethics is one of the most difficult leadership challenges faced by chief executives and leaders in the fire service today.[23] Such interactions can often have impacts at all levels of the department as well as for many years on the department and the governing agency. Joann B. Seghini, Ph.D., of Midvale City, Utah, describes what she, as mayor, sees as the fire chief's role and responsibility. It is a fine overview, but it does not address an important issue that all in the fire service face and have to deal with daily—the issue of politics and ethics (see Figure 6.1).

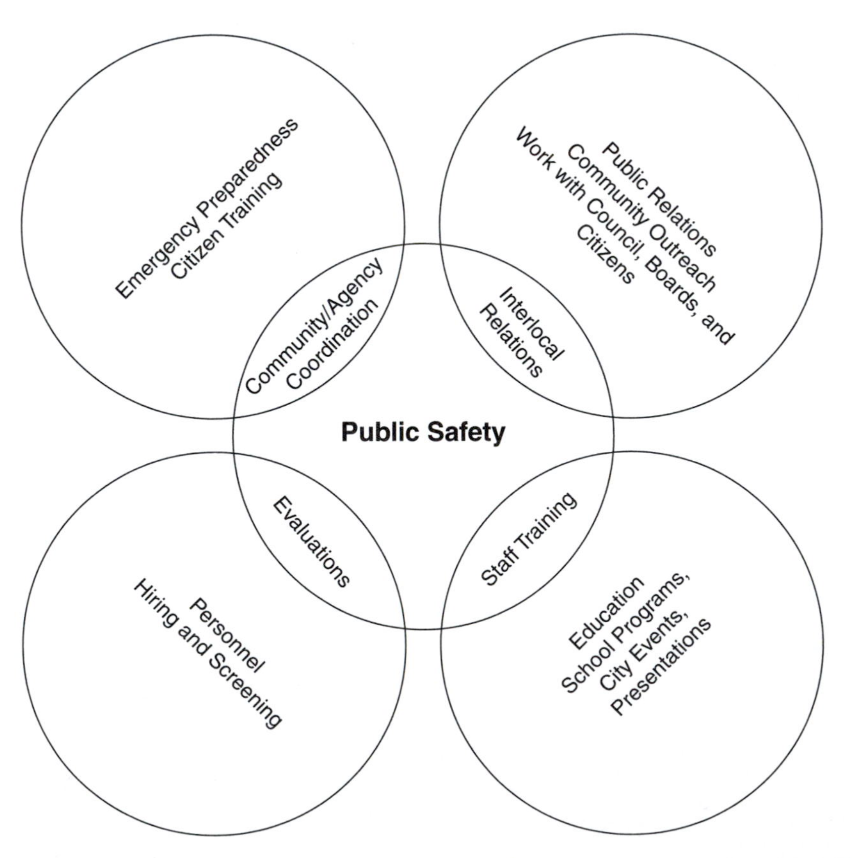

FIGURE 6.1 ◆ The Fire Chief: A Mayor's View *(Courtesy of Randy R. Bruegman,* The Chief Officer: A Symbol Is a Promise, *p. 87)*

The political process is an aspect of our job that is vital to the success of our organization, yet it is one that most of us learn only by on-the-job training. This process, if mishandled, can create ethical challenges for the individual and the organization; and, if misunderstood, can create enormous frustration and be one of the most difficult challenges leaders in the fire service are faced with today. There is not a great deal of academic attention or training about dealing with the political process prior to accepting the job as fire chief or the promotion to battalion chief or captain. Unfortunately, throughout the department, whether as a new recruit or a seasoned veteran, you will find yourself in situations that can turn from social to political very quickly. Politics often has a negative connotation for most people and, therefore, is often an avoided topic. Nevertheless, it is a real and an essential fact of life, especially for those of us who are fire chiefs, labor leaders, volunteer association presidents, or informal leaders within the department or community. Assuming greater leadership responsibility in the department comes with an understanding that the political landscape may determine the posture and position taken throughout your career. The political landscape that you operate in is often very dynamic and apt to change very quickly. Sometimes one has fire service personnel who think of themselves as politicians, to the point where it may be unclear whether their role is to suppress fires or to be on the phone with elected officials. On the other side, certain elected officials fancy themselves as would-be firefighters. They long to take a more hands-on approach in the operation. The fire chief and others in leadership roles in the department often feel caught in the middle. This dynamic creates a level of uncertainty not conducive to the chief executive doing his or her job and possibly unnecessary tension in the organization. Many people who become fire chiefs or who are promoted within the organization have had very little experience with politics and do not understand how the political process works. This can create a very frustrating situation for many new chief officers and others of influence in the department, as they attempt to learn that politics and political gamesmanship are a way of life for most of today's fire service professionals.

It is important to consider how ethics and politics are integrated into an organization and how the political structure incorporates the fire chief and those who hold positions of authority into the dealings of the government they work for. Chief fire officers or captains charged with heading a fire company have to come to grips with "political ethics" early in their careers, which means being clear about what you will and will not do in terms of organizational politics. What is acceptable to you as a professional? Is it all right for your subordinates to deal with elected officials behind your back? Are officials going directly to your subordinates without your knowledge? Is a strict chain of command followed by both? Whatever the case, you need to understand the ground rules because what is acceptable in one organization is not tolerated in another. Certain department heads and staff routinely talk to elected officials, and vice versa; in other organizations everything goes through the manager or mayor. Understanding an organization's political dynamics is critical, especially if you are the new fire chief, a division leader, the labor president, or a volunteer association leader. Determining your comfort level and what you are willing to accept, understanding the rules of the game, and identifying the key players in terms of roles, relationships, and responsibility are vital.

OUR PAST IS OFTEN A WINDOW TO OUR FUTURE

One of the best ways to grasp the political dynamics of any organization is to understand what occurred in the past. There were people in your position long before you arrived, and the reality is that much of what occurs today is built upon events or relationships that happened under someone else's watch—some good, some bad. Past organizational events or relationships often dictate the present ethical and political structures and impact the current organizational dynamic. These effects can be seen in how people deal with one another; the honesty and ethics of the organization; and the relationship dynamics between labor and management, between management and the elected officials, and between the department and the community. Organizational history is a window on the future when you face issues similar to those of your predecessors. The reactions and perceptions often occur the same way, and it can be extremely helpful to understand and determine how to approach and present an issue or in some cases avoid it. Knowing what your principles are and being able to live by them are the personal values you bring to the job, which ultimately drive how you conduct yourself and your interactions with others, and your professional success or failure.

This arena continually demands a defined framework of principles within which to operate in our role of leadership. When we speak of ethics, we speak of the principles that help to define proper behavior. Such principles do not always dictate a single course of action but instead provide a "measuring stick." Being able to choose among different options contributes to being a good public servant and an effective fire chief. Making good ethical decisions often requires the ability to make distinctions between competing choices. Many of those choices involve gray areas for leaders who frequently require that integrity be the best asset. Unfortunately, many in our profession have crossed the ethical line to maintain their position, enhance their political stock, or overlook questionable processes to maintain political and departmental harmony.

COMMISSION ON PROFESSIONAL CREDENTIALING CODE OF CONDUCT

In fulfilling his or her responsibilities, each individual participant in the Chief Fire Officer Program serves as a moral and ethical agent. Every action affects the health and well-being of individuals, organizations, and communities; therefore, participants must assess the consequences of their decisions and actions and accept responsibility for them. Chief fire officers (CFOs) must speak out and strive for the most moral and ethical course of action for themselves and the organizations they lead. Participants in the Chief Fire Officer Program are required to comply with this *Code of Professional Conduct* and the administration thereof. Noncompliance may be cause for termination from the program or revocation of the CFO designation.

Responsibilities to Individuals

The chief fire officer shall:

- Set an exemplary standard for subordinates and peers to follow.
- Be courteous and tactful in all interactions.
- Ensure the communication of rights, responsibilities, and information is upheld to foster informed decision making.
- Respect the customs and beliefs of others—consistent with the mission of the organization.
- Respect the confidentiality of information, except when it is in the public's interest or when there is a legal obligation to divulge such information.
- Promote competence and integrity among individuals associated with fire and emergency services.

Responsibilities to the Profession

The chief fire officer shall take a leadership role to ensure the fire service:

- Serves the public interest in a moral, ethical, and efficient manner.
- Strives to provide quality services as defined by the community and based upon accepted industry standards.
- Communicates truthfully and avoids misleading representation and raising unreasonable expectations in other persons or in the community as a whole.
- Uses sound management practices and makes efficient, effective, economical, and ethical use of resources.
- Promotes understanding of public protection, safety services, and issues.
- Conducts inter- and intraorganizational activities in a cooperative way that improves community well-being and safety.
- Develops and maintains the required level of physical and mental health to enhance and promote individual quality of life, which allows for the proper discharge of duties.
- Reports to the Commission on Chief Officer Designation when there are reasonable grounds to believe a member has violated this Code of Professional Conduct.

Responsibility to the Community and Society

The chief fire officer shall:

- Abide by the laws of all levels of government but may seek changes by lawful means when deemed appropriate.
- Contribute to improving the well-being and safety of the general population, including participation in educational programs, dialogue, and recommendations to enhance the quality of life and to improve fire and emergency services.
- Strive to identify and meet the needs of the community within the resources available and within the mission of the organization.
- Consider the effects of management policy decisions on the community and society and make recommendations based on these considerations.

Conflict of Interest

A conflict of interest exists when the chief fire officer uses his or her position, authority, or privileged information to:

- Obtain an improper benefit, tangible or otherwise, either directly or indirectly.
- Obtain an improper benefit for another.
- Make decisions that attempt to, or do, negate the effectiveness of the Chief Fire Officer Program.

The chief fire officer shall:

- Conduct all relationships in a way that assures management decisions are not compromised by a perceived or real conflict of interest.
- Disclose to the appropriate authority all direct or indirect personal or financial interests, appointments, or elections that might create a conflict of interest, whether real or perceived.
- Neither accept nor offer personal gifts or benefits with the expectation or appearance of influencing decisions.
- Refrain from using the chief officer credential to promote or endorse commercial products or services without the express written permission of the Commission on Professional Credentialing.

- Value ethics within the fire and emergency services. Most professions abide by a Code of Ethics or Conduct, which expresses their members' agreement as to what constitutes acceptable behavior.

The Code of Professional Conduct has been adopted by the Commission on Professional Credentialing to promote and maintain the highest standards of professional performance and personal conduct. Abiding by these standards is required for continued designation as a chief fire officer and serves notice to the pubic who entrusts its confidence in the abilities and integrity of the chief fire officers.[24] Although these were directed for the chief officer, the same tenets hold true for all who wear the uniform in the fire service.

Your leadership ethics defines you as a person. In the past several years, we have witnessed so many ethical misadventures of those in leadership roles that they have become commonplace, from fabrication of résumés to altering response data to meet the adopted local standard. What is always amazing in these situations is that when these people are exposed, they deny their actions or, in many cases, attempt to blame them on someone else. This is a comment on the integrity not only of the individual but also of the system in which he or she works. The individual, the system, and the industry as a whole are tainted by such experiences whenever an ethical misstep occurs within the fire service.

We are not unique in that regard. If a major league ballplayer is caught using performance-enhancing steroids, we all then suspect other ballplayers who are performing exceptionally well. We begin to question the validity of their accomplishments. When a local council member is indicted on charges of kickback, the community becomes suspect of those on the remaining governing body. What are they up to? Are they all corrupt? We cannot trust them anymore!

When a local fire department is caught in an indiscretion, we all come under the microscope. The breadth of media coverage dictates whether your department is judged by what someone else has done: for example, the fire department that confiscated illegal fireworks only to go back to the station and set them off; the department whose fire captain and company frequented a porn convention while on duty; the fire chief who lied on his résumé to get a job; or the multiple incidents of firefighter fatalities while driving under the influence of alcohol when responding to a call. When these stories are played on the local or national media, our entire industry is affected. As an example, when the news broke regarding the fire chief who had doctored his résumé in a major metropolitan city, many fire chiefs across the country received freedom of information requests for their résumés submitted at the time of application to verify the validity of what had been presented.

It is important for you as a leader and public servant to have an ethical guidepost. Similar to the Commission on Professional Credentialing CFO code of professional conduct, the *IAFC Code of Ethics* also provides such a framework. Both can be used when faced with competing choices. Both can provide a touchstone for issues that force leaders to challenge their ethical boundaries. These ethical principles are consistent with professional behavior of not only chief fire officers but also all in the fire service and include the following (see Figure 6.2):

The IAFC Code of Ethics

- Recognize that we serve in a position of public trust that imposes responsibility to use publicly-owned resources effectively and judiciously.
- Do not use a public position to obtain advantages or favors for friends, family, personal business ventures, or ourselves.

FIGURE 6.2 ◆ CFO Logo *(Courtesy of Center for Public Safety Excellence)*

- Use information gained from our positions only for the benefit of those we are entrusted to serve.
- Conduct our personal affairs in such a manner that we cannot be improperly influenced in the performance of our duty.
- Avoid situations whereby our decisions or influence may have an impact on personal financial interest.
- Seek no favor and accept no form of personal reward for influence or for official action.
- Engage in no outside employment or professional activities that may impair, or appear to impair, our primary responsibility as chief fire officials.
- Comply with all the laws and campaign rules in supporting political candidates and engaging in political activities that may impair professional performance.
- Handle all personnel manners on the basis of merit.
- Implement policies that are established by our elected officials and policy makers to the best of our abilities.
- Refrain from financial investments or business that conflicts with or is enhanced by our official position.
- Refrain from endorsing commercial products through quotations and use of photographs and testimonials for personal gain.
- Develop job descriptions and guidelines at the local level to produce behaviors [in keeping] with the Code of Ethics.
- Conduct training at the local level to inform and educate personnel about ethical conduct policies and procedures.
- Have systems in place to resolve ethical issues.
- Orient new employees to the organization's ethics program during new employee orientation.
- Review the ethics management program in management training experiences.

- Deliver accurate and timely information to the public and to elected policy makers to use when deciding critical issues.[25]

THE POLITICAL DYNAMIC

The credibility you establish at the local level and the integrity you display determine your ability to have an effective relationship with your elected officials. However, this alone is not enough. When you present an item for consideration to an elected body, it will be judged not only, hopefully, on its merit but also in many cases politically. Understanding the political nature of your position and how to interact within it is very important. Many proposals submitted to the elected body may be the right thing to do based upon good science and yet are not acted upon, not due to the objective evaluation but to the politics involved.

Most fire service leaders will have come through a typical career model of being evaluated and promoted based largely upon their training in emergencies. This model is fairly linear, fairly black and white (something is either right or wrong), and fairly objective, and we tend to like it. It is easy for us to make a plan, set an objective, develop a course and take action, reevaluate, and move on. The political arena, however, is very different. Elected officials often lean more toward the subjective or softer side of the issues. This is often a key element in their process of decision making, and it is extremely important for fire service leaders to understand it and learn to accept it. When you lose a battle because of a political versus an issue-based decision, you cannot take it personally; you have to be able to let it go and move on to the next issue. This is part of the process and in most cases was not meant as a personal attack on you or your department. When you are ethical in your dealings, you will live to fight another battle on another day.

Survival, though, depends upon how we get through those loaded political encounters. Continue to remind yourself of what the outcomes and the objectives are, and what you are trying to achieve in the political process. Remain focused on the issue and not on the people involved in the debate. If you do not, you will begin to focus on the personalities and the political discussions taking place and lose perspective as a leader in your organization. Many proposals get voted down, not because there was a lack of political support, but because the person presenting the issues mistook the tough questioning as a personal attack. In one case, instead of listening, the fire chief mounted a full assault on the council member, calling into question the ethical character of the council member. In this scenario you can become very vulnerable and lose your own credibility—and, if it is severe enough, your job. Always remain focused on the issue at hand and remember why you presented it in the first place. Remind yourself to listen to all the options stated. Though you are the fire service professional, there are times when politically motivated suggestions are offered, and some of them might be good ideas. You have to be willing to consider them. Getting through politically loaded encounters requires fire service leaders to be focused yet at the same time flexible, politically sensitive, and open to alternatives.

How can you help to create a positive political climate for yourself? The more you can deal with people face to face, the better off you will be. Relationships are established best with one-on-one encounters. Increase your own tolerance for ambiguity and, when dealing with elected officials or members of a political body, remember that they may not be used to handling the black-and-white issues of the fire service. Ambiguity is sometimes the saving grace for politicians, and they often choose it over the more definitive proposal. Become more receptive to politicians' ideas as they relate to your proposals.

Take an open, receptive stance when you bring issues forward. Take a broad view, one that not only incorporates the needs of your organization but also takes into account the city's or county's goals and objectives. Realize that buying a new ladder truck might not be the most important thing on the agenda of the city or county at a particular point in time. These officials may have more pressing needs or issues to address. Share your own goals with your elected body; hopefully, with understanding, they will begin to support what you are trying to do organizationally, both on a short- and a long-term basis. If they don't, it is a good indication that either they are out of step with the goals and objectives of the organization, that they do not understand, or in some cases that your credibility is so low that you need to look for a new job.

Remember that understanding does not always mean agreement. Understanding the political nature of the community and how it operates is very important not only to the fire chief but also to the labor president, the volunteer association, the leadership, and the organization as a whole. This does not mean you have to agree to the final decision on an issue. Yet you do need to realize why the decision was made and see that it is implemented. Sometimes a decision will be reached for totally political reasons that probably go against the grain of most fire service leaders, but this is the reality that most of us are faced with. In either case, whether you agree with the decision or not, it is your job to support it.

The late president of France, Charles de Gaulle, once said that *politics is too serious to be left to the politicians*. This is certainly the case in local government. Politics in itself is neither good nor bad. It largely depends on how it is used, the people involved, and their own motivations. The political climate that exists in a given community will create an environment of trust, distrust, or skepticism. Strong loyalties can form and strange bedfellows made as a result of the political dynamic. All, to some degree, will influence how the political process will work within and/or upon the organization. This is a reality you should not ignore if you wish to be effective, nor should you allow the art of politics to place you in a position that compromises your ethics.

THE TEN COMMANDMENTS OF POLITICAL SURVIVAL

Several years ago, George Protopapas published his version of the ten commandments of political engineering. These are paraphrased in respect to the role of today's fire service leadership. The ten commandments of political survival for fire service leaders provide a good road map that allows all of us in the fire service to deal with the politics of local government today.

1. *Never show animosity toward any elected official.* Sometimes elected officials will do or say something to make us look bad in public, either at a meeting with elected officials or in front of a constituent. Always be professional. Coolness under fire can go a long way toward creating a good relationship with the entire elected body.
2. *Know your budget thoroughly.* Elected officials often refer back to specific areas of the budget that you may not have looked at for the past several months. Questions on your budget often come when you least expect them. When you are presenting an item that has a big budget, has no cost ramifications, and has already been approved, you may be asked questions about something totally unrelated and you are expected to have an answer. So it is extremely important that you and your staff have a good understanding of everything in your budget and where it can be found at a moment's notice.

3. *If you have a proposal but know you do not have the votes to win, do not bring it to the governing body.* Give yourself time to lay the groundwork before you place it on the agenda.
4. *For senior fire service leadership,* stay out of political campaigns, especially in elections that will result in replacing people on your elected body. It is a lose-lose situation. If the incumbent you choose to support loses the election, you will have an opposing vote on many of your proposals from the new elected official. Conversely, if you support local candidates, they often think that you owe them a special favor, which can present ethical dilemmas later.
5. *Make a point of getting to know your elected officials better.* When attending conferences or other events together, get to know them personally. This can tell you a lot about what motivates them and is important to them as individuals, and not as council or board members.
6. *Encourage your elected officials and city and county managers to participate with you and your personnel in events,* if this is allowed by the political structure of your governing body. It is a great opportunity for them to learn more about the organization and its people and provides a way to share information.
7. *Conduct field trips for your elected officials.* In the fire service, we have a lot of impressive equipment and great people. A good way to create a positive relationship with our elected officials is to encourage them to spend some time with us doing what we do. Allowing them to ride along and participate with the organization in prearranged events will help these officials see what the fire service does on a day-to-day basis. This can offer a great opportunity for discussing with them issues of importance to the organization. Remember, once is not enough.
8. *Follow up every problem referred to you by an elected official,* even if it is a minor one. Do this immediately, return phone calls, and write to the official to confirm that the matter has been handled. Deal with the issue in concert with your city and county managers, if appropriate.
9. *Work with your elected officials individually* before the open business session. Often, being able to field questions from individual elected officials helps them to understand more fully the issue that you are bringing forward. It can also be a great opportunity for you to learn what they will be asking when the regular open business meeting begins. If you do this, you cannot do it for just one official; you must do it for all. Make sure that you do not offend any board members by leaving them out. Again, this has to be done in concert with, and with the approval of, your city and county managers if appropriate.
10. *Sometimes your recommendations are not going to pass;* you must recognize when that occurs. When it does, do not continue to push the agenda item and end up antagonizing your elected officials. Let it go and move on.[26]

Understanding the methods of political survival for fire chiefs and fire service leaders is essential. In discussing ethics and the political dynamics of our organization, leadership, ethics, and politics are interwoven into the fabric we call government. Chief fire officers should be dedicated to the concepts of effective and democratic forms of government in their response to elected officials and professional management. This is essential to achieving the objective of providing quality service. To do so, one must maintain a constructive, creative, and practical attitude toward the way local government works. We must understand and believe that each of us in our various roles and positions has a responsibility as a public servant. We can fulfill that responsibility only by being dedicated to the highest ideals of honor and integrity in our public, personal, and professional relationships. Our measure of success is the respect and confidence given to us by our elected officials, our peers, our employees, and the public.

While I was chief at Clackamas Fire District No. 1, we developed a code of conduct. It is something I have tried to live by, and I offer it to you as a starting point when you deal with the political and ethical issues as a member of the fire service.

- Do no harm.
- Place the safety and welfare of others above all concerns.
- Seek to help those who are in difficult circumstances.
- Provide service fairly and equitably to everyone.
- Assist and encourage those who are trying to better themselves in the fire service.
- Foster creativity and be open to the innovations that may improve the performance of our duties.
- Comply with the laws of local, state, and federal governments.
- Defend the Constitution.[27]

Probably no single action that leaders can take will affect our credibility more than ensuring that our behavior is ethical and professional.

SCORING POLITICAL POINTS

The fire service has been reorganized toward becoming a more progressive, all hazards response industry. The roles and responsibilities throughout our organizations have changed dramatically over the past 25 years. The biggest shift may have been in the political arena. Whereas the fire chief was once the top firefighter, today the chief is a CEO, content expert, facilitator, and politician. Although not elected, the position of fire chief stands out and can become highly politicized very quickly. Those in leadership positions must understand as much about the political process as about incident command. Whether in a small all-volunteer organization or a large career metropolitan department, you will work in a political environment. It may involve the volunteer association, the local union, citizen groups, and/or elected officials. You must have the expertise to blend your skills, ethics, values, and organizational needs into the political process in such a manner to be professionally successful, create organizational improvements, and have positive relationships throughout the political arena. The following guidelines for political survival for the fire chief and others in leadership roles are essential to long-term job success and, in some cases, survival:

- Understand that past events often dictate current situations and dynamics.
- Identify the players in the game.
- Learn the rules of the game; every community is different.
- Determine your own ethical bottom line.
- Develop a framework for making ethical decisions.
- Understand that the political process can be as important as the substance of the issue.
- Focus on the issue, not on the personalities.
- Do not take the rejection of your idea or concept personally.
- Perseverance is a must.
- The most important thing to remember is that fire chiefs and staff don't vote.

LEADERSHIP STRATEGIES FOR PERSONAL SUCCESS

The issue of ethics has a direct impact on the fire service middle manager or company officer.[28] The officer must make decisions, provide guidance and leadership, and set an example for subordinates in the context of what is right/wrong, acceptable/unacceptable, or good/bad. The ethical standards of the fire department are influenced by what society in general expects; what the local community believes is the job of the fire department; and numerous laws, codes, and other standards of behavior. Each department's ethics mirrors what is valued and rewarded, as well as what is not valued and therefore penalized. This internal value system or culture, a powerful regulator of ethics, can be a help or a hindrance to the company officer trying to do the right thing in all situations. It is not an easy job! Our personal and professional ethics are shaped by such differing sources as:

1. Our family's values and culture
2. Community attitudes, including conflicting social and economic expectations
3. The United States Constitution and Bill of Rights, and numerous laws, codes, and ordinances
4. Religious beliefs and teachings
5. Our life and work experience, including the standards, beliefs, and attitudes of peers, superiors, subordinates, and policy makers

We know ethics and standards change, causing ethical dilemmas for fire department officers. For example, it was not that long ago that the dumping of hazardous wastes was rather casual and considered "OK" if done economically and short distances from human activity. Now, with changing public attitudes, new legislation, and rigid restrictions, fire departments must be prepared to enforce the laws, mitigate unsafe storage and transport, and coordinate cleanups of hazardous materials spills. Ethical questions arise about the acceptable level of exposure to firefighters and other emergency personnel. When the AIDS epidemic first surfaced, it caused similar soul-searching as well as changes in policies and procedures. We realize that codes of ethics can serve as only rough guidelines for ethical decisions, because day-to-day ethical dilemmas cannot be specifically anticipated and solved with a formula. Rather, company officers must rely on numerous sources of guidance for decisions and anticipate gaps or organizational blind spots where no explicit guidelines exist to help with decisions.

We understand that mid-level managers or company officers play a difficult and vital role in managing and influencing the day-to-day operational ethics of the department. For example, they are primarily responsible for ensuring that the community gets its money's worth (return on investment) by the effective use of personnel, materials, and time on a day-to-day or shift-to-shift basis. As fire officers and public servants, with proper planning and foresight, we can avoid falling into ethical traps that would follow each of us throughout our fire service career or in some case put an end to it.

OUR OWN PERSONAL CODE OF CONDUCT

Based upon a very personal assessment of what is right or wrong, our code of conduct begins with early childhood and matures through our life experiences and the attitudes of people we respect, the confidence we have in our own standards, how often we have been supported or burned by past decisions, and so on. What is often described as a gut feeling about what is right has medical confirmation! Except for rare sociopaths, people receive a distinct message from the pit of their stomachs when either thinking about or doing what they believe to be wrong! The galvanic skin response, used in the controversial but common lie detector test, takes advantage of this basic physiological response to truth or falsehood. It is similar to the fight or flight reflex we have when in a stressful situation. Our body alerts us to our most basic options! Codes of ethics arose over centuries to provide specific guidelines to members of a profession, craft, or business. They are designed to inform in-group members of a common set of standards, encouraging them to live by those standards. Their purpose is to protect the integrity and reputation of the whole group by publishing what is considered to be correct or expected behavior, and an individual's obligations to the group. A well-known code of ethics is the Hippocratic oath of physicians, which states in part, "Above all, do no harm."

GUIDELINES AND RESOURCES

Many groups (and individuals) have a stake in decisions, and they may all have legitimate but contradictory needs. The fire officer's own personal ethics, standards, and integrity are always part of the picture because there are numerous courses of action that he or she can take in making a decision. There is seldom a clear, totally right or wrong answer to ethical dilemmas. A person must base a decision on community, organizational, and/or personal standards, and the clearer these are, the easier the decision! Which guidelines currently exist in your fire department to guide and support your decisions and also determine what gaps might exist? All fire departments, no matter how simple or sophisticated, have a unique jigsaw puzzle of values, standards, expectations, requirements, and legal mandates. These parts of the organizational whole directly reflect decisions that have been made in the past about what is right and wrong, proper and improper. Because some of these decisions have been made officially and some unofficially, based on the culture and values of the members of the department, it is difficult for a fire officer to sort out the formal from the informal, to clarify standards, and to arrive at ethically proper decisions. Presumably, the more formal the guidelines that exist in the department, the more help an officer has in making decisions. Not necessarily so! Sometimes past organizational decisions are not consistent but contradictory. Serious review of department documents and culture may only confuse the responsible officer. In the final analysis, the bottom line is that each person must weigh and honor existing guidelines but make decisions based on personal ethics, ideals, and convictions.

Unfortunately, there are no formulas for the gray areas or for the tough decisions. A major part of ethical behavior is accepting that you must do what you believe is right, given the specific situation, and be prepared to justify and defend your decision if necessary. An ethical decision will:

1. Honor formal department decisions and documents.
2. Not violate laws, rules, and the like.
3. Reflect reasonable and positive cultural characteristics.
4. Reflect the interest of all parties as much as possible.
5. Feel right!

Here are some tests of an ethical decision (organizational or personal):

1. Do I feel unembarrassed, unashamed, not guilty, not defensive?
2. Do I object to my decision being published openly?
3. Am I willing to risk criticism for my decision?
4. Could I justify and defend my decision to my greatest critics and enemies?
5. Have I considered all who would be affected by the decision and given them proper priority or weight?
6. Does it feel right in my gut—is my conscience clear?

Characteristics of an Ethical Organization

1. Senior management defines and clarifies standards, values, and ethics.
2. Senior management demonstrates a commitment to those ethics and expects a similar commitment of all members.
3. The organization supports and rewards ethical behavior and ethical solutions to problems.
4. The organization gives consideration to all stakeholders—the community, policy makers, employees, and special-interest groups.
5. The organization as a whole prides itself on its ideals and on striving toward them.

Characteristics of an Ethical Leader

1. Models ethical behavior after others
2. Attempts to balance personal ethics with those of the organization
3. Considers impact of decisions on all others who will be affected
4. Operates with integrity, honesty, and courage
5. Approaches ethics from a positive point of view, guided by his or her ethical compass or conscience.

> Men (and women) often stumble over the truth, and most manage to pick themselves up and hurry off as if nothing had happened.
>
> *Winston Churchill*

Trust and ethics in business are at an all-time low and for good reason. The last 20 years have seen larger and larger scandals in which the pressure to produce results quickly has pushed and often surpassed the bounds of acceptable behavior. Many feel there is no longer any business code of ethics and that businesspeople simply bow to profit. To some degree, of course, this is true. After all, business leaders are required to manage for maximum share price, not maximum ethics. Yet this does not mean they must throw values, community, and employees out the window. Often, long-term prospects of a business are linked to how it treats the human issues. Over time, employee retention, community support, regulation, and morale all reflect a company's ethical decisions. We are rarely taught how to consider, balance, and implement these "soft" issues in the bottom-line environment of business. Personal integrity and ethical behavior from people who occupy leadership positions throughout government and other civil service agencies have come into question since our government was formed. The public often has a stereotype of the typical civil service worker as lazy and unmotivated, who shows up to work only to obtain an easy paycheck and the generous benefits that go with it. Managers of these agencies often come from the rank and file and are often seen as less than accountable.

When a community votes a local government leader into office, the public trusts that the person will make decisions for the good of the people, not just some of the people. In local government, people expect their neighborhoods to be safe, their garbage to be picked up, their streets to be repaired, and their tax dollars to be spent wisely. There is a serious ethical dilemma when a government is so old and entrenched in its ways that it quits caring about—or even thinking about—the people who trust it to do what is right for the community. The approach to ethics is rooted in government, having been conceived in the nineteenth century by Jeremy Bentham and John Stuart Mill to help legislators determine which laws were morally best. They both suggested that ethical actions are those that provide the greatest balance of good over evil—the greatest good for the greatest number.

GOVERNMENT LEADERSHIP

The ethics and values of leaders are reflected in their actions and behaviors and subsequently in the actions and behaviors of those they lead (see Figure 6.3). Whether in education, business, or government, if organizational leaders do not act ethically, their followers will not have a model of higher standards to live up to. Wilcox and Ebbs wrote, "The role of leadership of the college or university is attributed to the president, who has the obligation of ethical and academic responsibility."[29] In business, this leadership role is played out by a CEO and in government by elected officials and their appointees. When an ethical role model is not presented to the people whom these leaders influence, who in turn must lead smaller groups, a community is created that is void of an ethical sense, and individuals are left to interpret on their own what the organizational standards may be.

Government officials have a duty to the taxpayers who elected them to put in place a leadership group that will place an emphasis on doing good for all instead of for special-interest groups. They must be surrounded by people who share a vision. In industry or on college campuses, when a new leader enters the organization, he or she brings along people with that shared vision. In government, that is not the case, although it could and should be. In private industry, leaders must be able to run businesses in competition with others, taking into account all of the factors involved in business competition—customer needs, economics, accepted business

FIGURE 6.3 ◆ U.S. Capitol *(Courtesy of Bill Koplitz/FEMA News Photo)*

standards and practices—and getting the most return on investment, just to name a few. It behooves these leaders to present a strong ethical model; their business success relies on it. However, in government entities, little, if any, of this is true. Public sector leaders are given annual budgets of taxpayer money to spend as they see fit, and few are held accountable for how that is done or how much value was received in return. To further complicate matters, monies are divided into many different accounts and budgets, making tracking their use even more difficult. Certain divisions within these entities serve only "inside" customers, not the taxpayers, and can launder funds originally earmarked for certain uses that would have benefited the taxpayers back into the general fund where government officials can use them for special-interest purposes.[30]

DEVELOPING AN ETHICS CODE

> The ultimate answer to ethical problems in government is honest people in a good ethical environment. No web of statute or regulation, however intricately conceived, can hope to deal with the myriad possible challenges to a [person]'s integrity or his [or her] devotion to the public interest.
>
> *John F. Kennedy, Message to Congress, April 27, 1961*

An *ethics code* is a framework for day-to-day actions and decision making by officeholders and an entire agency, depending on how the code is written. The fundamental premise of an ethics code is that it is easier for people to do the right thing when they know what it is.[31]

An agency usually has three objectives when adopting an ethics code: encouraging high standards of behavior by public officials, increasing public confidence in the institutions that serve the public, and assisting public officials with decision making.[32] It must be emphasized that achieving these goals requires a well-conceived process for both adopting and implementing the code.

Values-Based Versus Rule-Based Codes There are two types of ethics codes. One emphasizes rules ("don'ts"). Such codes often parallel, if not duplicate, state laws relating to ethics. The other type emphasizes values and the kinds of behaviors that demonstrate those values ("do's").[33] It is a commitment to uphold a standard of integrity and competence beyond that required by law.[34] An ethics code thus creates a set of *aspirations* for behavior, based on values associated with public service held by public servants and the communities they serve. The process of adopting and reviewing an agency's ethics code enables agency officials to clarify these values and link them with standards of conduct. Therefore, ethics codes complement ethics laws by going beyond the minimum ethical requirements established by ethics laws to define *how public officials act when they are at their best*.[35]

A values-based ethics code is a complement to ethics laws. An ethics code identifies those areas in which agency officials set their sights higher than the bare minimum requirements of the law. The values-based approach reflects the general distinction between the law and ethics. Fundamentally, ethics means obedience to the unenforceable.[36] Laws, of course, are enforceable—typically by those other than local agency officials. Obedience to the unenforceable requires self-regulation in light of ethical values.[37]

SAMPLE CODE OF ETHICS FOR ELECTED AND APPOINTED OFFICIALS AND EMPLOYEES

Declaration of Policy The proper operation of democratic government requires that public officials and employees be independent, impartial, and responsible to the people; that governmental decisions and policy be made in the proper channels of the governmental structure; that public office not be used for personal gain; and that the public have confidence in the integrity of its government. In recognition of these goals there is hereby established a code of ethics for all officials and employees of the agency, whether elected or appointed, paid or unpaid. The purpose of this code is to establish ethical standards of conduct for all such officials and employees by setting forth those actions that are incompatible with the best interests of the agency and by directing disclosure by such officials and employees of private financial or other interests in matters affecting the agency. The provisions and purpose of this code and such rules and regulations as may be established are hereby declared to be in the best interests of the agency.

Responsibilities of Public Office Public officials and employees are agents of public purpose and hold office for the benefit of the public. They are bound to uphold the Constitution of the United States and the Constitution of the state in which they reside and to carry out impartially the laws of the nation, state, and municipality and thus to foster respect for all government. They are bound to observe in their official acts the highest standards of morality and to discharge faithfully the duties of their office regardless of personal considerations, recognizing that the public interest must be their primary concern. Their conduct in both their official and private affairs should be above reproach.

Dedicated Service All officials and employees should be loyal to the objectives expressed by the electorate and the programs developed to attain those objectives. Appointed officials and employees should adhere to the rules of work and performance established as the standard for their positions by the appropriate authority. Officials and employees should not exceed their authority or breach the law or ask others to do so, and they should work in full cooperation with other public officials and employees unless prohibited from doing so by law or by officially recognized confidentiality of their work.

Fair and Equal Treatment

Interest in Appointments. Canvassing of members of the elected body, directly or indirectly, in order to obtain preferential consideration in connection with any appointment to the municipal service shall disqualify the candidate for appointment except with reference to positions filled by appointment by the elected body.

Use of Public Property. An official or employee shall not request or permit the use of agency-owned vehicles, equipment, materials, or property for personal convenience or profit, except when such services are available to the public generally or are provided as outlined in agency policy for the use of such elected or appointed official or employee in the conduct of official business.

Obligations to Citizens. An official or employee shall not grant any special consideration, treatment, or advantage to any citizen beyond that which is available to every other citizen.

Conflict of Interest Elected or appointed officials or employees shall not engage in any business or transaction or have a financial interest or other personal interest that is incompatible with the proper discharge of their official duties, is not in the public interest, or would tend to impair their independence or judgment or action in the performance of their official duties.

Incompatible Employment. Elected or appointed officials or employees shall not accept private employment when such employment is incompatible with the proper discharge of their official duties or would tend to impair their independence of judgment or action in the performance of their official duties.

Disclosure of Confidential Information. Elected or appointed officials or employees shall not disclose confidential information concerning the property, government, or affairs of the agency to advance their financial or other private interests without proper legal authorization.

Gifts and Favors. Elected or appointed officials or employees shall not accept any valuable gift, service, loan, or promise from any person or firm that to their knowledge is interested, directly or indirectly, in any manner whatsoever in business dealings with the agency; nor shall any such official or employee (1) accept any gift, favor, or item of value that may tend to influence him or her in the discharge of his or her duties, or (2) grant in the discharge of his or her duties any improper favor, service, or item of value.

Representing Private Interests Before Agency Departments or Court. An elected or appointed official or employee whose salary is paid in whole or in part by the agency shall not appear on behalf of private interests before any department of the agency. He or she shall not represent private interests in any action or proceeding against the interests of the agency in any litigation to which the agency is a party. An elected or appointed official may appear before an agency on behalf of constituents in the course of his or her duties as a representative of the electorate or in the performance of public or civic obligations. However, no elected or appointed official shall accept a retainer or compensation that is contingent upon a specific action by the agency.

Contracts with the Agency. An elected or appointed official or employee who has a substantial or controlling financial interest in any business entity, transaction, or contract with the agency, or in the sale of real estate, materials, supplies, or services to the agency, shall make known to the proper authority such interest in any matter on which he or she may be called to act in his or her official capacity. He or she shall refrain from voting upon or otherwise participating in the transaction or the making of such contract or sale. An elected or appointed official or employee shall not be deemed interested in any contract or purchase or sale of land or other item of value unless such contract or sale is approved, awarded, entered into, or authorized by him or her in his or her official capacity.

Disclosure of Interest in Legislation. An elected or appointed official or employee who has a financial or other private interest in any legislation shall disclose for the record of the elected body or other appropriate authority the nature and extent of such interest. Any other official or employee who has a financial or other private interest, and who participates in discussion with or gives an official opinion to the elected body, shall disclose for the record the nature and extent of such interest.

Applicability of Code. This code shall be operative in all instances covered by its provisions except when superseded by an applicable statutory or other charter provision of the agency and statutory or charter action is mandatory, or when the application of a statutory or charter provision is discretionary but determined to be more appropriate or desirable.

WHAT IS AN ETHICS AUDIT?

An ethics audit is one tool local agencies can use to assess the degree to which ethical standards influence decision making by both the agency and individuals within the agency. Ethics audits can serve either as reassurance that the agency's ethical house is in order or as an early warning of potential ethical blind spots that, if left unaddressed, could lead to embarrassment or worse down the road. There are three kinds of ethics audits:[38]

1. *The compliance audit* analyzes the degree to which one's ethics program meets the standards required by law and the degree to which both the organization's and individuals' behavior satisfies legal requirements.
2. *The cultural audit* explores how employees and other stakeholders feel about the organization's standards and behavior, including the perceived priorities and ethical effectiveness of individuals and subunits of the organization, as well as the organization as a whole.
3. *The systems audit* assesses the degree to which the ethical principles, guidelines, and processes are integrated within the organizational system.

Every agency should perform a compliance audit on a periodic basis as part of its minimum due diligence with respect to ethics laws. Although compliance with ethics laws is a floor and not a ceiling for ethical conduct, it is important to assure the agency is meeting minimum legal requirements for its practices. Not every agency needs or will benefit from a systems audit, which is most useful when an agency has reached a preliminary conclusion that increased attention to ethical issues would be beneficial. A cultural audit can sometimes assist the agency in reaching a preliminary conclusion that increased attention to ethical issues would be beneficial. Adopting a positive, values-based ethics code is one step a local agency can take to demonstrate its commitment to ethical conduct and decision making. Such codes can only make a difference, however, if the values expressed in them are actively reinforced on a daily basis. When such consistent reinforcement occurs, the agency is more likely to meet the aspirations in the code; when reinforcement does not occur, the likelihood of the code making a difference in the agency's behavior is significantly diminished. Ethics audits are one such way to measure the degree to which an agency is walking the talk.

KEY ETHICS LAW PRINCIPLES FOR PUBLIC SERVANTS

The following principles drive the ethics laws in many states.[39] If you find yourself in a situation that implicates one of these principles, talk with your immediate supervisor or agency counsel as soon as possible about the specifics of what the law does and does not allow.

Personal Financial Gain—Appearing to Influence Decisions

Public Officials

- Must disclose their financial interests.
- Must disqualify themselves from participating in decisions that may affect (positively or negatively) their financial interests.
- Cannot have an interest in a contract made by their agency.
- Cannot request, receive, or agree to receive anything of value or other advantages in exchange for a decision.
- Cannot influence agency decisions relating to potential prospective employers.
- May not acquire interests in property within redevelopment areas over which they have decision-making influence.

Personal Advantages and Perks Relating to Office

Public Officials

- Must disclose all gifts received of $_____ and may not receive gifts aggregating to over $_____ (each state and local jurisdiction may have different limits established) from a single source in a given year.
- Cannot receive compensation from third parties for speaking, writing an article, or attending a conference, if paid to attend by their employer.
- Cannot use public agency resources (money, travel expenses, staff time, and agency equipment) for personal or political purposes.
- Cannot participate in decisions that may affect (positively or negatively) their personal interests.
- Cannot accept free transportation from transportation companies.
- Cannot send mass mailings at public expense.
- Cannot make gifts of public resources or funds.
- Cannot receive loans over $___ from those within the agency or those who do business with the agency.

Fairness, Impartiality, and Open Government

Public Officials

- Cannot participate in decisions that will benefit their immediate family (spouse or dependent children).
- Cannot participate in quasi-judicial proceedings in which they have a strong bias with respect to the parties or facts.
- Cannot simultaneously hold certain public offices or engage in other outside activities that would subject them to conflicting loyalties.
- Cannot participate in entitlement proceedings—such as land use permits—involving campaign contributors (does not apply to elected bodies).
- Must conduct the public's business in open and publicized meetings, except for the limited circumstances when the law allows closed sessions.
- Must allow public inspection of documents and records generated by public agencies, except when nondisclosure is specifically authorized by law.
- Must disclose information about significant fundraising activities for legislative, governmental, or charitable purposes.

A Public Official's Conflict of Interest Checklist

Key Concepts

- A public agency's decision should be based solely on what best serves the public's interests.
- The law is aimed at the perception, as well as the reality, that a public official's personal interests may influence a decision. Even the temptation to act in one's own interest could lead to disqualification, or worse.
- Having a conflict of interest does not imply that you have done anything wrong; it just means you have financial or other disqualifying interests.
- Violating the conflict of interest laws could lead to monetary fines and criminal penalties for public officials. Do not take that risk.

Basic rule: A public official may not participate in a decision—including trying to influence a decision—if the official has financial or, in some cases, other strong personal interests in that decision. When an official has an interest in a contract, the official's agency may be prevented from even making the contract.

MAKING ETHICS A CORE VALUE

Examine Your Ethical Climate and Put Safeguards in Place Organizations are composed of cultures. Take a good close look at your culture. What are the norms of behavior? What is valued? Are employees rewarded for succeeding at any cost, or are they urged to be stewards of the organization's reputation as well as its assets? What pressures do they face to commit misconduct? What systemic problems exist that could encourage good people to make bad decisions?

Don't Just Print, Post, and Pray If you have a code of conduct or an ethics code, printing copies and posting them on the wall and bulletin board is not enough. Codes of conduct are an outgrowth of the department's missions, visions, and values. Thoughtful and effective corporate codes provide guidance for making ethical business decisions that balance conflicting interests. Codes of conduct need to be actual living documents encouraged and valued at the highest levels. Elected officials and senior executives have to set an example for the type of conduct they expect from others. Ethical lapses at the upper echelons of management tend to be perceived as tacit permission to choose the "path of least resistance" at lower levels. Senior management needs to hold itself to the highest standards of conduct before it can demand similar integrity from those at lower levels. Executives who refuse to tolerate misconduct among their peers and who actively seek to model high standards of honesty, transparency, and trustworthiness can best demonstrate the commitment to ethical conduct.

Publicly Commit to Being an Ethical Organization Go public. Departments that are open about their ethical standards and conduct seem to be more trustworthy than those who stay silent. Openly post your vision, values, and codes of conduct on your web sites for public viewing.

Talk with Employees at All Levels Often! Failure to communicate causes far more pain than smashing your thumb with a hammer. The sore thumb will heal; poor communication can be fatal.

In the 1980s, Tom Peters talked extensively about *managing by walking around* (MBWA). In the purest sense, MBWA is a way for supervisors and managers to best communicate their (task and ethical) expectations and requirements in daily, informal meetings with employees. These informal conversations give employees two sets of data: spoken information that is exchanged and inferred data that they glean from the more subtle communications that accompany a manager's words. Employees basically want two things: to know what is expected or required of them and to be successful. They also want to know "how they are doing" at this point in time.

Communicate the Following: Goals, Roles, Expectations, and Priorities

- *Goals.* When "wandering around" make certain that you remind people of the short-term and long-term goals of the job. They should see how their goals support the organization's mission and vision. This is also an excellent opportunity to tie goals to the code of conduct or code of ethics. Let employees know that how you accomplish a goal is just as important as accomplishing the goal itself. Cutting corners can hurt the department, its reputation, and eventually the individual employee.
- *Roles.* Let employees know how their piece of the job fits into the bigger picture. Remind them of their importance and value. Ensure that they understand their role as it relates to yours. Ensure that they understand what kind of conduct you expect.
- *Expectations.* Be certain that employees understand exactly what you expect. What has to be done? When? To what standards? How will it be evaluated? What should they do if they encounter any roadblocks or unanticipated changes? How do you want them to handle questions and/or "gray areas" in which expectations may be unclear or conflicting?
- *Priorities.* Remind your employees of the organization's operational priorities. If safety, quality, and customer service come first, for example, then make that clear to your employees. Be clear about what you expect them to do when they experience conflicts between any of these core values. Clarify what constitutes ethical conduct. Don't just assume your employees know where you stand. Ethics Resource Center Fellows Program research indicates that unless leaders clearly and consistently communicate their values, employees will assume they are neutral on the subject.

Build Ethical Conduct into Your Department Define your position as an ethical business. Provide employees and customers with a written pledge. These are our values, how we define what is right, fair, and good. We promise that all employees (at every level) of this organization will treat each other and customers accordingly. Train your employees on their ethical responsibilities. Teach people how to translate the pledge into specific actions that support the pledge and build trust. Provide support and guidance to employees. Take time to share what you have learned about how the pledge applies in particular cases within the organization. Measure your success. Implement simple systems to measure the effectiveness of this ethics initiative. Determine whether employees are living the pledge, and measure the differences it makes to your employees, your customers, and your organization's performance. Reward those employees who choose to live up to their promises, and deal with those who do not.

Choose to Live Your Corporate Values No rules and regulations manual, regardless of its thoroughness, can cover every contingency. And if one could be written that did cover all possibilities, it would occupy so much space and be so cumbersome to use that its covers would never be opened.

By equipping employees with department-supported values and empowering them to make decisions based on those values, you will free them to take action even when specific guidance is not readily available. You will also enjoy the peace of mind that comes from knowing your employees have common ground from which all decisions can be made.

Keep the Lines of Communication Open If you ask about what is going right, what is going wrong, and what makes employees uncomfortable in their jobs, you can usually identify pitfalls before you step into them. Communicate openly and honestly.

> Have the courage to say no. Have the courage to face the truth. Do the right thing because it is right. These are the magic keys to living your life with integrity.
>
> *W. Clement Stone*

ADVICE FROM THE EXPERTS

How have you installed ethics in your organization?

MICHAEL CHIARAMONTE

Chief Fire Inspector, Retired, Lynbrook Fire Department (New York): Through example.

BETT CLARK

Former Fire Chief, Bernalillo County Fire Department (New Mexico): Slowly, patiently, by actions rather than preaching.

KELVIN COCHRAN

Fire Chief, Atlanta Fire and Rescue (Georgia): As fire chief of the Shreveport Fire Department, I instill ethics beginning with inspiring a shared and compelling vision for the future and the benefits of shared organizational ethics in achieving the vision. Community and organizational ethics must be emphasized in the rules, regulations, policies, and procedures of the fire department. Those of greatest significance must be communicated on an ongoing basis by the chief and reinforced through accountability. The fire chief must proactively communicate high-risk, high-consequence ethical behavior to department members and then model the behavior in the community and department. Utilizing cable television broadcast internally to all fire stations, district meetings, and fire station visitation are methods I use frequently to reinforce ethical behavior.

CLIFF JONES

Fire Chief, Tempe Fire Department (Arizona): First by attempting to hire members for the department who are ethically oriented. Then by stressing ethical behaviors during the pre-academy orientation, during the recruit academy, and in the city-wide new employee orientation immediately following academy graduation. By stressing the importance of self-discipline, as many of the ethical problems we face boil down to issues of self-discipline. The first rule in our department's rules and regulations states, "All members of the Department are expected to be self-disciplined." On a continuing basis, through example setting, by myself, the command staff, and the company officers of the department. By encouraging members to have the courage to be ethical at all times through the notion of personal leadership, through appropriate and aggressive follow-up, when ethical conduct is in question, and progressive discipline as warranted.

BILL KILLEN

Retired Fire Chief and IAFC President 2005–2006: I try to instill ethics within my organization and throughout the plant by setting the example in ethical behavior. Showing respect to every member of the company and our clients at the plant is where ethics begins, particularly in those instances when my behavior may influence subordinates. It does not matter whether they work directly for me or for another division or one of our clients, I make every effort to demonstrate that I live by the principles and code of ethics of the IAFC.

KEVIN KING

U.S. Marine Corps Fire Service: Because we are a federal government agency, a lot of our ethics rules are thrust upon us legislatively, often to the detriment of running a high-performing organization. We have seen this in much of our procurement laws and rules. However, I do not think that you can instill ethical behavior legislatively. We have tried to instill ethics in many ways by focusing on our major customers. If the military forces cannot trust us to perform ethically while they are protecting our country, then we are letting down the people protecting our freedoms. I also believe that the peer process plays an important role in ethics. If we can get our people to understand the importance of ethics and how the public perceives our actions, then I think our people will learn the importance of strong ethical behavior. (These are opinions of Mr. King and do not reflect official policy of the Department of Defense or the United States Marine Corps.)

MARK LIGHT

Executive Director, International Association of Fire Chiefs: Ethics is instilled in organizations by what we say, do, and, most important, how we do it. We have a code of ethics, and rules and regulations, but ethics goes well beyond that. At times, someone in the organization will say something is not ethical or something is not right. It often sets us back when they say that, and yet when you stop and think about it, that tells us that ethics are alive and well in our organization. What better way to see that our staff embraces ethics than for them to openly question something among their peers? As a leader, we must show our ethics every day. We must be willing to question behavior and to discuss what is and is not appropriate. This often may go against popular opinion, but that is a part of leadership. The true test is seeing staff holding up a high level of ethics when no one is watching.

LORI MOORE-MERRELL

Assistant to the General President, International Association of Fire Fighters: Some would say that the fire service cannot give you what your parents did not; however, I disagree. Ethics or morals and values are areas that must be addressed with recruits in the very early stages and throughout the academy. Then it must be continually refreshed as it is "lived" daily in the stations on the street. Those who breach ethical behavior must be held responsible for their actions, must be dealt with quickly and clearly, and must not just be transferred to another station to cover the problem and hope it will go away. Additionally, officers must be held accountable for breach and inappropriate behavior of those working under them. Most, if not all, U.S. colleges and universities today have an honor code. If that honor code is broken for any reason, a student can be punished and may be expelled based on the magnitude of the breach. Inappropriate and unethical behavior is rarely tolerated. The fire service should maintain such a code. We are not looking at unintentional mistakes or ignorance here; however, we are addressing completely irresponsible and inappropriate behaviors for professional public servants, and leaders must know the difference. Unethical behavior by any firefighter must not be tolerated!

JAY REARDON

Fire Chief, Retired, Northbrook Fire Department (Illinois), currently President of the Mutual Aid Box Alarm System (MABAS): Ethics is a pretty broad subject and human dimension. Is ethics doing what is right only when everyone's looking or more important when nobody's looking? The answer is within one's ethical nature. Will giving only the ethically correct position always prove to produce the best outcome even if its result causes harm? Again, what's the correct answer, situational ethics? Perhaps the simplest answer on how to instill ethics in an organization is to understand yourself, your values, and your attitudes. Then, be strong enough to exercise them on and off duty. Find those worth mentoring for tomorrow and pass them on. People find with experience that the most powerful human emotion is shame. The fire service holds honesty, trust, and the "brotherhood" dear to its core reasons to belong. Unethical behavior is not tolerated within the fire service; shame within the work group is the fear that prevents that kind of behavior from surviving. Leaders must recognize opportunities to instill ethical behavior and have the commitment to act and reinforce the value in themselves and the organization.

GARY W. SMITH

Retired Fire Chief, Aptos–La Salva Fire Protection District (Aptos, California): Ethics are dictated by the actions and practices of the leadership and are promoted and made real by the team. The culture and status of the team is reflected by the actions of the group as shaped over the years of their existence. Instilling ethics into an organization that has been led and developed by unethical leaders is very difficult. People tend to be distrustful, and it becomes an organization of finger-pointing advocates for one direction or the other. The leader must work hard to focus the team on the importance of the new value system that the ethical leader wants to be a priority. The team may doubt the motive and be unsure or distrusting of the actions that the leader calls ethical. It will be tested constantly until the leaders within the group decide to accept that practice as "ethical" and worthy of accepting as a new part of the organizational culture.

STEVE WESTERMANN

Fire Chief, Central Jackson County Fire Protection District (Blue Springs, Missouri), and IAFC President 2007–2008: The biggest rule to follow here is to make sure you can walk the talk. Do it by example in every facet of your life. Training and seminars are great, but you have to live by what is right every second of the day. Begin to make sure that those who work directly for you understand the limits and abide by them.

There will always be questionable situations that arise. Discuss them openly with your staff or your crew. After discussion, if there is still a hint of a question, or it does not pass the "smell test," then err on the side of safety.

THOMAS WIECZOREK

Executive Director, Center for Public Safety Excellence, and Retired City Manager: One of the things that I tried as a city manager and have tried to include in the policies of the CPSE is prohibiting activities that bring public anger. Financial management must be good, reportable, and transparent. Employment policies must be written, enforced, and not be so cumbersome that they are viewed as "avoidable." Too often managers, chiefs, and others produce huge multivolume processes, procedures, and policies that even the most conscientious employee would have trouble comprehending. A good starting point is either the ICMA or IAFC Code of Ethics, which guides the actions of individuals and organizations. However, hanging it on the wall is not the answer. The leader must be willing to follow the production because the rank and file can spot a fraud.

JANET WILMOTH

Editor, *Fire Chief* magazine: In the magazine it's tough NOT to be ethical. Best I can offer is accountability and trust.

FRED WINDISCH

Fire Chief, Ponderosa Volunteer Fire Department (Houston, Texas): Ethics is as simple as approving expenditures that are efficient and effective. Ethical behavior is a constant that is demonstrated on a daily (hourly?) basis. A leader cannot play favoritism and must focus on the right decision versus the most popular decision. I attempt to demonstrate ethical behavior in everything we do. From daily customer interactions to assuring that we do not let our personal emotions interfere with decisions, we try to be ethical every day, every time. Setting positive examples of this behavior will work via osmosis!

Review Questions

1. What is your definition of ethics?
2. Does your department have an ethics code? If so, what was its origin?
3. Provide an overview of the ethical changes that have occurred during the past three decades.
4. Describe the ethical environment of your own department.
5. Has your department or jurisdiction ever dealt with improprieties of employees? If so, how was the situation handled?
6. Describe the potential consequences of not reporting an ethical violation or misconduct.
7. What extent does ethics play into the culture of your organization?
8. How does ethics impact the decision-making process?
9. Describe the political climate in your jurisdiction as it relates to ethical behavior.
10. What are the key elements of law that drive most ethics laws in most states?

References

1. Kenneth Majer, *Values-Based Leadership: A Revolutionary Approach to Business Success and Personal Prosperity* (San Diego, CA: Majer Communications).
2. "Quotable Quotes," *Chicago Tribute Magazine* (January 1, 1996), p. 17.
3. Al Gini, *Moral Leadership and Business Ethics* (Loyola University Chicago in Ethics & Leadership Working Papers) (College Park, MD:Academy of Leadership Press, 1996).
4. John Rawls, "Justice as Fairness: Political Not Metaphysical," *Philosophy and Public Affairs*, 4, no. 3 (1985), 223–251.
5. Pope Pius XI, "*Quadragesimo Anno* (On Reconstructing the Social Order,)" in David M. Byers, ed. *Justice in the Marketplace: A Collection of the Vatican and U.S. Catholic Bishops on Economic Policy*, 1891–1984 (Washington, DC: United States Catholic Conference, 1985), p. 61.
6. Henry Ford, Sr., quoted by Thomas Donaldson, *Corporations and Morality* (Upper Saldle River, NJ, Prentice-Hall, 1982), p. 57.
7. Ibid., p. 14.
8. R. Edward Freeman, *The Problem of the Two Realms,* speech, Lonola University, Chicago, spring 1992.
9. The Quotations Page, www. Quotationspage.com. 2007.
10. Thomas J. Peters, and Robert H. Waterman, *In Search of Excellence* (New York: Harper and Row, 1982), p. 245.
11. The Quotations Page, www. Quotationspage. com. 2007.
12. William James, *The Will to Believe* (New York, Dover Publications, 1956), pp. 1–31, 184–215.
13. John Gardner, *On Leadership* (New York: Free Press, 1990), p. 9.
14. Gail Sheehy, *America's Search for Leadership* (New York: William Morrow, 1988).
15. Howard S. Schwartz, "Narcissistic Project and Corporate Decay: The Case of General Motors" (*Business Ethics Quarterly*, 1, no. 3 1991), p. 250.
16. Ibid.
17. Robert Jackall, *Moral Mazes* (New York; Oxford University Press, 1988), p. 6.
18. 2003 National Business Ethics Survey (NBES), www.ethics.orglerc-publications/ethics-today.asp?aid-733
19. See this article on the Ethics Resource Center (ERC) web site for a list of works that contributed to the development of the Organizational Integrity approach.
20. Selections from *Ethics Today* Online, Volume 2: September 2003 to July 2004.
21. Josephson Michael, suggestions for formulating standardiraly and workable policie, www.americanresearchinstitution.org, 2002.
22. "Everyday Ethics for Local Officials, "Navigating the Perils of Public Service, Part One and Part Two," *Western City Magazine*, August 2005 and September 2005.
23. Randy R. Bruegman, *The Chief Officer: A Symbol Is a Promise* (Upper Saddle River, NJ: Prentice Hall, 2005).
24. Commission Fire Officer Designation (CFOD) Code of Conduct. Chartilly 2005
25. International Association of Fire Chiefs, Code of Ethics. FairFax, Feb. 2002.
26. George Protopapas, "Ten Commandments of Political Engineering," *County Engineers of California (CEAC) Newsletter*, 1993.
27. Clackamas Fire District No. 1, Oregon. *Milwabee,* Oregon. 2000.
28. Federal Emergency Management Agency, United States Fire Administration, National Fire Academy, *Leadership: Strategies for Personal Success, Ethics*, Student Manual, January 1994.
29. John R. Wilcox and Susan L. Ebbs, *The Leadership Compass: Values and Ethics in Higher Education* (San Francisco: Jossey-Bass, 1992).
30. Mitchell Crowden, *Lost Ethics in Public Sector Leadership*, California State

University Sacramento, Educational Leadership and Policy Studies, EDLP 225, Advanced Seminar: Higher Education Ethics, Dr. Rosemary Papalewis, September 29, 2004.

31. Carol W. Lewis, *The Ethics Challenge in Public Service: A Problem Solving Guide* (San Francisco: Jossey-Bass, 1991), p. 139.
32. J. S. Zimmerman, "Ethics in Local Government," *Management Information Service Report 8*, International City/County Management Association, August 1976.
33. Lewis, *The Ethics Challenge in Public Service,* p. 143.
34. Jane G. Kazman and Stephen J. Bonczek, *Ethics in Action: Leader's Guide* (Washington, D.C.: International City/County Management Association, 1999), p. 97.
35. The concept of ethics defining how local officials behave when they are "at their best" is a theme that runs throughout the city of Santa Clara's groundbreaking code of ethics and values. The city developed the code with the help of Dr. Thomas Shanks of the Markkula Center for Applied Ethics at the University of Santa Clara.
36. Early twentieth-century English jurist Fletcher Moulton, quoted in Rushworth M. Kidder, *How Good People Make Tough Choices* (New York: Simon & Schuster, 1995), p. 66.
37. Patricia L. Brousseau, "Ethical Dilemmas: Right versus Right," in *The Ethics Edge* (Washington, D.C.: International City/County Management Association, 1998), p. 38.
38. Frank J. Navran, *Ethics Audits: You Get What You Pay For* (Washington, DC: Ethics Resource Center, 1998) (www.ethics.org).
39. Institute for Local Self Government, *Key Ethics Law Principles for Public Servants*, California League of Cities.

Personnel Management: Building Your Team

Key Terms

cognitive styles, p. 296
coping styles, p. 295
IAFF, p. 313
Maslow's hierarchy, p. 290
participative management, p. 323
team performance model, p. 290

Objectives

After completing this chapter, you should be able to:

- Define the areas of importance to building a successful team.
- Conduct a SWOT analysis on your department.
- Explain why people often resist change.
- Describe how coping styles impact group dynamics.
- Describe the profile of each generation.
- Describe how morale can affect and impact an organization.
- Discuss the importance of the labor–management relationship.

BUILDING YOUR TEAM

The next time you go to a movie, stick around and watch the credits roll at the end. Try to visualize all of the people and the wide range of skills it took just to make that one movie! Movie production is a highly complex process, requiring the work and coordination of many people.

Teamwork is something most of us do almost every day. If you are a member of a family, have worked on a committee, or taken part in organized sports, you have participated in teamwork. Most of us recognize that teams are more creative and productive than individuals working alone.

Because teamwork is so important to success in the field of organizations, when you become part of one, it is essential that you learn how a group of individuals develops into a cohesive unit. The fire service and its delivery of service is built upon a strong foundation of teamwork. Without it, fire service employees place themselves at risk and cannot adequately serve the public.

Robert Fulghum wrote *All I Really Need to Know I Learned in Kindergarten.* An excerpt from his book sums up what teamwork is all about.

> Most of what I really need to know about how to live and what to do and how to be I learned in kindergarten. These are the little things I learned—share everything, play fair, don't hit people, put things back where you found them, clean up your own mess, don't take things that aren't yours, say you're sorry when you've hurt somebody, wash your hands before you eat, flush, warm cookies and cold milk are good for you, lead a balanced life, learn some, think some, draw, paint, sing and dance, play and work every day. Take a nap every afternoon. When you go out in the world watch for traffic, hold hands and stick together. Be aware of wonder. Remember the little seed in the plastic cup—the roots go down and the plant goes up. Nobody really knows how or why but we all like that. Goldfish, hamsters, white mice and even the little seed in the plastic cup—they all die and so do we.[1]

When I was growing up, I remember reading the children's book *Dick and Jane* and one of the first words I learned, may be the most important word of all: *look.* Everything you need to know is there somewhere if you take the time to look. The Golden Rule, love and basic sanitation, ecology, politics, and sane living. Think of what a better world it would be if we all, the whole world, had cookies and milk at 3:00 every afternoon and then laid down with our blankets for a nap. Or what if we had a global basic policy always to put things back where we found them and to clean up our messes. It is still true no matter how old you are, when you go out in the world, it is best to hold hands and stick together. Fulghum's commonsense approach is a poignant reminder of the importance of the basics in life.

WHOSE SIDE ARE THEY ON?

To be an effective leader at all levels of the organization, we must return to the basic premises that Fulghum writes about. It really is about holding hands and sticking together. But getting to that point with senior staff, company officers, and elected officials is often very difficult. Building your team can mean very different things, depending upon the culture of the organization, the level of trust within the organization, and the leadership talent you bring to the table. It is amazing that on the fire ground we can demonstrate one of the most effective displays of teamwork known to any modern organization, yet the same people sitting around a table in a work group can be one of the most dysfunctional teams ever assembled.

The art of leadership lies in the ability to blend the talents of the team to achieve great things together. If you are to be a successful leader and manager, building your team is a must. Assessing and changing the dynamics of existing groups within the organization are often not easy tasks, but many strategies are available to bring people and groups together toward organizational vision and objectives.

> In the world of corporate olympics, there are many different games to be entered. And there are common characteristics found in those teams that help them to compete and win, again and again. These characteristics—strength,

skill, discipline of the athlete—focus on individual excellence, coupled with the ability to work within a well-organized team.[2]

There are *six areas of importance in building an effective team* that must be understood and built upon if we are to be successful in our team building efforts.

1. *What is the level of trust in the organization?* For most organizations, the current level of trust is a window on past events that have created that trust level. It is important for the new chief executive to the company officer to understand the level of trust in the department and the reasons for it. Some organizations are so damaged by such things as poor labor–management relations, bad leadership, and poor work habits that to be effective, those in leadership positions will require a different approach. In such situations, leaders must spend a great deal of energy attending to the basic psychological needs of personnel and attacking the culture of the organization. The word *attacking* denotes the importance of addressing a culture of negativity and energy required to do so. A culture that promotes the status quo, where the level of trust is low and open communication is not fostered, can paralyze an organization.
2. *How willing are your personnel to be honest with each other?* It is a fascinating experience for a new fire chief, battalion chief, or company officer to watch how the dynamics work, or do not work, in the department, in the battalion, or at the company level; but, to experience this, you must be willing to observe and be open minded. During the first several months of any new assignment, you are trying to determine the strengths and weaknesses of the team you will be working with. Conducting a *strength/weakness opportunity threat (SWOT) analysis* may provide many insights and often fun surprises. Many people simply cannot tell the truth about where the organization as a whole stands or where their own division or company is in relation to the organization's missions and performance.

Begin your assessment with a look at the values and culture in play. Each organization has a distinctly different culture and value system. You can utilize several methods of analysis to gain a perspective on your organization. A useful and easy-to-use one that provides a broad view of the organization, the division, or the company is the SWOT assessment (SWOT: Strengths, Weaknesses, Opportunities, Threats). This can give you a quick overview of how others view your organization, division, or company and allows you to determine where you will need to take your team in the first 12 months. SWOT also identifies organizational barriers to improvement. For example, if one division leader cites a program as a strength and another division leader cites it as a weakness, this signals an opportunity and a barrier—poor communication. The SWOT assessment allows you to open an organizational dialogue on what needs to be focused on and built upon (see Table 7.1).

I. Describe the issue being evaluated in detail and explain how it impacts the community.

II. Situational analysis (where are we now?)

A. The Situational Environments

1. Demand trends. (What is the forecast demand for the service? Is it growing or declining? How, when, where, what, and why?)
2. Social and cultural factors of the department, the jurisdiction, the city, and/or county.
3. Demographics.
4. Economic and business conditions for services at this time and in the geographical area selected.
5. State of technology for the agency. (Is it high-tech, state-of-the-art? In short, how is technology affecting service delivery in a positive or negative manner?)
6. Politics. Are politics (current or otherwise) in any way affecting the situation for the service? What is the political landscape, both internally and externally to the department?
7. Laws and regulations. (What laws or regulations are having a positive/negative impact?)

TABLE 7.1 ◆ Analysis of the Organizational Situation

Analysis of the Organizational Situation (SWOT)

STRENGTHS	WEAKNESSES
OPPORTUNITIES	THREATS

Sample SWOT Outline

Source: Randy R. Bruegman, *The Chief Officer: A Symbol Is a Promise,* p. 67.

B. The Local Environment Factors

1. Financial environment. (How does the availability or unavailability of funds affect the situation?)
2. Government environment. (Is current legislative action in the federal, state, or local government likely to affect the agency?)
3. Media environment. (What's happening in the media? Does current publicity favor the agency?)
4. Special-interest environment. (Are there direct competitors or any influential groups that are likely to affect the agency?)

C. The Competitive Environments

1. Describe your main competitors (for the fire service this is often other departments in local government) and their services, plans, experience, know-how, and financial, human, and capital resources. (What are their strengths and weaknesses?)

D. The Organizational Environments

1. Describe services, experience, know-how, and financial, human, and capital resources of the agency you work for. Do you enjoy the favor of the employees? If so, why? What are the strengths and weaknesses of the system?

III. The Service Area

Describe your service area in detail by using demographics, geographic, lifestyle, or whatever segmentation tool is appropriate.

IV. Problems and Opportunities

State or restate each opportunity and indicate why it is, in fact, an opportunity. State or restate every problem. Indicate what you intend to do about each of them.

V. Objectives and Goals (Where do we want to be?)

State precisely the objectives and goals in terms of your strategic plan and the time needed to achieve each of them.

VI. Strategy

Consider alternatives for implementing the overall strategy.

VII. Summary

An interesting aspect to the SWOT analysis, while it can be applied to larger-scale organization assessment, is that it can also be used as an effective tool to evaluate a battalion or a single-company house.[3]

3. *Does your organization have the ability to communicate up, down, and sideways?* It is very useful to watch how organizations, especially when coming in from the outside, communicate throughout the department. This observation tells you a lot about the personalities involved, their leadership style(s), and the organizational culture. An organization of excellence has the

ability to communicate in all directions. If there are communication issues, determine whether they are symptomatic of the organization as a whole or whether they concern one person or one division. Organizations that only communicate upward may be reflective of a lack of leadership and are often being driven from below. Organizations that allow only top-down communications are autocratic, hierarchical ("I have five bugles; therefore, we'll do it my way"), and often very myopic. Organizations that only communicate horizontally or sideways are being driven largely by their mid-level managers. Their leadership probably has no clue to what is occurring throughout the organization, nor does the firefighter stationed at the firehouse have a grasp of what the organization is moving toward on a macro scale. This creates a substantial void in the organization and can be very disruptive as it promotes different groups going off and doing their own thing. The goal is to promote effective communication in all directions, using appropriate chain of command for the agency. So, the nature of the organization's communication provides a great field of observation that, as a leader and manager, you must continually pay attention to, no matter what level you happen to be in.

4. *Does the organization kill the messenger?* Trust in an organization means being able to be open and honest with one another. In some organizations, the bearer of bad news is the first one to go. This does not promote open communication and reflects poor organizational culture and dynamics. If the bearer of bad news is punished, the messages will stop coming. This is a good measuring stick for leaders at all levels of the department to observe for the first three to six months in a new position or assignment. You can tell a lot about the individuals you work with and interact with by how they handle adversity. If they have a tendency to kill the messenger, begin to work with them on the issue. If you do not, a serious situation may develop and you may find yourself squarely in the middle. All organizations experience this to some degree. In a healthy organization this often self-regulates when another employee not involved in the issue steps in and stops the frenzy with logic and reason. When the organizational culture is unhealthy, others will let the frenzy continue to the detriment of the individual and the organization as a whole. What kind of department is yours?

5. *Is the focus on problem solving or decision making?* Depending upon your rank in the department, watch from your vantage point. Is the department decisive or indecisive? How does it process information and solve problems, if at all? Some people, once they pin a gold badge on their uniform, find it easy to make all the decisions. Although the authority is there to do so, this is not a productive way to empower the organization and create a team. On the other hand, if you are the new chief or company officer, indecision is what you may find simply because people are trying to figure you out, just as you are trying to figure them out. Because they do not want to step over the line, they will wait until they can determine your leadership style and what you expect of them. When you begin to build your team, you want to avoid a workforce that does not speak up because they fear being ostracized or cut off from the information flow. If the majority of the workforce is unwilling to speak to the people who can help solve problems, the leaders and managers of the organization have an inadequate pool of information for making quality decisions. This situation, in turn, often fuels workers' perception that managers are untrustworthy or incompetent. Therefore, it is extremely important that organizationally, as a team, we begin to consider how to process information to solve the problems of the department.

6. *To what degree do personnel have a shared organizational mission, values, and visions?* This is one of the most overlooked issues in any organization, whether it is a fire department or a *Fortune* 500 company. The culture and expectations of the organization will often dictate its success. Shared values often start with the leader of the organization outlining his or her values and expectations of behavior to the group, but this is not a one-way communication. If they are truly to be organizational values, they must incorporate those of the workforce. Be willing to write them down and incorporate them in your organization's strategic planning process and budget documents, posting values, value statements, and codes of conduct on the wall. This begins to create a shared expectation of how people within the organization should act, and it helps to develop a clear understanding of what you, as a chief fire officer or company commander, expect. Create shared values for the organization through many different processes—labor–management committees, volunteer or association–management processes, strategic planning, project initiatives, team building, and community outreach programs. Values accepted and shared by the majority of the workforce help to drive the organization by providing a mechanism for developing shared vision, goals, and objectives that are essential for future organizational success.

A NEW BEGINNING

Your objective as you begin to build your team is to evolve to a new level of understanding, attitude, and value regarding the team, and doing so can create mixed emotions. Creating a team environment requires providing a safety net so your people feel comfortable enough to take risks, knowing that if they fail, you will help them achieve their goal. Successful team building often depends on what I call the "four P's": *purpose, picture, planning,* and *part.* Framing the purpose lets the action become real to the people involved; in many cases, no purpose exists until it has been clearly defined and understood. You can strengthen the sense of purpose by painting a picture of how the outcome will look and feel. One way is to design a step-by-step plan, including what the ultimate outcome will be. Remember, in the fire service, we are trained for fire ground operations, so we are a linear group. If we can visualize the goal, we are much more likely to achieve it. Teams are an essential part of the change process; therefore, defining what part everyone is to play is vital. Imagine being assigned to a football team and when you show up on game day, the coach picks 11 players and tells them, "Get out there and score some points." Who's playing what position? What about the plays? Who is good at what? Adequately defining the part each of the team members is to play is a factor often overlooked. Providing this framing experience is essential for evolving your team's culture and embarking on becoming a unified group.

As we build our team, reorganization may be required. Reorganization may go against basic values that many in our departments hold dear. The team performance model (see Figure 7.1) is reflective of the stages departments, divisions, and small groups go through during changing times.[4] Most teams are continually moving through these stages as new people are added or changed when new assignments are given or organizations' environmental factors shift. Many of the proposed changes may contradict the way members believe things should be done and, as such, they often resist change for the following reasons:

1. They are afraid of the unknown.
2. They think things are fine and do not understand the need for change.
3. They are inherently cynical about change, particularly if it is viewed as just the latest management technique.
4. They doubt there are effective means to accomplish the proposed reorganization.
5. They see conflicting goals in the organization (i.e., increasing resources to accomplish the change while cutting costs to remain viable).

One of the best ways to address these issues is through increasing communication. Detailed plans should be developed and communicated. Plans do change; understanding this, make sure you communicate changes and why they have occurred. Encourage forums within the organization that allow employees to express their ideas, concerns, and frustrations. Whenever change occurs, something ends and something new begins.

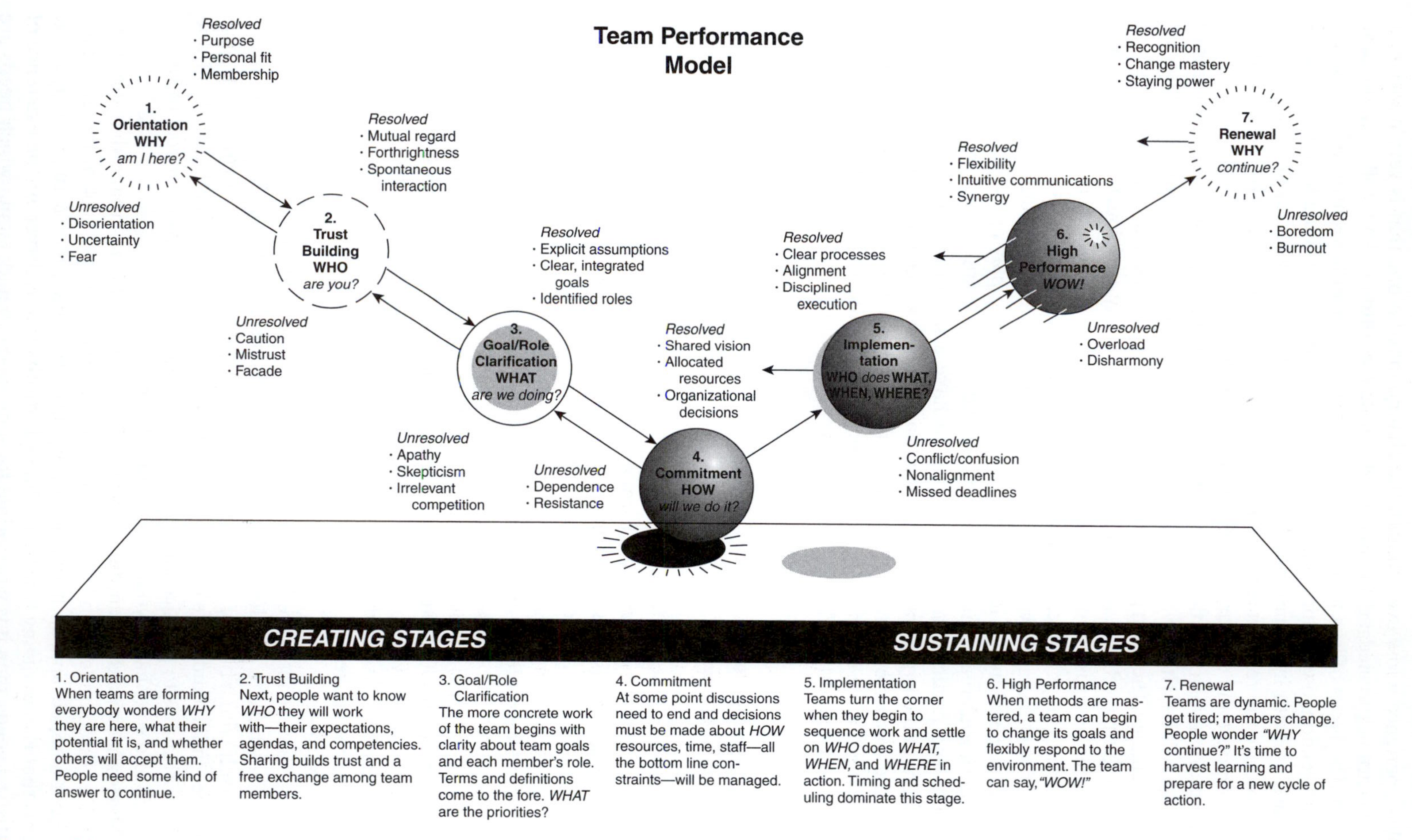

FIGURE 7.1 ◆ Team Performance Model
(Courtesy of Randy R. Bruegman, The Chief Officer: A Symbol Is a Promise, *p. 137)*

People experience a loss even though the change may be one that is very positive; if the loss is not acknowledged and managed, employees often will not follow in the new direction. The types of loss that employees experience include security, relationships, sense of direction, ownership, and territory. People are more willing to make changes if they are led with information than if they are simply told what to do. Information helps to reduce the resistance to change, and the response will likely be more positive with the support and encouragement from leadership.

team performance model
Allan Drexler and David Sibbet developed a comprehensive model of team performance that shows the predictable stages involved in both creating and sustaining teams. The *Drexler/Sibbet Team Performance™ Model (TPM)* illustrates team development as seven stages, four to create the team and three to describe levels of performance.

The more involved senior leaders are with their teams and each other, the easier the change will be. Those who create trusting relationships are more successful during periods of change. The **team performance model** reflects what organizations and groups go through during a change process. Teams often repeatedly go through the various levels of creating and sustaining performance as a team. As a leader, you must recognize your leadership helps shape the direction of the teams, and the process requires understanding of and empathy with the work and emotions of the individuals in the group.

Last, but certainly not least, give each person a part to play in the plan. The more stakeholders you create, the more likely you will be successful. People need a way to contribute and participate, and they want to have a meaningful role in relation to what others in the organization are doing. Each person's role must be defined in regard to the outcomes that you anticipate from a leadership perspective. Bringing every member of the organization into the process, defining roles, and providing clear direction and support will help to create successful teams.

BUILDING HIGH-PERFORMANCE TEAMS

As discussed previously, fire services can respond to an emergency demonstrating extremely effective teamwork yet form a very dysfunctional team at the conference table. The outcome is quite different when there is no purpose, motivation, understanding, leadership, and fellowship. What can you as a leader and manager do to develop high-performing teams that create opportunities, devise solutions for your most complex problems, and explore future ideas and concepts? It often comes down to your ability to assess and possibly change the existing group dynamics and then create an environment that allows group members to develop into high-performance teams. Chapter 3 introduced Abraham Maslow's ideas on motivation and people's needs. One of the key roles of any leader is to coach team members to achieve their best. As coach and leader you will typically help your team members to solve problems, make better decisions, and learn new skills or otherwise progress in their role or career.

While some leaders are fortunate enough to get formal training as coaches, many are not. They have to develop coaching skills for themselves. While challenging, if you arm yourself with a base of knowledge of what drives a team to perform at peak performance, understand some of the proven techniques that make teams successful, and find opportunities to practice and learn to trust your instincts, you can become a better coach and so enhance your team's performance.

Maslow's hierarchy
A theory in psychology that Abraham Maslow proposed in his 1943 paper, *A Theory of Human Motivation,* which he subsequently extended. His theory contends that as humans meet their basic needs, they seek to satisfy successively higher needs that occupy a set hierarchy.

Maslow's hierarchy identifies a series of levels of satisfaction that either motivate persons or discourage them from acting. Although many people look at the hierarchy as a stair-stepped approach that is continually escalating, it is actually a dynamic process that is undergoing constant change. If motivation is driven by the existence of unsatisfied needs, then it is worthwhile for a manager to understand which needs are more important for individual employees. In this regard, Abraham Maslow developed a model in which basic, low-level needs such as physiological requirements and safety must be satisfied before higher-level needs such as self-fulfillment are pursued. In this hierarchical model, when a need is mostly satisfied, it no longer motivates and the next higher need takes its place. Maslow's hierarchy of needs is shown in Table 7.2.

TABLE 7.2 ◆ Impacts That Affect Maslow's Hierarchy of Needs

Maslow's Hierarchy of Needs			
Needs	***Expression***	***Threat***	***Action***
Self-Fulfillment	Realizing potential Creativity Self-development of talents Self-projection of opportunities Individualism	Inflexible policies Narrow definitions of job tasks No interest in job enrichment ↓	↑ Trust Progressive climate Tolerating failure Management Development programs Empowerment
Ego Needs	Self-esteem mastery (confidence) (self-confidence) (independence) Reputation (recognition) (appreciation) (respect) (personal identify)	Job fragmentation Careless criticism Failure to praise or reward Centralization of authority Accoutrements of office Theory X style ↓	↑ Decision consultation Use of rewards and recognition Job enlargement and influence Subordinate feedback Visibility and exposure Use of goals
Social Needs	Belonging Association Acceptance Giving and receiving attention Friendship Love and be loved	Unfriendly action by management Office cliques and rejection Poor communication Ostracism ↓	↑ Participation in management Conduct morale and opinion surveys Communications and special bulletins Project management Theory Y style
Safety Needs	Protection against danger (arbitrariness) threat deprivation insecurity Unpredictability in environment	Arbitrary action by management Sudden changes Poor communication Lack of policy and procedure Favoritism Discrimination ↓ No benefits Lack of proper equipment	↑ Performance appraisal system Written policies and procedures Position guide Newsletters Bulletin boards Grievance process
Physiological Needs	Air, food, water, space, rest, and exercise Protection from the elements	Inadequate benefits Poor pay No break Cramped conditions Poor facilities Excessive hours ↓	↑ A sound and effective safety program Good working environment Work planning and scheduling Rest, lunch, and holiday periods Shorter work week

As individuals and group members, all persons can function at different levels in the organizational hierarchy, depending upon the actions of their superiors and/or other group members. A basic premise of Maslow's hierarchy is that people cannot move to the next level until they have satisfied their needs at their current level. Similarly, groups will not have an adequate self-concept, a critical factor in performance, until they have worked through the lower levels and reached the esteem or group fulfillment level (the top level in Maslow's hierarchy). No group can feel good about itself when it feels threatened. Each group must be given an opportunity to perform at its highest possible level. Nevertheless, no group functions at its highest possible level if it is pushed back, held back, or otherwise prevented from ascending to the next level. Study the groups under your command or the groups you are a part of, and watch the reaction(s) the next time there is a personnel action, a lack of communication, or an unwanted or misunderstood policy. These groups will often move up and down the hierarchy of needs based upon these forces and their environmental impacts on them and the organization (see Table 7.3).

Choose a couple of work groups or fire companies to evaluate.

1. Where would you place each group on the hierarchy?
2. What factors have placed them at the satisfaction level?
3. What actions could you as a department leader/manager take to make each group more effective?

We often think of Maslow's hierarchy of needs as a theoretical aspect of employee psychology but, in fact, it has a practical application to group dynamics. Here

TABLE 7.3 ◆ Group Hierarchy of Needs

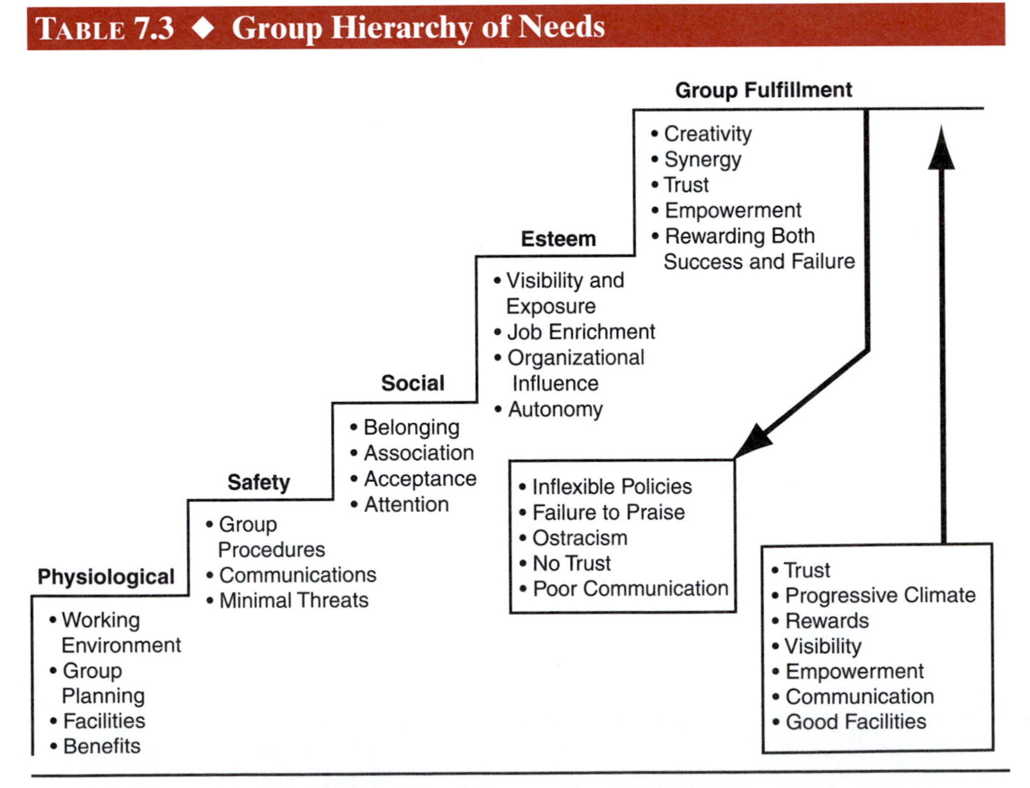

Source: Randy R. Bruegman, *The Chief Officer: A Symbol Is a Promise*, p. 143.

are some examples. Say you go through contract arbitration; the management proposals win and the firefighters' association loses. What happens to morale? Where do the employees immediately go on the hierarchy of needs? They go to the bottom, where they focus on little else than their own needs. You may have to cut your budget by 15 percent, lay off 10 firefighters, and close a fire station. Or say your budget is increased, and you will be able to open a new fire station, hire new firefighters, or purchase several pieces of equipment. Where do you think the majority of employees will be on the hierarchy, and what challenges will you, as a leader, face as a result of each scenario? Events can drive the entire organization, or different groups within it, to different levels within the hierarchy of needs. Understanding this from a leadership perspective will help you to determine what your challenges are and where you need to focus your time and energy (see Figure 7.2).[5]

The second assessment tool is the framework for high-performance programming. Take the same groups you used before, and place them in the framework model.

1. Where would you place each group in the model?
2. What factors have placed them there?
3. What actions would impact the group and move it to a higher level?

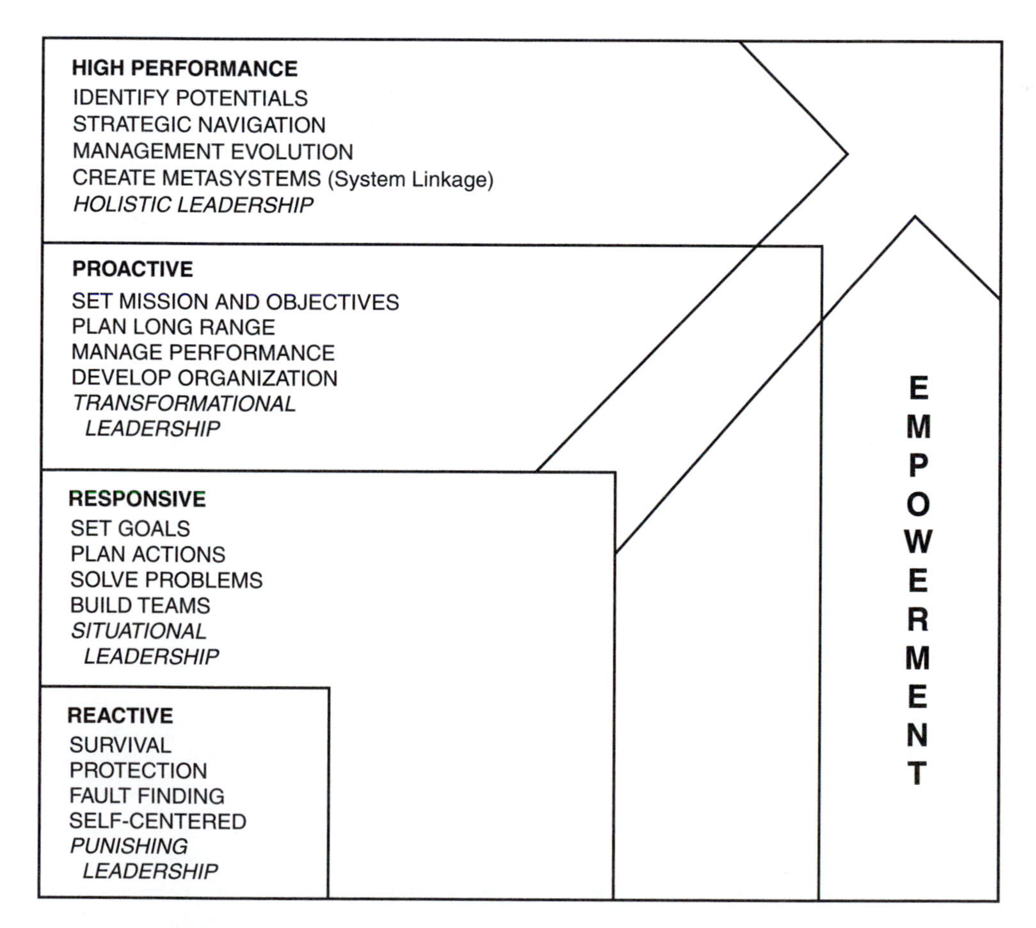

FIGURE 7.2 ◆ The Framework of the High-Performance Programming Model
(Courtesy of Randy R. Bruegman, The Chief Officer: A Symbol Is a Promise, *p. 144)*

CHARACTERISTICS OF A HIGH-PERFORMANCE TEAM

High-performance teams seldom exist by accident. Rather, they are an articulation of leadership that stems from adherence to basic principles such as trust, meaningful responsibility, commitment, high standards, and the willingness to confront inadequate performance and reward exceptional performance. In a group, several elements, both internal and external, contribute to the development of high-performance teams. These factors increase or decrease the ability of the team to perform as a group (see Figure 7.3).

Desmond Tutu has said, "I would not know how to be a human being at all, except I learned this from other human beings. We are made from a delicate network of relationships, of interdependence."[6] Is this not true within our departments or work groups as well? We often learn how to behave and act in the workplace by observing our peers' actions and reactions to situations.

Individual Needs Effective facilitation of groups necessitates an understanding of the needs of individuals in a group context. Everyone in a new group situation experiences the following three primary needs, although, individuals vary in the degree to which these needs are felt.

1. *Inclusion, Identity*. Every person in a group needs to develop a viable role to identify within the group. Until individuals experience congruence between what is expected and what can be delivered, they will be preoccupied, anxious, and not entirely able to pay attention to the group's external tasks. Instead, their emotional energy will be channeled into the personal issue of finding a fit in the group situation.
2. *Influence*. Each individual in a group needs a certain amount of influence and control. In a sense, this reflects a core conflict of human experience—the need to feel powerful and appropriately independent of others, while still retaining membership in a group.

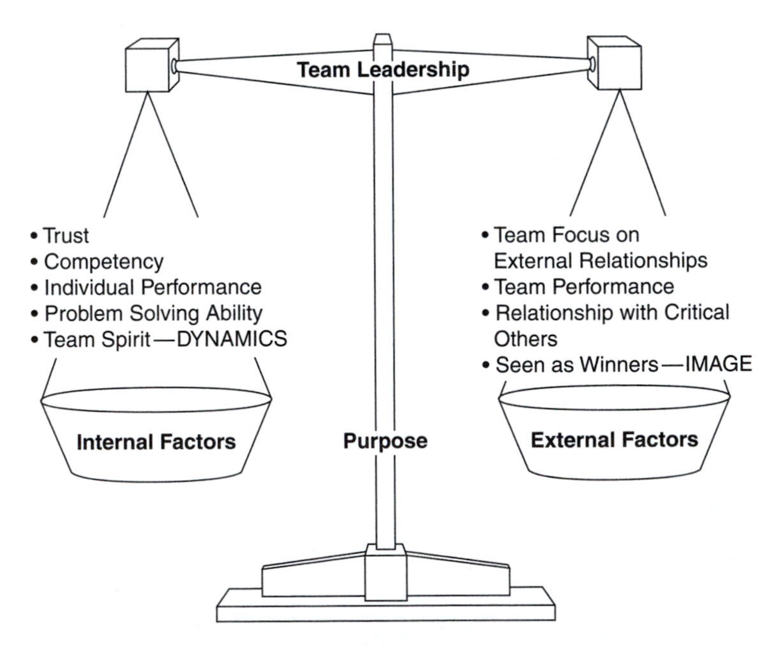

FIGURE 7.3 ◆ Team Leadership Provides Balance
(Courtesy of Randy R. Bruegman, The Chief Officer: A Symbol Is a Promise, *p. 145)*

3. *Acceptance.* Everyone in a new group needs to feel a degree of personal acceptance above and beyond the basic need for inclusion. This need reflects the other side of the core conflict—the need to belong in a deeper sense, to be truly accepted, and to achieve the basic security that comes with that.

Because these needs reflect the basic human needs for security, mastering of the environment, and acceptance, they can powerfully impact individuals. When these needs are not met, anxiety and preoccupation with the dynamics of the group often result, but when they are met, positive energy and engagement with the group result.

Every group member, including the leader, struggles to satisfy these needs. A group cannot fulfill its mission until the members have learned their needs can be met to some degree by belonging to the group. Feelings of comfort in a group situation can be expedited by focusing on the purpose of the group as well as by activities geared toward individuals getting to know one another. Although all individuals have similar needs in a group environment, people vary in how they cope in new group situations.

Coping Styles Groups can vary immensely in the kind of team dynamic they create, which often determines how successful they will be. The team dynamic often dictates whether the group actually develops into a high-performance team or gets stuck in one of the early stages of group development. Groups' differences are related to varying environments, leadership, and, most important, the complex interaction of the personalities of group members. Understanding emotional **coping styles** sheds some light on individual behaviors that impact group development.

coping styles
The concept of coping styles attempts to capture the variety of ways in which individuals try to cope with the demands of the environment. A common distinction is between "problem" and "emotion" focused strategies. There are many competing frameworks for describing and understanding coping styles. One important question is whether they genuinely capture the diversity of the stress experience.

Counterdependent type. Comfort in expressing assertive/aggressive, controlling feelings, but discomfort with and suppression of affection. Group members of this type cope with anxiety by actively testing the interpersonal environment. Behaviors might include challenging, controlling, competing with, and resisting other members, especially authority figures in order to sort out one's own identity.

Dependent type. Comfort in expressing affection and tender feelings, but discomfort with and suppression of aggression. Group members with this style tend to cope with anxiety by creating close relationships through supporting, helping, and developing alliances with others, and by leaning on authority in the attempt to sort out their own identity.

Logical thinker. Comfort with procedures, rules, and group processes that rule out the expression of either aggressive or loving emotions, based on discomfort with expressing emotions in a group context. Logical thinkers usually cope by withdrawing and passively feigning indifference, or by actively promoting nonemotional processes. The latter would include formal rules and procedures.

Every person has the capacity to use each of these styles, depending on the situation. However, most people develop a characteristic style that they lean on in group settings. Where would you place yourself? This model helps us understand why people have difficulty arriving at a consensus in a group situation. Members leaning toward any one of the styles are cognitively tuned to certain categories of interpersonal information. Thus individuals either do not understand where people of different styles are coming from and what their needs are, or they are threatened by different styles.

For example, counterdependent types are tuned toward identifying stronger and weaker members, those who can be controlled and those who are controlling, those who are winning and those who are losing. Counterdependent types tend to be threatened by too much affection. They attempt to create a group in which each member can test him- or herself against others and gradually determine a workable "pecking order."

Dependent types seek to determine other members who are warm and those who are cold, those who are helpful and those who are not, those who are supportive and those who are threatening. Dependent types tend to be threatened by conflict and by the possibility they will not be accepted or liked or will be overwhelmed by feelings of hostility. They attempt to create a group in which people can work closely together, are supportive, and like each other.

Logical thinkers generally attempt to identify other members who think clearly and those who are fuzzy, those who are accurate and those who are not, and those who are oriented toward procedures and those who are not. Logical thinkers tend to be threatened by emotionality in a group because of fear of being overwhelmed by either their own feelings of aggression or their positive feelings. Therefore, they try to create a group in which logic, structure, and procedure dominate, a setting in which feelings are irrelevant and can be legislated off the agenda.

Individuals bring styles learned in earlier cultural settings into new group experiences. Initially different styles are likely to cause communication problems and slow down group formation because each type, in an effort to create the perfect group world, threatens each of the other types. If the task of the group demands working together, members gradually learn through interaction what one another's biases are and how to accommodate them. However, if a group is trying to accomplish a common task while members are still having difficulty understanding and accepting each other's different emotional styles, not much will be effectively accomplished. Rather, energy will be drained by attempting to reduce interpersonal anxiety. On the other hand, variations in personal style can begin to aid the group in accomplishing its task. Initially groups need the energy and initiative often supplied by members who feel most comfortable being assertive. When external demands produce stress and frustration, the group needs the emotional "glue" often supplied by members who are most comfortable with tender, supportive feelings. The group needs the procedures, problem-solving processes, and structure supplied by members who need to create structure in order to feel most comfortable. One of the paradoxes of group formation is that the diversity of styles that is initially difficult to deal with becomes the strength of the group in later stages. Another map for viewing differences among group members is that of individual cognitive styles.

cognitive styles

The cognitive styles describe how the individual acquires knowledge (cognition) and how an individual processes information (conceptualization). The cognitive styles are related to mental behaviors, habitually applied by an individual to problem solving and generally to the way that information is obtained, sorted, and utilized. Cognitive styles are usually described as personality dimensions that influence attitudes, values, and social interaction.

Cognitive Styles As discussed in depth in Chapter 4, variations in **cognitive styles** arise because people have different assumptions about the nature of the universe, truth, human nature, and relationships. One of the most widely researched and used theories on psychological type/cognitive styles, the Myers-Briggs Type Indicator, provides an understanding of individual differences based on preferences on four models of personality.

1. *Extroversion/introversion*. This scale indicates where people tend to get their energy and focus their attention—on their own inner world of ideas and concepts, or on the outer world of people and things.
2. *Sensing/intuition*. This scale describes opposite ways of obtaining information from the environment. Sensing types trust and rely most on their five senses. Intuitive types trust and utilize their intuition, extrapolating meaning and relationships from all of their previous experiences.
3. *Thinking/feeling*. This scale refers to how people reach conclusions, make decisions, and form opinions. For thinking types, this is based on logical, objective alternatives based on values—how important something is, how much people care about it. Feeling types prefer to decide using values and/or personal beliefs.
4. *Judging/perceptive*. The final scale refers to the type of lifestyle individuals prefer to live and refers back to the previous two scales. Judging types tend to be more in the decision-making mode and prefer to live planned, orderly lives. Perceptive types spend more time taking in information and tend to like flexibility and spontaneity.

Like differences in coping styles, individuals enter new groups with a frame of reference based on their own cognitive style. When opposites work together, there can be great frustration if individuals do not have an understanding of innate differences in how others approach tasks and process. Disagreements are less irritating and indeed are enriching when it is recognized others are not being purposefully contrary but are approaching situations from a different frame of reference. Success in any enterprise demands a variety of types. The clearest vision often comes from an intuitive, the most practical realism from a sensor, the most incisive analysis from a thinker, and the greatest understanding of people from a feeler. Opposite types can complement each other in any joint undertaking. When two people approach a problem from opposite sides, each sees things not visible to the other. They tend to suggest different solutions. When people recognize and know how to draw on these differences, problem solving and decision making are greatly enhanced.[7]

TASK/PROCESS ORIENTATION: THE IMPORTANCE OF TASK AND PROCESS CHARACTERISTICS

Our entire industry is built around the concept of teamwork. From the time we enter the recruit academy and throughout our careers, the focus is on the team. Throughout this book we address individual leadership styles and the importance of understanding our own personal framework, our cognitive style, the integration into a team, and how our style meshes with others. In order for a group of individuals to become a team, they must achieve a unifying goal or accomplish a task. The term *group process* refers to how well members interact with each other to accomplish the task. Individuals differ in their needs to focus on the task or process and in their skills in each of these areas. Attention to both is critical to team success. Understanding the following task and process roles can help create positive group interaction. Think of these in your present position with respect to what happens on the fire ground, in your station, in your division, and in your department.

Task Roles

1. *Problem definition.* Defines the group's problem.
2. *Seeks information.* Requests factual information about group problems or procedures, or asks for clarification of suggestions.
3. *Gives information.* Offers facts or general information about group problems or procedures, or clarifies suggestions.
4. *Seeks opinions.* Asks for the opinions of others relevant to the discussion.
5. *Gives opinions.* States beliefs or opinions relevant to the discussion.
6. *Tests feasibility.* Questions reality; checks practicality of suggested solutions.

Process Roles

1. *Coordination.* Clarifies a recent statement and may relate it to another statement in such a way as to bring both together; reviews proposed alternatives.
2. *Mediating/harmonizing.* Intercedes in disputes/disagreements and attempts to reconcile them; highlights similarities in views.
3. *Orienting/facilitating.* Keeps the group on track, points out deviations from agreed-upon procedures, directs group discussion; helps the group process by suggesting other procedures to make the group more effective.
4. *Supporting/encouraging.* Expresses agreement with others' ideas; verbally supports others.

Not everyone likes the task or the process part of the bureaucracy in which we work. An excellent tool to determine team characteristics is a survey instrument developed by Richard Ross and Tom Isgar entitled *Building High Performance Teams.*[8] It is useful in figuring out who in the group is task oriented versus process oriented. Through measurement of four categories—innovator, achiever, organizer, and facilitator—a team leader can identify the strengths of each team member in order to get the most from the team.

Innovators Like New Ideas

- Brainstorming is great.
- Crazy ideas—the more, the better.
- Stimulate thinking by challenging the status quo.
- A better way is the only way.
- A risk-free atmosphere is a must.
- Time is not a big factor.

Achievers Like to Be Directed

- Goal-directed—when, where, and how.
- Can overcommit to too many projects.
- Need to get things done to feel good.
- Momentum is created by many projects.
- Action oriented: get things done; move on to the next project.
- Reports on the status to keep track of all the projects.
- Let's celebrate when we finish a project.

Organizers Like to Follow the Steps

- Resources are well defined.
- Preparation is a must.
- All steps must be followed.
- Time efficiency is critical to the organizer's organizational plan.

Facilitators Like Good Process and Consensus Building

- Clear expectation of the role/purpose of each member.
- Designs—how to make it happen.
- Sensitive to feelings of the team.

MEASURING YOUR LEADERSHIP STYLE

Ross and Isgar's *Building High Performance Teams* survey evaluates 60 items that describe characteristics of leaders in organizations. A variety of situations common to leadership have been included to cover a wide range of characteristics and, thereby, to provide you with meaningful information about yourself as a leader. You can also use this survey to help determine the leadership characteristics of your senior staff, your station officers, or your crew at the company level. This interesting exercise can be insightful and can help to open a meaningful dialogue in regard to the characteristics that help to shape how we process information and work with others. Each major category (e.g., "Innovation") is represented in 15 different situations. The four alternatives to each situation differ. Therefore, read all four alternatives before answering so that you can select the alternatives *most* and *least* characteristic of you. There are no right or wrong answers. The best answer is the one most descriptive of you. Therefore, answer honestly, because only realistic answers will provide useful information about yourself.

Instructions From each set of four alternatives, select the one most characteristic of you. Place the letter for that item on the scale at the point that reflects the degree to which that item is characteristic of you. Then, select the alternative least characteristic of you and place its letter at the appropriate point on the scale. Once you have found the most and least characteristic alternatives, enter the letters of the remaining alternatives within this range according to how characteristic each alternative is of you. Do not place alternatives at the same point on the scale (no ties). For example, on a given set of four items you might answer as follows:

1. a. Imaginative **b.** Sensitive
c. Slightly Competitive **d.** Rational

Completely Characteristic **Completely Uncharacteristic**

:__:__: D: C: A: __:__: B: __:__:__:

10 9 8 7 6 5 4 3 2 1 0

Place the letter of the "characteristic" on the scale above the number to reflect your order of importance.

BUILDING HIGH PERFORMANCE TEAMS: LEADERSHIP ROLES

1. a. Imaginative **b.** Sensitive
c. Slightly Competitive **d.** Rational

Completely Characteristic **Completely Uncharacteristic**

:__:__:__:__:__:__:__:__:__:__:__:

10 9 8 7 6 5 4 3 2 1 0

2. a. Prudent **b.** Listener
c. Pragmatic **d.** Risk-taker

Completely Characteristic **Completely Uncharacteristic**

:__:__:__:__:__:__:__:__:__:__:__:

10 9 8 7 6 5 4 3 2 1 0

3. a. Results-oriented **b.** Concerned for feelings of others
c. Creative **d.** Analytical

Completely Characteristic **Completely Uncharacteristic**

:__:__:__:__:__:__:__:__:__:__:__:

10 9 8 7 6 5 4 3 2 1 0

4. a. Situationally sensitive **b.** Hard Worker
c. Organized **d.** Innovative

Completely Characteristic **Completely Uncharacteristic**

:__:__:__:__:__:__:__:__:__:__:__:

10 9 8 7 6 5 4 3 2 1 0

5. Emphasizing

a. Getting Results

b. Human Interaction

c. Logical Thinking

d. Ideas and Innovation

Completely Characteristic **Completely Uncharacteristic**

10 9 8 7 6 5 4 3 2 1 0

6. Producing Results by

a. Getting People to Work Together

b. Doing My Part

c. Motivating Other People

d. Establishing Helpful Systems and Procedures

Completely Characteristic **Completely Uncharacteristic**

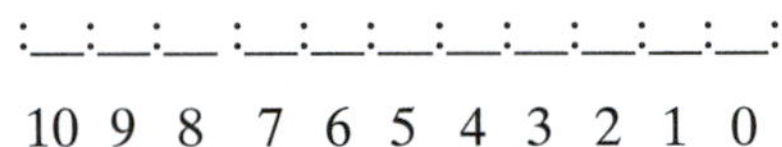

7. Tending to Focus on

a. What Is New That Could be Done

b. The Caliber of What Is Being

c. How the Work Is Being Done

d. What Is Being Done

Completely Characteristic **Completely Uncharacteristic**

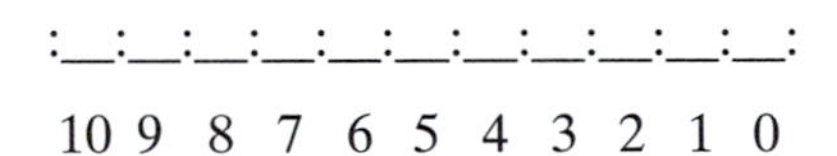

8. Being Described as

a. Highly Dedicated to My Work

b. People-oriented

c. Thoughtful and Precise

d. Charismatic and Enthusiastic

Completely Characteristic **Completely Uncharacteristic**

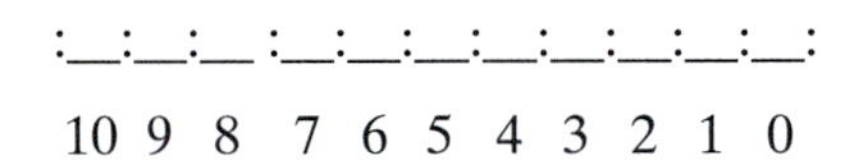

9. In Meetings, Tend to

a. Talk

b. Listen Impatiently

c. Listen for Underlying Motives and Potential Conflict

d. Listen Critically for What May Not Work

Completely Characteristic **Completely Uncharacteristic**

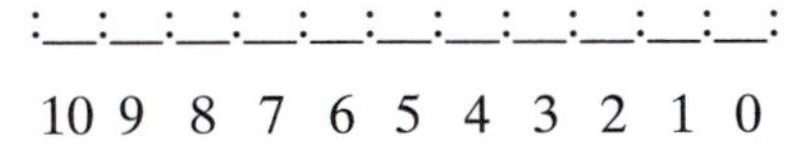

10. In a Work Group

a. Wanting to Keep People Working Together Harmoniously

b. Looking at What We Could Be Doing Differently

c. Wanting to Do the Job Efficiently

d. Like to Keep People Focused on the Task

Completely Characteristic **Completely Uncharacteristic**

:__:__:__:__:__:__:__:__:__:__:__:

10 9 8 7 6 5 4 3 2 1 0

11. Valuing Information That Is

a. About Technical Concerns

b. About Goals and Policies

c. About Opportunities

d. About Others' Beliefs on Issues

Completely Characteristic **Completely Uncharacteristic**

:__:__:__:__:__:__:__:__:__:__:__:

10 9 8 7 6 5 4 3 2 1 0

12. Using Free Time to

a. Think About New Methods, Procedures, Controls, Etc.

b. Seek Out Information from Others, Sift Rumors, Interact

c. Create a New Project or Revamp/Change Ongoing Work

d. Complete More Work

Completely Characteristic **Completely Uncharacteristic**

:__:__:__:__:__:__:__:__:__:__:__:

10 9 8 7 6 5 4 3 2 1 0

13. In Making a Decision

a. Look for the Right Answer

b. Use Group Input

c. Act Quickly

d. Consider Precedents

Completely Characteristic **Completely Uncharacteristic**

:__:__:__:__:__:__:__:__:__:__:__:

10 9 8 7 6 5 4 3 2 1 0

14. In a Conflict Situation

a. Use Conflict to Pressure Subordinates

b. Become Uncomfortable

c. Ignore it, or Fight Based on Rule and Policies

Completely Characteristic **Completely Uncharacteristic**

:__:__:__:__:__:__:__:__:__:__:__:

10 9 8 7 6 5 4 3 2 1 0

15. Excelling by

a. Getting Things Done

b. Accomplishing Things with Systems and Procedures

c. Creating a New Venture

d. Working with Groups to Accomplish Tasks or Get Agreement

Completely Characteristic **Completely Uncharacteristic**

:__:__:__:__:__:__:__:__:__:__:__:

10 9 8 7 6 5 4 3 2 1 0

LEADERSHIP STYLES

Scoring Key To score the inventory, merely transfer the rankings you assigned to the appropriate item number below, and then add the columns. If possible, pair up with another person to read off the rankings.

	Innovator	Facilitator	Achiever	Organizer
1.	a. ______	b. ______	c. ______	d. ______
2.	a. ______	b. ______	c. ______	d. ______
3.	a. ______	b. ______	c. ______	d. ______
4.	a. ______	b. ______	c. ______	d. ______
5.	a. ______	b. ______	c. ______	d. ______
6.	a. ______	b. ______	c. ______	d. ______
7.	a. ______	b. ______	c. ______	d. ______
8.	a. ______	b. ______	c. ______	d. ______
9.	a. ______	b. ______	c. ______	d. ______
10.	a. ______	b. ______	c. ______	d. ______
11.	a. ______	b. ______	c. ______	d. ______
12.	a. ______	b. ______	c. ______	d. ______
13.	a. ______	b. ______	c. ______	d. ______
14.	a. ______	b. ______	c. ______	d. ______
15.	a. ______	b. ______	c. ______	d. ______
	______ TOTAL	______ TOTAL	______ TOTAL	______ TOTAL

On effective teams, members have the variety of skills needed for performance of the tasks, for the ability to process, and to maintain the team as a viable group. It is usually understood members must have adequate technical skills to accomplish the task at hand. Just as important are the skills required to elicit knowledge and integrate information into effective decisions. Research on high-performance teams indicates they are aware of and consciously focus on both team functions. These factors reflect the basic human need for security and must be achieved in the development of any team. If you are a new group member, whether the fire chief or a new company officer, you will be challenged to satisfy those needs. When new groups are formed, they cannot fulfill their mission until the members learn to meet their own needs to some degree. The feeling of comfort in a new group situation can be enhanced by focusing on the purpose of the group and/or by providing activities designed to help the members gain an understanding of each other. As individuals join groups, they bring with them their own frame of reference based on their own styles and experiences. When opposites work together, there can often be frustration if the individuals involved do not understand the differences in the way others approach problems. Understanding that people approach situations from different perspectives often reduces disagreement and can actually become a great learning experience.

Teams who reach a high performance level do so through effective leadership and realizing the potential of their collective experience and expertise. Whatever methodology you use to get there, high-performance teams exhibit several common characteristics.

CHARACTERISTICS OF HIGH-PERFORMANCE TEAMS

Individual Level

- High personal commitment to the group
- High level of trust among members
- High involvement with the group, which inspires each individual's personal best
- High level of personal development
- Perceived fun and excitement for participation in the group

Group Level

- Internalized purpose and mission basis of action helps inspire individuals to achieve
- Results driven; keen focus on making a difference
- Effective utilization of members' skills and abilities (diversity of skills, points of view, and values)
- Open communication; effective norms for surfacing and working through differences
- High standards of excellence
- Smooth task and process flow within the group
- Open expressions of appreciation, recognition, and caring
- Optimal communication and exchange with the "outside world," other organizational stakeholders
- Willingness to experiment and try new ways of doing things; flexibility and versatility[9]

Potential often begins with gaining an understanding of the people in the group (see Figure 7.4). Team building starts with understanding how others process information

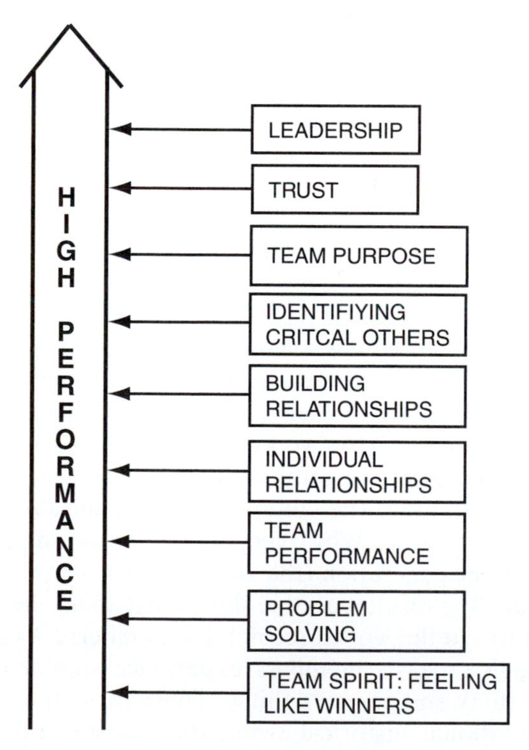

FIGURE 7.4 ◆ Components of a High-Performance Team
(Courtesy of Randy R. Bruegman, Exceeding Customer Expectations: Quality Concepts for the Fire Service, *p. 159)*

and what motivates them to participate. Organizations and groups, like individuals, need motivation to succeed. Understanding individual leadership and personality styles will be an important determinant of your ability to form your teams and motivate them to do great things.

A DIVERSIFIED WORKFORCE

Suppose in the next decade 40 to 50 percent of the workforce consists of women and minorities. Think about how much change this will bring to our industry, not only in the way we look but also in the way we approach our organizational problems, our industry, and our community. In addition, for the first time in the history of the fire service and our country, four and possibly five generations will be in the workforce at the same time. Each generation brings a unique set of values and job expectations, which may cause conflict in organizations because the values and expectations of the different generations can collide. Couple this with the increasing diversity of the workforce versus the traditional makeup of the fire service, and you can see we are about to enter a workforce revolution involving diversity, value structures, and employee expectations. All will challenge the leadership to blend this new mosaic. Team building will be critical in the immediate future.

Most research has been conducted and many books have been written regarding the generational differences existing today. The impacts of these differences on society reflect the shared characteristics and similar interests found in each generational group. The four generations in today's workforce are:

Veterans (traditionalists)	1922–1945
Baby Boomers	1946–1960
Generation X	1960–1980
Generation Y (millennials)	1980–2000
?	2000–

Why is this important to you as a leader or manager? Each generation brings certain values, traits, and expectations to the workforce. Depending upon the makeup of your workforce, the influence of each generation on your organization will make for a continual state of transition. As workers retire, new personnel are hired. As the organization grows in numbers or reduces its force, each of these events impacts the culture of the organization. The culture of the organization you manage today will be very different in 10 years. Why? Very simply, we are in the midst of a generational wave, in which one generation (the veterans) is leaving and a new one, yet to be named, is emerging. The result is a value shift. What about the next wave, when the baby boomers start to exit the workforce in large numbers? As each of these generations moves through time, its members experience similar events that help to shape the way they think and perceive things. These formative events are often significant national, emotional, historical events that become a permanent and defining memory for the group. Such experience helps to shape the basic characteristics, values, work ethic, and outlook of the generations involved. Although not everyone will fit neatly into any generational grouping, the profiles of these groups are surprisingly accurate.

A Profile of Five Generations

Veterans (traditionalists), 1922–1945	***Baby Boomers, 1945–1960***	***Generation X, 1960–1980***	***Generation Y (millennials), 1980—2000***	***2000–***
Also Known As				
Traditionalists	Boomers	Xers	Millennials	
Mature GIs	Postwar generation	Twenty-somethings	Generation Y	
World War II generation	Vietnam generation	Thirteeners	Generation 2001	
Silent generation	Sixties generation	Baby busters	Nintendo generation	
Seniors	Me generation	Post-boomers	Generation net	
Builder generation			Internet generation	
What Shaped Them				
Patriotism	Prosperity	Watergate, Nixon resigns	Computers	September 11
Families	Children in the spotlight	Latchkey kids	Oklahoma City bombing	War on terrorism
Great Depression	Television	Stagflation	*It Takes a Village*	War in Iraq
World War II	Suburbia	Single-parent home	TV talk shows	Madrid bombing
New Deal	Assassinations	MTV	Multiculturalism	London bombing
Korean War	Vietnam	AIDS	Girls' movement	*Columbia* disaster
Golden age of radio	Civil rights movement	Computers	McGuire and Sosa	
Silver screen	Cold war	Fall of Berlin Wall	*Challenger* explosion	
Rise of labor unions	Women's liberation	Wall Street frenzy	Persian Gulf War	
School and church	Space race	Glasnost, perestroika		
	Kent State killings	*Roe v. Wade*		
	Economic affluence			
	Education and technology			
Core Values				
Dedication/sacrifice	Optimism	Diversity	Optimism	
Honor	Team orientation	Thinking globally	Civic duty	
Adherence to rules	Personal gratification	Balance	Confident	
Hard work	Health and wellness	Technoliteracy	Achievement	
Law and order	Personal growth	Fun	Sociability	
Duty before pleasure	Youth	Informality	Morality	
Respect for authority	Work	Self-reliance	Street smarts	
Conformity	Involvement	Pragmatism	Diversity	
Frugality				

(Continued)

A Profile of Five Generations (*Continued*)

Veterans (traditionalists), 1922–1945	*Baby Boomers, 1945–1960*	*Generation X, 1960–1980*	*Generation Y (millennials), 1980—2000*	*2000–*
Their View on the World				
Outlook: Practical	Optimistic	Skeptical	Hopeful	
Work ethic: Dedicated	Driven	Balanced	Determined	
View of authority: Respectful	Love/hate	Unimpressed	Polite	
Leadership by: Hierarchy	Consensus	Competent	Pulling together	
Relationships: Personal sacrifice	Personal gratification	Reluctant to commit	Inclusive	
Turnoffs: Vulgarity	Political incorrectness	Cliché, hype	Promiscuity	

Sources: Adapted from Zemke et al. (2000) and McIntosh (1995).[10]

Defining Events

1930s	Great Depression
	Election of Franklin D. Roosevelt
1940s	Pearl Harbor
	D-Day
	Death of Franklin D. Roosevelt
	Victory in Europe (VE Day) and Victory over Japan (VJ Day)
	Hiroshima and Nagasaki atomic bombings
1950s	Korean War
	TV in every home
	McCarthy House Committee on Un-American Activities hearings
	Rock 'n' roll
	Salk polio vaccine introduced
1960s	Vietnam War
	Election of John F. Kennedy
	Civil rights movement
	Assassinations of John F. Kennedy and Martin Luther King, Jr.
	Moon landing
	Woodstock
1970s	Oil embargo
	Resignation of Richard Nixon
	First personal computers
	Women's rights movement
1980s	*Challenger* explosion
	Fall of Berlin Wall

	Killing of John Lennon
	Election of Ronald Reagan
1990s	Persian Gulf War
	Oklahoma City bombing
	Death of Princess Diana
	Clinton scandals
2000s	September 11
	War on Terrorism
	War in Iraq
	Columbia disaster
	Hurricane Katrina

WHAT DOES THIS MEAN TO THE FIRE SERVICE?

Although the workforce has changed rapidly in the United States, the current workforce of many fire departments still consists predominantly of Caucasian males. During the next 20 years, however, many changes are predicted to occur:

- The percentage of Caucasian males entering the workforce will drop dramatically.
- A majority of new firefighters will be women or minorities.
- The average age of all American workers will rise to 39.
- Approximately one-third of all workers will be 65 or older.
- A rising percentage of all workers will not have a high school diploma.
- The number of immigrants entering the workplace will grow faster than at any time since World War I.

As the dynamic of the workforce changes, so do the impacts on the fire service on several fronts. Some of the effects that may be felt by the fire service community include the following:

- There may be a disruption in the sense of continuity in many organizations as these workforce changes occur.
- New challenges will result in such areas as training and development; language barriers and marginal mechanical experience will force many organizations to adjust our traditional training methods.
- A sense of "culture shock" may occur among the traditional Caucasian male population of firefighters.

As the workforce transitions, so will the key training issues facing us today and in the future (see Figure 7.5): structural engineering; hazardous materials handling, containment, and abatement; chemical engineering; first aid/EMS procedures; regionalization and cooperation with other public agencies; personal safety (exposure to bodily fluids); physical fitness; technology integration; technical rescue; terrorism response; and disaster management.

In government, and especially in public safety, we face potential liability in almost everything we do. If we provide emergency medical services, we face liability if anything goes wrong. However, if we do not provide these services, then we may face liability anyway for no response. We are liable for injuries to the people in our departments and for living up to unfunded mandates. We continue to deal with the liabilities existing between legislated regulations and unlegislated guidelines. Every action in our organization poses some kind of risk. The fire service faces several key legal and safety issues,

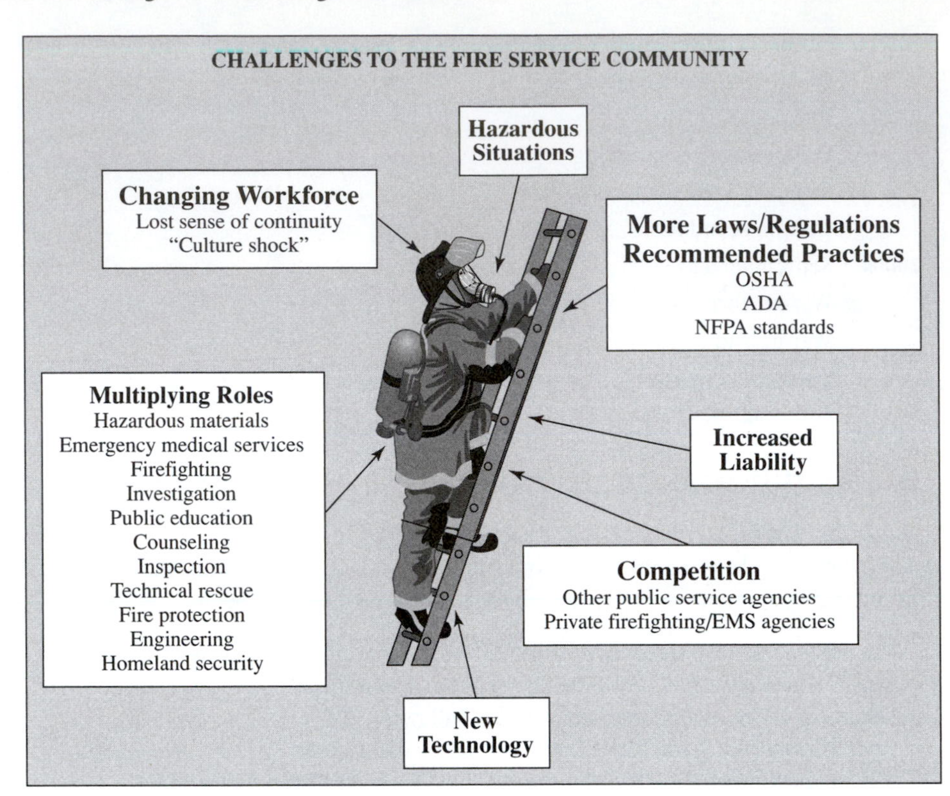

FIGURE 7.5 ◆ Challenges to the Fire Service Community
(Courtesy of Randy R. Bruegman, Exceeding Customer Expectations: Quality Concepts for the Five Service, *p. 160)*

including proliferating legislation on acceptable firefighting procedures, personal liability to lawsuits, and the amount and complexity of documentation required to document response to emergencies effectively. There has been an increased number of regulations relating to human resources. Just as the demographics of our firefighters shift, we are increasingly exposed to the need for compliance with federal and state legislation regarding employees. Some of the legislation that has had an effect on fire departments in the past 10 years includes the Americans with Disabilities Act, the Age Discrimination in Employment Act, and the Family Medical Leave Act (see Chapter 8). Several other key employee issues will continue into the foreseeable future, including:

- Concerns about crimes against firefighters
- Fear of exposure to infectious diseases
- Worries about exposure to hazardous and biological materials
- Increased costs of liability, disability insurance, and workers' compensation, which may impact the benefits provided
- Increased pension cost will force restructuring of many plans
- Higher wage expectations compared to the private sector
- Gender, race, and age equality issues
- Critical incident stress disorders

It is evident that to be successful in the future, our personnel strategies must involve all members to build consensus around these key employee issues.

BRIDGING THE GENERATIONS

In a recent survey conducted by the Society for Human Resource Management, 40 percent of human resource professionals have observed conflict among employees as a result of generational differences![11] In organizations with 500 or more employees, 58 percent of human resource professionals reported conflict between younger and older workers, largely due to differing perspectives on work ethics and work–life balance. These data suggests there is a huge potential for miscommunication, low morale, and poor productivity unless the generations learn to handle conflict successfully.[12] The generational issues are basic business issues that every industry will face in the next decade. Age has taken its place beside gender, race, and culture as a way that defines generational experiences whether you grew up during the Great Depression, witnessed the turbulent times of the 1960s, or grew up with the Internet and the latest explosion of technology.

When it comes to the workplace, you can see generational differences offer a complex variety of perspectives and talents. The truth is generational issues we face today in the fire service and in the world are important reflections of who we are to become. Yet they pose critical issues for every organization. This is not merely an issue of young versus old or what generation we belong to. We are truly living through an historic shift in our framework, in what our are and working life careers, and it is reflected in the unpredictability of today's private sector marketplace and the expectations of public sector performance. Each generation has matured in a very different world, and these distinct societal conditions have helped to define how we react and respond to life in general, to our workplace, and to our coworkers. Leaders and managers must have some understanding of what frames the generational issues for each. The challenge for leaders and managers is to bring together this diverse generational workforce and help shape it into a collaborative team. This will occur only when we can utilize the strengths that each brings to the table, with an understanding that when leaders and managers bridge the generations, they are building the future of their organizations.

- *The traditionalists,* born 1922–1945, are much more likely to attend symphonies than rock concerts and eat steak instead of tofu. They like consistency and uniformity and prefer detail and logic when approaching projects. Computers were foreign objects and once feared. Their values include hard work, dedication, sacrifice, respect for rules, duty before pleasure, and honor. As leaders, they use a directive style.
- *The boomers,* born 1945–1960, grew up with Elvis and the Beatles and watched the Vietnam War on television. They are good team players and typically are reluctant to go against peers. This group has values of optimism, teamwork, involvement, personal growth. As leaders, they are typically collegial and consensual.
- *The Gen Xers,* born 1960–1980, are the most recent group integrated into the workplace and are the first to be techno-literate. They loathe paying dues and want companies to be loyal to them. They view job portability as a necessity to acquiring skills and experience and will stay with a company only if conditions are right. They bring great adaptability and creativity to their work and as leaders are fair, competent, and straightforward. Their values include diversity, fun and informality, self-reliance, and pragmatism.
- *The Millennials,* born 1980–2000, are just beginning to enter the workforce and are a group that has dealt with a number of significant events in their young lives. Columbine, Jonestown, Clinton/Lewinski, and 9/11 are some of the major events that have impacted

the psyche of this generation. For this group, change is a part of daily life and they are multitasking masters. They want and prefer to handle multiple jobs at once. Forget about paying dues; this group wants learning opportunities and will combine teamwork with can-do attitudes. Values for the millennials include optimism, civic duty, confidence, achievement. and respect for differences.

RECRUITMENT

If you want to appeal to the *traditionalists* to join your company, do not limit job prospects to full-time employment. Send messages about family and home, and view age and experience as an asset, not a liability. Use good grammar in the interview, including *please* and *thank you*, and avoid profanity.

If the job you have is right for *boomers*, you will need to be sure to give credit for experience and accomplishments. Let them know they will make a difference, and depict your company's human side as well as its dynamic, leading-edge nature.

With *Gen Xers,* include the phrase "I want you to have a life" in the interview process. You must convince them their performance and ideas will be evaluated by merit, not by a person's years on the job. Don't BS; Gen Xers have a sixth sense about that. Stress how your company values technological skill and innovation.

The *millennials* want a clear picture of the way work ought to be. They will be a very demanding workforce and are used to getting what they want. They will be politically and socially oriented and will want an opportunity to influence pay and gender equity.

MANAGING

Traditionalists like to be up to date on the company and the projects at hand. It is important to articulate how they can contribute to long-term goals. A personal approach versus e-mail, voice mail, and faxes works better with this group. For incentives, traditional rewards are the best motivators.

The key to managing *boomers* is to treat them as unique and give them a lot of public recognition. They want to be involved in decisions and like consensus. This group has a strong work ethic, is willing to work long hours, and wants the chance to prove themselves.

Gen Xers want freedom from standard management practices, so it is wise to give this group a lot of elbow room. They want to handle multiple projects and are best left to prioritizing them themselves so they can feel in control of their work. This group is eager for opportunities to learn new skills; move them into new situations (sideways is fine with them). Constructive feedback works well. Make them feel like insiders.

The new kids on the block, the *millennials*, will require plenty of time to orient. Their personal goals are their priority, so learn what they are and try to weave them into the job. Watch out for potential conflict between this group and the Gen Xers.

RETENTION

Traditionalists are old-fashioned and will need technology training for many jobs in the future. Be respectful of their existing knowledge and tactful in helping them to learn new skills. Use respected leaders as mentors. "No news is good news" works when it comes to feedback with this group. They value authority and discipline, so unless you tell them something is wrong, they are happy to go along with the stated company objectives.

Boomers often need to develop their strategic planning skills and like projects as a means to learn these skills. Involve them in task forces and recognize efforts "above and beyond." Provide resources such as business books and training tapes, and encourage them to tap into these resources. Annual updates with lots of examples and documentation is a good evaluation tool with boomers. They love to communicate; this is the group that pushed and created the once-a-year appraisals. Boomers often clash with traditionalists, who are not into sharing of feelings and open communication.

It is critically important to *Gen Xers* to learn new skills, so the key to retaining this group is providing opportunities to do so. Keep training brief; they are self-learners, so highlight what is important and let them go. Coach and mentor them to figure things out for themselves. "What do you think is the best approach? How do you plan to solve this problem?" This group wants instant feedback. "Are you busy? I just want to know how I'm doing." Give straight feedback; everything else is viewed as unreliable.

Millennials are best matched with seasoned employees. Develop a mentor program for this group and provide plenty of training. It is too early to tell what kind of feedback will work with this group, but they will probably want it fast, individualized, and often with a technology component.[13]

Conflicting Career Goals

- Traditionalists—"Build a legacy."
- Baby Boomers—"Build a stellar career."
- Generation Xers—"Build a portable career."
- Generation Y Millennials—"Build parallel careers."

Views on Retirement

- Traditionalists—"Reward"
- Baby Boomers—"Retool"
- Generation Xers—"Renew"
- Generation Y Millennials—"Recycle"

Views Around Job Changing

- Traditionalists—"Job changing carries a stigma."
- Baby Boomers—"Job changing puts you behind."
- Generation Xers—"Job changing is necessary."
- Generation Y Millennials—"Job changing is part of my daily routine."

Views on Feedback

- Traditionalists—"No news is good news."
- Baby Boomers—"Feedback once a year, with lots of documentation!"
- Generation Xers—"Sorry to interrupt, but how am I doing?"
- Generation Y Millennials—"Feedback whenever I want it at the push of a button."

Views on Training

- Traditionalists—"I learned it the hard way; you can too!"
- Baby Boomers—"Train 'em too much and they'll leave."
- Generation Xers—"The more they learn, the more they stay."
- Generation Y Millennials—"Continuous learning is a way of life."

Balancing the Generations

- Traditionalists—"Support me in shifting the balance."
- Baby Boomers—"Help me balance everyone else and find meaning myself."
- Generation Xers—"Give me balance now, not when I'm 65."
- Generation Y Millennials—"Work isn't everything; I need flexibility so I can balance all my activities."

Recruiting the Generations

- Create cross members of Generation Xers, boomers, and Yers.
- When recruiting, focus on how the job benefits the applicant, not just the organization.
- Manage people as individuals.
- Set up programs for flexible or semiretired work.
- Set up continuous learning programs to retain workers.
- Challenge each generation with appropriate responsibilities.

Training Generations X and Y

- Accept them.
- Care.
- Hands off; be there.
- Talk, talk, talk.
- High input.
- Mentor.
- Exert authority from reason.
- Learn to move faster in making change.
- Convey the meaning of assignments.
- Be explicit about your expectations.

As we move into the future and try to balance the needs of all of our employees, the role of leadership will become increasingly complex. As we try to find the balance and develop strategies to accommodate the different learning styles and lifestyle desires of the workforce, our relationships with labor groups and volunteer associations can often become strained. The next 10 to 15 years will bring a cyclical relationship between labor and management as we work to find a balance. A cooperative labor–management partnership that deals with the changes in race, gender, ethnicity, worker generations, and shifts in our industry is a must. This will help us prepare to meet the challenges involved in recruiting, retaining, and training quality people; diversifying our workforce; and leading the revolution in the fire service. It will force many labor and management leaders to change their approach to each other. The leaders of today and those of the future must strive for empowerment and inclusion. For both management and labor, this means becoming more global in their perspectives and collaborative in their approach.

The next several years will provide a unique opportunity for labor–management to establish a new dynamic of cooperation, which will result in improved organizational outcomes. Since the early 1990s, the International Association of Fire Chiefs (IAFC) and the International Association of Fire Fighters (**IAFF**) have worked on a collaborative effort known as the *Labor/Management Fire Service Leadership Partnership.* Since 2000 they have cooperatively conducted Fire Service Leadership Partnership workshops throughout the country. These two-part workshops designed to promote cooperative labor–management relationships have paid dividends for many fire departments. Traditionally, they have been cosponsored by the state fire chief and firefighter associations. The workshops focus on four goals: starting productive communications, understanding the cooperative process, understanding the benefits of the cooperative process, and providing tools to enhance partnerships. The participants engage in interactive discussions and exercises with their labor–management partners to gain a better understanding of each other and the pressure each faces from respective superiors or constituents. These provide ways to understand the roles and relationships of labor and management and to create effective strategies at the local level. Hopefully, this will eliminate the conflicts of the past between fire chiefs and labor leaders, including votes of no confidence, picketing, and retribution from chief fire officers.

IAFF
International Association of Fire Fighters.

The relationship between volunteer firefighters and local fire chiefs has also changed dramatically, as have the roles and responsibilities of the volunteer fire service. The demands placed on volunteers have become increasingly time consuming and stressful, leading to a whole new dynamic for the local chief fire officer who leads and manages a volunteer organization. Many volunteer departments are evolving into combination departments, which have both career and volunteer firefighters to meet these demands and the growth of their communities. This too creates an interesting leadership opportunity for the future.

As we move into the next decade, our increasingly diversified workforce will have a dramatic impact on our relationships with local labor leaders, local volunteer association presidents, and labor and management in general. At the same time, we have to understand how the changing workforce will affect the day-to-day dynamics of our departments. A positive relationship between labor and management can provide a strong foundation to address appropriately many of these issues that are sure to emerge.

EMPLOYEE MORALE IS CRITICAL TO TEAM SUCCESS

One of the questions most frequently asked in organizations today is, "How can we improve morale?" Because morale affects every aspect of an organization, it is an important question to ask and to answer. Objectives related to your department's success—such as improving quality, productivity, and customer service; reducing turnover, absenteeism, and safety-related costs—are all influenced by employee morale. Therefore, keeping morale high should be a goal of everyone in the organization. Every employee has an impact on morale; it may be just in the station or company you are assigned to, but you have an impact. We have all worked with people who, no matter what you did for them or how great their day was, were negative about everything: the calls, the company, the chief, the department. On the flip side is the person who, no matter how messed up the day has been, is optimistic and elevates the mood of everyone in his or her presence. It is the classic analogy: is the glass half empty or is it half full? Leaders cannot change people who are distinctively negative about most things in life, but they can be fair, upbeat, and focused on the positives occurring within the department.

Let me share an example of this, which happened to me as a chief; if you are not observant of such behavior, you can find yourself becoming negative or angry due to someone else's attitude. We had a tremendous amount of positives occurring in the department: three new stations under construction; 14 new engines, ladders, and specialty vehicles placed in service within 18 months; 90 new people hired; approximately 60 promotions; a $6 million renovation project on existing facilities completed; labor contracts positively negotiated; and the department had received the highest rating from the citizens of any department in the city. Sounds like a lot of positive things were happening. Also a new policy was issued on physical fitness that changed the physical fitness time, a controversial change for some. The reaction from a firefighter, in an e-mail forwarded to me through command channels, was classic of the glass is half empty attitude: "Well, we can see the chief and administration have proven once again that they do not care about their people. This action takes away the supposed positive things that have been taking place and puts the entire department back in the basement in respect to morale." You can guess what my initial reaction was. Yet if you allow yourself time to process and not react, this negative act was an interesting and insightful response. It reminded me that for employees such as these, you can never do enough, as they will never allow themselves to leave the basement of negativity. With this said, every time I see this employee, I never treat him or her any different from any other employee. My hope is that one day I will get him or her to look at that glass as half full!

IMPROVING MORALE: WE MUST ASK THE RIGHT QUESTION

Most managers and human resource professionals start their quest to improve morale on the wrong foot. They doom their morale building efforts from the beginning by asking the wrong question. "We need to improve morale. What program would you recommend that does not cost much (or anything)?" The way the issue is framed reveals two serious flaws in perspective and it offers a clue why morale might be a problem in the first place.

Being "Penny Wise and Dollar Foolish" The fact that the request includes the qualifier "does not cost much (or anything)" reveals an interesting perspective on how important morale is in the first place. Not being willing to invest in employee morale, which so powerfully affects organizational success, is simply being "penny wise and dollar foolish." Approaching the issue of improving employee morale from the perspective of "we want to improve morale as it is critical to success, but we do not want to invest time and money in making it happen" makes as much sense as saying, "We want to deliver world-class customer service, but we do not want to invest in hiring the best employees or take the time and money to train them well."

You Will Not Solve the Problem with a Gimmick In many organizations the solution often comes in the form of a program, as if just the right event, award ceremony, or training program will make a lasting change in morale. It will not. Special events by themselves do not lead to high morale, nor do any quick-fix "solutions." In fact, when such events and programs contradict workers' daily experiences of not being respected, valued, or appreciated, these approaches have just the opposite effect. As an example, a local public sector workforce had been underfunded for some time, not only in wages and benefits but also too much work and not enough resources. Morale was low. Management's solution? Bring in a feel-good program and spend several hundred thousand dollars on getting the workforce energized toward improving morale. Do you think this worked? Not at all. The resulting employee perspective led to an even more cynical, distrustful, and disengaged workforce.

What leads to high morale is an intrinsically rewarding work experience whereby employees feel respected, valued, and appreciated; get to be players and not just hired hands; and get to make a difference. With such work experience, employees do not need to be bribed with goodies to make them want to come to work and do their best. Morale problems are often a result of a negative or dissatisfying work experience, whether due to the actual job itself, one's relationship with one's boss, not having adequate training, or the myriad of other factors that affect morale. If morale is a result of an unsatisfying work experience, then the answer is to change the work experience to make it rewarding. You do not create such work experience with one-time events or material perks. Holding an employee appreciation day, having dress-down Fridays, or giving employees company logo products do not create an intrinsically rewarding work experience. What does? Designing a work experience based on research regarding which organizational factors are the most important to the workforce, sound managerial practices, and meeting the human needs does lead to an inspired, engaged workforce.

FOUR POINTS TO GUIDE YOUR MORALE BUILDING EFFORTS

Gimmicks Are the Frosting, Not the Cake Although gimmicks are not the solution to improved morale, they do have a place in the overall approach. They are appropriate when done as part of a larger organizational effort. Organizations known for having a great workplace frequently put on a variety of fun events and special programs, and often shower employees with various rewards. These programs and perks work for them because they are an honest presentation of how management feels about and treats employees day in and day out. Leaders in these departments recognize that such programs and perks are the frosting on the cake; they are not the cake. They understand that the cake is a positive work experience. For these organizations, their rewards and programs are a congruent manifestation of the ongoing relationship between labor and management, and an extension of their employees' work experience.

It's the Little Things, and Every Little Thing Matters Morale is not improved by a one-time dramatic display of appreciation. Morale is improved, or damaged, one interaction at a time. Every time employees interact with their manager, it is a moment of truth. Every time they interact with their employer, whether in the form of a company-wide policy or communication, it is a moment of truth. Thus, instead of focusing on one-time events and displays of concern and appreciation, as leaders and managers, we need to "think small." Focus on these simple day-to-day encounters that might seem insignificant but will, through their cumulative effect, determine morale. It matters whether a manager notices the good things an employee does or notices just their mistakes. It matters whether a manager asks employees for their input before making a decision that impacts their daily work or just goes ahead and makes the change, expecting employees to "deal with it."

It matters whether managers get back to employees promptly about their requests or have to be repeatedly pursued for an answer. It matters whether managers say "thank you" or take it for granted when employees go the extra mile. It matters when employees recognize each other for a job well done. In short: everything matters, and we all have a part in making it matter.

It is important to understand this for two reasons. First, with many people today being overloaded with work, it is natural for managers to sprint through the day without taking time to consider the impact of their interactions with others. "Everything matters" helps them remember the importance of paying attention to each interaction and giving it their best. Second, because many people are reluctant to give their boss negative feedback, managers may not realize the negative impact of a mishandled moment of truth. Because they do not get feedback, they do not receive evidence that everything matters. Thus, by promoting an organizational culture in which everything matters, it encourages everyone to become more alert to, and mindful of, those little moments of truth each day brings and increases the odds that the outcome of each will be morale building.

Most of the Answers Are Within the Workforce . . . So Ask The answers to improving morale in most organizations will not come from the latest management book but are usually found within your workforce. Each department has a unique culture and set of problems causing diminished morale. No off-the-shelf, one-size-fits-all, quick-fix "solution" will address the unique challenges, culture, and needs facing your organization. Trying to force a prepackaged solution onto employees usually backfires. No one likes to have things forced on them; what they do like is to be involved in solving problems. Creating a "homegrown," customized solution for creating high morale obviously requires finding out the causative factors of why morale is low in the first place. Rather than guess what they are, ask. Just as important, make sure you do not ask unless you are honestly willing to address the issues that may surface. When employees are asked to give input and then nothing is done with it, the results are decreased morale, increased resentment, mistrust, and cynicism.

Doing this right also means involving employees in generating solutions. Because everything matters, the fact that you involve employees in generating solutions will win you a few "morale brownie points." Getting them involved shows you respect them. It taps into a basic employee need, the need to matter. Most employees want to be players, not just hired hands, and have an innate drive to solve problems, and these factors strongly impact morale.

Be Willing to Look in the Mirror If there is a morale problem, there is a leadership problem. The problem is, when things are not going well, it is human nature to look outside ourselves for the cause. If you are the chief, battalion chief, or captain, have you asked yourself, "What am I doing that might be contributing to—or even driving—low morale?" If you are contributing to low morale, chances are good no one has told you this. Bosses do not hear these things because most employees realize criticizing their boss is not exactly the fast track to success. In many cases, because bosses never hear about the many things they inadvertently do that diminish employee morale, they continue to do things that damage morale and wonder why turnover is high or employee relations issues plague their department. Because power often brings immunity from feedback, the leadership must actively seek out feedback if it is truly serious about improving morale. Those in leadership positions will need to ask for feedback and learn how to make it safe for people to respond honestly. Approaches and tools that can yield useful information include the many leadership assessment tools available, 360-degree survey tools, having human resources or an external consultant interview people you deal with, and executive coaching. (I have used the 360-degree survey for the past 10 years and found it to be a very useful tool. I do not require people to sign it and typically receive approximately a 70 percent return rate. A word of caution: you must be willing to hear some very truthful observations.) If you want to improve employee morale, remember that gimmicks are not the answer. They are the icing on the cake, not the cake. The cake is an intrinsically rewarding positive work experience. To find out how you can create one, ask. Then work together with your employees or coworkers to make it a reality.

By increasing awareness you will learn how you (if you are in a leadership position) impact your staff's morale or (if you are in a subordinate position) the morale of your coworkers. Taking the time to learn more about the factors that impact morale can dramatically improve your own effectiveness. It is important to remember our ability to cultivate high morale, high productivity, and high employee engagement. This is a responsibility we all have, no matter what our rank or position is in the department.

FOCUS ON WHAT YOU CAN CONTROL, NOT ON WHAT YOU CANNOT

Often when working in the firehouse you hear, "You should be talking with administration. They are the ones who need to hear this." Although they may be right, experience has shown an employee's supervisor, the company officer, or the battalion chief affects his or her performance and loyalty far more than does the CEO or the overall organization climate. Thus, even if the senior management team does not seem interested in improving morale, research indicates that company officers and battalion chiefs can make a huge difference. The key to both your effectiveness and your job satisfaction is to focus on the things you can control and influence, and practice letting go of those things totally out of your control. You have control over whether you take the time to learn what factors and practices affect morale. You have control over whether you make a conscious effort to do the things that make a difference and whether you engage in professional development to improve your supervisory skills. You also have control over whether you study how to become more influential, so that you can increase the odds that others will do their part to improve morale.

"The Big Three"

1. *Practice noticing when your people do something well.* Then tell them about it. Unfortunately, noticing good things does not come naturally. Noticing what is wrong is actually hard-wired into the human brain. Our survival was more closely linked to noticing what is wrong, that is, potential danger ("There is an accident up ahead") than to noticing what is right ("Isn't it a beautiful day today"). Thus, it takes conscious attention and daily discipline to offset this hard-wired tendency.
2. *Don't just talk at employees; listen to them.* Listen to their ideas about process improvements. Listen to their concerns. Listen to their opinions. This does not mean you agree, nor does it mean you have to act on every recommendation you hear. It does mean you respect employees as intelligent adults. Few things damage morale and an employee's respect for management more effectively than a know-it-all boss who does not value the ideas of the people in the trenches. Not listening to concerns also creates a "Why should I care about you, if you do not care about me?" attitude in employees. Conversely, managers who listen engender engagement and loyalty. Listening also cultivates respect because frontline employees know it is just

common sense that the people doing the job might have a few good ideas about how to do it better. Managers who fail to realize the importance of listening are often those who have lost the respect of their people.

3. *Practice showing more appreciation.* Several landmark studies over the last several decades have shown appreciation is the number-one motivator for employees. Managers who do not express appreciation not only miss out on this powerful motivator but also sow seeds of discontent and disengagement. Few things alienate workers more than when their hard work, going the extra mile, and showing initiative are taken for granted. Therefore, practice noticing when your workers do these things and then let them know you appreciate their efforts.

Engage Your Staff Tell your staff you are interested in improving morale, and you want to get their thoughts about what the team can do together to improve morale. It is important to emphasize "together," so it is clear this is a team effort, rather than a "wish list" workers get to make and managers are supposed to satisfy. Everybody needs to look at how to contribute. You will also want to be crystal clear that this is not a magical wish list. Not all ideas will be feasible, but all will be discussed and assessed. Most organizations with high morale engage employees in conversations about improving morale in a formal way, through employee focus groups and surveys. They make it a practice to keep connected into the voice of their customer (their employees). Despite how useful such formal approaches are, they do not take the place of one-to-one informal conversations that build strong manager–employee bonds. Often when only formal approaches are used, employees become skeptical of management's sincerity. They have witnessed far too many new initiatives that start with fanfare but end up crashing and burning. If this is the case in your department, you might want to take a lower-key approach by casually engaging staff in conversations about how things are going for them and asking for their insights and suggestions. Besides showing sincere intent, it also helps make such important two-way communication a regular part of their work experience, which in itself increases morale and engagement.

Ask for Feedback When you become a supervisor, your job is to bring out the best in your employees. To be effective you must be open to their feedback. Doing this not only provides you with useful information about how to manage each person more effectively but also keeps hurt feelings from festering and getting in the way of employees' working enthusiastically. It also communicates that you care about and respect them. As we know from personal experience, whether or not we believe our boss cares about and shows us respect has a huge impact on our morale and level of involvement in our organization.

If a leader tends to be confrontational or if staff is intimidated by the leader, they will be reluctant to give honest feedback. If you truly want honest feedback, you will need to prove it. Do this by checking in with your people every once in a while, rather than giving a one-time speech. Through repeated exposures, your workers will begin to believe you really do mean it. You also demonstrate your sincerity by graciously receiving the feedback rather than getting defensive. Doing so is a challenge for most of us. In fact, when the Franklin Covey organization compiled the results of 360-degree feedback surveys its clients had conducted over the past decade, the two items that received the lowest average score were "receives feedback without getting defensive" and "is open to constructive criticism." Even the great managers, on average, have a lot of room for improvement in this area. It is hard for most of us to respond nondefensively to feedback, especially if we disagree with the perception. You might want to get some coaching on how to respond productively. This increases the odds that you will eventually receive the kind of feedback you need to improve your ability to cultivate high morale and productivity. Regardless of where you are in your organization's hierarchy, you have tremendous influence on the people you work with and on the morale and productivity of the department. By engaging in these actions, you will improve both. To make these efforts more than just another "flavor of the month" or fad in the eyes of the workforce, this needs to become an integral part of the organizational leadership and managerial style, which is reflected in daily actions.[14]

EMPOWER YOUR TEAM

Empowerment refers to increasing the political, social, or economic strength of individuals or groups. It often involves the empowered person or group developing confidence in its own capacities. The methods to empower an individual or a group can take on many forms dependent upon the situation.

Sociology Sociological empowerment often addresses members of groups that social discrimination processes have excluded from decision-making processes through, for example, discrimination based on race, ethnicity, religion, gender, and so on. Note in particular the empowerment technique often associated with feminism: consciousness-raising.

Management In the sphere of management and organizational theory, empowerment often refers loosely to processes for giving subordinates (or workers generally) greater discretion and resources, distributing control in order to better serve both customers and the interests of employing organizations. (This use of the word appears somewhat at odds with other usage, which most often assumes the empowerment of groups and individuals to better serve their own interests.) One account of the history of workplace empowerment in the United States recalls the clash of management styles in railroad construction in the American West in the mid-nineteenth century. At that time "traditional" hierarchical East Coast models of control encountered individualistic pioneer workers, strongly supplemented by methods of efficiency-oriented "worker responsibility" brought to the scene by Chinese laborers. In this case, empowerment at the level of work teams or brigades achieved a notable (but short-lived) demonstrated superiority.

Economics In economic development, the empowerment approach focuses on mobilizing the self-help efforts of the poor, rather than providing them with social welfare.

Personal Development In the arena of personal development, empowerment forms an apogee of many a system of self-realization or identity (re-)formation. Realizing the impracticality of everyone attempting to exercise power over everyone else, empowerment advocates have adopted the word *empowerment* to offer the attractions of each power, but they generally constrain its individual exercise to potential and feel-good uses within the individual psyche.[15]

WHAT IS EMPOWERMENT?

There has been a lot of discussion about what empowerment is and what it is not. Empowerment is the process of enhancing the capability of individuals or groups to make choices and to transform these choices into desired actions (see Figure 7.6).

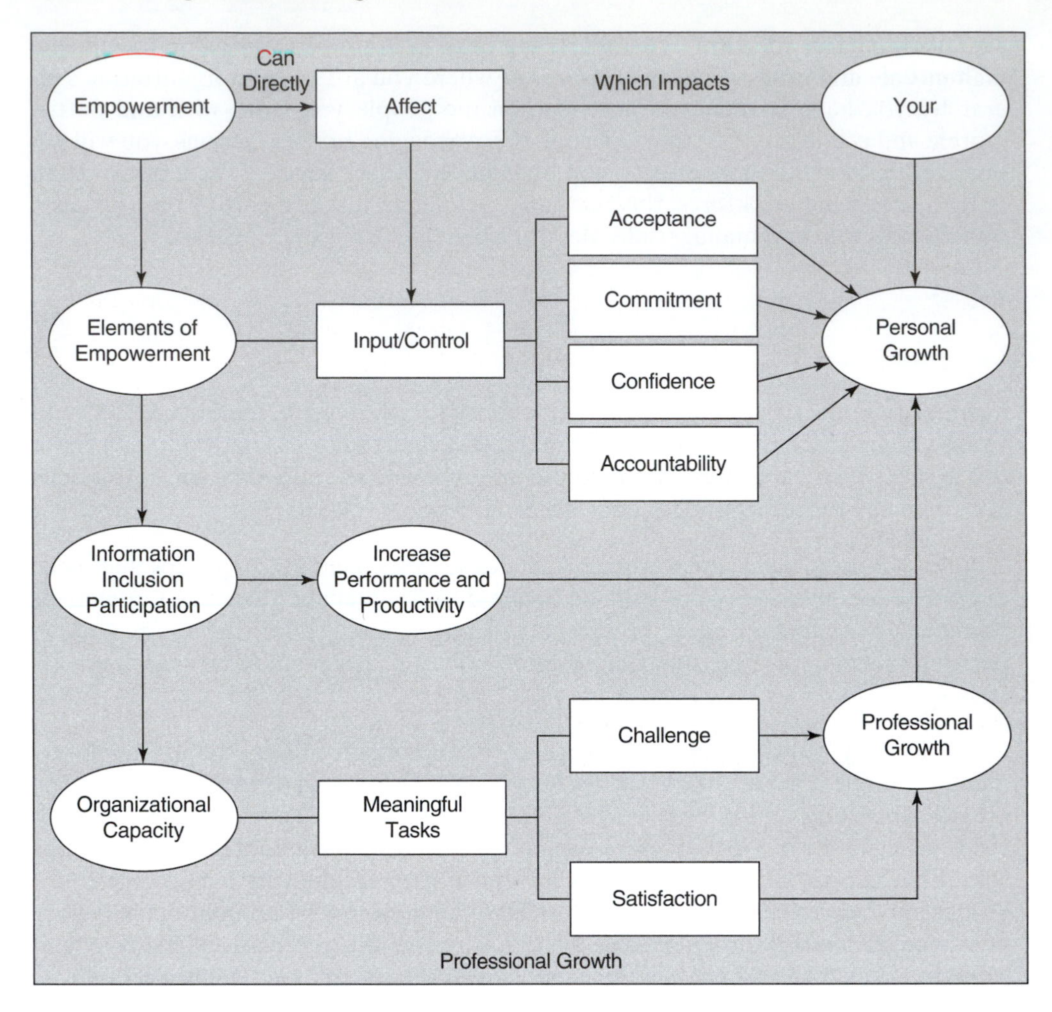

FIGURE 7.6 ◆ Elements of Empowerment
(Courtesy of Randy R. Bruegman)

Central to this process are actions that build individual and collective assets and improve the efficiency and fairness of the organizational and institutional context that govern the uses of these assets. Empowered people have freedom of choice and action, which in turn enables them to better influence the course of their lives and their decisions. There are thousands of examples of empowerment strategies used by people, local government, society, and the public sector. Although there is no single institutional model, empowerment experience indicates certain elements are almost always present when empowerment efforts are successful. These elements include access to information, inclusion, participation, accountability, and organizational capacity by promoting quality service (see Figure 7.6). Although these elements can be discussed separately, they are closely intertwined.

ACCESS TO INFORMATION

It is no secret employees are much better equipped to take advantage of the opportunities to exercise their own rights and become more accountable when they are armed with knowledge. Information is relevant if presented in forms that can be understood so that it is possible for people to take effective action. Therefore, information and knowledge dissemination cannot stop with the written word but also has to include a variety of methods—from discussions to telling our story, to debates, to many other forms of use of internal media—to ensure timely access to the information is provided and, as important, in a manner by which our employees can take advantage of it. Many institutional reform projects must include re-engineering our efforts to provide quality information and invest in the dissemination of such throughout the organization. Organizational knowledge not only plays an important role in connecting people together with the information needed to do their jobs but also articulates the strategic objectives of the organization.

INCLUSION AND PARTICIPATION

With an empowering approach to participation, we treat our people as coproducers with defined authority and control over decisions and resources at the lowest appropriate level. Such an approach begins to include more traditional groups in priority setting and decision making, allowing you to start to build a base for empowerment within the organization. However, an effort to sustain inclusion in a form of participation often requires changing our own rules so that we can create space for people not only to debate the issues but also to participate directly and indirectly in items such as strategic planning, budgeting, and, of course, the delivery of our basic services. As we know, participatory decision making is not always a harmonious process; therefore, be prepared to have some conflict resolution mechanisms in place to manage the disagreements.

ACCOUNTABILITY

Accountability refers to our ability as public officials to account for and answer to the public for policies, actions, and use of funds. As we have discussed throughout this book, ethics in leadership and integrity of our actions are paramount to our ability to be successful as individuals as well as organizations. Accountability can take many forms, and for many of us in local government, it starts with political accountability of our local representatives through representative elections. Those representatives ultimately appoint the higher positions in government, including fire chiefs who in turn hire staff to conduct the day-to-day business of providing fire and emergency medical services. Therefore, the linkage to our political accountability is direct and can be seen at the polls every four years.

We also have an administrative accountability of all government agencies through our own internal accountability measures whether it be ethics, financial, or just our normal standard operating procedures. There are external and administrative accountabilities through the local, state, and federal laws and ordinances we adhere to. There are also public and social accountabilities that hold governmental agencies accountable to their citizens whether through the electorate and/or some other means. Council or board meetings require us to be accountable and reinforce our need to be accountable to the citizens we serve. This is an important aspect of empowerment because we often talk about empowering our organizations, but we have to remember a critical element of empowerment rests with being accountable for our actions. The accountability of public resources at all levels must be assured through good fiscal management and local government oversight.

Today, government is challenged to offer services by providing competent fiscal management and contract accountability. Ensuring the most efficient and effective services are provided to the residents is a critical element of empowerment. However, some leaders do not really want their employees to empower change, which is unfortunate but true. These individuals live off the drama of the problems within the organizations. Some derive their energy from complaining about the way things are, how poor their staff is, whereas others just keep their staff operating at the lower possible levels so they can continue to engage in their power trips, helping them to feel superior to the people they work with. The problems with this are that you do not get very much done and you end up spending an enormous amount of energy dealing with issues that have very little relevance. Leaders can empower their organizations by doing the following:

1. *Ask your staff to bring you solutions rather than problems.* Many of our captains, battalion chiefs, and firefighters are also our best problem solvers. If you are going to empower your workforce, you need to engage them in the process of defining and bringing forward solutions. Frankly, many will not like this, some because they know you will ask them to figure out a solution for it.
2. *Ask your employees to move away from being risk averse to fixing things,* even when they are not broken. In the public sector sometimes the old adage, "if it's not broke, let's not fix it," is a cornerstone of much of what we do on a day-to-day basis and a foundation of the bureaucracy we work in. Unfortunately, it does not provide our organizations with leading edge solutions. Leaders and managers must be willing first to ask people to fix the things they know are not functioning properly and then look at how to improve existing services even when they are perceived to be working effectively and efficiently.
3. *Ask your staff to be the very best at what they do.* An old concept called the Peter Principle is when one is promoted one step beyond his or her capability. Do you have any of these folks within your own organization? We so often find these individuals were not the top performers in their previous classification, yet did well enough on the test to be promoted. The fact of the matter is if there were more accountability in the overall testing and evaluating process, these people might not have been promoted in the first place. The best way to ensure peak performance in the future is to make sure the best people get promoted. Future peak performance will rest with your staff and only be accomplished when they are doing top-level work in their current positions.
4. *Train your people to listen up.* Listen to what people are saying rather than just listening to them. When you listen up you hear the meaning behind the words being spoken, you hear what the person's real needs are and what may be stopping this person from getting the job done. Listening up takes practice, but it is worth the time and effort.

EMPLOYEE EMPOWERMENT: A CRUCIAL INGREDIENT IN PROMOTING QUALITY SERVICE

1. *Quality starts with people.* Any sound total quality management process, which will be discussed in Chapter 10, should be concerned with more than just the mechanical aspects of the change. Instead, it should focus on improving the more indirect value characteristics of the organization such as trust, responsibility, participation, harmony, and group affiliation. Empowerment is the most important concept in organizational quality, because employees must be empowered to make the necessary organizational changes. The concept of empowerment is based upon the beliefs that employees need the organization as much as the organization needs them, and leaders understand employees are their most valuable asset.
2. *Participative management is more than a management buzzword.* Research has shown there is a positive link between participation and satisfaction, motivation and performance. The self-managed work team is a new way of viewing the relationship of the worker-management-organization. Employee involvement teams, which consist of small groups of employees who

work on solving specific problems related to quality and productivity, represent one way of practicing participative management. Such teams have proved effective in resolving problems related to productivity and quality, as well as to improved employee morale and job satisfaction. Whatever the definition is, **participative management** requires responsibility and trust in the employees. It is important that we as managers recognize the potential of employees to identify and derive corrective actions to quality problems. However, if we refuse to act upon any of the team's recommendations, the team members faith in their ability to develop a quality program/process solution will have been destroyed, and the concept of an empowered team will not exist.

participative management
A management technique that allows nonmanagement personnel an opportunity to participate in decision making, provide input to management decisions, and/or execute management actions.

Critics often argue employees may be given the impression that quality and employee empowerment are just management buzzwords, and the decision-making process is still dominated at the top of the organizational hierarchy. In many organizations, this traditional labor division is the principal reason managers find it difficult to delegate responsibility. To some people, empowerment means more delegation in the form of indirect control. Some subordinates may view empowerment as abandonment of leadership responsibility or a means to shift workload.

3. *We are all in it together.* Workers affected by proposed changes must be involved in the decision to change or else they will likely fight progress. In an empowered organization, people should not expect to be told what to do but should *know* what to do. The primary role of management is to support and stimulate their people, cooperate to overcome cross functional barriers, and work to eliminate fear within their own work environment. However, many supervisors think empowerment may lead to them losing authority and ultimately their jobs. Therefore, it is logical that most of the resistance to empowerment comes from middle management. This resistance to change can be reduced by setting, measuring, and evaluating performance together with the team. In addition, managers often argue employees are unable to get the whole picture of the organization and are not all qualified to make decisions. Often work teams are unable to see the connection between process improvements and the overall strategy for the organization.

4. *Empower from the bottom up.* The most important concept of empowerment is to delegate responsibility to the lowest levels in the organization. The decision-making process should be to a high degree decentralized, and individuals or work designed teams should be responsible for a complete part of work processes. For instance, Saturn, a successful American car manufacturer, empowered its employees by turning assembly lines into dedicated process-oriented workstations managed solely by the work team. Even the design process involves a high degree of employee participation especially when directly linked to responsibility, and employees are allowed to make suggestions how to improve processes. The ultimate success of a quality program is based on its ownership by employees and their empowerment to make changes. It is crucially important that management value employee suggestions and manage accordingly. Naturally, workers directly involved in a process know best how to improve it. In an empowered organization, employees feel responsible beyond their own job, because they feel the responsibility to make the whole organization work better. Employee empowerment does not directly constitute the success of a quality program, because quality is always on the center stage in such a strategy. Employee empowerment is usually the result of an organization's strategy, focusing not only on how to improve cost, speed, and efficiency through quality improvements but also on the employee.

5. *Employees—the most important assets in organizations.* Empowered personnel have responsibility, a sense of ownership, satisfaction in accomplishments, power over what and how things are done, recognition for their ideas, and the knowledge they are important to the organization. Without productive employees, the organization is stagnant and can do little to improve. Empowerment works the best when employees need their organization as much as the organization needs them, "and the need is much more than a paycheck and benefit package."

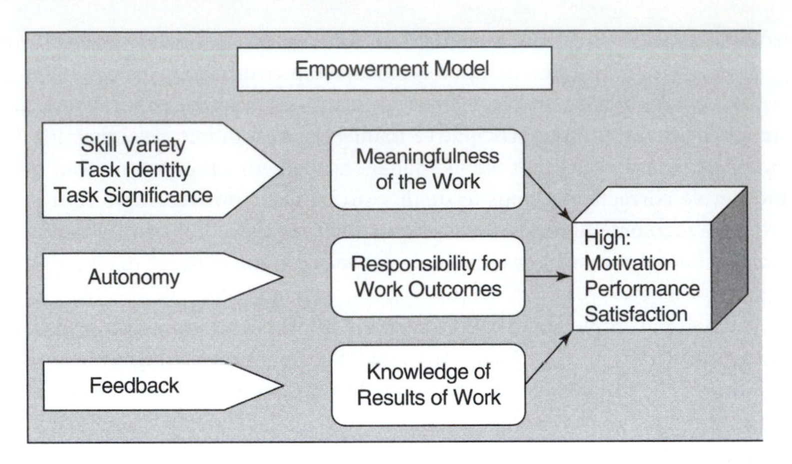

FIGURE 7.7 ◆ Empowerment Model
(Courtesy of Randy R. Bruegman)

The empowerment process is successful only when there is room for feedback and autonomy in the organizational culture (see Figure 7.7). Only in such a scenario is it possible to utilize the capabilities of your employees fully. The golden rule is "leaders should treat their employees the way they want their bosses to treat them."

6. *Treat your employees the way you want your customers to be treated.* The most critical success factor in quality is recognizing the importance of living up to customer expectations. It is important to understand there is a direct, positive correlation between satisfying internal customers and meeting external customers' needs. Employees who are not treated correctly cannot be expected to treat external customers differently. Internal satisfaction can be achieved through establishing a high degree of participative management, by decentralization of hierarchy power, and by creating a large degree of autonomy throughout the organization through the development of effective work groups. Because all these are based on the concept of employee empowerment, they will support the objectives of developing positive employee and customer relationships.

UNDERSTANDING THE PARADOX OF EMPOWERMENT

Although many in leadership roles talk about empowerment, often not many achieve it. Empowerment means letting go while taking control. These two actions seem contradictory, and indeed this paradox often traps many in leadership roles and undermines their attempts to provide transformation for their organizations and divisions and in their areas of responsibility. Those leaders who successfully embrace this paradox of empowerment understand their responsibility to create the right environment in which people flourish on behalf of the organization. By doing so, they can transform a department by intervening, sometimes in drastic ways, to change the way people work, relate, think, and feel. Yet they also realize when it is time to step back to let empowerment take root, grow, and become institutionalized within the organization.

Like any paradox, the paradox of empowerment is full of land mines, which many in leadership positions cannot accept: they cannot live with the contradictions of taking control and letting go. As such, many in leadership positions either advocate or continue to meddle in the operations. Those who understand the paradox often become the coaches and learn to cultivate the true aspects of an empowered organization. Although advocates talk empowerment with great enthusiasm, often spreading the word in a variety of ways, it is still often business as usual by either not establishing clear goals or not creating shared values to truly empower the organization. One can declare the goal is to be an empowered organization. Yet if nothing changes to transform an organization—no changes in the basic work process, no reorganization of teams, no reeducation or training—the department ultimately remains in the traditional hierarchy and silo mentality.

Those who speak of empowerment also may have difficulty letting go of their authority to make it work. These command and control freaks have a genuine fear of truly sharing power. Self-managed teams within their departments or divisions really do not fit their view of what their leadership role is, and from their vantage point, the empowered workforce appears to leave little role for the chief executive or the leader of the division. So, to compensate, they micromanage, demand immediate results, and often second-guess the process. People under their command often learn to delegate upward and never take ownership of the process, products, and tasks they have been assigned.

On one side are those who understand the paradox of empowerment and control. They have the ability to set a game plan, define the mission, establish goals and strategies, make changes in the infrastructure to create an empowered environment, and enable their people to be the best by providing the support to meet their needs. We call them the coaches, but the coaches do not play the game. They know the critical means of intervention and interference. Such people believe that it is human nature to seek work and responsibility, that people desire fulfilling productive jobs, and that in the right environment people will flourish on the job and exercise self-control and self-correction to serve the organization. On the other side are those who continue to hold negative views of human nature with their understanding that people cannot be trusted, do not like work, and take on very little responsibility. They truly do not trust their people, and without this trust they really cannot let go and are trapped in the paradox we speak of.[17]

Employee empowerment is more than a management buzzword and a textbook definition. It is a proven way of managing organizations toward a more complex and competitive future. Today providing quality service starts with engaging the people responsible for doing the work, who are often in the best position to provide meaningful insight to make it work better. Yet many in leadership positions argue they are unable to understand the global aspects of the organization. However, participative management has proven this not to be the case and has been very successful in fostering responsibility, motivation, and ownership in organizations with high autonomy and flexibility.

> Empowerment is epitomized in an ancient Chinese poem: Go to the people/Learn from them/Love them/Start with what they know/Build on what they have/But of the best leaders, when their task is accomplished, their work is done, the people will remark: We have done it ourselves.

ADVICE FROM THE EXPERTS

Management generational issues are becoming increasingly important. What have you done to address these issues?

MICHAEL CHIARAMONTE

Chief Fire Inspector, Retired, Lynbrook Fire Department (New York): Learn to understand the motives behind the various generations and adapt.

BETT CLARK

Former Fire Chief, Bernalillo County Fire and Rescue (New Mexico): Sit and discuss the values and issues throughout the organization. Promote the fostering of diversity among the crews and command staff. Our fire service must represent the community we serve. Young adults need the feeling of being individuals while belonging. We try to find common ground. This is not to say middle ground as I find that nothing is 50–50. We strive to meet individual needs while staying focused on the goals of the department. For example, my firefighters started to wear the "patch" of hair below their bottom lip. I made a deal with them, I would allow the patch for a period of time, and they would continue to do the quantitative fit test. I told them if there were no issues with the fit testing and no problems arose in the course of their duties, I would consider allowing the patch under the personal appearance and uniform policy. Well, eight months later, we have full compliance with fit testing and thus the "labret patch" as part of our policy . . . managing generational issues.

KELVIN COCHRAN

Fire Chief, Atlanta Fire Department (Georgia): Managing generational issues is an ongoing challenge within the Shreveport Fire Department. To manage the issues department-wide effectively, our leadership training for officers has focused on the topic on a year-to-year basis. Job orientation for the new generation of employees hired over the past 10 to 12 years has become critically important. Orientation and indoctrination of new hires establish expectations for every aspect of the job and its culture. The basic training program and the probationary firefighter checkoff program during the first year measure the new hires' ability and willingness to comply with those expectations. This generation of employees expects senior members and officers to comply with the same standards and expectations with which they are required to comply. They expect the leadership to set the example. They expect to be told what to do and require a balance of directive and supportive leadership behavior to perform successfully. Our officers are held accountable for creating this type of atmosphere at work. Recognition and rewards are also essential to managing generational differences. Officers must be empowered to grant certain rewards to Generation Xers and Generation Nexters. The Shreveport Fire Department has also implemented a special recognition program where all members can be granted commendations for outstanding performance.

CLIFF JONES

Fire Chief, Tempe Fire Department (Arizona): I have attempted to be adaptive to a degree. This is to understand that the future of the fire service rests with the new people who choose it as a career. I have continued to be demanding about the things that I feel are important to the service and the safety of the people in it. I have attempted to improve my insight into what motivates people in different generations, so that I might have a better understanding of how they operate and what to expect from them in the workplace.

BILL KILLEN

Retired Fire Chief and IAFC President 2005–2006: I spend a lot of time talking with my children, who have been in the workplace for more than 20 years, and asking them about their jobs, their supervisors, and what they think about the "generational" differences they have encountered. Their remarks have been candid, open, and most of all beneficial. I have tried the same with my grandsons, who have been in the workforce for a couple of years. The feedback from my children and grandchildren, as well as my observations of the workforce over the years, has enabled me to better understand some of the thoughts, habits, and values among the generations. Not so much early in my career, but more so in the last 15 years, I have developed a better understanding of the generational differences and how to create an effective working relationship with younger personnel.

KEVIN KING

U.S. Marine Corps Fire Service: I believe this is one of the greatest challenges for future leaders. I just do not think the "do it because I said so" is going to cut it anymore. One of the first things we have done is to try and understand what the generational differences are and how they affect our workplace. Second, we spend a lot more time telling our younger personnel why certain procedures and rules are important. If we can get them to understand the rationale behind the rule or procedure, then I believe we have a much better chance of getting their buy-in. Also, we have to get rid of procedures, processes, and traditions if they do not provide value. Our people are extremely busy, and we do not need to burden them with "busy work." (These are opinions of Mr. King and do not reflect official policy of the Department of Defense or the United States Marine Corps.)

MARK LIGHT

Executive Officer, International Association of Fire Chiefs: We continually address this issue by learning more about our workforce and how they expect to interact with others in the workplace. It is unique in that when we try to learn about how to work with Generation X and the Millennial generation, they often get offended that we would even have to do that. Yet today's workers are different and have different sets of expectations and goals. It is important for leaders to set the pace for addressing this with the frontline supervisors. This is often tough, as the frontline supervisors may see younger generations getting away with things they didn't, and they still have very fresh memories of what they were expected or not expected to do.

We must continue to change to embrace our workforce of the future. The unique thing about this is that, back in the 1970s and 1980s when many chief officers were entering the fire service, our predecessors said the same thing about us. Look how we turned out.

LORI MOORE-MERRELL

Assistant to the General President, International Association of Fire Fighters: Generational differences must be handled much like cultural diversity as that is how they are exhibited. There are varied cultural generations that have now entered the fire service. For example, there are the Generation Xers, who are hooked on electronic entertainment, and the Generation Me'ers, who are selfish and motivated only by money. Leaders must learn about what motivates the firefighters in their department and perhaps even change leadership styles based on those motivations. Remember, however, that there are still those motivated by family, work performance, pride, and honor, so do not change the nature of what the fire service should be; just change how we deal with those whose motivations are different from our own.

JAY REARDON

Fire Chief, Retired, Northbrook Fire Department (Illinois), currently President of the Mutual Aid Box Alarm System (MABAS): The way someone was "programmed" when born in the 1950s is tremendously different from that of someone born in the 1980s. You are what you were when! Values, attitudes, influences, experiences, and society wrote the individual's "program." You need to understand them! You need to understand you! Then you need to add their contribution capability to your contribution capability and move the pieces to the total sum. Negative contributions are the prejudices all bring to the table. Identify them, but do not allow them to taint the capability contribution calculation. The allowable variables within the capability contribution calculation include experience and maturity. They are often refined when the calculation is rounded off and consensus is reached.

GARY W. SMITH

Retired Fire Chief, Aptos–La Salva Fire Protection District (Aptos, California): A great book written by Warren Bennis, titled *Geeks and Geezers,* highlights the new era of leaders and does a lot of study on what causes change between generations. The background, education, environment, family experience, and leadership during the era of personal development (elementary school on to first job), have an impact on the type of leader/manager we will be when we become empowered. The new generation is concerned about balance in their lives; they want to support personal, family, and professional goals. They have a big concern for the future of our society and the need to deal with world challenges, and they are not easily inspired by leaders who want action strictly on the basis of "I am the boss." They change jobs and have been brought up in an era of quick-win logic. Personally, I have always enjoyed meeting the next generation, and I find great hope in their ability to manage and lead in the new world. It is like the most important aspect to change—a changing world needs new-generation leaders to take on the challenges that exist!

STEVE WESTERMANN

Fire Chief, Central Jackson County Fire Protection District (Blue Springs, Missouri), and IAFC President 2007–2008: Although we have no formalized program for this issue, we have done a couple of intentional informal actions. First, as a training tool, the management staff has picked a particular business or management book, given it as a reading assignment, and then picked it apart in discussion settings. We make sure that anytime someone runs across an article or newspaper clipping, we pass it around for review and comment. We have also found that we have a lot of 40- and 50-year-old employees who have now adopted children or are raising their grandchildren. If you are fortunate enough in your department, there is also what I will call more "mature Generation Xers" who have the ability to explain their perspectives aptly. Both of these groups have been helpful in assisting management staff in deciding how to handle individual situations.

THOMAS WIECZOREK

Executive Director, Center for Public Safety Excellence, and Retired City Manager: During my career as fire chief, city manager, and now executive director, I have tried to remain up to date and current on personnel issues. I have had the opportunity to hear some excellent speakers outline the pros and cons of the different generations. I remember seeing "demographics" on one agenda and thought it might offer a chance to catch up on some reading or, at worst, some sleep. It turned out that the individual not only looked at demographics but also supported them with advertising changes and concepts, along with other day-to-day research, that kept me focused and riveted to the point I bought both his book and tape for training. I brought the training back to the organization and tried to expose as many employees to it as possible. Obviously, some approached it with the same ideas I had in mind; most took back something. Education is key to any response.

JANET WILMOTH

Editorial Director, *Fire Chief* Magazine: Listen with an open mind.

FRED WINDISCH

Fire Chief, Ponderosa Volunteer Fire Department (Houston, Texas): There is clearly a difference with today's generation entering the workforce. It appears (to me) that people are less inclined (generally) to make personal commitments to success and instead place emphasis on "what is in it for me." Although this is not being critical and there are people who do not fit the above mold, I believe that leadership must be able to be flexible and not paint everyone the same color. Giving respect results in receiving respect, and when a person understands this, the entire organization wins.

Review Questions

1. What are the six areas of importance to building a successful team? Why are these important?
2. Conduct a SWOT analysis on your division or company. What recommendation for change would you propose?
3. What is the relationship of Maslow's hierarchy to group dynamics?
4. Explain why people resist change. How can you overcome this resistance?
5. Discuss the framework of the high-performance programming modeling.
6. How would you assess your department, division, and/or company in respect to the high performance team model?
7. Describe the characteristics and how to develop a high-performance team. Include the characteristics of a high-performance team.
8. Describe the task and process roles that must be addressed for effective team interaction.
9. How do coping styles impact group dynamics? Cite examples.
10. From a leadership perspective, how can cognitive styles be used to build a more effective team?
11. What is the difference between task roles and process roles in respect to teamwork?
12. Describe the profiles of each of the generations. Include the defining factors of each of the generations and discuss how to lead each of the generations.
13. Discuss how morale can affect an organization and the ability of the leader to lead. Include the four points to guide morale building efforts.

14. When looking at things you can and cannot control, what are the "big three"?
15. What changes do you envision for the fire service as a result of a diversifying workforce?
16. Why is the labor–management relationship so important to organizational success?
17. Discuss the importance and critical components of empowering your team. Describe what empowerment means.
18. Describe the crucial ingredients in empowering employees as it relates to total quality leadership.
19. What is empowered leadership?

References

1. Robert Fulghum, *All I Really Need to Know I Learned in Kindergarten: Uncommon Thoughts on Common Things* (New York: Ballantine, 1986).
2. Rosabeth Moss Kanter, *When Giants Learn to Dance*, (New York: Touchstone, 1989), p. 19.
3. Bruegman, *The Chief Officer*, pp. 154–156.
4. Ibid, p. 157.
5. John D. Adams, *Transforming Work: A Collection of Organizational Transformation Readings* (Alexandria, VA: Miles River Press, 1984), p. 227.
6. Quotos Fismon, www.quotosfismos.com, 2005.
7. Paper adapted from Edgar Schein, *Organizational Culture and Leadership* (San Francisco: Jossey-Bass, 1985).
8. Randy R. Bruegman, *The Chief Fire Officer: A Symbol Is a Promise,* (Upper Saddle River, NJ: Prentice Hall, 2005) p. 151–157.
9. Adams, *Transforming Work*, 1984, p. 227. Ron Zemke, Clair Raines, and Bob Filipczak, *Generations at Work* (New York: AMACOM, 2000). Gary McIntosh, "Three Generations: Riding the Wave of Change" Grand Rapids, MI: Fleming H. Rovell, 1995.
10. *Generation Gap*, http://library.thinkquest.org; *BridgeWorks,* www.generations.com; S. J. Adams, "General X," *Professional Safety* (January 2000), pp. 26–29; J. J. Salopek, "The Young and the Rest of Us," *Training and Development* (February 2000), pp. 26–30; M. L. Alch, "Get Ready for the Next Generation," *Training and Development* (February 2000), pp. 32–34; and W. Ruch, "How to Keep Gen X Employees from Becoming X-Employees," *Training and Development* (April 2000), pp. 40–43.
11. Linda Gravette, Ph.D., SPHR, *Managing Conflict Across Generations,* www.gravett.com/articles/04-12.htm
12. SNA, Suburban Newspapers of America, www.suburban-news.org/index.cfm?method=news.showNews&newsID=14 David Lee, *Human Nature at Work*, www.humannatureatwork.com
13. Wayne Baker, Ph.D., www.humaxnetworks.com/paradoxarticle.html
14. Ibid.

Managing Emergency Services

Key Terms

budget, p. 343
chain of command, p. 341
delegation, p. 342
line-item budget, p. 344
organization chart, p. 332
organizational design, p. 338
organizational structure, p. 337
performance appraisal, p. 368
performance budget, p. 346
planning, programming, and budgeting system (PPBS), p. 346
program budget, p. 346
span of control, p. 341
zero-based budget (ZBB), p. 348

Objectives

After completing this chapter, you should be able to:

- Explain the typical models of local governance and organization.
- Describe the various budget methods used at the federal, state, and local levels.
- Describe federal laws that influence labor relations.
- Explain differences among collective bargaining, mediation, and binding arbitration.
- Describe how to promote a healthy relationship between labor and management.
- Explain the current trends in personnel administration and human resource management development.
- Explain the importance of due process.

A key issue in accomplishing the goals identified in the planning process is structuring the work of the organization. Organizations are groups of people, with ideas and resources, working toward common goals. The purpose of the organizing function is to make the best use of the organization's resources to achieve organizational goals. Organizational structure is the formal decision-making framework by which job tasks are divided, grouped, and coordinated. Formalization, as an important aspect of structure, is the extent to which the units of the organization are explicitly

organization chart
A diagram of an organization's structure, showing the functions, departments, or positions of the organization and how they are related.

defined and its policies, procedures, and goals are clearly stated. The formal organization can be seen and represented in chart form. An **organization chart** displays its structure and shows job titles, lines of authority, and relationships among departments (see Figure 8.1)

LOCAL MODELS OF ORGANIZATION AND GOVERNANCE

Local governments use several governance models to provide services to their residents. Although variations exist, here is an overview of the most common ones utilized in the United States.

COUNCIL-MANAGER GOVERNMENT

The council-manager government is one of two main variations of representative municipal government in the United States. This system of government is used in the majority of American cities with populations over 12,000.

In the council-manager form of government, an elected city council (typically between 5 and 11 people) is responsible for making policy, passing ordinances, voting appropriations, and having overall supervisory authority in the city government. In such a government, the mayor performs strictly ceremonial duties or acts as a member and presiding officer of the council.

The council hires a city manager or administrator to be responsible for supervising government operations and implementing the policies adopted by the council. The manager serves the council, usually with a contract that specifies duties and responsibilities.

Municipal governments are usually administratively divided into several departments, depending on the size of the city. Though cities differ in the division of responsibility, the typical arrangement is to have departments handle the following roles:

1. Planning and Development
2. Public Works: Construction and maintenance of all city-owned or operated assets, including the water supply system, sewer, streets, storm water, snow removal, street signs, vehicles, buildings, land, and so on.
3. Parks and Recreation: Construction and maintenance of city parks, common areas, parkways, publicly owned lands, and the like. Also, operation of various recreation programs and facilities.
4. Police
5. Fire/Emergency Medical Service
6. Emergency Management (often found under the Fire Department)
7. Accounting/Finance
8. Human Resources: Incorporates human resources department for city workers.
9. Legal–Risk Management: Handles all legal matters including writing municipal bonds, verifying the city is in compliance with state and federal mandates, responding to citizen lawsuits such as those allegedly stemming from city actions or inactions.
10. Transportation: If the city has a municipal bus or light rail service, this function may be its own department or it may be folded into another of the departments listed here.
11. Information Technology Department: Supports computer systems used by city employees. May also be responsible for a city web site, phones, and other systems.
12. Housing Department

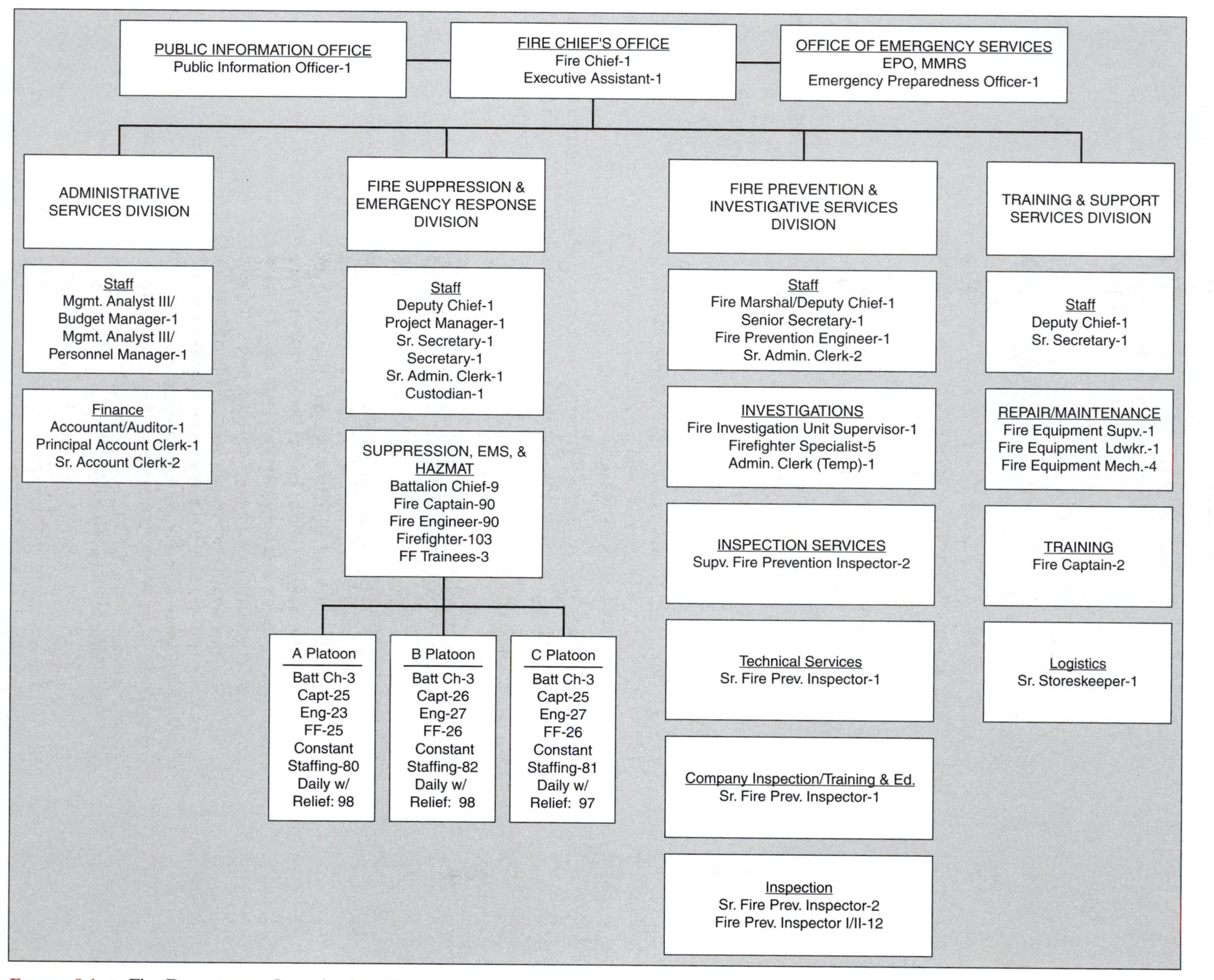

FIGURE 8.1 ◆ Fire Department Organization Chart

The council-manager system can be seen to place all power into the hands of the legislative branch. However, a city manager's role is similar to that of a corporate CEO in providing professional management to an organization. In a council-manager government, the city council appoints a city manager, makes major decisions, and wields representative power on behalf of its citizens.

City Manager A city manager is an official appointed as the administrative manager of a city in a council-manager form of city government. Some municipalities call this position the chief administrative officer.

Typical roles and responsibilities of a city manager include:

- Supervising the day-to-day operations of all city departments
- Supervising the department heads
- Preparing a draft city budget each year with options the council votes on
- Researching and making recommendations about topics of interest to the council
- Meeting with citizens and citizen groups to understand their needs better
- Providing executive leadership that encourages good performance by city workers
- Operating the city with a professional understanding of how all city functions operate together to their best effect

Typically, city managers have hire–fire authority over all city employees, though these decisions may require approval by the council and must comply with locally applicable civil service laws. This authority includes talent searches for "department heads" who are the managers of the city departments.

MAYOR-COUNCIL GOVERNMENT

The mayor-council government is the other variation of government most commonly used in representative municipal governments in the United States. The mayor-council variant can be divided into two types depending on the relationship between the legislative and executive branches.

Weak-Mayor Form In this form of the mayor-council government, the council possesses both legislative and executive authority. It may appoint officials, must approve mayoral nominations, and also exercises primary control over the municipal budget.

This form of government, most commonly used in small towns, is a variant of city commission government.

Strong-Mayor Form In the strong-mayor form of government, the mayor is given almost total administrative authority, with the power to appoint and dismiss department heads without council approval. Likewise, the mayor prepares and administers the budget, although that budget often must be approved by the city council.

In some strong-mayor governments, the mayor appoints a chief administrative officer (CAO) to supervise department heads, prepare the budget, and coordinate departments. This CAO is responsible only to the mayor.

CITY COMMISSION GOVERNMENT

City commission government as a form of municipal government was once common in the United States, but has fallen out of favor. Most cities formerly governed by commission having switched to the council-manager form. Some consider the city commission as a predecessor of the council-manager form.

In a city commission government, voters elect a small commission, typically from five to seven members, on a plurality-at-large basis. These commissioners constitute the legislative body of the city and as a group are responsible for taxation, appropriations, ordinances, and other general functions. Individual commissioners are assigned responsibility for a specific aspect of municipal affairs, such as public works, finance, or public safety. One commissioner is designated to function as chairperson or mayor, but this is largely a procedural or ceremonial designation and typically does not involve significant additional powers beyond that exercised by the other commissioners. As such, this form of government blends legislative and executive branch functions in the same body. Used mainly in the South Florida region, this form of government is in many ways similar to the weak-mayor form of mayor-council government.

TOWN

A town is a residential community of people ranging from a few hundred to several thousands, although it may be applied loosely even to huge metropolitan areas. Generally, a town is thought of as larger than a village but smaller than a city. The words *city* and *village* came into English from Latin via French. *Town* and *borough* (also *burrow, burgh, bury,* etc.) are native English and Scottish words.

In the United States, the meaning of the term *town* varies from state to state. In some states, a town is an incorporated municipality, that is, one with a charter received from the state, similar to a city. In others, a town is unincorporated.

The types of municipalities in U.S. states include cities, towns, boroughs, villages, and townships, although most states do not have all five types. Many states do not use the term *town* for incorporated municipalities. In some states, for example, Wisconsin, *town* is used in the same way that civil township is used elsewhere. In other states, such as Michigan, the term *town* has no official meaning and is simply used informally to refer to a populated place, whether incorporated or not.

In the six New England states, a town is a municipality and a more important unit than the county. In Connecticut, Rhode Island, and 7 out of 14 counties in Massachusetts, in fact, counties exist only as map divisions and have no legal functions; in the other four states, counties are primarily judicial districts, with other functions primarily in New Hampshire and Vermont. In all six states, towns perform functions that would be county functions in most other states. The defining feature of a New England town, as opposed to a city, is that a town meeting and a board of selectmen serve as the main form of government for a town, whereas cities are run by a mayor and a city council.

In New York, a town is similarly a division of the county, but with less importance than in New England. Of some significance is the fact that, in New York, a town provides a closer level of governance than its enclosing county, providing almost all municipal services to unincorporated areas, called hamlets, and selected services to incorporated areas, called villages. In New York, a town typically contains a number of such hamlets and villages. However, due to the independent nature of incorporated villages, they may exist in two towns or even two counties. Everyone in New York State who does not live in an Indian reservation or a city lives in a town and possibly in one of the town's hamlets or villages. (Some other states have similar entities called townships.) In New York, *town* is essentially short for *township.*

In Pennsylvania, only one municipality is incorporated as a town: Bloomsburg. Most of the rest of the state is incorporated as townships (there are also boroughs and cities), which function in much the same way as do the towns of New York or New England, although they may have different forms of government.

In Virginia, a town is an incorporated municipality similar to a city (though with a smaller required minimum population). Although by Virginia law cities are independent of counties, towns are contained within a county.

In Nevada, a town has a form of government but is not considered to be incorporated. It generally provides a limited range of services, such as land use planning and recreation, while leaving most services to the county. Many communities have found this "semi-incorporated" status attractive; the state has only 20 incorporated cities and towns as large as Paradise (186,020 in the 2000 Census), home of the Las Vegas strip. Most county seats are also towns, not cities.

In California, where the term *village* is not used, *town* usually refers to a community that is unincorporated, regardless of size. Because of this, some towns are larger than small cities, and any settlement with a name may be called a town, even though it may be only a relatively small grouping of buildings. Unincorporated communities, even large ones, are usually not referred to as cities.

Town Meeting A town meeting refers to a meeting whereby an entire geographic area is invited to participate in a gathering, often for a political or administrative purpose. It may be to obtain community suggestions or feedback on public policies from government officials or to cast legally binding votes on budgets and policy.

In the United States, a town meeting refers to a form of local government practiced in the New England region but is rare elsewhere. Typically conducted by New England towns, *town meeting* can also refer to meetings of other governmental bodies, such as school districts or water districts. Although the uses and laws vary from state to state, the general form is for residents of the town or school district to gather once a year and act as a legislative body, voting on operating budgets, laws, and other matters for the community's operation over the following 12 months.

This term's usage in the English language can also cause confusion. Town meeting is both an event, as in "Hagerstown had its town meeting last Tuesday," and an entity, as in "Last Tuesday, Town Meeting decided to repave Grantland Road." Starting with Jimmy Carter's presidential campaign in 1976, town meeting has also been used as a label for any moderated discussion group in which a large audience is invited, as in "John Kerry held a town meeting with voters to discuss issues in the upcoming election." To avoid confusion, this sort of event is often called a "town hall meeting."

Board of Selectmen The Board of Selectmen is commonly the executive arm of town governments in New England. The board may consist of three or five members with staggered terms.

In most New England towns, the adult population gathers annually in a town meeting to act as the local legislature, approving budgets and laws. Day-to-day operations were originally left to individual oversight, but when towns became too large for individuals to handle such workloads, they would elect an executive board of, literally, select(ed) men to run things for them.

These men had charge of the day-to-day operations; selectmen were important in legislating policies central to a community's police force, highway supervisors, pound keepers, field drivers, and other officials. However, the larger towns grew, the more power would be distributed among other elected boards, such as fire wardens and police departments. For example, population increases led to the need for actual police departments, of which selectmen typically became the commissioners. The advent of tarred roads and automobile traffic led to a need for full-time highway maintainers and plowers, leaving selectmen to serve as supervisors of streets and ways.

The term *selectman* is generally considered gender neutral and is usually applied even to female board members, although *selectwoman* is also used. Some towns have changed the official designation to the gender-neutral *Select Board*.

The function of the Board of Selectmen differs from state to state and even within a given state, depending on the type of governance under which a town operates. A selectman is almost always a part-time position that pays only a token salary.

The basic functions consist of calling town meetings, calling elections, appointing employees, setting certain fees, overseeing certain volunteer and appointed bodies, and creating basic regulations.

In larger towns, most of the selectmen's traditional powers are entrusted to a full-time town administrator or town manager. In some towns, the Board of Selectmen acts more like a city council, but retains the historic name.

In certain places, the head of the Board of Selectmen is the First Selectman, who historically has served as the chief administrative officer of the town. Sometimes this is a part-time position, with larger towns hiring a town manager, and sometimes the First Selectman exercises the powers typically associated with mayors.[1]

ORGANIZATIONAL STRUCTURE

Even though there are vast differences among organizations, their many similarities enable them to be classified. One widely used classification is the mechanistic versus organic form of **organizational structure** developed by Tom Burns and G. M. Stalker in their study of electronics firms in the United Kingdom.[2] The *mechanistic structure* is the traditional or classical design, common in many medium-size and large organizations. Mechanistic organizations are somewhat rigid in that they consist of very clearly delineated jobs, have a well-defined hierarchical structure, and rely heavily on the formal chain of command for control.

organization structure Framework that permits an organization to accomplish its mission, strategy, goals, and objectives; defines the functions of each portion of the organization; and establishes the communication and control lines associated with the operation of the organization; visualized in an organizational chart.

Bureaucratic organizations, with their emphasis on formalization, are the primary form of mechanistic structures, characterized by a rational, goal-directed hierarchy, impersonal decision making, formal controls, subdivision into managerial positions, and specialization of labor. Bureaucratic organizations consist of hierarchies with many levels of management. As such, people can become relatively confined to their own area of specialization. Bureaucracies are driven by a top-down or command and control approach in which managers provide considerable direction and have considerable control over others. Other features of the bureaucratic organization include functional division of labor and work specialization.

The *organic structure* is a more flexible, more adaptable, and much more participative form of governance, and one less concerned with a clearly defined structure. Organic organizations are open to the environment in order to capitalize upon new opportunities. They have a flat structure with only one or two levels of management. Flat organizations emphasize a decentralized approach to management that encourages high employee involvement in decisions. The purpose of this structure is to create independent small businesses or enterprises that can rapidly respond to customers' needs or changes in the business environment. The supervisors tend to have a more personal relationship with their employees. Rensis Likert has conducted extensive research on a nonbureaucratic organization design referred to as System 4 (participative-democratic), in which management and employees interact in a friendly environment characterized by mutual confidence and trust.[3]

INFORMAL ORGANIZATION

The informal organization is the network, unrelated to the department's formal authority structure, of social interactions among its employees. It is the personal and social relationships that arise spontaneously as people associate with one another in the work environment. This informal organization often affects the formal organization. The informal organization can pressure group members to conform to the expectations of the informal group that conflict with those of the formal organization. Recognizing the existence of informal groups, identifying the leaders within these groups, and using the knowledge of the groups to work effectively with them toward the departmental mission are important. The informal organization can make the formal organization more effective by providing support to management, stability to the environment, and useful communication channels.

ORGANIZATIONAL DESIGN

Designing an organization involves developing an organizational structure that will enable the department to achieve its goals most effectively (see Figure 8.2).

organizational design The determination of the organizational structure that is most appropriate for the strategy, people, technology, and tasks of the organization.

Fire department **organizational design** is normally based on an organizational structure, with traditionally functional and divisional hierarchy, which uses cross-functional teams for specific projects. Functions or divisions are often used to arrange traditional organizations. In both the public and private sectors, a variety of organizational structures can be found.

In a *functional organization,* authority is determined by the relationships between group functions and activities. Functional structures group similar or related occupational specialties or processes together under the familiar headings of finance, manufacturing, marketing, accounts receivable, research, surgery, and photo finishing. Economy is achieved through specialization. However, the organization risks losing sight of its overall interests as different divisions pursue their own goals and program objectives.

In a *divisional organization,* corporate divisions operate as relatively autonomous businesses under the larger corporate umbrella. In a conglomerate organization, divisions may be unrelated. Divisional structures are made up of self-contained

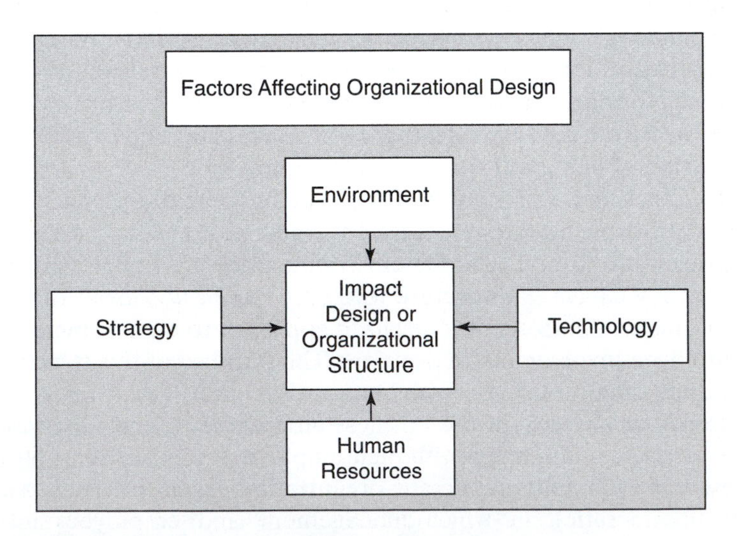

FIGURE 8.2 ◆ Factors Affecting Organizational Design

strategic business units that each produce a single product. For example, General Motors' divisions include Buick Chevrolet, Pontiac, GMC, Hummer, SAAB, Saturn, and Cadillac. A central headquarters, focusing on results, coordinates and controls the activities and provides support services among divisions. Functional departments accomplish division goals; however, a weakness is the tendency to duplicate activities among divisions. To overcome this weakness, cross-functional teams are formed and team members may report to two or more managers. Cross-functional structures utilize functional and divisional chains of command simultaneously in the same part of the organization, commonly for one-of-a-kind projects. Such a structure is used to develop a new product/service, to ensure the continuing success of a program or service to which several divisions of the department directly contribute, and/or to solve a difficult problem. By superimposing a project structure upon the functional structure, a cross-functional team allows the organization to take advantage of a diverse talent pool from many parts of the organization.

Boundaryless organizations have come to describe many of the business organizations of today and likely of the future. A boundaryless organization is the opposite of a bureaucracy with numerous barriers and divisions. In contrast, the organization without boundaries offers interaction and networking among professionals inside and outside the organization. This organizational model is fluid and highly adaptive because the structure of business is ever-changing. Professionals inside the organization form networks and links and emphasize collaboration on projects. Business relationships are informal, and people come together when they share a common need or problem. Employees are grouped by competencies centered on technology, information, and expertise.

The boundaryless organization has developed primarily due to the widespread dissemination of information and the presence of information technology. It is also difficult to distinguish between internal and external practitioners, and such distinctions offer little meaning in today's environment.

The Changing Role of Employees To be successful in the new boundaryless world of business requires a person to have strong team skills. It is important for employees to feel at ease with the free-form work structure and a work environment that may border on chaos. The tremendous networking and linking that occur change the role of employees to that of consultants. Employees no longer work in isolation but as part of a team on broad, company-wide projects, such as quality management, just-in-time methods, lean production, and supply chain management. Strategic alliances and collaborative arrangements, often between competitors and vendors, are other facets of the boundaryless organization.

Because technology plays a major role as a communication medium in the boundaryless organization, much work is done from a distance via e-mail, phone, text messaging, and fax. Less work is done in traditional face-to-face settings. Virtual collaboration makes it easier to use the expertise of a broader range of individuals. With telecommuting, international employees are more easily made a part of all business processes. Employees often like the freedom that boundaryless work offers them, particularly with virtual teams and more flexible work plans, arrangements, and schedules.

Boundaries and organizational affiliations are also blurring as large organizations team up with small businesses and consultants, as well as with other informal networks of groups, professional organizations, and businesses. The emphasis is on expertise, not location or affiliation. Employees may be part of multiple networks and organizations in the new workplace. Because they change roles and affiliations, the responsibility for training, education, and development now rests with the employee and not specifically with the organization.

Even in a boundaryless organization, some bureaucratic activities must take place. These include creating and maintaining a common task and group climate to focus groups and teams on the tasks at hand and on overall strategies. As organizations restructure, these boundary-spanning bureaucratic activities change as well.[4]

Contingency organizations are those that utilize the most appropriate organizational structure for each situation depending upon technology, organizational size, goals, strategy, environmental stability, and employees' characteristics. Mechanistic organizations are best suited to repetitive operations and stable environments, whereas organic organizations are best suited to an environment that is in a constant state of change.

ORGANIZING FUNCTIONS OF WORK

The organizing functions of work deal with all the activities that result in the formal assignment of tasks and authority and a coordination of effort toward a specific goal or task. The supervisor staffs the work unit, trains employees, secures resources, and empowers the work group to become a productive team. The nature and scope of the work needed to accomplish the organization's objectives determine work classification and work unit design. Division of labor, or work specialization, is the degree to which tasks in an organization are divided into separate jobs. Work process requirements and employee skill level determine the degree of specialization. Once jobs have been classified through work specialization, they are grouped to coordinate the common tasks. This is the basis on which work or individuals are grouped into manageable units. There are four traditional methods for grouping work activities.

- *Organized by the functions* to be performed. These functions reflect the nature of the business. The advantage of this type of grouping is obtaining efficiencies from consolidating similar specialties and people with common skills, knowledge, and orientations together in common units.
- *Organized by the programs/services* needed to provide a particular program/service under one executive. For instance, most fire departments are structured around groups such as administration, fire suppression, fire prevention, training, support services, and EMS.
- *Organized by geographical regions.* The foundation of our delivery system is built upon geographic districts based upon station locations in our community.
- *Organized by the process* on the basis of program, service, or customer need. Each process requires particular skills and offers a basis for homogeneous categorizing of work activities. A resident who calls 9-1-1 for a response would first engage in the 9-1-1 PSAP, then go through the dispatch process to receive an emergency response from multiple agencies and specialists. These services may each be administered by different departments.[5]

AUTHORITY

The organizational structure provides the framework for the formal distribution of authority. Formalization is the degree to which tasks are standardized and rules and regulations govern employee behavior. It influences the amount of discretion an employee has over his or her job. In an organization with high degrees of formalization, job descriptions and policies provide clear direction. Where formalization is low, employees have a great deal of freedom in deciding how to conduct their work. Within the same organization, different departments may have different degrees of formalization. Authority is the legitimate power of a supervisor to direct subordinates to take action within the scope of the supervisor's position.

Forms of Authority

Three forms of authority are line authority, staff authority, and team authority.

Line authority is direct supervisory authority from superior to subordinate. Authority flows in a direct chain of command from the top of the department to the bottom. **Chain of command** is an unbroken line of reporting relationships that extends through the entire organization defining the formal decision-making structure. It helps employees know to whom they are accountable and whom to go to with a problem. Unity of command within the chain states that each person in an organization takes orders from and reports to only one person. This helps prevent conflicting demands being placed on employees by more than one manager. However, the trend toward employee empowerment, fueled by advances in technology and changes in design from downsizing and re-engineering, has forced many departments to realign their organizational lines of reporting or span of control. **Span of control** refers to the number of employees who should be placed under the direction of one manager. Spans within effective organizations vary greatly. The actual number depends on the amount of complexity and the level of specialization. In general, a wide span of control is possible with better-trained, more experienced, and committed employees.

chain of command
A series of command, control, executive, or management positions in hierarchical order of authority.

span of control
The spatial, temporal, or resource limits accorded to a member of management.

Staff authority is more limited authority or advice. This authority is based on expertise that usually involves advising staff members in making decisions but does not make final decisions. Staff authority coordinates and provides technical assistance or advice. Examples include accounting, human resources, information technology, research, public relations, and legal services.

Team authority is granted to committees or work teams involved in an organization's daily operations. Work teams are groups of operating employees empowered to plan and organize their own work and to perform their work with a minimum of supervision. Team-based structures organize separate functions into a group based on one overall objective. Empowered employees create their own schedules, design their own processes, and are held responsible for their outcomes. This facilitates efficiencies in work process and the ability to detect and react to changes in the environment.

Centralization Versus Decentralization

Centralization is the degree to which decision making is concentrated in top management's hands. Decentralization is the extent to which decision-making authority is pushed down the organization structure and shared with many lower-level employees. Centralized organizations have more levels of management with narrow spans of control. Employees are not free to make many, if any, decisions. Decentralized organizations have fewer levels of management with wide spans of control giving employees more freedom of action. The current trend is toward broadening decentralization and employee empowerment. As the public intensifies the need for public organizations to be responsive, the employees at the lower levels and closest to customers become extremely important. They offer an excellent source of knowledge and implement changes that directly impact performance.[6]

Delegation

It is impractical and undesirable for those in supervisory positions to handle all decisions of the department directly. In order to meet the organization's

delegation
The act of empowering someone to act for another.

goals, to focus on objectives, and to ensure all work is accomplished, supervisors must delegate authority. Authority is the legitimate power of a supervisor to direct subordinates to take action within the scope of the supervisor's position. By extension, this power, or a part thereof, is delegated and used in the name of a supervisor. **Delegation** is the downward transfer of formal authority from superior to subordinate. The employee is empowered to act for the supervisor, while the supervisor remains accountable for the outcome. Delegation of authority is a person-to-person relationship requiring trust, commitment, and contracting between the supervisor and the employee. The supervisor assists in developing employees in order to strengthen the organization. He or she gives up the authority to make decisions that are best made by subordinates. This means the supervisor allows subordinates the freedom to make mistakes and learn from them. He or she does not supervise subordinates' decision making but allows them the opportunity to develop their own skills. The supervisor lets subordinates know he or she is willing to help but not willing to do their jobs. The supervisor is not convinced the best way for employees to learn is by telling them how to solve a problem, which results in these subordinates becoming dependent on the supervisor. Instead, he or she allows employees the opportunity to achieve and be credited for it.

The first step in delegating is to identify what should and should not be delegated. The supervisor should delegate tasks a subordinate can perform better or tasks providing valuable experience for subordinates to accomplish successfully. *Effective leadership and management require the ability to delegate work in a manner conducive to both the supervisor and the subordinate.*

- Establish the objectives of the delegation, specifying the task needing to be accomplished, and decide who should accomplish it.
- Meet with the chosen subordinate to describe the task and to ask the subordinate to devise a plan of action. As Andrew Carnegie once said, "The secret of success is not in doing your own work but in recognizing the right person to do it." For delegation to be successful, trust between the supervisor and employee fulfills the commitment.
- Review the objectives of the task as well as the subordinate's plan of action, any potential obstacles, and ways to avoid or deal with these obstacles. The supervisor should clarify and solicit feedback as to the employee's understanding, including the desired results (what, not how), guidelines, available resources, and consequences (good and bad). Delegation is similar to contracting between the supervisor and employee regarding how and when the work will be completed. The standards and time frames are discussed and agreed upon. The employee should know exactly what is expected and how the task will be evaluated.
- Monitor the progress of the delegation and make adjustments in response to unforeseen problems.
- Accept the completed task and acknowledge the subordinate's efforts.

An organization's most valuable resource is its people. By empowering employees who perform delegated jobs with the authority to manage those jobs, supervisors free themselves to manage more effectively. Successfully training future supervisors means delegating authority, which gives employees the concrete skills, experience, and confidence to develop themselves for higher positions. Delegation provides better managers and a higher degree of efficiency. Collective effort, resulting in the organization's growth, is dependent on delegation of authority.

Responsibility and Accountability Equally important to authority is the idea that when an employee is given responsibility for a job, he or she must also be given the degree of authority necessary to carry it out. For effective delegation, the authority granted to an employee must equal the assigned responsibility. Upon accepting the delegated task, the employee has incurred an obligation to perform the assigned work and to utilize the granted authority properly. Responsibility is the obligation to do assigned tasks. The individual employee is responsible for being proficient at his or her job; the supervisor is responsible for what employees do or fail to do, as well as for the resources under his or her control. Responsibility is an integral part of a supervisor's job.

Responsibilities fall into two categories: individual and organizational. Employees have individual responsibilities to be proficient in their job and for their actions. Nobody gives or delegates individual responsibilities. Employees assume them when they accept a position in the organization. Organizational responsibilities refer to collective organizational accountability and include how well departments perform their work. For example, the supervisor is responsible for all the tasks assigned to his or her department, as directed by the manager. Someone who is responsible for a task is liable or accountable to a superior for the outcome, and accountability flows upward in the organization. All are held accountable for their personal and individual conduct. Accountability is the reckoning, wherein a person answers for his or her actions and accepts the consequences, good or bad. Accountability establishes reasons, motives, and importance for actions in the eyes of managers and employees alike and is the final act in the establishment of one's credibility. It is important to remember accountability results in rewards for good performance, as well as discipline for poor performance.[7]

ORIGINS AND HISTORICAL CONTEXT OF BUDGETING

One of the most important areas for most fire agencies is budgeting. Depending upon the conditions at any given time in history, budgets have emphasized financial control, managerial improvements, planning, or all three. In looking back, it is difficult to imagine government without budgeting. However, unlike many other American institutions or practices, the concept of budgeting, as it is known today, was not introduced to the United States by the early colonists. Rather, it developed in the latter part of the nineteenth century.

At the turn of that century, the United States was "the only great nation without a budget system."[8] Congress raised and voted on the money needed to operate the national government in a more or less haphazard manner. The same was true of states and local governments, with the exception that more state and local variations existed due to their different governing agencies. In part, this legislative dominance was due to the notion of separation of powers and the fear of a strong executive deriving from American colonial experience. The result was budgeting, to the extent it can be said to have existed in the sense known today, was the preserve of the legislative branch. This pattern was dominant at all levels of government in the United States. However, it was the cities that sparked reform. Patronage-based political machines of the nineteenth century may have fostered the development of budgeting as it is now known. In the last decade of the nineteenth century, budgeting was defined as[9] "a valuation of receipts and expenditures or a public balance sheet, and as a legislative act establishing and authorizing certain kinds and amounts of expenditures and taxation."[10]

The idea of a **budget** as a control mechanism had been developing since the 1830s, but gained momentum after the Civil War with the growth of cities and the expansion of municipal services. By the end of the 1890s, there were three basic forms of municipal budgeting. Some cities simply used a tax levy, an approach disliked by reformers due to the lack of control through inattention to the expenditure side of budgeting, coupled with dominance by the city council. A second approach was a tax levy accompanied

budget
A financial plan for the coordination of resources and programs of the organization that articulates quantitative allocations of resources for specific activities and lists both the proposed expenditures and the expected revenue resources.

by detailed appropriations. Missing, of course, were details regarding revenue estimates. Still others used a tax levy but preceded it with detailed estimates of receipts and expenditures, a practice that found favor with business and middle-class reformers. However, city councils were not legally bound to adhere to these estimates.

Extremely important and influential in promoting municipal finance reform were activists in the rising professions of accounting, administration, and social work. Chief among these was the New York Bureau of Municipal Research, created in 1907, which highlighted poor fiscal procedures resulting in inefficiency. It worked to inject uniformity and responsibility in governmental finance. It deciphered how government operated and consequently recommended how to improve it. The New York Bureau of Municipal Research seized on the business corporation as the ideal model for bureaucratic organization and encouraged using scientific management concepts to promote planning, specialization, quantitative measurement, and standardization—the ingredients of efficiency. As such, the seeds were planted for training, in public administration, to produce a skilled pool of administrative technicians who could manage public services capably without regard to politics.

BUDGET REFORMS AND INNOVATIONS

The public administration movement with its scientific management ethic succeeded in creating widespread dissatisfaction with the budgetary methods of the time. Budgetary practices at most levels of government in the United States were dominated by the legislative branch of government. As such, departmental budget estimates at the local level, in most states, and the national government, were submitted directly to the legislative body. Seldom were supporting data included with estimates, and requests were usually lump sum. In fact, spending requests were not related to revenue projections or overall spending. At that time, a lack of standardization existed in accounting, and departments bargained with legislative appropriations committees directly or with local city council members. Little central oversight of departmental spending existed.

LINE-ITEM BUDGETING

line-item budget
Lists of revenue sources and proposed expenditures for budget cycle; most common budget system used in North America.

Prior to line-item budgeting, most budgets were lump sum. **Line-item budgets** simply listed categories (line items) of expenditure, such as salary, overtime pay, postage, gasoline, office supplies, and so forth. These objects of expenditure could be collapsed into broad categories such as personnel, operating, and capital expenses. Statutory or administrative controls could be imposed on the transfer of funds from one line item to another, or between broad categories of expenditure. In addition, another level of oversight was established with the legislative body's broad discretion in how detailed it appropriated funds. Line-item budgets, being relatively easy to use and understand, were attractive to legislative officials. Politically they were attractive because they did not focus explicit attention on substantive policy issues or choices, and line-item budgets had a number of appealing features. They allowed central control over inputs, or money, before they were used; were uniform, comprehensive, and exact; allowed routines to be established; and provided multiple opportunities for control to occur, such as in purchasing and hiring staff. They also allowed for budget cutting. With line items came central budget offices perpetuating the idea of control. Line items were to bring about more control over public spending. Even during the emphasis on line-item budgeting, some authorities had a more expansive vision for budgeting (see Figure 8.3).

	FY2003 ACTUALS	FY2004 ACTUALS	FY2005 AMENDED BUDGET	FY2006 PROPOSED BUDGET
160000-Fire Department				
10101-Genral Fund				
51101-Permanent Salaries	17,406,679	18,096,021	19,738,100	21 ,978,221
51102-Fringe	2,072,836	2,165,858	2,422,100	2,783,581
51103-Employee Leave Payoff	375,862	321,919	310,400	349,900
51201-Non-Permanent Salary	22,123	15,475	49,620	0
51202-Non-Permanent Fringe	1,697	1,184	3,000	0
51301-Overtime	2,568,055	1,795,012	1,617,900	1,265,700
51401-Premium Pay	343,079	417,805	511,400	1,056,276
51402-Relocation Payment	0	8,000	0	0
52601-Worker's Compensation	783,600	1,250,100	1,671,300	1,646,000
52901-Recurring Vehicle Allow.	11,310	10,740	10,500	13,000
53302-Prof Svcs/Consulting	7,767	23,800	0	9,000
53303-Public Relations & Infor.	2,230	11,080	0	0
53304-Prof Svcs (Non-Consulti	120,475	86,416	180,600	179,100
53401-Hazardous Waste Mana	2,220	4,816	500	500
53402-Specialized Services /T	19,747	21,725	11,100	13,600
54101-Utilities	240,758	236,414	307,500	307,500
54241-Landscaping & Grounds	282	7,399	5.000	2,500
54301-O/S Repair, Maint & Ser	7,481	6,017	8,000	8,000
54302-O/S Repair & Maint-Oth	0	0	0	6,400
54303-Service Contracts-Offic	700	8,660	3,000	9,900
54304-O/S Repair & Maint.—Ve	121,246	194,031	140,000	130,000
54305-O/S Repair & Maint.—Eq	34,614	30,539	78,200	78,200
54411-Space Rentals	9,504	11,568	10,000	10,000
54421-Equipment Rentals—Ex	644	2,558	0	0
55501-Printing & Binding—O/S	12.592	10.154	2,500	2,500
55801-Training	5,930	80,915	170,300	222,100
55803-Travel & Conference	7,723	13,598	2,600	11,700
55804-Misc. Subsistence Expe	18,553	5,557	1,000	1,000
55805-Mileage Reimbursemen	2,697	863	1,800	900
56101-Clothing & Personal Su	512,282	194,395	228,300	262,800
56102-Office Equipment-Unde	267	1,568	200	200
56105-Small Tools For Field O	5,888	8,088	1 7,400	17,400
56106-Postage	477	759	1,600	1,600
56107-Office Supplies	19,495	42,810	27,100	27,100
56108-Photographic Supplies	259	1,906	2,100	1,200
56109-Office Equipment Renta	0	0	0	0
56110-Computer Software	286	683	500	600

FIGURE 8.3 ◆ Fire Department Line-Item Budget

PERFORMANCE BUDGETING

The New York Bureau of Municipal Research had proposed early in its work that city budgets be on a unit cost basis and show work done as well as work proposed rather than just line items. The bureau was not successful in implementing the idea; it took 30 more years for these ideas to receive significant attention again. This came with the report of the first Hoover Commission (the Commission on Organization of the Executive Branch of Government) in 1949. This commission recommended budget information for the federal government to be structured in terms of activities rather than line items, and performance measurements were provided along with performance reports. With the expansion of governmental activity during the New Deal, attention began to shift to management efficiency, particularly at the national level. (Interestingly, the effects of this period on the state governments were often the opposite of the experience of the federal government. The states—due to inelastic taxes and debt restrictions—experienced constraints and did not expand as did the federal government during this period.)

An awareness of the need for efficiency and interest in management improvement were also evident at the local level. With the heightened interest in management came the concept of the **performance budget**. Performance budgeting emphasized the things government does rather than the things it buys. Therefore, it shifted attention from the means of accomplishment to the accomplishment itself. The Municipal Finance Officers Association presented a model accounting classification, emphasizing activity classifications within functions, in 1939 and again in the 1940s and 1950s. This had considerable influence on municipal budget practice.

performance budget
Type of budget categorized by function or activity; each activity is funded based on projected performance; similar to program budgets or outcome-based budgets.

Governmental expansion during the New Deal and World War II led to wider interest in performance budgeting in order to more efficiently use financial resources focused on activities and outputs, which could be identified and measured. Performance budgeting encountered a number of problems, particularly at the national level. Budget estimates were[11] "no more meaningful than those in line-item budgets."[12] Work measurement of governmental service presented problems and was imprecise. Inputs could be easily measured but not outputs. In addition,[13] "performance budgeting lacked the tools to deal with long-range problems."[14] Studies of state budget practices in the late 1960s and early 1970s found a great deal of variation in adoption of key techniques of performance budgeting. Workload statistics were introduced that usually gave some indication of the volume of governmental activity but failed to relate work to cost or performance[15].

program budget
Budget system used to categorize funds by program or activity.

planning, programming, and budgeting system (PPBS)
Budget system that provides a framework consistent with the organization's goals, objectives, policies, and strategies for making decisions about current and future programs using three interrelated phases: planning, programming, and budgeting; developed in the 1970s to coordinate planning, program development, and budget processes; used currently in industry and by the U.S. Department of Defense.

PROGRAM BUDGETING

The next budget reform phase emphasized program budgeting and planning. **Program budgeting**, however, suffered from a[16] "severe identification crisis in the budgetary literature."[17] Namely, it was used synonymously with performance budgeting and the planning, programming, and budgeting system budget reform. Program budgeting was more forward looking, whereas performance budgeting tended to focus on what had been already accomplished. The key elements of program budgeting include long-range planning, goal setting, program identification, and quantitative analysis such as cost-benefit analysis and performance analysis (see Figure 8.4).

PLANNING, PROGRAMMING, AND BUDGETING SYSTEM (PPBS)

Program budgeting was one of the key components of the **planning, programming, and budgeting system (PPBS)** reform that began in the 1960s. This budgeting system,

Program Outcome Measures	*Weight*	*FY2001/2002 Adopted*	*FY2002/2003 Adopted*	*FY2003/2004 Adopted*
◆ 9-1-1 and seven-digit emergency phone lines are answered within an average (mean) of ten seconds.	5			
—Seconds		10.00	10.00	10.00
◆ Emergency police calls are processed and dispatched within an average (mean) of 60 seconds.	5			
—Seconds		60.00	60.00	60.00
◆ Emergency fire and EMS calls are processed and dispatched within an average (mean) of 60 seconds.	5			
—Seconds		60.00	60.00	60.00
◆ The aggregate department performance index is at 100.	5			
—Performance Index		100.00	100.00	100.00
◆The Budget/Cost Ratio (planned cost divided by actual cost) is at 1.0.	4			
—Ratio		1.00	1.00	1.00
◆ All requests for property or evidence are completed within mandated laws and policies.	4			
—Percentage of Requests		100.00%	100.00%	100.00%
◆ Coordination of all recruitment, selection, and training for new sworn Public Safety personnel is provided, with at least 80% of those who enter the training program successfully completing probation (over a three-year rolling average).	3			
—Percentage of Personnel		80.00%	80.00%	80.00%
◆ 90% of Internal Affairs investigations are completed within 120 days without an appeal of the findings (over a three-year rolling average).	4			
—Percentage of Investigations		90.00%	90.00%	90.00%
—Number		120.00	120.00	120.00
◆ A satisfaction rating of 90% is achieved for Administrative and Technical Services.	4			
—Rating		90.00%	90.00%	90.00%
◆ All requests for information are processed within mandated guidelines so that California Department of Justice audit ratings reflect 95% accuracy.	3			
—Rating		95.00%	95.00%	95.00%

FIGURE 8.4 ◆ Program Performance Budget

which received its impetus during the Johnson administration from by Robert McNamara in the Defense Department, quickly surpassed the notion of performance budgeting. Interestingly, the impetus for PPBS did not come from budgeters but from three other sectors: economics, data sciences, and planning. PPBS was an attempt to use rational means of[18] "fusing planning processes, programming efforts, and the budget system."[19] Although little was new among these components, the uniqueness was in the attempt to combine the various elements. Planning was to be used to determine goals and programs to help achieve them. Programming would assist in administering efforts to accomplish goals efficiently. Budgeting would provide financial estimates of resources needed by agencies to execute the plans. PPBS failed to live up to its potential in the federal government, although elements of this practice remain in budget frameworks of several federal agencies.

ZERO-BASED BUDGETING

zero-based budget (ZBB) Budget type whereby all expenditures must be justified at the beginning of each new budget cycle, as opposed to simply explaining the amounts requested that are in excess of the previous cycle's funding. During ZBB planning, it is assumed that there is zero money available to operate the organization or program; then the contribution that the organization or program makes to the jurisdiction must be justified.

The basic concept for **zero-based budgeting (ZBB)** is to start with zero and build the budget. It requires the operation to take a fresh look at every expenditure without preconceived notions. There are several advantages to zero-based budgeting. It better equips management to make decisions when comparing actual program performance to the budget. Zero-based budgeting most often gives a better estimate of revenue projections and helps create a model for spending by breaking the habit of budgeting nonessential costs simply because they were incurred the prior year. The disadvantages of zero-based budgeting are that it is time consuming, and some categories in the budget are best estimated based on historical data because they are difficult to calculate from zero. For example, the cost of general supplies may best be determined by examining existing data for historical usage combined with the projected rate of inflation.

Interestingly, if PPBS was used as its proponents advocated, the concept included a zero-based budget (ZBB) idea because program analysis would be applied to all programs, old and new. It had been tried for a year in 1962 without much success in the Department of Agriculture. However, the concept did not gain national prominence until Peter Pyhrr wrote in the *Harvard Business Review* in 1970 about Texas Instruments' experience with it. In 1972, then Governor Jimmy Carter read the article and the rest is history. Carter introduced the idea to the Georgia state government and then to Washington when he was elected president in 1976.

Theoretically, ZBB required programs to be justified over and over again so that the traditional base budget, which received little or no scrutiny in traditional budget processes, was no longer sacrosanct. The concept called for identification of decision units, decision packages, ranking of units within packages, and evaluation of alternative spending levels for units in the packages. Managers were to provide estimates of different levels of funding, such as below current levels of support, maintenance of the current level, or a higher level of support with explanation of the impact of such alternative funding levels upon their program. The uniqueness of ZBB was in the formatting of information and the redefinition of budget base to include decrements and not just increases in funding. ZBB at the national level lasted only as long as the Carter administration and was quickly abandoned by the Reagan administration upon taking office. The concept was adapted and used by a number of local governments.[20]

No "major budgeting system dominated the 1980s public sector landscape."[21] Rather, the late 1970s and 1980s witnessed significant changes in American politics, as a period of reexamination of public support for government and significant demographic changes. With this came public preferences for lower taxes, reduced support for government social services, and more emphasis on personal responsibility. "Doing more with less" became a popular expression, which many in government found to be a mandate, if not the motto, of the era. Tax and spending limitation proposals gained popularity after a successful initiative (Proposition 13) in California in 1978. Nineteen states added some form of limit on themselves after 1978. By 1994, 43 states limited local taxes or expenditures in some fashion. The 1980s and the 1990s can be characterized as a period of tax resistance if not outright revolt.

BUDGET REFORM IN THE 1990s: THE "NEW" PERFORMANCE BUDGETING

In the 1990s, considerable attention had refocused upon performance budgeting. These concepts did not suddenly spring upon the scene, but rather became the object of renewed attention and interest as fiscal constraints persisted in many governmental entities. Several things converged to refocus attention on performance budgeting. One influence was the 1992 publication of *Reinventing Government, How the Entrepreneurial Spirit Is Transforming the Public Sector,* by David Osborne and Ted Gaebler.[22] They noted, among other things, what was widely known—Americans are cynical about their government.

As one remedy, they proposed a results-oriented budget system (although the term was not unique to Osborne and Gaebler). The idea was to hold governments accountable for results rather than focus upon inputs as traditional budgets and management did. Cost savings and entrepreneurial spirit would be rewarded. A long-term view would be facilitated in terms of strategy, costs, and planning for programs. Gaebler, a former city manager turned consultant, and Osborne, a writer and consultant, were exposed to a wide variety of governments. From this experience they focused on examples they found of entrepreneurial management in government. A local government example they cited as a "performance leader" was Sunnyvale, California, with its focus on outcomes rather than inputs.

Interestingly, as is often the case, efforts to improve performance and increase the economy in the public sector had been underway before *Reinventing Government* was published. At the federal level, for example, the Chief Financial Officers Act of 1990 (P.L. 101-576) required the development and reporting of systematic measures of performance for 23 of the larger federal agencies. The Governmental Accounting Standards Board had examined the use of service efforts and accomplishments (SEA) reporting for state and local government entities. Beginning in the early 1980s, it had encouraged governments to report not only financial data in budgets and financial reports but also information about service quality and outcomes (see Figure 8.4).

FEDERAL REFORM EFFORTS IN THE 1990s

One thing that popular ideas, such as reinventing government, do is get the attention of the media and political leaders. Drawing on this attention, Congress passed the Government Performance and Results Act (GPRA) of 1993 (P.L. 103-62). This act required federal agencies to prepare strategic plans by 1997, to prepare annual performance plans starting with fiscal year 1999, and to submit an annual program performance report to the president and Congress comparing actual performance with their plans beginning in the year 2000.

STATE REFORM EFFORTS IN THE 1990s

State interest in the "new" performance budgeting in the 1990s is evident in a number of ways. The National Governors' Association (NGA), for example, published *An Action Agenda to Redesign State Government* in 1993, which called for creating performance-based state government with measurable goals, such as benchmarks and performance measures, in order to move to performance budgeting. A year later, the National Conference of State Legislatures (NCSL) published a study entitled *The Performance Budget Revisited: A Report on State Budget Reform*. Together these two studies found interest in performance budgeting and measurement in a number of states. The rediscovery of, or renewed interest in, performance budgeting at the state and national levels appears to be drawing on a continuous theme in American budgeting—the need to make resource allocation decisions based on more than the inputs used to carry out public programs. The national and state interest in the new performance budgeting draws upon a common source: the cynicism and loss of confidence in government in the United States.

LOCAL REFORM EFFORTS IN THE 1990s

Some local governments have responded by becoming more entrepreneurial in their approach, such as Sunnyvale, California, the inspiration for the Government Performance and Results Act. A council-manager city of about 130,000 people located south of San Francisco, Sunnyvale is unique in its application of performance measurement and budgeting at the local government level. Its general plan looks 5 to 20 years into the future. The plan comprises seven elements and 20 subelements that set goals and policies for the city. Its resource allocation plan is a 10-year budget to implement the general plan. Each year the annual budget is a performance budget targeting specific service objectives and productivity measures linked to the larger plan. Therefore, its budget is a service-oriented document rather than the traditional line-item, input-oriented budget. In many ways, this city appears to be the embodiment of the contemporary interest in performance budgeting.[23]

PROSPECTS FOR BUDGETING IN THE TWENTY-FIRST CENTURY

Budgeting in the United States has experienced at least five phases, starting with control at the turn of the twentieth century, moving to management in the New Deal and post–World War II period, to planning in the 1960s, to prioritization in the 1970s and 1980s, and to accountability in the 1990s. Budget reform appears to be alive and well in the United States. The federal government and many state governments are continuing to experiment with program and performance information, as are local governments. Professional organizations, such as the Governmental Finance Officers Association, continue to nurture change and advancement in budget presentation and financial reporting. Local governments are a focal point once more for budget innovation and change (see Table 8.1).

FISCAL MANAGEMENT

As budgets are prepared at the local level, there are several factors to be aware of. Two agencies that have a great deal of impact on what happens at the local level are the Government Accounting Standards Board and the Generally Accepted Accounting Principles. Today, in local governments, there has been significant restructuring in respect to the general accounting practices and the regulations and guidelines applied at the local level by these two entities.

TABLE 8.1 ◆ Budget Reform Stages

Period	*Budget Idea*	*Emphasis*
Early 1900s	Line-item budget Executive budget	Control
1950s	Performance budget	Management Economy and efficiency
1960s	Planning Programming Budgeting System	Planning Evaluation Effectiveness
1970s and 1980s	Zero Base Budgeting Tasked Base Budgeting Strategic Planning	Planning Prioritization Budget reduction
1990s	New Performance Budget	Accountability Efficiency and economy management

Source: Charlie Tyer and Jennifer Willand, *Public Budgeting in America, A Twentieth Century Retrospective,* www.iopa.sc.edu/publication/Budgeting_in_America.htm

What Is the GASB? The *Governmental Accounting Standards Board (GASB)* is an independent, private sector not-for-profit organization that establishes and improves standards of financial accounting and reporting for U.S. state and local governments. Governments and the accounting industry recognize the GASB as the official source of *Generally Accepted Accounting Principles (GAAP)* for state and local governments.

In line with its mission, the GASB issues standards that result in useful information for users of financial reports (for example, owners of municipal bonds and members of citizen groups), and guides and educates the public, including issuers, auditors, and users about the implications of those financial reports.

GASB standards help constituents to determine the ability of their government to provide services and repay its debt. These standards also help government officials demonstrate accountability to constituents, including their stewardship over public resources. GASB standards help to ensure that those who finance government or who participate in the financing process have access to relevant, reliable, and understandable information that assists them to make better, more informed decisions.

The GASB works to accomplish its mission by issuing standards that:

- Improve the decision-usefulness of financial reports
- Foster reliable, relevant, and consistent information
- Recognize the unique and distinguishing characteristics of the governmental environment
- Improve constituent understanding of the information contained in financial reports
- Are accompanied by helpful and understandable implementation guidance

Government financial reports are used by a number of stakeholders including citizens and citizen groups; state, county, and local legislative and oversight officials; and creditors and persons involved in the municipal bond industry.

The GASB is not a federal agency. The federal government does not fund GASB, and its standards are not federal laws or rules. The GASB has no enforcement authority to require governments to comply with its standards. However, compliance with the GASB's standards is enforced through the audit process, when auditors render opinions on the fairness of presentations in conformity with GAAP, and through the laws of individual states, many of which require local governments to prepare GAAP basis financial statements. In addition, the municipal bond industry prefers that governments issuing debt prepare their financial statements on a GAAP basis.

On June 30, 1999, the GASB published comprehensive changes in state and local government financial reporting. GASB Statement No. 34, *Basic Financial Statements—and Management's Discussion and Analysis—for State and Local Governments,* provides a new focus of reporting public finance in the United States.

Under the new standard, anyone with an interest in public finance—citizens, the media, bond raters, creditors, legislators, and others—will have more and easier-to-understand information about their governments. Among the major innovations of Statement 34, governments are required to do the following:

- Report on the overall state of the government's financial health, not just its individual funds.
- Provide the most complete information available about the cost of delivering services to their citizens.
- Include for the first time information about the government's infrastructure assets, such as bridges, roads, and storm sewers.
- Prepare an introductory narrative section analyzing the government's financial performance.[24]

In 2004, the Governmental Accounting Standards Board released GASB Statement No. 45, *Concerning Health and Other Non-Pension Benefits,* also called other postemployment benefits, for retired public employees. These retiree health care programs are by far the most costly.

The GASB 45 was intended to bring governmental accounting standards more in line with private company standards. Although GASB has no power to change how governments fund retiree health, pension, and other benefits, it does govern the rules auditors must follow in providing options on the reliability of governmental financial statements. This new accounting rule increases the amount of quality information included in government financial reports, particularly concerning retiree health care and other retiree benefits.

State and local governments must take a series of steps that include quantifying the unfunded liabilities associated with retiree health benefits. Assessment results must be reported in governmental audits and updated regularly. Government financial statements then list an actuarially determined amount known as an annual required contribution.

In regard to health care, this contribution includes the normal costs—the amount to be set aside to fund future retiree health benefits earned in the current year—and unfunded liability costs—the amount needed to pay off existing unfunded retiree health liabilities over a period of no more than 30 years.[25]

What Is GAAP? Every industry has certain rules and guidelines to follow, and the accounting field is no different. Before you get involved in accounting, you must know and follow these guidelines and principles, known as Generally Accepted Accounting Principles (GAAP). If these rules and guidelines are followed on a regular basis, everything is fine. But when someone goes off course and decides to go by his or her own principles, accounting fraud develops.

GAAP is composed of many standards, interpretations, and opinions developed by the Financial Accounting Standards Board (FASB), the American Institute of Certified Public Accountants (AICPA), and the Securities and Exchange Commission (SEC). FASB is the group that developed the standards and guidelines that make up

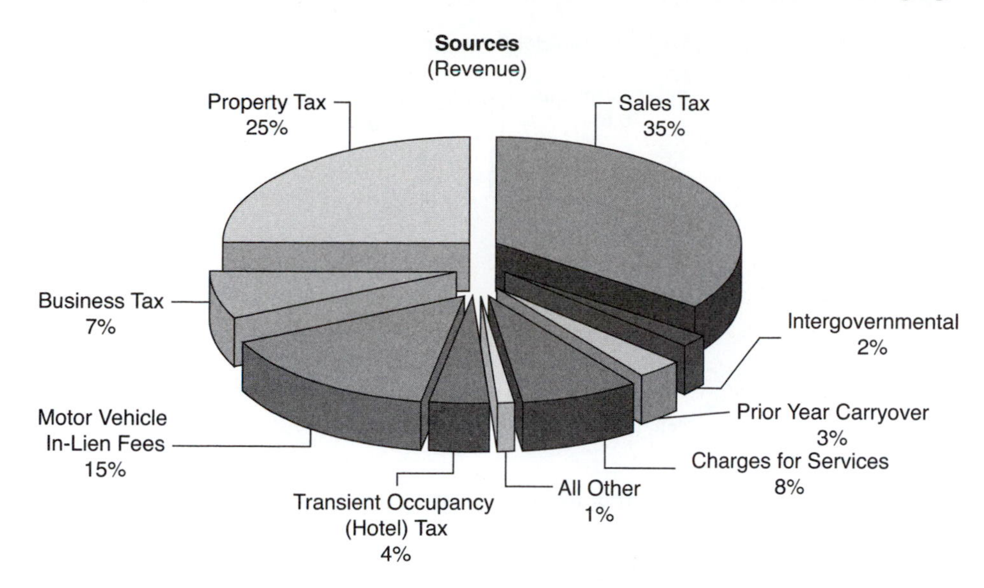

FIGURE 8.5 ◆ Sample of Municipal Fire Department Revenue Pie Chart

GAAP and put them into practice. AICPA enforces GAAP for all accountants and CPAs. It makes sure that anyone who is a CPA, CMA, or bookkeeper upholds the standards as set forth in the bylaws of the FASB. SEC is a government agency that makes sure all securities markets are handled correctly and properly by GAAP.[26]

Local Fiscal Management Regardless of the type of budget process (a line-item program, budgeting management by objectives), there must always be a balance between the revenues brought in annually and the expenses articulated in the form of a budget document. Revenue streams will be different among cities, counties, fire districts, and volunteer fire agencies. Each can derive its revenue from different sources. The examples in Figures 8.5 and 8.6 show the breakdown of revenue and expenditures for a municipal fire department.

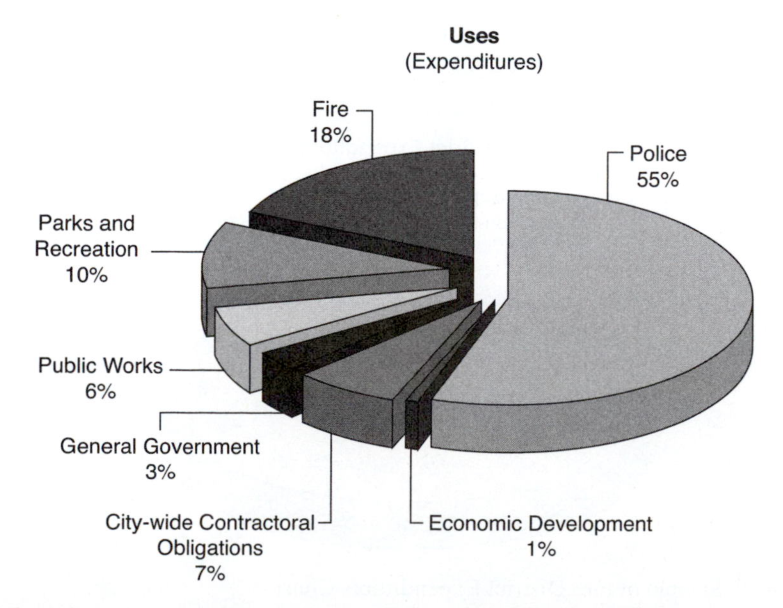

FIGURE 8.6 ◆ Sample of Municipal Fire Department Expenditures Pie Chart

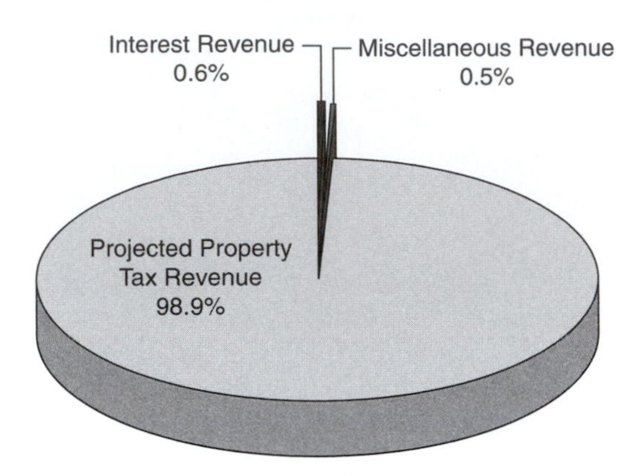

FIGURE 8.7 ◆ Sample of Fire District Revenue Chart

Cities often have a greater number of multiple revenue streams, such as property tax, sales tax, local impact fees, and fees for service, which help to create a larger pool of revenue from which to draw. However, at the same time, most cities have a much more diverse service delivery than that of special districts and/or volunteer fire departments. For special districts, the revenue is typically generated from property tax. In certain cases this is referred to as *ad valorem* tax, depending on the state. In many cases, capital expenditures are done with bond proceeds that are approved by the voters for specific capital projects. To pay the bonds, revenue is largely generated primarily from taxes on properties or parcels within the jurisdiction (see Figures 8.7 and 8.8).

Volunteer organizations often derive their money from a variety of sources including general fund support from local governments. A common revenue stream is a specific tax for the volunteer department based upon a percentage of the property tax. In many cases, volunteer departments are very effective in raising their own money to support their efforts. Two factors are of importance in basic fiscal management: the revenue–expense ratio of your organization and understanding the various revenue streams that fund the budget.

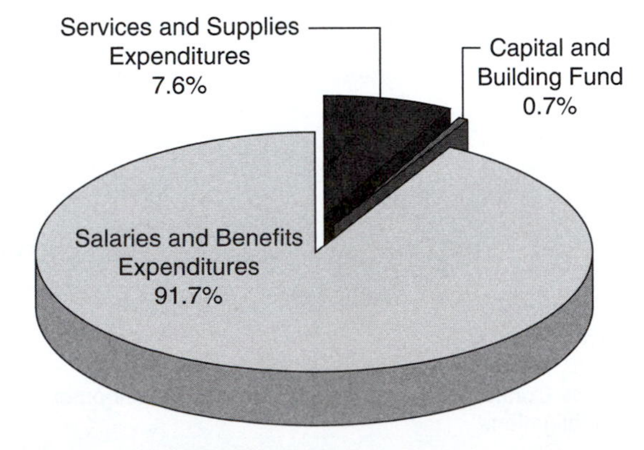

FIGURE 8.8 ◆ Sample of Fire District Expenditures Chart

EMPLOYMENT LAW

Fire departments are faced with a variety of employment laws concerning equal employment opportunity or fair and equitable treatment of any applicant. The challenges many organizations face are how best to implement the steps necessary to live up to the letter and the spirit of civil rights laws.

BASIS FOR EMPLOYMENT LAW

Employment law has its basis in federal and state laws as well as in how the courts have interpreted these laws (case law). A body of law called common law has its basis in general principles of fairness. The fire service manager must have a working knowledge of the laws, rules, and regulations and familiarity with decisions rendered by federal and state courts in order to avoid problems of unfair (or illegal) treatment of employees.

EQUAL EMPLOYMENT OPPORTUNITY

The days have long past when employers can manage their workforces any way they wish. Federal and state laws have been passed prohibiting discrimination against employees in the areas of race, national origin, color, creed, gender, age, disability, military experience, religion, and status with regard to public assistance. The term *discrimination* is used in many ways. *Webster's Collegiate Dictionary* defines it as "the act, practice, or an instance of discriminating categorically rather than individually" and "prejudiced or prejudicial outlook, action, or treatment." Employers must choose (discriminate) among applicants for a job (or promotion) on the basis of each candidate's qualifications. However, *discrimination* is also used to mean unequal treatment of members of a particular group when compared to members of another group. Individuals who are covered under equal employment laws are referred to as "members of a protected class." To implement laws barring discrimination, several regulatory agencies have developed guidelines and regulations. Several are listed here.

Equal Employment Laws and Regulations

Equal Pay Act of 1963 The equal pay provisions of the Fair Labor Standards Act as amended in 1963, 1968, and 1972 prohibit wage differentials based on gender for those who are employed in the same establishment in jobs that require equal skill, effort, and responsibility that are performed under similar working conditions.

1964 Civil Rights Act, Title VII This federal law is the keystone of federal equal employment opportunity legislation. Two important provisions of the act are as follows:

> Section 703(a): It shall be unlawful employment practice for an employer (1) to fail or refuse to hire or to discharge any individual or otherwise to discriminate against any individual with respect to compensation, terms, conditions, or privileges of employment, because of such individual's race, color, religion, sex, or national origin; or (2) to limit, segregate, or classify employees in any way that would deprive or tend to deprive any individual of employment opportunities or otherwise adversely affect employee status because of such individual's race, color, religion, sex, or national origin.

> Section 704(b): It shall be an unlawful employment practice for an employer, labor organization, or employment agency to print or publish or cause to be printed or published any notice or advertisement relating to employment by such an employment agency indicating any preference, limitations, specification, or discrimination based on race, color, religion, sex, or national origin.

Age Discrimination in Employment Act of 1967 The Age Discrimination in Employment Act (ADEA) of 1967, amended in 1978 and 1986, makes it illegal for an employer to discriminate in compensation, terms, conditions, or privileges because of an individual's age. The later amendments first raised the mandatory retirement age to 70, then eliminated it completely. Under the 1986 amendments, a seven-year exemption for public employers was included to allow them to adjust to mandatory retirements for police and fire officers.

Older Workers Benefit Protection Act of 1990 Another new piece of legislation is the Older Workers Benefit Protection Act (OWBPA). This law restricts an employer's ability to obtain waivers of age discrimination claims when terminating or laying off older employees. The OWBPA also allows employers to offer "voluntary retirement incentive programs" in conjunction with the ADEA. These retirement packages may be offered only to older employees without being held to be discriminatory.

Pregnancy Discrimination Act of 1978 In 1978, the Pregnancy Discrimination Act (PDA) was passed as an amendment to the Civil Rights Act of 1964. This act requires that women employees "affected by pregnancy, child birth, or related medical conditions will be treated the same for all employment-related purposes." The major impact of the act was to change maternity leave policies and employee benefit systems. Under the PDA, pregnancy must be treated just like any other medical condition. The same provisions regarding disability insurance and leaves of absence must apply to pregnant employees as to all other workers. A 1986 Supreme Court decision did allow states to pass laws requiring employers to give pregnant employees maternity leave and reinstate them in their same jobs at their same pay once they return to work.

Americans with Disabilities Act of 1990 President George H. W. Bush signed the Americans with Disabilities Act of 1990 (ADA). The employment provisions of the ADA were phased in between 1992 and 1994. Employers with 25 or more workers were required to comply with the ADA provisions on July 26, 1992. Employers with 15 or more workers were required to comply with the ADA on July 26, 1994. The equal access measures took effect in 1992. The ADA prohibits employers with 25 or more employees from discriminating against qualified employees with disabilities. The law defines a "qualified disabled person" as a person who, with reasonable accommodation, can perform the essential functions of the position. The ADA defines "reasonable accommodation" as steps an employer can take to help a person with a disability to perform the job including, for example, providing a reader for a blind employee. An employer does not have an obligation to accommodate an employee with a disability if the accommodations would cause the employer "undue hardship." The ADA also forbids preemployment physicals. An employer may require a job applicant to have a medical examination only if the examination tests for job-related functions, is a business necessity, and is required only after the employer has made an offer of employment to the job applicant.

National Labor Relations Act of 1935 There is also a federal National Labor Relations Act (NLRA). The NLRA prohibits employers from interfering with the rights of employees to form unions or engage in union activity.

Immigration Reform and Control Act of 1986 Because of concerns about the employment and other rights of illegal aliens in the United States, the Immigration Reform and Control Act was passed in 1986. Under this act, employers are prohibited from knowingly employing illegal aliens, with substantial fines possible for offenses. The act requires employers to examine the credentials of prospective applicants, and new employees must sign verification forms about their eligibility to work legally in the United States.

Veterans' Readjustment Assistance Act of 1974 The Veterans' Readjustment Assistance Act of 1974 provides that employees who leave their jobs to serve in the armed forces are entitled to reinstatement when their tour of duty is completed. The basic requirements for statutory coverage are that the veteran must have completed a tour of duty and must apply for a job within 90 days of release from active duty or within one year of hospitalization. If the veteran meets these requirements, he or she must be returned to his or her former position or an equal position, unless circumstances have changed so drastically that this would be impossible. After reemployment, the veteran may not be discharged except for just cause. It is important to note that whenever you are dealing with civil rights or legislative requirements, you must check local regulations. Many municipalities have added requirements for employers that may not be present in state or federal laws.

Employment Selection and Testing

Employment Application and Employee Interview An employer may not ask obviously illegal questions about a prospective employee on job applications, such as questions about an applicant's color, race, creed, religion, sex, national origin, and disability. An employer may not under any circumstances require an employee to submit to a polygraph, voice stress test, or any other test that purports to measure "truth."

Uniform Guidelines on Employment Selection Procedures To implement the provisions of the Civil Rights Act of 1964 and the interpretations of it based upon court decisions, the Equal Employment Opportunity Commission (EEOC) and other federal agencies developed their own compliance guidelines and regulations, each having a slightly different set of rules and expectations. In 1978, the major government agencies involved agreed upon a set of uniform guidelines. These guidelines affect almost all phases of employment practices, including hiring, promotions, recruiting, demotion, performance appraisals, training, labor union membership requirements, and license and certification requirements.

A case involving hiring practices is *Zamlen v. City of Cleveland*, 906 F.2d 209 (6th Cir. 1990). Only 21 females out of nearly 300 female applicants scored high enough on the written and physical portions of the exam to be placed on the eligibility list. Even then they were ranked far too low to have any reasonable prospects of being hired. Ultimately, a group of female applicants brought suit against the city challenging the rank ordering and the written and physical selection examinations. The trial court concluded that the possibility of hiring unqualified firefighters posed such a significant risk to the public that it was appropriate to hold the city to a lighter burden of proving that the employment criteria were job related. The court concluded the examinations had been appropriately validated and were job related. Finally, the trial court ruled there was sufficient correlation between higher test scores and job performance to warrant a rank-order hiring procedure. The purpose of fair employment laws is to achieve equality of employment opportunities through eradication of employment barriers that discriminate on the basis of race, gender, religion, and other protected classifications.

Implementing Equal Opportunity

Sex Discrimination/Discrimination in Job Assignments and Conditions Title VII of the Civil Rights Act prohibits discrimination in employment on the basis of sex. As with racial discrimination, it has taken a series of court decisions and EEOC rulings to determine exactly how broad this prohibition is.

Height/Weight Restrictions Many cases involving the use of height and weight restrictions were actually discrimination cases designed to bar women and certain races from jobs. Citing height and weight requirements as a substitute for strength may be determined to be discriminatory. The courts have generally found if an employer has a strength or agility requirement, the employer must test for it and not use a height/weight restriction.

Sexual Harassment As a general matter, sexual harassment refers to actions that are sexually directed and unwanted, and subject the worker to adverse employment conditions. Employees are protected against sexual harassment by Title VII, as well as state human rights legislation. The courts currently identify two different types of sexual harassment. The first, *quid pro quo* sexual harassment, involves the request by a supervisor or employer for sexual favors in return for favorable treatment in the workplace. The second, *hostile work environment,* usually involves an atmosphere in the workplace that creates or tolerates sex-based jokes or pictures, touching, questions regarding an employee's sex life, and the like.

Sexual harassment is a serious matter, and how supervisors and managers treat this issue is critical. If complaints are not handled with seriousness and sensitivity, the probability is high that they will be resolved outside the department by legal means. Violations of civil rights law carry heavy financial penalties to the violator. A supervisor or manager must be prepared to respond appropriately to both complaints/allegations and confirmed incidents. The supervisor, at a minimum, must conduct a careful and thorough interview with the complainant, interview the alleged "harasser," and consult with the executive management of the political subdivision to determine what action will be taken. It is absolutely essential that supervisors act "promptly and appropriately" upon receipt of a complaint of sexual harassment or upon the discovery of any type of conduct that would be construed as sexual harassment. The courts have consistently held employers liable for the actions of their employees if the harassment is not immediately remedied upon notification.

Pregnancy Discrimination The 1978 Pregnancy Discrimination Act has resulted in many employers changing their employee benefits policies. The major effect of this law has required employers to treat leaves of absence for pregnant employees in the same manner as other personal or medical leaves.

Conviction and Arrest Records Should this issue arise, legal advice should be sought. Most jurisdictions have established policies on methods of dealing with candidates who have a criminal record. Felony convictions are generally not acceptable for firefighter appointment.

Firefighter Selection

Job-Related Validation Strategy Validity simply means the applicant's test score actually is a predictor of the person's ability to perform the job for which he or she is being hired. Courts will look closely at tests administered to prospective candidates to ensure validity is present. In March 1975 the Supreme Court case of *Albermarle Paper v. Moody*, 425 U.S. 405, reaffirmed the idea that any real "test" used for selecting or promoting employees must be a valid predictor or performance measure for a particular job. The burden of proof for showing the test score is valid falls on the employer. It is obvious that selecting the appropriate test for firefighter selection is of the utmost importance. Many tests meeting the required criteria (and may already have been "court tested") are available commercially. Departments should not consider writing their own tests unless a person trained in test construction is available on staff and there is a means to validate it.

Personality and Psychological Tests Psychological tests attempt to measure personality characteristics. The Rorschach ("ink blot") test, the Thematic Apperception Test (TAT), and the Minnesota Multiphasic Personality Inventory (MMPI) are some examples. These types of tests are difficult to relate to jobs firefighters are required to perform and may be determined to be invalid if challenged in the courts. The main difficulty is connecting specific personality traits to specific job requirements.

Selection Interviews Selection is subjective, and courts view subjective decision making with some suspicion because, without some method of control, an employer may indulge in a preference for one class of applicants. Attorneys often recommend the following controls to reduce EEO concerns with the interview process: identify specific job criteria to be looked for in the interview, put those criteria in writing, and provide a review process for those difficult or controversial decisions.

Types of References Background information can be obtained from academic, prior work, and personal references given by the applicant. Personal references are often of minimal value. The applicant usually will not give a negative source as a personal reference. Investigating academic and prior work references is worthwhile, as information on the application form may be incorrect. Academic records will usually not be released without the applicant's written permission. Many employers are hesitant to give negative information about a former employee, because doing so may increase their liability.

Probation Periods Most fire departments require a new employee to complete some type of probationary period. Usually, this period is used to evaluate the suitability of the employee and requires that he or she complete some education requirements (First Responder, Firefighter I, etc.). Again, objective criteria must be used in determining what the employee is expected to accomplish during this period. Some examples include attendance at a specific number of training sessions, responses to fire calls, and successful completion of specific education and skills requirements. The most often used probationary period for career and volunteer firefighters is one year. At the successful completion of the probationary period, the firefighters become vested in the organization and are subject to all personnel procedures and employee rights. Probationary periods may be extended if personnel procedures or employee contract language establishes a process for extending probation. Usually the probationary extension period is no longer than six months.

EMPLOYEE AND LABOR RELATIONS

Fire chiefs are often caught between the bargaining interests of personnel and the political and economic realities of the world in which they work. The same issue sometimes arises on a national level. For example, the 1992 national controversy over the minimum size of initial fire attack crews resulted in the International Association of Fire Chiefs opting for local control on staffing levels, while the International Association of Fire Fighters pushed for a specific national staffing standard. This is a classic example of the dichotomy of interests between management and labor. Over time and through significant dialogue and work, both supported the passage of NFPA 1710 and **NFPA** (National Fire Protection Agency) 1720 in 2000. Ongoing communication between labor and management results in the best overall team attitude and a willingness to work out the differences.

NFPA
National Fire Protection Agency

Whereas the earliest fire departments in the United States were all volunteer, eventually the paid personnel in some places began to organize, and in 1918 the International Association of Fire Fighters (IAFF) was formed, with members in both Canada and the United States. The IAFF was chartered by the American Federation of Labor on February 28, 1918. This union represents over 70 percent of the career firefighters in the country, with more than 250,000 members, predominantly in the United States, and 2,100 "locals." An overview of the union movement is helpful when attempting to understand contemporary labor relations.[27]

At the turn of the century, unions were usually considered by the courts to be conspiracies in violation of the antitrust provisions of the Sherman Act of 1890, and injunctions against them were summarily issued. In the early period of the act, unions were often considered to cause artificial restrictions on wages and were violations against the act. The first legislative attempt to give legitimate unionism a chance to succeed came in 1932 in the form of the Norris–La Guardia Act. The new philosophy in this act implied the public interest would be served with union organizations, and collective bargaining would ensure proper rights for both employers and employees. Long and costly labor conflicts ensued with strikes and boycotts. New labor legislation was urgently needed to promote effective collective bargaining. Three years after the Norris–La Guardia Act was passed, the National Labor Relations Act (NLRA), or the Wagner Act, was passed. It guaranteed the rights of all (private) workers to organize and to bargain collectively. The National Labor Relations Board (NLRB) was created as a federal agency to ensure industrial harmony by administering and enforcing the Wagner Act. The agency could certify unions as the bargaining representative to oversee negotiations and break deadlocks.

Union membership quadrupled from 4 million to 16 million between 1935 and 1948. Legislation was called for to create a balance more favorable to business and industry because of the additional power gained by the unions. This legislation was known as the 1947 Labor Management Relations Act, also referred to as the Taft-Hartley Act. The then separate American Federation of Labor and Congress of Industrial

Organizations both opposed the Taft-Hartley Act, even though it ultimately placed both the management and the unions on a more equal basis. The Landrum-Griffin Act of 1959 detailed regulations in certain phases of labor relations, required additional accountability from employers, and put new restrictions on the activities of unions. All of the significant labor relations legislation excluded the public employee, so renewed emphasis was placed on passing legislation at the state level to provide public employees with recognition and limited collective bargaining rights. By the 1960s, the federal government began to recognize the rights of federal employees.

By 1962, public workers could join unions and bargain collectively. However, public employees in federal, state, and local governments are not included in the jurisdiction of the National Labor Relations Act. The 1978 Civil Service Reform Act, Title VII, created the Federal Labor Relations Authority (FLRA) to mediate and clarify issues. However, federal employees still may not strike or bargain collectively with respect to wages and benefits. Unions formed in the early nineteenth century did well during prosperous times but were abandoned when hard times arrived.

With the growth of membership in the 1930s, this pattern began to change. Mass production industries during the Great Depression felt a growth in permanent trade union membership, as did the traditional craft unions. This trend continued until the end of World War II, when union growth began to slow down. By 1950, union strength was concentrated in manufacturing, mining, transportation, and public utilities. Roughly 85 percent of the blue-collar workers in the United States were organized. From 1953 to the early 1960s, the economy began to slow down while technological changes in manufacturing occurred rapidly, and employment in manufacturing and production industries declined. Blue-collar employment declined, while white-collar employment within manufacturing increased due to the introduction of electronic and computer controls.

The growth of government unions and government employment in general also increased, partly because of executive orders by Presidents Kennedy and Nixon. Organizing and bargaining became easier at the federal level with the organization and development of the civil rights movement in the United States. In 1974, one in four workers in both the private and public sectors were members of labor unions. Two quite different scenarios have emerged during this time frame. By 2004, only 8.2 percent of workers in the private sector were members of a union. In contrast the union membership rate among public sector workers increased to 37.1 percent over the same period.[28]

SOCIETAL TRENDS

There are several factors in our national climate influencing employee and labor relations. Undoubtedly, national economics and the availability of jobs, especially in those occupations traditionally unionized, affect labor relations and union membership. Other factors in the national climate also influence employee and labor relations. There is special concern for worker health and safety, as well as a need to safeguard the environment. These concerns have brought changes in legislation and additional federal and state legislation. For example, special efforts have been made to protect workers from toxic materials in the workplace. The "right to know" laws have protected so-called whistle-blowers and have minimized the accidental discharge of, and/or exposure to, hazardous materials. Perhaps the strongest influences on labor relations will be the triple phenomena of large increases in the number of professional and technical workers, continuing growth of service industries, and the increasing percentages of women and minorities in the workforce.

Tomorrow's fire service should be culturally and technically prepared to deal with future fire service challenges. The fire service should provide the community with a respected level of emergency response, prevention, and preparedness awareness. The fire service has an opportunity to support its communities with strong leadership to address the many social and technical challenges facing its communities in the twenty-first century. It has been said that militant employee organizations rise because of negative actions of management. The availability of positive human resource development techniques is helping to generate an enlightened management. This, in turn, is facilitating the emergence of an enlightened and empowered workforce and improvement of labor and management relations.

"RIGHT-TO-WORK" STATES

Some states have enacted legislation that supports union missions, primarily through the establishment of laws mandating union membership for workers in unionized establishments or favoring union membership. The "shops" provide union security and are termed:

- Closed shop: job applicants must join the union.
- Open shop: employees have a choice regarding the need to join the union.
- Union shop: all workers must join following a probationary period.
- Preferential shop: union members have job preference.
- Agency shop: if worker is not a union member, a "service fee" must be paid to the union for its representation.
- Checkoff: union dues are deducted from the paycheck and transmitted directly to the union.

As permitted by federal legislation (Taft-Hartley Act), 22 states have passed laws to prohibit compulsory union membership. These "right-to-work" states allow only for the checkoff and not for the other forms of union security listed.

ORGANIZED LABOR ACTIONS

Employees of the federal government are prohibited from striking by statute, and most states also prohibit public safety employees from striking. Public sentiment and political reaction to public safety strikes are usually not good for future contract improvements for the firefighter and create a public relations nightmare for the department. Other types of job actions include sick-outs, work slowdowns, picketing, work-to-rule, and modifications of these. As an alternative to the strike option, and as a type of trade-off for statutory prohibition against striking, many states have passed binding arbitration laws, forcing both labor and management to accept a final ruling regarding an arbitration finding by a third party.

LABOR RELATIONS IN VOLUNTEER AND NONUNIONIZED DEPARTMENTS

It is important to keep in mind that "labor relations" exist in all types of fire departments, including those with all-volunteer members, combination departments, and nonunionized departments. Full-time, paid members of combination departments may be unionized and follow typical contract agreements. At this time, union affiliation is not offered to volunteer or paid-call firefighters. Nonunionized firefighters may work under the same type of agreement or formal understanding but do not have the impact of the larger affiliation on labor activity that unionized firefighters have. In many volunteer and part-paid departments, the grievance procedures, negotiation process, and other personnel policies and procedures may be less formalized than in union-represented agencies. Combination departments present special managerial challenges regarding labor relations. In some combination departments, where the paid presence is strong, often a climate is created that is negative to the volunteers. The opposite may exist if the department is predominantly volunteer. Departments that believe and show both career and volunteer members that they are highly valuable resources and all share a mutual respect for each other's contributions will have a labor relations climate that is often found to be mutually supportive between volunteer and career members.[29]

COLLECTIVE BARGAINING

The term *collective bargaining* was developed from legal terminology focused on labor law of the 1930s and 1940s. At the turn of the twentieth century, the Industrial Revolution began to boom. More employees became victimized by unfair labor practices. Union representation had been held in abeyance by the Sherman Act (1890). Federal courts held that the antitrust protection offered by the Sherman Act prohibited union representation of multiple labor groups. In 1932, the Norris–La Guardia Act revisited the Sherman Act and determined it was in the public's best interest to allow employees to bargain collectively. Unions began to flourish, but business and industry started developing "sweetheart" contracts with company unions. These mutually beneficial labor agreements left the rank and file worker without work. Strikes, boycotts, and other costly labor conflicts caused the federal government to develop the National Labor Relations Act (Wagner Act) in 1935. The act created the National Labor Relations Board, which established procedures for collective bargaining. The right to representation and the working conditions that were subject to negotiation became governed by specific written procedures and guidelines. In 1947 the Taft-Hartley Act further defined union and management rights. Right-to-work laws were tempered with restrictions on strikes.

The Landrum-Griffin Act, Civil Service Reform Act, Federal Labor Relations Authority, Fair Labor Standards Act, and, most recently, the Americans with Disabilities Act have further defined the responsibilities of labor and management. The collective bargaining process affords all employees, regardless of age, sex, race, or disability, a better-defined set of legal standards to operate within. When labor and management discover and use the best methods of negotiation to solve disputes productively, federal and state labor laws will have accomplished their purpose. Today, many examples of effective labor and management relationships prove the value of efficient negotiation skills. Over the last 10 to 20 years, arbitration awards and court decisions have further defined management and employee rights. In some cases management and employee rights have been specifically written into the employee–employer contract. Management rights usually include the right to:

- direct the workforce;
- hire, promote, transfer, and assign without interference;
- suspend, demote, discharge for cause, or take other disciplinary action;
- take action necessary to maintain a department's efficiency; and
- make reasonable rules and regulations.

SETTLING DISPUTES

Management should always function clearly within the parameters of management rights. When questioned by labor regarding a contract issue or a problem with a policy as it relates to the current scope of coverage, it is best to communicate openly with the labor representatives and solve the dispute. Employee–employer contracts become overburdened with detailed processes, procedures, and operational limitations because management and labor lose trust and are forced to protect their "turf" with written limitations. When labor relations fail, the attorneys and arbitrators are kept busy solving the problems that normal negotiation circumstances would usually solve (with good labor–management relations). Breaking down labor–management mistrust requires patient and persistent leadership skills. The team feeling is developed when labor experiences leadership that openly communicates the entire game plan and a motivation for progress. Leadership integrity and credibility are enhanced as the organization accomplishes the game plan and the community returns its good feeling for the organization efforts. The positive reactions are rewarding as leadership continues to make opportunities available for team success.

When agreement simply cannot be reached by the negotiating parties, several methods are available to reach a conclusion. These fall into three general categories: (1) use of a third party, (2) power plays (legal and illegal) by the union, and (3) power plays (legal and illegal) by management. The use of a neutral third party, called arbitration or mediation, may come in one of several forms. Compulsory *binding arbitration*, often considered the alternative to a strike, has the arbitrator (or arbitration board) considering the issue and reaching a decision that is binding on both sides. This form of arbitration typically is used with public employees who are forbidden to strike; obviously, it constrains the disputing parties. "Final" or "best offer" binding arbitration has the arbitrator selecting one or the other of the two parties' final, best offers. Knowledge of impending binding arbitration may prompt the disputing sides to renegotiate on their own, thus maintaining some control over the outcome. The form of arbitration lacking the final power of binding arbitration is called fact-finding arbitration. This version uses a fact-finding individual or panel to examine the issue and suggest a solution. Neither side is bound by the suggestion, but often the weight of outside opinion forces the parties to settle.

The mildest form of third-party involvement is called mediation. In this process a neutral mediator works with each side, either separately or together, in an attempt to get the parties to resolve the issue. The mediator acts as a neutral third party whose sole purpose is to open communication between the negotiating parties. Each party states its bottom-line requests, and the mediator attempts to find a way to develop an agreement. If third-party attempts to reconcile differences fail, either side may try to convince the other through a show of power. These actions by a union may take the form of picket, boycott, work slowdown, public petitions, work-to-rule, sick-out, or strike, where permissible. They are designed, of course, to call the public's attention to the situation. Slowdowns and work-to-rule actions may resemble some aspects of a strike, with firefighters responding to emergency calls but doing little else. Sick-outs have been used by fire unions but may be countered by the enforcement of medical checks by management.

Actions by management designed either to counter the power plays of the union or to demonstrate its own power may take several forms. These include no retroactive pay, impasse, last and best offer, compulsory contract agreement, salary freeze, and so on. Disciplinary action, injunctions, fines, and jail sentences may be imposed on labor union representatives and employees who participate in illegal labor activities. For example, contracts that have compulsory arbitration also will make striking or picketing illegal labor activities. Although extreme actions may bring a contractual impasse to a quicker end, the long-term repercussions will not prove worthwhile. Avoiding bitter confrontations will better serve the department and the community.

POWER AND MUTUAL GAINS BARGAINING

During the negotiation process, power can be exercised with several motives. Controlling power results in one party using organizational powers to force the weaker party to obey. This tactic, referred to as *win-lose* or *zero-sum* theory, often results in solving a small, immediate concern only to fuel the weaker party's motivation to prepare for a much larger "war." The more advisable use of power involves a cooperative spirit that strives for win-win or non-zero-sum agreements. The logic behind the *win-win* theory is the belief that collective agreement results in a better solution. The collective interests of both labor and management can be melded together for a better, more productive net result.

The term *mutual gains bargaining* refers to negotiations that are satisfying to all parties. Contract bargaining between the firefighter association and municipal management using mutual gains bargaining should result in both parties gaining; thus, service to the community should be enhanced. The *good of the community* becomes more of a real issue than an excuse for agreement with one side or the other (employee versus employer). Organizational goals and objectives become a guiding principle in the mutual gains agreement. Future labor agreements will be focused on mutual gains bargaining issues. The need to become more productive and accountable to the public is a reality in the twenty-first century.

THE GENERAL PROCESS OF NEGOTIATING

The process of negotiation between two parties requires communication with the intent to settle a given matter or issue. In order to initiate negotiations constructively, each party first should communicate its goals (desired outcomes). Then each party should review the goals of the other, and the first discussion should be about those issues that both sides agree upon. It also is appropriate to indicate issues that are totally unacceptable. Negotiators must learn how to say "no" reasonably, without intimidation or emotional outbursts. Those issues that are unresolved (the issues that fall between yes and absolutely no) should be the first issues to be negotiated. Compromise, middle ground, or alternate options may be found during the negotiation discussions.

When dealing with controversial issues (one party says absolutely no) the negotiator should typically expect four stages in the negotiation process: conflict, containment, accommodation, and cooperation. The conflict stage can be damaging to long-term relations. When feelings have been vented adequately and the clashing parties realize that continued conflict will bring further harm, the containment stage is reached. The accommodation stage begins when the parties settle some of the side issues, even though there is still disagreement regarding the entire solution. During the cooperation stage, labor and management begin to believe each side has truthfully expressed all feelings, motives, and bottom-line needs; and both sides start to focus on agreement on the major issues. A cooperative spirit allows both sides to save face and develop a compromise solution.

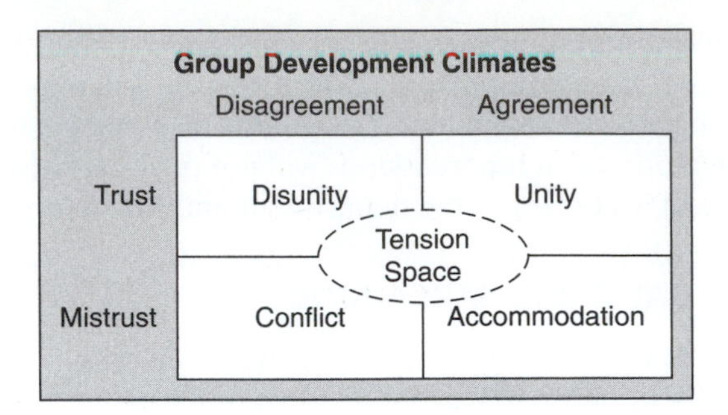

FIGURE 8.9 ◆ Group Development Climates (*Courtesy of Randy R. Bruegman*)

A similar concept of negotiation logic is depicted graphically in Figure 8.9. The four quadrants of group development climates portrayed in the illustration are defined as follows:

1. *Unity.* Both parties agree and trust each other.
2. *Accommodation.* The parties agree over certain issues but do not trust each other.
3. *Conflict.* The parties neither agree nor trust each other.
4. *Disunity.* The parties trust each other but are in disagreement over certain issues.

In organizational life, "professional" disagreements often arise. If these disagreements are discussed rationally, better decisions can be reached. If either or both parties begin to show or perceive mistrust, the climate moves to the conflict quadrant. Remaining in the disunity quadrant for long periods is difficult, because mistrust develops rapidly. Groups in the conflict quadrant can move to the accommodation quadrant, because this movement does not require trust. Trust is easy to lose but difficult to develop. Typically there is no movement directly from conflict to disunity or unity, because each of those quadrants requires the reestablishment of trust. The center area, labeled tension space, changes in volume and can be increased by either side through precipitous acts, thoughtless words, disregard of goodwill efforts, rumors, ego feeding, personalization of arguments, and the like. Increases in tension space push the climate into the conflict quadrant, so efforts to reduce the volume of tension space are well advised.

Principled Negotiation One need not be a professional negotiator to experience bargaining. Life is full of successful and unsuccessful attempts to reach agreement with others in business, pleasure, and ordinary living. Negotiations range from deciding where to have dinner to buying a car. Most of us recognize two types of negotiations: hard or soft. Hard negotiators are tough minded and do not give in until they win their way. Soft negotiators do give in, perhaps too quickly and easily, and thus lose in the bargaining process. Many negotiations take place in a win-lose, zero-sum environment, in which the big struggle is to emerge a winner. However, there is another form of negotiating, *principled negotiation*, that can be both hard and soft because it is hard on merits and soft on people. It advocates deciding issues on merit, rather than on what the parties indicate they will or will not do. It emphasizes mutual gains and looks for results based on fair standards. It works best when both sides understand its concepts. The basic sense of principled negotiating is expressed in Table 8.2.

TABLE 8.2 ◆ Problem/Solution

Problem		***Solution***
Positional Bargaining: Which Game Should You Play?		***Change the Game—Negotiations on the Merits***
Soft	***Hard***	***Principled***
Participants are friends.	Participants are adversaries.	Participants are problem solvers.
The goal is agreement.	The goal is victory.	The goal is wise outcome reached efficiently and amicably.
Make concessions to cultivate the relationship.	Demand concessions as a condition of the relationship.	Separate the people from the problem.
Be soft on the people and the problem.	Be hard on the people and the problem.	Be soft on the people and hard on the problem.
Trust others.	Mistrust others.	Proceed independent of trust.
Change your position easily.	Dig in to your position.	Focus on interests, not positions.
Make offers.	Make threats.	Explore interests.
Disclose your bottom line.	Mislead as to your bottom line.	Avoid having a bottom line.
Accept one-sided losses to reach agreement.	Demand one-sided gains as the price of agreement.	Invent options for mutual gain.
Search for the answer: the one they will accept.	Search for the answer: the one you will accept.	Develop multiple options to choose from; decide later.
Insist on agreement.	Insist on your position.	Insist on using objective criteria.
Try to avoid a contest of wills.	Try to win a contest of wills.	Try to reach a result based on standards independent of will.
Yield to pressure.	Apply pressure.	Reason and be open to reason; yield to principle, not pressure.

Principled negotiating, or bargaining, uses six guidelines:

1. Don't bargain over positions.
2. Separate the people from the problem.
3. Focus on interests, not positions.
4. Invent options for mutual gain.
5. Insist on using objective criteria.
6. Develop your best alternative to the negotiated agreement you want to reach.

Maintaining the bargaining process so that cooperative relationships result demands negotiators who can do the following:

1. Deal well with differences.
2. Disentangle relationship issues from substantive ones.
3. Be constructive.
4. Balance emotions with reason.
5. Learn how the other side sees things.
6. Listen well and consult before deciding.
7. Be wholly trustworthy but not wholly trusting.
8. Persuade, not coerce.
9. Deal seriously with those with whom we differ.[30]

PERFORMANCE APPRAISAL METHODS

performance appraisal
The process of evaluating an individual's performance by comparing it to existing standards or objectives.

Performance appraisals take many forms. Essays are written narratives assessing an employee's strengths, weaknesses, past performance, and potential, and provide recommendations for improvement. Several other types of performance appraisal methods are common in both the public and private sectors.

- *Comparative standards or multiperson comparison.* This relative, as opposed to absolute, method compares one employee's performance with that of one or more others.
  - In *group rank ordering*, the supervisor places employees into a particular classification such as top one-fifth and second one-fifth. If a supervisor has 10 employees, only two could be in the top fifth, and two must be assigned to the bottom fifth.
  - In *individual ranking*, the supervisor lists employees from highest to lowest. The difference between the top two employees is assumed equivalent to the difference between the bottom two employees.
  - In *paired comparison*, the supervisor compares each employee with every other employee in the group and rates each as either the superior or weaker of the pair. After all comparisons are made, each employee is assigned a summary or ranking based on the number of superior scores received.
- *Critical behaviors.* The supervisor's attention is focused on specific or critical behaviors that separate effective from ineffective performance.
- *Graphic rating scale.* This method lists a set of performance factors, such as job knowledge, work quality, and cooperation, that the supervisor uses to rate employee performance using an incremental scale.
- *Behaviorally Anchored Rating Scales (BARS).* BARS combine elements from critical incident and graphic rating scale approaches. The supervisor rates employees according to items on a numerical scale.
- *Management by objectives (MBO).* MBO evaluates how well an employee has accomplished objectives determined to be critical in job performance. This method aligns objectives with quantitative performance measures such as sales, profits, and zero-defect units produced.
- *360-degree feedback.* This multisource feedback method provides a comprehensive perspective of employee performance by utilizing feedback from the full circle of people with whom the employee interacts: supervisors, subordinates, and coworkers. It is effective for career coaching and identifying strengths and weaknesses.

APPRAISAL JUDGMENTS

The performance appraisal represents a legal communication to an employee and should be supported by objective reasoning. Appraisal judgments can be objective or subjective. Objective factors are observable and results are measurable. They include quantities, errors, and attendance records. Subjective factors are opinions, which are difficult to measure or document and invite bias or charges of bias. Included as subjective factors are personality traits, such as dependability, initiative, and perseverance. To remain objective, the supervisor double-checks ratings to be sure he or she does not favor one employee or make unsupported judgments. Supported judgments have documented critical incidents of employee performance to illustrate ratings.

OBJECTIVES OF PERFORMANCE APPRAISALS

Performance appraisals seek to meet specific objectives, including tell and sell, tell and listen, problem solving, and mixed model. *Tell and sell,* is used for purely evaluative purposes. The supervisor coaches by telling the employee the evaluation and then persuading the employee to follow recommendations for improvement. *Tell and listen* is evaluative in nature. The supervisor coaches by telling the employee the evaluation and then listens to the employee's reactions to the evaluation in a nonjudgmental manner. *Problem solving*, developmental in nature, involves counseling. It is used for employee development purposes. The supervisor does not offer evaluation but lets the employee decide his or her weak areas and works with the employee to develop an action plan for improvement. The *mixed model* combines coaching and counseling and is used for both evaluative and development purposes. The supervisor begins the appraisal with a problem-solving session and concludes with a more directive tell-and-sell approach.

AGENDA FOR THE PERFORMANCE APPRAISAL CONFERENCE

Prior to the appraisal conference, the subordinate and the supervisor should prepare written evaluations. The supervisor follows the steps in the appraisal conference.

Step 1. Communicate the Evaluation Begin with a quick review of the rating. Ratings should include results of performance, as well as motivation, effort, and cooperation. Give enough attention to what went right. Start with the success the person has had during the evaluation process and then emphasize where, how, and why improvement is needed.

Step 2. Resolve Any Misunderstandings Share the discussion with the subordinate. Give the subordinate the opportunity to respond to your appraisal. Listen without interruption. React to the subordinate's feedback nondefensively. Use concrete examples to describe performance. Acknowledge that you are receptive to the employee's point of view. Make sure the subordinate understands your complaint before proceeding further.

Step 3. Seek Acceptance of the Rating Ask the subordinate, "Can you accept this rating?"

Step 4. Identify Areas for Improvement One of the goals of performance appraisal is to provide a means of giving feedback to a subordinate concerning ways to improve job performance. Discuss goals toward which the subordinate might work. Agree on specific targets to meet in the period ahead. Review what you can do to help.

Step 5. Secure Commitment to Future Goals Try to focus the employee on the future rather than on the past.

CONDUCTING THE EVALUATION

There are several ways to conduct a performance evaluation, each with its own primary purpose. Evaluations may be conducted in order to improve employee performance and productivity, or to determine whether some reward in salary, status, and benefits should be given. Evaluation also ascertains whether a remedial or corrective program is warranted. It is difficult for a performance evaluation to serve more than one purpose and conclude with the employee feeling that he or she received fair and honest treatment. A performance evaluation should help the employee improve job performance. The appraisal, describing both strengths and weaknesses observed by the appraiser over a set time period, should lead to improved performance. A critical evaluation should be explained clearly to the employee, along with a self-improvement plan that outlines how the employee can improve to acceptable levels. Department members must know and believe that an evaluation is based on the duties of the position and provides an accurate measurement of the member's performance in the major dimensions of the job. The appraisal period must be a predetermined period of time and reflect the entire time-frame evaluated.

Whatever the instrument or method used, the employee needs evidence that the method and the person conducting the evaluation are fair, honest, unbiased, and objective. Employees often make their own judgments about who works hard, is loyal, and so forth, and if evaluations do not reflect this, the evaluations and the evaluators lose credibility. There are a variety of reasons why an evaluator might be incapable of an accurate appraisal. The evaluator may be biased or prejudiced, or may act on friendship. Evaluators may not want to evaluate fellow employees negatively and thus "score" everyone high. The evaluators may not know what the performance standards require, or they may not have been trained to use the evaluation system. Evaluators who have not appraised performance over the stipulated period may inappropriately make judgments based on only one or two incidents.

Performance appraisals can survive a court test if there is (1) personal knowledge and contact with the employee being appraised; (2) a review process that prevents the supervisor performing the appraisal from being able to affect the career of the individual; (3) formal evaluation criteria; and (4) absence of adverse impact (substantial underrepresentation of protected group members) as determined in the cases *Brito v. Zia Company*, 478 F2d 1200 (1972) and *U.S. v. City of Chicago*, 549 F2d 415 (1977) Cert denied 434 U.S. 875 (1977). Employees should receive assistance to improve their performance over a sufficient period of time. Lacking this assistance and corrective action, the value of the performance evaluation is severely diminished.

EVALUATION METHODS AND INSTRUMENTS

The evaluation process is one of three critical elements of the organizational performance cycle. The first element is a job analysis that describes each personnel classification (job responsibility and quality expectations). The second element is performance standards that describe the job expectations and the quality of the accomplished job. The third element is the evaluation system that identifies how successful the individual and the team have been in accomplishing desired results (job standards, mission, goals, and objectives). In other words, the job analysis describes the employees, the performance standards describe the quantity and quality of job expectations, and the evaluation gives a description of measured accomplishment for the employee who is attempting to achieve desired results (measured in part by conformance to job standards). The entire process can be termed the organizational performance cycle. The management strategy for implementing an evaluation process should be properly planned within an organization. The evaluation process should be focused on measuring progress toward accomplishing desired results (as targeted by the goals and objectives) and personal ability to meet job standards. A number of key points in the development of a good job evaluation process are summarized as follows:

- The evaluation process should be focused on measurable job functions (as established in the job analysis job description) and preestablished job standards to which the employee evaluated is expected to perform.
- The evaluation form should be developed so that the measurement criteria are designed clearly, in order for the employees involved with the evaluation process to understand how to use the forms. Each performance factor should be described clearly in a reference guide. This is especially important for the evaluation measurement expectations. The evaluation system should be described clearly so that all evaluators know the expectations.
- The evaluation system should be packaged as a program with the evaluation reference guide. The entire program should be presented to the management team (all supervisors). Training is essential in order to prevent misguided use of the evaluation process. The evaluation program should be included in the ongoing management training program. All new supervisors must be well prepared to perform their first job evaluation.
- The evaluation system should encourage written follow-up for measured evaluation factors, so that the evaluator can record feedback on observations of job performance and judge employee performance.
- The evaluation form should include a number of important organizational components, summarized as follows:
  - Name, position, date of evaluation.
  - Performance factor description and measurement criteria. Each factor should be followed by space for evaluator comments (a notation on the evaluation form for extra comments on a separate evaluation comments sheet is recommended).
- The evaluation form should include an opportunity to give a summary of employee performance.
- A section of the evaluation form should include an opportunity for the employee to comment.
- Signature blocks should be included to show who performed the evaluation and who received it. A section stating that the employee refused to sign may be provided if the evaluation procedure allows the employee that choice.
- The evaluation form definitely should include a section for the next level (or next two levels) of management personnel to acknowledge and approve the evaluation. This gives management the opportunity to correct evaluator error and understand the strengths and weaknesses of the team.

Foremost in the decision to include certain employee evaluation measurement factors in the departmental evaluation process is the concept of job relatedness. The evaluation factors should relate back to job requirements (identified in the job analysis and documented job standards). The federal Equal Employment Opportunity Commission (EEOC) guidelines stress the importance of valid appraisal processes that reference specific job situations when measuring job-related employee performance. This is especially important when the evaluation performance becomes the subject of a personnel action such as promotion, demotion, disciplinary action, or termination.[31] The bottom line regarding an evaluation process is that the measurement criteria be reliable and valid. The evaluation system should provide an ongoing, reliable, and predictable method (as used by all evaluators) and have job-related validity.

PROMOTIONAL PROCESS

Today many fire departments use assessment processes for promotion. Although there are many different approaches to assessment centers, most share some common testing elements.

Overview of Assessment Processes Assessment processes are designed to evaluate specific performances. Unlike the promotional tests previously used in the fire service, assessment processes focus on behaviors rather than simply knowledge. The process attempts to measure what you can do rather than what you know. The characteristics of assessment processes follow:

- Assessment processes evaluate performance rather than knowledge. Successful performance requires technical knowledge, but it is based on what the participant does rather than what the participant knows.
- Assessment processes are based on the principle that "behavior predicts behavior." In other words, a participant who successfully performs the behaviors in a process will probably be able to perform them in the position being tested for.
- Job analysis guides the development of the assessment process. Each valid assessment process is based on a thorough analysis of the identified job minimum qualifications. The dimensions, behaviors, and tasks included in the assessment reflect those required in the position.
- An assessment process is a systematic process for measuring the identified behaviors and dimensions.
- Preparation for an assessment process requires practice and professional development. An assessment process is more than regurgitating knowledge; it is evaluation of a person's ability to apply knowledge. Unlike a written test, if a person has not taken the time to develop those skills and behaviors evaluated in the assessment process, reading and studying will not be of any value.

Dimensions A dimension is an attribute or quality necessary to perform a task or a job. For any given job in the fire service, specific dimensions are the foundation for being able to perform the job successfully.

During the development of an assessment process, the specific dimensions required for the position are developed. Based on those dimensions, the components are identified that will be used in the assessment. It is essential that every promotional candidate have clear understanding of the dimensions for the job being sought. This understanding makes preparation for the assessment more straightforward. Dimensions are different than personality traits. Personality traits such as honesty, initiative, and so on, are embedded in each dimension, and those traits become part of how each person applies the dimension. For example, an important personality trait is respect for others. This trait is a key ingredient in such dimensions as interpersonal relations and decision making. Many different dimensions are used in evaluation processes. However, 11 are commonly found in fire service assessment processes.

1. Oral communication: The ability to communicate orders and information to others clearly and concisely, and the ability to listen to messages others are sending.
2. Written communication: The ability to create, complete, and disseminate written materials such as reports, memos, and letters.
3. Problem analysis: The ability to break a problem into individual parts and identify the relationships between the parts and the problem.

4. Interpersonal relations: The ability to create and maintain positive working relationships with other people. The relations should foster honesty, trust, and open communication.
5. Delegation: The ability to identify tasks that can be given to others and the ability to communicate expectations and specifics about the tasks.
6. Decision making: The ability to identify solutions to problems, evaluate the pros and cons of each solution, select the best solution, and then communicate that solution to others.
7. Decisiveness: The readiness to make decisions and stay committed to a course of action.
8. Organization: The ability to efficiently establish a plan of action for self and subordinates to achieve a specific goal. This includes effective use of personnel and resources.
9. Technical expertise: The skills necessary to perform technical operations safely, effectively, and efficiently. For example, as a company officer and battalion chief, technical expertise is required in fire ground and emergency operations.
10. Evaluation: The ability to assess personnel performance or progress toward a goal objectively.
11. Time management: The ability to effectively and efficiently use time as a resource in accomplishing assigned tasks and duties.[32]

Components of the Assessment Process

Oral Presentation Candidates prepare then deliver short talks on assigned topics. These may be instructing the staff of a business on the safe use of fire extinguishers, a talk to kids about fire safety in the home, or a presentation to senior staff on a new deployment concept.

Written Presentation Candidates prepare a memo or report, per instructions. These instructions may be to prepare a report to the chief regarding a complaint from another city department.

Role Play Candidates interact with a subordinate with some hidden problem requiring interpersonal skills, supervisory skill, knowledge of department policy, and so on. The subordinate might be starting to miss work or be under some suspicion of improper activity.

Tactical Scenario Candidates respond to real-time, unfolding, difficult tactical scenarios that test their ability to think quickly and accurately. Such scenarios include emergency events, often with one or more usual circumstances present.

Group Problem Solve Candidates work together on an assigned problem, or prepare an agenda or plan a public relations event. Often, the written exercise mentioned above will form the foundation of this common exercise.

Structured Interview Questions Candidates all respond to a uniform set of questions such as why they want this promotion, what they view as contemporary issues in the fire service, and the like.

DISCIPLINARY PROBLEMS

Effective discipline can help to eliminate ineffective employee behavior. An employee should be disciplined when he or she chooses to break the rules or is not willing to perform the job to standards. Discipline is corrective actions taken by a supervisor when an employee does not abide by organizational rules and standards. Common categories of disciplinary problems are attendance, poor performance, or misconduct. *Attendance* problems include unexcused absence, chronic absenteeism, unexcused or excessive tardiness, and leaving without permission. *Poor performance* includes failure to complete work assignments, producing substandard products or services, and failure to meet established production requirements. *Misconduct* includes theft, falsifying employment application, willfully damaging organizational property, and falsifying work records.

PROGRESSIVE DISCIPLINE

Disciplinary treatment in most organizations is progressive, whereby the organization attempts to correct the employee's behavior by imposing increasingly severe penalties for each infraction. The usual steps are verbal warning, written warning, suspension without pay, and termination of employment.

Employees accept fair, equitable, and consistent discipline. Positive, progressive, *hot stove* approaches work best.

Immediacy. The more quickly the discipline follows the offense, the more likely the discipline will be associated with the offense rather than with the dispenser of the discipline.

Warning. It is more likely that disciplinary action will be interpreted as fair when employees receive clear warnings that a given violation will lead to a known discipline.

Consistency. Fair treatment demands that disciplinary action be consistent.

Impersonal nature. Penalties should be connected to the behavior (violation) and not to the personality (person) of the violator.

DISCIPLINARY ACTION

Before conducting a discipline discussion, the supervisor should be able to do the following:

- Describe the incident by answering, who? what? when? how? where? witnesses? why?
- Refer to the policy or procedure that was violated.
- Determine whether the employee was previously notified of the correct operating procedure and be able to provide documentation, if it exists.
- Know whether the employee has been disciplined previously.
- Provide documentation of verbal counseling, if possible.
- Determine whether other employees have violated the same policy/procedure and what discipline, if any, they received.

In discipline discussions with an employee, the supervisor points out the unsatisfactory behavior, explains the need for and purpose of the rule or practice that is being violated, and expresses confidence in the employee's willingness and ability to make the necessary changes in behavior. During a discipline discussion the supervisor should be objective in reviewing the situation and give the employee specific examples of the behavior that is causing the problem. The employee should be allowed an opportunity to present his or her own case. The supervisor needs to make sure the employee has a clear understanding of the consequences of his or her behavior. The supervisor and the employee should agree on specific recommendations for correcting the performance.[33]

DISCIPLINE

Discipline is often thought of as the application of punishment for the infraction of a rule. Children are disciplined at home or in school for misbehaving. Some managers carry that same logic forward when supervising their employees. The results of such a simplistic view of discipline can be devastating for individual employees and the team as a whole. Discipline should be viewed as a positive experience. Leaders develop discipline through quality training; clear and understandable standards of operation; team-supported policies and procedures; and constant, trusted communication between supervisors and employees. *Disciplinary action* is used to correct an employee's acknowledged violation of the standards of operation. A proper level of discipline should be in place before punitive disciplinary action is taken. Employees cannot be expected to comply with a standard of operation or behavior that has not been properly explained. Discipline as an organizational function associated with controlling and teaching involves goal setting, team building, peer pressure, conflict resolution, employee assistance, due process, grievance processes, and intervention strategies.

Purposes of Discipline Personnel who *make a mistake* must have their error pointed out and corrective action clearly described by an immediate supervisor. This can be accomplished through an honest and thoughtful communication between the supervisor and employee. If the mistake was deliberate, repeated, or a serious violation of known standards, the supervisor may decide that a stronger disciplinary action is needed. The first level of disciplinary action begins with an oral reprimand. Disciplinary action may progress to a written reprimand, time off, demotion, and, possibly, termination. Most disciplinary problems can be managed through a step-by-step process called *progressive discipline*.

Progressive discipline starts with an official recognition of an employee weakness or failure to behave accordingly to known standards of operation. Simple violations, such as a beginning problem associated with attendance or an inability to complete assigned duties properly, will receive an informal inquiry and discussion by the supervisor. Failure to remedy the performance problem will be reflected in the employee evaluation. The supervisor should attempt to find the root of the problem (potentially a personal problem at home or with other employees). Failure to comply with the standards of operation should be dealt with fairly and firmly. Supervisor support and understanding also should be tempered with appropriate disciplinary action. Oral reprimands, followed by written reprimands and days off without pay, are the disciplinary measures that should accompany the supervisor's serious attention to the continued employee problem.

In most cases, an employee performance problem can be corrected with a low or moderate level of disciplinary action. Severe disciplinary action, such as demotion, long-term leave without pay, or termination, should be considered when previous progressive discipline has failed. When the violation of job standards is obviously improper, previous warnings have been given, or violations of law have been committed (such as theft), harsher disciplinary actions also are necessary. Employees who have drug and/or alcohol abuse problems present special disciplinary problems. The employer may consider a "last-chance" agreement (according to the jurisdictional personnel policy and procedures on drug and alcohol abuse) with the employee, whereby the employee is given an opportunity to seek help for substance abuse in exchange for an agreement that future violation may result in immediate termination (employer-requested substance abuse testing also may be a part of the last-chance agreement). It is always best to seek the advice of an attorney with a background in the legal mandates for employee rights in your state. Last-chance agreements should be carefully worded and developed to be clearly understood and comprehensive for both the employer and employee.

The primary purpose of disciplinary action is to correct employee behavior with an appropriately planned level of punitive action. Employees who understand their behavioral problems and sincerely desire to correct mistakes should be dealt with less stringently (with less punitive action) than those employees who repeatedly violate job standards or commit serious violations of personnel rules and regulations, or local, state, or federal laws. Fire officers should not discipline an employee to prove a point to the rest of the organization. The disciplinary process is a highly personal experience between the employee and supervisor. Fair, honest, and firm management of disciplinary problems (without any need to show "who's the boss") usually results in a positive and respected understanding between the employee and supervisor. There is satisfaction in knowing where one stands and in the hope for a better future relationship, given certain specific changes in behavior.

The Application of Discipline It is important to keep in mind the two major factors relative to the process of disciplining: the goal of improving the employee's performance to a satisfactory level, and the application of an appropriate level of discipline that follows the spirit and letter of due process. Experienced managers have developed guidelines that may be helpful in disciplinary cases:

- Always thoroughly investigate second-party observations that may require a disciplinary action toward the suspected employee. There are usually three sides to a story: the stories of both parties and what really happened.
- Provide immediate feedback to employees regarding what they did and why discipline is being applied. Avoid personal attacks or confrontational attitudes when providing feedback to the employees. Listen closely to employees' feelings.
- Be sure everyone understands organizational policy.
- Be consistent and fair. Apply policy with the same sense of fairness for all employees. Take time to think before making a judgment against an employee. A day or two of personal reflection or discussion with a trusted peer or higher authority provides an excellent opportunity to develop a fair and effective solution.
- Watch for the common problems of tardiness, absenteeism, alcohol and drug abuse, and slipping performance. Impaired decision-making ability, changes in mood or habits, and preoccupation often are signs of personal problems that may develop into significant disciplinary concerns. An employee assistance program provides an excellent opportunity for troubled employees to seek professional support to solve personal problems. When following due process, each step of a progressive disciplinary program must follow legal protocols.

DOCUMENTATION

The documentation of employee behavior starts with a note in the supervisor's personal log or a note indicating job performance on the employee evaluation and grows in formality to a written suspension or termination notice. In all cases, there are rules to be followed.

The following documentation checklist should help the supervisor keep more accurate and complete records:

- Was the documentation done promptly?
- Does the report include date, time, and location?

- Was the action or behavior exhibited recorded adequately?
- Were the persons or work products involved indicated?
- Were the specific performance standards that were violated or exceeded listed?
- Were the rules or regulations that were violated or exceeded listed?
- What was the consequence of the action or behavior?
- Is the record objective?
- Was the supervisor's response to the action or behavior recorded?
- Were the employee's reactions to the supervisor's efforts recorded?

Formal written disciplinary proceedings resulting in a suspension of more than three days should be evaluated thoroughly by personnel staff and legal advisers for correctness. In many states, written reprimands involving suspension for more than a few days are subject to legal appeal. In all cases, the supervisor should investigate the facts thoroughly before enforcing a formal suspension notice. The employee's appeal rights and right to hear charges according to personnel procedures must be made clear in the written notice to suspend. The reasons for the suspension should be clearly tied to specifically recognized violations of personnel procedure or other formally expected job performance (city procedures, laws, or other documented policy). The punishment associated with the written notice should properly fit the level of violation of disciplinary standards. Valid charges must be proven consistent with other personnel practices and job expectations for comparable employee experiences and related performance expectations. Progressive discipline is always a good defense for appropriately applied suspension or termination notices.

THE DISCIPLINARY INTERVIEW

The disciplinary interview is important not only at the beginning of a disciplinary process but also if the problem continues. A good supervisor should have the ability to communicate supervisory concerns and elicit employee feedback so that a positive and productive resolution to the problem can be reached:

- Face the employee in order to hear and see him or her.
- Determine employee personal prejudices in advance.
- Listen closely and encourage employee feedback.
- Encourage the employee to explain the situation.
- Repeat what the employee has said to be sure you understand.
- Watch for body language (on both your part and the employee's).
- Do not try to have the last word.
- Consider the facts and take time before making final judgment.

During this initial interview, it is vital for the employee to understand the following points:

- Corrective action must be taken for unacceptable work behavior.
- Assistance is available through training programs or an employee assistance program if emotional or other personal problems are contributing to the unacceptable performance.
- Further steps may be taken if the employee's performance does not improve in a reasonable time.
- Drug and/or alcohol problems should be addressed with a good understanding of departmental policy regarding rehabilitation plans and last-chance agreements.

It is important that the supervisor be specific about employee behavior when discussing job performance; be consistent in the evaluation of all employees; judge job performance and encourage professional help for personal problems; listen to personal problems as a good friend, but refer the employee to professional support for solutions (the employee assistance program provides a great opportunity for professional help in solving personal problem); and be firm and clear about what improvement is expected in job performance.

At the end of the interview develop a method of following up on the initial session with ongoing support to overcome the personnel problem and to improve the performance weaknesses.

GENERAL GUIDELINES

Once the decision has been made to discipline, several general guidelines should be followed if the action is to be successful.

- Give the employee warning. No disciplinary action should come as a surprise.
- Assure the privacy of the action. Do not embarrass an employee in front of his or her peers.
- Get all the facts before judging.
- Be prompt. Do not put it off.
- Strive to be objective. Know your personal prejudices.
- Know the limits of your authority.
- Follow up to assure success.

DUE PROCESS

All of us enjoy certain rights under the law, and any process applied to us in a disciplinary setting must comply with the law. If any part of a disciplinary or grievance process is not conducted as specified in departmental procedures and applicable laws, the disciplinary action will not be entirely valid. Due process is defined by applicable federal, state, and local laws, and civil service regulations. Should questions arise, the personnel appeal process can lead to arbitration through the personnel commission or a judicial review. In its publication on discipline, the International City/County Management Association presents a listing of some issues to be considered.

1. *Was the employee warned that the particular act or behavior would result in discipline?* The warning can be given orally or in writing, but it is more effective if it is a written statement contained in a department's work rules. It is a rule that must be given to all employees, and in a disciplinary hearing, it must be established that the specific employee was knowledgeable of the rule. There are some exceptions to this. For example, some actions are so clearly wrong that it is not necessary that a specific rule be violated. Examples would be stealing or threatening a supervisor or other employees.
2. *Was the broken rule or order reasonably related to the orderly, safe, or efficient operation of the department?* Appellant bodies will overturn disciplinary actions if they can see no relationship at all to the job functions. An example of this might be a dress code that has no reasonable relationship to the work being performed.
3. *Has the department applied the rules, regulations, or work orders in a consistent fashion and without discrimination?* An employee cannot be disciplined if it comes to light that the rule or order has never been applied but was utilized only for that particular employee. On the other hand, evidence of an infraction that was not caught or disciplined is not automatically sufficient to overturn discipline.
4. *Did management, before administering discipline, make an effort to discover whether the rule or order had been broken or disobeyed?* The key to this question is the word *before*.

Discipline will be overturned if management acted prematurely or in anger. A word of advice: back off, cool down, investigate.

5. *Was the investigation fair and objective?* Get the facts as to who, when, and where. Give the employee a "day in court." Let the employee explain his or her side before disciplining.
6. *Did the person making the disciplinary decision have sufficient evidence that the employee committed the act(s)?* Reasonable proof is required and will vary from case to case. Except in a court of law, the evidence does not always have to establish guilt beyond a reasonable doubt. If the disciplinary action is a discharge, however, the evidence will have to be substantial and documented. A rule of thumb is to avoid relying on rumor or unsupported accusations.
7. *Is the discipline related to the seriousness of the offense and to the employee's record?* This is the most important question and the one that probably leads to the overturn or reduction of more disciplinary actions than all of the other six. As a rule of thumb, minor offenses result in milder disciplinary action. However, discipline is progressive and the record of the employee is relevant. If he or she has been disciplined for an offense in the past, it is logical that discipline will be more severe for a second or third offense. Most significant is the record of the department in other cases of this nature. What has been your record in treating this type of offense? If the offense is serious in your department and you have treated it as such in the past, chances are good that your consistent discipline will be sustained. In summary, consider the nature of the offense, the record of the employee, and your practice in dealing with this type of offense.

Additional key questions that must be addressed before disciplinary action is taken include the following:

- Are there inconsistencies in the documentation?
- Are there extenuating circumstances?
- Are the conclusions drawn from facts?
- Were adequate help and training provided?
- Was the worker given every opportunity to hear and discuss the charges?

Did the worker have representation by an attorney and/or the union at all hearings? Has the case been discussed with the union? With a management attorney? Are there political overtones to the issue? Has any media reaction been anticipated? What will be the reaction of other members?

APPEAL PROCESS

Any significant disciplinary action that involves the reduction of an employee's *property interest* (significant amounts of compensation, usually over five to ten days of pay) in continuation of his or her employment, demotion, or termination requires an appeal process. Most municipalities have appeal rights clearly defined in a personnel resolution adopted by the city council or in the labor agreement (employee contract). The State of California Supreme Court defined a permanent employee (not probationary) right to appeal in the 1975 landmark case *Skelly v. State Personnel Board* S.F. No. 23241, Supreme Court of California 1975. The provisions of *Skelly* are relevant to today's disciplinary process in California and are duplicated in many other states. The case held that a permanent civil service employee held a property interest in his or her job that is protected by state and federal constitutional due process provisions. The *Skelly* decision was a precedent-setting case by which all governmental agencies must abide. Probationary, reserve, and volunteer personnel are not subject to the *Skelly* decision requirements. If a labor agreement is in place, the appeal process usually is spelled out, assuming it is an issue covered by the local contract. If a member feels aggrieved over an issue that is protected by federal or state laws, then appeal could be made to an agency or court at the appropriate level.

Another landmark case involving appeal process rights is *Cleveland Board of Education v. Laudermill*, 470 U.S. 532, 105 S.Ct 1487, 84 L.Ed. 2nd 494 (1985). Here the Supreme Court ruled that when a hearing is conducted after termination, the employer must afford the terminated employee all hearing rights, including the opportunity to refute the charges. The case (actually two cases—a security guard and a bus mechanic) went to the U.S. Supreme Court following a Civil Service Commission appeal. An example of a federal agency hearing an appeal, in this case the National Labor Relations Board, is *NLRB v. Weingarten* 485.2D1135 (1974). The NLRB ruled that an employee who is a member of a union has the right to have union representation at any interview that could result in disciplinary action. The concept of "just cause" is found in many labor agreements, typically within a context that states discipline can be administered only for just cause. Because what constitutes just cause can be debated, the phrase can be a trouble spot. Seven tests for just cause help describe its basic element:

1. Notice. Did the employer give to the employee forewarning or foreknowledge of the possible or probable consequences of the employee's disciplinary conduct?
2. Reasonable rule or order. Was the employer's rule or managerial order reasonably related to (a) the orderly, efficient, and safe operation of the employer's business, and (b) the performance that the employer might properly expect of the employee?
3. Investigation. Did the employer, before administering the discipline, make an effort to discover whether the employee did in fact violate or disobey a rule or order of management?
4. Fair investigation. Was the employer's investigation conducted fairly and objectively?
5. Proof. At the investigation, was there substantial evidence or proof that the employee was guilty as charged?
6. Equal treatment. Has the employer applied its rules, orders, and penalties evenhandedly and without discrimination to all employees?
7. Penalty. Was the degree of discipline administered reasonably related to (a) the seriousness of the employee's proven offense and (b) the record of the employee in his or her service with the employer?

GRIEVANCE PROCEDURES

In most jurisdictions, the processing of disciplinary actions, including employee rights and appeal processes, is detailed in a specific section of the city personnel resolution or employee labor contract. The grievance procedure generally is located in another section of those same documents. It is established to give the employees the right to a fair, just, and orderly review of specific employer activities. Issues that could be subject to grievances include changes in work environment, changes in rules and regulations, perceptions of unjust management or supervisory activities, and perceived violations or misinterpretation of the personnel rules and regulations or labor agreement. Grievance procedures include formal and informal proceedings starting with an informal discussion regarding the employee's concerns. The process formally requires a written description of the employee's grievance concerns. An established time frame for a written response by management (usually 15 days) is required. The grievance procedure allows for appeal through the chain of command to the fire chief and eventually the city's personnel director (usually the city manager) or the fire board (fire districts). Some jurisdictions include the opportunity for arbitration when settling a grievance other than administratively. Most grievance issues are solved through constructive communications and positive relations. Listening to the employee's issues can result in a better understanding of what the real problem is. The solution to the problem usually can be mediated by using the insights of all involved to develop a new position. Management need not sacrifice important concerns, and the employee's needs should be valued.

DISCIPLINE AND VOLUNTEERS

The concepts of discipline and disciplinary action should apply to volunteer firefighters as well as to career members, although the due process and property interests appeal rights usually are not legally mandatory for volunteer firefighters. When officers of volunteer departments hesitate to apply disciplinary steps to volunteers, they simply encourage inappropriate and nonproductive behavior. There are differences between volunteer and career firefighters relative to discipline in respect to property rights. Volunteers typically are disciplined for serious infractions by being placed on "inactive" status for a stipulated time period. Although different from placing a career firefighter on suspension without pay, this is viewed by volunteers as a significant deterrent.

EMPLOYEE ASSISTANCE PROGRAMS

Modern personnel management focuses on the wise use of human resources and has concern for individuals. Expert programs to assist employees and volunteers who are experiencing problems affecting job performance and their lives are often used. The employee assistance program (EAP) is available for use by employees without an acknowledgment by the organization's officers or staff. Professional counselors help employees with personnel problems, marital problems, drug or alcohol abuse, and other mental and physical problems. In addition to expert support offered through the employee assistance program, a peer support group may be organized. Peer groups, with support from a professional counselor, are helpful for employees who request group support for organizational problem counseling. An employee may desire peer feedback on how to deal with a personnel problem or a distasteful relationship with a supervisor. The peer group should be well trained and be prepared to offer constructive advice. Peer groups also are useful for critical incident stress debriefings. Many times a group replay of a horrendous employee experience is helpful for releasing stress and emotional concerns. Participation in these programs is entirely confidential and a notation should not be placed in the personnel folder unless attendance at the EAP was mandated by disciplinary action. Employee assistance programs provide a useful resource for managers who realize that professional help can improve a worker's nonconforming behavior.[34]

Today's fire service leader must embrace a much larger community interest than the fire service has recognized historically. Not only must we learn how to deal with zoning and land use issues, private sector business interests, and neighborhood safety concerns, but we have entered an era that includes regulatory mandates that necessitate coordination among law, health, planning, and public works interests. The fire service leader of today must now employ the same planning tools and business logic as in the contracts and agreements developed by the private sector. The who, what, why, where, and how of management must be developed with an understanding of the political realities and the multidisciplinary functions now embedded in almost every aspect of what we do on a day-to-day basis. To do so, we must coordinate with other governmental organizations to develop successful and creative innovations that lead to safe and healthy communities. We must have a clear understanding of how we manage our emergency services today, and how we will plan for what needs to occur in the future.

Our role as fire service leaders is to provide the policy makers (governing board) with the correct information so they can make quality policy decisions and develop strategic methods for spending the taxpayers' money. Effective and well-directed communication is critical for a successful overall coordination. The ability to manage any organization depends largely on the quantity of the resources that are available to accomplish the desired result and the quality of our leadership and management. In addition, the legal and regulatory requirements in which fire organizations operate, the parameters of the organizational foundation, and the budgetary challenges affect our strategic planning and our vision on how we are to manage our own future. Understanding the dynamics of how to manage a fire service organization and clarity of vision are critical if we are to be successful in designing our own future. At all levels of the organization, we must understand the budget development; how funds are allocated; and how general fund, enterprise fund, capital fund, and asset replacement funds are utilized in our organizations. We must have a clear understanding of the liabilities and regulatory mandates that not only have an ongoing cost but also require us often to change our operational procedures and organizational structures to meet those mandates. The connection among strategies, mission, goals, objectives, action plans, values, and visions take into account everything we have discussed in this chapter. The relationships among budgeting organizational design, labor management relations, discipline, and how we treat our people are all pieces of the puzzle of how we lead and manage our organizations.

Allan Kaye, formerly of Apple Computers, was once quoted as saying, "The best way to predict the future is to invent it." We can only do so through effective strategies built upon a clear understanding of how our organizations work.

ADVICE FROM THE EXPERTS

Organization design and structure have changed quite a lot in your career. What does the future hold?

MICHAEL CHIARAMONTE

Chief Fire Inspector, Retired, Lynbrook Fire Department (New York): It holds promise if we let it.

BETT CLARK

Former Fire Chief, Bernalillo County Fire and Rescue (New Mexico): Whatever the upcoming leaders' vision is of their future.

KELVIN COCHRAN

Fire Chief, Atlanta Fire Department (Georgia): Fire department organizational structures and design will continue to be hierarchal with a pyramid configuration. As departments grow in size and complexity or downsize due to reduced activity or funding, the levels of rank and divisions of labor change. The greatest change in fire department organization structure will be theoretical and intrinsic. Fire departments are currently transitioning from classical organizations to human relations organizations, to human resource organizations. The traits of a human resource organization include formal and informal communications; chain of command flexibility; emphasis on employee participation; work to meet group and individual needs; decentralized decision making; authority granted based on knowledge and competence; informal organizations recognized; seeking to develop each member; entrepreneurial opportunities and systems of meaning rewards and equitable discipline.

CLIFF JONES

Fire Chief, Tempe Fire Department (Arizona): The future holds a framework of increased demand compounded by an inability to generate adequate numbers of command and support staff. Organization design and structure has been impacted in many departments by financial constraints in recent years. Organizations have become flatter, and more has been expected of command staff members. At the same time more has been added to the list of responsibilities of many fire departments including emergency management and weapons of mass destruction. This situation places greater emphasis on work teams, especially from a labor–management perspective. Work teams—properly led, managed, and utilized—can accomplish much and drive change from a grassroots level by empowering employees to make contributions to the continuous improvements of their departments.

BILL KILLEN

Retired Fire Chief and IAFC President 2005–2006: Flatter organizational structures; increased empowerment of individuals; and greater trust in the abilities, skills, and knowledge of employees. Increased automation and use of information technology will have significant impacts on the structure of organizations and how fire and emergency services are trained, administered, and organized.

KEVIN KING

U.S. Marine Corps Fire Service: I believe the future will continue to be driven by budgetary pressures. Governments continue to seek cost reductions, and thus we need to look for ways to provide our services with fewer personnel, because this is by far our biggest cost driver. I believe we will be forced to focus our efforts on educating the public to take more care of itself, rather than having so much reliance on the fire service. I believe we will see more efforts in public education and prevention with a corresponding trade-off in operational reductions. However, on the operations side, we will need to ramp up quickly when the big incidents take place. This requires greater use of mutual aid, partnerships, consolidations, and other efforts that allow for rapid deployment of emergency resources to large incidents. (These are opinions of Mr. King and do not reflect official policy of the Department of Defense or the United States Marine Corps.)

MARK LIGHT

Executive Director, International Association of Fire Chiefs: I believe we will see more self-directed work teams. Fire departments are very hierarchical, and I believe we will see less hierarchy and more emphasis on functions and programs. In the past, to make a contribution, you had to be at least a captain before anyone would listen to you. In today's world and the future, firefighters are entering the service with a tremendous amount of talent and skills and are not afraid of bringing these forward. As such, I believe we will see firefighters leading major programs in the department, and in fact, they will progress through the ranks without taking the traditional approach of riding on a fire engine. This will be difficult to manage later in their careers, when you have a very bright officer who has never commanded an incident and is suddenly placed back in an operational role. The department must be aware of this and make every opportunity available for nontraditional progressions to get operational experience in new and different ways, such as through virtual training.

LORI MOORE-MERRELL

Assistant to the General President, International Association of Fire Fighters: Everyone from risk managers to human relations professionals to city attorneys have their fingers in the fire service. This is incredibly frustrating for fire service leaders and rightfully so. Many times chiefs are thwarted in their leadership and/or management roles because these "outsiders" make decisions for them or tell them what decision they must make when the "outsiders" rarely understand the stakes at hand. In our litigious society, this trend is likely to continue. Therefore, it is important to educate the "outsiders" about the actual role of firefighters as public servants. Mini-academies are perfect ways to facilitate this education. Fire service leaders should also educate themselves on the information necessary to engage these "outsiders" using their terms. Perhaps increased understanding of opposing roles will lead to better and less frustrating decisions in the future.

JAY REARDON

Fire Chief, Retired, Northbrook Fire Department (Illinois), currently President of the Mutual Aid Box Alarm System (MABAS): Organizational design will continue to flatten out and stovepipes are doomed for the wrecking ball. Total involvement in organizational access and comment as an employment responsibility, but military structure and function on the incident scene with operational deployments. No debates as "everyone" has been involved in designing, molding, and understanding deployment operations. Centrally coordinate the involvement and planning process, but decentralize deployment and execution of operations.

GARY W. SMITH

Retired Fire Chief, Aptos–La Salva Fire Protection District (Aptos, California): I still value many of the pearls of wisdom about leading and managing that came from the leaders and managers of the past. I find that the messages developed by Maslow, McGregor, Taylor, Fayol, Herzberg, Deming, and many more noteworthy leadership and management advisers still have value, just as the words and thoughts of Plato and Socrates have been valued over the centuries by philosophers and leaders of today. Some things are valuable forever, but we should always strive to understand more about what makes for success. Now I read and learn a great deal from the current futurists and leaders of our time. I like to understand the dynamics of what is going on today and work on visions that would improve operations. We certainly have many areas to work on! From disaster response, terrorism, and pandemics, to the wildland fire threat, and from economic challenges to a new energy policy to drive our country: these are many challenges to address in the future. We must learn, have vision, and work together with a spirit of "can do" logic.

STEVE WESTERMANN

Fire Chief, Central Jackson County Fire Protection District (Blue Springs, Missouri), and IAFC President 2007–2008: If you want to be responsive to your customers (internal and external), I think three things will need to occur. First, develop more regionalization/consolidation in order to remove duplicative costs and overhead while increasing the amount of resources available. Second, I believe we will need to keep our organizations as flat as possible in order to be as responsive to our own people. Third, with a flat organization you push the authority to make decisions and act on those decisions down to the customer contact point.

THOMAS WIECZOREK

Executive Director, Center for Public Safety Excellence, and Retired City Manager: I believe flatter, flexible, fair, and futuristic would summarize my answer. Flatter in terms of fewer "ranks" and steps. The leader and the troops need to be closer and work together because having multiple ranks in between leads to distant leadership and suspicion. Flexible in the area that do we need 30 categories in every contract that spell out in excruciating detail what an individual can do? Rather, should they be flexible and be cross trained to handle a variety of situations? I have been in departments where firefighters "fight fire and nothing else." Unfortunately, those are the ones who make headlines and who cannot continue. We need to look at how we can produce the best outcome for taxpayers as well as employees, and it will take flexibility on all sides. Fair in terms of not rewarding creativity and flexibility with disdain or short-term solutions; futuristic in terms of not being locked to the traditions of the past. We have to respect and know the traditions and history or we are bound to repeat it. The key is to learn and become the force of the future. Information is the key. If one doubts it, look at the growth of the Internet each day and the amount of data that can be reached from your PDA or home computer.

JANET WILMOTH

Editorial Director, *Fire Chief* magazine: I think we can learn from looking at other powerful organizations of the past and present. The Mafia and gangs of today (seriously!). No matter how large an organization, people want a sense of purpose and belonging. I believe as fire departments evolve into emergency response agencies—capable of responding to all hazards—a sense of belonging and pride in what employees do will become even more important. The riskier the work, the higher the demand for a support network.

FRED WINDISCH

Fire Chief, Ponderosa Volunteer Fire Department (Houston, Texas): I believe that the chief officer of tomorrow must have a formal education that addresses change models and must be willing to change within. The days of promoting through the officer ranks because they pass tests must come to an end.

Review Questions

1. Describe what is meant by "managers often get the type of labor relations they deserve."
2. Distinguish between the technical functions of personnel administration and the purposes of human resource

development as each relates to employee and labor relations.

3. Identify four trends in contemporary society that cause modifications to the labor–management climate and the expectations of each constituency.
4. List four important federal laws and decisions that influence labor relations, as well as any state and local regulations in your area that significantly affect labor relations.
5. Describe the roles of the International Association of Fire Fighters, other relevant national labor groups, and the professional and affiliated organizations that are concerned with labor–management relations and the rights of emergency workers.
6. Outline a technique or program to foster productive and cooperative relationships between management and worker groups.
7. Define the differences between right-to-work states and other states, relative to labor–management relations.
8. Select a prominent trend in our society and describe how it has influenced the labor–management climate and the goals of managers and members.
9. Identify and describe a law from your own state that affects, positively or negatively, fire department labor relations.
10. Describe a program that exists between labor and management that can be improved with cooperation.
11. Describe the most productive form of labor relations in a combination department.
12. Explain the meaning and process of collective bargaining, mediation, and binding arbitration.
13. List three examples of soft and hard positional bargaining, and principled bargaining.
14. Describe how the operating styles of personnel managers and labor leaders involved in bargaining are determined by their personal definitions of power.
15. Describe the guidelines for initiating, negotiating, and maintaining cooperative labor–management relationships that are part of principled bargaining.
16. List five ways personnel managers may strengthen productive and harmonious relationships between organized labor and management.
17. Describe one employee evaluation form, technique, or program that fosters the development of desired knowledge, skills, and attitudes.
18. Explain the concept of discipline as it applies both to the operations of a fire department and to the need for human resource development within that same organization.
19. Compare any differences in the application of discipline to volunteer members and career members.
20. Describe the importance of due process as it relates to official charges and discipline.
21. Describe two important court decisions related to the disciplining of career employees in fire departments.
22. Describe how discipline can be a positive function of human resource development as well as a necessity in personnel administration.

References

1. http://en.wikipedia.org/wiki/
2. Tom Burns and G. M. Stalker, *Management of Innovation* (London: Tavistock Publications, 1961), p. 19.
3. Rensis Likert, *The Human Organization: Its Management and Value* (New York: McGraw-Hill, 1967), pp. 4–10.

4. www.referenceforbusiness.com/small.Bo-Co/Boundaryless.htm
5. http://ollie.dcccd.edu/mgmt1374/book_contents/3organizing/org_processing/org_process.htm
6. http://ollie.dcccd.edu/mgmt1374/book_contents/3organizing/pwr_auth/power.htm
7. http://ollie.dcccd.edu/mgmt1374/book_contents/3organizing/deleg/delegate.htm
8. A. E. Buck, *Public Budgeting: A Discussion of Budgetary Practice in the National, State and Local Governments of the United States* (New York: Harper and Brothers, 1929).
9. Charlie Tyer and Jennifer Willand, *Public Budgeting in America, a Twentieth Century Retrospective*, www.iopa.sc.edu/publication/Budgeting_in_America.htm
10. M. J. Schiesl, *The Politics of Efficiency: Municipal Administration and Reform in America 1800–1920* (Berkeley: University of California Press, 1977).
11. Tyer and Willand, *Public Budgeting in America.*
12. G. J. Miller, "Productivity and the Budget Process," *Budgeting: Formulation and Execution*, ed. J. Rabin et al. (Athens, GA: Carl Vinson Institute of Government, University of Georgia, 1976).
13. Tyer and Willand, *Public Budgeting in America.*
14. Miller, "Productivity and the Budget Process."
15. Tyer and Willand, *Public Budgeting in America.*
16. Ibid.
17. Miller, "Productivity and the Budget Process."
18. Tyer and Willand, *Public Budgeting in America.*
19. Miller, "Productivity and the Budget Process."
20. Tyer and Willand, *Public Budgeting in America.*
21. A. C. Hyde, *Government Budgeting: Theory, Process, Politics*, 2nd ed. (Brooks-Grove Publishing Co., 2nd Edition, 1992).
22. David Osborne and Ted Gaebler, *Reinventing Government: How the Entrepreneurial Sprit Is Transforming the Public Sector* (Reading, MA: Addison-Wesley, 1992).
23. Tyer and Willand, *Public Budgeting in America.*
24. http://www.gasb.org/news/nr63099.html
25. http://www.gasb.org/statement45
26. http://beginnerguide.com/accounting/gaap/gaap-overview.php
27. United States Fire Administration, National Fire Academy, Degrees at a Distance Program, *Personnel Management for the Fire Service Course Guide.*
28. United States Bureau of Labor Statistics, www.BLS.gov.Relims/Union
29. United States Fire Administration, National Fire Academy, *Unit 11: Employee and Labor Relations.*
30. United States Fire Administration, National Fire Academy, *Unit 12: Collective Bargarining.*
31. http://ollie.dcccd.edu/mgmt1374/book_contents/5controlling/evaltg/evaluate.htm
32. Ed Kirtley, M.Ed, *Preparing for Fire Department Assessment Centers,* www.osufst.org/resources.php
33. http://ollie.dcccd.edu/mgmt1374/book_contents/5controlling/discplg/discipline.htm
34. United States Fire Administration, National Fire Academy, *Unit 8: Discipline*.

CHAPTER 9

Analytical Approaches to Public Fire Protection

Key Terms

advanced life support, p. 439
alarm processing time, p. 406
basic life support, p. 438
criterion/criteria, p. 423
critical infrastructures, p. 396
effective response force, p. 409
FESSAM, p. 392
fibrillation, p. 412
fire flow required, p. 408
fire management area, p. 429
GIS, p. 427
hazard, p. 429
high hazard risks, p. 428
Insurance Services Office, p. 412
level of service, p. 428
low risk, p. 427
mitigation, p. 394
moderate risk, p. 427
MOU, p. 421
NFPA 1710, p. 433
NFPA 1720, p. 441
performance indicator, p. 423
PSAP, p. 406
purpose, p. 433
response reliability, p. 410
response time, p. 434
risk, p. 430
standard operating procedures, p. 441
standards of response coverage, p. 427
terrorism, p. 397
total response time, p. 406
travel time, p. 406
turnout time, p. 406
Utstein criteria, p. 411
vulnerability, p. 430

Objectives

After completing this chapter, you should be able to:

- Define the history of resource deployment.
- Explain the term *standards of coverage*.
- Explain the national strategy for homeland security.
- Describe what probability and consequence are and why they are important in the development of community risk management.
- Discuss the insurance service rating schedule.
- Describe the CFAI accreditation process.
- Describe NFPA 1710 and 1720.

THE HISTORY OF RESOURCE DEPLOYMENT

When Benjamin Franklin founded the Union Fire Company in Philadelphia, the placement of the station was probably dictated more by the donation of a piece of property or the availability of a building in which to house the equipment than consideration for response time and coverage. Under Franklin's persistence, a group of 30 men came together to form the Union Fire Company on December 7, 1736.

Their equipment included leather buckets, with strong bags and baskets (for packing and transporting goods), which were to be brought to every fire. The group met monthly to talk about fire prevention and firefighting methods. Homeowners were mandated to have leather firefighting buckets in their houses (see Figure 9.1).

Other men who desired to join the Union were urged to form their own companies so the city would be better protected. Within a short span of time, Philadelphians witnessed the birth of the Heart-in-Hand, the Britannia, the Fellowship, as well as several other fire companies.[1] The early days of the fire service in the United States was not built so much around response time but around the availability of personnel who could pull the hose carts and staff the bucket brigade. Station location was based more upon the limitations of the available personnel and the means for hauling heavy equipment.

As discussed in Chapter 1, the history of the evolution of the fire service, with the conflagrations of the late 1800s and the creation of full-time fire departments, a systematic methodology began to develop by which fire resources were distributed throughout the community. Whereas fire stations had originally been staffed based upon a neighborhood or the location of volunteers, new stations were added due to community growth. The concept of multiple fire stations and the need to space them sufficiently apart so that the overall community was protected began to occur. Because the response at the time was based upon the use of horses to haul the equipment, it was natural to look at the capacity of the horse teams to arrive at an emergency in a relatively short period of time. A good team of fire horses could haul a steamer approximately 1½ miles in five minutes.

FIGURE 9.1 ◆ Leather Bucket

It is interesting to note that this still has application today when we talk of response times to emergencies of the twenty-first century, although the method of how we get there and the equipment we arrive with are much different.

With the conflagrations of the 1800s, it was evident that the American fire service needed a more precise system to evaluate the effectiveness of its fire suppression efforts. Creation of the National Board of Fire Underwriters had an immediate impact on the fire service, as the Underwriters established an evaluation system based on science, past practice, and experience. This evaluation system had a significant impact on the formation of the fire service during this period of time and has helped to guide the application of response and deployment for over a hundred years. For example, the work that was done to create fire stream hydraulics is based on very specific studies and considerable data of the era. The study of national fire losses resulted in fire flow figures developed for various construction types. Based on the design of the community rating schedule, the concepts incorporated into the system used assumptions and methodologies to prevent conflagrations from occurring. While the grading schedule has evolved (covered in more detail later in this chapter), other methodologies have emerged in the fire service that have assisted in the development and application of a more comprehensive integrated risk management system.

THE DEVELOPMENT OF NATIONAL STANDARDS OF FIRE COVER

The world was about to go to war, and elected officials were scrambling to prepare for the onslaught that was sure to take place. The populace was being prepared for everything including terrorizing attacks, which had never before been seen. The year was not 2005; rather it was pre–World War II, and the preparations taking place were occurring in England. The German invasions of countries on the European and African continents had been viewed from afar but with alarming consequences. England knew it was to be next. What steps could be taken to protect the populace and the assets of the country? How would emergency services be positioned and prepared for terrorizing attacks from the air, potentially from within as acts of espionage, or from the sea, which had traditionally provided a natural block to invaders?

In 1936, Great Britain was very conscious that war with Nazi Germany was extremely likely and potentially imminent. In 1936 the Riverdale Committee formed to contemplate the effects of aerial attacks upon the United Kingdom, particularly in light of similar events in 1930 during the Spanish Civil War and the Japanese attacks on China. In these conflicts, deliberate aerial assaults on undefended towns and cities had occurred for the first time for no other reasons than to terrorize the occupants, break the morale of the nation, and overwhelm the civilian infrastructure. All this served as a backdrop as the United Kingdom was at war in 1939 and, by 1940, was under serious and sustained air assault.

The national standards recommended by the 1936 Riverdale Committee were further refined in postwar years. As a result of a comprehensive review in 1958, the development of international standards of response of cover was completed for Great Britain. The result has provided a mechanism for the British fire service over the course of the past 70 years to adopt a series of standards of cover dealing with a wide variety of conditions ranging from rural to urban settings.[2] This concept has evolved over time, with the most recent edition unveiled in a May 2004 report entitled *Integrated Risk Management Planning: The National Document*.

The new term, *integrated risk management planning (IRMP)*, seeks to move the traditional standards of cover approach one more step further in protecting the safety of not only citizens but also firefighters. It attempts to create a strategic methodology that not only answers the fire call but also seeks to intervene *before* the alarm is sounded. The additional component is addressing prevention in code enforcement and mitigation. The new concepts also anticipate a chemical, biological, radiological, nuclear, or explosive (CBRNE) attack occurring on British soil. How ironic considering the 2005 subway and bus bombings that took place in London. Perhaps the consequence of this new methodology was the ease and relative calm that appeared to prevail in the wake of those terrorist attacks. The emergency response to those events can be directly attributable to the new IRMP, which requires agencies not only to plan the response for anticipated situations before they occur but also to design standardized response to defined incidents on a national basis.

Think of this in comparison to the variations of response, training, and equipment existing in the American fire service. Local performance guided by various state training requirements and over 30,000 units of government has resulted in a kaleidoscope of response and capability. This body of knowledge has become the basis for the work that has been done by the Center for Public Safety Excellence (CPSE), the entity that oversees the fire service agency accreditation project. This research has led to the development of an American-based standard of response coverage document, which is being utilized today by many jurisdictions in the United States and Canada. In addition, we can see influence of this work in aspects of NFPA 1710 and 1720 in respect to response, reporting, and firefighter safety.

The history of deployment analysis from the 1920s until 1968 indicates little was done in respect to examining the response methodology within the American fire service other than updating the Board of Fire Underwriters (ISO) grading schedule. In 1968, the Rand Institute developed a research project to study the variables of response patterns and the community impact on fire station location. This included the review of the factors of both time and distance. The Rand studies were complex and often difficult for local government and fire service personnel not only to understand but also to articulate in a meaningful way to their local elected officials. Although several academic principles were expressed in these studies, at the time this research had little widespread impact either on the operational fire service or in the context of the insurance industry application at the local level.

During this same time frame, the International City/County Management Association (ICMA) began a series of exchanges with the insurance industry regarding the concerns expressed. ICMA's position was that the insurance industry criteria were antiquated and inconsistent with contemporary issues facing the local governmental fire service. It is from these discussions that a fire station location package was developed utilizing actual street networks and grids. The station location package, developed by the Public Technology Incorporated (PTI) organization, was reflective of the Rand studies methodology and was first made available to local governments in 1971. Although the system required a larger amount of computing data not readily available at the time, many communities subscribed to the service and conducted individual studies.

In the mid-1980s, the International Association of Fire Chiefs (IAFC) approached the ICMA regarding the creation of a more comprehensive approach to the evaluation of the local fire service. *In 1986, the IAFC signed a memorandum of understanding with the ICMA for the development of a concept of fire department self-assessment.* Over the course of the next decade, a working group of more than 250 people created a development process, resulting in the 1997 signing of an agreement between the IAFC and the ICMA to form the Center for Public Safety Excellence (CPSE), at that time known as the Commission on Fire Accreditation International (CFAI). With the publication of the first edition of the *Fire and Emergency Service Self-Assessment Manual* (**FESSAM**) and the implementation of the process, numerous fire agencies began to develop documentation for their departments in order to achieve fire agency accreditation. The development of this process has resulted in a significant amount of research in the methodology and application of resource deployment and how to define community risk.[3] The influence of standard of response and integrated risk management can also be seen in several aspects of NFPA 1710 and 1720 and the ISO grading schedule.

FESSAM
Fire and Emergency Service Self-Assessment Manual.

COMMUNITY RISK MANAGEMENT

The methods traditionally used by fire departments to assess risks and prepare to respond are applicable to all hazards.[4] Each community and/or jurisdiction should assess all of its risks and determine its ability to respond to them. Considerations in such an assessment follow:

1. *Life safety.* What is the hazard to life? What events threaten injury or death? What is the likelihood that the hazard will cause death or injury?
2. *Responder risk.* What is the potential risk to responders? What can be done to protect responders from harm? What are reasonable risks for firefighters to take? What steps or actions can be taken that would mitigate an event should it occur and at what cost?
3. *Property loss.* What is the community potential for property damage or loss?

LIFE SAFETY

The most important aspect of any risk assessment is assessing potential harm to individuals. Humans are constantly making choices, either conscious or unconscious, about risks they are exposed to as part of their everyday lives. Likewise, communities must make choices on how they will respond to the potential life safety risk in their jurisdiction. There are many examples of fire risk assessments evaluating things that pose a threat to life. For example, there are important life-safety concerns to address in hospitals, nursing homes, and other nonambulatory facilities. As a result, building and fire codes for such facilities are more restrictive than those for other types of facilities. Once a life-safety threat is identified, new codes are developed or existing codes modified accordingly. Local jurisdictions will not always be able to identify all the life-safety threats facing them and often use historical events or past instances in the adoption of new codes. Whereas an obvious risk will result in definitive action, the more subtle threats may or may not receive immediate attention. It is the responsibility of the fire service to serve as an expert, raising public awareness and enabling local policy makers to make informed decisions. Because the role of fire departments has expanded, there is now more to be done than just fight fires. But no matter what job is assigned to the fire department, it always has a life-safety element: there are lives to be saved. Think of the types of service your agency provides: EMS, special rescue, hazardous materials response, response to terrorist acts, and whatever else falls within the realm of emergency response. Regardless of the emergency, the fire service's first priority is to save lives.

Risk assessments related to life safety can be divided into two types: individual (or small-group) risks and potential mass-casualty events. Medical emergencies, vehicle accidents, and even residential fires present risks to individuals or small groups of people. These constitute the majority of incidents in most communities. Fire departments should be well prepared to respond to and function at these types of incidents.

Most fire departments handling EMS report that 50 to 80 percent of their responses are related to medical treatment. How does your department assess risks with respect to EMS? Has there been an evaluation of potential calls and a plan for the response? In many cases, the system has evolved without much forethought or planning. The choice for most fire station locations was because either the land was affordable or they allowed equipment to reach fires in accordance with insurance standards. However, because the same forethought does not usually enter into the equation when adding EMS response, this challenges the paradigms of any organization and indicates the importance of risk assessing and planning an integrated emergency response.

Although some medical calls are not easy to anticipate, many present an evident risk. Senior-living centers are a growth industry due to the general aging population but are a predictable source of increased call volume. Interstate highways, rural roads, and high-incident intersections are predictable locations of accidents and subsequent injuries, yet many departments do not calculate their impact on the deployment matrix. The ability to identify the major causes of EMS incidents in a community has become part of the risk assessment for departments. One challenge that fire departments must address is the ability to access and evaluate available data. Oftentimes, crash data are readily collected through police agencies; more often these data are not shared with the fire department.

Single-family residence hazards related to fire are an acknowledged risk, which most fire departments are staffed and equipped to address. Individual and small-group risk incidents often define the makeup and services provided by the local fire department. Although all communities should be able to handle these types of incidents with the resources they have, many cannot. Of course, there are things that communities can do with respect to building and fire codes enforcement, fire alarms, automatic sprinklers, and education.

Even though many fire departments have a base understanding of the common risks found in their community, incidents involving multiple or mass casualties, which present a much greater challenge to local departments, are often discounted because of the "it will not happen here" mentality. Many communities choose to ignore the possibility of these incidents until they happen and then fumble through the incident. Although it is not possible for all communities to prepare for these massive events or to have sufficient local resources necessary to mitigate all possible events, they should at least analyze the possibilities and have a response plan in place. These events can be humanmade or natural disasters. Incidents can be weather related, accidental, or terrorist acts. Since the events of September 11, 2001, virtually all communities have an awareness of the potential for terrorist incidents. The inclusion of critical infrastructure vulnerability and key resource analysis has now become part of risk assessment. Hurricane Katrina, on August 29, 2005, and Hurricane Rita, on September 24, 2005, once again displayed our vulnerability to respond effectively to large-scale natural disasters.

RESPONDER RISK

One risk to a community that is often ignored is the risk to response personnel. Think of the possible negative impact on a community if a firefighter is severely injured or, worse, killed in the line of duty. The effect on the department and the community is tremendous. Obviously, adding to the intangible damage done, there is tremendous financial and operational fallout. Departments must evaluate the risks in the community, plan how to respond to the risks, and ensure that firefighters or first responders have the necessary training and equipment to respond safely and effectively. It is not appropriate to send employees into situations for which they are not fully prepared.

Work that has been done in the United Kingdom resulting in the expansion of the traditional standards of cover to an integrated risk management plan approach showed that without mitigation and prevention, the safety of the firefighter and citizen would plateau. It is through aggressive prevention and **mitigation** that safety can truly be impacted. Once communities identify particular risks and hazards, they must prepare their employees accordingly. It is not acceptable to ignore firefighter safety when considering community risk management.

mitigation
Activities designed to reduce or eliminate risks to persons or property or to lessen the actual or potential effects or consequences of an incident. Mitigation measures may be implemented prior to, during, or after an incident. Mitigation measures are often developed in accordance with lessons learned from prior incidents. Mitigation involves ongoing actions to reduce exposure to, probability of, or potential loss from hazards. Measures may include zoning and building codes, floodplain buyouts, and analysis of hazard-related data to determine where it is safe to build or locate temporary facilities. Mitigation can include efforts to educate governments, businesses, and the public on measures they can take to reduce loss and injury.

The events of September 11, 2001, have raised the level of the public's awareness of the dangers first responders face. Aside from the obvious deaths and injuries that occurred on that day and during the immediate search and rescue action, there have been longer-term health impacts and detrimental effects on the psychological welfare of the workforce. Such situations have a negative impact on the departments and the jurisdictions involved and often have a long-term effect on service levels, recovery efforts, and the economic health of the community.

Recognizing the unusual and unprecedented nature of these types of events, it would have been difficult (at least up until that point in time) to have anticipated the magnitude of 9/11 and made necessary and reasonable preparations. However, many risks can be anticipated, and consideration given to the safety of responders must become engrained in our thought process in how we manage emergency incidents.

The development of standards such as the National Fire Protection Association (NFPA) 1500 (Standard of Fire Department Occupational Safety and Health Program), NFPA 1710 (Standard for the Organization and Deployment of Fire Suppression Operations, Emergency Medical Operations, and Special Operations to the Public by Career Fire Departments), NFPA 1720 (Standard for the Organization and Deployment of Fire Suppression Operations, Emergency Medical Operations, and Special Operations to the Public by Volunteer Departments), and related standards are designed to reduce the risk of harm to firefighters and other responders. As risks are identified, threats to responders must be addressed to minimize the impact. The analysis of risks faced by firefighters and first responders must become part of the community-base model in respect to the effectiveness of the actions taken, programs instituted, and other risk reduction efforts that have been implemented.

The Integrated Risk Management Planning Model, released in 2004 by the United Kingdom, clearly states the risk to responders reaches a plateau that will not be further impacted unless mitigation and prevention are made priorities before incidents occur. Failure to conduct and implement preplans and risk assessment strategies can be evidenced by the outcome of efforts during Hurricane Katrina. The Federal Highway Administration, utilizing data from asset management research conducted in Australia and New Zealand, has found that for every $1 spent on prevention, $4 can be saved from reconstruction on roadways. Hurricane Katrina's multibillion-dollar cleanup will far surpass earlier dollar estimates of preventative measures such as the strengthening of levies and other proactive steps.

PROPERTY LOSS

Much has been written about assessing the risk related to fire threats in terms of potential damage and fire department deployment. The Insurance Services Office's (ISO) Fire Suppression Rating Schedule, NFPA Standards 1710 and 1720, and the Center for Public Safety Excellence (CPSE) assess fire risks and community response capabilities. Although each has a slightly different methodology, collectively they focus attention on the need for a comprehensive analysis in respect to community risk and deployment of resources. However, the risks to a community go beyond the immediate result of a disastrous fire or other catastrophic event. The community lives with the results long after the fire is extinguished or the incident is over. What happens when a building burns or another significant event occurs in your community? There is often the loss of jobs, the loss of tax revenue, the negative impact on neighborhoods where burned-out or damaged structures remain, the commitment of resources that follows any significant loss, and the psychological damage to those involved that cannot be easily measured.

A significant risk that has emerged in many communities is wildfires crossing into suburban and urban areas known as the urban interface. These fires have become modern-day conflagrations, as growth has interfaced with wildland and forested areas. There have been many instances, since the 1960s, of communities burning down as a result of urban interface fires. The impact of global warming and climatic changes has been noted as a contributing factor to this growing problem. As codes, prevention, and suppression efforts have improved, the frequency of such burnouts has declined. Yet many communities have continued to build in urban interface areas and continue to experience the incidence of heavy losses. Although such hazards are easy to identify, they are often more difficult to prevent due to the freedom of choice and lack of zoning and planned development. Many communities are constantly making efforts to reduce the risk of fire occurrence and improve response capabilities if a fire does occur.

Protecting against the risks to buildings is a major component of building codes. Are buildings constructed to withstand a fire? Is there built-in fire protection? Is the building segregated to keep the fire relatively small and within the capabilities of the fire department? Codes have been developed to cope with earthquakes, hurricanes, and other natural disasters. These codes are often geographically specific; California would be more prepared for earthquakes and Florida more prepared for hurricanes. The adoption of national (or international) codes in protecting communities is a critical element of community risk management. Some people still believe it is the insurance companies that exert the most influence on code compliance. However, the fire marshal and local building official remain the lead professionals in most localities for identifying and protecting against hazards in the community.

Communities expect safe buildings. Of course, single-family residences, which are subjected to much less stringent code compliance, are not always safe. There is a strong belief in this country that "a man's home is his castle"; therefore, only minimal governmental control over what goes on in private residences, including the monitoring of fire safety, is allowed. This may be a contributing factor as to why more than 80 percent of loss of life occurs in single- and multifamily dwellings.

RISK AND PLANNING

Today, the identification of community risks is not enough. Risk assessment is effective only when planning takes place as a result. This planning needs to occur on two levels: strategic and operational. Strategic planning involves a conscious effort to identify risks and design the community response to the risks. As the risks are identified, local governing bodies make policy choices concerning issues such as standards of coverage and level of service. Deployment policies are based upon risk and the community's expectation of service. There can be decisions to prevent or mitigate, respond or react, or just plain ignore. Laws, codes, regulations, and standards are designed to eliminate or minimize threats. If the prevention and mitigation efforts do not work, a response is necessary. For a safe and adequate response, sending sufficient resources to the scene of an emergency in a defined period of time to affect a positive outcome is the goal. With the recent increase in terrorist acts (and threats of such acts), governments at all levels are taking action to prevent or mitigate such acts. For example, airports and mass transportation providers are increasing security and changing the way they do business in the interest of providing a safe environment for air travel. Some of their actions are mandated by the federal government and others result from national, state, or local initiatives. Today risk planning and community risk assessment has taken on a broader meaning using more precise measurement to determine success.

THE PHYSICAL PROTECTION OF CRITICAL INFRASTRUCTURES AND KEY ASSETS

On July 16, 2002, President Bush issued the *National Strategy for Homeland Security*, a strategy for mobilizing and organizing our nation to secure its homeland from terrorist attacks.[5] It communicates a comprehensive approach "based on the principles of shared responsibility and partnership with Congress, state and local governments, the private sector, and the American people"—truly a national effort, not merely a federal one. The *National Strategy for Homeland Security* defines homeland security and identifies a strategic framework based on three national objectives. In order of priority, these are (1) preventing terrorist attacks within the United States, (2) reducing America's vulnerability to terrorism, and (3) minimizing the damage and recovering from attacks. The *National Strategy for Homeland Security* aligns homeland security efforts into six critical mission areas:

Homeland Security Critical Mission Areas

Intelligence and warning
Border and transportation security
Domestic counterterrorism
Protecting critical infrastructures and key assets
Defending against catastrophic terrorism
Emergency preparedness and response

critical infrastructures
Systems and assets, whether physical or virtual, so vital to the United States that the incapacity or destruction of such systems and assets would have a debilitating impact on security, national economic security, national public health or safety, or any combination of those matters.

The *National Strategy for the Physical Protection of Critical Infrastructures and Key Assets* (the *Strategy*) takes steps to reduce the nation's vulnerability by protecting its **critical infrastructures** and key assets from physical attack. It identifies a clear set of national goals and objectives and outlines the guiding principles toward efforts to secure the infrastructures and assets vital to national security, governance, public health and safety, the economy, and public confidence. It also provides a unifying organizational structure and identifies specific initiatives to drive national protection priorities and resource allocation processes. Most important, it provides a foundation for building and fostering the cooperative environment in which government, industry, and private citizens can perform their respective protection responsibilities more effectively and efficiently. This *Strategy* recognizes the many important steps that public and private entities across the country have taken in response to the World Trade Center and Pentagon attacks on September 11, 2001, to improve the security of their critical facilities, systems, and functions. This *Strategy* provides direction to the federal departments and agencies that have a role in critical infrastructure and key asset protection. It also suggests steps that state and local governments, private sector entities, and concerned citizens across America can take to enhance the collective infrastructure and asset security.

> The United States will forge an unprecedented level of cooperation throughout all levels of government, with private industry and institutions, and with the American people to protect our critical infrastructure and key assets from terrorist attack.
>
> *National Strategy for Homeland Security*

A NEW MISSION

Prior to September 11, the normal model used for risk assessment by the fire service referred to risk, hazards, and property value. Simply put, a community looked at the different risks it faced, normal natural threats, and what properties would be impacted. What properties would be crucial to the continued ability of government to meet the basic necessities of food, shelter, and water? A key question to identify was the frequency that these natural forces impacted the community. The final step was to look at how much monetary value or employment value would be lost given the different scenarios.

September 11 and the threat of **terrorism** not only changed the way government looked at its vulnerability but also changed the equation by which all prior models based their decision-making process. Terrorism events cannot be tracked with any frequency—rather they are designed to occur wherever and whenever. Terrorism events also targeted what would cause maximum loss of life in a single event or have the greatest psychological impact, such as the destruction of a national monument. In other words, the usual method of risk, hazard, and valuation was reversed with a key for terrorists being the value to create terror in the populace.

terrorism
Any activity that (1) involves an act that (a) is dangerous to human life or potentially destructive of critical infrastructure or key resources; and (b) is a violation of the criminal laws of the United States or of any state or other subdivision of the United States; and (2) appears to be intended (a) to intimidate or coerce a civilian population; (b) to influence the policy of a government by intimidation or coercion; or (c) to affect the conduct of a government by mass destruction, assassination, or kidnapping.

The September 11 attacks on the World Trade Center and the Pentagon demonstrated our national-level physical vulnerability to the threat posed by an enemy focused on mass-destruction terrorism. Given these realities, it is imperative to develop a comprehensive national approach to physical protection. Protecting America's critical infrastructures and key assets represents an enormous challenge. Our nation's critical infrastructures and key assets are a highly complex, heterogeneous, and interdependent mix of facilities, systems, and functions that are vulnerable to a wide variety of threats. Their sheer numbers, pervasiveness, and interconnected nature create an almost infinite array of targets for terrorist exploitation. To be effective, the national protection strategy must be based on a thorough understanding of these complexities as we build and implement a focused action plan.

HOMELAND SECURITY AND INFRASTRUCTURE PROTECTION: A SHARED RESPONSIBILITY

Protecting America's critical infrastructures and key assets calls for a transition to an important new national cooperative paradigm. The basic tenets of *homeland* security are fundamentally different from the historically defined tenets of *national* security. Historically, securing the United States entailed the projection of force outside its borders. We protected ourselves by "keeping our neighborhood safe" in the global, geopolitical sense. The capability and responsibility to perform this mission rested largely with the federal government.

The emergence of international terrorism within our borders has moved the frontline of domestic security to Main Street, U.S.A. Faced with the realities of the September 11 attacks, the mission of protecting our homeland now entails "keeping our neighborhood safe" in the most literal sense. Acting alone, the federal government lacks the comprehensive tools and competencies required to deliver the most effective protection and response for most homeland security threats. To combat the threat terrorism poses to our critical infrastructures and key assets, the homeland security strategy integrates the resources and capabilities of state and local communities and private sector entities that compose our national critical infrastructure sectors.

> Homeland security is a concerted national effort to prevent terrorist attacks within the United States, reduce America's vulnerability to terrorism, and minimize the damage and recover from attacks that do occur.
>
> *National Strategy for Homeland Security*

THE SIGNIFICANCE OF CRITICAL INFRASTRUCTURES AND KEY ASSETS

America's critical infrastructure sectors provide the goods and services that contribute to a strong national defense and thriving economy. Moreover, their continued reliability, robustness, and resiliency create a sense of confidence and form an important part of our national identity and strategic purpose. They also frame our way of life and enable Americans to enjoy one of the highest overall standards of living of any country in the world. When we flip a switch, we expect light. When we pick up a phone, we expect a dial tone. When we turn on a tap, we expect drinkable water. Electricity, telecommunications, and clean water are only a few of the critical infrastructure services we take for granted. They have become so basic in our daily lives that we notice them only when service is disrupted. When disruption occurs, we expect reasonable explanations and speedy restoration of service. The *National Strategy for Homeland Security* categorizes our critical infrastructures into the following sectors.

Critical Infrastructure Sectors

Agriculture
Food
Water
Public health
Emergency services
Government
Defense industrial base
Information and telecommunications
Energy
Transportation
Banking and finance
Chemical industry and hazardous materials
Postal and shipping

Critical infrastructures are "systems and assets, whether physical or virtual, so vital to the United States that the incapacity or destruction of such systems and assets would have a debilitating impact on security, national economic security, national public health or safety, or any combination of those matters."

USA Patriot Act

Together these industries provide the following.

Production and delivery of essential goods and services. Critical infrastructure sectors such as agriculture, food, and water, along with public health and emergency services, provide the essential goods and services Americans depend on to survive. Energy, transportation, banking and financial services, chemical manufacturing, postal services, and shipping sustain the nation's economy and make possible and available a continuous array of goods and services.

Interconnectedness and operability. Information and telecommunications infrastructures connect and increasingly control the operations of other critical infrastructures.

Public safety and security. Government institutions guarantee national security, freedom, and governance, as well as services that make up the nation's public safety net. The highly sophisticated and complex facilities, systems, and functions that compose the critical infrastructures consist of human, capital, physical infrastructure, and cyber systems. Each encompasses a series of key areas, in turn, essential to the operation of the critical infrastructures in which they function.

The Importance of Key Assets Key assets represent individual targets whose destruction could cause large-scale injury, death, and destruction of property, and/or could profoundly damage national prestige and confidence. Such assets and activities may not be vital to the continuity of critical services on a national scale, but an attack on any one of them could produce, in the worst case, significant loss of life and/or public health and safety consequences. This category includes such facilities as nuclear power plants, dams, and hazardous materials storage facilities. Other key assets are symbolically equated with traditional American values and institutions or U.S. political and economic power.

National symbols, icons, monuments, and historical attractions preserve history, honor achievements, represent the natural grandeur of the United States, and celebrate the American ideals and way of life. Such an attack would not or might not impact the ability to govern but, rather, it would cause fear and panic in the populace. The effects of this fear would cascade into other impacts with the ultimate goal of financially debilitating the country. The airline industry and many tourist destinations are only now beginning to see pre-9/11 levels of attendance.

Challenges to Protecting Critical Infrastructures and Key Assets: The New Frontlines Our technologically sophisticated society and institutions present a wide array of potential targets for terrorist exploitation. Private industry owns and operates approximately 85 percent of the critical infrastructures and key assets. Facility operators have always been responsible for protecting their physical assets against unauthorized intruders. The unique characteristics of critical infrastructures and key assets, their continuing (often rapid) evolution, and the significant impediments complicating their protection will require an unprecedented level of key public and private sector cooperation and coordination. The United States has more than 87,000 jurisdictions of local governance. The challenge is to develop a coordinated and complementary system that reinforces protection efforts rather than duplicates them and that meets mutually identified essential requirements.

NATIONAL RESILIENCE: SUSTAINING PROTECTION FOR THE LONG TERM

The nation's critical infrastructures are generally robust and resilient. These attributes result from decades of experience gained from responding to natural disasters, such as hurricanes and floods and the deliberate acts of malicious individuals. The critical infrastructure sectors have learned from each disruption and applied those lessons to improve their protection, response, and recovery operations. Over time, residents of communities in areas that are persistently subjected to natural disasters become accustomed to what to expect when one occurs. Institutions and residents in such areas grow to understand the nature of catastrophic events, as well as their roles and responsibilities in managing the aftereffects. They are also familiar with and rely on trusted community systems and resources in place to support protection, response, and recovery efforts. As a result, they have confidence in their communities' abilities to contend with the aftermath of disasters and their ability to learn from each event (see Table 9.1).

TABLE 9.1 ◆ The Protection Challenge

Agriculture and Food	1,912,000 farms; 87,000 food-processing plants
Water	1,800 federal reservoirs; 1,600 municipal wastewater facilities
Public Health	5,800 registered hospitals
Emergency Services	87,000 U.S. localities
Defense Industrial Base	250,000 firms in 215 distinct industries
Telecommunications	2 billion miles of cable
Energy	
Electricity	2,800 power plants
Oil and Natural Gas	300,000 producing sites
Transportation	
Aviation	5,000 public airports
Passenger Rail and Railroads	120,000 miles of major railroads
Highways, Trucking, and Busing	590,000 highway bridges
Pipelines	2 million miles of pipelines
Maritime	300 inland/coastal ports
Mass Transit	500 major urban public transit operators
Banking and Finance	26,600 FDIC insured institutions
Chemical Industry and Hazardous Materials	66,000 chemical plants
Postal and Shipping	137 million delivery sites
Key Assets	
National Monuments and Icons	5,800 historic buildings
Nuclear Power Plants	104 commercial nuclear power plants
Dams	80,000 dams
Government Facilities	3,000 government owned/operated facilities
Commercial Assets	460 skyscrapers

These are approximate figures.

Our challenge at the local level is to identify, build upon, and apply the lessons learned from the September 11 attacks and natural disasters, such as Hurricane Katrina, to anticipate and protect against future events that may impact the critical infrastructures and key assets. Can you identify the potential threat areas and risks in your own community and determine which preventive action is necessary and those for which response preparation needs to be made? In many communities, there still exists a mind-set of "we've always done it this way" or "it won't happen here." Fire service leaders must continue to challenge this complacency.

PROBABILITY AND CONSEQUENCES

As noted, jurisdictions assess risks based upon the potential frequency (probability of occurring) and the potential damage, should it even occur. For example, a terrorist act has a low probability, but the damage, should such an act occur, and the psychological impact to the country are potentially very high. This same outlook, regarding risk assessment, can also be applied to natural disasters. For example, tornadoes generally do not hit the same communities every year, but if they do strike, the damage can be great. Conversely, medical emergencies happen every day. The overall potential damage from medical emergencies to the community as a whole is not nearly as significant as that from a tornado or other natural disaster (though these individual incidents greatly affect those requiring the service). Organizations should be able to compare the potential frequency and potential damage of events that may affect their community. Communities of all sizes need to conduct this type of analysis. Often there is a knee-jerk reaction to a disaster. Tornado sirens are installed after a tornado devastates an area, fire codes get tougher after a major fire, Congress appropriates more funding for the war on terrorism after the World Trade Center and Pentagon attacks. There are many ways to look at risk management. Simply, it comes down to the chances of an event occurring and the damage that could result from the event (see Table 9.2).

For example, structure fires are relatively frequent in comparison to other types of incidents in a community, and the subsequent dollar loss, loss of irreplaceable items, and loss of business or jobs make the consequences of such fires high. In some communities, activation of automatic fire alarms is high probability with low consequence. Tornadoes and hurricanes may be infrequent but represent a large potential loss to life and property. Comparatively, a Dumpster fire may be high probability but have little consequence outside the fire response. With an understanding of the different levels of probability and consequences, proper strategic planning in respect to risk management can take place. If the only threat in your community is Dumpster fires, there might not be much need for a fire department. If structure fires are a daily event, the fire department should be appropriately staffed. Begin thinking of the threats matrix in your own community. What is your call volume, breakdown of calls, risk factors, critical and key assets, and current standards of cover or integrated risk management plan?

TABLE 9.2 ◆ Probability Matrix

High Probability High Consequence	High Probability Low Consequence
Low Probability High Consequence	Low Probability Low Consequence

COMMUNITY RESPONSE TO RISKS

If it is easy to identify risks, probabilities, and outcomes, then why are all communities not prepared? First of all, many communities and fire departments may not assess risks. Some communities and/or fire departments are not proactive but wait for problems to occur before taking any action. Others proceed as if nothing of significance will happen in their communities (just as individuals do, because fires happen to others, not us!). Some communities do not view risk assessment as a priority, just as some fire departments are virtually inactive in fire prevention programs. If a community is good at assessing the risks it faces, then why might it still not be prepared when something goes wrong? Certainly fire professionals and other risk managers can identify potential problems. Even if risks are identified, many communities do not take the necessary steps to invest in the needed resources to address the issue. Why? Because communities have to make choices, just as individuals do. The choices are often based upon financial considerations. The community may elect not to spend funds at the level needed to address a particular problem properly. This can be a conscious choice, or it can result from a situation in which the community does not have a sufficient tax base to fully address all the potential risks identified. Equally dangerous is the development of a "one size fits all" response. The level of deployed resources—both physical and mechanical—to handle a high-rise fire is and should be different than that for a 2,000-square-foot residential structure. The resources necessary to respond to a Dumpster fire should be different from the response to a five-story, wooden-frame warehouse fire in an old industrial neighborhood.

Another reason for lack of action may be local, state, or federal politics. Politics is about influence, decisions, and relationships. Response to risks may be based upon policy decisions that are the responsibility of the elected officials. Community risk management may not be a critical issue to those responsible for setting policy and budget plans. Again, think of how you make decisions as an individual regarding risk analysis; policy makers do the same thing. They consciously choose based upon their willingness or ability to pay and their perception of the risk factors in play based upon their knowledge of the situation. An example of this, on both the personal and community levels, is providing flood insurance. You may never need it, but when it does flood, it is too late to do anything about it. What risks are addressed in your community and which others are ignored? Why do you think this is so? The challenge to community risk management does not lie solely in the work necessary to assess the probabilities of an emergency event in a community, but in the political arena as well. As with so much of what is done for emergency response, it is the policy makers who determine the level of service and, in many instances, the policy makers may not be aware of the community risks. Only through active participation in the analysis of community risk can fire service agencies affect policy through education and strategic planning.

COMMUNITY RISK ASSESSMENT—FIRE SUPPRESSION

The evaluation of fire risks must take into account the frequency and severity of fires and other significant incidents.[6] Determining risk by analyzing the real-world factors in the service area is essential to the development of a workable five-year emergency services strategic plan. The risk assessment divided into four quadrants imposes different requirements for commitment of resources in each area.

The relationships between probability and consequence determine the needed distribution and concentration of resources. *Distribution is the number of resources*

TABLE 9.3 ◆ Probability and Consequence Matrix

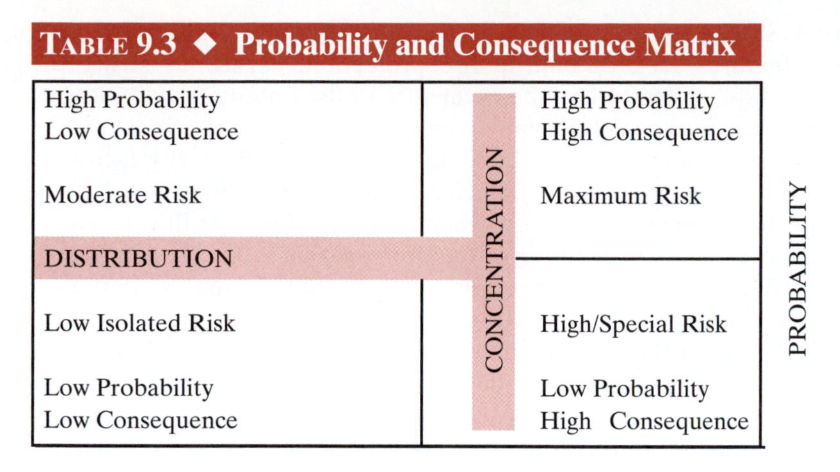

placed throughout a community. Concentration is the number of resources needed in a given area in the community. This varies depending on many factors including the number of events (calls) for service; the risk factors of the area; the availability, reliability, and time of arrival of secondary responding units; and so on. A challenge for the fire chief and local government is balancing the distribution and concentration of resources in a community to achieve the best results. If the *distribution* (fire stations and companies) is too low in comparison to *concentration* (resources, staffing), the outcomes will not be successful (see Table 9.3).

STANDARDS OF RESPONSE COVERAGE: INTEGRATED RISK MANAGEMENT PLANNING

A critical element in the assessment of any emergency service delivery system is the ability to provide adequate resources for anticipated fire combat situations, medical emergencies, and other anticipated events. Each emergency requires a variable amount of staffing and resources to be effective. Properly trained and equipped fire companies must arrive, deploy, and mitigate the event within specific time frames if successful emergency event strategies and tactical objectives are to be met.

Each event—fire, rescue operation, major medical emergency, disaster response, and other situations—requires varying and unique levels of resources. For example, controlling a fire before it has reached its maximum intensity requires a rapid deployment of personnel and equipment in a given time frame. The higher the risk, the more resources needed. More resources are required for the rescue of persons trapped within a high-risk building with a high-occupant load than for a low-risk building with a low-occupant load. More resources are required to control fires in large, heavily loaded structures than in small buildings with limited contents. Creating a level of service requires making decisions regarding the distribution and concentration of resources in relation to the potential demand placed upon them by the level of risk in the community. Each quadrant of Table 9.3 creates different requirements in the community for the commitment of resources.

The objective is to have a distribution of resources that is able to reach a majority of events, regardless of how insignificant they are, over the jurisdiction the companies protect in the service-level objectives established by their communities. Many factors make up the risk level that would indicate the need for a higher concentration of resources, such as the inability of occupants to take self-preserving actions, construction features, lack of built-in fire protection, hazardous structures, lack of needed fire flow, and nature of the occupancy or its contents. Evaluation of such factors leads to the number of personnel needed to conduct the critical tasks necessary to contain the event in an acceptable time frame. Although all risk factors have some common thread, the rationale of placing an occupancy within any risk assessment category is to assume the worst for structural protection. Fire flow is one such factor used as a risk assessment criterion or requirement that is based on defining the problem that will occur if the occupancy is totally involved, creating the maximum demand upon fire suppression services. The level of service provided by an agency should be based on the agency's ability to cope with the various types and sizes of emergencies that they can reasonably expect, after conducting a risk assessment. This process starts with looking at the most common community risk, the potential fire problem, target hazards, critical infrastructure, and a review of historic call data.

RISK ASSESSMENT MODEL

The risk assessment model incorporates the various elements of risk to the relationship among the community as a whole, the frequency of events that occur, the severity of potential losses, and the usual distribution of risks. Overall, the community is likely to have a wide range of potential risks; and, yes, there will be an inverse relationship between risk and frequency. In short, the daily event is usually the routine and results in minimal losses, whereas the significant events are less frequent. As we move up the chart toward the highest risk levels, the events are less frequent. If the risk management system is working in the community, a catastrophic loss should be an extraordinary event. Community-based management involves trying to keep routine emergencies from becoming serious loss situations. It is accomplished only when a comprehensive standard of response coverage integrated risk management plan has been developed, which provides the necessary resources for community risk (see Table 9.4).

TABLE 9.4 ◆ Community Risk Assessment Model

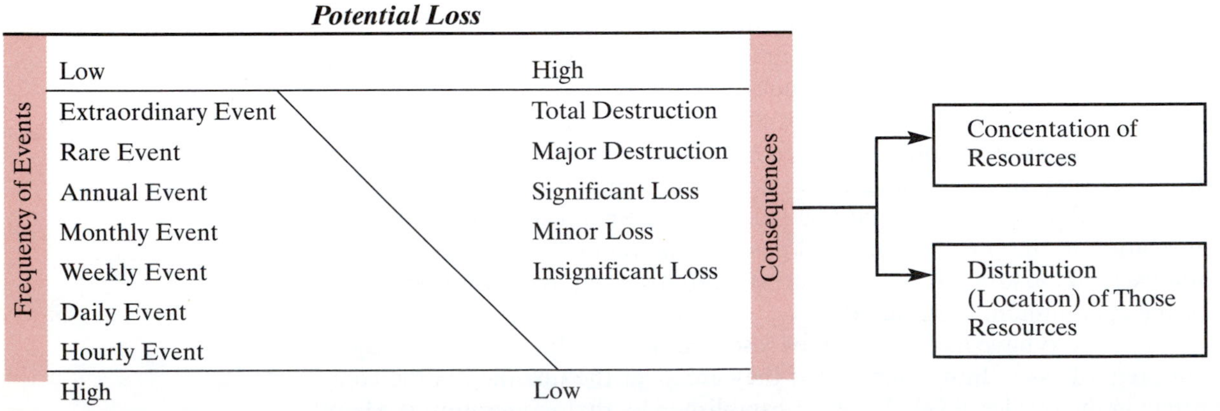

Source: Center for Public Safety Excellence.

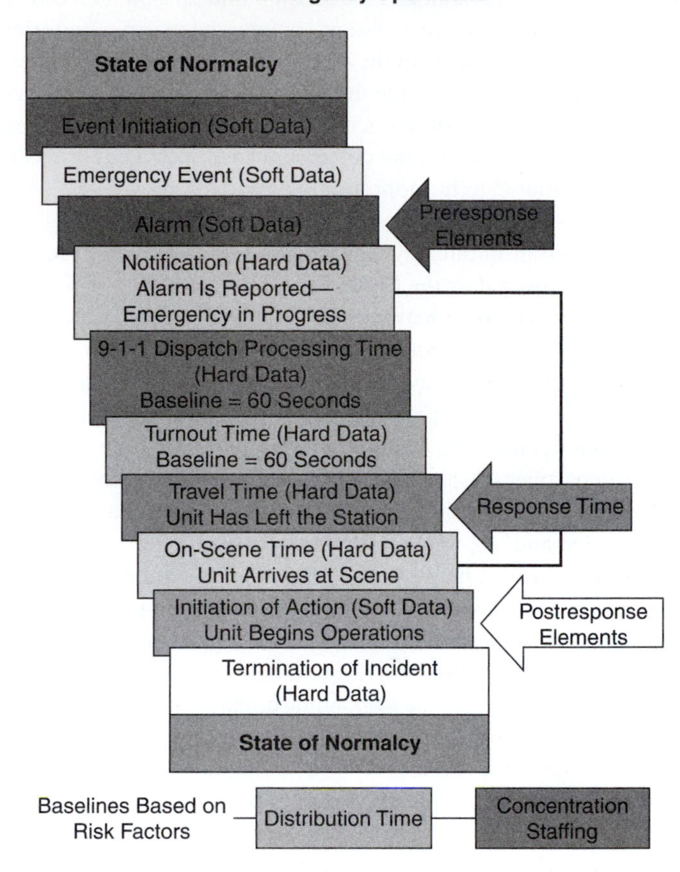

FIGURE 9.2 ◆ Cascade of Events Associated with Emergency Operations
(Courtesy of Randy R. Bruegman)

THE CASCADE OF EVENTS

In every emergency there is a sequence of events that are critical elements in respect to time and evaluation of the response system. Known as the cascade of events, it occurs on every emergency call[7] (see Figure 9.2).

The response performance continuum is composed of the following:

- *Event initiation point.* The point at which factors occur that may ultimately result in an activation of the emergency response system. Precipitating factors can occur seconds, minutes, hours, or even days before a point of awareness is reached. An example is the patient who ignores chest discomfort for days until it reaches a critical point at which he or she makes the decision to seek assistance.
- *Emergency event awareness.* The point at which a human being or technologic sentinel (i.e., smoke alarm, infrared heat detector, etc.) becomes aware that conditions exist requiring an activation of the emergency response system.
- *Alarm.* The point at which awareness triggers notification of the emergency response system. An example of this time point is the transmittal of a local or central alarm to a public safety answering point. Again, it is difficult to determine the time interval during

PSAP
Public safety answering point.

dispatch processing time
The portion of a fire department's response time that begins when the dispatcher receives an alarm and ends when the dispatcher assigns the proper companies to respond to the emergency. Also referred to as *alarm processing time.*

alarm processing time
The elapsed time from the receipt of an alarm by the dispatch center and the notification of specific fire companies that are to respond. May also be defined as dispatch time.

turnout time
The portion of response time when fire companies are donning personal protective clothing and boarding their apparatus. The time begins once the companies have been notified to respond and ends when they begin travel time (vehicle is moving).

which this process occurs with any degree of reliability. The *alarm transmission interval* lies between the awareness point and the alarm point. This interval can be significant, as when the alarm is transmitted to a distant commercial alarm monitoring organization, which then retransmits the alarm to the local 9-1-1 dispatch facility. When there is an automatic transmission of the signal, the fire department gains valuable time in controlling the event. An example of this situation occurs in many jurisdictions when 9-1-1 is called from a hard-wired monitored smoke detector, which often goes to a central answering point and is then rerouted to the appropriate dispatch center.

- *Notification.* The point at which an alarm is received by the public safety answering point (**PSAP**). This transmittal may take the form of electronic or mechanical notification received and answered by the PSAP.
- ***9-1-1 dispatch time.*** The time between the first ring of the 9-1-1 telephone at the dispatch center and the time the computer-aided dispatch (CAD) operator activates station and/or company alerting devices (also referred to as the **alarm processing time**). This can, if necessary, be broken down into two additional parameters: *call taker interval* (the interval from the first ring of the 9-1-1 telephone until the call taker transfers the call to the fire department dispatcher) and *dispatcher interval* (the interval from the time when the call taker transfers the call to the dispatcher until the dispatcher/CAD operator activates station and/or company alerting devices).
- ***Turnout time.*** The interval between the activation of station and/or company alerting devices and the time when the responding crew are aboard the apparatus and the apparatus is beginning to roll toward the call, as noted by the mobile computer terminal by voice to dispatch that the company is responding.
- ***Travel time.*** The point at which the responding apparatus signals the dispatch center that it is responding to the alarm and ends when the responding unit notifies the dispatcher that it has arrived on scene (again, via voice or mobile computer terminal notification).
- *On-scene time.* The point at which the responding unit arrives on the scene of the emergency.
- *Initiation of action.* The point at which operations to mitigate the event begin. This may include size-up, resource deployment, and so forth.
- *Termination of incident.* The point at which the unit(s) have completed the assignment and are available to respond to another request for service.
- ***Total response time.*** Alarm processing time plus turnout time plus travel time.

EVALUATING FIRE SUPPRESSION CAPABILITIES

travel time The portion of response time that is utilized by responding companies to drive to the scene of the emergency. Travel time begins when assigned fire companies actually begin to drive to the emergency and ends when they arrive.

Firefighters encounter a wide variety of conditions at each fire.[8] Some fires will be at an early stage, and others may already have spread throughout the building. This variation in conditions complicates attempts to compare fire department capability. A common reference point must be used so that the comparisons are made under equal conditions. In the area of fire suppression, service-level objectives are intended to prevent the flashover point, a particular point of a fire's growth that makes a significant shift in its threat to life and property. Fire suppression tasks required at a typical fire scene can vary a great deal. What fire companies must do, simultaneously and quickly, if they are to save lives and limit property damage is to arrive within a short period of time with adequate resources to do the job. Matching the arrival of resources within a specific time period is the objective of developing a comprehensive standard of cover integrated risk management plan.

THE STAGES OF FIRE GROWTH

Virtually all structure fires progress through a series of identifiable stages.

Stage 1. Ignition Stage. The ignition of a fuel source takes place. Ignition may be caused by any number of factors, from natural occurrences such as lightning to premeditated arson.

Stage 2. Flame Stage. The fuel initially ignited is consumed. If the fire is not terminated in this stage, it will progress to the smoldering stage or go directly to flashover.

Stage 3. Smoldering Stage. The fuel continues to heat until enough heat is generated for actual flames to become visible. It is during this stage that large volumes of smoke are produced and most fire deaths occur. Temperatures rise throughout this stage to over 1,000 degrees Fahrenheit in confined spaces, creating the hazard of back draft or smoke explosion. This stage can vary in time from a few minutes to several hours. When sufficient oxygen is present, the fire will progress to the free-burning phase.

Stage 4. Free-Burning or "Flashover" Stage. The fire becomes free burning and continues to burn until it has consumed all contents of the room of fire origin, including furnishings, wall and floor coverings, and other combustible contents. Research into the flashover phenomenon has yielded criteria that precisely measure when flashover occurs; however, any exact scientific measurement in the field is extremely difficult. Observable events that would indicate a flashover are "total room involvement" and "free burning." These indicators are clearly observable by firefighting personnel and the public and can be easily recorded and retrieved for future evaluation. Both scientific tests and field observations have shown that when flashover is experienced, it has a direct impact on fire protection and the ability of the emergency services system to control the fire.

total response time
The total elapsed time from the point of notification to a responding fire company and the arrival of that unit at the scene. Total response time equals notification, plus alarm processing/dispatch time plus turnout time plus travel time.

1. Flashover occurs at a temperature between 1,000 and 1,200 degrees Fahrenheit. These temperatures are well above the ignition points of all common combustibles in residences, businesses, and industries. When this temperature range is reached, all combustibles are immediately ignited. Human survival after this point is highly improbable without specialized protective equipment.
2. At the point of flashover, lethal fire gases such as carbon monoxide, hydrogen sulfide, and cyanide increase explosively. People exposed to these gases, even when not directly exposed to the fire, have drastically reduced chances of survival.
3. Flashover can occur within a relatively short period of time. Precisely controlled scientific tests indicate that it can happen in as little as two minutes from the flame stage. On the other hand, field observations of actual fires indicate that total room involvement can take as long as 20 minutes or more. There is no way to ascertain the time to flashover because it is not possible to determine when a fire started. We can nevertheless draw a correlation between flashover and the entire fire protection system. As suggested previously, the number of times that fires are controlled before flashover depends on the entire fire protection system and is not solely dependent on emergency response forces. Built-in fire protection, public education, extinguishment by citizens, and even the rate of consumption of fuel by the fire are all factors that affect flashover. Even when fires are not extinguished by firefighting forces, these personnel often provide other services, ranging from smoke removal to the restoration of built-in control

TABLE 9.5 ◆ Pre- and Postflashover

Preflashover	*Postflashover*
Limited to one room	May spread beyond one room
Requires smaller attack line	Requires larger, more attack lines
Search and rescue is easier	Compounds search and rescue
Initial assignment can handle	Requires additional companies

systems. The key point is that all components of the fire protection system, from public education to built-in fire protection to manual fire suppression, must be maintained and the performance of each evaluated.

Flashover is a critical stage of fire growth, as it creates a quantum jump in the rate of combustion, and necessitates a significantly greater amount of water to reduce the burning material below its ignition temperature. A fire that has reached flashover often indicates it is too late to save anyone in the room of origin, and a greater number of firefighters are required to handle the larger hose streams needed to extinguish the fire. A postflashover fire burns hotter and moves faster, compounding the search-and-rescue problems in the remainder of the structure, at the same time more firefighters are needed for fire attack (see Table 9.5).

The Significance of Flashover Staffing and equipment needs can be reasonably predicted for different risk levels and fire stages. The correlation of staffing and equipment needs in accordance with the stage of growth is one of the key determinant factors in the development of response coverage (see Figure 9.3). The goal is to maintain and strategically locate enough firefighters and equipment so that a minimum acceptable response force can reach a reasonable number of fire scenes before flashover and provide much needed care in critical medical emergencies.

SCENE OPERATIONS

fire flow required (or estimated)
The quantity of water that should be available for a period of two to three hours at a minimum pressure of 20 psi in a water distribution system. The estimated amount of water is calculated for individual structural conditions based upon a structure's size, construction type, fire loading, compartmentation, installation of built-in fire protection devices, and relationship to other exposed structures. May be compared to the available fire flow to determine relative ability to control specific problems.

The combination of property and life risk determines the fire ground tasks that must be accomplished to minimize loss. These factors, although interrelated, can be separated into two basic types: fire flow and life safety. Fire flow tasks are related to getting water on the fire; life-safety tasks are related to finding injured/ill persons and providing definitive emergency medical care, or trapped victims and removing them from the building. The **fire flow required (or estimated)** is based on the building: its size, structural material, distance from other buildings, horizontal and vertical openness (lack of partitions); and its contents, type, density, and potential energy (BTUs per pound). Life-safety tasks are based upon the number of patients in an emergency medical incident, or occupants, in a fire situation, their location (e.g., a low-rise versus high-rise), their status (awake versus asleep), and their ability to take self-preservation action. For example, ambulatory adults need less assistance than nonambulatory ones. The elderly and small children always require more assistance.

The key to a fire department's success at an emergency incident is coordinated teamwork, regardless of whether the tasks are all fire flow–related or a combination of fire flow, rescue, and life safety. A fire in an occupied residential single- or multifamily structure requires a minimum of eight tasks to be conducted simultaneously in order to stop the loss of civilian lives, stop further property loss, and minimize the

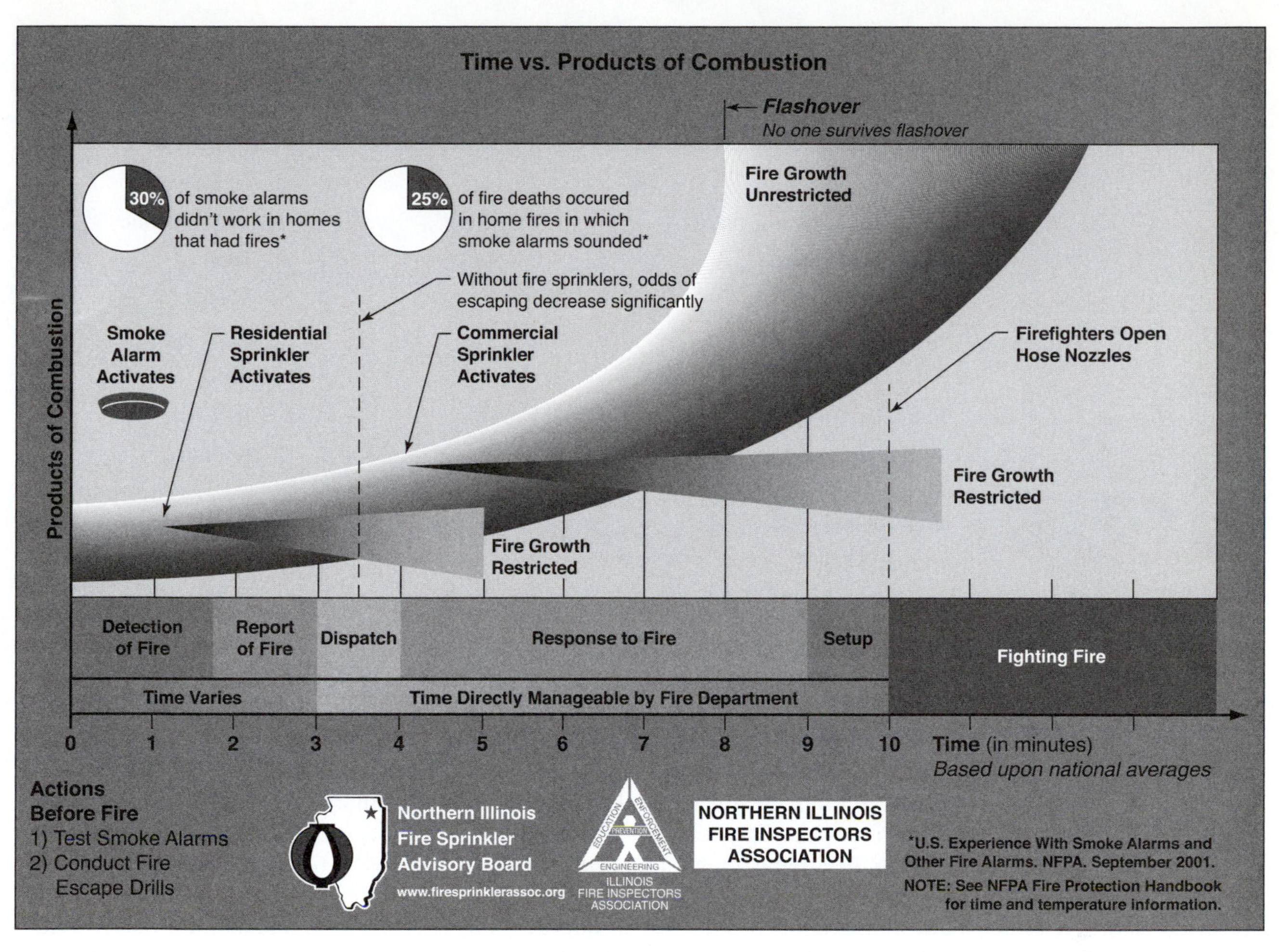

FIGURE 9.3 ◆ Time Versus Products of Combustion

risks to the firefighter. The number and type of tasks needing simultaneous action dictates the minimum number of firefighters needed at different types of emergencies. Table 9.6 is an example of these tasks, which usually are performed simultaneously in the majority of fire responses to our most prevalent risk: single- and multifamily dwellings. These tasks usually occur within the first 5 to 15 minutes of fire ground operations.

WHAT IS AN EFFECTIVE RESPONSE FORCE?

An **effective response force** is defined as the minimum amount of staffing and equipment that must reach a specific emergency in time to mitigate the emergency safely. Considering that the fire department cannot hold fire or other risks to zero or successfully resuscitate every patient, our response objective should find a balance among effectiveness, efficiency, and reliability that will keep community risk at a reasonable level and, at the same time, yield the maximum life and property savings. Any study should include a comprehensive review of safety for the first responder or firefighter.

effective response force
The minimum amount of staffing and equipment that must reach a specific emergency zone location within a maximum prescribed travel or driving time and is capable of initial fire suppression, EMS, and/or mitigation.

TABLE 9.6 ◆ Minimum Tasks Necessary at a Moderate-Rise Structural Fire

Task	*Number of Firefighters*	*Company Assigned*
Attack Line	2	1st Engine
Rapid Intervention Team	2	Truck/Engine
Search and Rescue	2	Truck
Ventilation	2	Truck
Backup Linc	2	2nd Engine
Safety Officer	1	Assigned
Pump Operator	1	1st Engine
Aerial Operator (optional depending on the incident)	1	Assigned
Water Supply	1	2nd Engine/WT
Command Officer	1	Battalion Chief
Command Aid (optional depending on the incident)	1	Assigned
Total Personnel	14/16	

response reliability
The probability that the required amount of staffing and apparatus that is regularly assigned will be available when a fire or emergency call is received; that is, the percentage of time that all response units are available for a dispatch. This is a function of the average amount of time that a fire unit is unavailable for a dispatch because it is already committed to another response. When a response unit is unavailable, the response time to an emergency in its first-due area will be longer, because a more distant unit will have to respond to the call. Response reliability is a statement of the probability that an effective response force may not be provided when a call is received.

Response Reliability **Response reliability** is defined as the probability that the required amount of staffing and apparatus will be available when a fire or emergency call is received. The department's response reliability would be 100 percent if every piece of fire department apparatus were available every time an emergency call were received. In reality, there are times when a call is received for a particular company but that company is already at another call. This requires a substitute (second-due) company to be assigned from another station, which often cannot respond within the established service-level objectives. As the number of emergency calls per day increases, so does the probability that a needed piece of apparatus will already be busy when a call is received. Consequently, the department's response reliability for that company and subsequently the entire system decreases. The size of the area that a station covers, the number of calls, the types of calls, and the population density all affect response reliability. The more densely populated, the more likely a second-due call will occur. An analysis of current response data can reveal variations in the response reliability among stations. The optimal way to track response reliability would be to analyze the total call volume for a particular fire management area and then track the number of double and triple calls to assess what the true response reliability is for that given area and the companies assigned to respond into the area.

EVALUATING EMS CAPABILITIES

From an emergency medical perspective, the service-level objective typically is a six-minute time frame, as brain damage is very likely at six minutes without oxygen.[9] However, in a cardiac arrest situation, survivability dramatically decreases

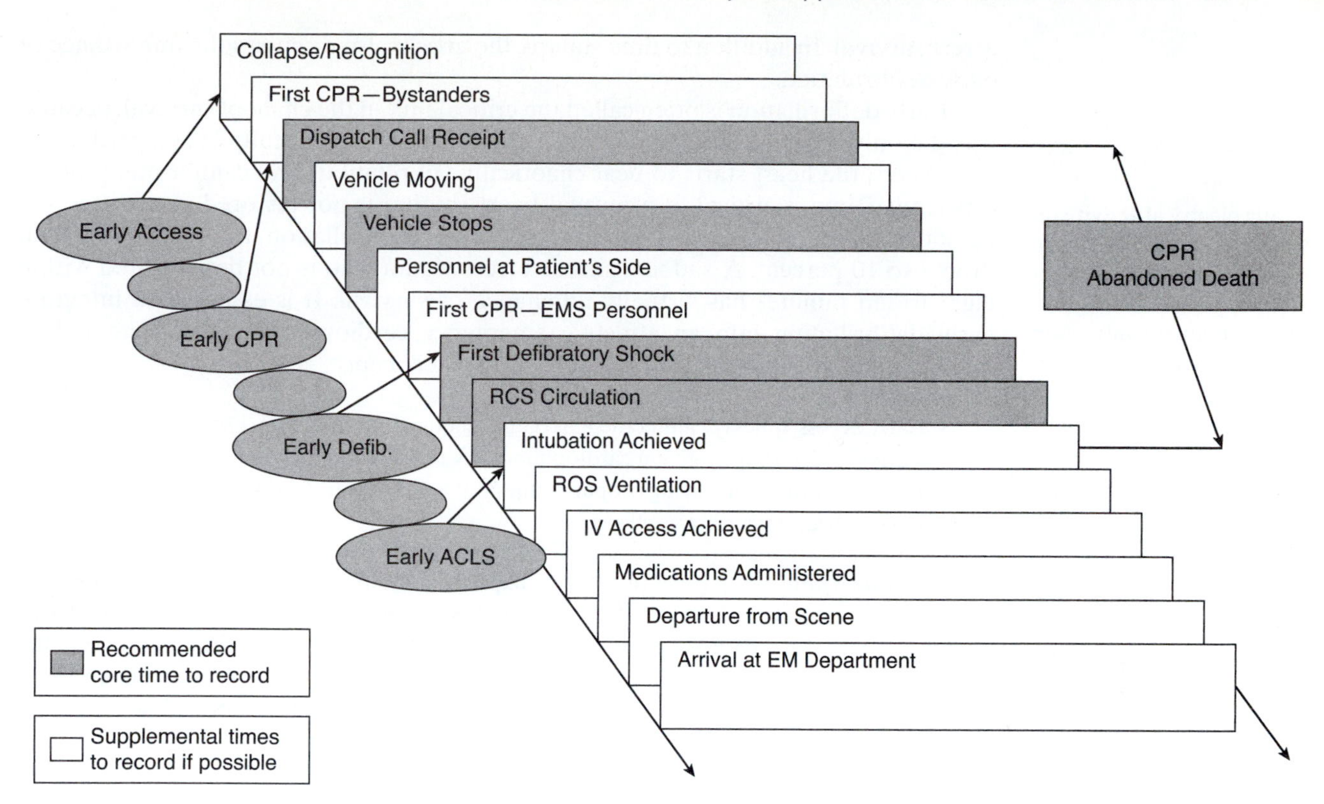

FIGURE 9.4 ◆ Events Associated with Cardiac Arrest Resuscitation Attempts
(Courtesy of Randy R. Bruegman)

beyond four minutes without appropriate intervention. Following a careful review of current medical literature related to emergency cardiac care, the American Heart Association, in concert with clinicians, administrators, and researchers, published a series of guidelines in the *Journal of the American Medical Association* (*JAMA*).

The research recommends using the **Utstein criteria** for outcomes research and capture of the following time stamps/points in the cascade of events in an EMS call that should be tracked (see Figure 9.4):

Utstein criteria
A recommended guideline for uniform reporting of data from out-of-hospital cardiac arrest.

- Time call for help is received
- Time emergency vehicle is dispatched
- Time first response vehicle stopped at scene
- Time cardiac arrest is confirmed (if EMDs began phone-directed CPR, time should be noted)
- Time bystander CPR is initiated
- Time of first defibrillation shock
- Time each treatment is provided
- Time patient arrived at Emergency Department
- Time death is declared

Recording the time stamps recommended by the *JAMA* will enable departments to present a more accurate picture of how changes in the EMS system affect cardiac arrest survival. In addition to time stamps, the article also discusses the importance of early defibrillation.

fibrillation
Erratic electrical activity of the heart. Atrial fibrillation, or erratic electrical activity of the upper chambers, causes irregular contraction of the lower chamber and palpitations, characterized by the feeling of galloping horses. Although palpitations can be disconcerting, atrial fibrillation is generally not a serious condition. Ventricular fibrillation results in inefficient pumping of the lower chamber and can cause death if it is not treated immediately.

Early defibrillation is often called the critical link in the chain of survival, because it is the only way to treat most sudden cardiac arrests successfully. When cardiac arrest occurs, the heart starts to beat chaotically (**fibrillation**) and cannot pump blood efficiently. Time is critical. If a normal heart rhythm is not restored in minutes, the person will die. In fact, for every minute without defibrillation, the odds of survival drop 7 to 10 percent. A sudden cardiac arrest victim who is not defibrillated within eight to ten minutes has virtually no chance of survival. It is essential to integrate early defibrillation into an effective emergency cardiovascular care system. This means employing the four-part chain of survival concept.

- Early access: quickly calling the emergency medical services (9-1-1) system
- Early CPR: promptly giving cardiopulmonary resuscitation when needed
- Early defibrillation: having proper equipment and being trained to use it when indicated
- Early advanced cardiovascular care[10]

An important evaluation point, not captured by many agencies, is the time that crews reach the patient's side. Often the clock stops when the vehicle arrives or stops at the address. Key to a successful outcome is this last piece of time, which occurs from the moment wheels stop to the point the patient is actually contacted. In high-rise communities or other large complexes, this time period can be substantial and most certainly affects the outcome of the intervention.

EVALUATION OF FIRE DEPARTMENTS AS A FACTOR IN PROPERTY INSURANCE RATING

Insurance Services Office (ISO)
A national organization that evaluates public fire protection and provides rating information to insurance companies. Insurers use this rating to set basic premiums for fire insurance.

Since the middle of the nineteenth century, U.S. property insurance companies have funded initiatives aimed at fire mitigation. In today's battle against fire losses, one of the insurance industry's most important tools is the Public Protection Classification (PPC™) program administered by the **Insurance Services Office (ISO).** ISO's PPC program evaluates a community's public fire protection capability and assigns a protection-class rating from 1 to 10. Class 1 represents exemplary fire protection; Class 10 means that the area's fire suppression program does not meet ISO's minimum criteria. Insurance companies use information from the PPC program to help market, underwrite, and price homeowners, business owners, and other types of commercial property insurance. Communities rely on the program to help plan, budget, and justify improvements or changes in their fire protection.

ISO (see Figure 9.5) is a leading source of information, products, and services related to property and liability risk. Working with communities, fire officials, and insurance companies, ISO maintains information on more than 45,000 fire protection areas. For a broad spectrum of commercial and personal lines of insurance, ISO also provides statistical, actuarial, underwriting, and claims information and analyses; property repair and replacement cost estimations; catastrophe modeling; policy language; information about specific locations and communities; fraud identification tools; and data processing services. ISO serves the insurance industry, regulators, and other government agencies.

FIGURE 9.5 ◆ ISO Logo
(Courtesy of Insurance Services Office)

EARLY HISTORY OF EVALUATING FIRE DEPARTMENTS

In the 1800s, insurers' primary reason for wanting to evaluate fire departments was concern over the threat of conflagration. At the turn of the twentieth century, the central areas of most cities consisted of wood-frame and ordinary wood-joisted masonry buildings, closely spaced, usually with heights of three to eight stories. Surrounding a core of retail and office buildings were buildings for wholesalers, jobbers, and various manufacturers. Because businesses, their employees, and their customers depended on public transportation, most commercial and manufacturing concerns were concentrated around the central part of the city, with only the very large or noxious industries located elsewhere.

At that time, piped water-supply systems were fairly new. Because of the cost of construction, systems were often inadequate to supply water for fighting fires in areas of predominantly combustible construction. Certain fire departments had a small number of steam-powered pumpers to boost water pressure for fire streams, but the majority of fire companies had only hose-carrying wagons. So, for fighting fires, communities relied mainly on hose streams supplied directly from relatively low-pressure hydrant systems. Other issues compounded this problem and included wooden water mains that could not support higher pressures (or that provided little pressure because of leaks), different hydrant types, and nonstandardization of threads on hydrants and appliances. With the combination of closely spaced combustible construction, buildings taller than effective hydrant hose streams could reach, and often inadequate water systems, a number of cities suffered major conflagrations.

In 1889, the National Board of Fire Underwriters (NBFU) hired a former fire department officer to examine the vulnerabilities of large jurisdictions to the threat of conflagration. The resulting reports included assessments of the needs of the associated fire departments and fire facilities. At first, the NBFU's representative used his own judgment to evaluate the available public fire protection. Later, to provide a more consistent evaluation, an NBFU committee developed a public fire protection rating schedule, which provided for five classes of protection. For each class, the schedule gave specifications for the water system and the fire department.

In 1904, the Baltimore conflagration dramatically focused attention on the susceptibility of cities to major fires. The possibility of another conflagration posed a serious threat to the financial stability of insurance companies. Cities were experiencing rapid growth, subject to little if any advance planning and controlled by almost no building or zoning laws. It was apparent to the insurers that, for purposes of planning and underwriting, they needed uniform information on the fire-loss characteristics of cities across the country. After the Baltimore conflagration, the National Board of Fire Underwriters assembled an engineering staff to survey the fire conditions of the major cities in the United States. Its reports covered water supply, fire department, fire alarm, fire-loss statistics, police department, building, electrical, and explosive and flammable substances laws and their enforcement, heating fuels used, temperatures, winds, width and condition of streets, location and description of structures, description of the most conflagration-prone blocks, and overall potential and probability of conflagrations in the central business districts and similar information.

First, the reports included recommendations for improvement of all major features of the municipalities' public fire protection. They also provided detailed underwriting information about the probability of major fire losses and especially about the potential of a conflagration in each of the cities visited. A second important purpose of the reports was to encourage each community to make improvements in structural conditions, in the control of occupancy hazards, and in the overall ability to control large fires. To further that purpose, the NBFU made a follow-up visit to each city within one or two years to evaluate progress on recommended improvements, which it then reported to its member insurance companies.

In preparing the reports, the NBFU made considerable efforts to analyze the reliability as well as the adequacy of the firefighting facilities. Equipment malfunctions were frequent and replacement or repair time was lengthy. Considering the poor structural conditions, a small fire could quickly grow into a conflagration if a fire alarm were delayed because of a malfunction or if the firefighting capabilities were reduced because of apparatus or water system failure.

In 1905, the NBFU developed a model building code for adoption by cities whose officials wanted to control the structural hazards of their developing municipalities. Also, in October 1905, the NBFU drafted a report on San Francisco. It read:

> Not only is the hazard extreme within the congested value district, but it is augmented by the presence of a compact surrounding great-height, large-area frame residential district, itself unmanageable from a fire-fighting standpoint by reason of adverse conditions introduced by topography. In fact, San Francisco has violated all underwriting traditions and precedent by not burning up. That it has not done so is largely due to the vigilance of the fire department, which cannot be relied upon indefinitely to stave off the inevitable.

Unfortunately, six months later, the San Francisco Earthquake caused the greatest conflagration in American history. The fire burned for two days, consuming 4.7 square miles and destroying 25,000 buildings.

A PERMANENT FIRE PROTECTION GRADING SYSTEM

The NBFU's original intent was to make only one complete survey of each major city. However, insurers and the communities both received the service so well that the organization decided in 1909 to make the program permanent. To quantify the information from the public fire protection reports, the NBFU then began developing a numerical classification system, representing the various features described in the reports. In 1916, the organization published its *Schedule for Grading Cities and Towns of the United States with Reference to Their Fire Defenses and Physical Conditions.*

The 1916 schedule evaluated both the adequacy and reliability of the water supply system, as well as the adequacy of the fire department and condition of its apparatus. The schedule established benchmark conditions for a "well-protected" municipality and assigned deficiency points when a municipality could not meet the benchmark. The schedule considered the features in Table 9.7.

The schedule called for additional deficiency points for unusual climatic conditions and for a large difference between the evaluations of the major features of water supply and fire department. Those points recognize the fact that even the best fire department will be less than fully effective if it has an inadequate water supply. Similarly, even a superior water supply will be less than fully effective if the fire department lacks the equipment or personnel to use the water. The emphasis in the 1916 schedule was on the ability to provide public fire protection to the central business district. The structural conditions section of the schedule contained an extensive construction analysis of the buildings in the central business district. The 1916 schedule evaluated a total of 236 items and subitems. Because of the extensive engineering evaluation required for the grading system, it soon became the means for considering public fire protection in the fire insurance rating procedure for individual properties.

KEEPING THE SCHEDULE UP TO DATE

After 1916, fire departments gradually changed from horse-drawn to motorized apparatus. To keep pace with the changes, the NBFU issued new editions of the schedule in 1922 and 1930.

During the 1930s, increased private and public use of automobiles, buses, and trucks resulted in a gradual development of commercial, warehouse, and light-manufacturing buildings in outlying parts of the cities. The 1942 edition of the schedule recognized the change by placing more emphasis on protection for areas beyond the

TABLE 9.7 ◆ ISO Schedule Features

	Percentage of Grading	*Deficiency Points*
Water Supply	34%	1,700
Fire Department	30%	1,500
Fire Alarm	11%	550
Police	1%	50
Building Laws	4%	200
Hazards	6%	300
Structural Conditions	14%	700
Total	100%	5,000

central business district. After World War II, the development of commercial buildings outside the central business district accelerated, and smaller mercantile businesses also moved out of the central business districts to escape increasing traffic congestion. The 1956 edition of the schedule and the 1964 amendments further recognized the growing importance of the areas of a city beyond the old central business district.

During the 1950s and 1960s many cities improved the capacity of their water systems in response to municipal growth and a general increase in the domestic use of water. Those improvements had a beneficial effect on the adequacy and, in some places, the reliability of systems for providing fire protection. Another social improvement that became a common fixture in homes and businesses was the telephone. This development considerably reduced the percentage of fires reported through street fire-alarm boxes. To accommodate the automobile in the central business districts, many cities removed older and less desirable buildings to provide parking space. The general trend to modernization led to removal and replacement of many other buildings. Fortunately, the buildings removed were usually of weaker construction, and their removal reduced the conflagration hazard. Today, few cities have extensive areas with a probability of serious conflagration. The riots that occurred in many older cities in the late 1960s dramatically illustrated the change. In no case, despite multiple simultaneous fires and often severely depleted firefighting resources, did a fire spread beyond the block of origin.

ISO TAKES RESPONSIBILITY FOR MUNICIPAL GRADING

In 1971, several national insurance rating and statistical organizations and a variety of regional and state fire rating bureaus consolidated their operations within ISO. As part of the consolidation, ISO became responsible for conducting municipal fire grading activities on behalf of the industry.

ISO's approach to grading recognized the rapid changes taking place in cities. The 1974 edition of the schedule applied by ISO evaluated central business districts the same as any other commercial districts in a city. To determine the necessary firefighting facilities for each city, the schedule determined an average needed fire flow for the actual built-up areas in the city. (Older schedules looked at a city's population to determine necessary facilities.) The schedule also reflected a change in emphasis—from conflagrations and the firefighting facilities needed to control them, to a concern for fires in individual buildings.

THE FIRE SUPPRESSION RATING SCHEDULE TODAY

In 1980, ISO introduced a new version of the schedule, known as the *Fire Suppression Rating Schedule* (FSRS), as the basis for its public fire protection classification activities countrywide. This version represented a significant departure from past practices in that it began referencing the relevant national standards promulgated by the National Fire Protection Association and the American Water Works Association. Generally, as these professional standards changed, the FSRS would prospectively recognize the most up-to-date criteria in its evaluations, without the need for continuous republication.

Using the FSRS, ISO develops a classification (ranging from Class 1 to Class 10) for each community. The classification represents the average class of fire protection for small to moderate-size buildings, which generally compose the vast majority of all buildings in nearly all cities. The classification is derived by totaling points assigned to various categories of performance involving alarm and dispatch (10 percent weight), fire department operations (50 percent weight), and water supply (40 percent weight). In 2001, a revision to the FSRS included criteria for the recognition of superior fire departments operating with limited water supplies (class 8B).

The following paragraphs explain the major items considered in a PPC evaluation of a community. For a complete description of the criteria used by ISO, visit www.isomitigation.com.

Items 410–414: Telephone Service. The schedule gives credit for each telephone line provided for fire department emergency and business service, up to the number needed. The number of lines needed depends upon the number of calls received. The schedule also gives credit for fire department listings (both emergency and business numbers) in convenient and conspicuous locations in local telephone directories.

Items 420–422: Operators. The schedule gives credit for each fire alarm operator, up to the number needed. The number of operators needed depends upon the total number of calls and the method of operation.

Items 430–432: Dispatch Circuits. Fire departments need adequate means for notifying personnel of the location of fires. The schedule gives credit for the availability and reliability of an alerting system to notify firefighters expected to respond. (Separate rules apply for single-station departments with on-duty personnel receiving alarms directly at the station.)

Items 510–513: Engine Companies. The schedule gives credit for each in-service pumper, up to the number needed. The number of pumpers needed depends on building fire flows, response distances, and method of operation. The amount of credit for each engine company depends on the equipment the company carries. Equivalent tools are recognized for some of these equipment items. A list of these tools can be found on the ISO web site at www.isomitigation.com/ppc/3000/ppp3007.html.

Items 520–523: Reserve Pumpers. The schedule gives credit for pumpers in reserve.

Items 530–532: Pump Capacity. The schedule gives credit for pump capacity of in-service pumpers, reserve pumpers, and pumps on other apparatus, up to the needed pump capacity. The needed pump capacity depends upon the fifth-largest needed fire flow for the community, not to exceed 3,500 gallons per minute (gpm).

Items 540–549: Ladder and Service Companies. The schedule gives credit for each in-service ladder and service company, up to the number of needed companies. The number of needed companies depends upon the height of buildings, the number of buildings with needed fire flows greater than 3,500 gpm, the response distances, and the method of operation. The schedule also gives credit for ladders, tools, and ladder-truck equipment normally carried on in-service apparatus for operations such as forcible entry, ventilation, salvage, and overhaul. Equivalent tools are recognized for some of these equipment items. A list of these tools can also be found on the ISO web site at www.isomitigation.com/ppc/3000/ppc3007.html.

Items 550–553: Reserve Ladder and Service Trucks. The schedule gives credit for ladder and service trucks in reserve.

Items 560–561: Distribution of Companies. The schedule gives credit for the area of the city within satisfactory first-alarm response distance of pumpers, ladder companies, and service companies.

Items 570–571: Company Personnel. The schedule gives credit for the number of personnel responding to first alarms.

Items 580–581: Training. The schedule gives credit for facilities used to train individual firefighters and companies; training at fire stations; training of officers, drivers, and recruits; and prefire planning inspections.

Items 610–616: Water Supply. The schedule gives credit for the available water supply at representative locations in the city. In every community the water supply works, the water distribution system, or the spacing of fire hydrants may limit the adequacy of the water supply. For areas where fire hydrants are not available, the schedule includes criteria for recognizing alternative water supplies provided by fire departments. For a fire department to receive recognition for an alternative water supply, it must provide 250 gpm of uninterrupted fire flow for a minimum of two hours (30,000 gallons of water).

The schedule allows credit for suction points, such as rivers, canals, lakes, wells, and cisterns. To be recognized as a water source such a suction point must have enough available water to satisfy the needed fire flow during freezing weather, floods, and the 50-year drought. (This is a drought with a 2 percent chance of happening in any one year). There must be an all-weather access road, and the fire department must have permission to use the water.

Items 620–621: Hydrants—Size, Type, and Installation. The schedule gives credit for the number of satisfactory hydrants installed.

Items 630–631: Inspection and Condition of Hydrants. The schedule gives credit for the frequency and completeness of hydrant inspections and for the condition of the hydrants.

Items 700–701: Total Credit and Divergence. This item develops a community's PPC by summarizing the credits developed in the "Receiving and Handling Fire Alarms," "Fire Department," and "Water Supply" sections of the FSRS. An inadequate water supply can limit the effectiveness of a fine fire department, and a poorly equipped and trained fire department cannot effectively use a plentiful water supply. Therefore, a community's preliminary FSRS score is subject to modification by a divergence factor, which recognizes any disparity in the effectiveness of the fire department and the water supply.

Items 800–802: Class 8B. Class 8B is for communities that provide superior fire protection services and fire alarm facilities but lack the water supply required for a PPC of Class 8 or better. To be eligible for Class 8B, a community must meet the fundamental requirements for a classification better than Class 9. The community must have an adequate number of well-organized and properly trained firefighters, reliable fire-alarm facilities, reliable fire apparatus with proper equipment, adequate fire station facilities, and operational records.

However, instead of providing a minimum fire flow of 250 gpm for two hours, the fire department must deliver an uninterrupted fire flow of 200 gpm for 20 minutes beginning within five minutes of the first-arriving engine company. The department must be able to provide the minimum fire flow to at least 85 percent of the built-upon areas of the community and score well in the "Receiving and Handling Fire Alarms" and the "Fire Department" sections of the FSRS.

Items 810–812: Class 9. Class 9 is for fire departments that lack a water supply for fire suppression meeting minimum criteria (250 gpm for two hours) and that have minimal fire suppression apparatus and equipment.

Individual Property Fire Suppression The Fire Suppression Rating Schedule provides separate rules for rating very large unsprinklered buildings that have a needed fire flow greater than 3,500 gpm. For such buildings, ISO determines the PPC by comparing the available protection with the protection needed for each building.

Items 1000–1003: Evaluation of Fire Department Companies. The schedule gives credit for each in-service engine and ladder company, automatic-aid engine and ladder company, reserve pumper and ladder truck, and outside-aid engine and ladder company up to the number of needed engine and ladder companies. The number of needed companies depends upon the needed fire flow for the subject building.

Items 1100–1101: Water Supply System. The schedule gives credit for the available water supply for the subject building. The water supply works, the water distribution system, or the spacing of fire hydrants may limit the adequacy of the supply system.

Items 1200–1211: Classification for an Individual Property. This item develops a PPC for a specific building by considering the credit for fire department companies or the credit for water supply system, whichever is lower. The PPC for the subject building is the same as for the city as a whole unless the PPC for the building is lower. In such cases, the poorer class (but not less than Class 9 if the city is Class 9 or better) applies to the subject building.

DETERMINING THE PPC FOR A COMMUNITY

In sum, ISO's evaluation of a community's fire suppression system includes a review of the dispatch center, fire department, and the water supply infrastructure. A community's strengths and/or weaknesses relative to specific criteria in each of those categories will determine the community's public protection classification. Communities can have different combinations of strengths and weaknesses yet still receive the same PPC. Therefore, the PPC number alone does not fully describe all the features and capabilities of an individual fire department.

Generally, the classification numbers suggest the following:

- Classes 1 through 8 indicate a fire suppression system with a creditable dispatch center, fire department, and water supply.
- Class 8B recognizes a superior level of fire protection in an area lacking a creditable water supply system. Such an area would otherwise be Class 9.
- Class 9 indicates a fire suppression system that includes a creditable dispatch center and fire department but no creditable water supply.
- Class 10 indicates the area's fire suppression program does not meet minimum criteria for recognition.

For many jurisdictions, ISO publishes a "split class," such as 6/9. In such jurisdictions, all properties within 1,000 feet of a water supply (usually a fire hydrant) and within five road miles of a fire station are eligible for the first class (Class 6 in the example). Properties more than 1,000 feet from a water supply (usually a fire hydrant) but within five road miles of a fire station are eligible for Class 9. All properties more than five road miles from a fire station are Class 10.

THE EFFECT OF PPC ON INSURANCE PREMIUMS

ISO provides insurance companies with public protection classifications and associated details, including fire station locations, response area boundaries, the location of hydrants, and other water supply details. But because insurance companies, not ISO, establish the premiums they charge to policyholders, it is difficult to generalize how an improvement (or deterioration) in PPC will affect individual policies, if at all.

However, ISO's studies have consistently shown that, on average, communities with superior fire protection have lower fire losses than communities whose fire protection services are not as comprehensive. Consequently, PPC plays an important role in the underwriting process at most insurance companies. In fact, virtually all U.S. insurers, including the largest ones, rely on PPC information for one or more of the following reasons: to identify opportunities for writing new business, to help achieve a reasonable concentration of property risks, and to decide on offers of coverage, deductibles, and pricing for individual homes and businesses.

Here are some general guidelines to help communities understand the benefits of improved PPC ratings for residences and businesses.

- PPC may affect availability and/or pricing for a variety of personal and commercial insurance coverages, including homeowners, mobile home, fine-arts floaters, and commercial property insurance (including business interruption).
- Assuming all other factors are equal, the price of property insurance in a community with a good PPC is lower than in a community with a poor PPC.
- The greater the change in PPC class, the more likely the premiums for individual policies will be affected, especially for commercial properties. Even when savings on a percentage basis appear to be small, the total savings for a community as a whole can be significant. A few percentage points of reduction can add up to substantial dollar savings, especially when extended over a period of years.
- Even without an overall improvement in the community's PPC class, individual policies can be affected significantly by the construction of new firehouses, the availability of recognized water sources, or the execution of recognized automatic aid agreements. How insurers apply these factors depends on the proximity of the insured property to the improvements.

HOW COMMUNITIES RELY ON THE PPC PROGRAM

The PPC program offers economic benefits—in the form of lower insurance premiums—for communities that invest in their firefighting services. And the program helps fire departments and other public officials as they plan for, budget, and justify improvements.

According to a 2003 survey of municipal fire officials conducted for ISO by Opinion Research Corporation, nearly 90 percent of fire chiefs and other officials interviewed said that in planning for, budgeting for, or justifying improvements or changes in their community's public fire protection, the effect of such changes on the PPC is important. Sixty-four percent of the respondents reported they had actually used the program that way in the last three years, and 76 percent said they planned to do so in the next three years.

Finally, many fire departments take advantage of ISO's program and information by accessing ISO's "Fire Chief On-line" web site. This secured web site offers information and features to help local officials improve their PPC. For example, the site contains an electronic version of the ISO Community Outreach Program questionnaire to facilitate faster ISO awareness of the latest local improvements in fire departments, water supply, and alarm systems. In addition, fire department officials can directly access valuable summary reports from ISO on specific commercial buildings in their jurisdictions that include details on needed fire flow, construction of walls, floors and roof, occupancy classification, and the existence and adequacy of fire sprinkler systems. Oftentimes, this information can be utilized in preparation of prefire plans on the respective commercial properties within the chief's jurisdiction. And, the site also contains an interactive electronic map of the respective fire department's response area—displaying fire station locations, response boundaries, and hydrants—all provided to allow for efficient and convenient updating of the protection details on file with ISO.

SUMMARY

Fire is the leading cause of loss for personal and commercial property insurance. There is a definite correlation between improved fire protection, as measured by the PPC program, and reduced losses. Insurance companies have recognized this correlation for more than a century. By offering substantial economic benefits to communities that invest in their public protection, the PPC program encourages improvements and helps communities plan for, budget, and justify expenditures that reduce property damage from fires; and better fire protection saves lives. In communities all over the country the PPC program helps fire departments do their most important job.[11]

THE CFAI

The *Commission on Fire Accreditation International (CFAI)* is a nonprofit entity formed in October 1988 through a Memorandum of Understanding (**MOU**) between the International City/County Management Association (ICMA) and the International Association of Fire Chiefs (IAFC). In 2006, the corporate name was changed to the *Center for Public Safety Excellence (CPSE)*. The CPSE oversees both the Commission on Fire Accreditation International (CFAI) and the Commission on Professional Credentialing (CPC) (see Figure 9.6).

MOU
Memorandum of understanding.

The mission of the CFAI is to assist fire and emergency service agencies throughout the world in achieving excellence through self-assessment and accreditation in order to provide continuous quality improvement and enhancement of service delivery to their communities. The CFAI provides a comprehensive system of fire and emergency service evaluation. In addition, it helps local government executives evaluate expenditures that are directly related to improved or expanded service delivery in the community and provides a nationally accepted set of criteria by which communities can judge the level and quality of fire, EMS, and other services they provide. The CFAI also offers a measurement tool to use in gauging the effectiveness of the fire and emergency service agency. Decisions on accreditation, general organizational operation, and special programs and activities for the CFAI are made by a commission that also oversees revisions to the accreditation model and self-assessment process, training, education, research, and development, as well as other issues related

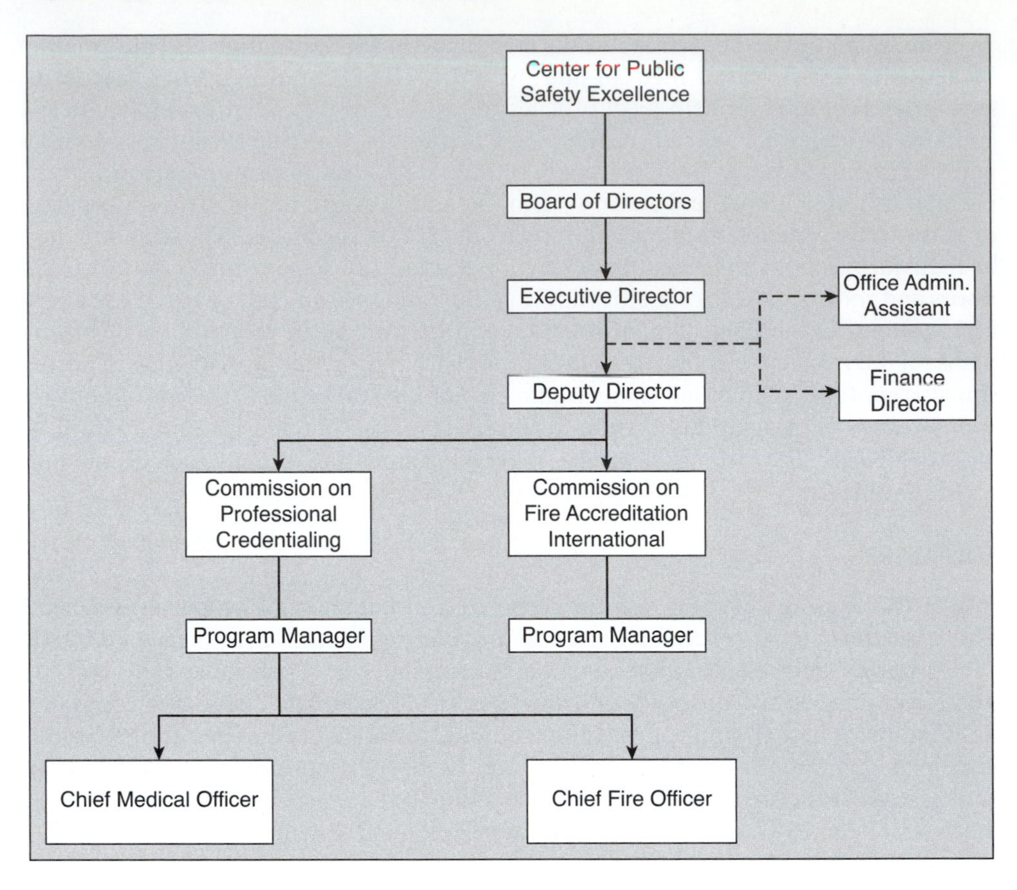

FIGURE 9.6 ◆ CPSE Organizational Chart

to self- and peer assessment. Eleven CFAI commissioners represent a variety of fire service stakeholders (see Figure 9.7).

Evaluating fire and emergency service agencies is critical today for all governmental functions. By doing so, we provide for periodic organizational evaluation to ensure effectiveness, help to manage change, and raise the level of professionalism within the

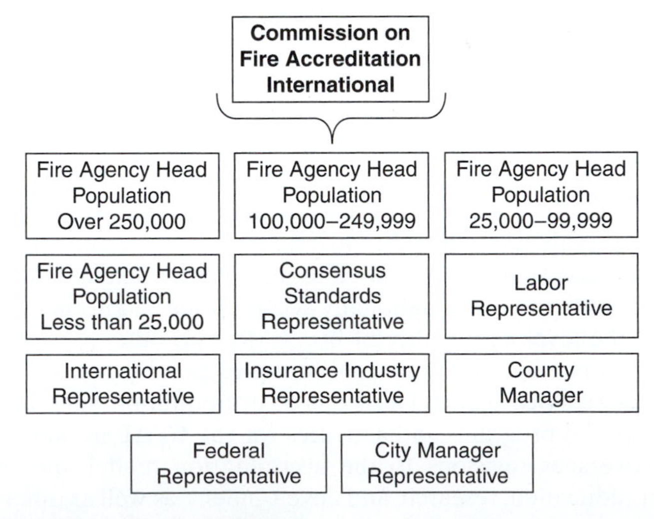

FIGURE 9.7 ◆ CFAI Organizational Chart

organization and the profession. The CFAI offers an excellent self-assessment process for fire and emergency service agencies that is designed to answer three very basic questions:

1. Is the organization effective?
2. Are the goals, objectives, and mission of the organization being achieved?
3. What are the reasons for the success of the organization?

Program evaluation focuses on identifying the efficiency and significance of the organization's activities. It seeks to answer these questions:

- Is the organization producing benefits that justify the expenditures?
- Are there better ways to achieve the organizational objectives?
- Are performance measures used to quantify the effectiveness of departmental programs?
- Are the goals and objectives being achieved with the appropriate allocation of resources?
- Will the achievement of the goals contribute to providing a quality and stated level of service for the area being protected?
- To what extent are the activities of the organization making an impact on the stated purpose (mission) of the organization?

The accreditation model includes 10 categories used in the evaluation of departmental performance: governance and administration, assessment and planning, goals and objectives, financial resources, programs, physical resources, human resources, training and competency, essential resources, and external systems relationships.

PERFORMANCE INDICATORS

Within each category there are criterion and performance indicators. A **criterion** is a measure or a set of measures or an index on which a judgment may be based to evaluate performance. Criterion defines a major area within each category of the self-assessment. A performance indicator is the desired level of achievement toward a given objective, defining the ability to demonstrate doing a particular task as specified in the accreditation process. There are 244 **performance indicators** in the self-assessment process, 77 of which are core competencies. Here is an example of criterion and related performance indicators from Category 1 Governance and Administration.

criterion/criteria
A measure or index (or set of measures or indexes) on which a judgment or decision may be based to evaluate performance or define a major area within each category. *Criterion* is singular, the plural is *criteria.*

performance indicators
The desired level of achievement toward a given objective and the ability to demonstrate doing a particular task as specified in the accreditation process.

Criterion 1A: Governing Body The governing board and/or agency manager has been legally established to provide general policies to guide the agency, approved programs and services, and appropriated financial resources.

Performance Indicators

1A.1 The agency has been legally established.

1A.2 The governing authority having jurisdiction over the fire service organization or agency periodically reviews and approves programs and ensures compliance with basic governmental as well as agency policies.

1A.3 There is a method that utilizes qualifications and credentials to select the agency's chief fire officer.

1A.4 The governing body approves the administrative structure that carries out the agency's mission.

1A.5 The governing body has policies to preclude individual participation of governing board members and staff in actions involving possible conflict of interest.

1A.6 There is a communication process between the governing body and the administrative structure of the agency.

1A.7 The role and composition of various policy making, planning and special purpose bodies are defined in a governing body organization chart for the authority having jurisdiction.

OBJECTIVE OF THE CFAI PROGRAM

The objective of this program is to provide an accreditation system that will improve the ability of a fire agency and the community to recognize and understand their respective community risks and associated emergency protection needs. The system has proven to improve the local fire agency's resources and emergency service delivery systems. The program is designed to improve the quality of life in the communities served by the fire and emergency service organizations and to give recognition for quality service, and is a mechanism in the design of a plan for improvement.

The CFAI program defines a model accreditation system that is credible, realistic, usable, and achievable. It is designed to be used by fire service agencies, city administrators, city/county managers, and elected officials to evaluate community fire risks using state-of-the-art practices. Those communities that have utilized the self-assessment process are successful in the development of policies that reduce fire/EMS risks and gain results in improved delivery of services. This program, a voluntary system, reflects a unique combination of fire protection engineering standards, community values, and an inventory of commonly used acceptable practices. *The project involves a "systems" model instead of a singular, stand-alone "pass or fail" document.* The ultimate goal of the program is to provide an accreditation process that will improve the ability of communities to recognize and understand their respective fire and life safety risks, provide a balance of public and private involvement in reducing the risk, and improve the quality of life for citizens within the communities using this model.

The primary purposes of a fire service agency are to prevent fires from starting; prevent the loss of life and property when a fire does occur; and provide means to evaluate a variety of other locally related services, such as emergency medical response, specialized rescue service, and public education efforts, to satisfy the needs of its jurisdiction and citizens. The traditional responsibility of a fire service agency has been to prevent fires and to suppress them if they should occur. Modern-day, progressive fire agencies (whether career, volunteer, or combination due to changes in society's demands) have expanded their services beyond basic prevention and suppression of fires to other services (see Table 9.8).

TABLE 9.8 ◆ Fire Agencies' Services

Emergency medical services	Hazardous materials response teams
Highly technical rescue services	Repair and maintenance facilities
Communications operations	Site plan review services
Building plan review processes	Public education programs
Weed abatement programs	Fire and arson investigation
Fire and arson investigation	Pre-emergency planning
Basic and advanced training programs	Preparation and response to terrorist events and natural disasters
Human relations program	

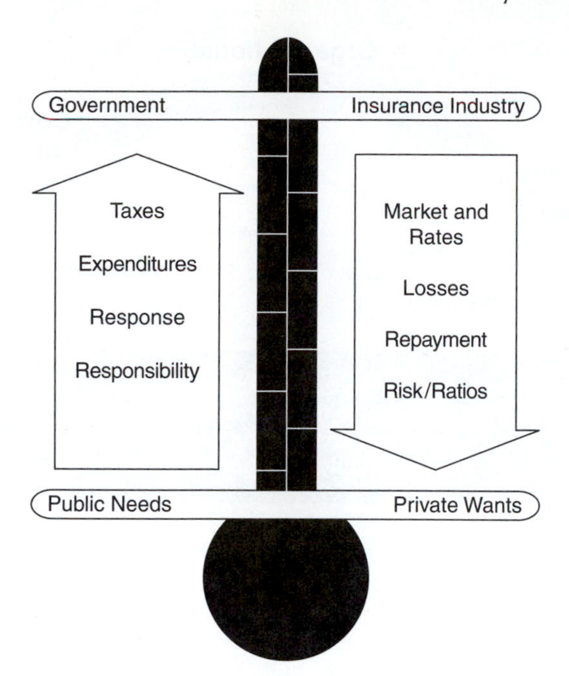

FIGURE 9.8 ◆ Government/Insurance Industry

EFFECTS ON LEVELS OF SERVICE

The resources devoted to controlling the community's risk factors are a combination of public and private funding sources. The protection of life and property from fire can be categorized as both a public responsibility and a community expectation. The relationship of government and the insurance industry in relation to the effects of private and public expenditures for fire protection is often overlooked. Whenever the public sector invests in higher levels of service, the risk to the insurance sector is reduced. Conversely, when the public sector decreases resources, the insurance sector and the public are at increased risk (see Figure 9.8). In every community the element of fire losses is three dimensional: life safety, responder risk, and property loss. The insurance industry focuses on the property loss aspect. Today's fire service organizations must focus on each dimension, or the true impact of risk cannot be quantified. The bridge between the public and private sector relationships is an important link when evaluating the effectiveness of a fire service organization (see Figure 9.8).

Like all other government agencies, the fire service is not an isolated organization within a government structure. What other departments do within a community dramatically affects the capability of the agency to accomplish its purposes. For example, the fire service agency should have input into the policy-making process to assure that other government operations do not adversely affect the fire service and conversely that the fire service does not adversely affect other civic operations. To evaluate the emergency response delivery system, the chief executive must assess the interrelationships of all the facets of fire protection and emergency medical services and the impact of actions taken on the ability of the local agency to deliver the stated services. Fire service organizations have an array of criteria that ranges from subjective requirements for measurements to mandated levels of performance. This CFAI model is highly dependent upon the dialogue between the fire service and community

FIGURE 9.9 ◆ Organizational Realities

leaders in establishing and maintaining programs, resources, and public safety policy. Often, elected officials and appointed fire officials in both fields find themselves at opposite ends of the pendulum in justifying or assessing resource allocation to reduce the fire losses, prepare for major emergencies, or provide for an expanded emergency service mission (see Figure 9.9).

Integrated risk modeling can help balance the scale in respect to interrelated issues that are taken into account when developing a deployment strategy. It is easy to understand the dilemma that many communities face today. The resolution of the community fire problem and other general service demands almost always involves a commitment of financial resources, either public or private. Although this can be viewed as either time or money, in all cases, it can be reflected in the community acceptance or rejection of specific practices. The issues facing community leadership must balance all the variables: level of risk, level of commitment, performance, and costs. If the level of risk is low and the commitment of resources is high, the system may not be efficient in terms of cost. If the level of risk is high and the commitment of resources is inadequate, the system may be cost efficient but dangerously ineffective. The objective of both fire and community leaders should be to find the balance of resource commitment to effectively control those risks for the community.

ASSESSING BENEFIT

Elected officials can consider a variety of factors in assessing the services provided by the fire agency. Among the indicators are fire losses; response times; percentage of time actually used in providing the principal services of the department such as fire prevention, inspection, pre-emergency planning, disaster response, and training and education of personnel; the number of complaints registered against the agency by citizens; the nature and value of fire losses; comparisons of data concerning the agency with that of agencies in other similar cities; and overall productivity of department personnel in principal fire department activities. In order for elected officials to understand the operations of their fire agency better, they must consider the local factors and peculiarities that affect the agency's ability to provide service. Among these items are the various federal and state laws and agreements governing its operations; the geographical area served by the agency and its unique characteristics, both natural and humanmade; the water distribution system serving the agency; staffing levels; the degree of fire prevention efforts, the relationship between the fire agency and other governmental departments; the type and condition of equipment used by the agency; and the various safety and personnel factors bearing on the day-to-day operation of the organization.

standards of response coverage
A written statement that combines service-level objectives with staffing levels to define how and when a fire agency's resources will respond to a call for service.

GIS
Geographical information system.

CREATING AND EVALUATING AN INTEGRATED RISK MANAGEMENT PLAN: STANDARDS OF RESPONSE

Standards of response coverage can be defined as those written policies and procedures that establish the distribution and concentration of fixed and mobile resources of an organization. The accreditation process requires an agency to have an adopted system that defines risk categories within the response area of that jurisdiction based upon that risk and the anticipated workload. As such, a written policy statement should be developed that defines the service-level objectives for emergencies such as fires, medical calls, and other anticipated emergencies. An original model developed by the CFAI was based upon building occupancy. Because of the range of complex issues faced today by the fire service, it was necessary to expand the model to include critical infrastructure, key assets, national disaster modeling, integration of deployment strategies, and the use of geographical information systems (**GIS**).

low risk
Small commercial structures that are remote from other buildings; detached residential garages and outbuildings.

moderate risk
Built-up areas of average size, where the risk of life loss or damage to the property in the event of a fire in a single occupancy is usually limited to that occupancy. In certain areas, such as small apartment complexes, the risk of death or injury may be relatively high. The moderate/typical risks are often the greatest factor in determining fire station locations and staffing due to the frequency of emergencies in this category. To assure an equitable response and to provide adequate initial attack/rescue capability to the majority of incidents, the typical risk is often used in determining needed resources.

RISK ASSESSMENT MODEL

The risk assessment model assumes that in every community the ratio of risks will be different. In general, it is anticipated that in most communities the vast majority of the risk will fall into the **low risk** and **moderate risk** categories with smaller percentages being distributed among the other probability quadrants. The majority of fire service concern should be directed toward the development of fire defense strategies for occupancies that fall into the high-probability and/or high-consequence categories. The development of an organizational strategy to achieve the desired service-level objectives is very important to the credibility of any fire organization (see Figure 9.10). The risk levels in one community may be based on structural conditions only. However, many other risks must be assessed as well. For example, an agency that has wildland firefighting responsibilities may define its risk on the basis of topography, geography, fuel cover, and weather conditions. An area with an urban-wildland interface may have a risk assessment that combines structural conditions with the existing or potential natural fuel load.

In order to translate the efforts of a fire service agency into terms that the public and policy makers in the community can evaluate, they must be defined in terms that

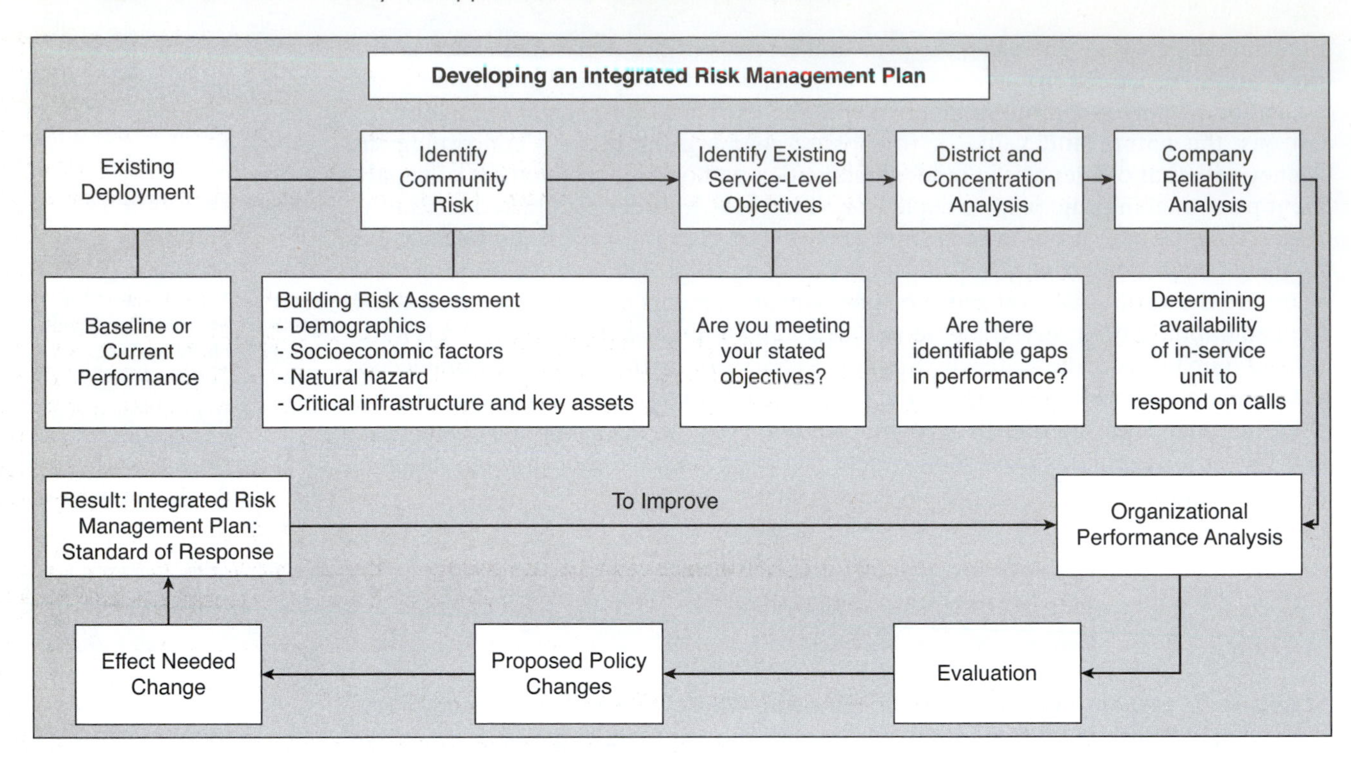

FIGURE 9.10 ◆ Developing an Integrated Risk Management Plan

level of service
The resources needed to meet the stated service-level objective(s). Level of service is defined only in terms of what resources are provided and not in terms of effectiveness or quality.

high hazard risk
Built-up areas of substantial size with a concentration of property presenting a substantial risk of life loss, a severe financial impact on the community, or unusual potential damage to property in the event of fire.

can be measured. A common term used in the evaluation of the fire service is **level of service**, which is defined as:

> The resources needed to meet stated service-level objectives. Level of service is defined only in terms of what is provided and not in terms of effectiveness or quality.

As the magnitude of emergencies ranges from small to catastrophic, the requirements for resources can vary greatly. An area of **high hazard risk** could require a timely deployment of more fire companies for several reasons. More resources are required for the possible rescue of persons trapped within a high-risk building with a high occupant load as compared to a low-risk building with a low occupant load. More resources are required to control fires in large, heavily loaded structures than are needed for small buildings with limited contents. Therefore, creating a level of service consists of the analysis made regarding the distribution and concentration of resources needed in relation to the potential demand placed upon them by level of risk.

STANDARDS OF MEASUREMENT

The following basic principles are considered to be the factors in developing a level of service. Developing a risk assessment consists of eight key elements.

Fire Flow. The amount of water required to control the emergency, which is based on contents and combustible materials.

Probability. The likelihood that a particular event will occur within a given period of time. An event that occurs daily is highly probable. An event that occurs only once a century is very unlikely. Probability is an estimate of how often an event will occur.

Consequence. Life safety (risk to the lives of occupants from life-threatening situations that include both fire and EMS), economic impact (the losses of property, income, or irreplaceable assets), and responder risk (risk to the emergency responders who are called to handle the incident).

Occupancy Risk Assessment. An assessment of the relative risk to life and property resulting from a fire inherent in a specific occupancy or in generic occupancy classes.

Fire Management Areas. An area used to assist in defining the management of risk is typically defined in terms of geographical boundaries, referred to as **fire management areas**. The zones are often representative of a station's first-in district.

fire management areas
A geographic area of a jurisdiction that is classified according to one or more risk categories. The size and classification of a fire analysis area is usually based upon either a specific area or a building.

Community Profile. The overall profile of the community based on the unique mixture of demographics, socioeconomic factors, occupancy risk, fire management areas, and the level of services currently provided.

Distribution. The station and resource locations needed to assure rapid response deployment to minimize and terminate emergencies. Distribution is measured by the percentage of the jurisdiction covered by the first-due units within adopted public policy, normally a service-level objective. Policies shall include benchmarks for intervention such as arrival prior to or at flashover and arrival at EMS incidents prior to brain death in cardiac arrest. From risk assessment and benchmark comparisons, the jurisdiction will use critical task analysis to identify needed resource distribution and staffing patterns based on the hazard, risk, value, call volume, and public policy direction.

Concentration. The spacing of multiple resources arranged so that an initial effective response force can arrive on scene within sufficient time frames to mobilize and likely stop the escalation of an emergency in a specific risk category. Such an initial response may stop the escalation of the emergency, even in high-risk areas. An initial effective response force is not necessarily the total number of units or personnel needed if the emergency escalated to the maximum potential. For example, if a building is preplanned for a worst-case fire flow of 4,000 gpm, it is possible that the jurisdiction plans an initial effective response force to provide the resources necessary to contain the fire to a reasonably sized compartment of origin. Additional alarms or units from farther away could be planned on, including mutual aid. If one level of risk is predominant in a fire management area or a community, then the initial effective response force should be planned for the predominant type of risk found and the historical response data, including fire loss and call demand.

DEVELOPING STANDARDS OF RESPONSE COVERAGE

hazard
Something that is potentially dangerous or harmful, often the root cause of an unwanted outcome.

To clearly define standards of coverage, agencies should have a statement of policy regarding how risks are categorized within the context of their own jurisdiction. Standards of response coverage must include an element of time—the maximum prescribed travel that indicates the level of service that is anticipated. Usually this is

risk
Exposure to a hazard based on the probability of an outcome when combined with a given situation with a specific vulnerability. The level of risk can be described as the probability of a specified loss over a given period of time. All structures, for example, are subject to destruction by fire; however, individual structures vary considerably as to the possibility of loss as a result of their construction, contents, and built-in protection.

benchmark
A benchmark is defined as a standard from which something can be judged. A benchmark is the best practice found in a specific service and/or product and will help define what superior performance is of a product, service, or process.

vulnerability
A measure of adverse consequence that might occur to a structure as a result of exposure to an uncontrolled fire. It is usually expressed as an indication of the difference between a level of risk and a level of service. For example, if a building has a calculated fire flow of 5,000 gpm and the level of service can deliver only 3,000 gpm, the structure is vulnerable to total loss unless the fire is controlled at the compartment level. Vulnerability is increased as the size and complexity of a risk exceeds the resources available to contain a fire to a limited level.

referred to in the service-level objective statements of the agency and is expressed in terms of a specific response by a specific period of time. Many factors will impact times, such as urbanization of the jurisdiction, population density, frequency and nature of service demand, available and adequate road network, and availability of resources for deployment.

RISK ASSESSMENT

The purpose of risk assessment is not only to evaluate risks and hazards in a fire department's response area but also to provide a basic methodology to evaluate existing response coverage. The process begins with the identification of community hazards and risks. **Hazard** is defined as a source of potential danger or an adverse condition. **Risk** is defined as the possibility of loss or injury; the exposure to the chance of loss; the combination of the probability of an event; and the significance of the consequence (impact) of the event: Risk = Probability × Impact. An evaluation system that is currently under development by the CFAI and several partner organizations groups risks into three separate categories that include property, life, and critical infrastructure. Each category contains subsets of risks/hazards relevant in determining overall community risk and vulnerability.

Once the details of risks/hazards are known for a community, then the community can design deployment of resources (or other activities, e.g., smoke detectors, public education, disaster planning, building/fire code amendments, etc.) either to manage the known risks or to respond and mitigate the emergency when an adverse risk event occurs. (Fires and medical emergencies are adverse risk events, as are natural or other disasters.)

Department leaders must provide sufficient information to the elected officials to determine (1) what resources to commit to risk management (prevention/preplanning/preparation), (2) what resources to commit to response/mitigation, and (3) what level of risk to accept. These concepts are built upon the existing basic infrastructure, the response capability, and the current level of community preparedness.

The information compiled regarding community risks/hazards and the resources committed for risk management and/or response/mitigation of risk events that occur can be analyzed and an overall community **vulnerability** score calculated. Integrated risk management evaluation will assist communities to achieve this objective (see Figure 9.11).

CFAI BENCHMARKS AND BASELINES OF PERFORMANCE

To establish a common benchmark/baseline for the purposes of evaluating response time criteria, the CFAI has the following times that should be incorporated into the development of an integrated risk management plan.

Response Time

- **A.** *Notification/9-1-1 dispatch processing time:* 60 seconds, 90 percent of the time is benchmark; 90 seconds, 90 percent of the time is the baseline.
- **B.** *Turnout time:* 60 seconds, 90 percent of the time is benchmark; 90 seconds, 90 percent of the time is the baseline.
- **C.** *Travel time:* Based on criteria for the different risk categories and within guidelines provided for service area and/or population density.

Total response time = *A* + *B* + *C*

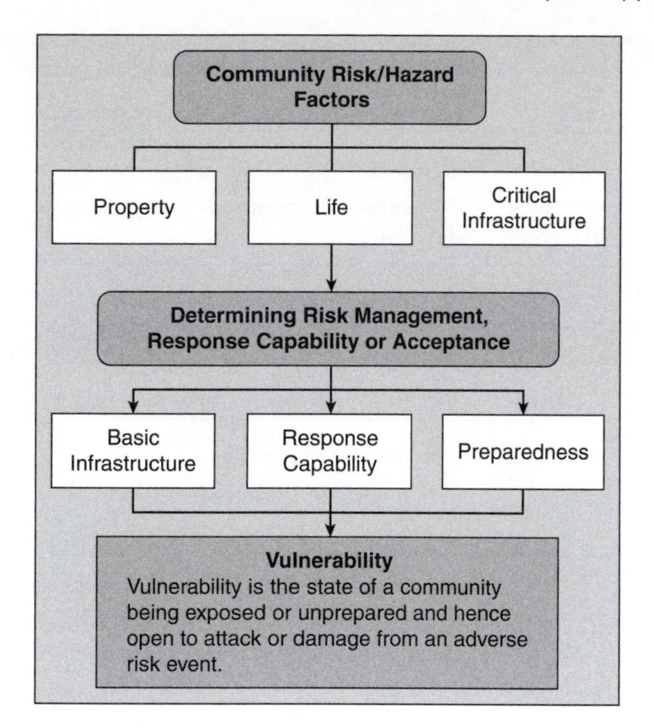

FIGURE 9.11 ◆ Community Risk Factors

Service Area/Definitions and Travel Time Benchmarks and Baselines *Metropolitan* designation means an incorporated or unincorporated area with a population of over 200,000 people and/or a population density over 3,000 people per square mile (see Table 9.9).

Urban designation means an incorporated or unincorporated area with a population of over 30,000 people and/or a population density over 2,000 people per square mile (see Table 9.10).

TABLE 9.9 ◆ Metropolitan Benchmark/Baseline

	1st Unit	*2nd Unit*	*Balance of a 1st Alarm*	*Performance*
Benchmark	4 minutes	8 minutes	8 minutes	90%
Baseline—70%	5 minutes/ 20 seconds	10 minutes/ 40 seconds	10 minutes/40 seconds	90%

TABLE 9.10 ◆ Urban Benchmark/Baseline

	1st Unit	*2nd Unit*	*Balance of a 1st Alarm*	*Performance*
Benchmark	4 minutes	8 minutes	8 minutes	90%
Baseline—70%	5 minutes/ 20 seconds	10 minutes/ 40 seconds	10 minutes/40 seconds	90%

TABLE 9.11 ◆ Suburban Benchmark/Baseline

	1st Unit	*2nd Unit*	*Balance of a 1st Alarm*	*Performance*
Benchmark	5 minutes	8 minutes	10 minutes	90%
Baseline—70%	6 minutes/ 50 seconds	10 minutes/ 40 seconds	13 minutes	90%

TABLE 9.12 ◆ Rural Benchmark/Baseline

	1st Unit	*2nd Unit*	*Balance of a 1st Alarm*	*Performance*
Benchmark	10 minutes	14 minutes	14 minutes	90%
Baseline—70%	13 minutes	18 minutes/ 20 seconds	18 minutes/20 seconds	90%

Suburban designation means an incorporated or unincorporated area with a population of 10,000 to 29,999 and/or any area with a population density of 1,000 to 2,000 people per square mile (see Table 9.11).

Rural designation means unincorporated or incorporated areas with total population less than 10,000 people or with a population density of less than 1,000 people per square mile (see Table 9.12)

Wilderness designation means any rural area not readily accessible by public or private maintained road (see Table 9.13)

The criteria listed here provide a range of performance within each category, from the target benchmark to a lesser baseline of performance at the 70 percentile of the benchmark. The analysis developed in the three areas of call processing, turnout time, and travel time should demonstrate the agency's own baseline performance falls within the ranges provided in the chart. These baselines should be evaluated annually to determine quality of service. Strategies used to reduce each element may well be part of other accreditation criteria. During the risk assessment process, an agency may find there is a mix of service areas as described previously within the jurisdiction. It is appropriate for an agency to have multiple service-level objectives based upon the nature of the area. For example, a predominantly metropolitan community with a large population base may also have service areas that are rural in nature and population density. An agency deserving of accredited agency status would have created geographic planning areas, evaluated each area for hazards and risks, and established appropriate response time and standard of cover deployment objectives based upon the results. "One size fits all" does not apply to developing an integrated risk management plan—standards of coverage.[12]

TABLE 9.13 ◆ Wilderness Benchmark/Baseline

	1st Unit	*2nd Unit*	*Balance of a 1st Alarm*	*Performance*
Benchmark	N/A			
Baseline—70%	N/A			

Other service-level objective statements identify response levels to identified target risks; critical infrastructure and key assets; critical tasks that staffing plans must meet; and desirable levels of distribution, concentration, and reliability.

INTEGRATION, REPORTING, AND POLICY DECISIONS

The final integrated risk management plan should provide a clear, comprehensive picture of what has been found and what recommendations will be necessary for future planning and implementation. The document with the use of graphs and mapping-based displays should foster informed policy discussion. The key points presented should be existing baseline of current performance, historical performance, identification of community risk factors, current service-level objectives, critical task analysis for anticipated events, company distribution and concentration analysis, company reliability analysis, revised service-level objective statements with cost-benefit analysis, and recommendations.

In public presentations care must be taken to inform elected officials and the public as to the current levels of services and the proposed service-level objectives. Ultimately, the final deployment plan will be determined based upon the desire of the residents served, policy makers, values, community expectations, and economics.

NFPA 1710 AND 1720

ORIGIN AND DEVELOPMENT OF NFPA 1710

The development of this benchmark standard, adopted in 2000, was the result of a considerable amount of work over several years by the technical committee members appointed by NFPA as representatives of several fire and governmental organizations. In the case of this standard, their work is the first organized approach to developing a standard defining levels of service, deployment capabilities, and staffing levels for those "substantially" career fire departments. Research work and empirical studies in North America were used by the committee as a basis for developing response times and resource capabilities for those services being provided, as identified by the fire department. **NFPA 1710** provides the user with a template for developing an implementation plan in respect to the standard.[13] The NFPA 1710 standard set forth in concise terms the recommended resource requirements for fires, emergencies, and other incidents. It requires the emergency response organization to evaluate its performance and report it to the authority with jurisdiction. The *scope* and **purpose** help to define what the standard does and what it covers. In both cases, the standard defines the minimum acceptable requirements, while still allowing more stringent or more comprehensive ones if a community so decides.

NFPA 1710
Standard for the Organization and Deployment of Fire Suppression Operations, Emergency Medical Operations, and Special Operations to the Public by Career Fire Departments.

purpose
That which is expected to be achieved if the organization is successful in completing its mission. It can be expressed in either qualitative or quantitative terms, within the parameters to be able to objectively verify them. That which we hope to create, accomplish, or change with a view toward influencing the solution to a problem.

With respect to scope, these minimum requirements are related to how fire, EMS, and special operations are organized and deployed in departments that are substantially career. The minimum requirements address these organizations' objectives as well as their functions. Not surprisingly, the standard emphasizes three key areas of a successful operation: service delivery, capabilities, and resources. The standard sets forth the minimum criteria related to the effectiveness and the efficiency of public entities that provide fire suppression, emergency medical service, and special operations. Both effectiveness and efficiency are specifically related to protecting the public and fire department responders.

When adopted, the concern by many in local government was that the NFPA 1710 standard may create a number of legal implications. The one issue that seemed to generate the most concern during the standard's development was whether jurisdictions could be held liable for failing to comply with the standard. The courts have traditionally been reluctant to hold cities, towns, and fire departments liable for the consequences of their discretionary decisions related to fire department resource allocations. This reluctance is the product of the age-old common law doctrine of *sovereign immunity*, under which courts in the English common law system traditionally held that the "king can do no wrong."

In some recent cases, the courts have found exceptions to this rule, and over the past several years, court cases have been based upon a jurisdiction failure to fund or staff its fire department adequately. Courts often take into account standards in determining whether such liability is properly imposed upon public fire departments and their municipalities. Many NFPA mandates have been enacted into law at the federal, state, provincial, and local levels. Some have argued that because jurisdictions having authority are not required to automatically enact a particular NFPA standard, violation of an NFPA standard does not automatically give rise to a finding of liability against a jurisdiction that has not adopted the standard. However, one should be mindful that courts frequently rely upon NFPA standards to determine the *industry standard* for fire protection and safety measures. Judicial reliance on NFPA doctrines is more frequently found in common law negligence claims. To prevail in a common law negligence claim, the plaintiff must show that the defendant owed a duty of care to the plaintiff, that the defendant breached this duty of care, and that this breach was the cause of the plaintiff's injury.

The NFPA 1710 standard creates a common template for evaluating performance. The organizational statement of the standard sets forth the minimum information required concerning what the organization does, how it is structured, and the staffing required to achieve its objectives. Service delivery objectives found in the standard are specific requirements for deployment, staffing, response times, and the necessary support systems. These support systems include safety and health, communications, incident command, pre-incident planning, and training. There are three time components defined in the NFPA 1710 standard relating to emergency response system performance. All three components must be measured and documented by departments in their quadrennial report. The components and definitions are as follows:

- *Call receipt and processing time.* The interval between receipt of the emergency alarm at the public safety answering point to the moment where sufficient information is known to the dispatcher and applicable units are notified of the emergency. The maximum time for this component is specified in NFPA 1221.
- *Turnout time.* The interval between acknowledgments of notification of the emergency by the units to the beginning point of response time.
- ***Response time.*** The time that begins when units are en route to the emergency incident (wheels rolling) and ends when units arrive on scene (wheels stopped at the address). This time component is referred to as *travel time* in the CFAI process.

response time
The total amount of time that elapses from the time that a communications center receives an alarm until the responding unit is on the scene of the emergency and prepared to control the situation. Response time is composed of several elements.

Figure 9.12 illustrates the NFPA 1710 response time elements that have been assembled for fire suppression incidents and EMS response.

The purpose of department evaluation and reporting is to measure and document compliance with the NFPA 1710 standard. According to the standard, a department must perform an annual evaluation of service based on actual response data. Using these data, the department must prepare and submit a quadrennial written report to

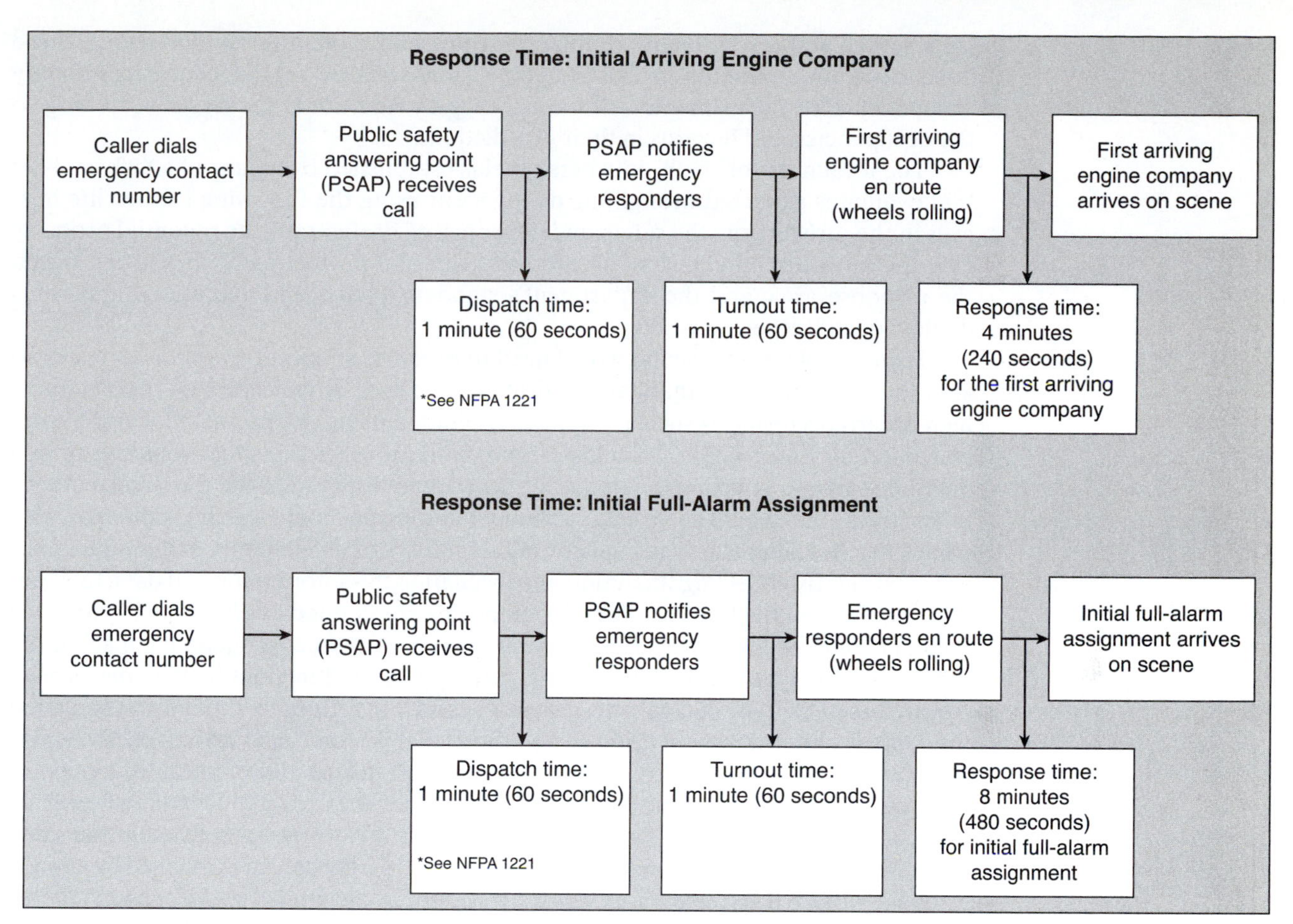

FIGURE 9.12 ◆ Response Time: Initial Arriving Engine Company
(Courtesy of Randy R. Bruegman)

the governmental authority noting compliance or noncompliance with the standard. The report must explain any deficiencies, the consequences of the deficiencies, and the offer of improvements on how the department plans to become compliant. The NFPA 1710 standard acknowledges, where fire and EMS response are concerned, many communities have decided to work together across geographic and municipal boundaries. The standard recognizes the existence and usefulness of mutual-aid pacts in their various forms. It stipulates that these *agreements must be in writing and must address certain specific concerns:* liability, disability retirements, service costs, staffing, and equipment.

The standard requires that personnel from multiple organizations working at the scene of an incident must have common procedures in order to ensure safe operations. This can occur only if the personnel train together and are familiar with procedures and equipment used to control an incident. Finally, a crucial component of effective intercommunity response is adequate communication. The NFPA 1710 standard mandates that these organizations be equipped with compatible communications equipment for that purpose.

Fire Suppression Fire departments must be capable of establishing the following functions at each structural fire: incident command, water supply, attack lines, backup lines, search and rescue teams, ventilation teams, and rapid intervention teams. These benchmark requirements are based upon a 2,000-square-foot detached single-family occupancy. Fire departments will have to deploy additional resources according to the occupancies and hazards in their jurisdictions.

The total number of on-duty personnel is established by means of a task analysis that evaluates expected firefighting deployment using the following factors: life hazards in the jurisdiction; the safety and efficiency of firefighters; the potential property loss; the nature, configuration, hazards, and internal protection of properties within the response area; and the department's standard tactics and evolutions, apparatus deployed, and expected results.

For example, a jurisdiction would need to evaluate all locations within its response area to determine those that have tactical hazards such as concentrated fire potential; high hazard occupancies such as schools, hospitals, nursing homes, manufacturing complexes, refineries, or high-rise buildings; geographical restrictions that could result in a delayed response affecting the frequency, severity, and spread of the fire occurrence; or other factors that would necessitate additional staffing per company and additional companies for the initial alarm assignment, additional alarm assignments, and simultaneous emergencies. By collecting, analyzing, and evaluating this information and data, total on-duty staffing can then be established. Companies are defined as either engine or truck (ladder) companies or specialized apparatus, such as rescue or squad companies, depending on the type of apparatus and the fire suppression functions that are performed. Regardless of the type of company, each must consist of a group of trained and equipped firefighters, under the supervision of an officer, that operates and arrives on the emergency scene with one piece of fire apparatus. The standard allows for an exception in those instances when multiple apparatus are used to make up a company. However, such exceptions require that these multipiece companies are always dispatched at the same time and arrive together, are continuously operated together, and are managed by a single company officer. *The standard recognizes and clarifies the limited use of such multipiece companies.* Examples include the use of a fire department personnel vehicle if the apparatus does not have adequate seating; an engine and water tanker, such as those used in some suburban and rural response, where a water supply (hydrant or natural water bodies) is not available; and an engine and an EMS unit (ambulance or rescue).

The standard's intent for fire suppression is to have the first-due engine capable of arriving within its response area consistently within four minutes, 90 percent of the time. The "and/or" criterion is intended to recognize the effects of simultaneous emergencies, training, or other occurrences that take one or more companies out of service and not to relieve a department of its responsibility to plan for overall deployment of resources by location to satisfy the four-minute criterion. The fire department shall have the capability to deploy an initial full-alarm assignment within the eight-minute response time. The number of people required falls between 14 and 17 depending on whether or not an aerial is used, both pumpers are being used to provide attack and backup lines, and a safety officer is required. The following is a list of functions defined in the standard and the number of personnel required to be deployed to perform these functions.

- Incident command shall be established by the deployed supervisory chief officer outside the hazard area for the overall coordination and direction of the initial full-alarm assignment. A minimum of one supervisory chief officer shall be dedicated to this task.
- The supervisory chief officers shall have staff aides deployed to them for purposes of incident management and accountability at emergency incidents. A minimum of one individual shall be dedicated to this task for each supervisory chief.

- A safety officer shall be dispatched to an initial alarm assignment when significant risks to firefighters are present and shall be deployed to all emergencies that go beyond an initial full-alarm assignment to ensure that the health and safety systems are established at the emergency incident. A minimum of one individual shall be dedicated to this task.
- An uninterrupted water supply of a minimum 400 gpm for 30 minutes shall be established. Supply line(s) shall be maintained by an operator who shall remain with each fire apparatus supplying the water flow to ensure uninterrupted water flow application.
- An effective water flow application rate shall be established: 300 gpm from two hand lines, one of which shall be an attack line with a minimum of 150 gpm and one of which shall be a backup line with a minimum of 150 gpm. Attack and backup lines shall be operated by a minimum of two personnel each to effectively and safely maintain the line.
- One support person shall be provided for each attack and backup lines deployed to accomplish hydrant hookup and assist in line lays, utility control, and forcible entry. This individual shall be permitted to be assigned as a member of the initial rapid intervention team (IRIT) if the fire department determines that the support person can abandon his or her tasks without placing any personnel in jeopardy.
- A minimum of one search and rescue team shall be part of an initial interior attack. Each search and rescue team shall consist of a minimum of two personnel.
- A minimum of one ventilation team shall be part of an initial interior attack. Each ventilation team shall consist of a minimum of two personnel to perform structure ventilation in coordination with the primary interior attack.
- If an aerial device is used in operations, one person shall function as an aerial operator who shall remain at the primary control of the aerial device at all times.
- An initial rapid intervention team (IRIT) shall be established consisting of a minimum of two properly equipped and trained personnel. When an incident escalates beyond the initial full-alarm assignment, or when there is significant risk to firefighters due to the magnitude of the incident, the incident commander shall upgrade the IRIT to a full rapid intervention team (RIT) that consists of four dedicated, fully equipped, and trained firefighters.
- The fire department shall have the capability for additional alarm assignments that can provide for additional personnel and additional services including the application of water to the fire; engagement in search and rescue, forcible entry, ventilation, and preservation of property; accountability for personnel; and provision of support activities for those situations that are beyond the capability of the initial alarm assignment.

Figure 9.13 illustrates these staffing and task requirements for the response to the benchmark room-and-contents fire in a 2,000-square-foot single-family occupancy without a basement and with no exposures (adjacent buildings). In an urban environment with high-population-density dwellings in very close proximity (multifamily occupancies, industrial areas, high-occupancy institutions, including hospitals and schools), the fire department's response capability must be enhanced with additional apparatus, personnel, and resources for the initial alarm assignment.

When jurisdictions use *quint* apparatus, the standard requires that these units be designated as either an engine or a truck company, unless the jurisdiction expects multiple roles (i.e., engine and truck company functions) to be performed simultaneously from this unit. If such is the case, the standard requires that additional staffing, beyond the four-member minimum, be provided for such companies.

The NFPA 1710 standard does require all fire departments to provide a basic level of EMS. The standard calls for the department to be capable of responding to emergency

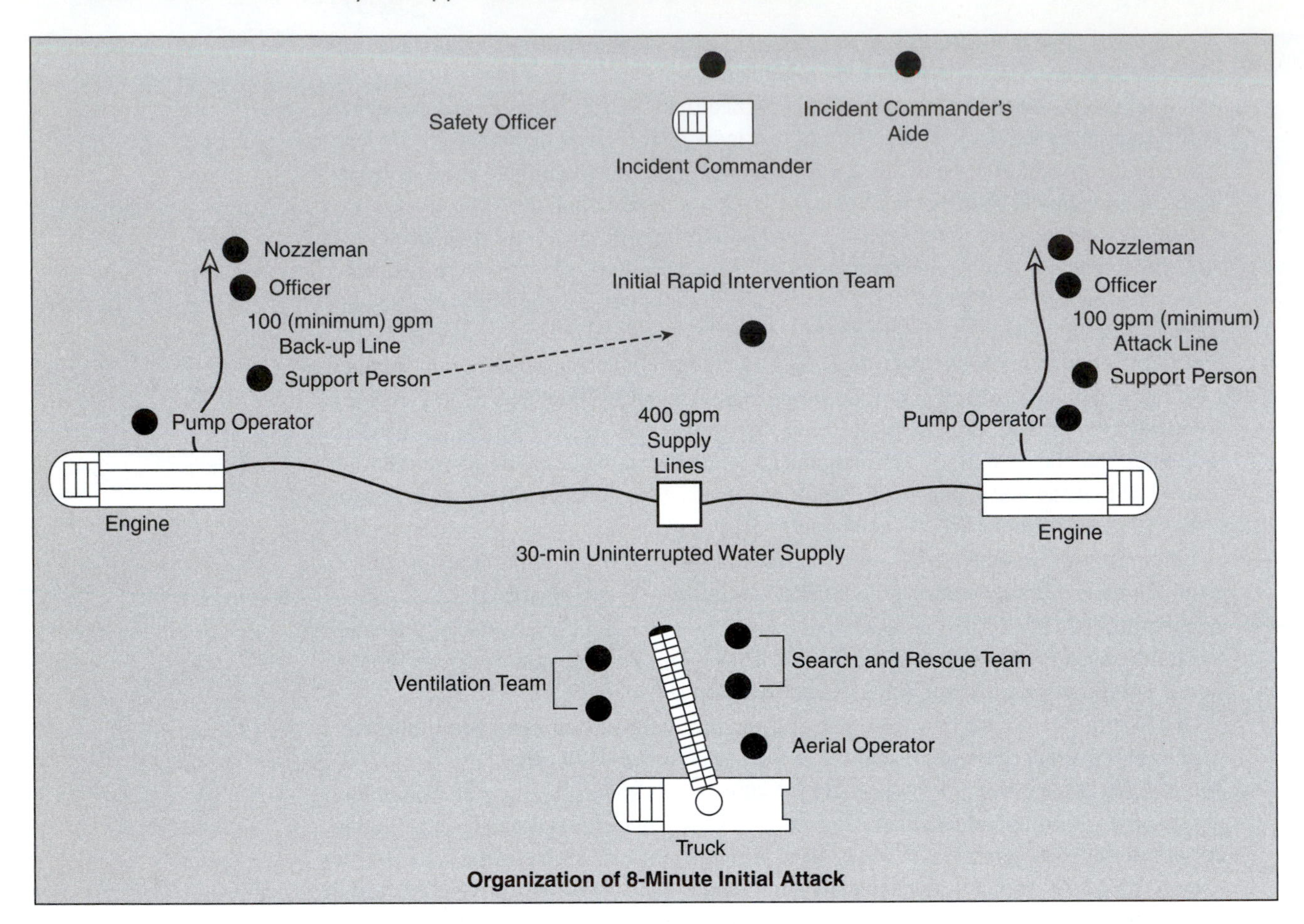

FIGURE 9.13 ◆ Organization of 8-Minute Initial Attack
(Courtesy of Randy R. Bruegman)

medical incidents at the first responder level with automatic external defibrillators (AED). If the department chooses to provide EMS at a higher level, the standard sets forth operational requirements for that service as well. Fire departments that provide EMS at any level must establish in their organizational statements the criteria for the types of incidents to which they will respond. The established level of EMS provision must be recognized and the department must allocate the necessary resources to adequately provide the services required by the local jurisdiction and expected by the citizens. Necessary resources include personnel and equipment. Where EMS beyond the first responder level is provided by an entity other than the fire department, the higher-level provider must adhere to minimum staffing, deployment, and response criteria recommended by the fire department according to the requirements in the NFPA 1710 standard. These operational requirements must be set forth in both the fire department's organizational statement and any contract or other agreement between the jurisdictional authority and the EMS agency or private company.

basic life support
A primary level of prehospital care that includes the recognition of life-threatening conditions and the application of simple emergency lifesaving procedures, including the use of adjunctive equipment, aimed at supporting life.

There are three levels of EMS provision recognized in the NFPA 1710 standard: **first responder with AED, basic life support** (BLS), and **advanced life support** (ALS). The standard also recognizes EMS transport as a service that may be provided by the fire department. It is not a requirement that a fire department provide

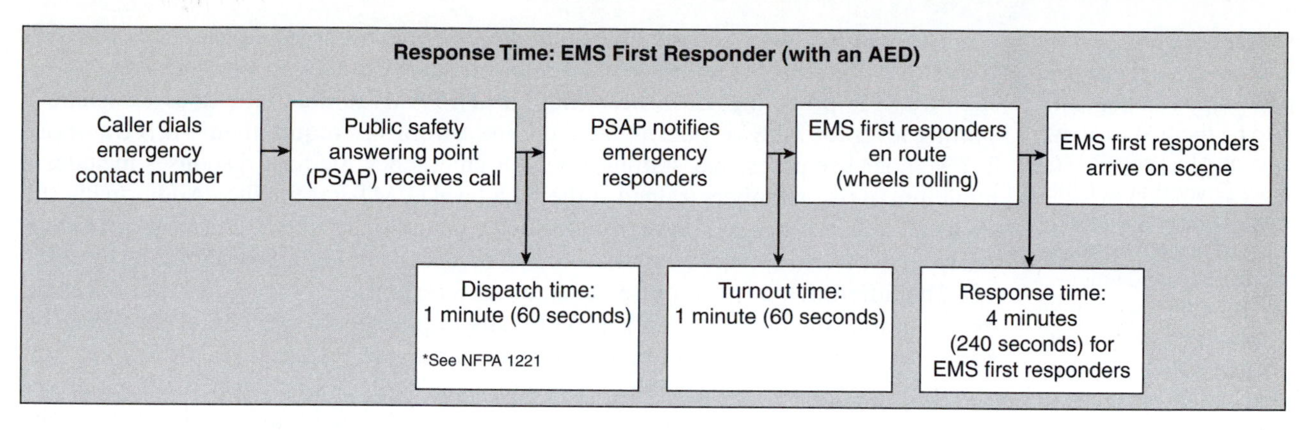

FIGURE 9.14 ◆ Response Time: EMS First Responder (with an AED)
(Courtesy of Randy R. Bruegman)

all levels of EMS service beyond first responder. However, the standard establishes operational requirements for each level that is provided by a department. For each level, operational requirements are set forth as follows:

1. ***First Responder.*** A fire department must appropriately train all response personnel at the first responder with AED capability level, and personnel must arrive within a four-minute response time frame to 90 percent of all emergency medical incidents (see Figure 9.14). The number of personnel must be sufficient to assure adequate care capability and member safety.
2. ***BLS.*** A fire department that provides BLS beyond the first responder level shall adhere to staffing and training requirements as set forth by the state or provincial licensing agency. The department must also deploy sufficient mobile resources to arrive within a four-minute response time frame for 90 percent of all incidents.
3. ***ALS.*** A fire department that provides ALS beyond the first responder and BLS levels shall adhere to staffing and training requirements as set forth by the state or provincial licensing agency. The department must also deploy sufficient mobile resources to arrive within an eight-minute response time frame for 90 percent of all incidents (see Figure 9.15).

advanced life support
A sophisticated level of pre-hospital care that builds upon basic life support procedures and includes the use of invasive techniques such as advanced airway management, cardiac monitoring and defibrillation, intravenous therapy, and the administration of specified medications to save a patient's life.

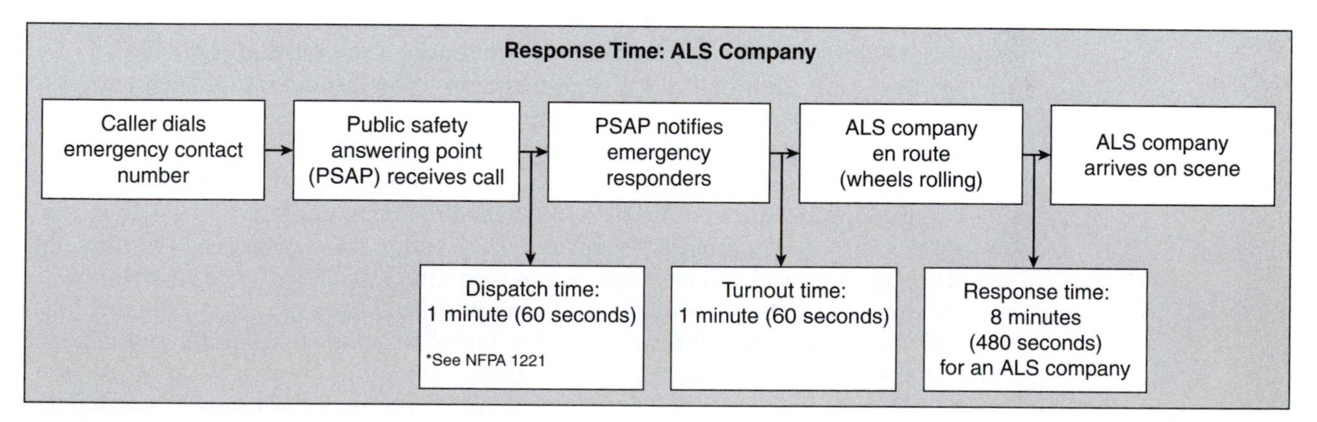

FIGURE 9.15 ◆ Response Time: ALS Company
(Courtesy of Randy R. Bruegman)

first responder

A term used for the person who is trained and/or certified to be the first to arrive at a scene of a specific type of emergency, that is, EMS or hazardous materials. Local and nongovernmental police, fire, and emergency personnel who, in the early stages of an incident, are responsible for the protection and preservation of life, property, evidence, and the environment, including emergency response providers as defined in section 2 of the Homeland Security Act of 2002 (6 U.S.C. 101); as well as emergency management, public health, clinical care, public works, and other skilled support personnel (such as equipment operators) who provide immediate support services during prevention, response, and recovery operations. First responders may include personnel from federal, state, local, tribal, or nongovernmental organizations.

The NFPA 1710 standard states that staffing and training requirements for both BLS and ALS transport units are to be determined by the state or provincial agency responsible for providing EMS licensing. The NFPA 1710 standard does designate a staffing level for ALS response that is different from the requirement ALS transport. The standard requires that staffing for ALS emergency medical responses includes a minimum of two members trained at the ALS level (EMT-Paramedic). Additionally, the standard requires that ALS responses include a minimum of two BLS trained providers. All response personnel are to arrive within the response time frame established for ALS.

The different staffing requirements for ALS responses is based on experience and expert consensus that time-critical ALS calls require more personnel resources on scene for assessment and initiation of care than those required for BLS level incidents or transport. Additionally, the American Heart Association has long-established guidelines for response to the most time-critical incident—cardiac arrest. The NFPA 1710 standard clearly requires the establishment of a quality management program as a basic function of EMS provision. The purpose of this program is to ensure adequate response capability and quality patient care. All quality reviews of both BLS and ALS services must be documented. An additional quality management requirement of the standard is that the department must provide a mechanism for immediate communication with the overseeing EMS system supervisor and physician (medical director).

Special, ARFF, Marine, and Wildland Operations The fire department is required to define formally the types of special operations that it is required or expected to perform in an emergency or other incident. These types of special operations include, but are not limited to, hazardous materials response, confined space response, technical rescue, high-angle rescue, and water rescue. Regardless of the fire department's defined special operation capability, all firefighters who provide emergency response must be trained to the first responder operations level for both hazardous materials and confined space responses. Likewise, all fire departments must define their response capability to natural disasters, terrorism incidents, large-scale emergencies, and mass-casualty events. Where fire departments have established that they will provide response beyond first responder level for hazardous materials or confined space emergencies, they are required to ensure that all members involved in this level of response be trained to the levels specified in the standard. The fire department must also determine the availability of resources outside the fire department through federal, state, provincial, or local assistance or private contractors who are deployed to emergencies and other incidents and the procedures for initiating such outside response. The fire department must also limit the level of response to special operation emergencies to the level for which it has staffed, trained, and equipped its personnel. Additionally, it must have the capacity to initiate a rapid intervention crew during any and all special operations responses. The NFPA 1710 standard requires that airport fire departments be organized to ensure that their response capabilities to nonaircraft incidents (non-airframe structural fires and EMS emergencies) within the department's response jurisdiction are identical to nonaircraft rescue firefighting fire department capabilities.

Marine firefighting was recognized as a specialized fire suppression support function by the NFPA 1710 standard, with the initial response to a land-based incident provided by land-based fire suppression forces as well as shipboard crews and sea-based response provided initially by the shipboard crew. Land-based forces staffing and deployment criteria are recognized by the standard and require the jurisdiction that responds to such emergencies to address the resource allocation to such events.

The U.S. Coast Guard or other legal authority regulates sea-based crew staffing and deployment capability with jurisdiction over navigable waterways.

The NFPA 1710 standard recognizes that many, if not most, fire departments must respond to either wildland or wildland/urban interface fires. Accordingly, the fire department must address the service delivery for such occurrences. The standard specifies the minimum wildland staffing for defined wildland companies, as well as engine and truck companies that respond to wildland or urban interface/wildland emergencies. Likewise, deployment requirements for a wildland initial direct attack are specified.

A system is a functionally related group of components. These are areas where a set of needs or requirements work closely together and are interrelated to achieve a key result. The NFPA 1710 standard addresses five of these systems.

- *Safety and health.* Each organization must have an occupational safety and health program meeting the requirements of NFPA 1500, Standard on Fire Department Occupational Safety and Health Program.
- *Incident management.* Each organization must have in place an incident management system designed to handle expected incidents. The system must be in accordance with NFPA 1561, Standard on Emergency Services Incident Management System.
- *Training.* Each organization must ensure members are trained to execute all responsibilities consistent with its organizational statement. This training must be accomplished using a programmatic approach that includes a policy.
- *Communications.* Each organization must have a communications system characterized by reliability; promptness; and **standard operating procedures,** terminology, and protocols. Departments must also comply with all the requirements set forth in NFPA 1221, Standard for the Installation, Maintenance, and Use of Emergency Services Communications Systems.
- *Pre-incident planning.* Safe and effective operations are grounded in identifying key and high-hazard targets. The standard requires departments to develop operational requirements to gather information regarding these locations.

standard operating procedures
A term used to describe written direction provided to personnel in a manual format. Similar to the general operating guideline but may be more specific requiring specific actions.

Together these five systems help to ensure that emergency responders have the essential tools, information, procedures, and safeguards to operate effectively and efficiently.

ORIGIN AND DEVELOPMENT OF NFPA 1720

NFPA 1720
Standard for the Organization and Deployment of Fire Suppression Operations, Emergency Medical Operations, and Special Operations to the Public by Volunteer Fire Departments.

In 2001, the first edition of **NFPA 1720** was known as the standard for the organization and deployment of fire suppression, emergency medical operations, and special operations to the public by volunteer fire departments. The development of this benchmark standard was the result of a considerable amount of work and time by the technical committee members and the organizations they represented. This standard was the first organized approach to defining levels of service, deployment capabilities, and staffing levels for substantially volunteer fire departments.[14] This standard was developed to identify minimum requirements relating to the organization and deployment of fire suppression operations and emergency medical operations in volunteer fire departments. Approximately three out of every four fire departments in the United States are volunteer; therefore, this standard, as well as related practices (accreditation, certification, etc.), has a profound effect on the direction of the volunteer fire service. However, NFPA 1720 does not include such important issues as fire prevention, community education, fire investigations, support services, personnel management, and budgeting. The components of the standard (see Tables 9.14, 9.15, and 9.16) provide a common basis for the operation of substantially volunteer departments.

TABLE 9.14 ◆ Components of NFPA 1720

4.1	Suppression operations, adequate personnel, equipment, sufficient/efficient/effective resources.
4.1.1	Organization, operations, deployment, written R&R, SOPs, orders.
4.1.2	Develop community risk management plan (storage, use, and transportation of hazardous materials).
4.1.3	Procedures clearly state succession of command responsibility.
4.1.4	Organized into company units and response teams and have appropriate apparatus and equipment.
4.1.5	Identify minimum staffing required for safe and efficient operations.
4.1.6	Maintain standard reports for responses; including location, nature of incident and operations performed, members responding.
4.1.7	Mutual-aid response and agreements; predetermined locations; regulate dispatch of companies, response groups, command to fire and other responses.
4.1.8	Number and type of units assigned to respond by risk analysis and prefire plans.
4.2.1	Incident command assignments.
4.2.1.1	Assumption and identification communicated to all responding units.
4.2.1.2	Incident command shall be responsible for coordination and direction of all activities.
4.2.1.3	Incident commander ensures accountability system is used immediately.
4.2.1.4	Company officer/crew leader aware of identity, location, and activity of every member assigned.
4.2.1.5	Each member of a company aware of the identify of the company officer/crew leader.
4.2.1.6	Orders to members, verbal at incident transmitted through company officer/crew leader.
4.2.2.1	Upon assembling necessary resources, shall have the capability to safely initiate attack within two minutes, 90 percent of the time.
4.2.2.2	Initial attack operations shall be organized so that at least four members are assembled before an interior attack is made.
4.2.2.2.1	Hazardous materials incidents shall have two people work as a team.
4.2.2.2.2	Outside hazardous materials area, two people present for assistance on rescue, one person may be engaged in other activities.
4.2.2.2.3	No assignment can be made if abandoning the critical tasks of rescue.
4.2.2.3	Initial attack can occur if life-threatening situation is imminent.
4.2.2.4	Fire department has capability for sustained operations, suppression, search, rescue, forcible entry, ventilation, preservation of property, accountability, rapid intervention team, support activities.
4.3.1	Mutual aid, automatic aid, fire protection agreements in writing and liability issues, injury, death, retirement, cost, and authorization for support services.
4.3.2	Procedures for training for personnel within all agreements for response or support.
4.3.3	Companies responding to mutual-aid incidents shall be equipped with communications equipment to communicate with various officers.
4.4.1	EMS capability (people, equipment, resources) initially on arrival; automatic or mutual aid.
4.4.1.1	Do you deliver emergency medical services?
4.4.1.2	Maintain clear document of rule, responsibilities, function, and objectives for delivering EMS.

(Continued)

TABLE 9.14 ◆ Components of NFPA 1720 (*Continued*)

4.4.2	Basis treatment levels within system —First responder —Basic life support —Advanced life support
4.4.3.1	Fire basic functions 1. ___ First responder with AED 2. ___ Basic life support 3. ___ ALS response 4. ___ Patient transport 5. ___ Assurance of response quality management
4.4.3.2	Department shall be involved in any or all functions identified in 4.4.3.1 through 4.4.3.1.5.
4.5.1	Fire department has a quality management program.
4.5.2	All responders and basic life support programs are reviewed and documented.
4.5.3	Advance life support systems have named medical director with responsibility to oversee and ensure quality medical care within state law.
4.5.4	Advance life support system provides immediate communication with supervisor and medical oversight.
4.6.1	Special operations consist of: ___ sufficient people ___ equipment ___ resources ___ initial deployment and subsequent deployments
4.6.1.1	Involved in special operations response.
4.6.2	Have adopted plan for operations response specifying role and responsibilities for hazardous materials and trained to NFPA 472.
4.6.3	All members who respond beyond first responder level for hazardous materials are trained to NFPA 472.
4.6.4	Fire department shall have capacity to implement rapid intervention crew during all special operations to meet NFPA 1500.
4.6.5	If higher level emergency response is needed beyond department, department has determined availability and capability of other responders.
5.1	Health and safety provided in accordance with NFPA 1500.
5.2.1	Incident management system provided in accordance with NFPA 1561.
5.2.2	Use of incident command designed for structures, wildland, hazardous materials, EMS, and other types of incidents.
5.3	Department has training program for competency; effective, efficient, and safe.
5.4.1	Department has reliable communication system to facilitate prompt delivery of suppression, EMS, and special operations.
5.4.2	All communication facilities, equipment, staffing, and operating procedures comply with NFPA 1221.
5.4.3	Radio communications have standard protocols and terminology.
5.4.3.1	Radio terminology is in compliance with NFPA 1561.
5.5	Department does pre-incident planning and especially for target hazards.

Table 9.15 ◆ Staffing and Response Time

Demand Zone	*Demographics*	*Staffing and Response Time*	*Percentage*
Special Risks	AHJ	AHJ	90
Urban	>1,000 people/mi.^{2}	15/9	90
Suburban	500–1,000 people/mi.^{2}	10/10	80
Rural	<500 people/mi.^{2}	6/14	80
Remote*	Travel distance ≥ 8 mi.	4	90

*Upon assembling the necessary resources at the emergency scene, the fire department should have the capability to safely commence an initial attack within two minutes 90 percent of the time.

Table 9.16 ◆ Alarm Time Line

Alarm Time Line at Communications Centers

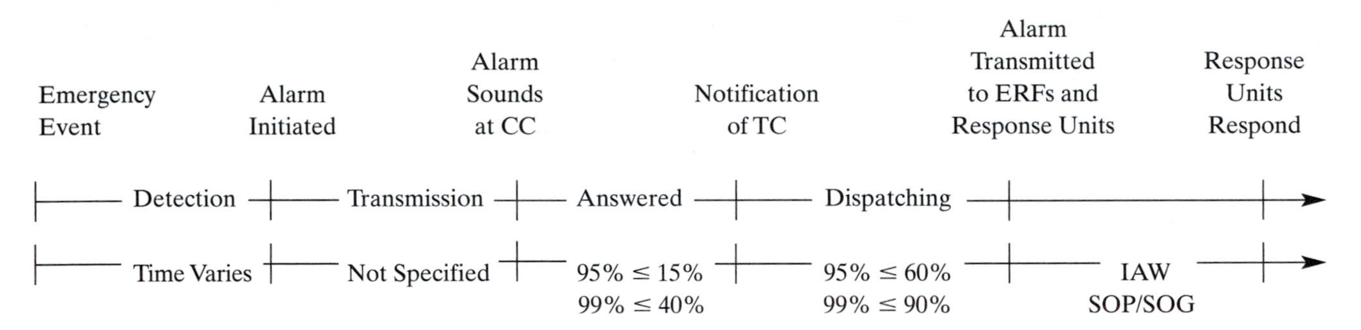

Notes: 1. *Alarm Sounds* means audible or visual annunciation, or both.
2. *TC* stands for Telecommunicator.
3. *CC* stands for Communications Center.
4. *IAW* stands for in accordance with.

Alarm Time Line Where Primary PSAP Is Communications Center

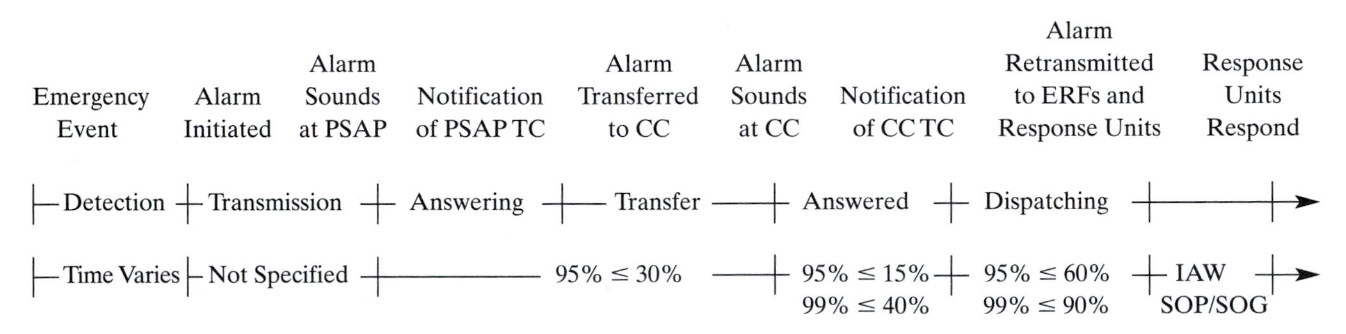

Notes: 1. *Alarm Sounds* means audible or visual annunciation, or both.
2. *TC* stands for Telecommunicator.
3. *CC* stands for Communications Center.
4. *PSAP* stands for Public Safety Answering Point.
5. *IAW* stands for in accordance with.

The challenge for today's fire service leader is to have the ability to analyze a community's risk factor(s) and propose the appropriate deployment level for the community. As we have realized over the course of the past 25 years, to be effective, a deployment resource model must be built on sufficient, accurate, and complete data no matter what methodology or model is chosen. In the fire service one of the difficulties we are faced with is the selection of the right tool from the toolbox to correct the problem. Sometimes the tools we use determine the way the problem or the situation is addressed. There is an old saying that if you give a child a hammer, everything in the house looks like a nail. In today's fire service, the "one size fits all" or one solution for every fire department in respect to the response delivery system is simply not feasible. The fact is that the risk analysis should be problem driven instead of method driven, which is to say that today's leadership must start with the identification of the problem and then seek appropriate methods to address the issue, rather than start with a methodology and look for a problem to apply it to!

In this chapter several methods have been outlined from which to choose to begin risk analysis of your community, which will assist in the development of a comprehensive deployment package. The underlying theme in all of the discussion is that the development of an integrated risk management model must take into account all the factors of the community risk level(s) and the response needed regardless of the system chosen. What we have seen in the past decade is a convergence of thought. If you overlay the grading schedule of the ISO, the CFAI accreditation model, and NFPA 1710 and 1720, you begin to see threads of commonality. As we become more adept and proficient at risk analysis in the United States, we will see more of the convergence of thought and application in respect to how we analyze risk in this country and how we deploy our resources to minimize the effects of emergencies in each of our jurisdictions.

ADVICE FROM THE EXPERTS

Community risk management has evolved over the past two decades. What have you done in your own organization to address this issue?

MICHAEL CHIARAMONTE

Chief Fire Inspector, Retired, Lynbrook Fire Department (New York): New York is behind in this area.

BETT CLARK

Former Fire Chief, Bernalillo County Fire and Rescue (New Mexico): We completed risk assessments as part of our emergency operations plan in a quad-county area. We then went further to identify threats and threat-specific responses, achievable levels of mitigation, and mutual-aid needs. Our risks have evolved to include terrorism that we never dreamed of dealing with 20 or even 10 years ago. We set our budget's goals to deal with the real risks that may occur in our region, such as a growing older population and untapped prevention education opportunities to reach this population. As we outreached to them we witnessed a true decline in elderly fire deaths. We are a dry region with the Rio Grande River and lots of farming lands that are irrigated by the river. Fifteen years ago we responded to an average number of a dozen drownings per year in the river and ditches. We saw a real need to manage this risk better; therefore, a Ditch Safety Task Force was created. Political leaders were lobbied to enact legislation for mandatory security around the ditches, safer water flow rates, and a swim pass program. The swim pass program allows youngsters to go to any local fire station and receive a day swim pass to any public pool. We have not had a drowning in our area for the last six years.

KELVIN COCHRAN

Fire Chief, Atlanta Fire Department (Georgia): To enhance community risk assessment in the city of Atlanta, the fire department has worked to improve our relationship with the local Office of Homeland Security and Emergency Preparedness. For years its mission has been community risk assessment but did not include formal consultation and partnering with other emergency response agencies. Currently all community risks have been evaluated, probability assessments have been conducted, and contingency plans developed. Frequent exercises, tabletops, and meetings with area emergency response agencies have strengthened relationships and enhanced community risk management. Preparedness was tested significantly in managing the impacts of Hurricanes Katrina and Rita on northwest Louisiana.

CLIFF JONES

Fire Chief, Tempe Fire Department (Arizona): We have utilized the framework provided by the Center for Public Safety Excellence for fire risk. We have divided the community into management districts for fire department purposes, and those districts are comanaged by a fire captain and a fire inspector for risk management issues. Additionally, we have worked through a FEMA program called Project Impact (no longer an active program for FEMA), which facilitated a disaster risk assessment for the community that we keep current. Finally, we have recently completed a community risk assessment for weapons of mass destruction.

BILL KILLEN

Retired Fire Chief and IAFC President, 2005–2006: In my previous organization I was able to influence and implement the adoption of the Center for Public Safety Excellence Self-Assessment program, which was ideal for evaluating community risks and developing standards of cover. Completion of the community risk management led to accreditation of several Army, Navy, and Air Force fire departments. I have not been able to do much in my current organization due to the unique and compact community we serve. Risk management in our community is unique in that we do not have any residential, mercantile, medical, educational, or recreational occupancies and all production facilities, maintenance shops, and key administrative buildings are totally sprinklered. The products of the facility are explosives for munitions used by the Department of Defense, and community risk management has been in place for more than 60 years.

KEVIN KING

U.S. Marine Corps Fire Service: One of the most important efforts we undertook was to focus on what risks our communities face. Do we really have major fire risks at most of our military installations? Generally we do not, at least by historical standards. However, we certainly have potential terrorist risks, natural disaster risks, and multicasualty risks. As we have identified these risks, we have then tried to apply resources toward the risks. One of our more recent community risk efforts is in the area of critical infrastructure protection. We are now trying to identify those critical infrastructure assets and apply appropriate mitigation strategies to those assets. (These are opinions of Mr. King and do not reflect official policy of the Department of Defense or the United States Marine Corps.)

MARK LIGHT

Executive Officer, International Association of Fire Chiefs: Departments continue to look at the risk in their communities and to develop plans to address that risk. Computer programs have given us tools that make it easier to quantify the risk. However, the biggest risk that most departments face is the residential areas in their response areas.

The biggest thing I see is fire departments getting the communities involved in what they do. The fire department is one of the few government buildings that are in every community within a municipality. Fire departments must become a regular part of their community. Company officers should know the civic organization leaders in their response area. Community watch groups should be very comfortable with stopping by the fire station and talking with the firefighters. The fire stations should be a safe haven for anyone in the community and truly make citizens feel that this is their fire station and that the firefighters care about them and their families. I continue to see more interaction at the community level, which will go a long way toward addressing the risks in every community. Involvement breeds familiarity, and familiarity lets us learn about the risks in the community and how to best deliver the service based on the risk.

LORI MOORE-MERRELL

Assistant to the General President, International Association of Fire Fighters: Whether low, medium, or high hazards, recognized risks are in every community. Continuing inspections and public education are essential. However, just as important are identification and preplanning for target hazards. One means by which fire departments may effectively engage in risk management is by going through fire department accreditation.

JAY REARDON

Fire Chief, Retired, Northbrook Fire Department (Illinois), Currently President of the Mutual Aid Box Alarm System (MABAS): The risk management road map is based on accreditation principles—a continuous, ongoing process that is all risk based and never is brought to closure. Community perception and expectations will not tolerate complacency and stagnation.

GARY W. SMITH

Retired Fire Chief, Aptos–La Salva Fire Protection District (Aptos, California): The most effective tool that I have used to address the fire, hazardous materials, and emergency medical risks within my community is the standards of response coverage (SORC). This tool, supported and promoted by the IAFC and the ICMA, is an excellent way to organize and monitor the use of local resources to accomplish the response goals that the community can afford. The SORC process clearly defines the local ability to address the risks that exist, showing the value of using new resources wisely, or the direct (measurable) effect of reducing resources, and the distribution and concentration of response efforts for the risks that exist within the community. The computer-supported geo-based information is a quick way to show the effect of fire resource placement.

STEVE WESTERMANN

Fire Chief, Central Jackson County Fire Protection District (Blue Springs, Missouri), and IAFC President 2007–2008: One of the first and most important realizations is that no matter how big you are, there are potential situations that you cannot resolve with your own resources. We are a suburban department in a UASI metropolitan area. As a region, we have conducted a risk assessment and then, once the risk was evaluated, made regional decisions on how the funds could best be utilized for the region. For instance, one of the major risks that was recognized is our regional capabilities for major structural collapses. We wanted to create a team that could handle our regional needs before a USAR team could be on the ground. We allocated sufficient funds and created four specialized rescue teams using six response units spread around the metropolitan area. We jointly decided what equipment was needed and then ordered the equipment as one purchaser. Training is being conducted in the same way.

THOMAS WIECZOREK

Executive Director, Center for Public Safety Excellence, and Retired City Manager: Before 9/11, I had sought to move all of our agencies through some form of accreditation process. The goal of this process was to ensure that for every dollar spent, we had looked at all options and opportunities and could say that we were providing good stewardship of those dollars. After 9/11 this became the focal point for blending our agencies and departments. All of our agencies undertook Incident Command (or Incident Management) training. It was interesting watching the public works, police, fire, and building staff sit at tables and work through IC situations. While suspicious at first, many great ideas began flowing, and the organizations began working together. We wrote an integrated risk management plan that did not assign incident command to an individual or department. We had the opportunity during a flood situation to test it and, from the test, garnered ideas on how to improve. It was amazing to see the different disciplines come up with ideas to save others work (like putting plastic and geo-textile plastic under manhole grates to keep storm water from flooding some areas and save on sandbagging). It was also interesting how little crisis management had to take place as a result. I think next time will be still easier. When I retired we were working on becoming interoperable—moving all disciplines to an 800 MHz system so that communication could take place throughout our organization and beyond.

JANET WILMOTH

Editorial Director, *Fire Chief* magazine: N/A

FRED WINDISCH

Fire Chief, Ponderosa Volunteer Fire Department (Houston, Texas): The primary change we have accomplished is the continual focus on personal safety related to fires that destroy property. It is not worth it to risk your life for a parking lot or an insurance claim. Risk management clearly should be more formalized in each and every emergency service agency, and we must understand that our lives are more important than buildings. By the same token in EMS, our lives must be protected from the various diseases and pathogens that are increasing every day. We have the methods and the desire; the hard part is changing the fire service culture of risking ourselves for low-gain situations.

Review Questions

1. Explain how history has evolved to create the current standards of cover practice the fire service has embraced.
2. Describe integrated risk management planning and how it affects the fire service.
3. Describe the three considerations made in a community risk assessment.
4. Discuss the Insurance Services Office's (ISO) Fire Suppression Rating Schedule, NFPA Standards 1710 and 1720, and the Commission on Fire Accreditation International (CFAI) bases for assessing fire risks and community response capabilities.
5. Explain the impact of the *National Strategy for Homeland Security* on the fire service. Include discussion on the strategic framework based on three national objectives and the six critical mission areas.
6. Define the difference between homeland security and national security.
7. Describe how the probability/consequence matrix works. Give an example for each quadrant.
8. Explain standards of coverage.
9. Describe the response performance continuum; discuss each of the components.
10. Discuss the various components of fire protection systems. Explain why it is essential to have the various levels of fire protection.
11. Describe the Utstein criteria. Discuss how these criteria play a role in response zones for EMS response units.
12. Explain the purpose of ISO. Include an overview of how ISO has evolved.
13. Explain the purpose of the CFAI accreditation process. Describe the components and the purpose behind each component.
14. Describe the various risk categories. Include discussion on the integrated risk management plan and the development of a risk assessment.

References

1. www.ushistory.org/franklin/philadelphia/fire.htm
2. Fire Brigades Union, the National Document, *Planning to Make a Real Difference* (Birmingham, England: FBU National IRMP Department, Folium Group Ltd., 2004).
3. Commission on Fire Accreditation International, *Creating and Evaluating Standards of Response Cover for Fire Departments*, 4th ed. (CD-ROM).
4. *Community Risk Management*, National Fire Academy, "Analytical Approaches to Public Fire Protection," pp. 6–9.
5. The White House, *The National Strategy for the Physical Protection of Critical Infrastructures and Key Assets*, February 2003.
6. Commission on Fire Accreditation International, *Self-Assessment Manual*, 6th ed.
7. Commission on Fire Accreditation International, *Creating and Evaluating Standards of Response Coverage for Fire Departments*, 4th ed. (Chantilly, VA:)
8. Ibid.
9. "Ensuring Effectiveness of Community Wide Emergency Cardiac Care," *JAMA*, 268, no. 16 (1992).
10. www.americanheart.org/presenter.jhtml?identifier+4550
11. Dennis Gage, *Evaluation of Fire Departments as a Factor in Property Insurance Rating*, © ISO Properties, Inc., 2005.
12. Commission on Fire Accreditation International, *7th Edition*, 2005.
13. National Fire Protection Association, *NFPA 1710*, 2001.
14. National Fire Protection Association, *NFPA 1720*, 2004.

CHAPTER 10

Quality of the Fire Service

Key Terms

baseline data, p. 476
benchmark, p. 470
continuous improvement, p. 455
customer service, p. 476
cycle time, p. 473
effectiveness, p. 482
efficiency, p. 482
plan-do-check-act (PDCA) cycle, p. 468
quality, p. 487
quality control, p. 490
quality improvement, p. 451
Six Sigma, p. 451
zero defects, p. 457

Objectives

After completing this chapter, you should be able to:

- Describe the principles of quality improvement.
- Explain what Six Sigma is.
- Explain how to use benchmarking to improve performance.
- Define performance measures, baselines, and benchmarks.
- Describe the obstacles to implementing quality in the fire service.
- Analyze how your department is using quality concepts to improve performance.

QUALITY IS JUST GOOD BUSINESS!

In 1993, I had the opportunity to represent the International Association of Fire Chiefs in Orlando, Florida, at Motorola's Total Customer Satisfaction (TCS) Showcase for its Land/Mobile Products sector. This was, in part, one of its corporate efforts toward achieving Six Sigma quality. The 20 teams chosen as finalists were showcased at this event and had competed with more than 900 other teams worldwide from this particular sector within Motorola during the past year. The winning team had the honor of being one of 20 teams attending and presenting its TCS problem-solving process corporate-wide. This entire event was indicative of the way Motorola had institutionalized the quality improvement process within its organization, calling it "total customer satisfaction." Companies that have implemented the quality improvement process (QIP) understand that an organizational commitment must be made to institutionalize the process if it is to succeed.

One team at the TCS Showcase was a great example of how the focus on **quality improvement** has an impact on daily operations. This group had identified that its facility was classified as a major source of air emissions for volatile organic materials and nitrous oxides by the Environmental Protection Agency (EPA). Under new EPA guidelines, such a classification would restrict future operations at this facility. The goal was simple: reduce emissions from more than 96 tons per year to less than 25 tons per year. This team, having applied the quality improvement process it was trained to do, involved all operations that contributed to these emissions to find solutions. The solutions included purchasing new equipment, reducing the use of chemicals, and changing processes to include new technology. Within 12 months the team's efforts resulted in a 75 percent reduction in emissions from nearly 100 tons per year to only 23.5 tons per year. This level met the new EPA standards without restricting the operation of the facility. Not only did these actions save the company a tremendous amount of money and allow it to continue on a normal production schedule, but they also institutionalized the quality process within its organization. It clearly demonstrates why this process works. In less than 12 months, six people were able to focus on identifying the problems, examine possible solutions, set a direction to correct the problems, and evaluate their success.

quality improvement
The actions taken to increase the value to the customer by improving the effectiveness and efficiency of processes and activities throughout the organizational structure.

WHY SIX SIGMA?

For Motorola, the originator of **Six Sigma** (see Figure 10.1), the answer to the question "Why Six Sigma?" was simple: survival. Motorola came to Six Sigma because it was being consistently beaten in the competitive marketplace by foreign firms that were able to produce higher-quality products at a lower cost. When a Japanese firm took over a Motorola factory that manufactured Quasar television sets in the

Six Sigma
This is the statistical measurement for Motorola's quality initiative. It represents no more than 3.4 errors per million opportunities.

FIGURE 10.1 ◆ Six Sigma (*Courtesy of Randy R. Bruegman*)

United States in the 1970s, it promptly set about making drastic changes in the factory's operation. Under Japanese management, the factory was soon producing TV sets with one-twentieth the number of defects it had produced under Motorola management. It did this using the same workforce, technology, and designs, making it clear that the problem was Motorola's management. Eventually, even Motorola's own executives had to admit "our quality stinks." Finally, in the mid-1980s, Motorola decided to take quality seriously. Motorola's CEO at the time, Bob Galvin, started the company on the quality path known as Six Sigma and became a business icon largely as a result of what he accomplished in quality at Motorola. Today, Motorola is known worldwide as a quality leader and profit leader. After Motorola won the Malcolm Baldrige National Quality Award in 1988, the secret of its success became public knowledge and the Six Sigma revolution was on.[1]

Question: How would you as a fire chief, battalion commander, or company officer like to be able to create a team that could boast a 75 percent improvement in a service you provide or in a process within the organization?

I left the TCS event with a realization that the public sector and the fire service cannot continue to approach our business as if we are not competing with anyone else. The fire service is in continuous competition at the local level to obtain sufficient tax dollars and budgets to operate our departments. We often underestimate the powerful forces occurring within the United States focused on the reinvention of government. Today, most in the fire service realize that any improvement process we hope to implement will be successful only when labor and management work together. For many fire service organizations, this means stepping into a new framework of cooperation. If we are to achieve our maximum potential, we must focus on outcome through quality measurements. Let us not forget that less than three decades ago the American corporate world was struggling to compete with the quality of products being produced outside the United States. Today, not only is our country competitive, but in many sectors it is leading the field.

THE IMPACT OF QUALITY

The impact of quality initiatives will be recognized only when we commit to the use of continuous improvement processes and focus on customer service and satisfaction. Prior to World War II, a gentleman named W. Edwards Deming (Chapter 3) approached our corporate leaders in the automobile industry with the concept of quality. He was rebuffed, largely because at that time the auto industry and corporate America had little competition. Most products sold in this country were made in America and, at the time, the quality of foreign-made goods was less than desirable. However, after the devastation and subsequent economic collapse Japan suffered as a result of World War II, Deming was asked to take his concepts of quality to Japan to help rebuild a country. Although it took 20 years, the decision by corporate America not to embrace the concept of quality had a significant effect on the economic fortunes of many American business sectors in the 1970s and 1980s. Japan, in a short period of time, began producing products with very high quality, more cheaply, and it began to capture significant market share in transportation and electronics. Today, while we have seen a turnaround in the quality of goods being produced in America, the market once enjoyed by U.S. companies has been lost. The realization that quality and customer satisfaction are woven into the fabric of today's customer expectations is an important fact that must also be integrated into the services provided by the public sector.

An objective for the fire service is to ensure it does not travel down the same road many U.S. corporations journeyed after World War II. When the leaders of any industry (public or private) become so complacent that they think they have "cornered the market," significant vulnerability is often the result. The corporate landscape is littered with many examples of this. AT&T, Bell Telephone, U.S. automakers, IBM, and General Electric (to name a few) had a disproportionate share of the market. This often leads organizations or industrial segments to evolve to an unhealthy compliance level. We in the fire service cannot be lulled into thinking no competition exists for the services provided. Although we may be the only local fire department, we compete for funding each and every day with other departments within the jurisdiction we serve.

So, how do we prove what we do is worthy of the investments made by the community? One of the first hurdles we must face in the fire service is to overcome the conventional wisdom that quality is not an important issue within our profession. Look at what is happening today in the public sector regarding many of the *reinventing government* initiatives. Many citizens have expressed considerable anticipation and expectation that the public sector will recognize and correct the bureaucratic redundancies found in government. They believe we must set a new course to run government more efficiently and must focus on the needs of the customers, not on the needs of the people who work in the system. This perspective often plays itself out in local elections for tax support, which have become increasingly more difficult to pass.

We have entered an era of rapid change within society and the fire service, which will require us to become more adept at quantifying our roles and impacts as service providers to the community. The change is already reflected in a higher level of community expectation from departments and their chief executives and the local fire stations to provide better service with fewer dollars. At the same time, there should be a higher level of understanding from the elected leadership and the citizenry at large about how their departments are using tax dollars to provide emergency response and related services to their communities. Quality management is a means of institutionalizing a process within each of our organizations. Quality management will help to provide the means to measure performance, facilitate needed internal change, and provide a higher level of service to our customers. There is a great deal of wisdom to be learned from one of the great entrepreneurs of the twentieth century, Walt Disney. From the moment you walk into Disney World or Disneyland, it is evident that the whole focus is to provide a quality experience to the customers. These parks' commitment to providing a high-quality product has been institutionalized by a commitment to train their people to "exceed the customers' expectations," their attention to detail, and their customer service. This is not by accident but by design. In an interview with Walt Disney in the early 1950s, as he was sharing his vision of what Disneyland would be and citing his attention to doing things right the first time, he emphasized that "quality is just good business" (see Figure 10.2).

What a great theme for the fire service: "exceed the customers' expectations." For many departments, the ability to meet the core mission will depend on the quality of services, the ability to minimize costs and maximize performance through measuring how and what is done, and how well it is done. When the fire service can do this, the competition for local economic resources becomes much easier (see Figure 10.3)

WHAT QUALITY IS NOT

One very important point to note, as quality is designed into the fire and emergency medical services delivery package, is that your quality initiative project must not become

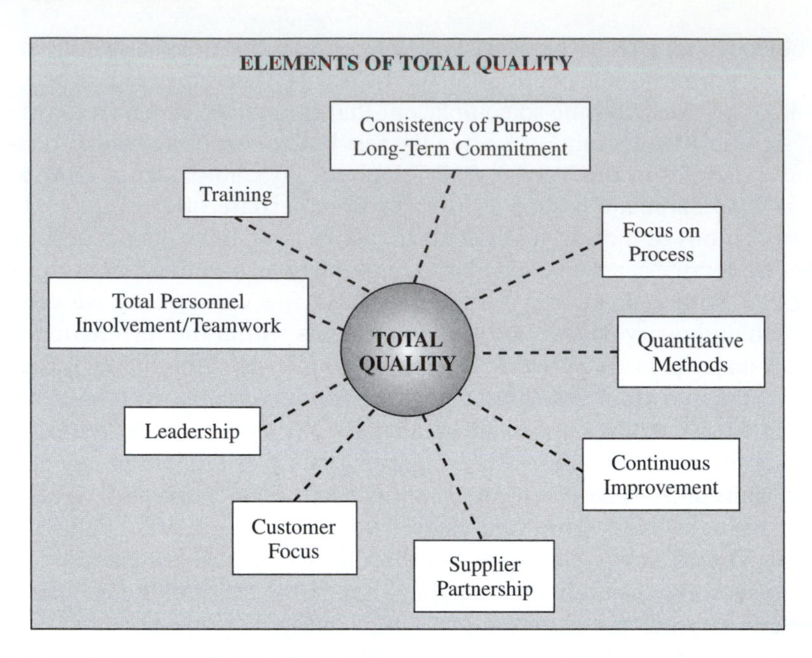

FIGURE 10.2 ◆ Elements of Total Quality (*Courtesy of Randy R. Bruegman* Exceeding Customer Expectations: Quality Concepts for the Fire Service, *p. xii*)

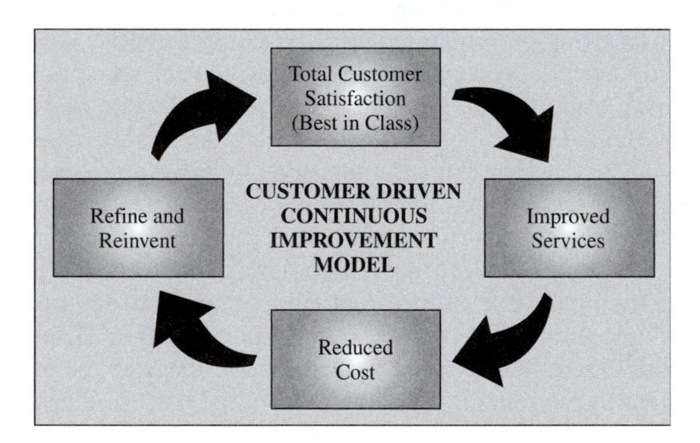

FIGURE 10.3 ◆ Customer Driven Continuous Improvement Model (*Courtesy of Randy R. Bruegman,* Exceeding Customer Expectations: Quality Concepts for the Fire Service, *p. xiii*)

so cumbersome as to create a bureaucracy unto itself. Quality is essential for any organizational success and competitive advantage in the future. It is the cornerstone of good customer service. However, just having a quality program is not synonymous with providing good quality service. As you begin to design your quality initiative, here are some factors to keep in mind:

1. Quality processes have little to do with emotion and everything to do with the facts at hand.
2. Quality improvement can be used as an organizational fad for short-term results, for winning quality awards, or as a marketing tool. This will not ensure a quality product. Make sure quality improvement is undertaken for the right reasons: to find better ways to meet your objectives and still maintain core effectiveness.

Obstacles to Implementing Total Quality
- The hope for instant success
- Looking elsewhere for solutions
- Saying "Our problems are different..."
- Continuing only to an "acceptable level of quality"
- Believing that employees cause all the problems
- Relying on quality control departments
- False starts with no organization-wide commitment

FIGURE 10.4 ◆ Obstacles to Implementing Total Quality (*Courtesy of Randy R. Bruegman*)

3. A good quality program does not identify and create new outside relationships. It means you strengthen the ones you have through good customer service and quality management techniques. Quality management is not about radical organizational transformation. Instead, it is about introducing new concepts, ideas, and paradigms (frameworks) into your organization to begin to think differently and approach processes in a cause-and-effect relationship.
4. Quality programs can create a paradox within your organization. On the one hand, you are trying to standardize the process to pursue consistent improvement, while on the other you are generating an inverse relationship toward the entrepreneurship and creativity needed in public sector organizations today.
5. Because quality can develop its own cumbersome bureaucracy, focus on the result and not on the creation of a whole separate area in your organization to implement this process.
6. Quality should not focus on minimum standards but on **continuous improvement**, no matter how small the improvements. Incremental improvements lead to the opportunity to create exceptional performance.
7. Quality as a philosophy, when integrated into daily activity, prompts fundamental concepts and process-improving organizational performance (see Figure 10.4). Think of all the activities you do on a daily basis, such as responding to calls, inspections, prefire plans, and public education. In any of these processes, is there an opportunity to improve them? In most organizations the answer is yes, but they often lack the means by which all levels of the organization can play a part in improving the systems.

continuous improvement
Perhaps the central tenet of the quality revolution is that improvement must be continuous. The goal of quality improvement is zero defects. Continuous improvement (CI) is the cornerstone of a lasting competitive advantage.

CHALLENGES TO THE FIREFIGHTING COMMUNITY

One of the hallmarks of the firefighting community has always been a strong sense of heritage, continuity, and commitment. Our competition with other divisions, departments, and units of government for our share of the available tax dollar has never been more extreme. Most communities today are faced with a growing list of community needs. Many federal programs have been delegated to the state and local governments, such as new regulatory requirements, new service changes, and terrorism response. This has resulted in several key financial issues for local and state public sector agencies today:

- Rise of competition from private sector to provide many governmental services
- Increasing sophistication and cost of equipment
- Rising need for specialized equipment and training
- Constrained public budgets
- Increasing regulation and standardization
- New service demands and unfunded mandates

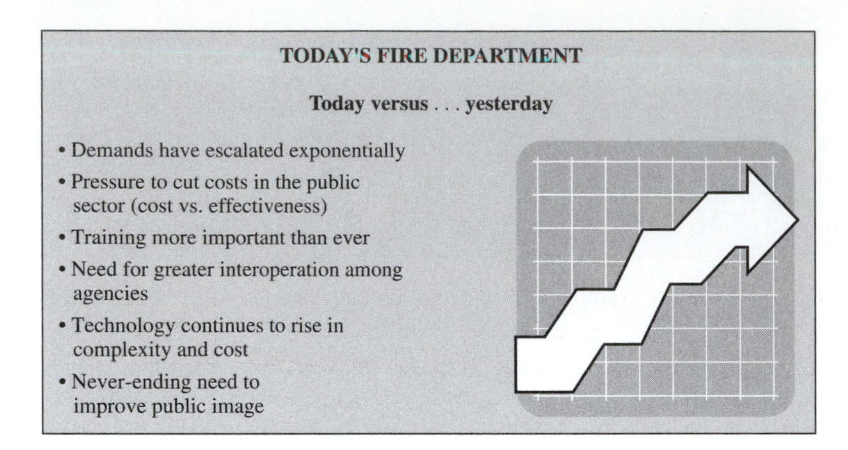

FIGURE 10.5 ◆ Today's Fire Department (*Courtesy of Randy R. Bruegman,* Exceeding Customer Expectations: Quality Concepts for the Fire Service, p. xvi)

TODAY'S FIRE DEPARTMENT

What effect do these changes have on fire departments, chief executives, and company officers? While the demands placed on fire departments have escalated, in many cases the resources have remained static (see Figure 10.5). A good example is the provision of emergency medical services. For most departments, what started 30 years ago as a very minor part of our duties now constitutes the highest percent of all service calls. With the aging population and the state of health care in the United States, the likelihood this will continue to increase is guaranteed. EMS consumes a large percentage of our resources and has dramatically increased our need for training and certification. Yet this is just one example of many services we have undertaken, developed, and implemented in the last three decades that have created impacts on the organization's ability to provide a quality level of service.

Pressures on cost cutting and cost containment in the public sector have been strong for more than a decade. The economic slowdown in recent years and the events of September 11, 2001, have magnified the lack of resources for the fire service. Even with more federal grant money available and an improving economy, it is safe to assume local constituencies will not immediately respond by approving larger city and county budgets for the fire service. Most would agree this type of environment can result in a great deal of anxiety and frustration. We face situations in which we are not comfortable with any of the choices available. It seems clear no matter how hard we have worked in the past, we will need to work smarter in the future. A point in our favor is that as fire and emergency service providers, we are used to handling tough jobs. Challenges are a part of what we do while facing the kinds of pressures that can be overwhelming if we are not prepared.

THE NEED FOR QUALITY IMPROVEMENT

We will face these challenges and adapt to the future through planning, training, teamwork, and a focus on a quality level of service. A surgical team faces similar pressures to that of a fire company (see Figure 10.6). It holds people's lives in its hands. The surgical team is exposed to huge liabilities and must contend with a staggering array of regulations, procedures, and guidelines. One tiny mistake can lead to a tragic

FIGURE 10.6 ◆ Surgical Team (*Courtesy of Randy R. Bruegman,* Exceeding Customer Expectations: Quality Concepts for the Fire Service, *p. xvii*)

loss of human life. The difference between the surgical team and a fire company is that the loss of life can be one of our own people.

So how do those on surgical teams do it? The answer is through quality improvement. Going into every surgery, they know they cannot afford to make a single mistake, so they plan for every possible contingency. They visualize the worst-possible scenarios and assume they might happen. Their goal is to function effortlessly as a team for optimum results and to prevent tragedy. Surgical teams operate on a single principle: everything must be done perfectly. Every operation must be performed exactly by the book. This is what quality improvement is all about: **zero defects**. It is important to any business. Motorola's goal of Six Sigma quality (3.4 defects per million) sounds almost impossible, yet it has helped to transform its corporate culture and viability. If you are headed to the hospital for surgery, Six Sigma quality from your doctor starts to sound like an absolute minimum. What this means is striving for error-free production or administrative functions 99.9999999998 percent of the time, resulting in only two defects per billion opportunities. Modified to accommodate realistic variations in any process results in *3.4 defects per million opportunities, or Six Sigma quality*.

zero defects

A situation that exists when all quality characteristics are produced within design specifications. This concept is reflected in the attitude that defects can be prevented, especially if more attention is given to the task at hand. It is the theme that most embodies the concept of "do it right the first time."

Doing the Math

6 Sigma = 3.4 defects per million
5 Sigma = 230 defects per million
4 Sigma = 6,210 defects per million
3 Sigma = 66,800 defects per million
2 Sigma = 308,000 defects per million
1 Sigma = 690,000 defects per million

So what does this really mean? To give you an example, what would you get from a supplier with 99.73 percent quality?[2]

- At least 54,000 wrong drug prescriptions each year
- More than 40,500 newborn babies dropped by doctors/nurses each year
- Unsafe drinking water about two hours each month

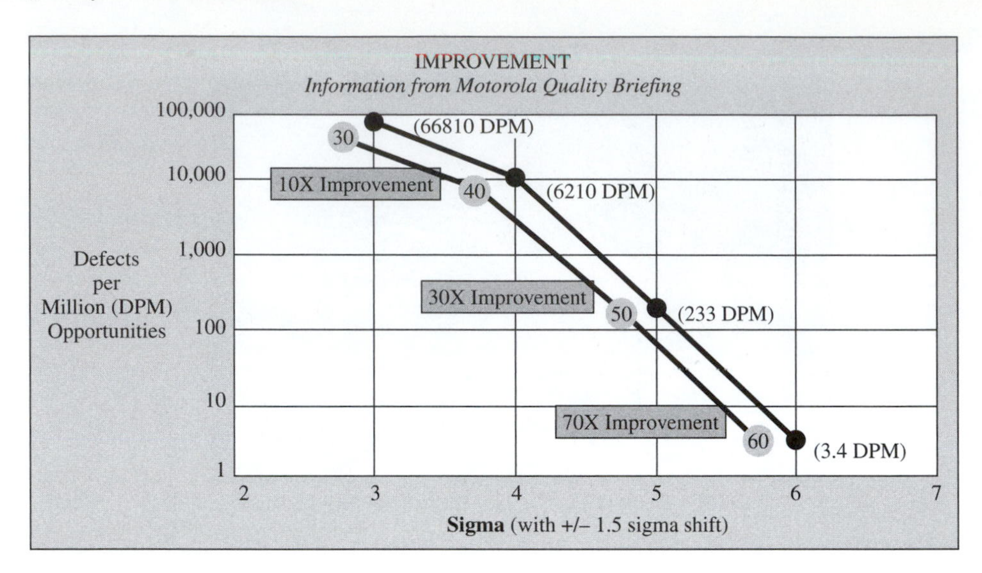

FIGURE 10.7 ◆ Improvement Information from Motorola Quality Briefing (*Courtesy of Randy R. Bruegman,* Exceeding Customer Expectations: Quality Concepts for the Fire Service, *p. 51*)

- No telephone service or television transmission for nearly 27 minutes each week
- Five short or long landings at O'Hare Airport each day
- Nearly 1,350 incorrect surgical operations per week
- Approximately 54,000 lost articles of mail per hour

Now, compare this with the effects of good quality, or what you get from a Six Sigma supplier (99.9999999998 percent), or 3.4 defects per million opportunities.[3]

- One wrong drug prescription in 25 years
- Three newborn babies dropped by doctors/nurses in 100 years
- Unsafe drinking water 1 second every 16 years
- No telephone service or television transmission for nearly 6 seconds in 100 years
- One short or long landing in 10 years in all the airports in the United States
- One incorrect surgical operation in 20 years
- About 35 lost articles of mail per year

Figures 10.7 and 10.8 show how quality improvement within an organization is used to achieve Six Sigma quality, and more important, what it actually means from a customer viewpoint. So where is your organization from a quality standpoint? Internally, is your organization a Six Sigma? Is it 90 percent effective, 95 percent effective, or 99.9 percent effective? How does it measure up with other organizations its size and type? How do you determine how effective your delivery of service is, and how does it compare with others offering similar services? The same holds true for the fire service. If your house is on fire or your child is having a medical emergency, do you expect Six Sigma quality?

WHAT DOES THIS MEAN TO YOU?

If you were to take a picture of where your fire department is today, what would it depict? A top-flight surgical team or ____ (*you fill in the blank*). The directions our industry will take in the future are directly connected to the rising demands placed on

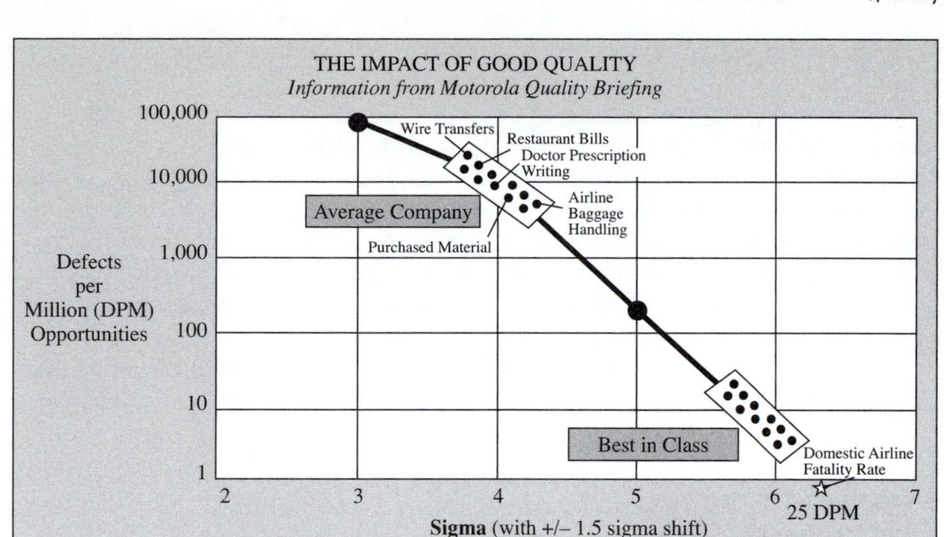

FIGURE 10.8 ◆ The Impact of Poor Quality (*Courtesy of Randy R. Bruegman,* Exceeding Customer Expectations: Quality Concepts for the Fire Service, *p. 52*)

all public service agencies. The value of a quality improvement program is well stated, but the real question for many in leadership and management positions will be, "What's in it for me?" Maybe you are convinced that a quality improvement program can offer value to your department. However, what will it do for you? The greatest benefit quality improvement can offer, whether you are a chief officer or a company officer, is the positive effects it will have on the people in your organization and the service it provides (see Figure 10.9). Because it is based on teamwork, commitment,

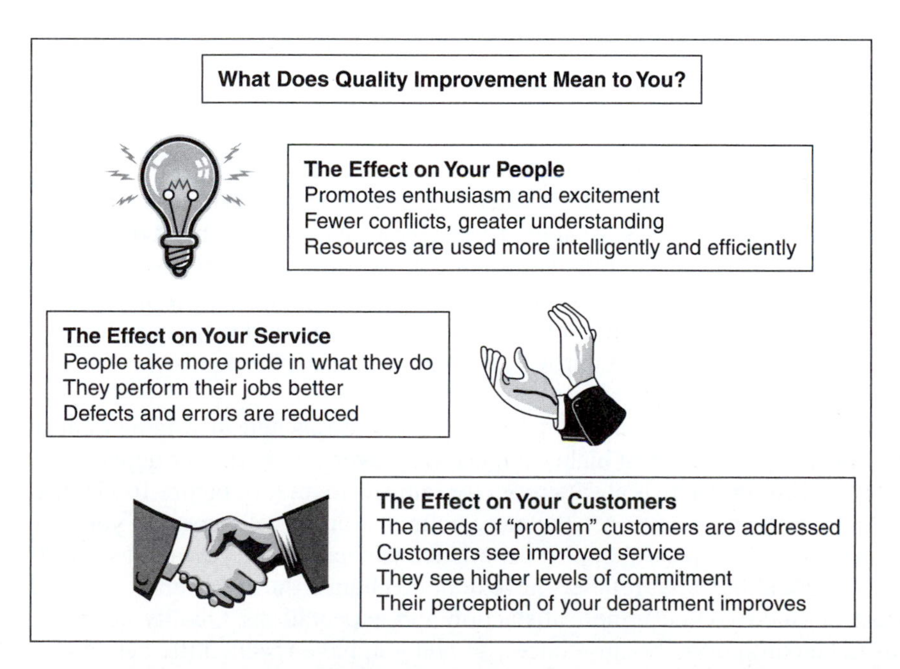

FIGURE 10.9 ◆ What Does Quality Improvement Mean to You? (*Courtesy of Randy R. Bruegman*)

and input, a good quality improvement program can build enthusiasm, excitement, and involvement within your organization.

A good quality improvement program can also help identify your "problem" customers and deal with them more effectively. Your problem customers may include developers, labor groups, elected officials, city managers, community groups, or ethnic groups. In many cases, the fire service does not realize the problem because we fail to ask the customer and fail to measure our performance. We face many conflicting demands from a variety of different sources. A good quality improvement program offers a way to manage those demands and match our resources to our customers' expectations. So, what is in it for you? Hopefully, your department or company will run more efficiently, perform better than ever before, gain a high level of public esteem and support, measure its performance, continually improve, and strive for zero defects.

THE ELEMENTS OF QUALITY IMPROVEMENT

If any organization hopes to move toward zero defects, it must address the vital elements of quality improvement: cross-functional teaming, identification of customer expectations, careful measuring and mapping, and a commitment to continuous quality.

Cross-Functional Teaming (No Silos). Cross-functional teaming means everyone involved in a particular process is connected directly with everyone else involved. By eliminating the silos, those imaginary walls existing between departments and command structures, the people who actually perform a job are the people in control. Nobody can spot a problem in a process better than the person who actually performs the process; yet when organizational silos exist, critical information is often not communicated.

Identification of Customer Expectations. Successful organizations have developed the mechanisms to determine what their customer expectations are (see Figure 10.10). What is the definition of quality? The answer is, whatever the customer wants. Only your customers can define quality. Have you ever asked them?

Careful Measuring and Mapping. Because quality is something the customer defines, you have to be able to measure what you do in the customer's terms. If a customer needs low cycle time (such as response times to emergencies), then you must determine how the customer thinks of cycle time and then compare that to what you currently provide. By putting those elements together—cross-functional teaming (no organizational silos), identification of customer expectations, measuring, and mapping—you lay the foundation to progress toward improving quality in your department.

Commitment to Continuous Quality. When Motorola reached Six Sigma quality consistently, the company set even higher targets. Wherever you find your department on the quality continuum, there is always room for improvement (see Figure 10.11). The cycle is continuous: the better you get, the more your customers will expect of you. For example, think about the typical shopping experience you had in a department store 20 years ago, and think about what retailers have done to change the experience. You can see the effect of quality on customer satisfaction and expectations. Quality improvement needs to be continuous, because once you feel you have reached the best-possible level of quality, your quality will have only one place to go . . . down.

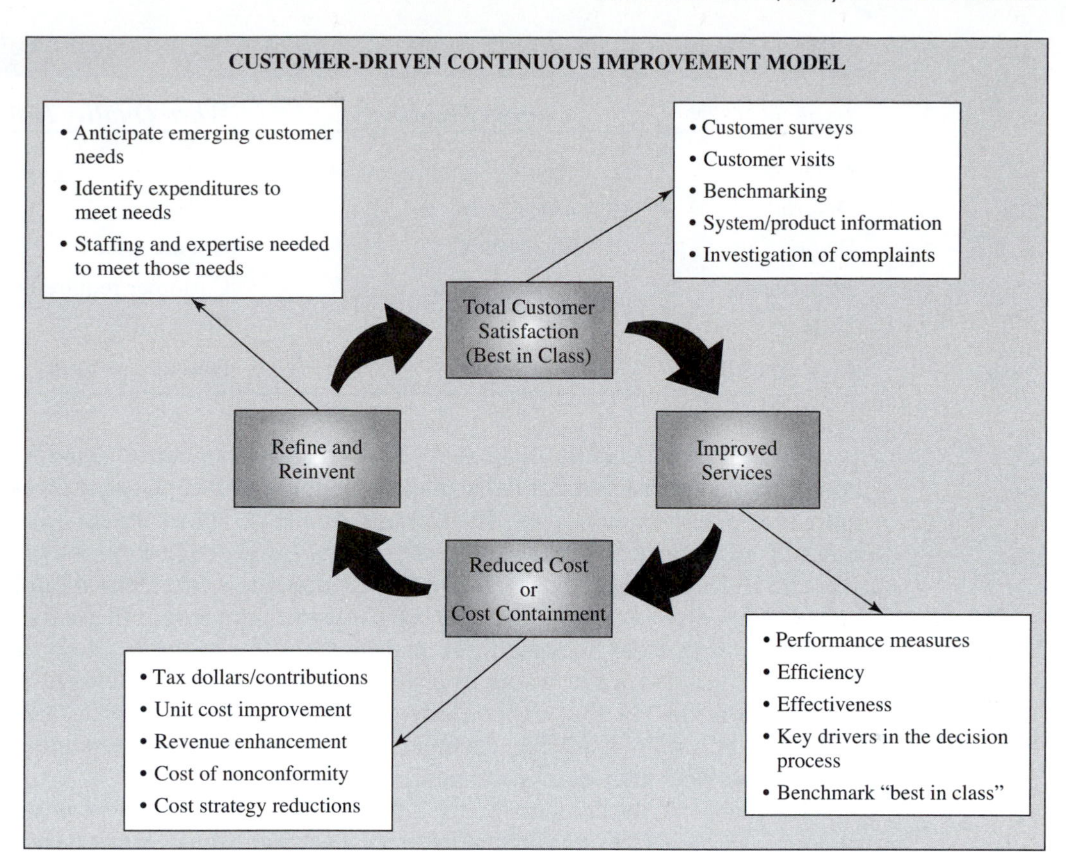

FIGURE 10.10 ◆ Customer-Driven Continuous Improvement Model (*Courtesy of Randy R. Bruegman,* Exceeding Customer Expectations: Quality Concepts for the Fire Service, *p. xix*)

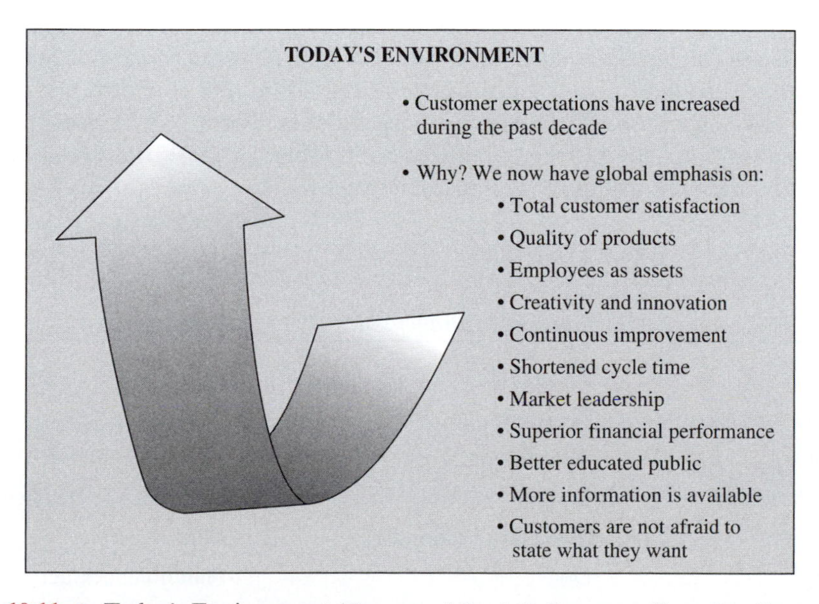

FIGURE 10.11 ◆ Today's Environment (*Courtesy of Randy R. Bruegman,* Exceeding Customer Expectations: Quality Concepts for the Fire Service, *p. xviii*)

TABLE 10.1 ◆ What Quality Is About

Issue	*Current Mind-Set*	*When Quality Is the Focus*
Mistakes	Who?	Why?
Defects	Fix	Prevent
Errors	Inevitable	Not accepted
Performance	Cost/schedule	Customer requirements
Training	Cost	Investment
Measurement	Justify	Find the problem

Most of us have heard the quality buzzwords in a number of organizational fix-it books. "Is quality just another fad in management literature?" Quality is really nothing more than plain common sense. The concepts of quality improvement are not a collection of confusing jargon meant to make you look and sound impressive (see Tables 10.1 and 10.2). Instead, they are simple, honest, down-to-earth ideas about how to help departments, divisions, and individual fire companies run more efficiently.

Firefighting today is very different from the time when a group of local men hitched a horse to a water wagon and simply doused a blaze with water delivered by a bucket brigade. For that matter, today's firefighter is also very different from the firefighter of just 20 years ago. Changes have swept across all segments of business, including services provided by the public sector.

The demands placed upon the fire service have expanded. Because a firefighter is no longer just a person who fights fire, technical expertise is required in a wide range of areas, including structural engineering, chemistry, hazardous materials handling and documentation, emergency medical procedures, and a variety of other specialty topics. Training has become an important line item in fire department budgets at a time when it is difficult to satisfy the numerous other budgetary requirements.

Firefighters also face a number of other challenges. Exposure to blood and bodily fluids raises concerns about blood-borne diseases. Recent criminal attacks on fire crews as they respond to emergencies have raised additional concerns and have significantly affected morale. Large-scale emergencies such as the first bombing of the World Trade Center, the horrific events of September 11, 2001, the Oklahoma City federal building, and various natural disasters such as Hurricane Katrina have shown

TABLE 10.2 ◆ Total Quality

Issue	*Current Management*	*Total Quality*
Vender choice	Price alone	Price/quality (best value)
Change	Resisted	Way of life
Technology	Automation	Empower/innovate
Information flow	Vertical	Horizontal/vertical/circular
Leaders	Manager or enforcer	Coach/visionary
Workers	Do	Plan/do/check/act
Performance	Standards	Efficiency/effectiveness
Customer service	What we define	What the customer defines

clearly how critical cooperation is among public agencies. Tearing down the traditional barriers and competition among agencies to deliver quality services is a clear expectation of the public today.

One issue that particularly impacts the firefighting community is that, although firefighting equipment has never been inexpensive or simple, the growth of fire service technology has been exponential during the past 20 years. Will the cost of cutting-edge technology spiral out of control until only the largest, best-funded departments can afford it? Will smaller departments make do with older, outdated equipment? This kind of "stair stepping" of technology distribution is common in any market in which the technology is advancing rapidly, but it is especially painful in our industry because the lives at stake include not only the civilian population but also those of our firefighters.

Wrapped around all these issues is the never-ending need to maintain and improve a fire department's image in the community (for our industry to articulate our values) and our performance to our customers. A continuous improvement program should achieve a rigorous process to get a handle on these challenges, make them manageable, and provide a program to prepare the organization for constant change and positive outcomes. The purpose of this chapter is not to teach participants how to build the ultimate quality program. Instead, it is to raise consciousness, demonstrate the need, explain the benefits, and present a few of the techniques to show a quality program is feasible and achievable in your own department. Development of a quality program within your organization will help you create a road map to exceed your customers' expectations, both today and in the future.

INTRODUCTION TO TOTAL QUALITY LEADERSHIP

We have entered a very dynamic era of change within government and the fire service. This change is reflected in a higher level of expectation at all levels of government to provide a broader range of services more effectively, often with fewer dollars. The total commitment of fire service teams to the communities they serve is demonstrated by programs designed to reduce the likelihood and severity of crisis (see Figure 10.12). They provide rapid and professional help when residents face an emergency.

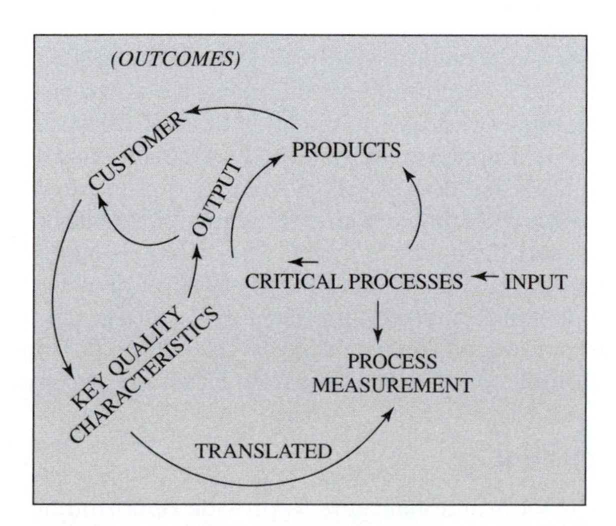

FIGURE 10.12 ◆ Outcomes (*Courtesy of Randy R. Bruegman,* Exceeding Customer Expectations: Quality Concepts for the Fire Service, *p. xx*)

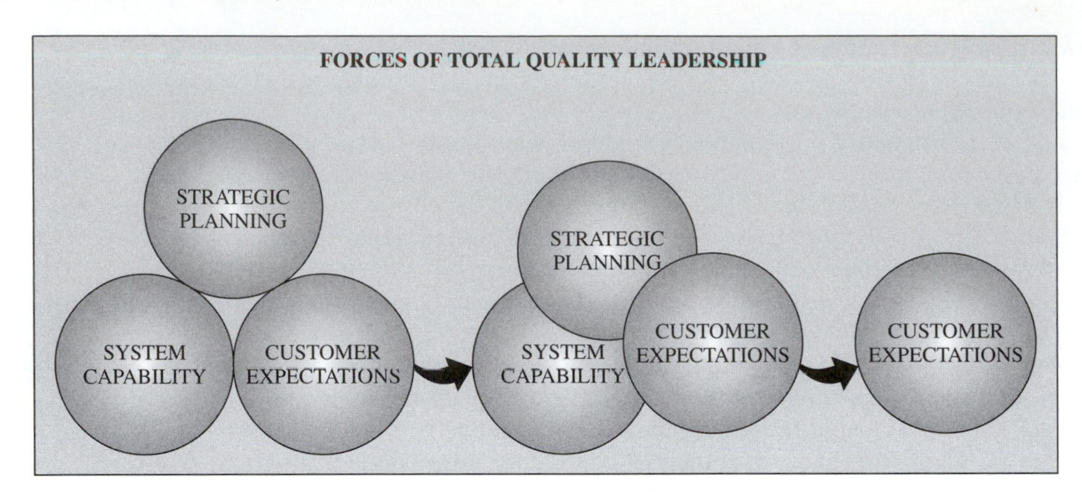

FIGURE 10.13 ◆ Forces of Total Quality Leadership (*Courtesy of Randy R. Bruegman,* Exceeding Customer Expectations: Quality Concepts for the Fire Service, *p. 2*)

Within the fire service, we have begun to develop a corporate culture that maximizes participation of employees through the use of committees or other specific work groups. These committees research and recommend proposals to staff, and they ultimately control and implement many of their own programs. During the last two decades, the movement to empower leadership and management in all levels within an organization has become evident. Although more employees are involved today than before, true change will come only when a process within an organization that measures the ability to perform becomes institutionalized. Quality improvement processes have been effective in doing this. Any quality improvement process must be based on the ability to provide personalized and efficient community services. Efforts made to provide quality services will provide clarity for the distinct areas of service. First, as our programs improve, the expectations of customers will also rise, providing continuous challenge to meet them (see Figure 10.13).

Second, as quality initiatives help to improve our services and processes, they provide a means to measure our performance, which will allow us to meet the rising needs of our customers in the future.

The fire service has a quality workforce that has been meeting these changing needs; however, to reach our maximum potential, we need to provide our employees with the tools, processes, attitudes, and direction that will allow them to be successful. A quality improvement process based on the mission and values of an organization can provide these tools and help us take advantage of and create organizational opportunities in the future. As fire service leaders, one of the obstacles to be overcome from an organizational standpoint is to get our personnel to look past the four walls of the firehouse. They must understand the world around us is revolving, changing, and placing new demands upon the fire chief, the captain, the lieutenant, the firefighter, and the organization. Quality initiatives can help us accomplish this objective, as they provide a means to frame the issue from a customer perspective.

QUALITY AND CHANGE

The future is for leaders and managers to provide opportunities for organizations and people to explore, innovate, and meet the challenges present in the future. As we

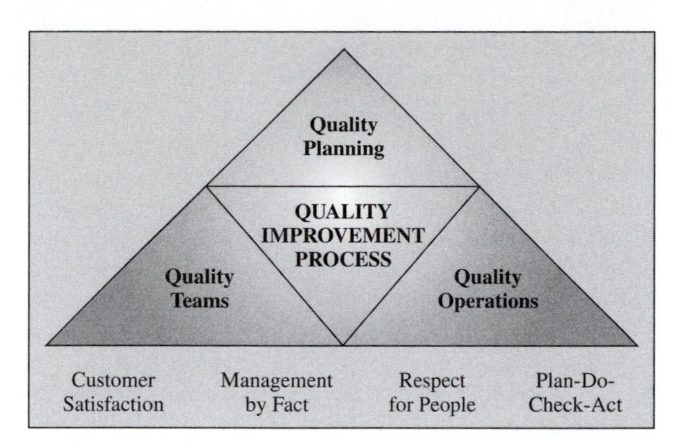

FIGURE 10.14 ◆ Quality Planning (*Courtesy of Randy R. Bruegman,* Exceeding Customer Expectations: Quality Concepts for the Fire Service, *p. 5*)

navigate organizational change and become the architects of the future, we must understand that this is not only about process but also about people. The quality improvement process is a way of ensuring customer satisfaction through the involvement of our employees by learning how to deliver quality services to our customers reliably. The goal is to improve internal and external customer satisfaction through total quality control. The quality improvement process is how we work toward this goal and has three components: quality teams, quality planning, and quality operations. (see Figure 10.14).

1. Quality initiatives engender an environment in which employees work together toward improving the quality of services our organizations provide; enhancing the quality of work life through meaningful participation; developing the skills and abilities of the employees involved; and promoting communication, networking, and teamwork.
2. Quality planning targets our organizational efforts and resources on identified priority issues toward increasing performance levels, improving communication, and attaining broad participation in the development and attainment of organizational goals.
3. Quality operations require the application of the plan-do-check-act (PDCA) philosophy to activities necessary to exceed the needs and expectations of our customers. The goals are to maintain gains achieved through improvement projects, achieve consistency in operations as well as results, clarify individual contributions toward achieving customer satisfaction, and improve daily services and operations.

Four Principles of Total Quality Improvement

1. *Customer satisfaction.* It is vital not only to satisfy the needs and reasonable expectations of customers but also to develop an organizational attitude that puts the needs of the customer first.
2. *Management by fact.* All employees, not just managers, should collect objective data and make decisions based on this information.
3. *Respect for people.* Each of us should recognize the capacity of other employees for self-motivation and creative thought.
4. *PDCA (plan-do-check-act).* A work philosophy emphasizing four key steps of any activity proven to assist in improving performance (see Figure 10.15).

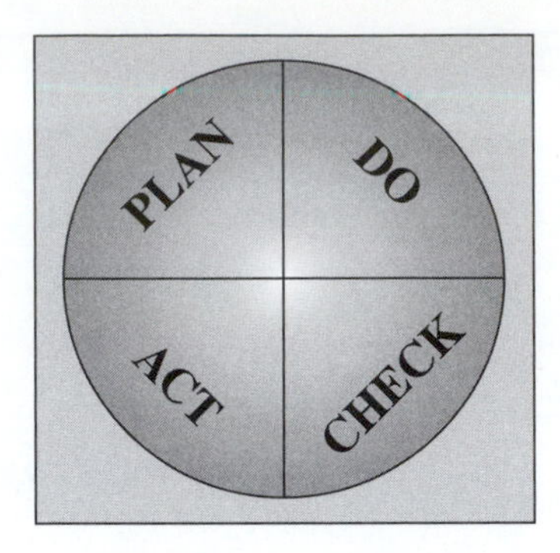

FIGURE 10.15 ◆ PDCA (*Courtesy of Randy R. Bruegman,* Exceeding Customer Expectations: Quality Concepts for the Fire Service, *p. 6*)

Plan What to Do

- Assess needs and prioritize.
- Collect information and data.
- Determine existing standards and requirements.
- Plan improvement.

Do It

- Educate everyone.
- Design controls and measurements.
- Implement changes and improvements.
- Collect data.

Check What You Did

- Clarify knowledge.
- Study effects.
- Report findings.
- Support effort of others.
- Recognize and celebrate.

Act to Prevent Errors and Improve Service

- Act on what was learned.
- Reassess.
- Hold on to the improvements made.

Customer Satisfaction. What is customer satisfaction? It is the heart of the quality improvement process (see Figure 10.16). We all serve both internal and external customers. Our customers deserve to be treated with the same care and attention we would wish to receive.

Here is the customer satisfaction philosophy:

1. We are in a long-term relationship with our customers.
2. We must help our customers identify and satisfy their needs and wants.
3. Their needs and wants are bound to change and evolve.
4. We must keep the lines of communication open between ourselves and our customers.

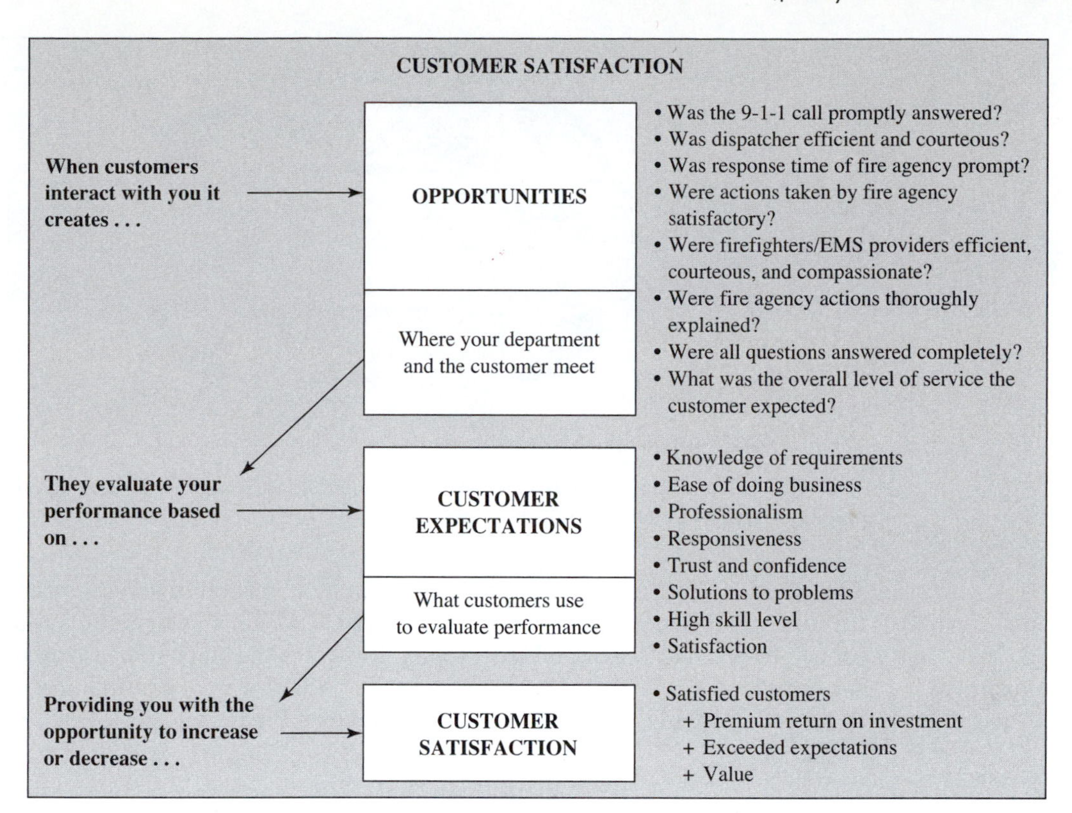

FIGURE 10.16 ◆ Customer Satisfaction (*Courtesy of Randy R. Bruegman,* Exceeding Customer Expectations: Quality Concepts for the Fire Service, *p. 7*)

The requirements for total customer satisfaction include the following:

Top-Down Commitment and Involvement

- Set the example.
- Be active in and support the quality improvement in every aspect of the department.

Measurement System to Track Progress

- Micro level: Baby steps—Many small steps pay big dividends.
- Macro level: Big picture—Assess overall organizational measurements.

Tough Goal Setting (reach out)

- Benchmark the "best in class" (know what and who the best is!).
- Audit results often (develop performance-based measurements).

Provide the Required Education

- The "why."
- The "how to."

Spread the Success Stories Throughout the Organization and Share the Improvement Gained with All

- Recognize and reward the team.

PDCA cycle

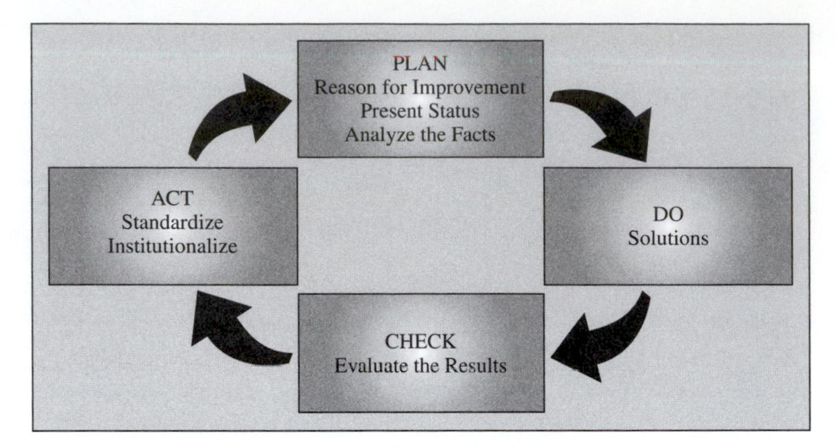

FIGURE 10.17 ◆ Plan, Act, Do, Check (*Courtesy of Randy R. Bruegman,* Exceeding Customer Expectations: Quality Concepts for the Fire Service, *p. 8*)

plan-do-check-act (PDCA) cycle
The systematic approach used to guide managers to quality. Also known as the Shewhart or Deming cycle. It is a scientific method useful for organizational decision making. The initials refer to specific phases that occur during this process: plan, do, check, act.

The Quality Picture. The quality improvement problem-solving process applies the four underlying principles of quality (**PDCA**) to a systematic, data-based approach to problem solving. This structure illustrates the steps to be taken by a team in the improvement process. It also provides a standard way of communicating the steps to be undertaken in any improvement model (see Figure 10.17).

Plan

- **Reason for improvement:** To identify a problem area and define the need to work on it.
- **Present status:** To define a problem and set a target for improvement. At this point, the team is collecting and interpreting data.
- **Analyze the facts:** The team's identification and verification of the root causes of the problem. The analysis step is designed to help teams focus on root causes rather than on symptoms of problems. It is critical to confirm the root cause before proceeding to the solutions step.
- **Solutions:** To confirm the problem and its root cause have been decreased and the target for improvement has been met.

Check

- **Results:** Have the team's solutions correct the identified root cause(s) of the problem?

Act

- **Standardize and institutionalize:** To prevent the problem and its root from recurring. Once the data in the results section of the story indicate solutions have been successful, the team begins to standardize its system for improvement.

The PDCA process helps to *focus on future goals* and allows the team to review its success and address any remaining issues. This structure is a management tool that reinforces the four principles of quality.

The Benefits of a Quality Improvement Process

For Your Employees

- Provides an opportunity for all employees to participate in the implementation and attainment of agency goals
- Encourages decision making at the most appropriate level in the organization

- Improves communication
- Promotes teamwork
- Recognizes the value of each employee by recognizing individual contributions
- Develops the leadership skills necessary to effectively manage
- Increases employee motivation and job involvement at all levels
- Uses the expertise and knowledge of the employees who actually do the work
- Enhances quality of services, efficiency, and productivity
- Develops a people-building philosophy

For Your Customer (see Figures 10.18 and 10.19)

- Enhances customer satisfaction
- Increases responsiveness to your customer

CUSTOMER-DRIVEN QUALITY

The result of any satisfaction measurement process must have three parts:

1. Information from customers.
2. A lens or framework to interpret this information and focus it on the right internal activities.
3. An understanding of how internal activities affect the customer.

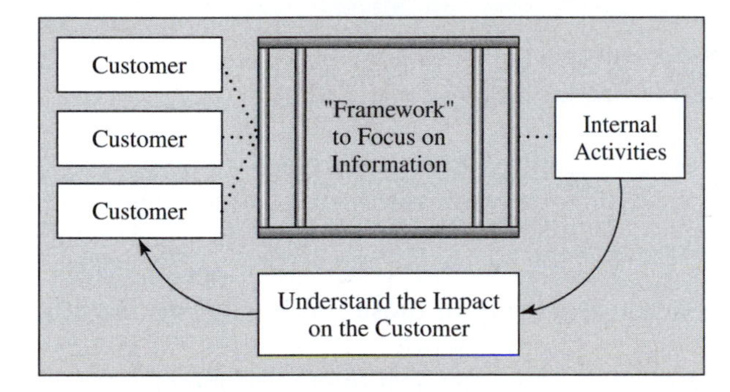

FIGURE 10.18 ◆ Customer-Driven Quality (*Courtesy of Randy R. Bruegman,* Exceeding Customer Expectations: Quality Concepts for the Fire Service, *p. 9*)

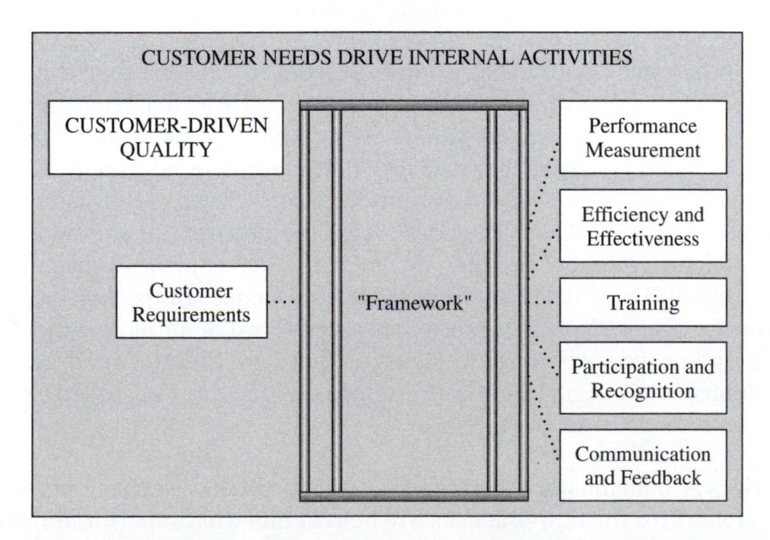

FIGURE 10.19 ◆ Customer Needs Drive Internal Activities (*Courtesy of Randy R. Bruegman,* Exceeding Customer Expectations: Quality Concepts for the Fire Service, *p. 10*)

- Improves services
- Identifies new services to be provided
- Promotes timely, efficient, and accurate service
- Improves communication
- Enhances the customer relationship
- Helps control costs
- Provides for decision making at the most appropriate level

For Your Organization

- Use the skills and knowledge of your people to generate recommendations that will lead to well-informed decision making
- Encourages decision making at the most appropriate level
- Improves communication and teamwork
- Reduces costs
- Increases innovation and creativity
- Reduces errors and improves safety
- Increases motivation
- Develops improved work relationships
- Increases acceptance of new ideas

BENCHMARKING FOR TOP PERFORMANCE—INTRODUCTION

It has been stated that competitive athletic activities would be a terrible waste of time if no one kept score. No one really wants to watch a lot of physical effort between two groups of individuals unless they can enjoy the excitement of finding out who did the best job of playing the game. In the fire service the opposing sides are often the demand for services and the delivery of those services versus the efficiency and effectiveness of those services. Keeping score in these cases is a matter of measuring the level of fire service effort as waged against the fire problem, EMS, inspections, or any service we deliver. In this context, the use of two different terms in keeping score is important. A foundational step in improving your quality of service is the development of two score cards: one is *baseline*; the other is *benchmark*.

Baseline. A baseline is defined as a database from which something can be judged. Baselines consist of the current level of performance at which a department, process, or function is operating. A baseline can be created only by the agency that is delivering the activities and programs, because it is a measurement internal to the organization. For example, a baseline very important for a fire service agency is its experience in responding to alarms. Over a period of years the department should have a database clearly identifying how fast the agency can receive, process, dispatch, and respond to an alarm. These data are the real-life experiences of the organization and a baseline for the agency. In fact, when we stop and think of all the pieces of a service we provide or a process we utilize to measure, there are literally hundreds of important data elements. We begin by choosing to measure the one that will promote value-added service to the customer(s).

benchmark
A benchmark is defined as a standard from which something can be judged. A benchmark is the best practice found in a specific service and/or product and will help define what superior performance is of a product, service, or process.

Benchmark. A **benchmark** is defined as a standard from which something can be judged. Searching for the best practices will help define what superior performance is. Therefore, a benchmark is the best performance that can be found by a department

TABLE 10.3 ◆ Types of Baselines/Benchmarks

Type	*Definition*	*Examples*	*Advantages*	*Disadvantages*
Internal Baselines	Similar activities in different stations, divisions, within the department	• Area covered • Alarm process time • Turnout time • Travel time • Training/exercises • Workload • Violations per inspection • Staffing	• Data easy to collect • "Easy" to evaluate • Can make quick changes	• Limited focus • Internal bias • No external comparison
External Benchmarks Active (can lead to)	Other departments, agencies in other cities of similar size, makeup	• Will provide target for improved performance in identified areas in need of improvement	• Can establish location of origin • Relevant information • Comparable practices/ process • Relatively easy to evaluate	• Data collection difficult • Sometimes unwilling to share • Takes time to collect • Takes time to analyze
State of the Art	Organizations recognized as having state-of-the-art products/ services/processes	• Identified by associations, journals, periodicals, etc. • Agencies involved in self-assessment • Accredited agencies	• High potential for discovering innovative practices • Development of professional networks • Access to relevant databases • Stimulating results	• Difficulty transferring practices • Some information not transferable • Time-consuming • May not have the resources to implement

or others performing similar services or functions. A benchmark is a standard against which the organization can assess itself to see whether its baseline is achieving an acceptable level of performance in the context of industry standards. In the case of the alarm processing area, a benchmark to comparable fire organizations might be the ability to receive, process, and dispatch apparatus and personnel in 60 seconds (see Table 10.3)

Now, if we go back to the comparison of the two concepts, an organization may have a baseline of 90 seconds for alarm processing, whereas the benchmark is 60 seconds. Therefore, the agency has identified an opportunity for improvement to reduce alarm processing time by 30 seconds.

Another good example of the use of baselines and benchmarks would be the travel time of apparatus. A benchmark goal might be that the organization wants to achieve a travel time of less than four minutes to 90 percent of its incidents. The baseline may indicate the organization is able to achieve this in only 75 percent of its incidents. The difference between the baseline and the benchmark can become a performance improvement goal for the organization (see Figure 10.20).

One of the most important aspects of quality improvement is measurement. Companies that have recognized the importance of quality realize they cannot afford to go by their own perceptions about what their customers want or by how well they are providing a service. The process of benchmarking your service(s) against others allows you to evaluate your performance as compared to that of others (see Figure 10.21).

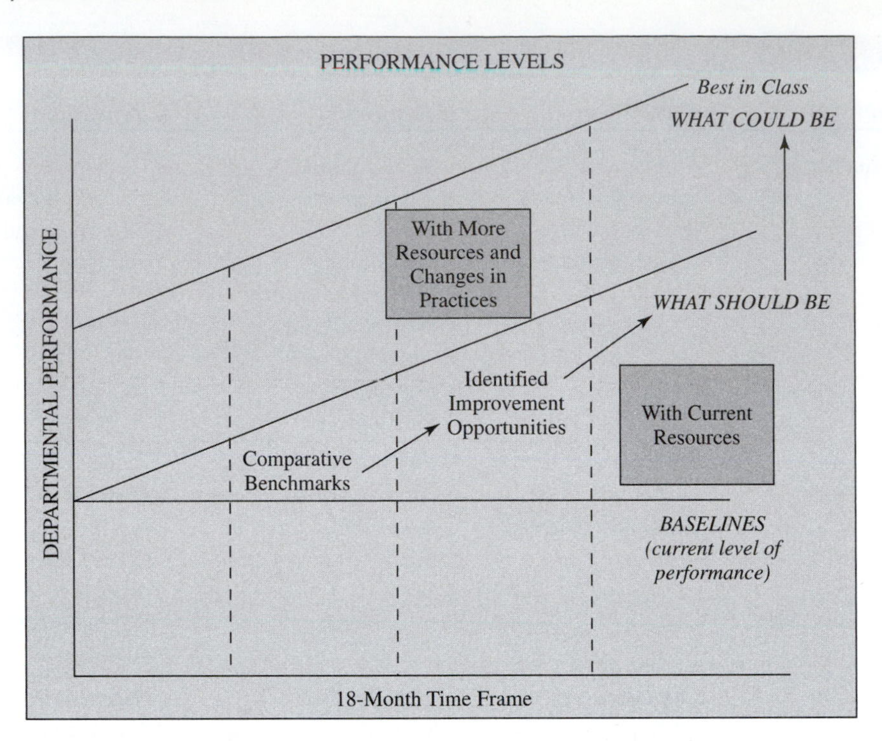

FIGURE 10.20 ◆ Performance Levels (*Courtesy of Randy R. Bruegman,* Exceeding Customer Expectations: Quality Concepts for the Fire Service, *p. 49*)

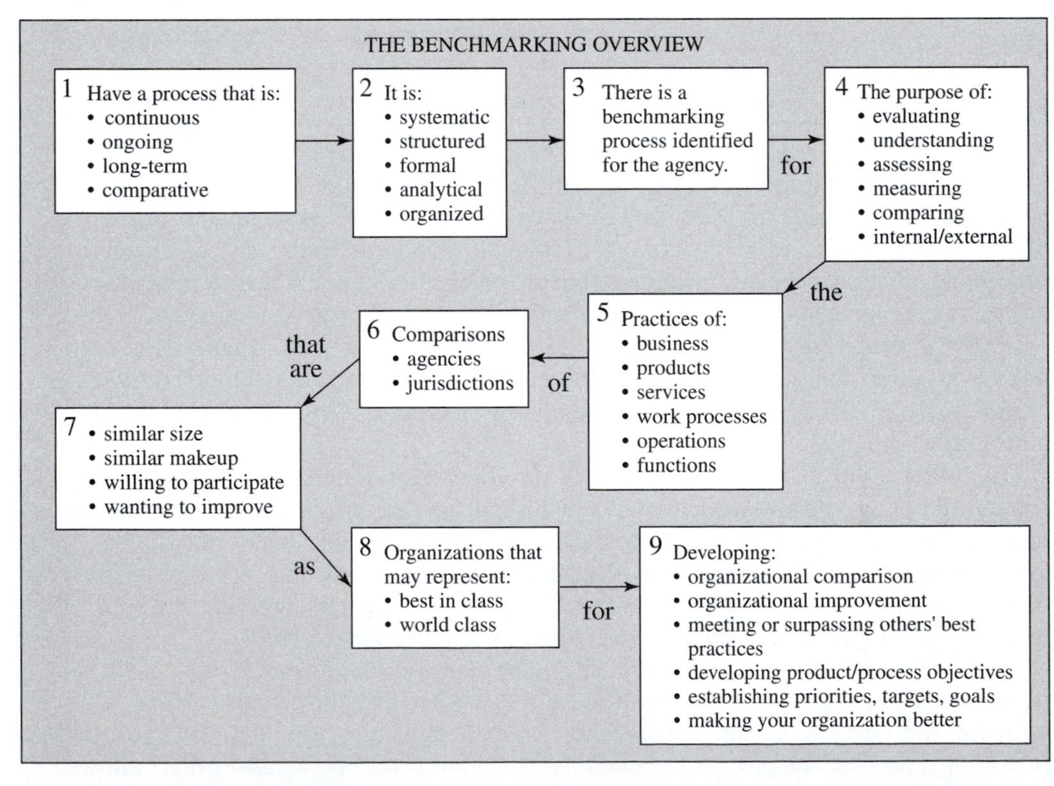

FIGURE 10.21 ◆ The Benchmarking Overview (*Courtesy of Randy R. Bruegman,* Exceeding Customer Expectations: Quality Concepts for the Fire Service, *p. 46*)

Any improvement in quality depends heavily on using the right measurements. If you improve something your customers do not particularly want or care about, then you may have been wasting time and effort that should have been spent improving what the customer does care about. Remember, it is about value added to the customer, and one of the key measurements used to improve quality is cycle time.

CYCLE TIME

cycle time
The moment from the time a customer first requests service to the time that all aspects of the service experience are completed.

Cycle time is a term often used when discussing quality improvement. It is defined as the moment from the time a customer first requests service to the time that all aspects of the service experience are completed. Total cycle time includes time for all escaping defects, and so one of the imperatives of reducing total cycle time is to do it right the first time. Reducing overall cycle time raises the customer's perception of quality and, at the same time, drives up productivity. A common measurement in the fire service is turnout time, the time interval between the notification of an emergency call to the rollout time of the fire station. Now, consider the customer whose house is on fire. Does the customer care how long our turnout time is? It depends. In one sense, the customer does care because the shorter the turnout time, the less time it takes to get to the customer's home. However, if two companies respond, which one will the homeowner be happier to see—the one with shorter turnout time or the one that arrives first? It is not difficult to answer the question. Turnout time by itself is not important, but the customer is more interested in total cycle time. And what is total cycle time? Think about what would happen if your home caught on fire. Total cycle time is the time it takes a fire company to reach you. This is definitely important. On the other hand, if Company A arrives in three minutes, and Company B arrives in four minutes, but Company B quickly deploys and is fighting the fire while Company A is still hooking up equipment, then the customer is happier with Company B. The cascade of events from the CFAI accreditation model is an excellent example of total cycle time (see Figure 10.22).

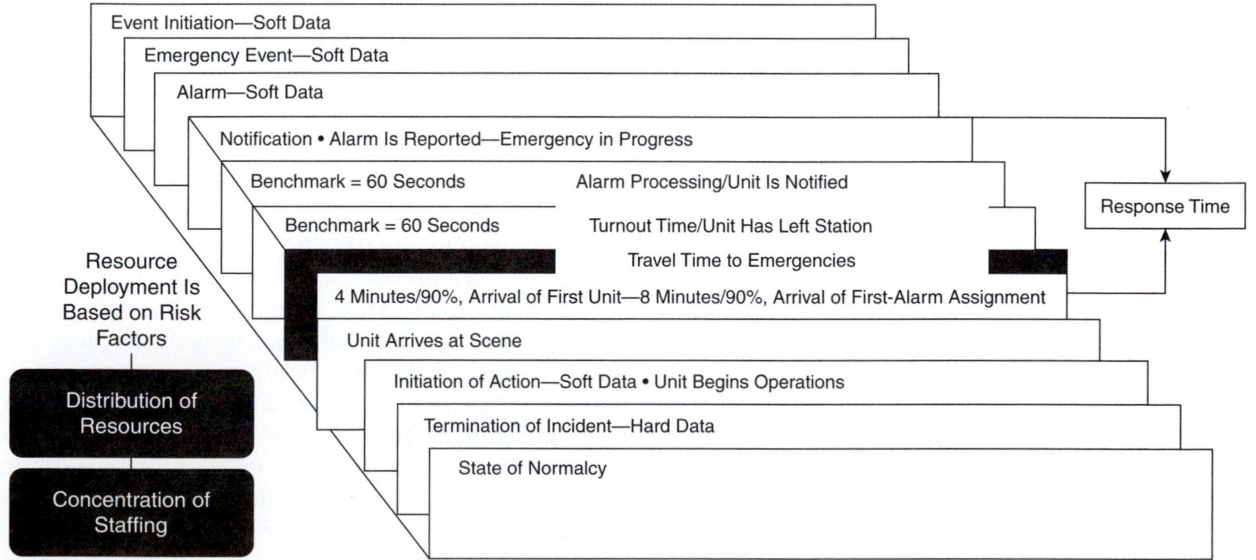

FIGURE 10.22 ◆ Cascade of Events (*Courtesy of Randy R. Bruegman*)

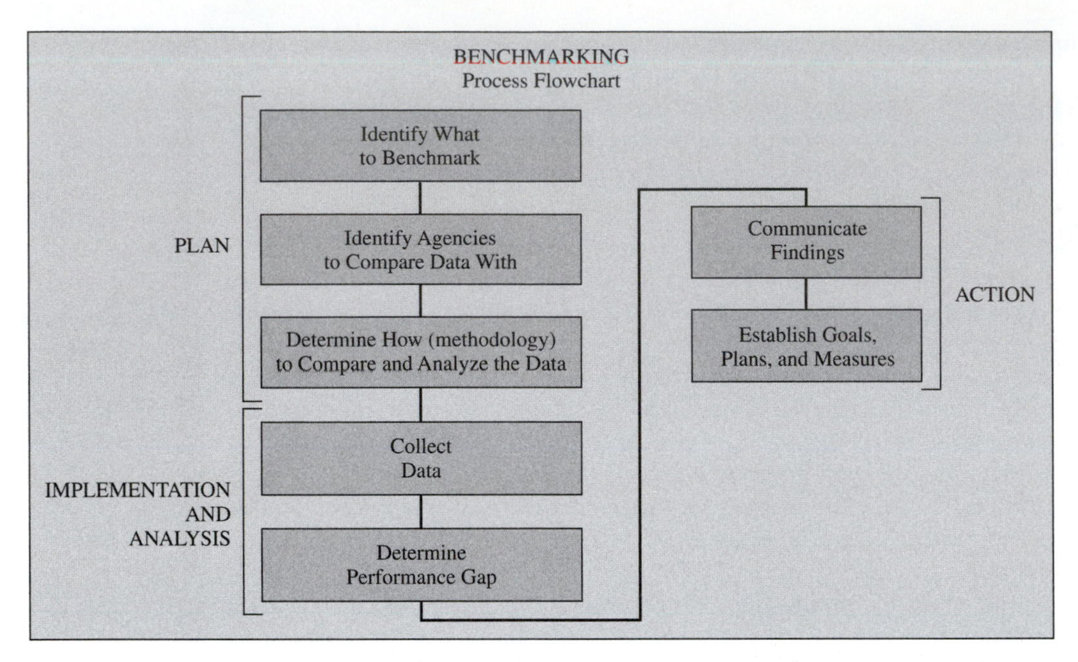

FIGURE 10.23 ◆ Benchmarking Process Flowchart (*Courtesy of Randy R. Bruegman,* Exceeding Customer Expectations: Quality Concepts for the Fire Service, *p. 49*)

Although we have talked about only one kind of cycle time as it relates to a fire department, there are many others. We have cycle times for complaints, plan reviews, data processing, inspections, correcting code violations, and many others. One of the best ways to improve key measurements, such as cycle time, is through benchmarking. It is an idea as simple as the phrase "Let's not reinvent the wheel." If one fire department develops a good way to reduce cycle time, it makes sense that it should pass the technique along to everyone else. The result is everyone shares a set of "best practices," which leads to higher customer satisfaction.

The basic idea behind benchmarking is two or more organizations of similar size and makeup work together to build a common set of performance measurements. Then they exchange information on their own performance and make comparisons on what has worked successfully and what has not (see Figure 10.23). As more organizations are added to this process, a performance curve begins to appear. Some organizations are at the bottom and some are at the top. The key practices and processes of the best performers are compared to those of the worst, and before long some important lessons begin to emerge. Effective benchmarking usually results in a series of "aha!" moments for underperformers, as they discover how their counterparts have successfully resolved problems in their work processes. It is not generally a matter of only the worst performers profiting from the process, either. Usually, even the best performers can learn from others, as they inquire about the processes they use.

Benchmarking can be defined as the search for the best practices that lead to superior performance. Benchmarking is the process of continuously comparing and measuring an organization with similar departments to gain information that will help the organization take appropriate action to improve its performance. Benchmarking can be used as a tool to improve organizational efficiency and effectiveness by identifying customer needs; selecting processes that are key to accomplishing organizational success; collecting quantifiable data with the ability to compare the internal

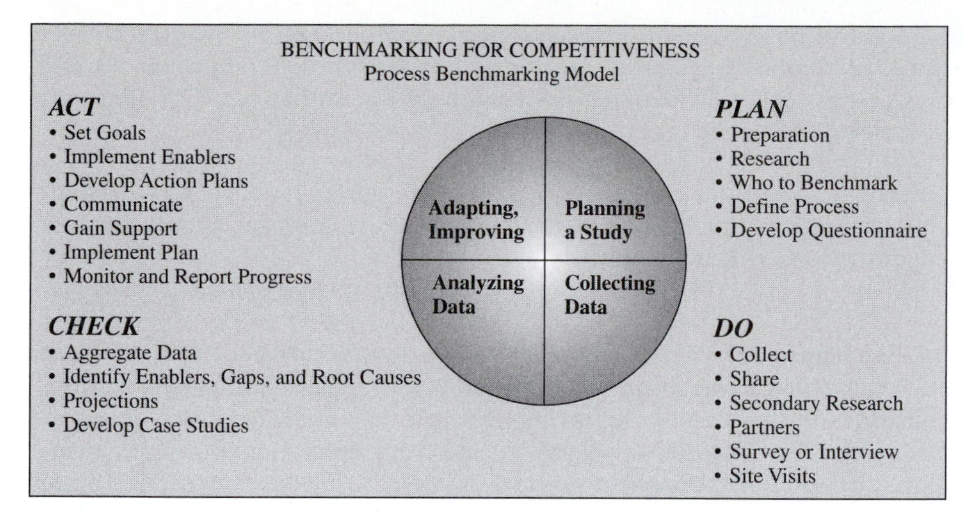

FIGURE 10.24 ◆ Benchmarking for Competitiveness *(Courtesy of Randy R. Bruegman,* Exceeding Customer Expectations: Quality Concepts for the Fire Service, *p. 50)*

baseline against benchmarking statistics; providing insights into new ways of doing business and delivering service; assisting in analyzing both internal and external data collection; identifying those items that enable consistent superior performance; assisting us in learning the best practices being used by others who have achieved superior performance; and assisting in developing goals, objectives, strategies, tactics, and organizational action plans aimed at making our departments the best in the market they serve. It is a natural element of the plan-do-check-act (PDCA) component of the quality improvement process (see Figure 10.24).

Benchmarking is not visiting another organization to see how it operates, searching for a cure-all for an organization's insufficiencies, merely copying what others are doing, or a stand-alone exercise. It is not quick and easy. According to Osborne and Gaebler (1992) in *Reinventing Government*, "If you don't measure results you can't determine success from failure."[4] Benchmarking has become one of the power tools used by government to improve [its] performance. Benefits of benchmarking include the following:

- Is a great way of improving effectiveness and efficiency
- Helps to accelerate the change process
- Assists in setting and achieving higher performance goals
- Provides the bridge to overcome the non-inventive here (NIH) traditional aspects found in many fire departments today
- Provides a means to meet customer expectations better

One has to look no further than two of our Malcolm Baldrige Award winners in the private sector, Motorola and Xerox, to understand just how important benchmarking is in the effort to stay competitive and maintain "best in class" within their respective fields. Much of what the private sector has experienced with quality improvement during the last two decades the public sector will face during the next 10 to 15 years. Much of what the fire service has to deal with today is cultural change; and from a leadership perspective, measurement will help drive positive cultural change. Measurement and rewards must be consistent with the change objectives you have established for your department, and one must remember that what you measure is what you get. People need to experience success, and to do so, there must be some form of measurement. In many cases our programs, processes, and delivery methods often must be changed to achieve the measured operational goals. It is for those reasons that benchmarking has been a critical component of the success of many private and public organizations during the last 25 years.

> Benchmarking is the continuous process of measuring products, services, and practices against the toughest competitors or those companies recognized as industry leaders (best in class).
>
> *The Xerox Corporation*

Using benchmarking as a learning tool, we observe what the best departments are doing and find a way to apply it within our own agencies. Ultimately, benchmarking provides the information to set our performance goals for attaining organizational leadership through development and implementation of action plans to achieve this position. It is a conscious process of measuring how our service stacks up against others we consider world class. If benchmarking is to be successful within any organization, top management has to recognize clearly the need for change and improvement. It has to have demonstrated its willingness to make the tough decisions needed to maintain competitiveness and success.

This is why in the private sector benchmarking has become such a critical management tool, a key element in the quality improvement process, and fuel for the race for excellence. Benchmarking was first popularized by Xerox when it realized the Japanese were selling copiers for less than it cost Xerox to make them. A team of engineers was then sent to Japan; the engineers returned with benchmarking information that shook Xerox, both literally and figuratively. A 1993 interview given by Robert Camp, who has written two books on benchmarking, provides an interesting insight into benchmarking. He said the toughest aspect about implementing a benchmarking project is getting people to understand there may be others who can do things better. The most important component of benchmarking, according to Camp, is preparation (see Figure 10.25). Benchmarking will focus on cycle time and **customer service**. Faster, better, cheaper, and with good customer service is what customers are demanding from us today.

customer service
Providing a product or service that meets the needs of a group of people, called customers, who use and consume the product or service.

One of the best ways to get started is by asking the following about possible benchmarking opportunities.

- What services are provided to customers?
- What programs/services account for customer satisfaction?
- What is the most critical factor to the organization's success (e.g., citizen/customer satisfaction, cost recovery ratio, response time, etc.)?
- What programs/services are causing the most trouble (e.g., not performing to expectations)?
- What specific problems (operational) have been identified in the organization?
- Where are the "competitive" pressures being felt in the organization?
- What are the major costs (or cost "drivers") in the organization?
- Which programs/services represent the highest percentage of cost?
- Which programs/services have the greatest room for improvement?
- Which programs/services have the greatest effect (or potential) for differentiating the organization from others in the public sector?
- What performance measures should be used to begin creating the **baseline data** that will be used in the benchmarking process?

baseline data
Internal information about a process that is collected repeatedly over a period of time prior to introducing a change.

The Japanese word *dantotsu* means "striving to be the best." To be the "best in class" you must continually search for practices that will lead your organization to top

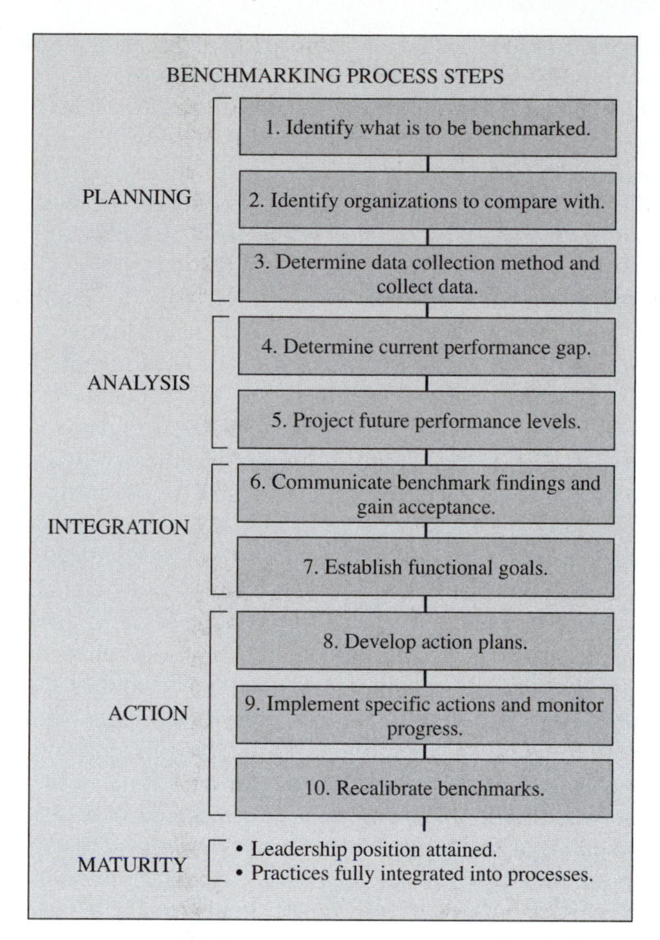

FIGURE 10.25 ◆ Benchmarking Process Steps (*Courtesy of Randy R. Bruegman,* Exceeding Customer Expectations: Quality Concepts for the Fire Service, *p. 53*)

performance. Benchmarking is about a continuous process of measuring your services and practices against organizations recognized as industry leaders. The purpose is simple: learn from others and then use this knowledge to make your organization better. How can you do it faster, better, cheaper, while at the same time providing quality customer service? It is impossible to accomplish this in a vacuum.

PERFORMANCE MEASUREMENTS

Performance measurements provide a means of defining program service levels both at the operational level and at the strategic level. Whether measuring fire suppression, fire education, arson investigation, or any other fire service program, performance measures can provide clarity of mission. Additionally, performance measurement systems provide a rational methodology to report program accomplishments to managers, customers, and policy makers. The International City/County Management Association (ICMA) has been keenly interested in productivity and measuring performance for more than two decades.

Operational performance measures are used by managers to plan and control programs at the operational level, while strategic performance measures provide guidance to both managers and policy makers who have to make decisions from a more global (big picture) perspective. Performance measures provide a means to clarify programs in understandable terms to citizens, customers, fire managers, and firefighters. These terms are typically formulated as inputs, outputs, and outcomes. Program costs can be calculated by evaluating the efficiency, effectiveness, and equity of the program. Although costs are not always quantitative—that is, measured in dollars and cents—there is often a tendency to consider only the financial costs. Equity and effectiveness costs typically must be measured in qualitative terms, which are much harder to measure and justify because they are based mostly on a set of values or assumptions about what is in the best interest of the public. Despite the difficulty in costing qualitative measures, it is extremely important to give it a best effort. Performance measures also offer opportunities to improve the services of a program.

Leading-edge organizations, whether public or private, use performance measurement to gain insight into, and make judgments about, the effectiveness and efficiency of their programs, processes, and people. These best-in-class organizations choose what indicators to use to measure their progress in meeting strategic goals and objectives, gather and analyze performance data, and then use the data to drive improvements and successfully translate strategy into action. Information collected about the programs offered can be utilized to evaluate outcome performance for customers and how well these programs and services are meeting the strategic objectives of the organization and the community. Evaluations based on predetermined performance measures can then be used to support requests for additional or reallocation of resources. Data can also help in analyzing how efficiently current resources are being utilized. The same data can be used to help identify both strengths and weaknesses in the program, thus supporting decisions to modify a program or sometimes aid in the decision to end a program. Although an evaluation may suggest a program should be ended, a good performance measurement system can provide early warnings of a program weakness, which can lead to changes to improve the service(s).

The levels of public services provided by any jurisdiction are often political-based issues that require political/policy decisions. The strongest, most comprehensive, and most understandable performance measurement systems do not change this fact, nor should they. Political leaders (city council members, fire district board members, and policy makers) are elected to make decisions about the allocation of resources. Fire service managers can, and should, play a role in developing performance measurement systems that can meet their communities' objectives in the best way possible. In this sense, managers and leaders in the fire service are public safety policy entrepreneurs who are constantly searching for opportunities to implement creative and innovative fire service programs. These programs must meet the needs of their customers and simultaneously provide for the overall public safety concerns of the community before it is reasonable to expect they will be funded.

The data provided by a good performance measurement system can be an effective tool in influencing political decisions. However, performance measures do not make decisions or replace people. They are intended to provide a systematic management approach that provides better data and evaluation opportunities, which are then used to make important programmatic decisions. For example, a well-planned public education program for community youth based on sound research and analysis and supported by a clear mission statement for the program can make the difference between gaining and not gaining community support. Support can be translated into budget dollars and staff to implement the program, and performance measurement results can be used to substantiate how well the program is meeting its objectives.

It is possible to have good operational performance measures for every fire department program without having a set of programs that is integrated into the strategic objectives of the fire department or the whole community. Operational performance measures are needed and are especially helpful to program managers. However, a holistic approach to strategic planning provides a set of programs complementary to the strategic mission of the organization, which identifies the most appropriate level of service for each program. Performance measurement systems succeed when the organizations' strategic and business performance measures are related to—that is, are in alignment with—overall organizational goals. Strategic performance measures are also needed because they address the community's strategic plan in a more comprehensive way than do operational performance measures. Yet, fire departments should have a strategic plan that provides direction and guidance for the development of fire service programs that are complementary to the objectives of the community. For example, if the community is concerned about its youth, then youth education programs are very important. On the other hand, if the community is primarily a resort town or retirement village, other services may take precedence.

PERFORMANCE MEASUREMENT DIMENSIONS

Several dimensions important to the development of an effective performance measurement system include inputs, outputs, outcomes, efficiency, and effectiveness (quality).

Inputs Inputs are the resources required to perform a program, deliver a service, or produce a product at some desired level. Inputs include dollars, staffing (additional personnel and time allocation of existing personnel), equipment, supplies, and other tangible goods or commodities.[5] Inputs can also include demand characteristics of a program based on target populations.[6] Cost analysis requires a review of all inputs whether they are direct or indirect. The indirect inputs tend to cause the greatest amount of anxiety among program managers and elected officials who are trying to cope with sensitive resource allocation decisions. Yet indirect costs (sometimes caused by unintended consequences) are important and, if overlooked, can cause a program to fail due to lack of adequate resources or other organizational support. Accurate inventory and analysis of inputs are critical before realistic cost projections and cost comparisons can be completed. A thorough understanding of program inputs is essential to the development of meaningful performance measures. Yet, often some of the simplest items are not accounted for in a program. An accurate description and accounting of program inputs also provide an opportunity for program managers and other stakeholders to have meaningful dialogue about the allocation of these scarce resources. Program managers who are struggling to stretch already overextended staffs can do an even better job of time management when inputs are accurately and honestly identified. Inputs in fire service programs are articulated in such things as firefighters per thousand population, the number and type of apparatus required to be dispatched to handle a specific type of incident adequately, and the number of personnel assigned to a piece of apparatus.

Outputs Outputs are the services provided or products produced by the program.[7] Outputs refer to the activity of an organization and are generally internal in nature.

Outputs generally address how much activity is generated within a program. Examples of activities are the number of complaints answered, the number of responses to an event, or the number of personnel required to complete a job. Outputs reflect how busy an agency is. Outputs are important to measure efficiency. In the most simplistic terms, efficiency is measured by dividing outputs by the number of inputs.[8] Understanding outputs is required before improvements can be made in the way something is being done or to determine the most right way to do something. However, outputs are not indicators of whether the activity is the right thing to do. Outputs are indicators of only how efficiently dollars are being spent, how efficiently staff time is being used, and how efficiently supplies are being allocated to accomplish the stated activities. The fire service, much like other organizations, tends to rely heavily on outputs as a measure for defending a program or requesting supplemental resources.

Output reports written to describe the amount of activity within a program are ubiquitous. The number of fires dispatched, the number of EMS incidents dispatched, the number of inspections conducted, the number of public education events held are all output reports. Each example above has a common theme; that is, the number of something is tallied. In other words, it is easy to categorize and count what was done, but critical questions must be asked. For example, when working with public education programs, should the program manager count the number of participants or the number of educational offerings (i.e., presentations)? This becomes an important question because the resulting numbers can be drastically different. A single focus on outputs can misrepresent the true performance of a program. The number of miles driven, feet of hose laid, and feet of ground ladders used at fires are each examples of measuring outputs. The number of apparatus dispatched to an incident is an output measure. Most departments count only dispatches regardless of the number of trucks dispatched. However, some fire departments report each apparatus dispatched on an incident.

Outcomes Outcomes are measures of effectiveness or the quality of a program, service, or product. Outcomes describe results and whether the program goals are being met.[9] An outcome measure is "the end result that is anticipated or desired (such as the community having clean streets or reduced incidents of crimes or fires)."[10] However, outcome measures should be developed with the user in mind. Managers need operational outcome measures useful to them in planning and controlling their programs at the operational level. Strategic outcome measures are very important to those persons responsible for developing, guiding, and evaluating the performance outcomes in relation to the overall mission of the organization and the community. Operational outcome performance measures and strategic outcome performance measures are also "end" outcome performance measures as defined previously.

Mathematicians and budget analysts try to describe outcomes in terms that can be explained by some numerical formula using ratios, ordinal scales, or some other technique that reduces the subjectivity in the analysis of outcome measures. While the objective is to eliminate subjectivity completely, we must take into account those intangibles such as the "public good," "quality of life," and "political considerations," which also must be addressed and are not easily quantifiable using precise mathematical formulas. Elected officials are typically most interested in outcome measures. Politicians want to know what the "bottom line" is for a given service or program, and they prefer to have the information in simple, easy-to-understand language that can be delivered in 30-second sound bites. This admirable objective is not always a realistic expectation. The fire service manager is often caught in the dilemma to provide sound analytic work in an informational format that satisfies the elected officials.

Outcome measures, like outputs, can also be used to measure efficiency. Simply divide outcomes by inputs. However, the real strength of good outcome measures lies with the story they tell about the quality of the service or the effectiveness of the program. Outcome measures go beyond measuring mere activity. Where outputs can measure whether something is being done, outcome measures can measure whether the right thing is being done. Although this looks like a subtle difference, it is an extremely important distinction. Outcome measures can help to bring clarity to what a program is supposed to accomplish; and, when clearly articulated, outcome measures can be used to steer training and education programs internally and future resource allocation. Additionally, outcome measures can be the focus of reports, which communicate how well the goals of the program are being met internally and externally.

Inside the organization, reports based on outcome measures give managers and workers important responses to how well they are doing at meeting their stated goals. Outside the organization, the same reports provide stakeholders and policy makers with feedback on how effective their decisions were and how well scarce resources are being used to meet customer needs. Additionally, these reports communicate how well the same resources are meeting the overall mission of the organization and community. Whereas output performance measures measure the number of apparatus dispatched to a fire, outcome performance measures evaluate how effective firefighters were after their arrival on the scene of an incident. If it takes 15 minutes to hook up to the available water supply and the house burns down, the output measures were not affected because only the resources dispatched and their arrival time were measured. However, the outcome was a disaster. An example of an operational performance outcome measure is the number of fires kept to the room of origin after arrival of the fire department. A fire department may want to define this outcome measure further by describing what fire department resources must arrive before it is reasonable to measure the outcome. This requires clearly stated goals and objectives that accurately reflect the fire department's expectations. These expectations should then be provided to the members of the department in the form of operating procedures.

One example of a strategic outcome performance measure is the reduction in dollars of fire loss in the community. This may be the result of an active fire prevention program or the implementation of a sprinkler ordinance. These programs, although very different from each other, contribute to the overall strategic mission of the organization. The number of youths taught how to call the fire department through 9-1-1 is an output measure by itself. However, if the number of youths who call 9-1-1 to report real emergencies increases, it is a positive outcome directly correlated to the fire department's educational program. This example demonstrates how an outcome measure may be simple to state, but collecting accurate and meaningful data may be a veritable challenge. For example, how do the current data compare with the number of calls received before the intervention program was implemented?

The same approach can be taken in training programs, apparatus maintenance programs, or facilities maintenance programs. In the classroom, inputs include the classroom, supplies, teacher(s), resources necessary to get the firefighters to class, and other resources required to conduct the class. Outputs are the number of firefighters who attended or the content of the material presented by the class facilitator. However, without outcome measures the department cannot verify wheather anyone learned anything. One way to measure outcomes is through testing in the classroom. Yet, many will argue classroom tests do not represent the real world. Thus, other ways of more accurately measuring the real outcome of the class are needed. Examples include the use of preceptor programs (such as those used in paramedic programs), peer review programs in the field, or critiques after incidents.

efficiency
Capacity to produce desired results with a minimum expenditure of time, energy, money, or materials.

Efficiency **Efficiency** is a measure that compares the cost of something, in terms of resources used, to the production of something, in terms of service, products, energy expended, or some other input. Efficiency is output (or sometimes outcomes) divided by inputs, thus providing a unit-cost ratio.[11] It is important to note that if an organization uses outputs instead of outcomes to measure efficiency, a lower unit-cost ratio . . . may achieve the result at the expense of the quality (i.e., outcome) of the service. Efficiency is an easily understood concept that has an important impact on decision makers from all sectors in society. Efforts to improve efficiency sometimes manifest themselves in downsizing, layoffs, and mergers in the private sector to reduce costs and improve margins of profit. Decision makers in the public sector typically look for ways to reduce taxes and government spending. This often results in efforts to privatize public services or contract services with the private sector, create public/private partnerships, or reduce staffing without reducing the levels of service.

Reductions in costs of inputs without reductions in outputs improve efficiency. Additionally, reductions in costs without commensurate reductions in service (usually measured in outputs) are also an improvement in efficiency. It is important to note efficiency must not be confused with effectiveness. Efficiency addresses only the cost to do something. Efficiency measures are important for several reasons. First, resources are scarce; therefore, decisions are required to allocate scarce resources based on some set of criteria, which is used to set priorities. Cost is certainly one criterion for making a decision, and for many policy makers it is their most important criterion. Second, managers must frequently be able to demonstrate they are getting the maximum outputs and outcomes from their already limited inputs before decision makers are willing to give them more resources.

Third, efficiency measures provide the persons doing the work feedback on how well they are utilizing their resources. Continuous improvement efforts in many organizations are about being more efficient. Fourth, in a free-market economy, the more efficient a private firm is, the more competitive it can be, thus improving its profits. The public sector faces similar challenges. The public sector is challenged to be more competitive (usually defined as doing the job cheaper, i.e., more efficiently) through competitive bidding with the private sector. Finally, efficiency measures are important to the public sector because public managers are responsible for eliminating wasteful spending of tax dollars, and citizens are holding them accountable for doing their job. Efficiency measures provide public managers with a means to report their achievements to everyone in the system from those who do the work, to those who receive the service, and ultimately to those who pay their taxes.

Because the fire service is generally part of the public domain, fire departments are not immune from the challenges to be more efficient presented by elected officials and by the public. Fire departments cannot ignore these challenges but must meet them realistically and responsibly. Operational efficiencies can most often be achieved through implementation of sound management practices and resourceful leadership. Creativity and innovation can be critical elements enhancing the development of sound performance measurement systems.

effectiveness
Is a measure of achieving the desired result.

Effectiveness **Effectiveness** is a measure of achieving the desired result and does not necessarily mean at the lowest cost. Effectiveness measures address whether the right thing is being accomplished, with consideration given to the quality of the service or product produced. Effectiveness measures rely on clearly stated program goals and objectives. Otherwise, most discussions about what a program is doing and whether it should be doing it usually regress into debates over the very subjective and personal values of those involved in the discussion. Therefore, effectiveness measures should be written after such discussion has taken place. Program goals and objectives are then decided after a clear understanding of what the program is intended to accomplish, preferably based on a consensus of the stakeholders. Some organizations use other terms in place of effectiveness. Effectiveness is one of the most important concepts in performance measurement, because it focuses on the quality of the service or program and is a clearer reflection of the purpose or scope of the program than is the measure of efficiency.

Effectiveness raises issues of customer expectations. While most people want to achieve their expectations at the lowest possible cost, they usually are not willing to affect outcomes severely just to achieve reduced costs. Effectiveness is also an important concept that must be included in any discussion regarding outsourcing, privatizing, downsizing, or reducing levels of service. The performance measures used to evaluate the effectiveness of the program in meeting its strategic organizational objectives must be clear. Sometimes it may make more sense to eliminate a program altogether rather than to cut the costs (by reducing the inputs) below the level that provides for a minimum level of effective service delivery.

Fire protection is surprisingly difficult to measure well. The main purpose of fire protection is to reduce loss of life and property, and it is difficult to measure or even estimate what tragedies have been averted as a direct result of education and prevention programs. There are two key measurement strategies: (1) measuring losses that occur, how they change over time in light of outside explanatory factors such as socioeconomic conditions and environment, and how the losses compare to those in other like municipalities, and (2) measuring intermediate fire protection efforts known to contribute to the desired goals.[12] Building inspections without the staff or resources to conduct adequate follow-up to assure compliance may not be an effective use of resources. Offering emergency medical service without the ability to support continuing education for EMTs and paramedics reduces the effectiveness of the EMS program. Lack of personal protective clothing hinders the fire department's ability to deliver effective fire suppression. Severely cutting or eliminating training in a fire department may save the department money, but the outcome may be an even greater reduction in the effectiveness of firefighters' capability during emergencies.

Fire departments are required to make routine budget cuts regularly with the caveat that the cuts must not reduce service levels. Often it is possible to accomplish these two conflicting objectives but only in the short term. However, the long-term consequences can be detrimental. This is why performance measurements are essential tools for fire service managers and leaders to understand the effectiveness of existing efforts. When the current efforts are not effective, cutting the program resources may not have serious consequences. In fact, cutting resources may even be an acceptable way to improve overall organizational efficiency without an immediate negative impact on outcomes. Therefore, it is imperative that the effectiveness of a program is understood and clearly communicated to all stakeholders.

LESSONS LEARNED

Jim Theurer, senior policy analyst in Minnesota, offers some lessons learned when establishing performance measures. Theurer shares seven "pitfalls" of performance measurement systems.

1. *Data by themselves have no meaning.* The data collected in performance measurement systems must be set within a particular context. There are at least two ways to accomplish this: (a) performance measures must be integrated into the long-term goals of the community,

and (b) comparisons of the community to other communities through benchmarking must be done to demonstrate the outcomes achieved.

2. *A lack of commitment from leaders.* There must be a strong commitment from leaders to move toward measuring performance and not just collecting data.
3. *Employees must have the capacity to develop measures*, or they will use whatever "measures" that are already available.
4. *The performance measurement system must not be used as punishment.* If measurement focuses on negative accountability, managers and employees will seek to avoid accountability when things go wrong.
5. *A performance measurement system should* provide information to policy makers and managers so they can make better decisions. The system will fail if the stakeholders are not able or willing to use the data and reports provided by the system.
6. *For many governments, the ultimate aim of management* based on performance measures is to integrate program performance and outcome information with the budget process. This is not an easy task, but one that must be tackled if the fire department expects to enjoy continued funding, even for some of its most critical programs, such as EMS and fire suppression.
7. *The system should be flexible* and provide for the diversity of objectives and expectations. Uniformity of forms and processes can be detrimental to the main purpose of the performance measurement system, which is to "provide reliable and valid information on performance."

Theurer provides some valuable lessons for managers and policy makers. Avoiding these pitfalls will improve the probability of the performance measurement system's success.[13]

THE COMMISSION ON FIRE ACCREDITATION INTERNATIONAL (CFAI) BENCHMARKING SURVEY

An early effort in the accreditation process was the creation of benchmarks for use within the fire service as a means of performance measurement. The process of collecting and analyzing information to be used as performance measurements and to compare the department with other organizations is a key component of quality improvement. The CFAI identified several areas within an assessment of a department that are indicative of performance and industry trends. These were categorized by size and type of organization. The collection and establishment of these important benchmarks will be dependent on the input of the fire service agency that you choose to compare it with and are relevant for discussion if you embark on a comparative survey for your organization. A sample of the survey is included in Appendix A.

In addition, workload performance measures can provide valuable information in respect to the cost (efficiency) and the outcomes (effectiveness) on a variety of fire service programs.

Sample Workload Performance Measurements

Workload	*Efficiency*	*Effectiveness*
Number of requests getting processed	Cost/processed request	Number of requests concluded/outcome

Evaluate Criteria The goal is to select and/or develop as complete a set of workload, efficiency, and effectiveness measures for each program activity as possible.

Prevention

Community Education

Workload	*Efficiency*	*Effectiveness*
Number of applicants for training	Cost per training graduate	Number of certified graduates per geographic districts
Number of participants in training		Number of certified graduates per 1,000 population

Workload	*Efficiency*	*Effectiveness*
Number of public education campaigns	Resources expended on campaign	Behavior/fire loss changed for specified occupancies

Workload	*Efficiency*	*Effectiveness*
Number of injury prevention programs presented	Resources expended on programs	Behavior change/ mitigation preventable injuries

Workload	*Efficiency*	*Effectiveness*
Number of targeted campaigns	Resources expended on campaign	Behavior change in responses to household and small business

Code Administration

Workload	*Efficiency*	*Effectiveness*
Number of inspections performed	Number of staff hours per inspection	Percentage of inspections done accurately
Number of code consultations	Number of person-hours per code enforcement	Percentage of inspections completed
Number of plan reviews	Number of person-hours per plan review	Percentage of violations resolved on fire reinspection
Number of permit/ license applications processed	Number of person-hours per permit license/ application processed	Percentage of recurring violations per inspection
		Percentage of appeals sustained
		Percentage of fires with cause determined
		Percentage of plan reviews, inspections, etc., done within target cycle time

Code Development

Workload	*Efficiency*	*Effectiveness*
Number of code sections evaluated by fire safety research reports and number of code section amendments proposed	Staff hours per code amendments submitted to governing body	Number of presented code updates adopted by governing body
		Entire code reviewed and updated for relevancy within required cycle time

Investigations

Workload	*Efficiency*	*Effectiveness*
Number of investigations	Number of person-hours per investigation Cost per investigation	Percentage of accurate investigations
Number of arson investigations	Number of investigations per investigator	Percentage of undetermined cause prosecution rate
Number of court-related work		Percentage indicted (information communicated effectively to relevant actors)
Number of investigations by company officers Number of advanced investigations		

Response

Disaster

Workload	*Efficiency*	*Effectiveness*
Number of times disaster plan activated	Amount of assistance per occurrence	Percentage of compliance with disaster plan per occurrence
Number of people, households, businesses, critical facilities involved		Percentage of disaster plan determined to be relevant

Fire

Workload	*Efficiency*	*Effectiveness*
Number of runs per company	Cost per run and cost per incident by type of incidents per firefighter	Fire loss per incident

Workload	Efficiency	Effectiveness
Number of fire incidents by type		Fire loss per assessed value Loss ratio value versus loss Response time Percentage of loss after arrival Percentage of times fire response unit not available

EMS

Workload	*Efficiency*	*Effectiveness*
Number of requests	Cost per incident	Percentage of calls correctly triaged
Number of responses		Percentage of responses that reduce the severity of EMS incidents through medical interventions Percentage of customer feedback rating service "good" to "excellent"

Hazardous Materials

Workload	*Efficiency*	*Effectiveness*
Number of calls	Cost per incident	Average time to bring incident under control
Number of incidents		Percentage of calls requiring specialized hazardous materials response

Specialty Rescue

Workload	*Efficiency*	*Effectiveness*
Number of specialty rescue	Cost per rescue	Time per rescue Time from alert through setup
Number of victims served		Percentage of viable victims rescued

quality
A characteristic or the value of a product or service from the perspective of the user. The extent to which a product or service meets or exceeds customer requirements and expectations. Good quality does not necessarily mean high quality. It means a predictable degree of uniformity and dependability at low cost, with a quality suited to the market.

MEASURING CUSTOMER SATISFACTION

Virtually all organizations compete to some degree on the basis of service. It is difficult to name even one industry for which service matters are unimportant. Indeed, as the 1990s unfolded, more and more executives in government, like their counterparts in the business sector, became keenly interested in service quality. Government executives now realize they must find a way to establish sustainable, technology-based, competitive advantages that will direct added attention and resources to value-added services.[14]

Quality is an attitude, a vision, a passion, and a core belief accomplished through a commitment to the way business is conducted. Quality uses a system-based approach to guide operations; it integrates quality improvement into the strategic and

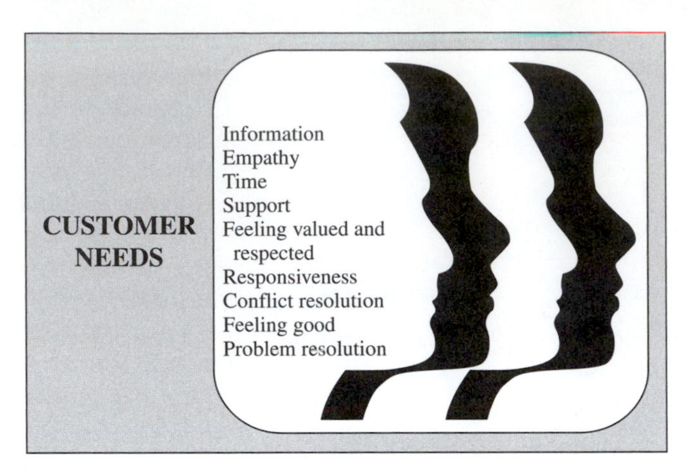

FIGURE 10.26 ◆ Customer Needs (*Courtesy of Randy R. Bruegman,* Exceeding Customer Expectations: Quality Concepts for the Fire Service, *p. 99*)

operational planning of a business. As such, quality focuses on both product and service quality, as well as the continuous improvement of both, and ensures that systems are in place for quality achievement and management to take place.

QUALITY CUSTOMER SERVICE

The goal of a quality customer service program is the customers should feel satisfied with the manner in which they are served, even when their specific wishes or demands are not met.

Think about the last time you had an interaction with someone in either a service or retail establishment. Did you walk away feeling as though you were someone special to that person? (See Figure 10.26.) This is really the heart of what quality customer service is all about: making people feel as though we have understood their problems even when we may disagree with their conclusions. This is also the heart of satisfying basic customer needs. In the fire service, we cannot become complacent, relying on the perceptions expressed in customer service surveys indicating we are doing an outstanding job. What they are telling us is that public perception toward the fire service is very positive. In fact, if you look at surveys conducted in different communities, not only within the United States but also internationally, you will find there is a high regard for fire service personnel. However, this does not always mean we are meeting customer expectations. It is critical, as we begin to design our future, to bridge the gap between this positive perception and the actual services we provide. We need to determine how to identify our customer needs and expectations and how we can meet or exceed them organizationally. In this chapter we discuss the customer-centered continuous improvement process, with a focus on the customer and not on the bureaucracy in which we work (see Figure 10.27). Such internal focus forces you to look at the ways your organization deals with meeting customer expectations. Several quality and continuous improvement checkups and surveys are provided that can be utilized to gain feedback from your customer base. What it all comes down to is an organizational attitude about how you treat your customers, identify their needs, provide for those needs, and analyze how well you are doing it. If you find you are not doing it very well, you must be willing to change your operation to be responsive to your customers.

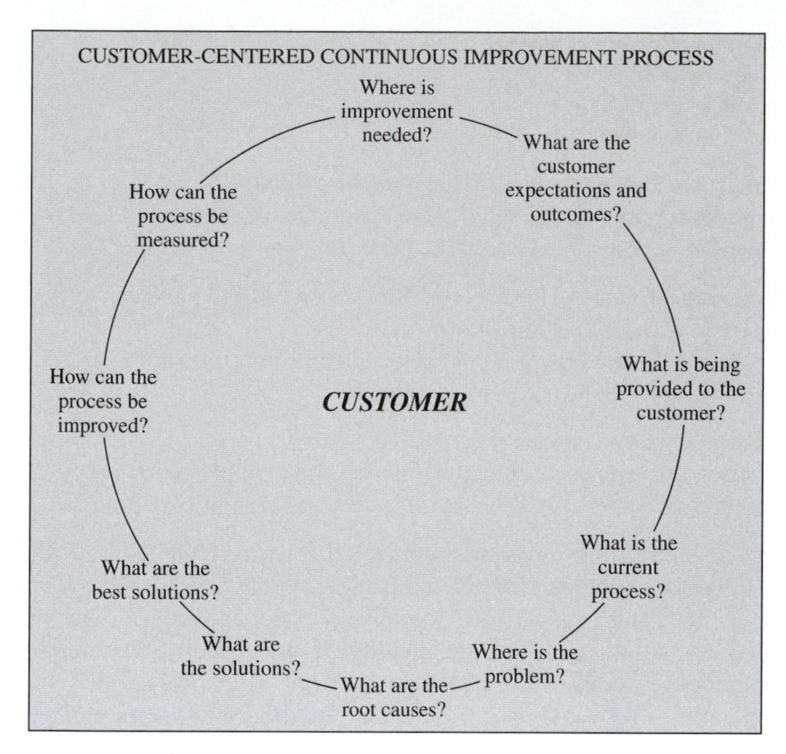

FIGURE 10.27 ◆ Customer-Centered Continuous Improvement Process (*Courtesy of Randy R. Bruegman,* Exceeding Customer Expectations: Quality Concepts for the Fire Service, *p. 101*)

Tenets of Continuous Improvement

- Nothing improves until it is measured.
- As soon as something is measured, it provides the basis for improvement.
- QUALITY = Meeting and Exceeding Customer Expectations

CUSTOMER-CENTERED CONTINUOUS IMPROVEMENT PROCESS

Step 1. Where is improvement needed?

- Identifies a focus area where improvement is needed.
- Generally selected because of some customer "feedback."

Step 2. Who is the customer? What are the customer's expectations? What is being provided?

- Identifies your customer, external or internal.
- Focuses on the customer's requirements and priorities.
- Assesses conformance to requirements.

Step 3. What is the current process?

- Creates a thorough understanding of your work process.
- Creates a map or flowchart illustrating sequences and stakeholders.
- Assesses conformance to requirements.

Step 4. Where is the problem?

- Specifies the problem area.
- Identifies difference between expected and actual performance.

Step 5. What are the root causes?
- Isolates reasons why problem exists.
- Provides guidance for appropriate solution.
- Selects the cause with the greatest impact.

Step 6. What are viable solutions? What is the best solution?
- Identifies several possible solutions to the problem.
- Selects the solution that best fits the need.

Step 7. How do we improve the process? How do we test and implement the plan?
- Designs a plan for improving the process.
- Develops actions and steps for implementing the new process.
- Test-drives the new process and modifies it.

Step 8. How do we measure the progress?
- Assesses effectiveness of continuous improvement implementation.
- Measures, documents, and communicates success of new work process.

There is an increasing focus on "quality" throughout the United States. When talking about quality, the common thread is meeting the needs of those who pay for and use the services and products provided by an organization. All types of industries have lowered costs and improved the quality of their operations and products by working to meet the needs of the people they serve. Many books have been written describing the philosophy and methods used in the quality movement. This chapter is not intended to be a substitute for those works, but rather it provides a brief overview of quality management principles. In concluding this chapter, it is important to review the work of three leaders who have shaped the discipline of quality improvement. These three pioneers of quality initiatives have all stressed the importance of management awareness and leadership in promoting quality.

W. Edwards Deming, who began his work in Japan in 1950 (see Chapter 3), was instrumental in building the Japanese industry into an economic world power. His strongly humanistic philosophy is based on the idea that problems in a production process are due to flaws in the design of the system, as opposed to being rooted in the motivation or professional commitment of the workforce. Under Deming's approach, quality is maintained and improved when leaders, managers, and the workforce understand and commit to constant customer satisfaction through continuous quality improvement.

Deming promoted the PDCA cycle and developed his famous 14 Points to transform management practices. *Joseph Juran*'s approach is based on the idea that the QI program must reflect the strong interdependency that exists among all of the operations within an organization's production processes. According to Juran, quality planning is the process of understanding what the customer needs and designing all aspects of a system that is able to meet those needs reliably. Designing a fire/EMS system to do anything less is wasteful because it does not meet customer needs. Once the system is put into operation, **quality control** is used to monitor performance constantly for compliance with the original design standards. If performance falls short of the standard, plans are put into action to deal quickly with the problem. Quality control puts the system back into a state of "control," that is, the way it was designed to operate. Quality improvement occurs when new, previously unobtained levels of performance—breakthrough performance—are achieved!

quality control

Measures output and identifies defects as they occur, with the intention of preventing them from reaching the customer. Although it is important to catch defects before they escape to customers, it is much more important to prevent them from happening in the first place.

Juran also proposed the idea of the "Vital Few and the Useful Many" that helps prioritize which QI projects should be undertaken. Any organization has a lengthy list of possible ideas for improvement. Because the resources to actually implement new ideas are limited, however, leaders must choose those vital few projects that will have the greatest impact on improving ability to meet customer needs. The criteria for selecting QI projects include potential impact on meeting customer needs, cutting waste, or marshaling the necessary resources required by the project. Juran also developed the idea of instituting a leadership group, or "quality council," consisting of the organization's senior executive staff. The quality council is typically charged with the responsibility for designing the overall strategy for quality planning, control, and improvement. Senior leadership involvement is a must because QI activities are as important as other management tasks (e.g., budgeting, human resource management, purchasing, and training), and leaders can integrate QI into every aspect of fire-based operations.

A third pioneer in promoting quality was *Philip Crosby*. He coined the phrase "quality is free," meaning that the absence or lack of quality is costly to an organization (e.g., in money spent on doing things wrong, over, or inefficiently). Conversely, spending money to improve quality (e.g., to reduce waste or improve efficiency) saves money in the long run. According to Crosby, ensuring quality should occur primarily at the design phase. Rather than spending time and money finding and fixing mistakes and errors, Crosby advocates organizational changes to encourage doing a job right the first time. Crosby challenges organizations to think of how processes can be designed or redesigned to reduce errors and defects to reach a goal of zero defects. Crosby believes managers' policies and actions indicate their commitment to quality. He also advocates a step-by-step approach for educating the entire workforce about quality principles, extensive measurement to document system failures, and formal programs to redesign faulty production processes.

The QI principles and methods of Deming, Juran, and Crosby provide a basic foundation for most QI efforts; however, at the end of the day, it all comes back to leadership. The fire service leader's role in promoting and developing quality initiatives begins with commitment. This commitment is a personal and organizational focus on the needs of the internal and external customer and consumer. Look within your own organization: Are there improvement opportunities to be found? Is your organization a Six Sigma or One Sigma organization? In many fire service organizations today the opportunity for significant improvement exists if we identify clear goals that define the expected outcome of our efforts, use facts and indicators to measure programs, include systematic cycles of planning execution and evaluation, concentrate on key processes as a route to achieve improved results, and focus on our customers and other stakeholders.

The most important results for achievement are improved quality of our services, improved outcomes of our efforts, and improved efficiencies of the resources we use. This is what quality is all about!

ADVICE FROM THE EXPERTS

Quality initiatives have become an important concept in both the public and private sectors. What have you done to implement quality improvement in your organization?

MICHAEL CHIARAMONTE

Chief Fire Inspector, Retired, Lynbrook Fire Department (New York): Other than stressing it, no formal policies have been written.

BETT CLARK

Former Fire Chief, Bernalillo County Fire and Rescue (New Mexico): First we had to define our mission. Then we set a minimum level of expectations from internal and external stakeholders. As we reviewed our minimum expectations, we decided to write goals for our five-year view of our future and how we could improve those expectations. When we review our quarterly program goals, we match them with our projected five-year vision. We do a realistic review of our goals every year. As we approach the completion of a goal, we encourage setting higher goals to encourage better quality in our services. We also take a real look at those areas in which we need to improve and create possible solutions. Some of our solutions: fire station engineering to improve turnout time, fire and EMS Clawson dispatching; staffing configurations based on seasonal weather events and projected fire behavior, and Citizen Address Response Enhancement (CARE) Program that we designed for better address response in rural areas.

KELVIN COCHRAN

Fire Chief, Atlanta Fire Department (Georgia): Fire departments and issues related to human resource management and emergency management are too great to lead and manage without the benefit of performance objectives, standards, and recognized models for establishing organizational excellence. Challenges abound, change is continuous, and the giants just keep on coming. While Chief at Shreveport, the Department developed a solid, ongoing, and realistic strategic planning team of members from every rank and division of labor to establish the direction and plan for the future of our department. Workshops are conducted two to three times a year. Our general planning team identifies and prioritizes the issues. Our executive planning team develops strategies for achieving our goals. Identifying nationally recognized standards for service levels and setting goals to achieve them also promotes organizational excellence. The Shreveport Fire Department obtained a Class 1 fire insurance rating in 1998 and is working to sustain it. The department established intermediate and long-term plans for EMS based on American Heart Association standards for the chain of survival and NFPA 1710, and is also in the process of accreditation through CFAI.

CLIFF JONES

Fire Chief, Tempe Fire Department (Arizona): Quality is addressed through a section of our department's five-year strategic plan and operational guide. This sets the stage for stressing the importance of quality in the department. Quality is not one person's responsibility but everyone's. Quality is one of the eight core values of the city organization stressed throughout the organization, including performance evaluations. Quality assurance is a major part of our emergency medical services program and is addressed through review of all patient encounter forms and through tape and chart reviews with the base station medical director. It is important that the notion of quality be on the minds of everyone in an organization and that it is reflected in the day-to-day work that people perform. I attempt to lead by example by doing quality work and those who report directly to me do as well. The critical need for quality is everywhere, from a chief officer managing a two-alarm fire to a fire mechanic doing a brake job on an aerial ladder truck.

BILL KILLEN

Retired Fire Chief and IAFC President 2005–2006: Total quality management is an integral part of the fire department mission and, in our organization, includes an annual performance and training plan for each employee. Fire department training is based on the National Fire Protection Association's Professional Firefighter Qualifications Standards and quality management concepts. The annual performance appraisal system includes an individual training plan that incorporates quality management principles. Each individual work plan is designed to assist the employee's professional development and growth, as well as obtain professional certifications required for his or her position.

KEVIN KING

U. S. Marine Corps Fire Service: It is interesting that you mention both the public and private sectors because in many ways I believe the private sector has forgotten about the customer today. Many of its efforts are focused on the "shareholder" and on "return on investment." However, we must continue to focus on our customers in the fire service because they are our shareholders, and our quality of service is their return on investment. We have worked at two major quality initiatives in the Marine Corps Fire Service. First is the focus on our customers, which is best exemplified by our "Protecting Those Who Defend America" motto. Our whole effort is to ensure our Marines and their families are protected at home so they can serve our country while deployed. Second, we have worked at becoming a full emergency service provider, so that whatever the emergency, we hopefully can provide relief to our customers. As we like to say, our most important call for service is the next one we receive. (These are opinions of Mr. King and do not reflect official policy of the Department of Defense or the United States Marine Corps.)

MARK LIGHT

Executive Director, International Association of Fire Chiefs: I see many fire departments that have adopted the flavor of the month. Yet in the long term, few fire departments have staying power on any of the quality programs such as TQM, Six Sigma, and so on. One of the best ways to look at quality and continuous improvement in a fire department is through the Center for Public Safety Excellence's Fire Department Accreditation Program. This program lets a fire department do a self-evaluation of its services and develop plans for improvement. This program is the only one that I feel will continue to push fire departments to look continually at quality. All fire departments should embrace this concept and set the program as one of their key goals for ensuring the future of their department.

LORI MOORE-MERRELL

Assistant to the General President, International Association of Fire Fighters: Quality is a concept that is difficult to grasp, but leaders should ascribe to it and recognize it when they see it. Measuring quality presents an entirely different dilemma, particularly for the fire service due to the lack of data. The first step in achieving quality performance in a fire department is to know how you perform today, and this cannot be done without data. Data collection is imperative to measuring and reporting performance. Once performance is noted, benchmarking best practices can be completed. Following benchmarking, leaders must compare performance to industry standards as well as community expectations. Once this is complete, the trek toward quality can begin. Continuously measuring performance in many aspects of a fire department allows comparison of a department to itself over time, noting changes and improvement. Measurement is a necessity according to quality experts because we do not improve what we do not measure. A process in place to assist fire service leaders with self-assessment and measurement is the accreditation process through the Commission on Fire Service Accreditation International (CFAI). For more information, visit www.CFAInet.org.

JAY REARDON

Fire Chief, Retired, Northbrook Fire Department (Illinois), currently President of the Mutual Aid and Box Alarm System (MABAS): (1) When customers call for help or direction, don't let them down. They expect our fire and civilian staff to become their advocates and ombudsmen when they are in search of an answer and cannot get one from another government agency, utility, and so forth. Example: A contractor cannot get the building inspector to return a phone call with a code question. The fire prevention technical services staff will connect with the building inspector and get the answer. Do not transfer the problem—make it yours, on the customers' behalf, and seek to resolve it for them. (2) Our mission statement reads, "We are professionals dedicated to preserving life, property, and a safe community through quality planning, education, enforcement, engineering, and emergency interventions." We are truly committed to the belief—our family helping yours by safely and diligently doing the right thing, at the right time, for the right reason. The key to it is its closure, "Do the right thing at the right time for the right reason." Translated into action, it provides the direction when a formal procedure or policy is lacking or absent. Example: A Northbrook engine company was returning from an incident when they noticed a couple of senior citizens with a bunch of luggage outside a three-story walk-up. The lieutenant stopped and asked whether they needed help. The seniors stated they had just been dropped off at home from the airport by a cab and were trying to figure out how to get their luggage up the stairs. The company carried the bags, but it is not in the job description; it is in their cultural attitude. Right thing—right time—right reason, exceed their expectations! (3) Everybody, including police officers, calls the fire department when they have a situation and cannot figure out what to do. They give you a new market and new customers. Think about accepting the challenge. If you don't, someone will fill the void, and it might be the biggest opportunity your business (fire department) was ever given. Smart organizations do not necessarily seek opportunities—they just see them and act.

GARY W. SMITH

Retired Fire Chief, Aptos–La Salva Fire Protection District (California): I am not sure what you are asking. I think that quality initiative is not a concept that can be adopted as a program or new management gimmick. It is more a value or a way of doing business. I feel that quality spins from the foundation of the organization. I have an acronym called SCORE: *s*afe, *c*lean, *o*rganized, and *r*eady for *e*verything. From that vantage point quality spins forward. It is a way of doing business that says we do not put "garbage out." Our systems work with a clear vision and an output that is ready for testing, always changing, and getting better because of our experiences and team-oriented evaluation. Engaging new knowledge, more network support, and more "doing" to master our service delivery.

STEVE WESTERMANN

Fire Chief, Central Jackson County Fire Protection District (Blue Springs, Missouri), IAFC President 2007–2008: (1) While not totally unique, we have created an EMS

QA/QI committee made up of subject matter experts, our medical director, and our EMS directors. They will review all life-threatening responses and evaluate the care given. If needed the committee will ask the responders on an individual basis to come in and discuss the response and if we could have improved the care given. If a trend is noted here or in our safety committee, we try to utilize our officer's meetings to refocus on the trend and develop possible solutions. Training sessions are then used to reinforce changes in procedures. (2) One of the other significant practices in our organization is the acceptance of the leadership/partnership program between labor and management. As an organization, we have tried to institutionalize the practice so that no programs or practices go without evaluation or, if problems arise, then discussion at the labor–management (L/M) arena. These discussions occur even if it is not what is typically thought of as a labor issue. If anything that affects our organization, our employees, and our culture comes up, it is worth discussing in an L/M setting. Although many chiefs may have trouble accepting this level of involvement, there can be huge dividends for the organization.

THOMAS WIECZOREK

Executive Director, Center for Public Safety Excellence, and Retired City Manager: As I stated earlier, we sought accreditation for all of our services, including fire in 2003 and police in 2006. We had written a draft accreditation model for public transportation and were called upon by the State of Michigan Department of Transportation to work on furthering the effort. We introduced the asset management processes in public works and public utilities; we were generating more outputs with better outcomes with less staff. As a result, we were asked to be a pilot for the state and had one of only three approved asset management plans for road transportation in the state. By working together on these processes, I would point out one small highlight. When we began developing our GIS system, hydrant maintenance had been spread across fire, DPW, and public utilities. Each year dozens of hydrants were often found frozen, nonworking, or with other problems. By working together and tracking repairs, defects, and other actions, if we had one hydrant each year that did not test appropriately it was a lot. In one case a homeowner complained to his council member that we never flushed the hydrant on his street, and he was sure it did not work. Not only could I call the council member back and tell him the dates, times, and individuals who flushed it, I could also report the flow rate, the pressure, and repair history (including painting)! The homeowner never called again!

JANET WILMOTH

Editorial Director, *Fire Chief* magazine: Accountability and empowerment are two keys to improving quality.

FRED WINDISCH

Fire Chief, Ponderosa Volunteer Fire Department (Houston, Texas): Quality improvement is a process. Why do we always have time to do it right the second time? We attempt to fix it the first time. At the same time, we operate some projects by the 80 percent rule, to assure we do not study something to death. There is some risk in this, but if the foundation is laid correctly and everyone knows that we can modify direction quickly, then we can implement change to better the organization.

Review Questions

1. What is Six Sigma? Cite examples.
2. Who are considered three pioneers of the quality movement?
3. What are the principles of quality improvement?
4. Diagram the continuous quality improvement model and describe the elements of total quality management.
5. Explain the abbreviation PDCA.
6. When defining what quality is not, describe what some of the factors of quality are.
7. Describe the obstacles of implementing a quality program in the fire service.
8. Describe the need for a quality improvement program.
9. The quality improvement process is how we work toward this goal and has three components: quality teams, quality planning, and quality operations. Describe these three components.
10. What is a baseline? Cite examples.
11. What is a benchmark? Cite examples.
12. What are performance measures? Cite examples.
13. What is the difference between efficiency and effectiveness?
14. Describe the common pitfalls of performance measures?
15. Explain customer satisfaction as it relates to the fire service. Describe the different components of a customer satisfaction program.
16. Discuss the benefits of a quality improvement program.
17. Explain benchmarking for top performance and how it should be used.
18. Explain the performance criteria as established by the CFAI and the importance of such criteria.
19. Explain each of the steps of the customer-driven continuous improvement process.
20. Provide an overview of the quality initiative undertaken in your department.

References

1. Thomas Pyzdek, *The Six Sigma Handbook,* 2nd ed. (New York: McGraw-Hill, 2003); Pyzdek, "The Six Sigma Revolution," www.qualityamerica.com.
2. Ibid.
3. Ibid.
4. David Osborne and Ted Gaebler, *Reinventing Government* (Reading, MA: Addison-Wesley, 1992), p. 147.
5. ICMA, Arizona Leader, Academy, Performance Modern Dimensions, 1997.
6. Ibid.
7. Ibid.
8. Ibid.
9. Ibid.
10. Ibid.
11. Ibid.
12. Hatry, Blair, Fisk, Greiner, Hall, and Schaenman, *How Effective Are Your Community Services? Procedures for measuring their Quality.* 2nd ed. ICMA the Urban Institute Press. 1992, p. 93.
13. Jim Theurer, "Seven Pitfalls to Avoid When Establishing Performance Measures," *Public Management,* 86, no. 7, 1998, pp. 22–24.
14. Parasuraman Zenthaml, and L. L. Berry, *Delivering Quality Service: Balancing Customer Perceptions and Foundations* (New York: Free Press, 1990), p. 1.
15. Ibid.

Community Disaster Planning

11 CHAPTER

Key Terms

area command (unified area command), p. 519
CERT, p. 525
CI/KR, p. 535
disaster, p. 498
DOJ, p. 526
emergency, p. 498
emergency operations center (EOC), p. 521
emergency operations plan (EOP), p. 500
emergency support function (ESF), p. 519
EMI, p. 526
ERT-A, p. 522
federal coordinating officer (FCO), p. 523
FEMA, p. 501
HSOC, p. 521
HSPD, p. 515
IIMG, p. 521
incident action plan (IAP), p. 532
incident command post (ICP), p. 519
incident command system (ICS), p. 500
incident commander (IC), p. 530
incident of national significance, p. 516
joint field office (JFO), p. 522
joint operations center (JOC), p. 522
JTTF, p. 522
National Flood Insurance Act, p. 503
National Incident Management System (NIMS), p. 516
NJTTF, p. 522
nongovernmental organization (NGO), p. 523
NRCC, p. 521
NSSE, p. 521
NVOAD, p. 524
PDA, p. 510
planning, p. 529
preparedness, p. 506
principal federal official (PFO), p. 523
recovery, p. 509
response, p. 508
RRCC, p. 522
SAC, p. 523
SCO, p. 523
senior federal official (SFO), p. 523
SFLEO, p. 523
SIOC, p. 522
Stafford Act, p. 501
standardization, p. 528
weapon of mass destruction (WMD), p. 501

Objectives

After completing this chapter, you should be able to:

- Describe the Stafford Act.
- Define the role that FEMA has in disaster response.
- Describe the four phases of emergency management.
- Describe the federal National Response Plan.

- Describe what an ESF is.
- Define the ICS system and each section.
- Define the six areas of homeland security.

WHAT IS A DISASTER?

While I was teaching a fire service management class recently, many participants were questioning why a disaster planning course is needed because fire and emergency departments routinely do pre-incident planning for small and large emergencies. After all, planning for an infrequent event such as an earthquake or terrorist attack will only take valuable training time away from the more pressing training issues facing the department. The fact is that all levels of training are necessary if we are to be effective in our response to a catastrophic event.

disaster
Any occurrence that inflicts widespread destruction or distress.

emergency
As defined by the Stafford Act, an emergency is "any occasion or instance for which, in the determination of the President, Federal assistance is needed to supplement State and local efforts and capabilities to save lives and to protect property and public health and safety, or to lessen or avert the threat of a catastrophe in any part of the United States."

Disasters are events that strike citizens in the course of their ordinary lives, disrupting them in extraordinary ways. They usually affect more than just one family or business, unlike the majority of daily emergencies handled by fire and emergency forces. In fact, most states legally define what a disaster is and what criteria an event must meet to be declared a disaster. These definitions vary from state to state, but generally they are built around the concept that during any disaster the resources of the local community are completely committed or exhausted, and the situation nonetheless is not under control. Thus, outside aid is required for the resolution of the crisis. A disaster is quite different from day-to-day emergencies. **Emergencies** are events requiring substantial resources from the community and surrounding jurisdictions, whose impacts are restricted and are very short term. Operational departments usually are well prepared to handle emergencies but not as well prepared to handle disasters. Emergencies, like disasters, usually have a legal definition and a set of formalized response criteria, yet are based upon basic planning assumptions and response procedures utilized daily (see Figure 11.1).

Some have argued there is little difference between an emergency and a disaster other than the requirement for mutual aid, such as with a large wildfire. This is a major misconception. Disaster involves every public service discipline, not just police supported by fire and EMS (as in a civil disturbance) or fire supported by public works and police (as in a major fire). Often a disaster stretches over a large geographical area, crossing several jurisdictional boundaries. It creates multiple problems, broad in scope and nature, requiring community-wide response and significant outside assistance. This assistance not only is governmental but also includes a wide variety of national, state, and local private agencies that play a critical role in support and recovery. Many similarities exist between disasters and many larger incidents that fire and emergency agencies respond to regularly. These similarities can provide a basis for reliable disaster planning and initial response. Large, wide-area fires often cross boundaries of multiple governments. Mass-casualty incidents produce an overflow of patients to medical facilities, taxing pre-hospital care and impacting the entire medical system. Support at hostage events often may require extended operations with many of the same safety concerns as disasters. There also are similarities among various types of disasters. Hurricanes, tornadoes, and windstorms produce similar damage, injuries, and problems. Floods and hurricanes create similar issues for resolution by local officials. Building collapses, whether caused by an F-4 tornado or a large explosion, result in similar injuries, rescue problems, and recovery issues.

FIGURE 11.1 ◆ New Orleans after Hurricane Katrina *(Courtesy of the Federal Emergency Management Agency)*

The only way to begin to prepare for the devastation that disasters create is through comprehensive disaster planning. In most states, the responsibility for this rests with a local government agency. Because a disaster generally affects the entire community, disaster planning must be community based. For many jurisdictions prior to September 11, 2001, this was often just a collection of police, fire, EMS, and public works response plans integrated into a single document. These plans were limited in scope, failed to integrate incident management, and were known to exist by only a few members in the local jurisdiction. Today, the plan must articulate a comprehensive outline of the responsibilities of every agency in solving the problems and meeting the challenges a disaster creates. It must identify the resources available in the community, identify existing weaknesses, and develop plans to overcome those weaknesses where shortfalls occur. Planning highlights areas in which local mutual aid can help or where requests for aid must be forwarded to an outside agency, that is, the state or federal government. Most important the plan must be trained on and utilized at all levels of the local jurisdiction.

THE RISK

As discussed in Chapter 9, each community is exposed to a set of unique risks. While there may be similarities with situations in other communities, the risk impact of a disaster and the risk level present will be specific for each jurisdiction. These risks form the foundation upon which disaster planning will occur. The risk from the same disaster can vary from community to community. A Category 4 hurricane will cause different problems for a coastal community with barrier islands than it will for an inland community in the mountains. As witnessed with Hurricane Katrina, tidal surge and subsequent levee breaks, along with high winds, devastated the city of New Orleans. In other areas of the Southeast, heavy rainfall and debris from high wind gusts caused small stream and ravine flash flooding, causing several deaths and substantial property loss.

In accomplishing thorough comprehensive disaster planning, organizations begin with a detailed hazard assessment. This assessment examines all potential threats and attempts to determine what the impact would be if any of these actually occurred. Such an assessment generally separates events into several types: natural hazards, such as storms, earth movement, epidemics, or drought; technological or human-made, such as building collapse, hazardous materials incidents, or mass transit wrecks; and terrorism, such as attacks on government facilities, disruption of normalcy, or mass poisonings. Some events fall into multiple categories, such as building collapses caused by terrorist attacks.

The hazard assessment evaluates the threat that a particular circumstance can create in a community and examines it from the point of view of normal public safety operations. The hazard assessment then expands the analysis to look at what happens when the event expands to have community-wide impacts. An incident can transition from a normal emergency handled within a community (and by its public safety forces with possible outside mutual aid with the support of other public sector providers) to a disaster that may threaten the entire region. To be effective in such large-scale events requires the efforts of many local governments and agencies to plan together, focus, and identify the areas requiring comprehensive disaster planning and the responsibility each agency will assume when an incident occurs. Lessons learned from drills and exercises can reveal the community's hazard vulnerability, giving direction to local agencies in improving response, mitigation preparedness, and recovery efforts.

THE PLAN

emergency operations plan (EOP)
The comprehensive plan maintained by various jurisdictional level for managing response to a catastrophic event.

incident command system (ICS)
A standardized on-scene emergency management construct specifically designed to provide for the adoption of an integrated organizational structure that reflects the complexity and demands of single or multiple incidents, without being hindered by jurisdictional boundaries. ICS is the combination of facilities, equipment, personnel, procedures, and communications operating with a common organizational structure, designed to aid in the management of resources during incidents.

To meet the goal of being prepared, communities must have comprehensive disaster plans. These plans will cover many facets of the disaster potential and historically have stood apart as individual plans. Today, these plans are integrated into the community **emergency operations plan (EOP)**. EOPs cover all four phases of disaster response (mitigation, preparedness, response, and recovery) and are not simply response plans. Although not a hard concept for fire/emergency personnel to understand, given the focus on operational planning in the **incident command system (ICS)**, the challenge is often the motivation to get it done, keep the plan current, and have key personnel in the local jurisdiction trained to the point they can actually perform as a team in a disaster.

The EOP is based on the hazard vulnerability assessment of the community as well as the capability review, which outlines the community's disaster resources. The plan is organized into a basic plan with supporting annexes, each of which may have one or more appendixes to support it. As the statement of how a community will prepare for, respond to, and recover from disasters, the plan should include all organizations, both private and public, that are expected to respond, the role they will fill, and how they will coordinate with other entities. The completion of a community emergency operations plan is not an end in itself but only the beginning. Having a plan does not guarantee the response in an actual disaster will be effective. This concept is exemplified in a comment by President Dwight D. Eisenhower that "plans are worthless, but planning is everything . . . keep yourself steeped in the character of the problem you may one day be called upon to solve, or help solve." Local jurisdictions must realize merely getting a plan on paper does not fulfill their disaster planning responsibilities. As we have witnessed so many times in the last decade, those communities that fail to plan and prepare also fail in their response to large-scale events. The results are disastrous on many levels and often mean a prolonged recovery period.

THE GOVERNMENT'S ROLE IN DISASTER RESPONSE

Although many disasters can be handled using the resources within a state, some require a larger response and recovery effort. This effort often includes the federal government because of its ability to draw resources from far beyond the area affected by the event. Congress has approved different types of federal aid to assist and support local and state governments during time of disaster. This aid ranges from assistance given by military bases to their surrounding communities to the specific authorities granted to **FEMA** by the Robert T. Stafford Disaster Relief and Emergency Assistance Act (Amended). This act, commonly referred to as the **Stafford Act**, provides the legal basis for the Federal Response Plan (FRP), which has been cosigned by 28 federal agencies and outlines the comprehensive response federal agencies will employ to meet the needs of an area affected by a covered disaster.

FEMA
Federal Emergency Management Agency.

Stafford Act
This statute authorizes the president to provide assistance to state and local governments, as well as some nonprofit entities and individual disaster victims, in the aftermath of presidential-declared emergencies and major disasters. Most of the Stafford Act authorities have been delegated to the director of FEMA pursuant to Executive Order 12148, as amended. Title II of the Stafford Act provides authority for a variety of federal disaster preparedness activities. Title III is comprised of the act's administrative provisions, while Titles IV and V of the act authorize programs for responding to major disasters and emergencies, respectively. Title VI contains authorities formerly in the Federal Civil Defense Act of 1950 for emergency preparedness and cross references the Defense Production Act to include "emergency preparedness" in the definition of "national defense."

The FRP is designed to meet the two levels of disaster emergencies as defined in the Stafford Act and as declared by the president. The first level (an emergency) is any event requiring federal assistance to supplement state and local efforts to save lives and protect property or to decrease the effects of a catastrophe. The second level (a major disaster) is any event causing damage of such severity and magnitude that the damage assistance outlined in the Stafford Act is necessary. FEMA can use its authority under the Stafford Act to respond to other types of incidents than natural disasters. This authority is also the foundation of federal response to the consequences of terrorism or **weapons of mass destruction (WMD)** incidents.

In every community there is the expectation that a local and state level of preparation will facilitate response and recovery efforts. During the last 20 years the many disasters we have experienced, both natural and human caused, have proven there is still much to accomplish in respect to disaster response. Experience has taught us that response to a disaster requires each component of the system to work in concert, or the system will become unbalanced and implode upon itself (see Figure 11.2).

Local community disaster planning outlines the community's understanding of its role in disaster response and the support it can expect from state and federal agencies. Key to this understanding is awareness of available resources and aid from beyond the traditional realm of local government and public safety. From community emergency response teams to industrial mutual-aid groups, disaster planning incorporates many elements into the community preparedness effort. This effort is what the citizens expect from government. It is how they will measure the effectiveness of every agency (federal, state, and local) if a disaster occurs. Think of the fallout after Hurricane Katrina. Federal response was criticized as being slow and uncoordinated.

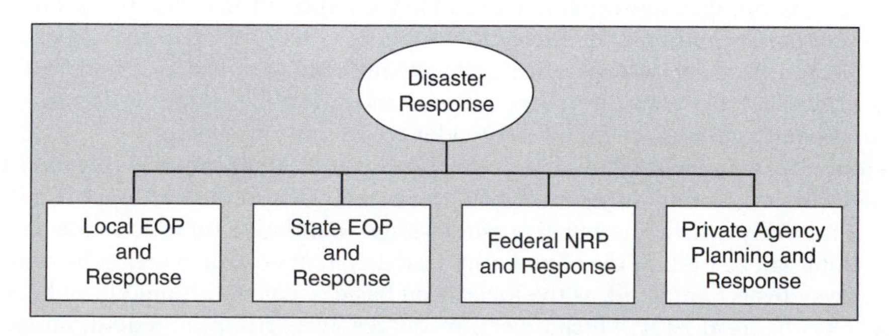

FIGURE 11.2 ◆ Disaster Response *(Courtesy of Randy R. Bruegman)*

weapon of mass destruction (WMD)
As defined in Title 18, U.S.C. § 2332a: (1) any explosive, incendiary, or poison gas, bomb, grenade, rocket having a propellant charge of more than 4 ounces, or missile having an explosive or incendiary charge of more than one-quarter ounce, or mine or similar device; (2) any weapon that is designed or intended to cause death or serious bodily injury through the release, dissemination, or impact of toxic or poisonous chemicals or their precursors; (3) any weapon involving a disease organism; or (4) any weapon that is designed to release radiation or radioactivity at a level dangerous to human life.

State response was questioned for failure to integrate with federal and local response, and the lack of a congruent effort between the state and federal response was evident in the national media. Locally, jurisdictions that have failed to plan and train are often simply overwhelmed, many not knowing how the state and federal EOPs and response system worked.

In addition to working on preparedness and mitigation issues, these programs focus on recovery. Good and reliable damage assessment programs are key postdisaster activities. This focus is predicated on the need to obtain a governor's declaration in the event of a disaster and request for a comparable presidential declaration as early as possible. Recent natural disasters and terrorism incidents have led to many changes in the federal programs that allow early and rapid assistance. The Stafford Act now allows deployment of federal resources before an event occurs and before a state governor requests assistance. With the noted deficiencies in response during Hurricane Katrina, we can anticipate even more change and possible structural realignment of the federal response system. Local fire and emergency personnel must keep abreast of such changes in the state and federal programs governing emergency and disaster assistance, as they often can impact local EOPs and the process for state and federal resource acquisition. However, it is clear these programs are not static, and the changes will continue to occur. Yet to be factored into reference materials and books are the lessons learned from the 2005 hurricane season. These lessons may only confirm what emergency planners have known for years or may also reveal new areas of concern. One thing has clearly emerged: to be effective, all levels (local, state, federal, private) must work in congruence, or the system suffers a disconnect, which results in a negative impact on the delivery of services in the field. Unfortunately, when this occurs more lives are lost, more property is damaged, and recovery takes longer.

PREPAREDNESS: A COMMUNITY EFFORT

The fire service always has been seen as a critical element of the community's response capability in major events and disasters. Because of this, in many jurisdictions the fire service has become the community preparedness or emergency management agency. The fire service as part of the community's overall preparedness is often the core component because it is the fastest responder with teams equipped to intervene in the types of problems resulting from a natural or human-made disaster. Yes, police may arrive first, but usually in limited numbers and without the appropriate equipment and training to impact the immediate rescues and mitigation often needed. And yes, public works agencies operate more heavy equipment than the fire service, but the fire service is the first-arriving team-based unit. This raises citizens' expectations that the fire service can and will effect a change in the outcome of the incident. Often, however, fire service plans are based only on those problems anticipated to have a fire component and lack attention to broader community-wide issues.

Good community-based disaster planning includes all agencies involved in service to its citizens. Community-based systems not only afford the best opportunity for intervening early in an event but also establish the foundation on which state and federal actions will be built. If the community-based system fails, then action by state and federal systems is hampered, as the focus is on basic issues that should have been addressed by the local EOP, which diverts resources away from the federal mission of support and recovery.

FEMA HISTORY

The Federal Emergency Management Agency (a former independent agency that became part of the new Department of Homeland Security in March 2003) is tasked with responding to, planning for, recovering from, and mitigating against disasters. FEMA can trace its beginnings to the Congressional Act of 1803, generally considered the first piece of disaster legislation, which provided assistance to a New Hampshire town following an extensive fire. In the century that followed, ad hoc legislation was passed more than one hundred times in response to hurricanes, earthquakes, floods, and other natural disasters.

By the 1930s, when the federal approach to problems became popular, the Reconstruction Finance Corporation was given authority to make disaster loans for repair and reconstruction of certain public facilities following an earthquake and later other types of disasters. In 1934 the Bureau of Public Roads was given authority to provide funding for highways and bridges damaged by natural disasters. The Flood Control Act, which gave the U.S. Army Corps of Engineers greater authority to implement flood control projects, was also passed. This piecemeal approach to disaster assistance was problematic, prompted legislation requiring greater cooperation among federal agencies, and authorized the president to coordinate these activities.

The 1960s and early 1970s brought massive disasters requiring major federal response and recover operations by the Federal Disaster Assistance Administration, established within the Department of Housing and Urban Development (HUD). Hurricane Carla struck in 1962, Hurricane Betsy in 1965, Hurricane Camille in 1969, and Hurricane Agnes in 1972. The Alaskan Earthquake hit in 1964, and the San Fernando Earthquake rocked southern California in 1971. These events served to focus attention on the issue of natural disasters and brought about increased legislation. In 1968, the **National Flood Insurance Act** offered new flood protection to homeowners, and in 1974 the Disaster Relief Act firmly established the process of presidential disaster declarations.

National Flood Insurance Act of 1968, as amended, 42 U.S.C. 4001 et seq.
This statute authorizes FEMA to administer the National Flood Insurance Program (NFIP). Under the NFIP, FEMA is authorized to provide flood insurance for commercial and residential structures that are built in communities that agree to adopt standards for the construction of buildings located within flood-prone areas of the communities.

However, emergency and disaster activities were still fragmented. When hazards associated with nuclear power plants and the transportation of hazardous substances were added to natural disasters, more than one hundred federal agencies were involved in some aspect of disasters, hazards, and emergencies. Many parallel programs and policies existed at the state and local levels, compounding the complexity of federal disaster relief efforts. The National Governor's Association sought to decrease the many agencies with which state and local governments were forced to work.

It asked President Jimmy Carter to centralize federal emergency functions. President Carter's 1979 executive order merged many of the separate disaster-related responsibilities into a new Federal Emergency Management Agency (FEMA). Among other agencies, FEMA absorbed the Federal Insurance Administration, the National Fire Prevention and Control Administration, the National Weather Service Community Preparedness Program, the Federal Preparedness Agency of the General Services Administration, and the Federal Disaster Assistance Administration activities from HUD. Civil defense responsibilities were also transferred to the new agency from the Defense Department's Defense Civil Preparedness Agency. Named as FEMA's first director, John Macy emphasized the similarities between natural hazards preparedness and the civil defense activities. FEMA began development of an integrated emergency management system with an all-hazards approach that included direction, control, and warning systems common to the full range of emergencies from small isolated events to the ultimate emergency—war.

Many unusual challenges faced the new agency in its first few years, emphasizing how complex emergency management can be. Early disasters and emergencies included the contamination of Love Canal, the Cuban refugee crisis, and the accident at the Three Mile Island nuclear power plant. Later, the Loma Prieta Earthquake in 1989 and Hurricane Andrew in 1992 focused major national attention on FEMA. In 1993 President Clinton nominated James L. Witt as the new FEMA director, who became the first agency director with experience as a state emergency manager. Witt initiated sweeping reforms that streamlined disaster relief and recovery operations, insisted on a new emphasis regarding preparedness and mitigation, and focused agency employees on customer service. The end of the Cold War also allowed Witt to redirect more of FEMA's limited resources from civil defense into disaster relief, recovery, and mitigation programs.

In 2001, President George W. Bush appointed Joe M. Allbaugh as the director of FEMA. Within months, the terrorist attacks of September 11 focused the agency on issues of national preparedness and homeland security, and tested the agency in unprecedented ways. The agency coordinated its activities with the newly formed Office of Homeland Security, and FEMA's Office of National Preparedness was given responsibility for helping to ensure the nation's first responders were trained and equipped to deal with weapons of mass destruction. Billions of dollars of new funding were directed to FEMA to help communities face the threat of terrorism. Just a few years past its twentieth anniversary, FEMA was actively directing its all-hazards approach to disasters toward homeland security issues. In March 2003, FEMA joined 22 other federal agencies, programs, and offices in becoming the Department of Homeland Security (DHS). The new department was headed by Secretary Tom Ridge until February 1, 2005, when he was replaced by Secretary Michael Chertoff. This consolidation brought a coordinated approach to emergencies and disasters, both natural and human-made. Mike Brown was the undersecretary of preparedness in charge of FEMA until Hurricane Katrina. Within a week, due to a perceived poor federal response, Brown resigned and was replaced with R. David Paulison, director of FEMA and former U.S. fire administrator.

Today, FEMA is one of four major branches of DHS. Approximately 2,500 full-time employees in the Emergency Preparedness and Response Directorate are supplemented by more than 5,000 standby disaster reservists.[1] FEMA's mission is to lead America to prepare for, prevent, respond to, and recover from disasters with a vision of "A Nation Prepared." At no time in its history has this vision been more important to the country than in the aftermath of September 11 and Hurricanes Katrina and Rita.

Since FEMA's establishment, many state and local organizations have changed the names of their organizations to include the words *emergency management*. The name change reflects the shift in orientation away from specialized preparedness for single hazards or narrowly defined categories of hazards and toward an all-hazards approach, natural and technological, to potential threats to life and property. As Congress and FEMA have been quick to point out, this change reflects not a reduction in security, but an increased emphasis on making the nation's emergency management capability responsive to any major emergency. This expansion into the all-hazards approach increased the need to further develop the concept of comprehensive emergency management (CEM). The concept consists of three interrelated components.

1. *All types of hazards.* The commonalities among all types of technological and natural disasters suggest strongly that many of the same management strategies can apply to all such emergencies.

2. *An emergency management partnership.* The burden of disaster management and the resources for it require a close working partnership among all levels of government (federal, regional, state, county, and local) and the private sector (business and industry, voluntary organizations, and the general public).
3. *An emergency life cycle.* Disasters do not just appear one day. Rather, they exist throughout time and have a life cycle of occurrence that must be matched by a series of management phases, which includes strategies to *mitigate* hazards, *prepare* for and *respond* to emergencies, and recover from their effects.

These three components form CEM. CEM suggests there are four phases of emergency management that must work together to protect a community: preparedness, response, recovery, and mitigation.

In combination, these four phases involve all the levels of government, nonprofit organizations, and the private sector. The specific roles and responsibilities of each government level and allied partner depend to some degree on the nature of the disaster and specific resource requirements. Whenever the government levels cooperate to address a specific function, communication is critical. In general, the flow of communication is between local and state levels and between state and federal levels. As the severity of an emergency increases and escalates to disaster proportions, the roles and responsibilities of local government change significantly. For minor emergencies, the local government is able to fully handle functions such as warning, communication, evacuation, sheltering, damage assessment, search and rescue, emergency medical services (EMS), provision of food and water, and actions needed to facilitate recovery or mitigate the effects of future similar emergencies. As the level of emergency increases, mutual-aid agreements with other communities may be activated (see Figure 11.3).[2]

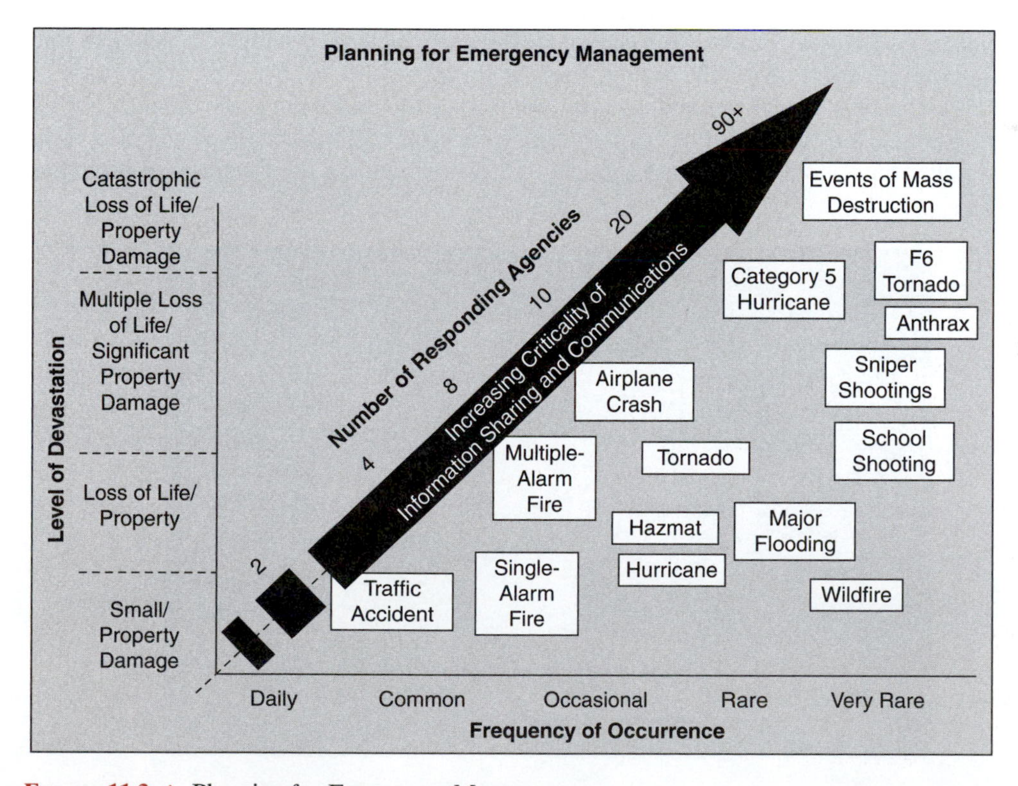

FIGURE 11.3 ◆ Planning for Emergency Management

If the emergency is severe enough to exhaust some or all local functional resources in the immediate operational area, state resources may be needed to protect life and property adequately. The role of local government will change in these functional areas. Coordination, communication, and reporting become critical to ensure efficient response whenever local and state resources work together. In a disaster, the state may act on information communicated by the local government (and its own analysis of the situation) to request federal assistance, some of which is available even without a declaration of disaster. The amount of local resources used in response to a major disaster may be quite small in proportion to the total resources applied. Yet the local role remains critical. It is local government that understands how local facilities are used, how local traffic moves, and how the local economy works. It is local government that can remain in touch with residents so they can voice their concerns. Even in the most serious disaster, local government is never insignificant; on the contrary, it is essential to success. In addition, the resources of nonprofit organizations such as the Red Cross and the private sector provide a number of immediate resources that, when integrated into overall response, can have a very positive impact on the outcome of the event.

preparedness
The range of deliberate, critical tasks and activities necessary to build, sustain, and improve the operational capability to prevent, protect against, respond to, and recover from domestic incidents. Preparedness is a continuous process involving efforts at all levels of government and between government and private sector and nongovernmental organizations to identify threats, determine vulnerabilities, and identify required resources.

PREPAREDNESS

Preparedness must be in place before an emergency; once disaster strikes, it is too late to prepare (see Figure 11.4). **Preparedness** includes measures that enable ready-responding forces and citizens of a locality to take prompt, appropriate action in the event of an emergency. Examples of preparedness measures a local government might take include preparing a comprehensive plan, conducting training to prepare local response forces to play their roles effectively, and negotiating mutual-aid agreements. Local government also should designate evacuation routes in advance and provide information to the public to help them protect themselves appropriately. FEMA's Family Preparedness Program supports state and local governments in their efforts to educate the public by providing opportunities, information, and also tools to help other organizations and agencies educate the public on disaster preparedness.

FIGURE 11.4 ◆ EOC *(Courtesy Mark Wolfe/FEMA News Photo)*

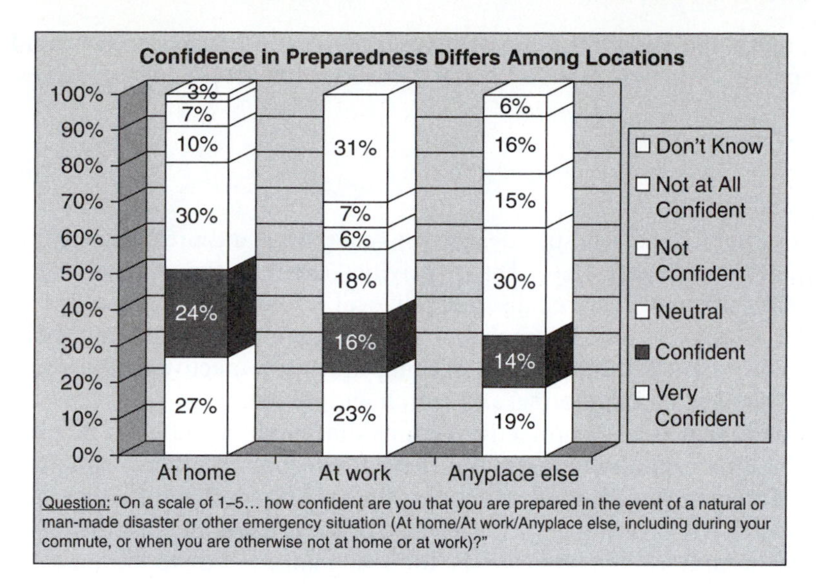

FIGURE 11.5 ◆ Confidence in Preparedness Differs

As we have seen in many of the recent disasters, the preparedness of individuals and families often does not occur. A survey conducted by the American Red Cross in 2003 provides an insight into our level of individual preparedness (see Figure 11.5).

Promoting citizen readiness is a critical element of a community's response. FEMA suggests the following four steps for personal and family safety:

1. Find out what could happen to you.
2. Create a disaster plan.
3. Complete a checklist.
4. Practice and maintain your plan.[3]

More information on preparedness materials can be found online at www.fema.gov and www.redcross.org. Other preparedness materials are available at these sites, as well as at www.ready.gov.

When preparing for a major disaster that could escalate to a presidential declared disaster, local government must refine its planning based on the best available sources of information. Federal sources may provide information on weather patterns, local geology, and dam stability. The job of local government is to collect this information, assess it, and do what it can to enhance preparedness. This is true whether the impending disaster is a major storm or hurricane heading for a local area, a nearby volcano that has become active, or an imminent earthquake suggested by geological data. To be effective, local government should conduct exercises, hold discussions among responding agencies, and/or take mitigation steps that could decrease the impact of a disaster. Local government may, depending on the nature of the emergency, publicize its evacuation plans to ensure citizens are able to leave the community by appropriate routes. It also should disseminate information to ensure that citizens are informed fully of actions they can and should take to protect their lives and property.

The state also will monitor information and remain in touch with the local operational area, preparing its own resources to act if need be. As part of its planning, the state also considers what federal resources might be available (even in an undeclared disaster) that could assist if the disaster exceeds the state's capacity to respond. It may even prepare agreements in advance that will allow it to access such resources readily.

RESPONSE

response
Activities that address the short-term, direct effects of an incident. Response includes immediate actions to save lives, protect property, and meet basic human needs. Response also includes the execution of emergency operations plans and of incident mitigation activities designed to limit the loss of life, personal injury, property damage, and other unfavorable outcomes.

Response is defined as activities designed to save lives and prevent harm to people and property (see Figure 11.6). These activities occur at the peak of the response phase in a presidential declared disaster. All local resources will be totally committed during the response phase. Examples of response activities include notifying public officials and responders; alerting and warning citizens; protective actions; evacuation; short-term feeding and sheltering; and search and rescue.

Response may include protective actions that might also be considered mitigation efforts. This overlap illustrates the close connection in practice among the four phases in emergency management. During the response phase of a presidential declared disaster (or one that may be declared), local government must clearly, efficiently, and accurately communicate the following information to the state: the extent of the forces it has committed to disaster response and the extent to which the severity of the disaster exceeds the capability of the forces, the exact nature of the unmet needs and what will be required to meet them, timelines for action, information needed, and other critical data. A major role of local government is to provide the state with critically needed information on the local operational area to help assess not only what has been damaged but also what the nature of the losses are and what the impact is to the viability of the community. Local government must communicate clearly its perception of needs based on its knowledge of the area, help state and federal forces maneuver through the area when they arrive, and help them interpret what they see. The state, in turn, must assess the information provided at the local level and determine when the disaster has exceeded state capacity. When this is the case, it must provide the federal government with the information it needs to decide whether a disaster should be declared.

FIGURE 11.6 ◆ Transporting Patient *(Courtesy of Win Henderson/FEMA News Photo)*

MEDIA COMMUNICATIONS

One area often overlooked and not adequately prepared for is media communications in the event of a disaster. Although we have become adept at reporting local routine emergencies in the event of a disaster media communications become critically important as we go from the local arena to a national stage. It is important for us to understand in those situations that we not only are speaking to our local residents but also are conveying key messages on a national and sometimes international basis. Understanding how we communicate is key not only in the words that we say but also in the tone and pitch we speak in, in the body language we portray, and in other nonverbal characteristics that each can deliver a very strong message. If one is called upon to deliver a news conference, whether it is a disaster or a local emergency, here are some key points in dealing with the media and ensuring the message is articulated effectively:

- Remember body language and eye contact are important.
- Control your pace and the pitch of your voice.
- Deliver a concise message.
- Know your audience.
- Anticipate the questions you may be asked.
- Review and know the content.
- Practice your delivery such as gestures and pauses used to emphasize a point.

IF YOU ARE THE SPOKESPERSON

Remember you are there to make the public understand what has happened and the sincerity of your delivery and the factualness of the information you provide will determine the level of credibility to those who are watching you on TV or listening to you over the radio. If you are the spokesperson, your purpose is to:

- Educate the public.
- Promote a positive organizational identity.
- Develop a positive media relationship.
- Influence the story by sharing your knowledge of the situation.
- Become a trusted source.
- Respond appropriately and effectively to media questions.

recovery
The development, coordination, and execution of service- and site-restoration plans for impacted communities and the reconstitution of government operations and services through individual, private sector, nongovernmental, and public assistance programs that identify needs and define resources; provide housing and promote restoration; address long-term care and treatment of affected persons; implement additional measures for community restoration; incorporate mitigation measures and techniques, as feasible; evaluate the incident to identify lessons learned; and develop initiatives to mitigate the effects of future incidents.

RECOVERY

Before publication of the Federal Response Plan (FRP), FEMA (especially the Federal Disaster Assistance Program) was oriented toward **recovery** activities (see Figure 11.7). Federal on-scene activities began after a presidential declaration of a major disaster or emergency and focused on providing individual and public assistance to disaster victims and affected governments. FEMA institutionalized the programmatic distinction between response and recovery in its field operations. Response focused on early lifesaving aspects of federal assistance to state and local governments, whereas recovery focused on grants, loans, or direct federal assistance to restore the affected area.

THE FEDERAL RESPONSE PLAN (FRP)

In the fall of 1989, the Stafford Act and FEMA were tested by Hurricane Hugo in South Carolina and the Loma Prieta Earthquake in California. Over the several years

FIGURE 11.7 ◆ Recovery Center *(Courtesy of Marvin Nauman/FEMA News Photo)*

prior to these large-scale disasters, there had been concern about potentially catastrophic earthquakes. This concern was the basis for the development of the FRP. As the plan developed, it was expanded to include all disasters declared under the Stafford Act, not only catastrophic earthquakes. The plan was first tested during Hurricane Andrew in 1992. Since July 1993, federal response has broadened to encompass early detection, monitoring, analysis of the situation, and predictive modeling.

The intent of the FRP was to be more proactive, short of actually delivering disaster assistance prior to an anticipated catastrophic event. New legislation and policies now allow and facilitate rapid deployment and, in some cases, prepositioning of federal resources to assist state and local governments in dealing with the consequences of a disaster or emergency. As witnessed with Hurricane Katrina, this system of deployment is still a work in progress. Because of the expanded role of the FRP in disaster assistance, states now need federal assistance to assess their needs after a disaster. A preliminary damage assessment (**PDA**) is no longer conducted solely to support a request for a presidential declared disaster; it also functions as a needs assessment that helps determine early response requirements. A new response and recovery directorate has taken the place of the previous recovery-oriented division. FEMA's organization now matches the reality of operations in the field during a disaster. Recovery includes both short-term activities that restore vital life support systems to minimum operating standards and long-term activities that return life to normalcy. Examples of such restorative activities include debris clearance, contamination control, disaster unemployment assistance, temporary housing, and facility restoration.

PDA
Preliminary damage assessment.

Decisions made during recovery will have a long-lasting impact on the community. The soundness of these measures will affect not only the tangible resources of a community but also the intangible ones, such as victims' beliefs that personal and community recovery are possible. The psychology of recovering from such events has not been thoroughly studied; however, it is suggested that a critical element in the recovery process begins with a belief that recovery is possible. The sooner cleanup begins and services resume, the sooner people feel a sense of accomplishment and return to normalcy, which promote quicker recovery from the disaster.

While recovery from a presidential declared disaster is underway, local government continues to play an important role in the response effort. Local resources maintain community relations, opening lines of communication with victims that enable them to understand current concerns and issues. Local government then can work with state and federal agencies to address these concerns through clearer communication, changes in priority, additional resources, or whatever is required. Federal and state agencies cooperate closely to determine that resources needed to aid recovery are applied efficiently in areas of need. They also impose substantial requirements for reporting and documenting how funds were spent. In a disaster it is difficult to pinpoint where the response phase ends and the recovery phase begins. The first steps toward recovery are taken when the situation begins to stabilize and, as such, response and recovery activities overlap in practice. As recovery begins, response efforts that are likely to continue may include any or all of the following: search and rescue efforts; medical treatment; restoration of transportation facilities, utilities, and sewage plants; control of looting, security problems, and civil disorder; and management of hazardous materials threats created by the disaster.

MITIGATION

During the recovery process, mitigation measures should be considered to reduce the community's vulnerability to similar disasters (see Figure 11.8). Throughout the recovery period, energy and commitment to protecting the community from future disasters are at their height, and the memory of factors that created vulnerability in the current disaster is fresh and clear. It is important to use this period to identify appropriate protection for the community. In fact, as a condition of receiving federal disaster assistance through the Presidential Disaster Declaration process, each state must develop a comprehensive hazard mitigation plan identifying mitigation opportunities in the declared disaster area as well as throughout the state.

Mitigation actually occurs during all phases of emergency management. It includes actions taken to prevent an emergency from occurring or to reduce the harm

FIGURE 11.8 ◆ Mitigation Center *(Courtesy of Adam Dubrowa/FEMA News Photo)*

that would result from an emergency should it occur. For example, an earthquake-prone area might retrofit buildings to increase their durability and safety in an earthquake; the local government might also pass codes requiring earthquake-resistant construction in new buildings. If an earthquake occurred and aftershocks were expected, immediate mitigation measures might include the demolition of a structure destabilized by the first earthquake that could fall and damage other structures or injure people. Similarly, fire-resistant construction is a mitigation measure that could prevent a life-threatening fire from spreading through a building. Warning devices, such as smoke detectors, are mitigation measures to reduce the likelihood that human life would be lost in the event of a fire. For potential flood areas long-term mitigation measures might include elevating structures or strengthening levees to protect them from high waters; more immediate mitigation might include sandbagging.

For presidential declared disasters, the local, state, and federal levels have a role in mitigation. As noted previously, once a disaster is declared, a mitigation plan is required. The plan will be effective only if the local area plays its part in taking effective mitigation steps, many of which can be accomplished only at the state and local levels. These include changing codes to require earthquake-resistant or flood-protective construction or rezoning a landslide area to prevent future building collapse/destruction.

Project Impact, introduced by FEMA in early 2000, was a program designed to build disaster-resistant communities through a combination of several concepts: partnerships within the community, hazard identification and hazard assessment, and mitigation. Once these initial goals have been met, Project Impact helps the community stay focused on them through the use of the media, publicity, and special events.

STORMREADY

Americans live in the most severe weather-prone country on Earth. Each year, Americans cope with an average of 10,000 thunderstorms, 2,500 floods, 1,000 tornadoes, as well as an average of six deadly hurricanes. Potentially deadly weather impacts every American. Communities can now rely on the National Weather Service's (NWS) StormReady program to help them guard against the ravages of Mother Nature.

Approximately 90 percent of all presidential declared disasters are weather related, leading to around 500 deaths per year and nearly $14 billion in damage. StormReady, a program started in 1999 in Tulsa, Oklahoma, helps arm America's communities with the communication and safety skills needed to save lives and property—before and during the event. StormReady aids community leaders and emergency managers in strengthening local safety programs. StormReady communities are better prepared to save lives from the onslaught of severe weather through better planning, education, and awareness. No community is storm proof, but StormReady can help communities save lives.

The NWS designed StormReady to help communities better prepare for and mitigate effects of extreme weather-related events. StormReady also aids in establishing a commitment to creating an infrastructure and systems that will save lives and protect property. Receiving StormReady recognition does not mean that a community is storm-proof, but StormReady communities will be better prepared when severe weather strikes. Although StormReady is a volunteer program separate from FEMA's Pre-Disaster Mitigation Program, the two programs complement each other by focusing on communication, mitigation, and community preparedness to save lives. Grant funding is not associated with being recognized as StormReady; however, Insurance Services Office (ISO) may provide *community ratings system (CRS)* points to StormReady communities, which may be applied toward lowering NFIP flood insurance rates. There is no fee for StormReady recognition; however, communities may need to upgrade their emergency preparedness infrastructure to qualify for StormReady status. National StormReady guidelines set minimum requirements for the program. Many local areas have specific weather-related needs that local NWS offices consider during weather emergency planning. As a result, StormReady allows the creation of local StormReady advisory boards that have the flexibility to create specific bylaws for their areas. Local boards also can modify national StormReady guidelines to meet their specific customer needs.[4]

THE INTEGRATED EMERGENCY MANAGEMENT SYSTEM (IEMS) CONCEPT

FEMA implemented the Integrated Emergency Management System (IEMS) concept endorsed by President Ronald Reagan and Congress in fiscal year 1984. The goal was to develop and maintain credible emergency management capability nationwide for all emergencies at all levels of government. Although, now that we are into the twenty-first century, the concept remains solid and has been a stabilizing force in disaster areas receiving needed FEMA assistance. The concept of an all-hazards approach to emergency management defined in CEM has been implemented by FEMA in its Integrated Emergency Management System (IEMS). This implementation is based on all the principles of CEM with the specific goals of:

- fostering a full federal, state, and local government partnership with provisions for flexibility at the various levels of government for achieving common national goals;
- emphasizing implementation of emergency management measures that are known to be effective;
- achieving more complete integration of emergency management planning into mainstream state and local policymaking and operational systems; and
- building on the foundation of existing emergency management plans, systems, and capabilities to broaden their applicability to the full spectrum of emergencies.

IEMS takes into account the fact that each community across the country has its own existing level of emergency management capabilities. Some jurisdictions already have in place the components of an effective emergency management system. IEMS will build upon these existing capabilities, providing incentives to improve, and further integrate localities and states into a national system (see Figure 11.9).[5]

In the IEMS approach, a community that has, to this point, done little toward developing emergency management activities will begin a process to develop emergency plans governed by national criteria specific enough to provide guidance but sufficiently general to allow flexibility for the local protection options communities believe are critical. The process begins with a comprehensive hazard assessment prepared by the community, possibly in conjunction with state and federal regional personnel, depending on the circumstances. It then proceeds through analysis of capability (identifying shortfalls of resources) and moves to the development of a generic operations plan with annexes for the emergency management functions and appendixes for the unique aspects of individual emergencies, the maintenance of capability, mitigation activities, emergency operations, and evaluation of such operations. The jurisdiction is then asked to prepare an EOP, followed by annual plan increments as the process proceeds. By following this process, a community can establish an IEMS with readiness to deal with both the common elements of preparedness and those requirements unique to individual emergencies.

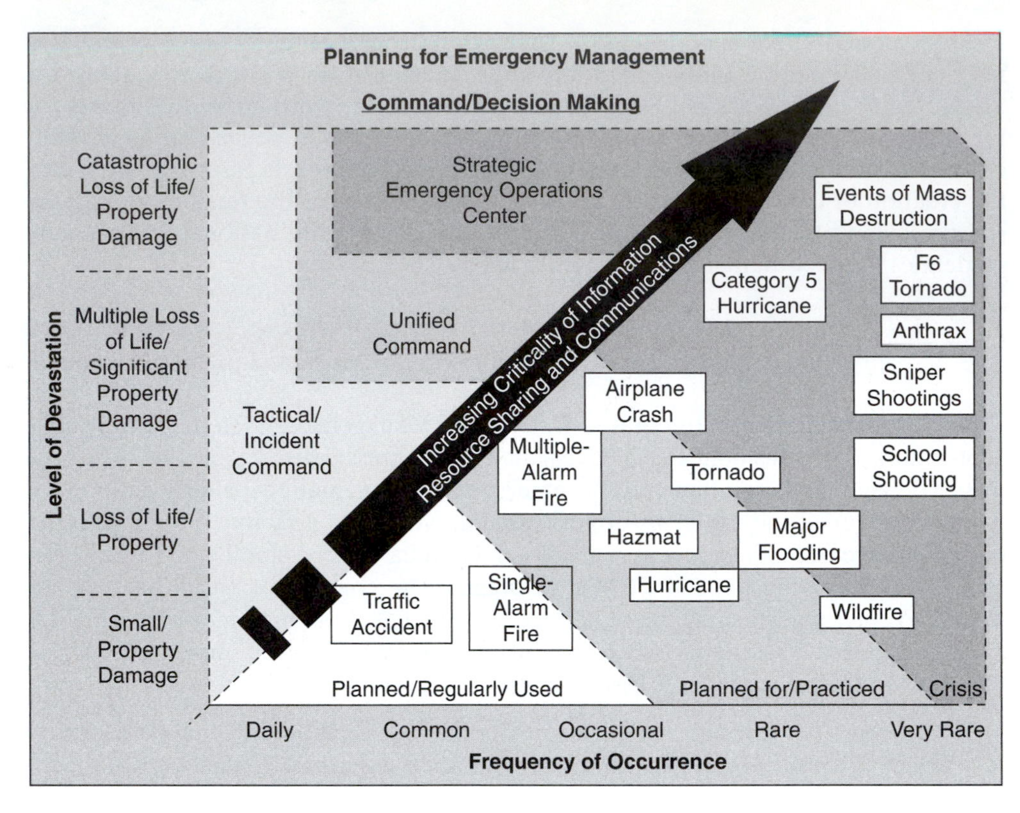

FIGURE 11.9 ◆ Planning for Emergency Management—Command/Decision Making

As such, hazard assessment is a major cornerstone of the IEMS philosophy and will require a detailed look at this critical process. It was evident during the 2005 hurricane season that those communities that undertook Project Impact and utilized IEMS were much better prepared to respond, recover, and mitigate the emergency.

HAZARD ASSESSMENT: WHY IS IT IMPORTANT?

Throughout the history of the United States, natural and technological disasters have destroyed lives, devastated property, and put large numbers of people at risk. Earthquakes, tornadoes, volcanic eruptions, floods, droughts, blizzards, fires, and hurricanes are but a few of the hazards threatening communities and challenging command fire officers. Various processes can lessen the negative impact of such incidents on citizens, the community, and individual property. Although no single process fits every community and every potential hazard, a carefully constructed community hazard assessment generally is considered a critical first step in any proactive hazard management plan.

Risk assessment is essential to the emergency planning process for a number of reasons:

- Most community response to perceived risk has been intuitive, not based on analytical assessment.
- Many communities are unaware of all the risks they face from various emergencies and disasters and are unprepared to respond to them.

- Mitigations selected to deal with the risks may not have been the most appropriate ones. In many cases, the mitigations require investment before the event, which is difficult to obtain. Communities would rather take their chances.
- Risk mitigation may not have kept pace with community, social, and technological changes.[6]
- Emergency managers have a clear challenge to heighten the public's attention and commitment to prepare for disaster before it occurs.

Community hazard assessment focuses the attention of the community on the present demand for protection from risk and on the long-term goals of the community to be relatively free from risk. Out of this process comes an opportunity for the community's fire service to acquire "natural publicity" if the process is open to the community. The spin-offs of an open planning process are substantial both for the fire service and for the community. A planning synergy is often developed because, as the community sets its priorities and the community leadership is asked to meet the expectations, leaders need to have political support to achieve the long-term goals for the community and to establish a larger framework that will cause the community to plan for the future.

Community hazard assessment assists in the budget process by making budget priorities logical and by creating political energy to support the budget process. With community hazard assessment, budgeting is a planned process instead of a reactive process. Community hazard assessment improves fire service quality, because fire service decisions focus on community risk priorities analyzed properly. These decisions follow from logical planning and the local budget process. Procedures are developed to handle critical risks that go beyond the capability of the community. Hazard assessment in any modern community is holistic, multidisciplinary, and community oriented. This means any properly considered hazard assessment calls upon a broad spectrum of the community for information. Decisions that affect a community's vulnerability to hazards are made daily by individuals, businesses, community leaders, and managers of emergency services. Effective decisions depend on the quantity, quality, accessibility, and usefulness of the hazard/risk information available to decision makers. The same is true for disaster preparedness. Every community has a different risk and exposure to natural and human-caused disasters. A comprehensive assessment of risk and hazard in a given jurisdiction provides the basic essential information for pre-event planning.

HAZARD IDENTIFICATION

A hazard is a natural or technological force or event that could cause or create a disaster. It would be ideal if communities were prepared for all types of hazards. However, in most cases this is not practical, as some may never occur. For example, it would be impractical to focus preparedness on hurricanes in the Midwest, as the probability of one occurring there is near zero. To begin planning a program to manage emergencies, the hazards that have the potential of affecting a jurisdiction must be identified, as well as how such an event may impact people, property, and structures that are in the jurisdiction served.

THE NATIONAL RESPONSE PLAN

In Homeland Security Presidential Directive (**HSPD**)-5, President Bush in 2003 directed the development of a new National Response Plan (NRP) to align federal coordination structures, capabilities, and resources into a unified, all-discipline, and all-hazards

HSPD
Homeland Security Presidential Directive.

approach to domestic incident management. This approach was unique in that it, for the first time, attempted to tie together a completed spectrum of incident management systems for disasters and other major emergencies. The objective is to improve coordination among federal, state, local, and tribal organizations to help save lives and protect America's communities by increasing the speed, effectiveness, and efficiency of response under a unified incident management system. The NRP represents a national framework in terms of both product and process. The NRP development process included extensive coordination with federal, state, local, and tribal agencies, nongovernmental organizations, private sector entities, and the first responder and emergency management communities across the country. The NRP incorporates best practices from a wide variety of incident management disciplines to include fire, rescue, emergency management, law enforcement, public works, and emergency medical services.

The NRP is built on the template of the **National Incident Management System (NIMS)**, which provides a consistent framework for incident management at all jurisdictional levels, regardless of the cause, size, or complexity of the incident. The activation of the NRP and its coordinating structures and protocols—either partially or fully—for specific incidents of national significance provides mechanisms for the coordination and implementation of a wide variety of incident management and emergency assistance activities. Included in these activities are federal support to state, local, and tribal authorities; interaction with nongovernmental, private donor, and private sector organizations; and the coordinated, direct exercise of federal authorities, when appropriate.

National Incident Management System (NIMS) A system mandated by HSPD-5 that provides a consistent, nationwide approach for federal, state, local, and tribal governments; the private sector; and NGOs to work effectively and efficiently together to prepare for, respond to, and recover from domestic incidents, regardless of cause, size, or complexity. To provide for interoperability and compatibility among federal, state, local, and tribal capabilities, the NIMS includes a core set of concepts, principles, and terminology. Provides for the systematic establishment of a complete, functional command organization for multiagency incidents that involve the U.S. government as well as state and local agencies.

INTRODUCTION TO THE NATIONAL RESPONSE PLAN (NRP)

The NRP

- Establishes a comprehensive, national, all-hazards approach to domestic incident management across a spectrum of activities.
- Is predicated on the National Incident Management System (NIMS). The NIMS is a nationwide template enabling government and nongovernmental responders to respond to all domestic incidents.
- Provides the structure and mechanisms for national-level policy and operational coordination for domestic incident management.
- Does not alter or impede the ability of federal, state, local, or tribal departments and agencies to perform their specific authorities.
- Assumes incidents are typically managed at the lowest possible geographic, organizational, and jurisdictional levels.

INCIDENTS OF NATIONAL SIGNIFICANCE

The NRP distinguishes between incidents that require Department of Homeland Security (DHS) coordination, termed **incidents of national significance**, and the majority of incidents occurring each year handled by responsible jurisdictions or agencies through other established authorities and existing plans.

Incidents of national significance are those high-impact events requiring a coordinated and effective response by an appropriate combination of federal, state, local, tribal, private sector, and nongovernmental entities in order to save lives, minimize damage, and provide the basis for long-term community recovery and mitigation activities.

incident of national significance Based on criteria established in HSPD-5 (paragraph 4), an actual or potential high-impact event that requires a coordinated and effective response by an appropriate combination of federal, state, local, tribal, nongovernmental, and/or private sector entities in order to save lives and minimize damage, and provide the basis for long-term community recovery and mitigation activities.

ROLES AND RESPONSIBILITIES

The NRP specifies the roles and responsibilities of the following parties (see Table 11.1).

TABLE 11.1 ◆ NRP Roles and Responsibilities

Governor	As a state's chief executive, the governor is responsible for the public safety and welfare of the people of the state or territory. The governor: • Is responsible for coordinating state resources to address the full spectrum of actions to prevent, prepare for, respond to, and recover from incidents in an all-hazards context to include terrorism, natural disasters, accidents, and other contingencies. • Under certain emergency conditions, typically has police powers to make, amend, and rescind orders and regulations. • Provides leadership and plays a key role in communicating to the public and in helping people, businesses, and organizations cope with the consequences of any type of declared emergency within state jurisdiction. • Encourages participation in mutual aid and implements authorities for the state to enter into mutual aid agreements with other states, tribes, and territories to facilitate resource-sharing. • Is the commander-in-chief of state military forces (National Guard when in State Active Duty or Title 32 Status and the authorized state militias). • Requests federal assistance when it becomes clear that state or tribal capabilities will be insufficient or have been exceeded or exhausted.
Local Chief Executive Officer	A mayor or city or county manager, as a jurisdiction's chief executive, is responsible for the public safety and welfare of the people of the jurisdiction. The local chief executive officer: • Is responsible for coordinating local resources to address the full spectrum of actions to prevent, prepare for, respond to, and recover from incidents involving all hazards including terrorism, natural disasters, accidents, and other contingencies. • Dependent upon state and local laws, has extraordinary powers to suspend local laws and ordinances, such as to establish a curfew, direct evacuations, and, in coordination with the local health authority, to order a quarantine. • Provides leadership and plays a key role in communicating to the public, in helping people, businesses, and organizations cope with the consequences of any type of domestic incident within the jurisdiction. • Negotiates and enters into mutual-aid agreements with other jurisdictions to facilitate resource sharing. • Requests state and, if necessary, federal assistance through the governor of the state when the jurisdiction's capabilities have been exceeded or exhausted.
Tribal Chief Executive Officer	The tribal chief executive officer is responsible for the public safety and welfare of the people of the tribe. The tribal chief executive officer, as authorized by tribal government: • Is responsible for coordinating tribal resources to address the full spectrum of actions to prevent, prepare for, respond to, and recover from incidents involving all hazards including terrorism, natural disasters, accidents, and other contingencies. • Has extraordinary powers to suspend tribal laws and ordinances, such as to establish a curfew, direct evacuations, and order a quarantine. • Provides leadership and plays a key role in communicating to the tribal nation, and in helping people, businesses, and organizations cope with the consequences of any type of domestic incident within the jurisdiction. • Negotiates and enters into mutual-aid agreements with other tribes/jurisdictions to facilitate resource sharing. • Can request state and federal assistance through the governor of the state when the tribe's capabilities have been exceeded or exhausted. • Can elect to deal directly with the federal government. (Although a state governor must request a presidential disaster declaration on behalf of the tribe under the Stafford Act, federal agencies can work directly with the tribe within existing authorities and resources.)

(continued)

TABLE 11.1 NRP Roles and Responsibilities *(continued)*

Secretary of Homeland Security	Pursuant to HSPD-5, the secretary of homeland security: • Is responsible for coordinating federal operations within the United States to prepare for, respond to, and recover from terrorist attacks, major disasters, and other emergencies. • Serves as the "principal federal official" for domestic incident management. The secretary is also responsible for coordinating federal resources utilized in response to or recovery from terrorist attacks, major disasters, or other emergencies if and when any of the following three conditions applies: • A federal department or agency acting under its own authority has requested DHS assistance. • The resources of state and local authorities are overwhelmed and federal assistance has been requested. • More than one federal department or agency has become substantially involved in responding to the incident. The secretary has been directed to assume incident management responsibilities by the president.
Attorney General	The attorney general is the chief law enforcement officer in the United States. In accordance with HSPD-5 and other relevant statutes and directives, the attorney general has lead responsibility for criminal investigations of terrorist acts or terrorist threats: • By individuals or groups inside the United States, or • Directed at U.S. citizens or institutions abroad. Generally acting through the Federal Bureau of Investigation (FBI), the attorney general—in cooperation with other federal departments and agencies engaged in activities to protect national security—coordinates the activities of the other members of the law enforcement community. Nothing in the NRP derogates the attorney general's status or responsibilities.
Secretary of Defense	DOD has significant resources that may be available to support the federal response to an incident of national significance. The secretary of defense authorizes Defense Support of Civil Authorities (DSCA) for domestic incidents as directed by the president or when consistent with military readiness operations and appropriate under the circumstances and the law. The secretary of defense retains command of military forces under DSCA, as with all other situations and operations. Nothing in the NRP impairs or otherwise affects the authority of the secretary of defense over the DOD.
Secretary of State	The secretary of state is responsible for coordinating international prevention, preparedness, response, and recovery activities relating to domestic incidents, and for the protection of U.S. citizens and U.S. interests overseas.
Nongovernmental Organizations (NGOs)	NGOs collaborate with first responders, governments at all levels, and other agencies and organizations providing relief services to sustain life, reduce physical and emotional distress, and promote recovery of disaster victims when assistance is not available from other sources.
Private Sector	DHS and NRP primary and support agencies coordinate with the private sector to effectively share information, form courses of action, and incorporate available resources to prevent, prepare for, respond to, and recover from incidents of national significance. The roles, responsibilities, and participation of the private sector during incidents of national significance vary based on the nature of the organization and the type and impact of the incident. Private sector organizations may be involved as: • An impacted organization or infrastructure: Private sector organizations may be affected by direct or indirect consequences of the incident. Examples of privately owned infrastructure

(continued)

	include transportation, telecommunications, private utilities, financial institutions, and hospitals. • A response resource: Private sector organizations may provide response resources (donated or compensated) during an incident including specialized teams, equipment, and advanced technologies. • A regulated and/or responsible party: Owners/operators of certain regulated facilities or hazardous operations may bear responsibilities under the law for preparing for and preventing incidents from occurring, and responding to an incident once it occurs. For example, federal regulations require owners/operators of Nuclear Regulatory Commission–regulated nuclear facilities to maintain emergency (incident) preparedness plans, procedures, and facilities and to perform assessments, prompt notifications, and training for a response to an incident. • A member of state/local emergency organizations: Private sector organizations may serve as an active partner in local and state emergency preparedness and response organizations and activities.
Citizen Involvement	Strong partnerships with citizen groups and organizations provide support for incident management prevention, preparedness, response, recovery, and mitigation. The U.S. Citizen Corps brings these groups together and focuses efforts of individuals through education, training, and volunteer service to help make communities safer, stronger, and better prepared to address the threats of terrorism, crime, public health issues, and disasters of all kinds.

Emergency Support Functions (ESFs) The **emergency support functions (ESFs)** serves as the coordination mechanism to provide assistance to state, local, and tribal governments or to federal departments and agencies conducting missions of primary federal responsibility (see Table 11.2). The ESF may be selectively activated for both Stafford Act and non–Stafford Act incidents and provides staffing for the incident management organizations.

emergency support function (ESF)
A grouping of government and certain private sector capabilities into an organizational structure to provide the support, resources, program implementation, and services that are most likely to be needed to save lives, protect property and the environment, restore essential services and critical infrastructure, and help victims and communities return to normal, when feasible, following domestic incidents. The ESFs serve as the primary operational-level mechanism to provide assistance to state, local, and tribal governments or to federal departments and agencies conducting missions of primary federal responsibility.

NRP Coordinating Structures The NRP coordinating structures used to manage incidents of national significance are defined on the following pages. *Pay particular attention to the abbreviations because in a disaster they can become their own language and must be understood by those at the local level.*

Incident Command Post (ICP). The field location at which the primary tactical-level, on-scene incident command functions are performed. The **incident command post (ICP)** may be co-located with the incident base or other incident facilities and is normally identified by a green rotating or flashing light.

Area Command (Unified Area Command). The **area command (unified area command)** is an organization established (1) to oversee the management of multiple incidents being handled by an ICS organization or (2) to oversee the management of large or multiple incidents to which several incident management teams have been assigned. Area command has the responsibility to set overall strategy and priorities, allocate critical resources according to priorities, ensure incidents are properly managed, and ensure objectives are met and strategies followed. Area command becomes unified area command when incidents are multijurisdictional. Area command may be established at an emergency operations center (EOC) facility or at some location other than an ICP.

incident command post (ICP)
The field location at which the primary tactical-level, on-scene incident command functions are performed. The ICP may be collocated with the incident base or other incident facilities and is normally identified by a green rotating or flashing light.

area command (unified area command)
An organization established to (1) oversee the management of multiple incidents that are each being handled by an ICS organization or (2) oversee the management of large or multiple incidents to which several incident management teams have been assigned. Area command has the responsibility to set overall strategy and priorities, allocate critical resources according to priorities, ensure that incidents are properly managed, and ensure that objectives are met and strategies followed. Area command becomes unified area command when incidents are multijurisdictional. Area command may be established at an EOC facility or at some location other than an ICP.

TABLE 11.2 ◆ Emergency Support Function's (ESF) Scope

ESF	*Scope*
ESF No. 1—Transportation	• Federal and civil transportation support • Transportation safety • Restoration/recovery of transportation infrastructure • Movement restrictions • Damage and impact assessment
ESF No. 2—Communications	• Coordination with telecommunication industry • Restoration/repair of telecommunications infrastructure • Protection, restoration, and sustainment of national cyber and information technology resources
ESF No. 3—Public Works and Engineering	• Infrastructure protection and emergency repair • Infrastructure restoration • Engineering services, construction management • Critical infrastructure liaison
ESF No. 4—Firefighting	• Firefighting activities on federal lands • Resource support to rural and urban firefighting operations
ESF No. 5—Emergency Management	• Coordination of incident management efforts • Issuance of mission assignments • Resource and human capital • Incident action planning • Financial management
ESF No. 6—Mass Care, Housing, and Human Services	• Mass care • Disaster housing • Human services
ESF No. 7—Resource Support	• Resource support (facility space, office equipment and supplies, contracting services, etc.)
ESF No. 8—Public Health and Medical Services	• Public health • Medical • Mental health services • Mortuary services
ESF No. 9—Urban Search and Rescue	• Lifesaving assistance • Urban search and rescue
ESF No.10—Oil and Hazardous Materials Response	• Environmental safety and short- and long-term cleanup
ESF No. 11—Agriculture and Natural Resources	• Nutrition assistance • Animal and plant disease/pest response • Food safety and security • Natural and cultural resources and historic properties protection and restoration
ESF No.12—Energy	• Energy infrastructure assessment, repair, and restoration • Energy industry utilities coordination • Energy forecast

(continued)

ESF	*Scope*
ESF No. 13—Public Safety and Security	• Facility and resource security • Security planning and technical and resource assistance • Public safety/security support • Support to access, traffic, and crowd control
ESF No. 14—Long-Term Community Recovery and Mitigation	• Social and economic community impact assessment • Long-term community recovery assistance to states, local governments, and the private sector • Mitigation analysis and program implementation
ESF No. 15—External Affairs	• Emergency public information and protective action guidance • Media and community relations • Congressional and international affairs • Tribal and insular affairs

Local Emergency Operations Center (EOC). The **local emergency operations center (EOC)** is the physical location at which the coordination of information and resources to support local incident management activities normally takes place.

State Emergency Operations Center (EOC). This is the physical location at which the coordination of information and resources to support state incident management activities normally takes place.

Homeland Security Operations Center (HSOC). The **HSOC** is the primary national hub for domestic incident management operational coordination and situational awareness. The HSOC is a standing 24/7 interagency facility fusing law enforcement, national intelligence, emergency response, and private sector reporting. The HSOC facilitates homeland security information sharing and operational coordination with other federal, state, local, tribal, and nongovernmental EOCs.

Interagency Incident Management Group (IIMG). The **IIMG** is a federal headquarters-level, multi-agency coordination entity that facilitates federal domestic incident management for incidents of national significance. The secretary of Homeland Security activates the IIMG based on the nature, severity, magnitude, and complexity of the threat or incident. The secretary of Homeland Security may activate the IIMG for high-profile, large-scale events that present high-probability targets, such as **NSSE**, and in heightened threat situations. The IIMG is comprised of senior representatives from DHS components, other federal departments and agencies, and nongovernmental organizations, as required. The IIMG membership is flexible and can be tailored or task-organized to provide the appropriate subject matter expertise required for the specific threat or incident.

National Response Coordination Center (NRCC). The **NRCC** is a multi-agency center that provides overall federal response coordination for incidents of national significance and emergency management program implementation. FEMA maintains the NRCC as a functional component of the HSOC in support of incident management operations. The NRCC monitors potential or developing incidents of national significance and supports the efforts of regional and field components. The NRCC resolves federal resource support conflicts and other implementation issues forwarded by the Joint Field Office (JFO). Those issues that cannot be resolved by the NRCC are referred to the IIMG.

emergency operations center (EOC)
The physical location at which the coordination of information and resources to support domestic incident management activities normally takes place. An EOC may be a temporary facility or may be located in a more central or permanently established facility, perhaps at a higher level of organization within a jurisdiction. EOCs may be organized by major functional disciplines (e.g., fire, law enforcement, and medical services), by jurisdiction (e.g., federal, state, regional, county, city, tribal), or by some combination thereof. A central location where those in authority congregate to allow for exchange of information and conduct face-to-face coordination in the making of decisions. The center, often referred to as the EOC, provides for centralized emergency management in major natural disasters and other emergencies.

HSOC
Homeland Security Operations Center.

IIMG
Interagency incident management group.

NSSE
National special security event.

NRCC
National Response Coordination Center.

RRCC
Regional Response Coordination Center.

ERT-A
Emergency response team—advance element.

SIOC
Strategic Information and Operations Center.

NJTTF
National Joint Terrorism Task Force.

JTTF
Joint terrorism task force.

JFO
Joint Field Office

joint field office (JFO)
A temporary federal facility established locally to provide a central point for federal, state, local, and tribal executives with responsibility for incident oversight, direction, and/or assistance to effectively coordinate protection, prevention, preparedness, response, and recovery actions. The JFO will combine the traditional functions of the JOC, the FEMA DFO, and the JIC within a single federal facility.

joint operations center (JOC)
The JOC is the focal point for all federal investigative law enforcement activities during a terrorist or potential terrorist incident or any other significant criminal incident, and is managed by the senior law enforcement official. The JOC becomes a component of the JFO when the NRP is activated.

Regional Response Coordination Center (RRCC). The **RRCC** is a standing facility operated by FEMA that is activated to coordinate regional response efforts, establish federal priorities, and implement local federal program support. The RRCC operates until a JFO is established in the field and/or the principal federal officer, federal coordinating officer, or federal resource coordinator can assume his or her NRP coordination responsibilities. The RRCC establishes communications with the affected state emergency management agency and the National Response Coordination Center (NRCC), coordinates deployment of the emergency response team—advance element (**ERT-A**) to field locations, assesses damage information, develops situation reports, and issues initial mission assignments.

Strategic Information and Operations Center (SIOC). The FBI **SIOC** is the focal point and operational control center for all federal intelligence, law enforcement, and investigative law enforcement activities related to domestic terrorist incidents or credible threats, including leading assigned investigations. The SIOC serves as an information clearinghouse to help collect, process, vet, and disseminate information relevant to law enforcement and criminal investigation efforts in a timely manner. The SIOC maintains direct connectivity with the HSOC and IIMG. The SIOC, located at FBI headquarters, supports the FBI's mission in leading efforts of the law enforcement community to detect, prevent, preempt, and disrupt terrorist attacks against the Unites States.

The SIOC houses the National Joint Terrorism Task Force (**NJTTF**). The mission of the NJTTF is to enhance communications, coordination, and cooperation among federal, state, local, and tribal agencies representing the intelligence, law enforcement, defense, diplomatic, public safety, and homeland security communities by providing a point of fusion for terrorism intelligence and by supporting Joint Terrorism Task Forces (**JTTFs**) throughout the United States.

Joint Field Office (JFO). The **JFO** is a temporary federal facility established locally to coordinate operational federal assistance activities to the affected jurisdiction(s) during incidents of national significance. The JFO is a multi-agency center that provides a central location for coordination of federal, state, local, tribal, nongovernmental, and private sector organizations with primary responsibility for threat response and incident support. The JFO enables the effective and efficient coordination of federal incident-related prevention, preparedness, response, and recovery actions.

The JFO utilizes the scalable organizational structure of the NIMS incident command system (ICS). The JFO organization adapts to the magnitude and complexity of the situation at hand, and incorporates the NIMS principles regarding span of control and organizational structure: management, operations, planning, logistics, and finance/administration. Although the JFO uses an ICS structure, the JFO does not manage on-scene operations. Instead, the JFO focuses on providing support to on-scene efforts and conducting broader support operations that may extend beyond the incident site.

Joint Operations Center (JOC). The **joint operations center (JOC)** branch is established by the senior federal law enforcement officer (SFLEO) (for example, the FBI SAC during terrorist incidents) to coordinate and direct law enforcement and criminal investigation activities related to the incident. The JOC branch ensures management and coordination of federal, state, local, and tribal investigative/law enforcement activities. The emphasis of the JOC is on prevention as well as intelligence collection, investigation, and prosecution of a criminal act. This emphasis includes managing unique tactical issues inherent to a crisis situation (e.g., a hostage situation or terrorist threat).

When this branch is included as part of the joint field office (JFO), it is responsible for coordinating the intelligence and information function (as described in NIMS), which includes information and operational security, and the collection, analysis, and distribution of all incident-related intelligence. Accordingly, the intelligence unit within the JOC branch serves as the interagency fusion center for all intelligence related to an incident.

Field-Level Organizational Structures: JFO Coordination Group The field-level organizational structures and teams deployed in response to an incident of national significance include the following potential members of the JFO Coordination Group.

Principal Federal Officer (PFO). The **principal federal officer (PFO)** is personally designated by the secretary of Homeland Security to facilitate federal support to the established incident command system (ICS) unified command structure and to coordinate overall federal incident management and assistance activities across the spectrum of prevention, preparedness, response, and recovery. The PFO ensures that incident management efforts are maximized through effective and efficient coordination. The PFO provides a primary point of contact and situational awareness locally for the secretary of Homeland Security.

Federal Coordinating Officer (FCO). The **federal coordinating officer (FCO)** manages and coordinates federal resource support activities related to Stafford Act disasters and emergencies. The FCO assists the unified command and/or the area command, and works closely with the principal federal officer (PFO), senior federal law enforcement official (SFLEO), and other **senior federal officials (SFOS)**.

In Stafford Act situations in which a PFO has not been assigned, the FCO provides overall coordination for the federal components of the JFO and works in partnership with the state coordinating officer (**SCO**) to determine and satisfy state and local assistance requirements.

Senior Federal Law Enforcement Official (SFLEO). The **SFLEO** is the senior law enforcement official from the agency with primary jurisdictional responsibility as directed by statute, presidential directive, existing federal policies, and/or the attorney general. The SFLEO directs intelligence/investigative law enforcement operations related to the incident and supports the law enforcement component of the unified command on-scene. In the event of a terrorist incident, this official will normally be the FBI senior agent-in-charge (**SAC**).[7]

principal federal official (PFO)
The federal official designated by the secretary of Homeland Security to act as his or her representative locally to oversee, coordinate, and execute the secretary's incident management responsibilities under HSPD-5 for incidents of national significance.

federal coordinating officer (FCO)
The federal officer who is appointed to manage federal resource support activities related to Stafford Act disasters and emergencies. The FCO is responsible for coordinating the timely delivery of federal disaster assistance resources and programs to the affected state and local governments, individual victims, and the private sector.

senior federal official (SFO)
An individual representing a federal department or agency with primary statutory responsibility for incident management. SFOs utilize existing authorities, expertise, and capabilities to aid in management of the incident working in coordination with other members of the JFO Coordination Group.

NONGOVERNMENTAL AND VOLUNTEER ORGANIZATIONS (NGO)

Nongovernmental organizations (NGOs) collaborate with first responders, governments at all levels, and other agencies and organizations providing relief services to sustain life, reduce physical and emotional distress, and promote recovery of disaster victims when assistance is not available from other sources.

SCO
State coordinating officer.

SFLEO
Senior federal law enforcement official.

SAC
Special agent-in-charge.

nongovernmental organization (NGO)
A nonprofit entity that is based on interests of its members, individuals, or institutions and that is not created by a government, but may work cooperatively with government. Such organizations serve a public purpose, not a private benefit. Examples of NGOs include faith-based charity organizations and the American Red Cross.

NVOAD
National Voluntary Organizations Active in Disaster.

The *National Voluntary Organizations Active in Disaster* (**NVOAD**) is a consortium of more than 30 recognized national organizations of volunteers active in disaster relief. Such entities provide significant capabilities to incident management and response efforts at all levels. For example, the wildlife rescue and rehabilitation activities conducted during a pollution emergency are often carried out by private, nonprofit organizations working with natural resource trustee agencies. The roles, responsibilities, and participation of the private sector during incidents of national significance vary based on the nature of the organization and the type and impact of the incident. For example, the *American Red Cross (ARC)* is an NGO that provides relief at the local level and also coordinates the Mass Care element of ESF. The ARC deserves attention due to its outreach, not only in the United States but worldwide.

For more than 122 years, the mission of the ARC has been to help people prevent, prepare for, and respond to emergencies. A humanitarian organization led by volunteers, guided by its congressional charter and the Fundamental Principles of the International Red Cross Movement, the ARC is woven into the fabric of our communities with 940 chapters nationwide. In fulfilling its mission, ARC seeks to empower Americans to take practical steps to make families, neighborhoods, schools and workplaces safer, healthier, and more resilient in the face of adversity. Through the "Together We Prepare" program, the ARC provides training for the public in community disaster preparedness and response, and lifesaving skills training (first aid and CPR). The program also encourages people to donate blood and volunteer to help build community preparedness.

Each year the American Red Cross responds immediately to more than 70,000 disasters, including house or apartment fires (the majority of disaster responses), hurricanes, floods, earthquakes, tornadoes, hazardous materials spills, transportation accidents, explosions, and other natural and human-made disasters. Although the American Red Cross is not a government agency, its authority to provide disaster relief was formalized when, in 1905, the Red Cross was chartered by Congress to "carry on a system of national and international relief in time of peace and apply the same in mitigating the sufferings caused by pestilence, famine, fire, floods, and other great national calamities, and to devise and carry on measures for preventing the same." The charter is not only a grant of power but also an imposition of duties and obligations to the nation, to disaster victims, and to the people who generously support the Red Cross's work with their donations.

Red Cross disaster relief focuses on meeting people's immediate emergency disaster-caused needs. When a disaster threatens or strikes, the Red Cross provides shelter, food, and health and mental health services to address basic human needs. In addition to these services, the core of Red Cross disaster relief is the assistance given to individuals and families affected by disaster to enable them to resume their normal daily activities independently. The Red Cross also feeds emergency workers, handles inquiries from concerned family members outside the disaster area, provides blood and blood products to disaster victims, and helps those affected by disaster to access other available resources. The mission of the American Red Cross Disaster Services is to ensure nationwide disaster planning, preparedness, community disaster education, mitigation, and response that will provide the American people with quality services delivered in a uniform, consistent, and responsive manner. The American Red Cross is an independent, humanitarian, voluntary organization, not a government agency. All Red Cross assistance is given free of charge, made possible by contributions of people's time, money, and skills.

CITIZEN INVOLVEMENT

Strong partnerships with citizen groups and organizations provide support for incident management prevention, preparedness, response, recovery, and mitigation.

PROGRAMS AND PARTNERS

After September 11, 2001, America witnessed a wellspring of selflessness and heroism. People in every corner of the country asked, "What can I do?" and "How can I help?" Citizen Corps was created to help all Americans answer these questions through public education and outreach, training, and volunteer service.

CITIZEN CORPS

The U.S. Citizen Corps works through a national network of state, local, and tribal Citizen Corps Councils, which bring together leaders from law enforcement, fire, emergency medical, emergency management, volunteer organizations, local elected officials, the private sector, and other community stakeholders. Other programs unaffiliated with Citizen Corps also provide organized citizen involvement opportunities in support of federal response to major disasters and events of national significance.

The *Citizen Corps* brings these groups together and focuses efforts of individuals through education, training, and volunteer service to help make communities safer, stronger, and better prepared to address the threats of terrorism, crime, public health issues, and disasters of all kinds. Local Citizen Corps Councils implement Citizen Corps programs, which include Community Emergency Response Teams (CERTs), Medical Reserve Corps, Fire Corps, and Neighborhood Watch, Volunteers in Police Service and the affiliate programs; provide opportunities for special skills and interests; develop targeted outreach for special-needs groups; and organize special projects and community events. Citizen Corps Affiliate Programs expand the resources and materials available to states and local communities through partnerships with programs and organizations that offer resources for public education, outreach, and training; represent volunteers interested in helping to make their communities safer; or offer volunteer service opportunities to support first responders, disaster relief activities, and community safety efforts.

The *Citizen Corps Guide for Local Officials* provides suggestions on how to form a Citizen Corps Council and how to get the programs and other activities started in your community. Everyone can contribute to help make communities safer through personal responsibility: developing a household preparedness plan and disaster supplies kits, observing home health and safety practices, implementing disaster mitigation measures, and participating in crime prevention and reporting. By engaging individuals in volunteer activities that support first responders, disaster relief groups, and community safety organizations, everyone can do something to support the response and recovery efforts at the local level.

CERT The Community Emergency Response Team (**CERT**) program educates people about disaster preparedness and trains them in basic disaster response skills. Using their training, CERT members can assist others in their neighborhood or workplace following an event and can take a more active role in preparing their community. The program is administered by DHS. The Community Emergency Response Team (CERT) program helps train people to be better prepared to respond to emergency

CERT
Community Emergency Response Team.

situations in their communities. When emergencies happen, CERT members can give critical support to first responders, provide immediate assistance to victims, and organize spontaneous volunteers at a disaster site. CERT members can also help with nonemergency projects that help improve the safety of the community. The CERT course is taught in the community by a trained team of first responders who have completed a CERT Train-the-Trainer course conducted by their state training office for emergency management, or FEMA's Emergency Management Institute (**EMI**), located in Emmitsburg, Maryland. CERT training includes disaster preparedness, disaster fire suppression, basic disaster medical operations, and light search and rescue operations.

EMI
Emergency Management Institute/FEMA.

Fire Corps The Fire Corps promotes the use of citizen advocates to enhance the capacity of resource-constrained fire and rescue departments at all levels: volunteer, combination, and career. Citizen advocates can assist local fire departments in a range of activities including fire safety outreach, youth programs, and administrative support. Fire Corps provides resources to assist fire and rescue departments in creating opportunities for citizen advocates and promotes citizen participation. Fire Corps is funded through DHS and is managed and implemented through a partnership among the National Volunteer Fire Council, the International Association of Fire Fighters, and the International Association of Fire Chiefs.

Medical Reserve Corps The Medical Reserve Corps (MRC) program strengthens communities by helping medical, public health, and other volunteers offer their expertise throughout the year as well as during local emergencies and other times of community need. MRC volunteers work in coordination with existing local emergency response programs and also supplement existing community public health initiatives, such as outreach and prevention, immunization programs, blood drives, case management, care planning, and other efforts. The MRC program is administered by Health and Human Services (HHS). The Medical Reserve Corps (MRC) program coordinates the skills of practicing and retired physicians, nurses, and other health professionals, as well as other citizens interested in health issues, who are eager to volunteer to address their community's ongoing public health needs and to help their community during large-scale emergency situations.

Neighborhood Watch An expanded Neighborhood Watch Program (NWP) incorporates terrorism awareness education into its existing crime prevention mission, while also serving as a way to bring residents together to focus on emergency preparedness and emergency response training. Funded by the Department of Justice (**DOJ**), Neighborhood Watch is administered by the National Sheriffs' Association. The Neighborhood Watch Program is a highly successful effort that has been in existence for more than 30 years in cities and counties across America. It provides a unique infrastructure that brings together local officials, law enforcement, and citizens to protect communities. Around the country, neighbors for three decades have banded together to create Neighborhood Watch groups. They understand that the active participation of neighborhood residents is a critical element in community safety—not through vigilantism but simply through a willingness to look out for suspicious activity in their neighborhood and report that activity to law enforcement and to each other. In doing so, residents can be of assistance in reclaiming high-crime neighborhoods, as well as making people throughout a community feel more secure and less fearful.

DOJ
Department of Justice.

In the aftermath of September 11, 2001, the need for strengthening and securing communities became even more critical, and Neighborhood Watch groups have taken on greater significance. In addition to serving a crime prevention role, Neighborhood Watch can also be used as the basis for bringing neighborhood residents together to focus on disaster preparedness as well as terrorism awareness; to focus on evacuation drills and exercises; and even to organize group training, such as the Community Emergency Response Team (CERT) training.

Volunteers in Police Service Volunteers in Police Service (VIPS) works to enhance the capacity of state and local law enforcement to utilize volunteers. VIPS serves as a gateway to resources and information for and about law enforcement volunteer programs. Funded by DOJ, VIPS is managed and implemented by the International Association of Chiefs of Police.

Since September 11, 2001, the demands on state and local law enforcement have increased dramatically. As a result, already limited resources are being stretched further at a time when our country needs every available officer in the field. Some local police departments are utilizing civilian volunteers to supplement their sworn force. These vital efforts are coordinated through VIPS. VIPS draws on the time and considerable talents of civilian volunteers and allows law enforcement professionals to perform their frontline duties better. The program provides resources to assist local law enforcement officials by incorporating community volunteers into the activities of the law enforcement agency, including a series of best practices to help state and local law enforcement design strategies to recruit, train, and utilize citizen volunteers in their departments.[8]

MANAGING THE NATIONAL RESPONSE PLAN

A basic premise of the NRP is that incidents are generally handled at the lowest jurisdictional level possible. Police, fire, public health and medical, emergency management, and other personnel are responsible for incident management at the local level. In some instances, a federal agency in the local area may act as a first responder and may provide direction or assistance consistent with its specific statutory authorities and responsibilities. In the vast majority of incidents, state and local resources and interstate mutual aid normally provide the first line of emergency response and incident management support.

ORGANIZATIONAL STRUCTURE

The national structure for incident management establishes a progression of coordination and communication from the local level to the regional and to the national headquarters level. The local incident command structures are responsible for directing on-scene emergency management and maintaining command and control of on-scene incident operations. The support and coordination components consist of multi-agency coordination centers/emergency operations centers (EOCs) and multi-agency coordination entities. Multi-agency coordination centers/EOCs provide central locations for operational information-sharing and resource coordination in support of on-scene efforts. Multi-agency coordination entities aid in establishing priorities among the incidents and associated resource allocations, resolving agency policy conflicts, and providing strategic guidance to support incident management activities.

NATIONAL INCIDENT MANAGEMENT SYSTEM (NIMS)

On February 28, 2003, the president issued Homeland Security Presidential Directive (HSPD)-5, which directs the secretary of Homeland Security to develop and administer a National Incident Management System (NIMS). According to HSPD-5:

> This system will provide a consistent nationwide approach for Federal, State, and local governments to work effectively and efficiently together to prepare for, respond to, and recover from domestic incidents, regardless of cause, size, or complexity. To provide for interoperability and compatibility among Federal, State, and local capabilities, the NIMS will include a core set of concepts, principles, terminology, and technologies covering the incident command system; multi-agency coordination systems; unified command; training; identification and management of resources (including systems for classifying types of resources); qualifications and certification; and the collection, tracking, and reporting of incident information and incident resources.

To provide this framework for interoperability and compatibility, the NIMS is based on an appropriate balance of flexibility and **standardization**.

standardization
A process by which a product or service is assessed against some standard, performance, or quality.

Flexibility. The NIMS provides a consistent, flexible, and adjustable national framework within which government and private entities at all levels can work together to manage domestic incidents, regardless of their cause, size, location, or complexity. This flexibility applies across all phases of incident management: prevention, preparedness, response, recovery, and mitigation.

Standardization. The NIMS provides a set of standardized organizational structures—such as the incident command system (ICS), multi-agency coordination systems, and public information systems—as well as requirements for processes, procedures, and systems designed to improve interoperability among jurisdictions and disciplines in various areas, including training; resource management; personnel qualification and certification; equipment certification; communications and information management; technology support; and continuous system improvement.

The NIMS integrates existing best practices into a consistent, nationwide approach to domestic incident management that is applicable at all jurisdictional levels and across functional disciplines in an all-hazards context. Six major components make up this systems approach.

1. *Command and Management.* NIMS standard incident command structures are based on three key organizational systems.
   a. *The ICS.* The ICS defines the operating characteristics, interactive management components, and structure of incident management and emergency response organizations engaged throughout the life cycle of an incident.
   b. *Multi-agency coordination systems.* These define the operating characteristics, interactive management components, and organizational structure of supporting incident management entities engaged at the federal, state, local, tribal, and regional levels through mutual-aid agreements and other assistance arrangements.
   c. *Public information systems.* These refer to processes, procedures, and systems for communicating timely and accurate information to the public during crisis or emergency situations.

2. *Preparedness.* Effective incident management begins with a host of preparedness activities conducted on a "steady-state" basis, well in advance of any potential incident. Preparedness involves an integrated combination of planning, training, exercises, personnel qualification and certification standards, equipment acquisition and certification standards, and publication management processes and activities.
   a. ***Planning***. Plans describe how personnel, equipment, and other resources are used to support incident management and emergency response activities. Plans provide mechanisms and systems for setting priorities, integrating multiple entities and functions, and ensuring that communications and other systems are available and integrated in support of a full spectrum of incident management requirements.
   b. *Training.* Training includes standard courses on multi-agency incident command and management, organizational structure, and operational procedures; discipline-specific and agency-specific incident management courses; and courses on the integration and use of supporting technologies.
   c. *Exercises.* Incident management organizations and personnel must participate in realistic exercises—including multidisciplinary, multijurisdictional, and multisector interaction—to improve integration and interoperability and optimize resource utilization during incident operations.
   d. *Personnel qualification and certification.* Qualification and certification activities are undertaken to identify and publish national-level standards and measure performance against these standards to ensure that incident management and emergency responder personnel are appropriately qualified and officially certified to perform NIMS-related functions.
   e. *Equipment acquisition and certification.* Incident management organizations and emergency responders at all levels rely on various types of equipment to perform mission essential tasks. A critical component of operational preparedness is the acquisition of equipment that will perform to certain standards, including the capability to be interoperable with similar equipment used by other jurisdictions.
   f. *Mutual aid.* Mutual-aid agreements are the means for one jurisdiction to provide resources, facilities, services, and other required support to another jurisdiction during an incident. Each jurisdiction should be party to a mutual-aid agreement with appropriate jurisdictions from which it expects to receive or to which it expects to provide assistance during an incident.
   g. *Publications management.* Publications management refers to form standardization, developing publication materials, administering publications—including establishing naming and numbering conventions, managing the publication and promulgation of documents, exercising control over sensitive documents, and revising publications when necessary.
3. *Resource management.* The NIMS defines standardized mechanisms and establishes requirements for processes to describe, inventory, mobilize, dispatch, track, and recover resources over the life cycle of an incident.
4. *Communications and information management.* The NIMS identifies the requirement for a standardized framework for communications, information management (collection, analysis, and dissemination), and information sharing at all levels of incident management. These elements are briefly described as follows:
   a. *Incident management communications.* Incident management organizations must ensure that effective, interoperable communications processes, procedures, and systems exist to support a wide variety of incident management activities across agencies and jurisdictions.

planning
The process of establishing objectives and suitable courses of action before taking action. Managerial function that determines in advance what organizations, subunits, or individuals should do and how they will do it.

b. *Information management.* Information management processes, procedures, and systems help ensure that information, including communications and data, flows efficiently through a commonly accepted architecture, supporting numerous agencies and jurisdictions responsible for managing or directing domestic incidents, those impacted by the incident, and those contributing resources to the incident management effort. Effective information management enhances incident management and response and helps ensure that crisis decision making is better informed.

5. *Supporting technologies.* Technology and technological systems provide supporting capabilities essential to implementing and continuously refining the NIMS. These include voice and data communications systems, information management systems (i.e., record keeping and resource tracking), and data display systems. Also included are specialized technologies that facilitate ongoing operations and incident management activities in situations that call for unique, technology-based capabilities.

6. *Ongoing management and maintenance.* This component establishes an activity to provide strategic direction for and oversight of the NIMS, supporting both routine review and the continuous refinement of the system and its components over the long term.[9]

In accordance with NIMS processes, resource and policy issues are addressed at the lowest organizational level practicable. If the issues cannot be resolved at this level, they are forwarded up to the next level for resolution.

THE INCIDENT COMMAND SYSTEM

The incident command system (ICS) is the combination of facilities, equipment, personnel, procedures, and communications operating within a common organizational structure, designed to aid in domestic incident management activities. It is used for a broad spectrum of emergencies, from small to complex incidents, both natural and human-made, including acts of catastrophic terrorism. ICS is used by all levels of government—federal, state, local, and tribal, as well as by many private sector and nongovernmental organizations. Some of the more important "transitional steps" necessary to apply ICS in a field incident environment include the following:

- recognizing and anticipating the requirement that organizational elements will be activated and taking the necessary steps to delegate authority as appropriate;
- establishing incident facilities as needed, strategically located, to support field operations;
- establishing the use of common terminology for organizational functional elements, position titles, facilities, and resources; and
- rapidly evolving from providing oral direction to the development of a written incident action plan.

incident commander (IC) The individual responsible for all incident activities, including the development of strategies and tactics and the ordering and release of resources. The IC has overall authority and responsibility for conducting incident operations and is responsible for the management of all incident operations at the incident site.

ICS ORGANIZATION

Functional Structure The ICS organization is composed of five major functional areas: command, operations, planning, logistics, and finance and administration. A sixth functional area, intelligence, may be established if deemed necessary by the **incident commander (IC)**, depending on the requirements of the situation at hand (see Figure 11.10).

Modular Extension The ICS organizational structure is modular, extending to incorporate all elements necessary for the type, size, scope, and complexity of a given incident. The IC structural organization builds from the top down; responsibility and

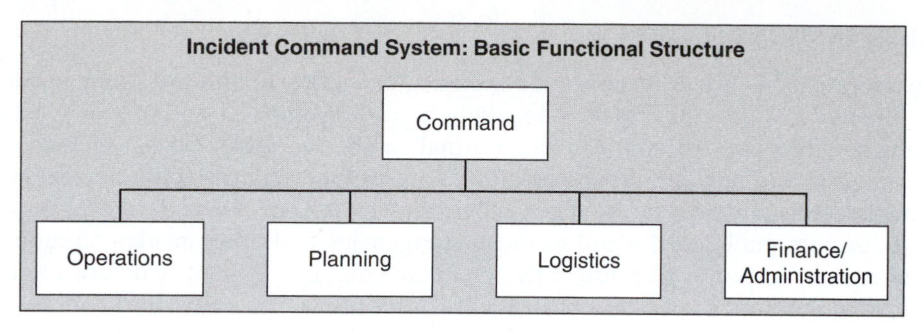

FIGURE 11.10 ◆ Command *(Courtesy of Randy R. Bruegman)*

performance begin with the incident command element and the IC. When the need arises, four separate sections can be used to organize the staff. Each of these may have several subordinate units, or branches, depending on the management requirements of the incident. If one individual can simultaneously manage all major functional areas, no further organization is required. If one or more of the functions requires independent management, an individual is assigned responsibility for that function.

The responding IC's initial management assignments are normally one or more section chiefs to manage the major ICS functional areas (operations, planning, logistics, and finance and administration). The section chiefs further delegate management authority for their areas as required. If a section chief sees the need, he or she may establish branches or units (depending on the section). Similarly, each functional unit leader further assigns individual tasks within the unit as needed.

This modular concept is based on the following considerations:

- developing the form of the organization to match the function or task to be performed;
- staffing only the functional elements that are required to perform the task, observing recommended span-of-control guidelines;
- performing the function of any nonactivated organizational element at the next highest level; and
- deactivating organizational elements no longer required.

Table 11.3 describes the distinctive title assigned to each element of the ICS organization at each corresponding level, as well as the leadership title corresponding to each individual element.

TABLE 11.3 ◆ ICS Organization

Organizational Element	*Leadership Position*
Incident Command	Incident Commander (IC)
Command Staff	Officer
Section	Section Chief
Branch	Branch Director
Divisions and Groups*	Supervisors
Unit**	Unit Leader

*The hierarchical term *supervisor* is used only in the operations section.

**Unit leader designations apply to the subunits of the planning, logistics, and finance/administration sections.

THE OPERATIONS SECTION

The operations section is responsible for managing tactical operations at the incident site directed toward reducing the immediate hazard, saving lives and property, establishing situation control, and restoring normal conditions. Incidents can include acts of terrorism, wildland and urban fires, floods, hazardous material spills, nuclear accidents, aircraft accidents, earthquakes, hurricanes, tornadoes, tropical storms, war-related disasters, public health and medical emergencies, and other incidents requiring an emergency response. Because of its functional unit management structure, the ICS is applicable across a spectrum of incidents differing in size, scope, and complexity. The types of agencies that could be included in the operations section include fire, law enforcement, public health, public works, and emergency services, working together as a unit or in combinations, depending on the situation. Many incidents may involve private individuals, companies, or nongovernmental organizations, some of which may be fully trained and qualified to participate as partners in the operations section. Incident operations can be organized and executed in many ways. The specific method selected will depend on the type of incident, agencies involved, and objectives and strategies of the incident management effort. The following discussion presents several different methods of organizing incident tactical operations. In some cases, a method will be selected to accommodate jurisdictional boundaries, and in other cases, the approach will be strictly functional. In still others, a mix of functional and geographical approaches may be appropriate. The ICS offers extensive flexibility in determining the appropriate approach using the factors described here. Figure 11.11 shows the primary organizational structure within the operations section.

incident action plan (IAP) An oral or written plan containing general objectives reflecting the overall strategy for managing an incident. It may include the identification of operational resources and assignments. It may also include attachments that provide direction and important information for management of the incident during one or more operational periods.

THE PLANNING SECTION

The planning section is responsible for collecting, evaluating, and disseminating tactical information pertaining to the incident. This section maintains information and intelligence on the current and forecasted situation, as well as the status of resources assigned to the incident. The planning section prepares and documents **incident action plans (IAPs)** and incident maps and gathers and disseminates information and intelligence

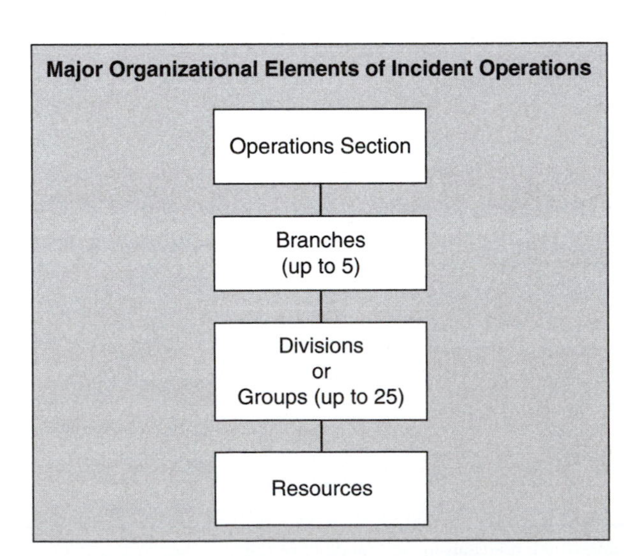

FIGURE 11.11 ◆ Major Organizational Elements of Incident Operations *(Courtesy of Randy R. Bruegman)*

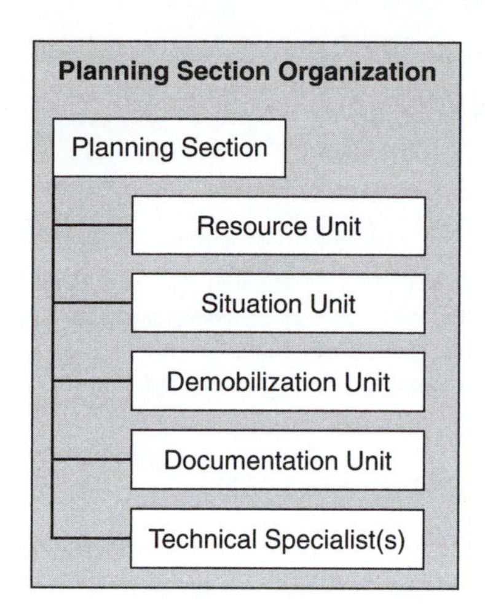

FIGURE 11.12 ◆ Planning Section Organization *(Courtesy of Randy R. Bruegman)*

critical to the incident. As shown in Figure 11.12, the planning section has four primary units and may include a number of technical specialists to assist in evaluating the situation and forecasting requirements for additional personnel and equipment.

THE LOGISTICS SECTION

The logistics section meets all support needs for the incident, including ordering resources through appropriate procurement authorities from off-incident locations. It also provides facilities, transportation, supplies, equipment maintenance and fueling, food service, communications, and medical services for incident personnel.

The logistics section is led by a section chief, who may also have a deputy. Having a deputy is encouraged when all designated units are established at an incident site. When the incident is very large or requires a number of facilities with large numbers of equipment, the logistics section can be divided into two branches (see Figures 11.13 and 11.14).

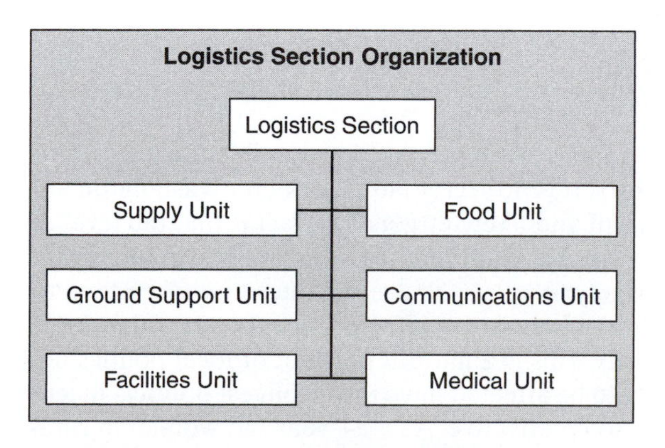

FIGURE 11.13 ◆ Logistics Organization *(Courtesy of Randy R. Bruegman)*

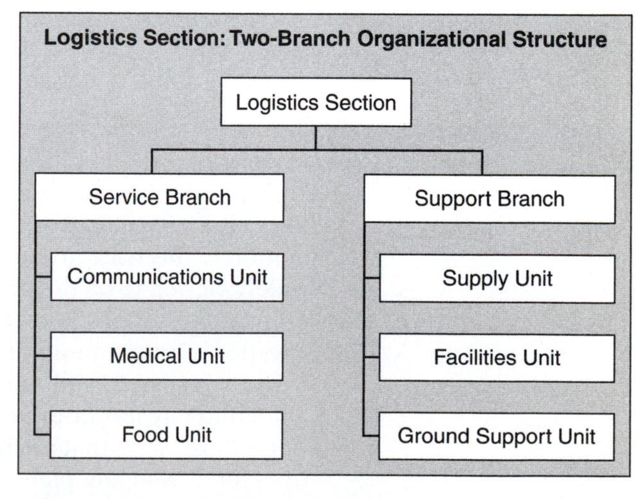

FIGURE 11.14 ◆ Logistics Structure *(Courtesy of Randy R. Bruegman)*

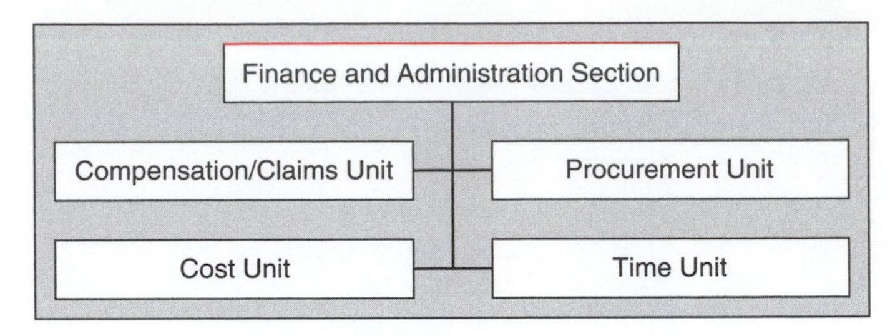

FIGURE 11.15 ◆ Finance/Administration Organization *(Courtesy of Randy R. Bruegman)*

THE FINANCE/ADMINISTRATION SECTION

When there is a specific need for financial, reimbursement (individual, agency, or department), and/or administrative services to support incident management activities, a finance/administration section is established. Under the ICS, not all agencies will require such assistance. In large, complex scenarios involving significant funding originating from multiple sources, the finance/administrative section is an essential part of the ICS. In addition to monitoring multiple sources of funds, the section chief must track and report to the IC the financial(s) as the incident progresses. This allows the IC to forecast the need for additional funds before operations are affected negatively. This is particularly important if significant operational assets are under contract from the private sector. The section chief may also need to monitor cost expenditures to ensure that statutory rules that apply are met. Close coordination with the planning section and the logistics section is also essential so that operational records can be reconciled with financial documents. Note that, in some cases, only one specific function may be required (e.g., cost analysis), which a technical specialist in the planning section could provide.

The finance/administration section chief will determine, given current and anticipated future requirements, the need for establishing specific subordinate units. In some of the functional areas (e.g., procurement), an actual unit need not be established if it would consist of only one person. In such a case, a procurement technical specialist would be assigned in the planning section instead. Because of the specialized nature of finance functions, the section chief should come from the agency that has the greatest requirement for this support. The section chief may have a deputy[10] (see Figure 11.15).

VARIATIONS IN PREPAREDNESS AT THE LOCAL LEVEL

There is a widespread variation in local terrorism preparedness and in the level of attention to this issue around the country. It is no secret New York and Washington are widely viewed to be the most significant and likely targets. The fact is that the level of preparedness is much lower in many smaller communities, which raises an important question. Those communities most likely to be at risk are the ones that often tend to be best prepared, and the places least likely to be affected are the ones that have tended to pay less attention to security. This is a natural product of local politics because those areas that are most likely to be affected have the strongest political incentives for developing plans for things to be effective. As a 94-year-old woman in rural Connecticut discovered (a woman who died of anthrax apparently transmitted through her mail), there are many issues of homeland security that pay no attention to the size of the community. You cannot assume that, simply because you do not live in a place presumed to be at risk, you are immune from the consequences of homeland security problems and terrorist attacks. Is this variation around the country in respect to preparedness due to the fact that many state and local governments have simply not taken the issue seriously?[11]

THE NATIONAL INFRASTRUCTURE PROTECTION PLAN

The *National Infrastructure Protection Plan (NIPP)* sets forth a comprehensive risk management framework and clearly defines critical infrastructure protection roles and responsibilities for the Department of Homeland Security; federal sector-specific agencies (SSAs); and other federal, state, local, tribal, and private sector security partners. The NIPP provides the coordinated approach that will be used to establish national priorities, goals, and requirements for infrastructure protection so that funding and resources are applied in the most effective manner.

The goal of the NIPP is to:

> Build a safer, more secure, and more resilient America by enhancing protection of the Nation's critical infrastructure and key resources **(CI/KR)** to prevent, deter, neutralize, or mitigate the effects of deliberate efforts by terrorists to destroy, incapacitate, or exploit them; and to strengthen national preparedness, timely response, and rapid recovery in the event of an attack, natural disaster, or other emergency.

CI/KR
Critical infrastructure/key resources.

THE GOAL

The NIPP goal requires meeting a series of objectives that includes understanding and sharing information about terrorist threats and other hazards, building security partnerships, implementing a long-term risk management program, and maximizing the efficient use of resources. Measuring progress toward achieving the NIPP goal requires that CI/KR security partners have:

- *coordinated risk-based CI/KR plans* and programs in place addressing known and potential threats and hazards;
- *structures and processes* that are flexible and adaptable to incorporate operational lessons learned and best practices, and quickly adapt to a changing threat or incident environment;
- *processes in place* to identify and address dependencies and interdependencies to allow for more timely and effective implementation of short-term protective actions and more rapid response and recovery; and
- *access to robust information-sharing networks* that include relevant intelligence and threat analysis and real-time incident reporting.

The NIPP risk management framework includes the following activities:

- *Set security goals.* Define specific outcomes, conditions, end points, and/or performance targets that collectively constitute an effective protective posture.
- *Identify assets, systems, networks, and functions.* Develop an inventory of the assets, systems, and networks, including those located outside the United States, that compose the

FIGURE 11.16 ◆ Protection—Manage Risks (*Source:* www.dhs.gov)

nation's CI/KR and the critical functionality therein; collect information pertinent to risk management that accounts for the fundamental characteristics of each sector.

- *Assess risks.* Determine risk by combining potential direct and indirect consequences of a terrorist attack or other hazards (including seasonal changes in consequences, dependencies, and interdependencies associated with each identified asset, system, or network), known vulnerabilities to various potential attack vectors, and general or specific threat information.
- *Prioritize.* Aggregate and analyze risk assessment results to develop a comprehensive picture of asset, system, and network risk; establish priorities based on risk and determine protection and business continuity initiatives that provide the greatest mitigation of risk.
- *Implement protective programs.* Select sector-appropriate protective actions or programs to reduce or manage the risk identified; secure the resources needed to address priorities.
- *Measure effectiveness.* Use metrics and other evaluation procedures at the national and sector levels to measure progress and assess the effectiveness of the national CI/KR protection program in improving protection, managing risk, and increasing resiliency (see Figure 11.16).

CI/KR protection is an ongoing and complex process. The NIPP provides the framework for the unprecedented cooperation needed to develop, implement, and maintain a coordinated national effort.

Achieving the NIPP goal requires actions to address a series of objectives that include understanding and sharing information about terrorist threats and other hazards, building security partnerships to share information and implement CI/KR protection programs, implementing a long-term risk management program, and maximizing efficient use of resources for CI/KR protection.

AUTHORITIES, ROLES, AND RESPONSIBILITIES

As outlined previously, the Homeland Security Act of 2002 provides the basis for Department of Homeland Security (DHS) responsibilities in the protection of the nation's CI/KR. The act assigns DHS the responsibility to develop a comprehensive national plan for securing CI/KR and for recommending "measures necessary to protect the key resources and critical infrastructure of the United States in coordination with other agencies of the federal government and in cooperation with state and local government agencies and authorities, the private sector, and other entities."

THE VALUE OF THE NIPP PLAN

The public–private partnership detailed in the NIPP provides the foundation for effective CI/KR protection, prevention, response, mitigation, and recovery. Government and private sector partners each bring core competencies that add value to the partnership and enhance the nation's CI/KR protective posture.

Many industries justify their CI/KR protection efforts based on corporate business needs. Government can support these industry efforts and assist in broad-scale CI/KR protection through activities such as the following:

- providing owners and operators with timely, analytical, accurate, and useful information on threats to CI/KR;
- ensuring industry is engaged as early as possible in the development and enhancement of risk management activities, approaches, and actions;
- ensuring industry is engaged as early as possible in the development and revision of sector-specific plans (SSPs) and in planning and other CI/KR protection initiatives;
- articulating to corporate leaders, publicly and privately, the business and national security benefits of investing in security measures that exceed individual business needs;
- creating an environment that uses incentives and encourages companies voluntarily to adopt sound security practices;
- working with industry to develop and prioritize key missions for each sector and enable their protection and/or restoration;
- providing support for research needed to enhance CI/KR protection efforts; and
- developing resources to engage in cross-sector interdependency studies through exercises, symposiums, training sessions, and computer modeling, which can enhance business continuity planning.

SECTOR PARTNERSHIP MODEL

The enormity and complexity of the nation's CI/KR, the diverse and dynamic nature of the actions required to protect CI/KR, and the uncertain nature of terrorist threats or disasters make the effective implementation of protection efforts challenging. To be effective, NIPP partners must be committed to sharing and protecting the information needed to achieve the NIPP goal and supporting objectives. DHS is responsible for the overall coordination of the NIPP partnership and information-sharing network.

The structure through which representative groups from federal, state, local, and tribal governments and the private sector can collaborate and develop consensus approaches to CI/KR protection consists of the following:

- The Private Sector Cross-Sector Council made up of the Partnership for Critical Infrastructure Security (see Figure 11.17)

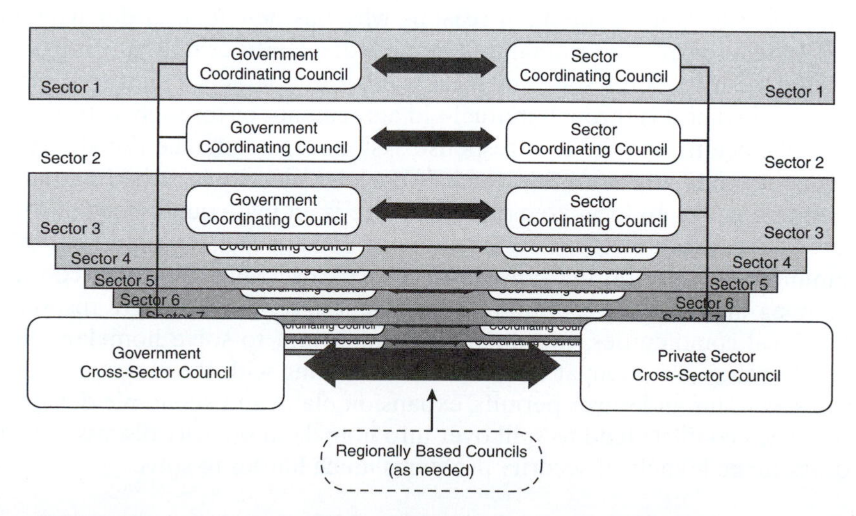

FIGURE 11.17 ◆ Regionally Based Concepts (*Source:* www.dhs.gov)

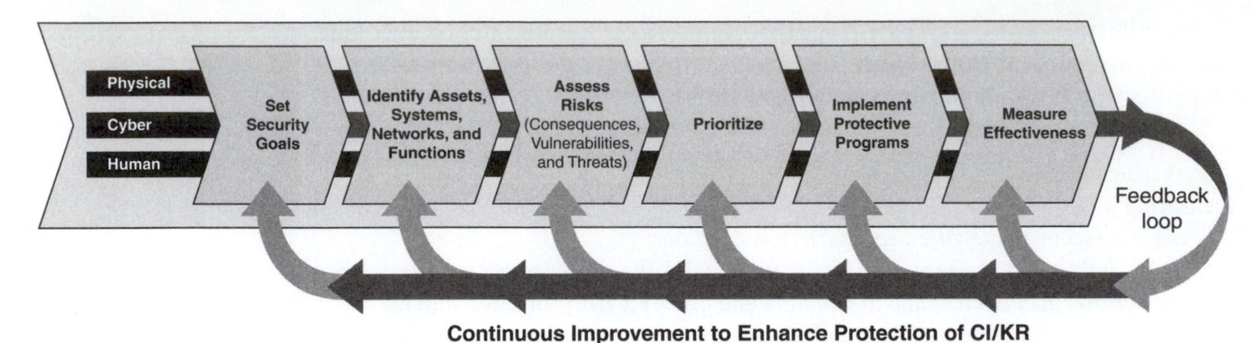

FIGURE 11.18 ◆ Continuous Improvement to Enhance Protection of CI/KR (*Source:* www.dhs.gov)

- The Government Cross-Sector Council, made up of two subcouncils: the NIPP Federal Senior Leadership Council and the State, Local, and Tribal Government Coordinating Council
- Individual Sector Coordinating Councils
- Government Coordinating Councils

RISK MANAGEMENT FRAMEWORK

The cornerstone of the NIPP is its risk management framework. This framework establishes the process for combining consequence, vulnerability, and threat information to produce a comprehensive, systematic, and rational assessment of national or sector-specific risk that drives CI/KR-protection activities (see Figure 11.18).

In the context of the NIPP, risk is the expected magnitude of loss (e.g., deaths, injuries, economic damage, loss of public confidence, or government capability) due to a terrorist attack, natural disaster, or other incident, along with the likelihood of such an event occurring and causing that loss.[12]

COORDINATION

Coordination is important; in many ways, it is the defining administrative strategy of homeland security. One of the main reasons why this new federal department was created is to secure better coordination among federal entities that may impact areas of homeland security. At the state and local levels, this is largely a matter of coordination as well, whether it is through mutual-aid agreements, or through a strategy to integrate public health into the first response system or to enhance the ability of fire departments to come to each other's aid. Homeland security is, at its foundation, an issue of coordination. In local governments there is an inadequate base of revenues and expertise to mount effective systems on our own. In fact, it would be foolish for all communities to develop the same level of expertise when thorough cooperative efforts among them would result in more efficient and more effective response systems. For local communities, coordination is important to solve homeland security problems, but they are often, at the same time, fighting with the same partners over such issues as water and sewer permits, expansion plans, and economic development strategies. Such conflicts tend to spill over into homeland security discussions and, in the process, make homeland security problems much harder to solve.

On the morning of September 11, we discovered just how important communication systems can be in a major event. As it turns out, the police department had better information about the condition of the World Trade Center towers than did the firefighters on the inside because the police department had a helicopter circling overhead. They helicopter radioed the information to police commanders, but the police commanders were not in touch with the fire commanders. The fire commanders in the lobbies of the World Trade Center did not have access to the TV coverage that the rest of the world was watching. Not only did they have difficulty communicating to the firefighters on upper floors, they also had trouble communicating among related agencies.

We have discovered that for homeland security to work, especially in cases of attack and terrorist events, effective communication is crucial. We also know there are many, many local communities around the country in which radio systems are not interoperable (see Figure 11.19).[13]

Police officers and fire officials often cannot talk to each other without going through a very complicated system through several emergency dispatch centers. There has been an effort in this country to solve this problem in respect to creating interoperability via technology and by defining radio spectrum by the *Federal Communications Commission (FCC)* for public safety[14] (see Figure 11.20).

Interoperability Techniques

	Method	**Fit**
LEVEL 6 Standards-Based Shared Systems	**Standards-Based Shared Systems**	**Best Long-Term Solution**
LEVEL 5 System-Specific Roaming	System-Specific Roaming	**Full-Featured, Wide Area**
LEVEL 4 Gateway (Console Patch)	Gateway (Console Patch)	**Short-Term System Modification**
LEVEL 3 Mutual-Aid Channels	Mutual-Aid Channels	**Simple Short-Term Solutions** (Easily deployed ↑ Time-consuming)
LEVEL 2 Talkaround	Talkaround	
LEVEL 1 Swap Radios	Swap Radios	

FIGURE 11.19 ◆ Interoperability Techniques

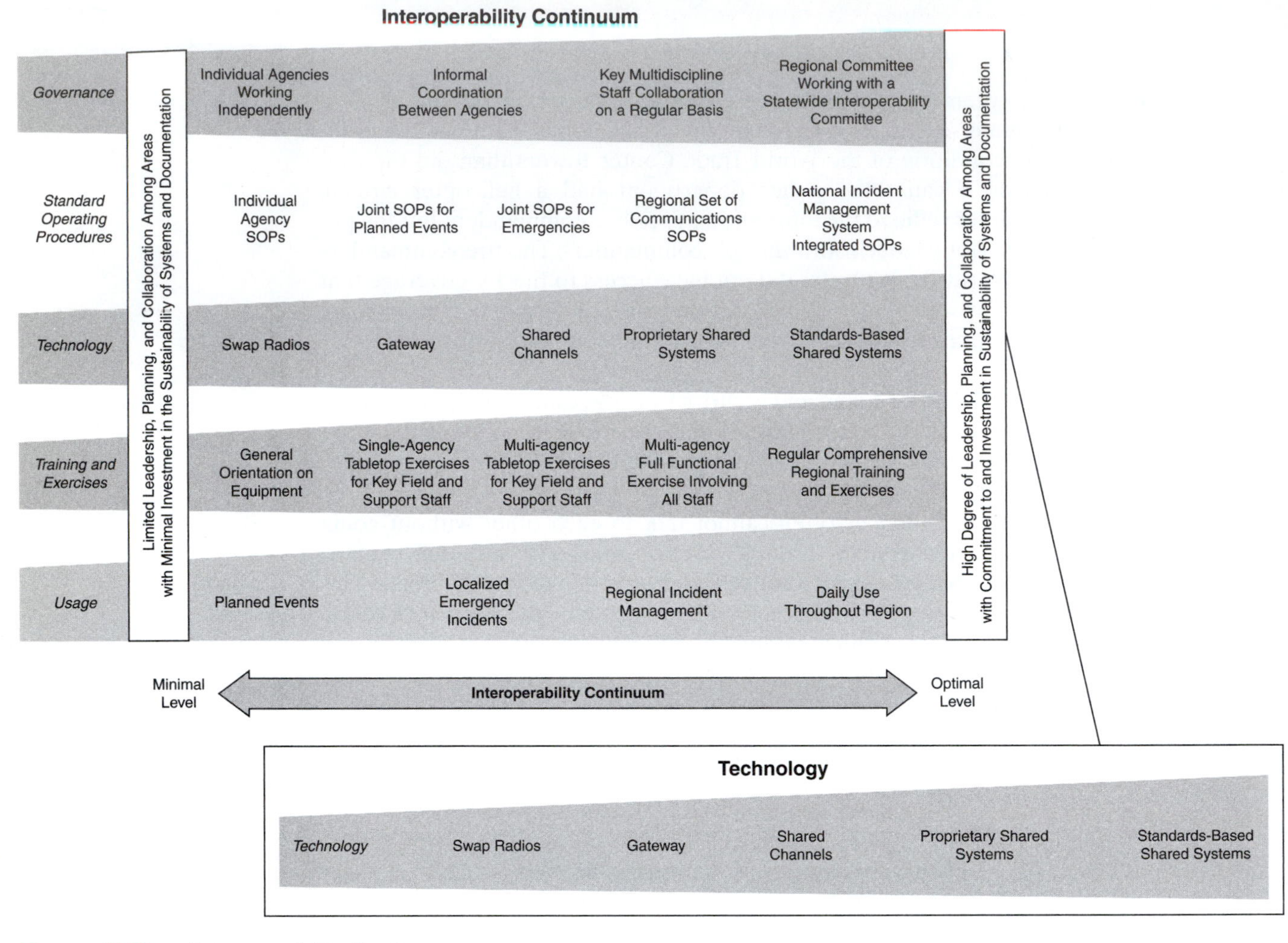

FIGURE 11.20 ◆ Interoperability Continuum

FUNDING

Many local government officials have stated, "If we really want to solve these problems, what we need is more money. We cannot come up with the cash ourselves; it must come from the federal government." While the actual flow of federal funds has been at a lower level than what was originally envisioned at the state and local levels, there is no doubt local government desperately needs additional funding to help purchase some of the communications and first responder equipment in order to be prepared in dealing with emergency response. The fact is simply adding more money into the system does not necessarily produce higher levels of homeland security. Additional funding may only perpetuate the problems already experienced unless more coordination at the local, state, and federal levels is a key element of future design. Federal aid systems will not solve all these problems. One way that may help solve this issue is to use the funds as incentives for local communities to do what they are otherwise not inclined to do. In short, encourage local governments to engage in the type of regional coordination required to prepare, respond, and mitigate such events because, at its core, homeland security is often an issue of coordination.

THE SIX MISSION AREAS OF HOMELAND SECURITY

The 2002 strategic plan for homeland security, issued by the Bush administration, illustrates the point that the intergovernmental partnership necessarily extends well beyond the response phase. In fact, state and local governments are critical players in each of the six major mission areas defined by this plan:

1. *Intelligence and warning.* Obtains early notice and advance intelligence of the threats that we have. Clearly, to be effective, information must flow from state and local governments about threats to their community, advising state and local governments of these threats so they can take real-time action.
2. *Border and transportation security* (airport, train station, or other mass transportation). Protects the nation's borders. State and local governments have a role to play and obviously are doing this.
3. *Domestic counterterrorism.* Generates intelligence information that states and local governments can often provide and use to help eliminate threats and gain advance warning.
4. *Protecting the critical infrastructure.* State and local governments own the infrastructure, and their investment and involvement are critical to national efforts to protect these assets against terrorist attacks. In the post 9/11 era, infrastructure has shifted in public debates from an asset needed to promote community economic development to become "critical national infrastructure." So a local bridge is no longer just a way to get to the store, it is a critical national infrastructure.
5. *Defense against catastrophic threats and bioterrorism.* Local government, specifically the fire service, plays a key role in this area. The initial response by the first responder will dictate the level of severity the incident escalates to.
6. *Emergency preparedness and response.*[15]

As discussed throughout this chapter, all disasters are local. If the initial response is poorly executed, the incident will likely cause more disruption to the community and impact the nation more severely.It is the responsibility of local government to build a strong foundation of response and emergency management based upon the likely events that may occur in their jurisdiction. Without this foundation, the support of regional, state, and federal resources may not be effectively utilized.

ADVICE FROM THE EXPERTS

How has the fire service changed in disaster management and response since September 11, 2001, and Hurricanes Katrina and Rita of 2005?

MICHAEL CHIARAMONTE

Chief Fire Inspector, Retired, Lynbrook Fire Department (New York): When I saw the actions after Katrina, I was thinking the things that were said in the numerous articles written after the event. You should never state an opinion on such things unless you investigate them. Upon investigation I found that Louisiana was ill prepared as much as they are now saying it was prepared. So now comes the tennis match of blame. In my opinion what happened, happened. Now is the time to put that all behind us, dig in, get to work, and solve the problems because things like this will happen again. The IAFC is working on a National Mutual Aid Task Force, which should speed things up a bit. This project has FEMA's blessing. We dragged our feet after 9/11 and all need to move quickly to identify what happened in this recent disaster, compare it with 9/11, and correct any problems as soon as possible. Forget blame—be a team. This is my wish.

BETT CLARK

Former Fire Chief, Bernalillo County Fire and Rescue (New Mexico): We have to think all-hazards planning. We have to think beyond our local response districts and move beyond the turf wars. We also must think about the next step. Where do I get assistance in mutual aid and funding? What could we realistically handle in three days if we were cut off from the rest of the state, the nation? What do we do after the disaster passes? How do we continue to provide service? How do we return back to normal? What do we do to assure the firefighters who are on duty for several days at a time? How do we handle communications? Do we work well with others: fire, law enforcement, civil response services, state EOM, and FEMA?

KELVIN COCHRAN

Fire Chief, Atlanta Fire Department (Georgia): Not only the fire service approach but also the global approach to emergency management has changed since September 11. We can no longer think of emergency management on a small local-level scale. Emergencies must be approached from the perspective of expanding from a local event to having the potential of an event of national consequence involving local, state, and federal resources. Since the establishment of the Department of Homeland Security (DHS), the National Response Plan, and the National Incident Management System, disaster management has been transformed. Hurricane Katrina was the first great test of DHS. To the misfortune of our nation and the people of the Gulf Coast region, we realized our emphasis on preparedness for human-made disasters caused us to underestimate the threat and consequences of natural disasters. Before we had a chance to lick our wounds, Hurricane Rita hit the Gulf Coast region again. Lessons learned revealed what the fire service has touted since the first World Trade Center bombing and the Oklahoma City bombing—the fire service needs to be adequately trained and equipped to respond to natural and human-made disasters to minimize the loss of lives and harm. The human needs of firefighters and their families must be planned and met during disasters to assure both their safety and that there will be first responders actually responding. Equally important, the fire service must have a role in policy-level decision making at the federal level, and fire service leaders must be appointed to key leadership roles within DHS.

CLIFF JONES

Fire Chief, Tempe Fire Department (Arizona): Disaster management and response have taken on heightened importance since the attacks of September 11, 2001, and the hurricanes of 2005. Other agencies have entered into the disaster management preparation and response arena because of demands of policy makers and because huge amounts of federal money have been made available through grants. The fire service finds itself stretched to meet all of the demands of disaster management and response. Although there is significant money available, sufficient staffing is not currently available to meet the demands of disaster situations and the expectations of policy makers. Again, the fire service finds itself taking on additional major program management responsibility with little or no additional resources. Enhanced cooperation in terms of day-to-day operations between the fire department and all other agencies associated with disaster management is both expected and demanded. Fire chiefs will need to move quickly to enhance their personal knowledge of the field of disaster management. Interoperability is high on the list of demands and much is to be done on this issue.

BILL KILLEN

Retired Fire Chief and IAFC President 2005–2006: It has not changed, not yet, but change is coming. The lessons learned from Hurricanes Katrina and Rita were nearly identical to the lessons learned from September 11, 2001, and the fire service repeated the same mistakes during these hurricanes. Change is forthcoming through the leadership of the International Association of Fire Chiefs' Mutual Aid System Task Force (MASTF). When MASTF is completed and implemented, it should provide a mechanism whereby interstate and intrastate mutual-aid response of emergency services will be seamless.

KEVIN KING

U.S. Marine Corps Fire Service: I think we have gotten better at sharing resources and managing large incidents since September 11; however, we clearly have a way to go, as evidenced by Katrina and Rita. I think we do a pretty good job today of mobilizing assets and managing incidents as long as the incident remains local or regional. However, when we get beyond that, and especially when we start getting federal assets involved, I believe we run into greater coordination and management issues. I believe the National Interagency Fire Center in Boise probably does the best job of managing incidents on a national scale, and we can learn a lot from its operations and procedures. (These are opinions of Mr. King and do not reflect official policy of the Department of Defense or the United States Marine Corps.)

MARK LIGHT

Executive Director, International Association of Fire Chiefs: Fire departments have realized emergency management and disaster preparedness are ways of life. In the past, emergency management was what a special part of government did. During tabletop drills or full-scale drills, departments would brush up and learn what their role was, and then promptly put the book back on the shelf and wait until the next drill. Both of these disasters brought planning and preparedness to the forefront. In many departments, it continues to be a struggle to get the rank-and-file firefighter to take preparation seriously, but through time, this will become as common as doing prefire plans.

LORI MOORE-MERRELL

Assistant to the General President, International Association of Fire Fighters: I will speak only to September 11 as much of the change that will come as a result of Katrina is yet to be seen. As a result of September 11, 2001, the fire service has undergone exponential change. Many dollars are available to local fire departments to train for the "big event." Funding is available for supply caches, for disaster trailers and other equipment to be stored "just in case." The focus on hazardous materials and weapons of mass destruction has multiplied. Much of this, however, has occurred to the detriment of regular day-to-day fire department operations. There is focus on the terrorism annex rather than maintaining basic infrastructure for emergency response. For example, some departments purchase new equipment that is stored for a disaster event and use antiquated and many times unsafe equipment on a daily basis. Preparation and planning for the big event is necessary; however, it must be considered in light of the events that occur daily. Fire service leaders must be prudent in weighing the excitement of disaster prep against the necessity of regular operational training and safe, efficient, and effective operations on a daily basis.

JAY REARDON

Fire Chief, Retired, Northbrook Fire Department (Illinois), currently President of the Mutual Aid Box Alarm System (MABAS): The media have molded/crafted societal expectations and perceptions (especially since Katrina). People, for now, believe government is responsible for everything—including irresponsible individual behavior. The fire service now needs to step up to the all-risk table—not just terrorism. Are we prepared to put our money where our mouths are? Will we become our own worst enemy? National mutual aid—are we willing to embrace it? Are volunteers willing to step up to the expectation plate? Will labor be willing to play on the same team as volunteers when society doesn't care when lives and property loss are the lead story? Will the fire service be able and willing to define itself for tomorrow's needs and reach unanimous concurrence on national doctrine? Will the fire service be able to gain the discipline it needs to survive and prosper?

GARY W. SMITH

Retired Fire Chief, Aptos–La Salva Fire Protection District (California): REACTION and awareness . . . September 11 woke the terrorism issue that had been discussed in many circles over the previous 20 years . . . when it finally occurred, we became aware and spent tons of money, with waste and mismanagement that follows panic. Then the hurricanes occurred; now, we recognize the limits of our response system. YES, Mother Nature can still kick our butts and always will. However, we can respond more effectively if we seriously consider the threats and work on the mutual-aid plans to support the disaster response that is needed to recover from a major hit from earthquake, fire, flood, weather, and so on. The expectation for the federal government to come in and fix everyone's local problems after a disaster is not realistic. We must learn to take on responsibility and to react and be aware of the challenges as well as the signs for the need to make the right moves with the resources available.

STEVE WESTERMANN

Fire Chief, Central Jackson County Fire Protection District (Blue Springs, Missouri), and IAFC President 2007–2008: I think the biggest change for the fire service is the realization that we are not just about fires. We are about emergencies and that means "all-hazard" emergencies. Whether we respond to a fire, a hurricane, an earthquake, or a tornado, we operate with the same basic command and control instincts. Your command structure has the same framework with slightly different designations depending on the task at hand. Because the fire service is all hazard, many fire chiefs or someone within the fire departments is well equipped to handle the EMD role in the community.

THOMAS WIECZOREK

Executive Director, Center for Public Safety Excellence, and Retired City Manager: I don't believe it has changed . . . yet. The fire service probably has the most credibility and opportunity to be viewed as the leader in emergency response. From GIS to incident command—the fire service has the opportunity to lead like never before.

JANET WILMOTH

Editorial Director, *Fire Chief* magazine: September 11, 2001, got the nation and the fire service's attention; the hurricanes were a test of what they learned. Some states have learned their lesson well and interstate training has been effective. Nationally, the fire service has failed.

FRED WINDISCH

Fire Chief, Ponderosa Volunteer Fire Department (Houston, Texas): The fire service has fully understood that we are our nation's FIRST, first responders. While we hope that the government will be coming soon, the fact of the matter is that *soon* is defined by the fire service as NOW. The fire service must develop a national mutual-aid response to address the needs of major disasters within five days. Good data are now available that demonstrate the direction the fire service must move toward in this regard. The pitfall is that some chief officers will be resistant to accepting an automated response. Hopefully, through education and assurances the product will be ASSISTANCE; those chief officers will understand that the help coming will not displace local resources, but it will enhance service levels that have been negatively affected (especially our own responders) by the disaster.

Review Questions

1. Explain the elements of planning. Include the importance of planning and how the federal government plays a role in disaster response.
2. Describe the role FEMA plays in disaster response. Include the history of FEMA and how it has evolved to its role today.
3. List and describe the four phases of emergency management.
4. Discuss the four steps for personal and family safety as recommended by FEMA in an emergency situation.
5. Describe the Federal Response Plan and the effect it has on emergency service providers.
6. Explain the Integrated Emergency Management System concept. Include the specific goals of the concept.
7. List the reasons risk assessment is essential to the emergency planning process.
8. Describe the three phases of hazard assessment.
9. Explain the National Response Plan and how it came into existence.

10. List each of the ESFs and describe their scope of responsibility.
11. Explain the role of nongovernmental and volunteer organizations in a disaster situation. Describe at least three recognized volunteer groups or programs.
12. List some of the more important "transitional steps" that are necessary to apply ICS in a field incident environment.
13. Describe each of the sections of the IC system.
14. Describe the six mission areas of homeland security.

References

1. www.FEMA.gov
2. Mike Worthington (Retired Corporate Vice President Global Safety and Security Solutions, Motorola), *Motorola Presentation to the IAFC Board,* 2002.
3. www.redcross/services/disaster
4. National Weather Service, *StormReady*, www.stormready.noaa.gov
5. Mike Worthington (Retired Corporate Vice President Global Safety and Security Solutions, Motorola), *Motorola Presentation to the IAFC Board,* 2002.
6. Thomas E. Drabek, "Managing the Emergency Response," *Public Administration Review* 45: 1985, 85–92.
7. National Response Plan (NRP) Course Summary (IS-800), December 2004. Washington, D.C.
8. www.citizencorps.gov
9. www.nimsonline.com
10. NIMS Document, Appendix A, www.fema.gov/nims/nims_compliance.shtm
11. Department of Homeland Security, *The Role of "Home" in Homeland Security* (Washington, DC: (Rockefeller Institute of Government, 2003).
12. Department of Homeland Security, *National Infrastructure Protection Plan (NIPP)* (Washington, DC: 2006).
13. Mike Worthington (Retired Corporate Vice President Global Safety and Security Solutions, Motorola), *Motorola Presentation to the IAFC Board,* 2002).
14. Department of Homeland Security, *Role of "Home" in Homeland Security*. (Washington, DC: 2003).
15. Ibid.

Shaping the Future

Key Terms

NIST, p. 567
swift water, p. 552
TSA, p. 554
urban search and rescue, p. 552
USAR, p. 552

In Malcolm Gladwell's *A Tipping Point: How Little Things Can Make a Big Difference,* he explores the concept of how ideas, products, messages, behaviors, industry shifts, and national perspectives spread similarly to a virus. Gladwell points out three characteristics that, when combined with one another, can have dramatic impacts on the world in which we live.

These three characteristics—(1) contagiousness, (2) the fact that little causes can have big effects, and (3) that change happens not gradually but at one dramatic moment—are the same three principles that define how measles moves through a grade-school classroom or the flu attacks every winter. Of the three, the third trait—the idea that epidemics can rise or fall in one dramatic moment—is the most important, because it is the principle that makes sense of the first two and that permits the greatest insight into why modern change happens the way it does. The name given to that one dramatic moment in an epidemic when everything can change all at once is the tipping point.[1]

So, as the fire service looks to shape its future, we have to begin to contemplate what our own tipping points may be, in the near future, that will establish the framework by which our industry will change and the roads our industry may travel as a result of these changes. We witnessed a tipping point on September 11, on many levels, for this country and for those of us in the fire service, because it brought many of us back to our core values and basic tenets, upon which not only this country but also the fire service were founded. These values are clearly articulated in the Declaration of Independence and the Bill of Rights, yet are specifically written in the American Constitution, which states, "We the People of the United States, in Order to form a more perfect Union, establish Justice, insure domestic Tranquility, provide for the common defence, promote the general Welfare, and secure the Blessings of Liberty to ourselves and our Posterity, do ordain and establish this Constitution for the United States of America."

Today, as we all work to ensure the general welfare of our citizens, the fire service has become a part of the nation's response system. In fact, as America's first responders we are an integral link to provide common defense of this country within the geographical borders of the United States. The events of September 11, 2001, clearly demonstrated that in the time of disaster or crisis, whatever the cause, the local fire service is the critical link in the first line of defense and the primary response to any such event. The fire service can no longer be looked upon just as a local resource but must be viewed as a national asset. With all of the disasters we have witnessed the last decade, each has clearly shown the need for a close relationship with the local fire service and the federal government toward improving our capabilities as the nation's first responders.

So, as we shape our future or it shapes us, what will be the catalyst that may create the tipping points that will move our industry in different directions? America's fire and emergency services reach every community throughout the nation covering urban, suburban, and rural neighborhoods. Nearly 1.1 million men and women make up the fire service. Almost 300,000 career firefighters and 800,000 volunteer firefighters serve in over 30,000 career/volunteer and combination fire departments across the United States.[2]

The fire service is the only entity that is locally situated, staffed, trained, and equipped to respond to all types of emergencies. As we know, our response to natural disasters such as earthquakes, floods, hurricanes, and tornadoes, as well as human-made catastrophes such as hazardous materials spills, arsons, and terrorism, have made America's fire service an all-hazard, all-risk response entity.

FIRE SUPPRESSION

Fire suppression (see Figure 12.1) is a traditional role of the fire service today. America's firefighters combat fires daily in residential, industrial, commercial, and wildland settings. While each scenario requires specialized training and equipment and a varying level of resource commitment, our communities have become more dependent on the fire service requiring firefighters to cross train, covering an increased variety of threats and hazards to which they respond on a daily basis.

In the near future, fire suppression will continue to be as it has been for the past century. With the amount of building stock currently in place, we can anticipate to be called upon to suppress fires manually into the foreseeable future. However, as this building stock ages and is replaced over the course of the next century and as we incorporate into new structures the technology of sprinklers, alarms, and enhanced building materials, the rate at which we will have to suppress fires manually will continue to decrease. Technology will also play a significant role in the way in which we suppress fires in the future. We are just beginning to see the impact that technology and its application in the field will have on the fire service. Items such as infrared thermal-imaging cameras, GIS tracking devices, compressed air foam systems, and high-pressure nozzles have all been introduced in the last two decades and are changing the way we attack and suppress fires manually. In the future technology will continue to evolve at a rapid pace and be refined so that our equipment will become lighter, easier to use, and more versatile.

FIGURE 12.1 ◆ Fire Suppression *(Courtesy of Jose A. Escobedo, Fresno Fire Department)*

EMERGENCY MEDICAL SERVICE

The majority of America's fire departments offer some level of EMS from basic life support to advance life support. Many fire departments require firefighters be trained to the emergency medical technician basic level, but many firefighters are trained to the paramedic level. This concept allows departments to provide lifesaving procedures such as advanced cardiac care and the administration of drugs. In the past two decades, EMS calls have outpaced fire suppression calls for most fire departments. In many jurisdictions, EMS calls account for a high percentage of the annual call volume. Emergency medical services of the future will be community-based health management systems that are fully integrated with the overall health system. We will have the ability to identify and modify illness and injury risk, provide acute illness and injury care and follow-up, and be significant contributors to the treatment of chronic conditions and community health monitoring. This will occur as the need to redistribute existing health care resources will force integration with health care providers and public health and safety agencies. Ultimately, it may improve community health and result in more appropriate use of acute health care resources; however, all EMS services will remain the public's medical safety net.

The *EMS Agenda for the Future: Implementation Guide* identifies three broad areas in which the fire service must be diligent in the future: building bridges, creating tools and resources, and developing infrastructure. Bridges will strengthen partnerships and result in new and enhanced relationships and partnerships among the many

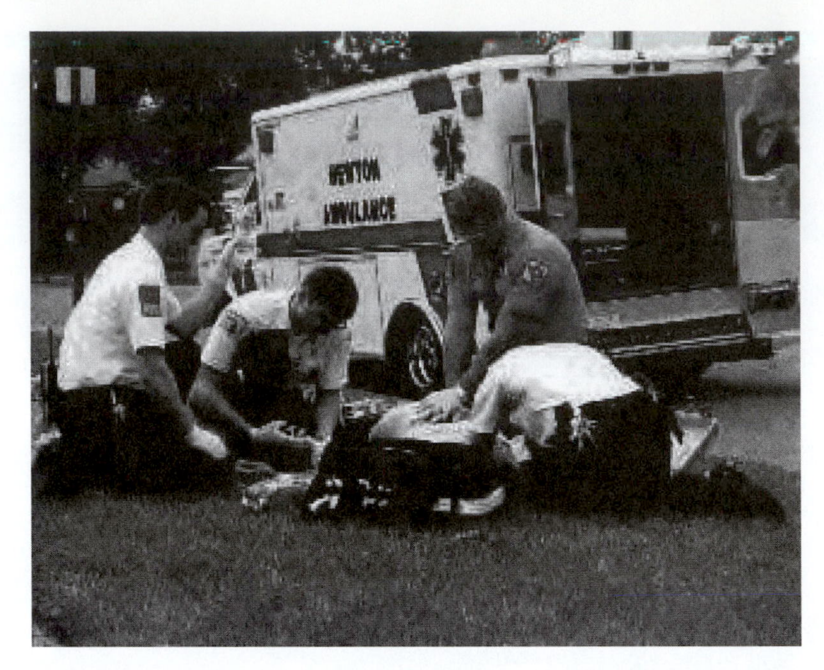

FIGURE 12.2 ◆ EMS Crew Providing Emergency Medical Service. *(Source: FEMA News Photo)*

agencies, organizations, and individuals with a stake in the future of EMS. These partnerships must seek diversified perspectives and invite enthusiastic participation if barriers on the path to the future are to be overcome. New tools and resources will enable progress and facilitate activities on a widespread basis. Improved infrastructure will add to the capacity of EMS to affect overall community health (see Figure 12.2).

Venturing toward the future vision for EMS is an ambitious undertaking. Activities must be initiated on national, state, and local levels. Ten "priority objectives" (short, intermediate, and long term) are proposed to be the initial focus:

- Develop collaborative strategies to identify and address community health and safety issues.
- Align the financial incentives of EMS and other health care providers and payers.
- Participate in community-based prevention efforts.
- Develop and pursue a national EMS research agenda.
- Pass EMS legislation in each state to support innovation and integration.
- Allocate adequate resources for medical direction.
- Develop information systems that link EMS across its continuum.
- Determine the costs and benefits of EMS to the community.
- Ensure nationwide availability of 9-1-1 as the emergency telephone number.
- Ensure calls for emergency help are automatically accompanied by location-identifying information.

The *EMS Agenda for the Future: Implementation Guide* is a tool for EMS providers, administrators, and medical directors; health care providers and administrators; public health and safety officials; local, state, and federal government officials; organization and community leaders; and all other entities and people with a potential interest in or influence on the structure or function of our nation's system for providing emergency medical care. The commitment to an accessible, reliable EMS system that contributes to the health of our communities will be a cornerstone of the future. The *EMS Agenda for the Future: Implementation Guide* is a call for action to join partnerships that will lead to the exploration of possibilities for the future of EMS—a crucial part of the health care system and the public's emergency medical safety net.[3]

In many communities, the current health care delivery system (including emergency rooms) and the emergency transport system are in crisis. The lack of health care for a large segment of the population has forced many to utilize their local EMS service as their primary medical provider. In many of the largest cities, EMS has become the universal health care program for the uninsured and the indigent population. The result has been overcrowded emergency rooms, EMS transport units out of service waiting to unload patients, and the disruption in many communities' EMS deployment modes. This trend is expected to continue into the foreseeable future and will require many EMS systems to evaluate alternative strategies of resource deployment to meet this growing demand for service.

SPECIALIZED RESPONSE

HAZARDOUS MATERIALS

Hazardous materials response (see Figure 12.3) requires specially trained hazmat technicians to have the basic knowledge of a chemists in addition to firefighting skills. Hazmat technicians train for weapons of mass destruction and preparedness in response. Hazmat incidents can range from a small incident, such as a fire in a private pool house where chlorine is stored, to a large train derailment that spills several thousand gallons of gasoline or some other hazardous or flammable substance. Trucks carrying potentially lethal cargo travel across the nation many times a day. The storage and use of hazardous chemicals have become commonplace in

FIGURE 12.3 ◆ Hazardous Materials Response Team at a Hazmat Incident *(Courtesy of Charles Tobias, Fresno Fire Department)*

FIGURE 12.4 ◆ USAR Rescue *(Courtesy of Sean Johnson, Fresno Fire Department)*

every community, resulting in the potential for a hazardous materials incident to be found in even the smallest communities. With the threat of terrorist attacks within the geographic borders of the United States, the entire concept of hazardous materials response has redefined itself and will continue to do so. In the future, hazardous materials teams will have to become specialized in response to explosive devices and potential biological exposure as they have been in the past to gasoline and oil spills in the street.

TECHNICAL RESPONSE

As the threat of terrorism has shaped our national response, it will also shape our response in the future at the local level for areas of technical and **urban search and rescue**. Technical rescue (see Figure 12.4) covers most of these emergencies: victims caught in flash floods, **swift water** rescue, window washers trapped 20 stories high, motorists trapped and crushed in overturned vehicles, and people trapped in collapsed buildings. Technical rescue may also include trench rescue, confined space rescue, scuba diving rescue, and open water operations rescue.

urban search and rescue
Operational activities that include locating, extricating, and providing on-site medical treatment to victims trapped in collapsed structures.

swift water
The rescue of persons trapped in rivers or flood control waterways.

USAR
Urban search and rescue.

The National Urban Search and Rescue Response System (**USAR**) was established under the authority of the Federal Emergency Management Agency (FEMA) in 1989. The USAR framework provides the structure for local emergency response personnel to integrate into a national disaster response. USAR involves the location of rescue, extrication, and initial medical stabilization of victims trapped in confined spaces. Structural collapse often is the cause of victims being trapped, but victims may also be trapped in transportation accidents, mining accidents, and collapsed trenches. Urban search and rescue is considered a multihazards plan, and it has proven to be very effective in a variety of emergencies and disasters including earthquakes, hurricanes, typhoons, storms, tornadoes, floods, dam failures, technological accidents, terrorist activities, and hazardous materials releases.

FIGURE 12.5 ◆ Wildland Firefighting Crew *(Courtesy of Timothy Henry, Fresno Fire Department)*

Our future will be shaped in this arena by our ability to link local resource and capability in technical rescue with that of the state and federal structured urban search and rescue teams. As with any emergency, if we can establish a sound initial response and lay the foundation for the next level of resources to respond and build upon what has been started at the local level, the result will be a more positive outcome.

WILDLAND FIREFIGHTING

One of the critical elements of response in the future will be that of wildland firefighting (see Figure 12.5). Most of the large wildland fires that have made the news in recent years have occurred in national parks and other federal lands. Many firefighters who are active in wildland firefighting are employees of the federal government, such as the U.S. Forest Service, Bureau of Land Management, National Park Service, and Bureau of Indian Affairs. Specialized firefighting teams, hot shots crews, smoke jumpers, aviation crews, and engine and hand crews are the backbone of wildland fire defense in this country. However, the use of local structural fire agencies to supplement federal deployment is common, and we know without their response the federal and state resources would in many cases not be sufficient to control the incident.

While the images of the national forest burning by the millions of acres throughout the United States is the most common perception of the wildland fire problem, large number of jurisdictions in the United States face some sort of wildland fire threat. As people continue to build in the wilderness areas, the threat of wildland/urban interface fires continues to grow, adding to the risks of communities located in forested areas. Fire departments traditionally trained to fight structure fires are now required to train and respond to wildland/urban interface fires. In some jurisdictions, structure departments are the first available responders for wildland fire events arriving often 24 to 48 hours before the federal system is able to bring in its wildland fire teams.

This will continue to be a driving force for the fire service for many years to come. Our ability to react quickly and with sufficient resources to confine these fires before they become conflagrations has been difficult, to say the least, in our recent history. Our challenge for the future is to revamp our deployment strategy with respect to wildland and urban interface fires. Part of this re-engineering will require more resources dedicated to the wildland/urban interface effort and the creation of more defined partnerships between federal response agencies and the state and local providers. This is critical if the fire service is to effectively utilize the resources available in a timely fashion, which can have the most impact on an incident while it is small. In addition, a reevaluation of our current first management plan in relation to fuel maintenance and reduction will need to occur. Without it, we will continue to experience a large wildland fire threat.

AIRCRAFT RESCUE FIREFIGHTING

Aircraft rescue and firefighting (see Figure 12.6) have risen to a new level as a result of heightened awareness for potential terrorist activity involving aircraft. Airport fire departments are responsible for aircraft rescue firefighting, emergency medical response, hazmat response, fire prevention and inspections, pre-emergency surveys, and integration with the federal agencies that are often found in airport facilities such as the Federal Aviation Administration (FAA) and the Transportation Security Administration (**TSA**). The FAA requires firefighters to be in place on the runway for declared aircraft emergencies within three minutes with a number of firefighters defined by the airport index, which depends on the size and number of aircraft landing at a given airport. In most of the airports in the United States, this is typically one to two firefighters. These firefighters must have knowledge of the type of aircraft arriving, reporting the number of passengers, location of emergency exits on the

TSA
Transportation Security Administration.

FIGURE 12.6 ◆ Aircraft Rescue *(Courtesy of Javier Lara, Fresno Fire Department)*

aircraft, location of critical systems on the aircraft, as well as a working knowledge of airport operations, which is a highly complex and very specialized area.

As we look into the future, one area that will change substantially is the number of firefighters who will be required to provide this service. Currently, there is a disconnect between what is known to be safe and effective operational parameters for the number of firefighters and the jobs that need to be done in the event of a downed aircraft. The number of firefighters and equipment type required by the FAA is based upon the number and type of planes departing and arriving from an airport facility. In the future, there must be some level of standard for minimum aircraft rescue and firefighting deployment. This standard should be not only built upon the size and number of aircraft arriving at and departing from a given facility but also built upon the critical tasks that the firefighters will be called upon to perform in the event of an aircraft emergency.

FEDERAL AND MILITARY FIRE PROTECTION

Federal and military fire protection plays a vital role in the protection of the infrastructure that protects this country. In the United States and throughout the world, many federal and military emergency services also provide a critical local resource for communities in and around the bases throughout the United States through mutual-aid agreement with their local fire service providers. In addition to fighting fires on the military bases and federal installations, federal firefighters provide EMS services, respond to hazardous materials situations, and provide inspection of military bases and other federal facilities. The federal fire service is charged with the protection of critical military infrastructure and requires a special knowledge and security clearance. A future challenge in this arena will include the consolidation of many military bases, which will challenge the Department of Defense emergency services to maintain sufficient resources on the bases they protect to be effective in their mission. In addition, this consolidation will also impact the fire protection and resources available to local communities as bases are closed and resources relocated.

INDUSTRIAL FIRE PROTECTION

As we have seen repeatedly in recent history, a fire in an industrial setting poses a very real risk not only to the facility but also to the community in which it is located. Industrial fire safety personnel are often private fire and emergency service responders who provide firefighting, hazardous materials, rescue, and EMS to fixed industrial sites. As with their federal and military counterparts, they often forge mutual-aid agreements with surrounding jurisdictions to ensure accurate protection of their facilities and the communities where they are located. They must have in place extensive preplanning and up-to-date Material Safety Data Sheets and evacuation plans, and often use the latest technology in fire protection systems to minimize the risk and extended damage caused by fire or chemical release. Because there are extensive government regulations from OSHA and the EPA as well as stringent insurance guidelines that industrial fire and safety agencies must follow, these response teams are often well equipped and well trained. As with their municipal counterparts these industries must deal with a number of potential threats that include hazardous materials, fire, confined space emergencies, and now terrorist events. As we look to the future, the partnerships between industrial fire and safety

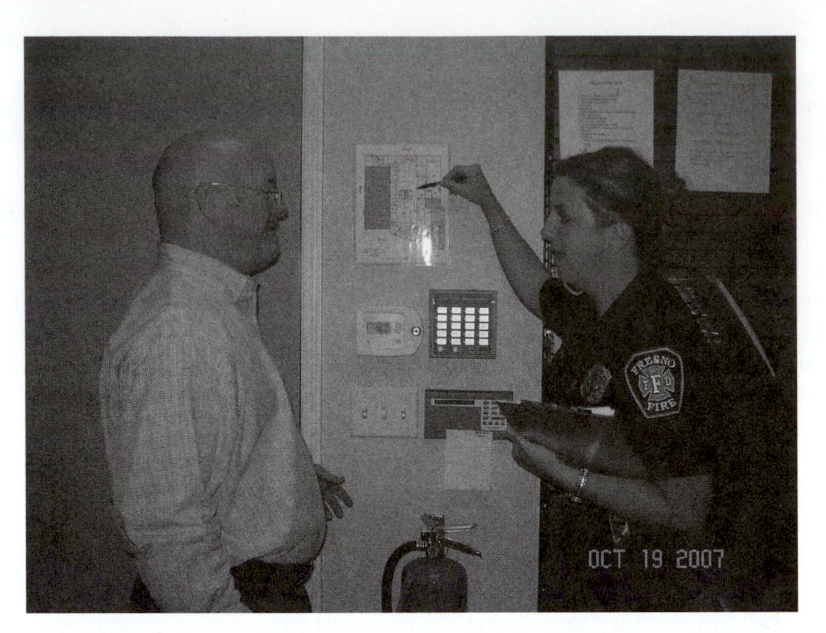

FIGURE 12.7 ◆ Code Enforcement by Fire Prevention Inspector *(Courtesy of Shandra Rodems, Fresno Fire Department)*

brigades and the local fire service provider will become even more important. Our ability to respond collectively and integrate our personnel in the event of a major emergency will be one of the challenges faced in this arena.

CODE ENFORCEMENT

In most jurisdictions, the fire department is charged with enforcing building and occupancy fire prevention codes (see Figure 12.7). This involves not only enforcing the laws as they relate to fire protection but also determining the existence of other conditions deemed to be hazardous to public health, safety, and welfare. Code enforcement officials conduct inspections to ensure the buildings are in compliance with local and state ordinances. The fire marshal's office within the department is usually responsible for this. The area of code enforcement may be one of the areas where we will see the biggest policy shifts in the future. Cost-effective technology is available and if a community chooses to utilize it, the risk factors of the community can be controlled through effective code enforcement. With the mandated use of sprinkler and alarm systems not only in commercial occupancies but also in residential occupancies, we can manage our fire risk in the future (see Tables 12.1, 12.2, and 12.3).

FIRE PREVENTION

Fire prevention through education (see Figure 12.8) will be another area that will see significant changes in the future. Most of today's departments have some type of fire prevention educational program. These outreach programs allow the fire department to educate the public regarding prevention of fire and life-safety issues. Programs include smoke alarm campaigns, fire safety for schoolchildren, public access defibrilla-

TABLE 12.1 ◆ Estimated Reduction in Civilian Deaths per Thousand Fires Due to Sprinklers by Property Use, 1989–1998 Structure Fires Reported to U.S. Fire Departments

Property Use	*Without Sprinklers*	*With Sprinklers*	*Percentage Reduction*
Public Assembly Properties	**0.8**	**0.0***	**100%**
Eating and Drinking Facilities	0.8	0.0*	100%
Educational Properties	**0.0***	**0.0***	**N/A**
Health Care Facilities	**4.9**	**1.2**	**75%**
Care of Aged Facilities	7.1	1.7	76%
Care of Sick Facilities	2.7	0.7	74%
Residential Properties	**9.4**	**2.1**	**78%**
One- and Two-Family Dwellings	9.7	4.7	51%
Apartments	8.2	1.6	81%
Hotels and Motels	9.1	0.8*	91%
Dormitories and Barracks	1.5	0.0*	100%
Stores and Offices	**1.0**	**0.3***	**74%**
Food or Beverage Store	1.2	0.0*	100%
Department Store	1.2*	0.0*	100%
General Office Building	0.6%	0.0*	100%
Industrial Facilities	**1.1**	**0.0***	**100%**
Manufacturing Facilities	**2.0**	**0.8**	**60%**
Storage Facilities	**1.0**	**0.0***	**100%**
Total	**7.6**	**1.1**	**86%**

*Based on fewer than two deaths per year in the 10-year period. Results may not be significant.

Note: These are national estimates of fires reported to U.S. municipal fire departments and so exclude fires reported only to federal and state agencies or industrial fire brigades. Fire statistics do not include proportional shares of fires with sprinkler status unknown or unreported.

Sources: Data from NFIRS and NFPA survey. Kimberly D. Rohr and John R. Hall, *U.S. Experience with Sprinklers and Other Fire Extinguishing Equipment* (Quincy, MA: Fire Analysis and Research Division, National Fire Protection Association, 2005).

tion, CPR, first-aid training for citizens, flu shot clinics that target at-risk populations, and citizen fire academies. These are a few examples of the many creative public education programs provided today. While many excellent programs deliver an effective fire safety message throughout many communities as a whole, the American fire service has been unable to measure adequately the effectiveness of our educational and outreach programs. The future of fire prevention, through educational programs, lies in our ability to measure the effectiveness of these programs so that resources are focused in those areas having the most impact in respect to the prevention of fires and the saving of lives. Currently, most jurisdictions simply do not have the data or the expertise to provide comprehensive performance analysis, which measures outcomes in the area of pubic education.

TABLE 12.2 ◆ Estimated Impact of Residential Sprinkler System in One- and Two-Family Dwellings

Impact of Sprinklers *Base of Comparison*	*Residential Sprinkler System and No Smoke Alarms*	*Residential Sprinkler System with Smoke Alarms*
A. Civilian Deaths		
1. Estimated reduction relative to death rate per thousand fires when no sprinklers or smoke alarms are present.	69%	83%
2. Estimated reduction relative to death rate per thousand fires when smoke alarms are present.	Not applicable	63%
B. Civilian Injuries		
1. Estimated reduction relative to injury rate per thousand fires when no sprinklers or smoke alarms are present.	46%	46%
2. Estimated reduction relative to injury per thousand fires when smoke alarms are present.	Not applicable	44%

Source: Rosalie T. Ruegg and Sieglinde K. Fuller, *A Benefit-Cost of Residential Fire Sprinkler Systems,* NBS Technical Note 1203, Gaithersburg, MD: U.S. Department of Commerce National Bureau of Standards, November 1984, Table 6.

TABLE 12.3[4] ◆ Estimated Reduction in Average Direct Property Damage per Fire Due to Sprinklers by Property Use, 1989–1998 Structure Fires Reported to U.S. Fire Departments

Property Use	*Without Sprinklers*	*With Sprinklers*	*Percentage Reduction*
Public Assembly Properties	**$21,600**	**$6,500**	**70%**
Eating and Drinking Establishments	$17,200	$5,900	66%
Educational Properties	**$13,900**	**$4,400**	**68%**
Health Care and Correctional Facilities	**$4,700**	**$1,700**	**64%**
Health Care Facilities*	$4,000	$1,600	59%
Residential Properties	**$9,400**	**$5,400**	**42%**
One- and Two-Family Dwellings	$9,600	$7,800	19%
All Apartments	$8,500	$4,400	49%
Apartments at least 7 stories tall	$3,200	$1,800	43%
All Hotels/Motels	$13,400	$5,900	56%
Hotels/Motels at least 7 stories tall	$13,400	$4,500	67%
Dormitories and Barracks	$7,400	$4,700	36%
Stores and Office	**$24,400**	**$12,200**	**50%**
Food or Beverage Stores	$21,000	$6,500	69%
Department Stores	$36,900	$14,900	60%
Offices	$22,700	$10,100	55%

(continued)

Property Use	*Without Sprinklers*	*With Sprinklers*	*Percentage Reduction*
All General Office Buildings	$23,100	$10,800	53%
General Office Buildings at least 7 stories tall	$27,700	$13,000	53%
Manufacturing Properties	**$50,200**	**$16,700**	**67%**
Food Product Manufacturers	$66,100	$23,300	655
Textile Product Manufacturers	$23,100	$12,000	48%
Footwear/Wearing Apparel Manufacturers	$137,500	$16,500	88%
Wood Product Manufacturers	$47,700	$14,100	70%
Chemical Product Manufacturers	$60,700	$24,900	59%
Metal Product Manufacturers	$45,800	$15,000	67%
Vehicle Assembly Manufacturers	$45,400	$21,600	52%
Other Manufacturers	$39,900	$15,400	61%

*Refers to care of aged and care of sick facilities only.

These are fires reported to U.S. municipal fire departments and so exclude fire reported only to federal or state agencies or industrial fire brigades. Fire statistics do not include proportional shares of fires with sprinkler status unknown or unreported. Direct property damage is estimated to the nearest hundred dollars.

Source: National estimates based on 1989–1998 NRIRS and NFPA Survey. Kimberly D. Rohr and John R. Hall, *U.S. Experience with Sprinklers and Other Fire Extinguishing Equipment* (Quincy, MA: Fire Analysis and Research Division, National Fire Protection Association, 2005).

FIGURE 12.8 ◆ Firefighter Teaching School-age Children About Fire Safety *(Source: Kerri Donis, Deputy Fire Chief, Fresno Fire Department)*

HUMAN BEHAVIORS, VALUES, AND IMPACTS ON FIRE AND LIFE SAFETY

A review of the literature indicates that countries experiencing the best fire statistics, at least in loss, simply put more effort into fire prevention education and the incorporation of building safety technology. This is apparent in Japan and Europe, which have developed a societal awareness of fire prevention. This is lacking in the United States. There is a willingness on the part of the individual and society as a whole to accept more restrictive codes and a higher degree of personal responsibility as it relates to their building environment. Built-in fire protection systems, along with comprehensive fire prevention educational programs that address attitudes about fire, have proven to be more effective in saving lives than trying to extinguish the fires once they start. Although it is difficult to change social attitudes and the importance of individual responsibility, the fire service also understands these are significant factors in the overall fire loss. As such, our educational efforts should be designed to influence these attitudes and to promote personal responsibilities.

The United States, while possessing a wealth of knowledge related to fire protection technologies, continues to have one of the worst fire loss records in the industrialized world. Research shows that human behavior and attitude play a significant role in determining the fire loss experience of the nation and its culture. Countries with lower fire loss strategies place a higher premium on their ability to prevent fires rather than their ability to extinguish them. Our challenge for the fire service in the future will be to overcome the mind-set that tolerates a cavalier attitude toward fire safety. In developing strategies for the future, the fire service must revisit and focus on the following:

- civilian fire deaths and injuries occurring in residences targeting specific age groups by need;
- funding of fire prevention activities that are effective and realistic;
- redefining the local fire prevention needs based on the need of individual communities;
- changing attitudes about the acceptability of fires in the civilian population and the fire service industry;
- increasing the cooperation of all government and industry agencies toward fire prevention;
- involvement in the engineering and enforcement of fire protection systems; and
- understanding the cultural diversity of individuals in our communities and finding better ways to communicate fire and life safety issues to them.

INTEGRATED RISK MANAGEMENT

In Chapter 9 we spent a considerable amount of effort evaluating the current analytical approaches to public fire protection: the accreditation model, the standards of response coverage concept, national standards such as NFPA 1710 and NFPA 1720, and the Insurance Services Office (ISO) grading schedule. Each has an impact on current strategies for how we define risk for the communities we serve. The future will see a shift that focuses on fire service management, which increases the fire service's level of sophistication through the use of technology and performance-oriented outcomes. The analysis will provide a more precise view of community risk factors and, subsequently, on how resources are deployed. A combination of technology and performance-based analysis will allow a local fire official and, ultimately, the local elected official to make decisions as they relate to community risk management through increased sophistication of our application of technology for evaluation purposes. Fire service leadership and local policy makers not only will become more adept at producing the desired service outcomes but also will be held accountable for the consequences of their decisions.

Such technologies as geographical information systems (GIS), Web-based information, and the linkage of information to other disciplines will allow us to collect statistical information and develop performance data in a way that we have been unable to do in the past. The ability of computer software and processing power has not been fully utilized as an analytical tool. The result has been the use of limited knowledge of many of the fundamental questions on fire service delivery to determine the balance of resources needed to achieve the greatest impact on fire and property loss and civilian and firefighter deaths and injuries. We have already experienced a tremendous growth in the means of delivering information to the user with the World Wide Web. Research via the Internet has allowed users to access information without regard to time and distance and via the emergence of extensive online document delivery. In addition, with the use of the Internet, we have been able to utilize software and access other computers, databases, and graphics without the need to purchase additional software or increase our computing power. This will continue to evolve and will allow the fire service the opportunity to access data by which it can fully maximize the risk management concept for our communities regardless of size or complexity. As computing powers have grown, so has the availability of software that can be run on a personal computer, which was previously impossible.

Geographic information systems (GIS) have become more user-friendly and integrated into many of the software programs available to the fire service. Such programs will assist in acquiring and developing base data that can be utilized for in-depth analysis. These data typically include street information; property and tax information; census data, which would include socioeconomic, demographic, and building stock information; and other locally obtained data such as infrastructure elements, for example, fire hydrants, standpipe connections, hazardous materials, structures, and other hazards. With this information at our fingertips, we can fully utilize and assess our ability to respond and control events that are likely to occur. As GIS has been embedded in other applications, such as computer-aided dispatch systems (CAD), our ability in computing power has grown, allowing us to extract a great number of data, which will ultimately yield a clear picture of community risk and how the local fire service provider will address the risk.

TECHNOLOGY

When we think of technology, many of us often just think of computers and computer software (see Figure 12.9). In the fire service we know technology takes on many forms, from the types of equipment utilized in the field on fire response to the medical equipment utilized on a daily basis when we are called to a medical emergency. As we look to the future, one of the significant shifts already occurring in the fire service is field impact technology. There has been a noticeable change over the past 20 years, evident at fire service conferences where equipment is displayed. Years ago there was little change from one year's conference to the next as the tools of our trade and the application of technology often looked the same. But during the last decade we have seen a

FIGURE 12.9 ◆ Dispatcher at Communications Center Utilizing New Technology *(Photo by Janeen Sanders)*

significant shift in the technology applications within the fire service. Much of this has been a result of the technology transfer that has occurred from the federal government. Exponential advancements in technology have become available to the fire service.

What does this mean for the future? Maybe the best way to understand the potential impact of technology on our profession in the future would be to ask, "What if?" For example, what if we could design a smart alarm system that had sensors throughout a building that could detect heat, smoke, and toxicity, and transfer information to the incoming officers via the mobile data computer or PDA? When firefighters arrived at the building that was on fire, they could tell exactly where the fire was, how much heat was being produced, the toxicity level of the atmosphere, and where the fire was actually spreading, even before they entered the building. They could track their progress of the fire through the use of technology so that the firefighters would know where to attack the fire and could determine whether the hose placement and/or sprinkler system were being effective. Would this change the method in which we deliver service today?

What if we had devices that could actually place firefighters in a high rise without having them climb 30 flights of stairs? What if we had a vertical hovercraft that would take groups of firefighters up the exterior of the building to allow them quick and easy access without the stress and strain of having to climb many flights of stairs? Such a device could be used to rescue victims of multistory buildings where our ladders could not reach and could assist with the reconnaissance of major events. What if a sprinkler system designed to use minimal water were so cost effective that it was installed in all new construction? Therefore, because of its low cost and easy installation, it gained widespread use in the retrofit of existing occupancies. What if, through the use of technology, with the placement of portable devices that firefighters could affix to the corners of the buildings, the measurement of the vibration of the burning building could determine when the roof and/or the building were ready to collapse? What if technology such as a visual infrared device could be utilized, and this same technology could be integrated as part of the SCBA mask for every firefighter instead of the handheld device being used today? The heads-up display of information would include real-time displays of outside environmental factors and also provide the firefighter's own body temperature, heart rate, and blood pressure. All this information would be sent back to the incident commander via wireless transmission so that he or she would know exactly what the firefighter was experiencing in the building and if it would be time to evacuate and/or change the field tactics.

It is difficult to grasp fully what technology will mean for the fire service in the next 25 years. Over the course of the last two decades, how has our industry changed as a result of technology? And just think how the incorporation of electronics has engrained itself in our industry, from the computers that we use in our offices to the electronics found throughout our fire apparatus. It is truly amazing when you stop to think about it. We can now respond on a call with the use of wireless technology that is available, access information while en route, display it on multiple monitors on the back of command vehicles, and calculate the potential spread of a fire using GIS-based technology. Surely technology will be one of the catalysts to create future tipping points for the fire service, whether it is home fire sprinkler technology, smoke or heat detection, or lighter and more versatile equipment for our firefighters. Technology will play an important part in shaping the future of the fire service.

FIREFIGHTER SAFETY

On March 10 and 11, 2004, more than 200 individuals assembled in Tampa, Florida, to focus on the troubling question of how to prevent line-of-duty firefighter deaths. Every year approximately 100 firefighters lose their lives in the line of duty in the United States, which equates to about 1 every 80 hours (see Figure 12.10). The inaugural National Firefighter Life Safety Summit convened to bring fire service leadership together to focus its attention on one of the most critical issues that faces the fire service. The last historic fire service gathering to have an impact on the line-of-duty

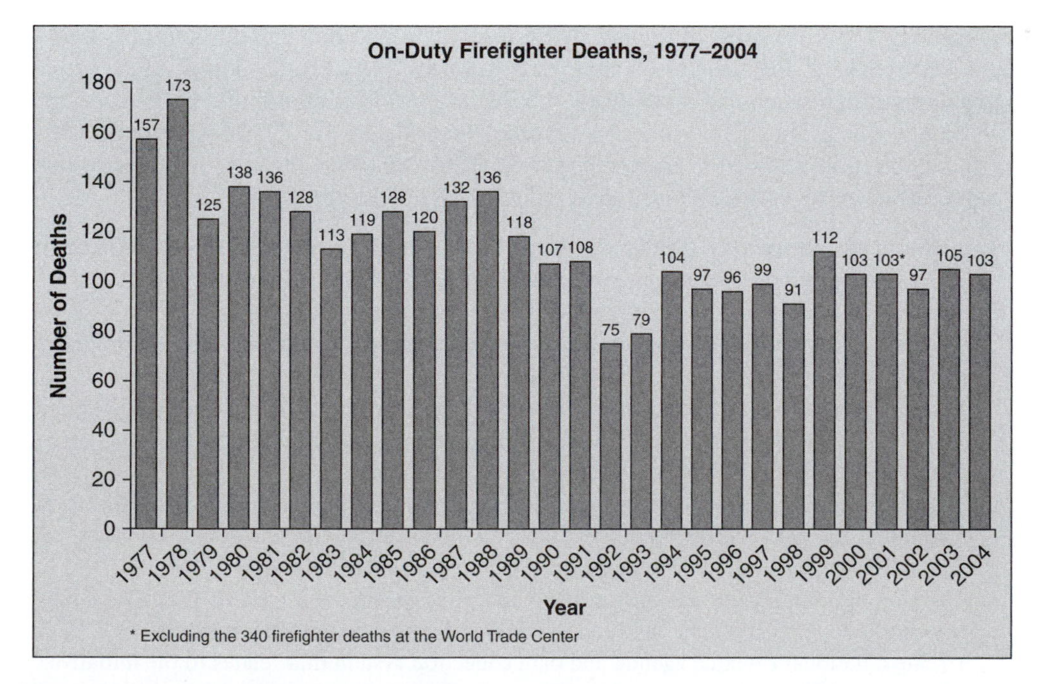

FIGURE 12.10 ◆ On-Duty Firefighter Deaths, 1977–2004

FIGURE 12.11 ◆ Everyone Goes Home *(Courtesy of the National Fallen Firefighters Foundation)*

death reduction was the original *America Burning* panel in the 1970s. However, since 1990, 99 firefighters on average have died each year in the line of duty. This does not include the 343 firefighters who were lost on September 11, 2001.

Of paramount importance in the future will be the ability to institutionalize firefighter safety in the fire industry. As Bill Manning wrote, "After 15 years of well-intentioned attempts to institutionalize firefighter safety, broad-base prescriptive measures have failed to push us past the 100-a-year threshold. It is not inappropriate to parallel the fire industry's situation with the aviation industry, whose leadership several years ago realized that the world's best regulation standards, research, and technology were not preventing and by themselves could not prevent aviation disasters. They came to the conclusion that performance- and attitude-based safety problems demand performance- and attitude-based safety solutions and that at the core are is a people thing. The "Everyone Goes Home" (see Figure 12.11) life-safety initiative is by far the most promising nationally based safety program that the fire service has rolled out in many years. The initiative includes 16 objectives.

1. Define and advocate the need for a cultural change within the fire service relating to safety, incorporating leadership, management, supervision, accountability, and personal responsibility.
2. Enhance the personal and organizational accountability for health and safety throughout the fire service.
3. Focus greater attention on the integration of risk management with incident management at all levels, including strategic, tactical, and planning responsibilities.
4. All firefighters must be empowered to stop unsafe practices.
5. Develop and implement national standards for training, qualifications, and certification (including regular recertification) that are equally applicable to all firefighters based on the duties they are expected to perform.
6. Develop and implement national medical and physical fitness standards that are equally applicable to all firefighters, based on the duties they are expected to perform.
7. Create a national research agenda and data collection system that relates to the initiatives.
8. Utilize available technology wherever it can produce higher levels of health and safety.

9. Thoroughly investigate all firefighter fatalities, injuries, and near misses.
10. Grant programs should support the implementation of safe practices and/or mandate safe practices as an eligibility requirement.
11. National standards for emergency response policies and procedures should be developed and championed.
12. National protocols for response to violent incidents should be developed and championed.
13. Firefighters and their families must have access to counseling and psychological support.
14. Public education must receive more resources and be championed as a critical fire and life-safety program.
15. Advocacy must be strengthened for the enforcement of codes and the installation of home fire sprinklers.
16. Safety must be a primary consideration in the design of apparatus and equipment.

Although these initiatives are not necessarily new, they are based on historic information and a set of fundamental truths. The fire service of the future will no longer accept that dying on the job is a normal way of doing business. Yes, the work is inherently dangerous, and no, the death toll for firefighters may never be zero. But firefighters are dying unnecessarily today and that must stop. The number of firefighters' injuries is also a significant concern.

Every individual in the fire service has to accept personal responsibility for his or her health and safety and the health and safety of his or her colleagues. Leaders and members of fire departments and fire service organizations must be accountable for themselves and for others. The work is inherently dangerous, but we must manage risks to function safely within an unsafe environment. Risk management will play a key role in reducing deaths. Risk management means identifying situations where predictable risks are likely to be encountered and then making decisions that will reduce, eliminate, or avoid them. While we are willing to sacrifice our lives, it should not be taken as an excuse to take unnecessary risks. Firefighters should not be losing their lives while trying to save property that is already lost.

The future will require mandatory training and qualification standards based on what duties individuals are expected to perform, no matter what their status is within the fire service or the type of organization. Standards must be clearly defined. A basic system of professional qualifications standards already exists, but applicability depends on too many different factors to be effective. Qualification and certification standards also must require continuing education, refresher courses, or some other training component and not be a certification for life that requires no additional action throughout a firefighter's time in service.

The future will require mandatory physical fitness standards. An increased emphasis on health and wellness is essential to reduce the number of deaths from heart attacks and other cardiovascular causes. Statistics show that this could be one area of significant reductions in loss of life, particularly in volunteer fire departments.[5] We can no longer accept placing out-of-shape firefighters with high medical risks in very physically demanding events and hope nothing catastrophic happens. We have enough statistics and history to know better.

NATIONAL MUTUAL-AID SYSTEM

When I was the president of the IAFC in the fall of 2002, the White House Homeland Security Office contacted us to provide a briefing on mutual aid. The discussion centered on the National Response Plan and the need to construct a system by which fire resources could be moved nationally in the event of a major catastrophe. At the time, we sent several chief officers from Illinois, California, Missouri, and Florida to provide the White House with an overview of how mutual aid works effectively on a local and statewide basis and to provide them with several templates of effective statewide systems by which to begin design of a national system. The test to the national preparedness was Hurricane Katrina, which caused most of those inside the fire service to recognize that such a system is not in place. As the Department of Homeland Security considers broad changes to the national response plan, including strengthening emergency operations, planning, and citizen protection capability, the National Response Plan has to be at the forefront of the discussions.

In September 2005, the president of the IAFC (2005–2006), Bill Killen, formed the national fire mutual-aid systems task force to develop a seamless process for interstate mobilization, integration, and utilization of fire service resources and assets in the event of a major disaster. This critical step will be the cornerstone to how the fire service will likely respond in the future to major disasters. Best practices that already exist in the fire service have been identified and will be studied as part of the effort to develop a national mutual-aid system. These studies include the California Office of Emergency Services, the Illinois Mutual Aid Box Alarm System, the Florida Fire Chiefs Association Mutual Aid Process, and the Kansas City Metropolitan Regional Response Compact. The principles that will guide this process in the future are built upon the following:

- Development of the system is predicated on a consensus process.
- The system design is not predetermined by any single system currently in existence.
- The best practices of existing systems will be identified and incorporated into the system design.
- The system must be adapted to both intrastate and interstate relationships.
- The system should be capable of supporting existing intrastate programs.
- The system must be capable of responding from an all-hazards perspective.
- The system is compatible with ongoing federal efforts with respect to NIMS (National Incident Management System) and the National Response Plan and is identified by the committee working on this project.

There are several attributes the final system should also display. They include acceptance by the fire service community and key affiliated allied organizations; acceptance from federal and state stakeholders on adopting, implementing, and utilizing the plan; the ability to enable state fire chiefs associations to adopt the plan and implement it for interstate systems request; participation that is voluntary and built upon existing mutual-aid programs; development by industry experts and petitioners; and the capability to expand the model to allied and affiliated organizations that have an emergency response and provide a supporting role in a disaster.

As we have seen with several other initiatives since September 11, once this national plan is complete, it may take several years for full implementation within the U.S. federal system of response. However, it will change the way in which emergency services will collectively respond to disasters in the not too distant future.

FIRE SERVICE RESEARCH

Fire service research will have a significant impact on what we do as a profession over the course of the next century. Laboratory studies of fire science, fire safety, engineering, building materials, computer-integrated construction practices, and structural, mechanical, and environmental engineering will all play a vital role in the future of the fire service. One agency that has been at the forefront of the fundamental fire science is the National Institute of Standards and Technology (**NIST**), which is an agency of the Commerce Department Technology Administration.

NIST
U.S. National Institute of Standards and Technology.

NIST was founded in 1901 as the nation's first federal physical science lab, and its mission is to promote U.S. innovative industrial competitiveness by advancing measurement science of standards and technologies to enhance economics, security, and the quality of life for Americans. Today, NIST operates eight laboratories from Gaithersburg, Maryland, and Boulder, Colorado, to conduct research on a wide variety of physical and engineering sciences. One of the laboratories is the Building of Fire Research Lab (BFRL).

BFRL works to improve quality and productivity in U.S. construction and to reduce human and economic loss due to fire, earthquakes, wind, and other hazards. The BFRL studies building materials, computer-integrated construction practices, fire science, fire safety engineering, and structural, mechanical, and environmental engineering. Products of the research conducted at the BFRL include measurements, test methods, performance criteria, and technical data that support innovation by industry and are incorporated into building and fire safety standards and codes. The primary research areas include high-performance construction materials and systems that enable scientific and technology-based innovation to modernize and enhance the performance of construction materials and systems. Fire loss reduction enables engineered fire safety for people, products, and facilities and enhanced firefighter effectiveness with a goal to reduce firefighter fatalities by 50 percent. Enhanced building performance provides a means to assure buildings work better through their useful lives. The homeland security component uses the lessons learned from the World Trade Center disaster to protect people and property better, to enhance the safety of fire and emergency responders, and to restore public confidence in the safety of high-rise buildings nationwide.

The BFRL is at the center of corporative research as well. BFRL provides outreach to many other private and public organizations on projects of mutual interest through cooperative research and development agreements. These agreements with other well-known fire research facilities such as Worcester Polytechnic Institute and Maryland Fire and Rescue Institute include the hosting of guest researchers, postdoctoral candidates, and collaboration with fire-based research worldwide.

An additional area of research has been promulgated through the funding of the Fire Act, which allows nonprofit associations to submit for noncompetitive grants to facilitate research into firefighter safety, training, and wellness, to name just a few of the topical areas. This has allowed associations such as the IAFC, IAFF, CPSE, ICMA, and many others to submit for grants for specified areas of research. Of note, the results of these research projects as they are completed and shared have stimulated new ideas and many collaborative efforts among the major associations, universities, and private vendors.

A new role that BFRL has undertaken resulted from the events of September 11. Under the National Construction Safety Team Act (NCST) signed into law in October 2002, NIST is authorized to investigate major building failures in the United States. The BFRL has been charged with this effort. The NIST investigations will establish the technical causes of building failures and evaluate the technical aspect of emergency response and evacuation procedures in the wake of such failures. The goal is to recommend improvements to the way in which buildings are designed, constructed, maintained, and used. Two major safety investigations of note include the

September 11, 2001, fire and collapse of the World Trade Center and the February 20, 2003, fire at the Station Night Club in West Warwick, Rhode Island. Such investigations will have direct impact on the fire service, as lessons learned from them can be immediately applied to related codes and standards in respect to building construction and response deployment.

It is evident that since September 11 the synergy of research created through NIST and BFRL has increased. This will promote only a greater collaborative worldwide research effort into fire service issues in areas such as enhanced building performance, fire loss reduction, and the use of high-performance construction materials and systems in homeland security. What will come of such research is yet to be known, but we can be sure that the impacts will have a positive effect on our ability as an industry to reduce the severity of property loss and life loss in this country.

NATIONAL FIRE SERVICE DOCTRINE

In 1989 the United States Marine Corps published a document that became the Corps doctrine, and it continues to be in use today. Titled *War Fighting,* it is presented in an easy-to-read format. All Marines are required to read the document and take its message to heart, as it has become a bedrock document for all operations of the Marine Corps.

The American fire service, as the United States Marine Corps was in 1989, is at a historic crossroads in its history. Our legacy of community-based fire protection has evolved to an all-hazards prevention and response agency. The historic mission of fighting fire has changed, and we now find ourselves on the frontlines of a worldwide war on terror, where the American homeland has become part of the battleground strategy. The federal government has become increasingly involved in creating policy and controlling the manner in which fire departments function. For the first time, the government is providing funding, and with the funding comes a new set of expectations for training, administration, and operational performance. American firefighters are now being deployed to all corners of the world, both friendly and hostile, and in doing so have served not only as first responders but also as diplomats for the American people.

One of the difficulties for the American fire service is the lack of a common doctrine for the over 30,000 existing U.S. fire agencies. While the diversity of these organizations and the communities in which they serve is one of our strengths, it also creates one of our greatest weaknesses, because there is no one doctrine that we all operate from. While there are a variety of standards that the fire service may choose to utilize and a number of state and local ordinances that must be followed and administered, a universal common view of the nature of the fire problem in the United States among its firefighting personnel is still lacking. The nature of the fire problem, the theory of firefighting, how we prepare for fire combat, our decision making, and command methodologies are not universal. The creation of a national fire service doctrine would guide every fire department and every firefighter in this country toward a philosophy of service and command and is a bold initiative. Yet, its time has come. Fire officers see pieces of the doctrine in many aspects of the fire industry today. As the fire service continues to evolve, one of these pieces to help drive and shape the future of the American fire service will be the creation of a national fire service doctrine with all the allied organizations representing the diverse stakeholders that make up the fire industry.

The fire service doctrine would not consist of procedures to be acquired in specific situations so much as it would set forth general guidance that requires judgment in its application. Therefore, an authoritative doctrine is not prescriptive; yet, it provides the framework by which, as a fire service, we are charged in ensuring the safety of our communities, and our fire service leadership must be true experts in the technical aspects of fire protection and emergency response. More than ever there is a critical need to take control of our destiny. To provide unification in describing who and what we are, we must define and live by a doctrine that outlines our core values, philosophies, and practices. A single industry doctrine would be a powerful force in shaping our future and would be a tipping point for the fire service in the twenty-first century.

THE FUTURE IS UP TO YOU

Many who start their fire service career are fortunate to have many good mentors and people willing to invest time and energy into seeing that these firefighters become all that they wished to be. Not everyone in the fire service has the opportunity to have this level of support. However, this does not preclude one from achieving one's own personal goals and in making a difference, not only within our profession but also for the firefighters whom we work with on a daily basis, and the community and residents whom we are called upon to serve many times a day.

During recruit academy graduation when the families are present to pin the badges on our new firefighters, I share with the families and friends in the audience that today is the start of great opportunity. An opportunity to enter a profession in which even the most disgruntled employees would say they love their job. You have the opportunity to be part of a very select group of individuals, each of whom knows how difficult it was to be chosen for this profession. Firefighters have the opportunity every day when they come to work, put on their uniforms, and pin on the badge to make a difference in someone's life. It could be to make the difference by the actions you take when an individual is having a significant medical problem. You could make a difference by the actions you take when responding to and combating fire, because of your knowledge, experience, and ability to do the job. You can make a difference by taking the time to share of yourself with the community through public education or just stopping to say hi and lend a smile to a child at the grocery store or simply when driving down the street.

Ours is a unique profession. As we anticipate what the future may be, the answer lies within each of you. You have an opportunity, as each generation has, to shape and influence the course and direction of our industry. You have the opportunity through your involvement with your department locally, regionally, and at the state level to impact what your department looks like, how you respond from a regional perspective, and what your state does in respect to the industry. It all comes back to your willingness to invest your time and your expertise to help shape your own future. So the destiny of our profession and your own destiny lie with you. Depending upon what you choose to do, it will dictate, in large part, the future of what the fire service will become. We truly are the architects of our destiny. With that comes a tremendous responsibility to prepare, to engage, and to make a commitment to make a difference each and every day when we come to work, not only for the citizens we serve and the firefighters we work with but also for the fire service as a profession.

ADVICE FROM THE EXPERTS

1. What do you see as the top five issues that will shape the future of the fire service over the next decade?
2. If you could give just one piece of advice to young fire officers, what would it be?

MICHAEL CHIARAMONTE

Chief Fire Inspector Retired, Lynbrook Fire Department (New York): (1) Generational management, disaster management, regionalization, consolidation, and public illness prevention education. (2) Study, observe, and listen to as much as you can. It may save your life or the life of a brother or sister someday.

BETT CLARK

Former Fire Chief, Bernalillo County Fire and Rescue (New Mexico): (1) Managing resources to meet community expectations within allotted budget. Defining the expectations of the community and the ability to meet those expectations within budget. Keeping fire service members safe in a changing world. Setting the professional standards for all fire service departments (professional does not necessarily mean paid). Changing the mentality of the fire service from reactive to proactive. (2) Start early and never forget the reason you joined the fire service.

KELVIN COCHRAN

Fire Chief, Atlanta Fire Department (Georgia): (1) *Leadership development*—Strengthen Executive Fire Officer and Chief Fire Officer Designation as national recognized criteria for developing, selecting, and appointing fire service leaders at the local, state, and federal level. Assimilate and institutionalize the IAFC *Officer Development Handbook* as a career development standard. *Prevention of line of duty deaths and injuries*—Continue to pursue changes necessary to reduce the trend of one hundred plus deaths and one hundred thousand injuries a year. *Improving the National Response Plan*—Developing a national mutual-aid system for fire service response to disasters; institutionalizing the NIMS; enhancing response to human-made and natural disasters; obtaining reliable radio communications for managing disasters. *Recruitment and retention*—Community demographics are changing, volunteers are increasingly more difficult to come by, and retirements are increasing yearly. National-level research to support local-level initiatives will be essential in the future. *Organizational change*—Fire service leaders must possess the leadership and management skills to direct the revolutionary changes occurring in the fire service. External and internal forces will continue to transform fire departments. Those who work to predict the future and plan for it will make the difference between struggling fire departments, good fire departments, and great fire departments. (2) Be a servant and seek to service at the highest level. Pursue the things that increase a fire officer's value. You will increase the quality and frequency of opportunities to contribute. Give the very best you have to give where you are. Though it is biblically based, it is absoulutely true as it relates to career development, advancement, and achievement: *The more lowly your service to others, the greater you are; to be the greatest be a servant; for those who think themselves great shall be disappointed and humbled, and those who humble themselves shall be promoted.*

CLIFF JONES

Fire Chief, Tempe Fire Department (Arizona): (1) Disaster management and mitigation demands and strategies; medical service demands from the aging baby boomer generation; financial constraints; leadership and management training requirements due to massive employee turnover; balancing staffing and deployment levels and labor costs and demands. (2) Take firefighter safety seriously and educate members under you about safety and enforce your expectations. Don't ever find yourself in a situation in which one of your members is seriously injured or worse, and it is determined that not adhering to a basic safety practice, such as wearing seat belts when fire apparatus is moving, probably would have saved a life.

BILL KILLEN

Retired Fire Chief and IAFC President 2005–2006: (1) Threats and acts of terrorism, changing ethnicity of America, regionalization of fire and emergency services, decline in numbers of volunteers in emergency services, national incident management, and national mutual-aid system. (2) Always remember to keep family first! Focus on your fire service goals and objectives, and the career successes will follow as long as you realize that fire and emergency services are a TEAM PROCESS.

KEVIN KING

U.S. Marine Corps Fire Service: (1) Applying effective information technology to the fire service; efficient use of personnel with an effort at reducing costs; improving safety and fitness of our personnel; implementing interoperable communications; mobilizing for large, multijurisdictional incidents. (2) Do not ever forget you are in this occupation to help people. Ours is a noble calling because we are there to help others in their time of need. It is not the money, the schedule, the trucks, or the camaraderie; it is the desire to serve and protect others. (These are opinions of Mr. King and do not reflect official policy of the Department of Defense or the United States Marine Corps.)

MARK LIGHT

Executive Director, International Association of Fire Chiefs: (1) Emergency management, terrorism response to major attacks in the homeland, development of future fire officers, technological advances in firefighter safety, and obtaining the revenue to stay on the cutting edge. (2) Take every opportunity afforded to you by your department, your officers, and your fire chief. The word "no" should be stricken from your vocabulary when it comes to things that can serve to develop you professionally. Take every experience, good and bad, and look at what you learned from it. Then expand on that. Over 10 years, you will have a tremendous amount of learning and a great book of things to guide you for the rest of your career. And finally, never, never, never give up.

LORI MOORE-MERRELL

Assistant to the General President, International Association of Fire Fighters: (1) Politics (federal, state, and local): fire service leaders and employees are public servants and thereby directly connected to politics. We must be involved in electing fire-service-friendly politicians and then be willing to work to educate these people on the needs of the fire service. Legal issues: tort liability, negligence law, municipalities losing sovereign immunity, stupid acts of leaders and employees will continue to steal precious time and resources from the real needs of the fire service. Terrorism threats: this is a sexy issue and will continue to drive funding and training. Funding (or lack of): will continue to limit necessary personnel, training, and equipment resources. Political correctness: tends to blur the lines between right and wrong. We must teach respect and maintain ethical behavior and the integrity of the fire service as a whole. (2) It is difficult to give only one piece of advice to someone in such an important position, so I will make a short list.

- Take your position seriously. You are responsible for the lives of others.
- Never stop learning. Stay proficient in your craft.
- Be assertive as a leader. Leave your department better than you found it.
- Challenge yourself and your comfort zone.
- Communicate and involve labor in all aspects of the department. You will find that the union can be your biggest ally and strongest advocate.
- Maintain your character including honor, pride, and integrity.

JAY REARDON

Fire Chief, Retired, Northbrook Fire Department (Illinois), currently President of the Mutual Aid Box Alarm System (MABAS): (1) Consensus and agreement on national doctrine for the fire service; all risk capabilities; technology and data based/metric decision making; a split between qualified and nonqualified chief fire officers; revenues and politically acceptable fire service standards; reinstill the sense of discipline to perform in today's world. (2) The only thing stopping you from being whatever it is you want to be is your ability and willingness. Set your career path on what you want for yourself and your family. Then program your brain to meet the challenge of willingness and your educational experience plan to meet the needed ability. Remember, no one will invest in you unless you invest in yourself. Finally, accept responsibility for yourself—it is no one else's fault for you not reaching your goals. Remember, you have ultimate control for your future—your destiny—you can simply quit!

GARY W. SMITH

Retired Fire Chief Aptos–La Salva Fire Protection District (California): (1) (a) My biggest concern is truly to understand and work with the community to understand the needs and desires. The customer connection between the fire department and the community occurs not only with our day-to-day visits by fire prevention and suppression service to schools, businesses, and homes. Community connection requires a much deeper value system by the fire organization, one that spins from leaders willing to join with other community leaders to address the entire picture of challenges, from housing to business challenges and from education to crime-related social concerns. We must engage as fellow leaders within the community and get to know each other, play off each other's talents, and work to make the entire community a better, safer place to live. (b) Organize and use CERT logic to engage neighborhood and personal preparedness. It is also the tool that could bring the fire service closer to its community. (c) The fire service has to come together on a national basis to deliver a unified message. We need to become much more involved as leaders to be considered in dealing with national public safety challenges. Our reputation and level of engagement has not kept up with the times. Even the fire plan for the White House puts the fire response second to the secret service and law enforcement response; the fire department may never be dispatched, police will take care of it! The same will occur with hazmat and rescue services; the law enforcement will go where the money and need show themselves, while the fire service stands by and whines. The comparison of the focus and viability of the fire service when compared to law enforcement is much the same as the strength and reputation of FEMA when compared to the Department of Homeland Security. One is developed and respected and the other is a helpless bureaucracy. Get it together and exhibit some national fire service leadership; stop infighting and develop a common vision that supports development of higher-end national involvement that will also improve local fire service respect and involvement (d) REALLY begin to plan community response with some reality of what local resources can afford; utilize the leadership, strength, and integrity to use the fire and EMS mitigations to reduce risk rather that to use traditional and ego-based logic by adding more engine and truck companies. Be true to the customer on the value of advanced emergency medical care, using the basic life support and the use of citizen-based defibrillators as a foundation for the emergency medical response system. In other words, use our resources wisely; stop quibbling about "more for less." If I need more, I will convince those who pay the bill and develop the way to figure out how to effectively "do more with more." (e) The only way we can succeed is to learn how to lead and succeed. It is simple: vision, knowledge, network and doing, always evaluating, and forecasting the future so that the vision changes to meet the next set of challenges. (2) Stay positive, engaged, and in sync with the times. Build yourself and your team to the talents and passions that put "fire in your belly" for doing!

STEVE WESTERMANN

Fire Chief, Central Jackson County Fire Protection District (Blue Springs, Missouri), and IAFC President 2007–2008: (1) Technology, terrorism tactics, volunteerism (or lack of it), economy, labor relations. (2) Be open to change. If it means changing yourself when appropriate, so be it (that doesn't necessarily mean to give up your value system). Learn to be flexible and adaptive when the environment around you is changing.

THOMAS WIECZOREK

Executive Director, Center for Public Safety Excellence, and Retired City Manager: (1) Integrated Risk Management Planning, GIS, emergency management, providing access to medical care, asset management. (2) Learn, learn, learn, and do not ever stop. At least once each year, go to a program offered by another discipline (for instance, police, public works, or water). As a city manager, I found that I often learned far more at the other conferences than I did at those sponsored by my own "group." I do not want to sound like I did not support those other organizations; it is just that we all have the tendency to be "comfortable." When I attended the other conferences, I found my beliefs and morals challenged. I believe it resulted in a more wide-eyed approach to problems, issues, and personnel.

JANET WILMOTH

Editorial Director, *Fire Chief* magazine: (1) Consolidation of departments, one more national disaster, standardization of SOPs, and emergency managers. (2) Keep your word. Trust is priceless.

FRED WINDISCH

Fire Chief, Ponderosa Volunteer Fire Department (Houston, Texas): (1) Volunteer recruitment, benefit reductions for career positions, less reliance on ISO ratings (cost reductions/reality checks), education of chief officers versus training, the explosion of combination departments. (2) Be a good person, love your fellow man, go to school, get that degree (even an associate's) because your learning capabilities will be with you forever.

References

1. Malcolm Gladwell, *The Tipping Point: How Little Things Can Make a Big Difference* (Boston: Little, Brown, 2000).
2. Michael J. Karter, Jr., *U.S. Fire Department Profile Through 2001* (Quincy, MA: National Fire Protection Association, 2002).
3. *EMS Agenda for the Future: Implementation Guide*, www.nhtsa.dot.gov/people/injury/ems/agenda/execsum.html
4. *USFA: Everyone Goes Home: The Firefighter Life Safety Summit*, www.usfa.fema.gov/about/administrator / 04-everyone-goes-home.html
5. Ibid.

Measuring Customer Satisfaction

One of the first places to start is to gain an understanding of the ways in which your organization is focused on meeting customer expectations and providing quality service. Some common characteristics correlate with the success of meeting customer needs. For each characteristic below, please rate to what extent you believe the statement is true about your own organization using the following scale (tenths can be used).

- 1—Not at all
- 2—To a small extent
- 3—To a moderate extent
- 4—To a great extent
- 5—All the time

A focus on the following seven key areas will help you evaluate your success in meeting customer needs.

Continuous Improvement

1. We invest in the development of innovation. ________
2. We take quick action to solve problems when they are identified. ________
3. We work continuously to improve the services we deliver. ________
4. We study the best practices of other fire agencies to learn how to do things better within our own organization. ________
5. We use cross-functional groups to reach shared organizational goals. ________

Total ________

22–25 = Excellent
19–21 = Good
16–18 = Fair
Less than 16 = Need help

Vision

1. Our customers' needs take precedence over serving our own internal needs. ________
2. Our aim is to do things right the first time. ________

(continued)

3. Our senior management team members demonstrate with their actions that customer satisfaction is important. ________
4. Our goal is to exceed the expectation of our customers in the things that matter most to them. ________
5. Our organization is totally committed to the concepts and ideals of good quality. ________
6. Customer focus is a major factor in determining who gets ahead within our organization. ________

Total ________

22–25 = Excellent
19–21 = Good
16–18 = Fair
Less than 16 = Need help

Eliminating Customer Problems

1. We look for ways to eliminate internal procedures and processes that do not create additional value for our customers. ________
2. We regularly ask the people we serve to give us feedback about our performance. ________
3. We have a system in place to monitor customer complaints. ________
4. We analyze customer complaints to identify problems. ________

Total ________

22–25 = Excellent
19–21 = Good
16–18 = Fair
Less than 16 = Need help

Communicating with Our Customers

1. Our customer's expectations of our organization are clearly understood. ________
2. Our senior managers clearly understand customer requirements. ________
3. Our senior and office staff has frequent contact with our customer base. ________
4. We know how our customers define quality. ________

Total ________

22–25 = Excellent
19–21 = Good
16–18 = Fair
Less than 16 = Need help

Customer Service

1. We try to resolve all customer complaints. ________
2. We make it easy for customers to complain to us about the services we deliver. ________

3. Our employees are encouraged to go above and beyond to serve our customers well. ________

Total ________

22–25 = Excellent
19–21 = Good
16–18 = Fair
Less than 16 = Need help

Empowering Our People

1. Employees are provided with the resources they need to do their job well. ________
2. Employees at all levels of the organization understand our services and how they are delivered. ________
3. Employees are empowered to use their judgment within all levels of the organization when decisive action is needed to make things right for a customer. ________
4. Employees at all levels are involved in making decisions about the future of the organization. ________
5. Employees are cross trained so that they can fill in a variety of roles within the organization. ________
6. Employees feel they are involved in the continuous improvement process. ________

Total ________

22–25 = Excellent
19–21 = Good
16–18 = Fair
Less than 16 = Need help

Customer Alignment

1. We strive to be a leader in our industry. ________
2. We know what attributes of our services our customers value the most. ________
3. We are engaged in consultative and partnership roles with our customers. ________

Total ________

22–25 = Excellent
19–21 = Good
16–18 = Fair
Less than 16 = Need help

Survey Total ________

140–155 = Excellent
124–139 = Good
108–123 = Fair
Less than 108 = Need help

QUALITY AND CONTINUOUS IMPROVEMENT CHECKUP

How is your quality effort measuring up? Are the following quality processes implemented currently in your organization? Place a check mark in the box next to the items being practiced.

- ☐ 1. Employees are empowered to act if quality standards are compromised.
- ☐ 2. Cooperation is emphasized within groups and between functional units.
- ☐ 3. Teams are encouraged to initiate projects they think are important.
- ☐ 4. Team members and employees are encouraged to participate in decision making.
- ☐ 5. Mistakes are treated as opportunities to improve and learn.
- ☐ 6. Credit is visibly given for the success of a project to the teams doing the work.
- ☐ 7. Teams are organized appropriately; overlaps and gaps in responsibilities are minimal.
- ☐ 8. Teams have appropriate systems and processes in place to help groups accomplish shared responsibilities.
- ☐ 9. Teams conduct business with minimal red tape.
- ☐ 10. Regular team meetings are held to share information and solve problems.
- ☐ 11. Team meetings are productive and do not waste the time of team members.
- ☐ 12. A training plan is in place that supports employees' training and development needs.
- ☐ 13. Employees and work teams are encouraged to sign up for quality training.
- ☐ 14. Management supports employees in using new quality tools.
- ☐ 15. Managers supply useful on-the-job coaching to help employees use new quality tools.
- ☐ 16. Problem solving in the work units relies on a team approach.
- ☐ 17. A useful problem-solving model/approach is used when teams or groups are working on problems.
- ☐ 18. Managers work with groups and teams to create processes that produce quality rather than check for quality later.
- ☐ 19. Work processes are analyzed and improved based on customer requirements.
- ☐ 20. Improvements are made proactively, not simply because a system or process threatens to collapse.
- ☐ 21. Root causes to problems are sought out.
- ☐ 22. Analytical data is used to solve problems.
- ☐ 23. Simple statistical tools are used to achieve fact-based decision making.
- ☐ 24. Measurements are used to monitor and change work procedures.

PROGRAM EVALUATION

One of the most important aspects of continuous quality improvement is the evaluation of the services provided to our customers. Included below are samples of quality of service surveys developed and used by fire agencies to survey their customer service provided.

EMERGENCY RESPONSES FOR A VARIETY OF SERVICES PROVIDED BY FIRE AGENCIES

The department will send a cover letter and a Quality of Service Survey card to recipients of the department's service based upon the following criteria:

> Every fifth emergency call excluding fatalities or activated fire alarms and every emergency call in which there was a fire and the fire loss was in excess of $500.

The cover letter and the survey card are normally sent within seven calendar days following the date of service. Once returned, staff would be responsible for reviewing all cards received and implementing any corrective actions as needed. If additional follow-up with the individual completing the card is requested, contact would be made immediately upon receipt of the card by a staff member. Tabulation of all responses received within a particular month would be included in the department's monthly report and subsequently in the department's annual report.

FIRE PREVENTION DEPARTMENT SERVICES

The department will send a cover letter and a Quality of Service Survey card to all customers who received a plan review and to every fifth customer who received a construction inspection and an annual or semiannual fire inspection. Follow-up will be as previously indicated.

Sample Cover Letter Accompanying All Surveys Mailed OR Distributed by Some Other Means

ABC Fire Department
123 Anywhere Street
ABCville, AB 12345

January 1, 2008

Dear Customer:

Recently, you had to use the services of the ABC Fire Department. In an effort to continually provide the best possible cost-effective service, I would like to know how well you feel the service was provided. I would appreciate a few minutes of your time to complete the enclosed Quality of Service Survey and mail it back to me for review. Space has been provided on the card to indicate a desire on your part to discuss the service received further. By marking this space, you will be contacted by me or another member of my command staff within several days of receiving your survey. Of course, if you would rather submit additional written information to my attention, please send it to the address on the survey and include the number written on the top of that card in all of your written correspondence. This number will permit cross-reference with our department records.

Your feedback, positive or negative, is important to the operations of the ABC Fire Department. I look forward to your reply. If you should have any questions or wish to contact me directly, please feel free to call me at 123-4567 during normal business hours.

Again, thank you for your assistance in this matter.

Sincerely,

Fire Chief Smith

Enclosure

Fire Suppression Customer Satisfaction Survey

"Quality is meeting our customers' needs and striving to exceed them."

	Outstanding	*Excellent*	*Average*	*Fair*	*Poor*
The 9-1-1 call was handled in a prompt, courteous, and competent manner.	☐	☐	☐	☐	☐
The firefighters' response time to the emergency was prompt.	☐	☐	☐	☐	☐
The firefighters' actions helped reduce property damage.	☐	☐	☐	☐	☐
The firefighters acted in a concerned, caring, and professional manner.	☐	☐	☐	☐	☐
The assistance provided after the fire.	☐	☐	☐	☐	☐
The overall experience with the services provided.	☐	☐	☐	☐	☐

	Yes	No
1. Did you have smoke detectors in your home before you called 9-1-1?	☐	☐
2. Do you have smoke detectors in your home at this time?	☐	☐
3. Was 9-1-1 called immediately upon discovering the possibility of a fire?	☐	☐
4. Was an attempt to extinguish the fire made before the arrival of firefighters?	☐	☐
5. Would a follow-up phone call from the fire department have been of assistance to you after the fire?	☐	☐

Do you have any suggestions about how we could improve services to you or any additional comments?

Alarm ____________ Date / / Station/Shift

Courtesy of Sarasota County Emergency Services

Emergency Medical Call Quality of Service Survey

ABC Fire Department
123 Anywhere Street
ABCville, AB 12345

Please help the ABC Fire Department to serve you better by completing this survey and returning it to the address above. Thank you!

Date: ____________ Run Number: ____________________

On a scale of 1 to 5, with 1 being poor and 5 being excellent, please rate the following aspects of your fire service experience.

1. Was the 9-1-1 call promptly answered?	1	2	3	4	5	N/A
2. Was the dispatcher efficient and courteous?	1	2	3	4	5	N/A
3. Was the response time of the fire department prompt?	1	2	3	4	5	N/A
4. Were the actions taken by the fire department satisfactory?	1	2	3	4	5	N/A
5. Were the firefighters/paramedics efficient?	1	2	3	4	5	N/A

6. Were the firefighters/paramedics courteous?	1	2	3	4	5	N/A
7. Were the firefighters/paramedics compassionate?	1	2	3	4	5	N/A
8. Were the fire department actions thoroughly explained to you?	1	2	3	4	5	N/A
9. Were all questions answered completely?	1	2	3	4	5	N/A
10. Was the overall level of service acceptable?	1	2	3	4	5	N/A

Name: ____________ Telephone No.: __________________

Please call to discuss this information in more detail. ______________________

Cardiac Customer Satisfaction Survey

"Quality is meeting our customers' needs and striving to exceed them."

Please circle the number of minutes you experienced chest pain before 9-1-1 was called.

< 15 45 75 2 hours > 4 hours

30 60 90 3 hours

Please circle the severity of your chest pain when paramedics first arrived.

0 1 2 3 4 5 6 7 8 9 10

None Moderate Pain Severe

Please circle the severity of your chest pain when paramedics delivered you to the emergency room.

0 1 2 3 4 5 6 7 8 9 10

None Moderate Pain Severe

	Outstanding	***Excellent***	***Average***	***Fair***	***Poor***
The 9-1-1 call was handled in a prompt, courteous, and competent manner.	☐	☐	☐	☐	☐
The 9-1-1 instructions given prior to arrival of the paramedics.	☐	☐	☐	☐	☐
The paramedic crew acted in a concerned and caring manner.	☐	☐	☐	☐	☐
The paramedic crew clearly explained the procedures performed.	☐	☐	☐	☐	☐
The paramedic crew and equipment presented in a professional manner.	☐	☐	☐	☐	☐
The overall quality of the care provided.	☐	☐	☐	☐	☐
The overall experience.	☐	☐	☐	☐	☐

Please comment on your overall experience with our service.

Alarm ______________ Date / / Station/Shift

Courtesy of Sarasota County Emergency Services wpdocs/m/Cardiac

Orthopedic Customer Satisfaction Survey

"Quality is meeting our customers' needs and striving to exceed them."

Please circle the severity of your pain when paramedics first arrived.

0 1 2 3 4 5 6 7 8 9 10

None | Moderate Pain | Severe

Please circle the severity of your pain when paramedics delivered you to the emergency room.

0 1 2 3 4 5 6 7 8 9 10

None | Moderate Pain | Severe

	Outstanding	*Excellent*	*Average*	*Fair*	*Poor*
The 9-1-1 call was handled in a prompt, courteous, and competent manner.	☐	☐	☐	☐	☐
The 9-1-1 instructions given prior to arrival of the paramedics.	☐	☐	☐	☐	☐
If you were lifted off the ground, the methods used by the paramedics to lift you.	☐	☐	☐	☐	☐
If your injuries were splinted to stabilize them, the methods used by the paramedics.	☐	☐	☐	☐	☐
The paramedic crew clearly explained the procedures performed.	☐	☐	☐	☐	☐
The ride to the hospital in the ambulance.	☐	☐	☐	☐	☐
The paramedic crew acted in a concerned,caring, and professional manner.	☐	☐	☐	☐	☐
The overall quality of the care provided.	☐	☐	☐	☐	☐
The overall experience.	☐	☐	☐	☐	☐

Please comment on your overall experience with our service.

Alarm ________________ Date / / Station/Shift

Courtesy of Sarasota County Emergency Services wpdocs/m/Orthopedic

Respiratory Customer Satisfaction Survey

"Quality is meeting our customers' needs and striving to exceed them."

Please circle the number of minutes you experienced shortness of breath before 9-1-1 was called.

< 15 45 75 2 hours > 4 hours

30 60 90 3 hours

Please circle the severity of your shortness of breath when paramedics first arrived.

0 1 2 3 4 5 6 7 8 9 10

None Moderate Pain Severe

Please circle the severity of your shortness of breath when paramedics delivered you to the emergency room.

0 1 2 3 4 5 6 7 8 9 10

None Moderate Pain Severe

	Outstanding	*Excellent*	*Average*	*Fair*	*Poor*
The 9-1-1 call was handled in a prompt, courteous, and competent manner.	☐	☐	☐	☐	☐
The 9-1-1 instructions given prior to arrival of the paramedics.	☐	☐	☐	☐	☐
The EMS crew acted in a concerned and caring manner.	☐	☐	☐	☐	☐
The EMS crew clearly explained the procedures performed.	☐	☐	☐	☐	☐
The EMS crew and equipment presented in a professional manner.	☐	☐	☐	☐	☐
The overall quality of the care provided.	☐	☐	☐	☐	☐
The overall experience.	☐	☐	☐	☐	☐

Please comment on your overall experience with our service.

Alarm ________________ Date / / Station/Shift

Courtesy of Sarasota County Emergency Services wpdocs/m/Respiratory

LOG #

Inquiry/Complaint Management Tracking Form

Date of Call: ________________ Time of Call: ______________________________

Name of Person Making Inquiry/Complaint: __________________________________

Address: ____

City: __________ State: __________ Zip Code: ________

Telephone Number (include area code): _____________________________________

Place and Time Person Can Be Contacted: ___________________________________

Information Received (Circle One): In Person By Phone By Mail Anonymously

Person Receiving Inquiry/Complaint:

Brief Summary of the Inquiry/Complaint (Attach Additional Sheet if Necessary):

Stop Here—Forward Form to the Executive Director's Office

Core Service (Circle All That Apply): Emergency Management EMS Lifeguards Other

Inquiry/Complaint Assigned to: _______________________ Date Assigned: _________

____

Disposition of Inquiry/Complaint (Attach Additional Sheet if Necessary):

Signature: _______________________ Date: ____________________________

Forward Completed Form to the Executive Director's Office

Courtesy of Sarasota County Emergency Services wpdocs/m/Inquiry

PUBLIC EDUCATION PROGRAM

Upon completion of a public education program, such as a station tour, block party, or extinguisher demonstration, a public education program evaluation form will be presented to each adult participant in the program. The individual will be asked to complete the survey and return it to the program presenter prior to leaving the program. In the case of a presentation to a school group, the evaluation form will be given to the teacher for completion. All completed evaluation forms will be forwarded to the Public Education Coordinator for review and follow-up.

Fire Safety Education Program Quality of Service Survey

ABC FIRE DEPARTMENT
123 ANYWHERE STREET
ABCville, AB 12345

Please help the ABC Fire Department to serve you better by completing this survey and returning it to the address above. Thank you!

Date of Event: ________________ Location of Event: ______________________

On a scale of 1 to 5, with 1 being poor and 5 being excellent, please rate the following aspects of your fire station tour.

1. Quality of fire safety instruction that took place during the event.	1	2	3	4	5	N/A
2. Quality of demonstrations conducted.	1	2	3	4	5	N/A
3. Quality of fire truck and/or ambulance tour or demonstration.	1	2	3	4	5	N/A
4. Quality of handout materials.	1	2	3	4	5	N/A
5. Appearance of firefighters.	1	2	3	4	5	N/A
6. Professionalism of firefighters.	1	2	3	4	5	N/A
7. Friendliness of firefighters.	1	2	3	4	5	N/A
8. Overall opinion of event.	1	2	3	4	5	N/A
9. Based on this event, would you plan on attending another fire safety education program in the future?	1	2	3	4	5	N/A
10. Based on this event, would you recommend that others attend such an event?	1	2	3	4	5	N/A

Name: ______________________________ Telephone No.: ________________

Please call to discuss this information in more detail. ______________________

Sample Fire Prevention Survey Letter

ABC Fire Department
123 Anywhere Street
ABCville, AB 12345

January 1, 2008

Dear Customer:

You recently had the opportunity to use the services of the ABC Fire Department's Fire Prevention Bureau. In an effort to continually provide the best possible cost-effective service, I would like to know how well the service was provided. Therefore, I would appreciate a few minutes of your time to complete the enclosed Service Survey card and then mail it back to me for review. Space has been provided on the card to indicate a desire on your part to discuss service received further. By marking this space, you will be contacted by me or

(continued)

another member of my command staff within several days of receiving your survey card back at our office. Of course, if you would rather submit additional written information to my attention, please send it to the address on the survey and include the number written on the top of that card in all your written correspondence. This number will permit cross-reference with our department records.

Your feedback, positive or negative, is important to the operations of the ABC Fire Department. Therefore, I look forward to your reply. If you should have any questions or wish to contact me directly, please feel free to call me at 123-4567 during normal business hours.

Again, thank you for your assistance in this matter.

Sincerely,

Fire Chief Smith

Enclosure

Sample "Station Tour" Evaluation

ABC Fire Department
123 Anywhere Street
ABCville, AB 12345

ABC Fire Department Fire Safety Education Program
"Station Tour" Evaluation

Group Name: ______________________________

Group Contact Person: ______________________________

Date of Station Tour: ______________________________

Location of Station Tour: ______________________________

On a scale of 1 to 5, with 1 being poor and 5 being excellent, please rate the following aspects of your recent fire station tour.

1. Quality of fire safety instruction that took place during the tour. 1 2 3 4 5 N/A
2. Quality of demonstrations conducted during the tour. 1 2 3 4 5 N/A
3. Quality of fire truck and/or ambulance tour or demonstration. 1 2 3 4 5 N/A
4. Quality of handout materials. 1 2 3 4 5 N/A
5. Promptness of firefighters. 1 2 3 4 5 N/A
6. Professionalism of firefighters. 1 2 3 4 5 N/A
7. Friendliness of firefighters. 1 2 3 4 5 N/A
8. Overall opinion of station tour. 1 2 3 4 5 N/A
9. Based on this station tour, do you plan on scheduling another tour in the future? ___ Yes ___ No
10. Based on this station tour, would you recommend other groups such as yours schedule a tour?

    ___ Yes ___ No

The reverse side of this evaluation provides ample space for your comments and suggestions. *All comments and suggestions are appreciated, even if negative.*

Please use that space to add any general comments and/or suggestions. Your opinions are helpful in improving our programs. Thank you!

Kindly return this evaluation to ABC Fire Department, 123 Anywhere Street, ABCville, AB 12345.

Sample Instructor Evaluation Form

ABC Fire Department
123 Anywhere Street
ABCville, AB 12345

Fire Safety Education Program
Instructor Evaluation Form

Dear Teacher:

Please take a few moments to complete the questions listed below. Your professional opinions and comments will help us improve our instruction techniques and overall program.

What is the name of the school where you teach? ________________________

What grade level do you teach at the above school? ______________________

On a scale of 1 to 5, with 1 being poor and 5 being excellent, please rate the following aspects of the Fire Safety Education Program.

1. Promptness of instructor	1	2	3	4	5	N/A
2. Preparedness of instructor	1	2	3	4	5	N/A
3. Teaching ability of instructor	1	2	3	4	5	N/A
4. Audio/visual material	1	2	3	4	5	N/A
5. Demonstrations	1	2	3	4	5	N/A
6. Handout materials	1	2	3	4	5	N/A
7. Length of program	1	2	3	4	5	N/A
8. Overall content of program	1	2	3	4	5	N/A

Please use the following space to make any comments or suggestions to help the department improve its instruction techniques or the program in general. Please use the back for additional comments, if needed. Please return the completed evaluation to the office.

Sample Public Education Survey Letter

ABC Fire Department
123 Anywhere Street
ABCville, AB 12345

January 1, 2008

Dear (Group Leader Name):

Thank you for taking the time to visit our fire station. We sincerely hope the tour was informative and enjoyable.

Enclosed with this letter is an evaluation form. In an effort to make our programs as educational and interesting as possible, your comments are greatly appreciated. Please

(continued)

take a few moments to complete the evaluation and mail it to the ABC Fire Department at the address listed below:

ABC Fire Department
123 Anywhere Street
ABCville, AB 12345
U.S.A.

Thank you for your time.

Sincerely,

Fire Chief Smith

Enclosure

Fire Prevention Bureau Quality of Service Survey

ABC Fire Department
123 Anywhere Street
ABCville, AB 12345

Please help the ABC Fire Department to serve you better by completing this survey and returning it to the address above. Thank you!

Date: ________________________

On a scale of 1 to 5, with 1 being poor and 5 being excellent, please rate the following aspects of your fire service experience.

1. Was the phone operator efficient and courteous?	1	2	3	4	5	N/A
2. Were return calls made promptly?	1	2	3	4	5	N/A
3. Was the inspection scheduled promptly?	1	2	3	4	5	N/A
4. Did the inspector explain code requirements of the ABC Fire Department?	1	2	3	4	5	N/A
5. Did the inspector explain code violations identified on the inspection report?	1	2	3	4	5	N/A
6. Were all questions answered completely?	1	2	3	4	5	N/A
7. Did you receive a copy of the inspection form?	1	2	3	4	5	N/A
8. Was the overall level of service acceptable?	1	2	3	4	5	N/A
9. If a permit was involved, were permit requirements explained?	1	2	3	4	5	N/A
10. Were your plans reviewed in an acceptable amount of time?	1	2	3	4	5	N/A

11. Indicate type of service received from the Fire Prevention Bureau.

    Annual or semiannual fire inspection _____
    Construction inspection _____
    Plan review _____
    Other (Please List) ________________________

Name: ________________________________ Telephone No.: ________________
Please call to discuss this information in more detail. ________________________

INSTRUCTIONS FOR COMPLETING THE BENCHMARKING SURVEY

GENERAL INFORMATION

The general information section should be completed, with attention to accurate contact information. Any questions about the information contained in the survey will be addressed to the individual named in the section.

DEMOGRAPHIC INFORMATION/DATA

This section seeks to gain background information about the department and the community served. The demographic information is important in conjunction with the statistical information so the benchmarks can be established and comparisons made for similar organizations.

Population Served—Using the most current census, list the population of the community or jurisdiction your agency serves.

Total Number of Personnel—This figure is the total number of personnel within your department. It is then broken down into segments including uniformed, paid, volunteer, civilian, paid on call, and sworn.

Total Number of Calls—This figure should be the total number of calls run by a jurisdiction within the last complete calendar or fiscal year. It is then broken down into suppression, EMS, and other types of incidents.

GEOGRAPHICAL DESIGNATIONS

Metropolitan designation means an incorporated or unincorporated area with a population of over 200,000 people and a population density over 3,000 people per square mile.

Urban designation means an incorporated or unincorporated area with a population of over 30,000 people and a population density over 2,000 people per square mile.

Suburban designation means an incorporated or unincorporated area with a population of 10,000 to 29,999 or any area with a population density of 1,000 to 2,000 people per square mile.

Rural designation means unincorporated or incorporated areas with total population less than 10,000 people, or with a population density of less than 1,000 people per square mile.

Wilderness designation means any rural area not readily accessible by public or private maintained road.

Square Miles in Jurisdiction—This figure is the total number of square miles within your service area or jurisdiction.

Minimum Staffing—This figure is the minimum number of personnel on duty each day. The minimum staffing for each type of apparatus is the minimum number of personnel required to place the unit in service or make the unit operational according to jurisdiction by local, state, or federal guidelines.

Income per Capita of Jurisdiction—This figure is the average income per person who resides within the jurisdiction. This information can usually be obtained through a community development department or finance office.

Type of Department—Mark the appropriate box for your department. If an appropriate category is not found, please fill in the type in the space provided.

Calls per Year—Check the appropriate boxes for your department.

Property Types in Department's Jurisdiction—Mark the appropriate boxes for your department.

Work Schedules—Mark the appropriate box for your department.

Shift Schedules—Mark the appropriate box for your department. If other, please fill in the type in the space provided.

Cultural Diversity—The cultural diversity chart is for the ethnic and gender breakdown of the department as a whole and the community served.

Number of Stations in Jurisdiction—This figure gives the number of stations within your service area or jurisdiction.

Total Department Budget—This figure is obtained from the total department budget for the last full calendar year or fiscal year.

Total Fire Loss (Annual)—This figure is obtained from the total direct fire loss, in dollars, within the jurisdiction for the last full calendar year or fiscal year.

Number of Activities Annually—Activity is defined as emergency responses, service calls, public education, plans, inspections, smoke detector installation/checks, hydrants.

Number of Firefighter Deaths and Injuries (Annual)—This figure is obtained from the total firefighter deaths and injuries for the last full calendar year or fiscal year.

Number of Civilian Deaths and Injuries (Annual)—This figure is obtained from the total civilian deaths and injuries for the last full calendar year or fiscal year.

Level of EMS Provided/Transport of Patients—Check the appropriate box for your department.

Average Alarm Processing Time–Alarm Processing Time—The elapsed time from the receipt of an alarm by the dispatch center and the notification of specific fire companies that are needed to respond. This figure is obtained by calculating the total amount of time for departments to process all alarms received in the last full calendar year or fiscal year. This number should then be divided by the total number of alarms received to calculate the average alarm processing time.

Average Turnout Time–Turnout Time—The portion of response time when fire or emergency service companies are donning personal protective clothing and boarding their apparatus. The time begins once the companies have been given their assignments and ends when they begin travel time. This figure is obtained by calculating the total number of minutes for departments to turn out for all alarms received in the last full calendar year or fiscal year. This number should then be divided by the total number of alarms received to calculate the average turnout time.

Average Response Time–Response Time—The total amount of time that elapses from the time that a communications center receives an alarm until the responding unit is on the scene of the emergency and prepared to control the situation. This figure is obtained by calculating the total number of minutes for departments to respond to all alarms received in the last full calendar year or fiscal year. This number should then be divided by the total number of alarms received to calculate the average response time.

Average Response Time for First-Arriving Unit—This figure is calculated the same as average response time; however, use only data related to first-arriving units.

Average Response Time for Second-Arriving Suppression Unit on Scene—This figure is calculated the same as average response time; however, use only data related to second-arriving suppression units.

Other Services Provided—Please list services provided by your department other than suppression.

General Information

Department Name:
Chief of Department:
Person Completing Survey:
Address:
Telephone: Fax: E-mail:
Date Completed:
Population Served:
Total Number of Personnel: ____ Paid: ___ Civilian: ___ Volunteer: ___ Paid on Call: _____
Total Number of Calls: ____ Suppression: _____ EMS: _____ Other: _____
Is your area urban, suburban, or rural?
Square miles in jurisdiction:
Minimum staff per shift: ____ for engines/pumpers: _____ for ladders/trucks: _____
for heavy rescue squads: _____ for ambulances: _____
Income per capita of jurisdiction:
Type of Department: ____Career ____Volunteer ____ Combination ____ Federal/Military
____ Industrial Fire Brigade ____ Other
Are there any companies that exceed:
1,000 calls per year? No. of companies _____
2,000 calls per year? No. of companies _____
3,000 calls per year? No. of companies _____
4,000 calls per year? No. of companies _____
Property types in department's jurisdiction (check all that apply):
___ Residential ___ Commercial ___ Light Industry ___ Heavy Industry
___ Marine ___ Wildland ___ Agriculture/Farming ___ Rural
___ Urban ___ Suburban
Which of the following work schedules (hours per week) does your agency use for operation/line personnel?
__ 40 __ 42 __ 48 __ 52 __ 56 __ 72
Which of the following best describes your shift schedule?
____ 24 hours on, 48 hours off
____ 24 hours on, 24 hours off
____3 days, 3 nights, 3 off (10-hour days, 14-hour nights)
____ 3 days, 3 nights, 3 off with a "Kelly" day
____ 8-hour shifts
____ Other (specify) _____
What is the percentage breakdown of the following for your agency?

	Female	Male	Community
African American	______	______	_________
American Indian	______	______	_________
Asian/Pacific Islander	______	______	_________
White	______	______	_________
Hispanic	______	______	_________

(continued)

Data

Please provide figures for the following:

Number of stations in jurisdiction:
Total department budget:
Total fire loss (annual):
Number of activities annually:
Number of firefighter deaths (annual):
Number of firefighter injuries (annual):
Number of civilian fire deaths (annual):
Number of civilian fire injuries (annual):
What level of EMS does the department provide?

- ____ First Responder
- ____ Basic Life Support
- ____ Advanced Life Support
- ____ Does Not Provide EMS

Does your department transport patients? ____ Yes ____ No
Average alarm processing time:
Average turnout time:
Average response time:
Average response time for first-arriving unit:
Average response time for second-arriving suppression unit on scene:
What other services are provided by the department? (hazmat, rescue, etc.)

WMD Response Assets Guide

DOD

Name	*Response Role*	*Size/Strength*	*General Capabilities*	*Anticipated Arrival*	*Location*
U. S. Department of Defense	Oversee America's armed forces.		Use of military assets in domestic situations is strictly governed by a variety of laws. One such law, the Domestic Preparedness initiative formed under the Defense Authorization Bill (PL 104-201, September 23, 1996), is commonly called the Nunn-Lugar-Domenici legislation. It originally provided funding for the DOD to enhance the capability of federal, state, and local emergency responders in incidents involving nuclear, biological, and chemical terrorism. DOD transferred primary responsibility for the Nunn-Lugar-Domenici Domestic Preparedness Program to the Department of Justice on October 1, 2000. In 2003, the Justice Department passed that authority to the Department of Homeland Security. DOD was heavily involved in relief efforts following Hurricane Katrina in August 2005. Approximately 70,000 military personnel aided civilian authorities in recovery efforts. In the wake of the disaster, the Bush administration proposed giving DOD a greater role in responding to domestic emergencies. Changes are not under consideration.		

AFRAT

Name	*Response Role*	*Size/Strength*	*General Capabilities*	*Anticipated Arrival*	*Location*
Air Force Radiation Assessment Team	Provide on-site detection, identification, and quantification of any ionizing radiation hazard.	Personnel include health physicists, bio-environmental engineers with expertise in industrial	Risk assessment and consequence management services. In the event of a nuclear weapon accident or related event, AFRAT can provide equipment	AFRAT is capable of deploying to any location worldwide within 48 hours.	

		Size/Strength	General Capabilities		
		hygiene and environmental quality, bioenvironmental engineering technicians, radio analytical laboratory technicians, and a radio chemist. AFRAT has Nuclear Incident Response Teams 1 and 2, a Radiation Reconnaissance Laboratory, and a Field Laboratory for Assessment of Radiation Exposure.	and consultation about health physics, industrial hygiene, and environmental quality to the incident commander.		

CBIRF

Name	*Response Role*	*Size/Strength*	*General Capabilities*	*Anticipated Arrival*	*Location*
U.S. Marine Corps Chemical Biological Incident Response Force	Consequence management and force protection for chemical, biological, radiological, nuclear, and high-yield explosive incidents. In 2001, CBIRF Marines were in Washington, D.C., to support the response to the anthrax attacks. In February 2004, the force returned to the nation's capital in response to the ricin attack on the Senate Office Building.	More than 450 Marines, sailors, and civilians spanning approximately 50 diverse military occupational specialties. Following the September 11 attacks, the Marine Corps reactivated the Fourth Marine Expeditionary Brigade as an antiterrorism force; it now includes CBIRF.	CBIRF provides a rapid-response force for CBRNE incidents; consequence management support in military and industrial agent identification; Advanced Life Support; casualty reconnaissance, extrication, and triage; personnel decontamination; medical treatment; low-level radiation detection; and stabilization for incident site management, including ordnance disposal, security, and patient evacuation. CBIRF also conducts research and development on consequence management using civilian and military technology.	A 120-member initial-response force is prepared to respond to a no-notice WMD incident in one hour by maintaining an on-call unit at all times. AFn additional 200-member force is available on a four-hour response basis.	

(continued)

CBIRF *(continued)*

Name	Response Role	Size/Strength	General Capabilities	Anticipated Arrival	Location
			CBIRF routinely deploys for National Special Security Events and is prepared to provide agent detection and identification, casualty search and rescue, personnel decontamination, and emergency medical care to stabilize contaminated victims.		

CBRNE

Name	Response Role	Size/Strength	General Capabilities	Anticipated Arrival	Location
U.S. Army 20th Support Command (Chemical, Biological, Radiological, Nuclear, and High Yield Explosive)	Activated on October 16, 2004, the 20th Support Command (CBRNE) provides a robust and scalable response force to meet the full spectrum of CBRNE and WMD threats faced by the nation and the Armed Forces worldwide.		The 20th Support Command (CBRNE) provides crisis response support both within the continental United States and overseas. The command leverages technology and sanctuary reach-back capabilities able to link subject-matter experts in America's defense, scientific, and technological communities with deployed elements and first responders. The organization provides training, readiness, technical expertise, and assistance to select National Reserve component units and, when requested, provides CBRNE technical expertise and assistance in support of National Reserve component units. The command possesses a deployable chemical and biological analytical capability to provide timely, more		U.S. Army Aberdeen Proving Ground, Maryland

accurate analysis of unknown samples, and a near real-time chemical/biological monitoring capability.

Subordinate elements include the 52nd Ordnance Group (EOD), the 71st Ordnance Group (EOD), and the 111th Ordnance Group (EOD), all with subordinate EOD battalions and companies regionally headquartered throughout the United States. Numerous elements have deployed in support of both Operations Iraqi Freedom and Enduring Freedom.

The 20th Support Command (CBRNE) plans to activate a chemical brigade headquarters and then place the two existing chemical battalions, the 110th Chemical Battalion (Technical Escort) and the 22nd Chemical Battalion (Technical Escort), as well as additional CBRNE capable units, under the command and control of this brigade.

A subordinate unit, the 22nd Chemical Battalion (Technical Escort), formerly the U.S. Army Technical Escort Unit, or TEU, has been headquartered at the Edgewood Area of Aberdeen Proving Ground. The 22nd Chemical Battalion (Technical Escort) also has regular, hands-on experience with hazardous

(continued)

CBRNE *(continued)*

Name	Response Role	Size/Strength	General Capabilities	Anticipated Arrival	Location
			materials and chemical warfare agents in both escort and ongoing remediation cleanup operations in the Untied States.		

EOD

Name	Response Role	Size/Strength	General Capabilities	Anticipated Arrival	Location
Explosive Ordnance Disposal	Explosive ordnance disposal.	There are 39 U.S. Army 52nd EOD Group units across the country and numerous other EOD units in National Guard units. The 52nd EOD Group is headquartered at Fort Gillem, Georgia.	Provides expertise on explosive ordnance disposal and explosive device components. EOD members and other specially trained EOD operations in special-mission units are the primary experts called on by the FBI for access and device disablement operations. EOD Response Teams can evaluate; render safe; and remove conventional, nuclear, chemical, biological, or improvised explosive devices. Many units have been deployed in support of the global war on terrorism.		

ESOH

Name	Response Role	Size/Strength	General Capabilities	Anticipated Arrival	Location
Environment Safety Occupational Health Service Center	Provide around-the-clock response for both routine and urgent situations pertaining to ESOH matters.		Provide rapid contact to multiple subject-matter experts, such as toxicologists, epidemiologists, virologists, microbiologists, industrial hygienists, agronomists, health physicists, chemists, biochemists, and environmental and chemical engineers.	Available to respond 24 hours a day.	Based at the Air Force Institute for Environment, Safety, and Occupational Health Risk Analysis, Brooks City-Base, in San Antonio, Texas

JTF-CS

Name	Response Role	Size/Strength	General Capabilities	Anticipated Arrival	Location
Joint Task Force Civil Support	JTF-CS plan integrates DOD support in the designated lead federal agency for domestic CBRNE consequence management operations. When directed by the commander of U.S. Northern Command, or USNORTHCOM, JTF-CS deploys to the incident site and executes timely and effective command and control of designated DOD forces to provide support to civil authorities to save lives, prevent injury, and provide temporary critical life support.		Planning and command and control for federal military responders in CBRNE situations. Communications capabilities include long-range, secure tactical satellite voice and data telecommunications equipment, as well as cellular and secure/unsecured Internet access.	Command assessment element and/or liaison officers can respond soon after an incident. The deployment of the main body, at the direction of the commander of USNORTHCOM and on the authority of the secretary of defense, would occur only after a governor requests federal assistance from the president and after the president issues a disaster declaration.	

MARS

Name	Response Role	Size/Strength	General Capabilities	Anticipated Arrival	Location
Military Affiliate Radio System	Provide emergency communication to military, federal, and local civil agencies as warranted. Thousands of trained volunteers maintain a nationwide system of HF and VHF radio communication.	MARS networks exist in all 50 states and territories plus selected overseas locations in Europe, with approximately 5,000 volunteer radio operators in all military services.		Initial response within an hour or less after notification. MARS may not provide support to civil agencies unless specifically requested.	

USAMRIID

Name	***Response Role***	***Size/Strength***	***General Capabilities***	***Anticipated Arrival***	***Location***
U.S. Army Medical Research Institute of Infectious Diseases	USAMRIID is capable of deploying an Aero-medical Isolation Team consisting of physicians, nurses, medical assistants, and laboratory technicians. Team members are specially trained to care for and transport a limited number of patients with diseases caused by biological agents requiring high containment.				Fort Detrick, Maryland

WMD-CST

Name	***Response Role***	***Size/Strength***	***General Capabilities***	***Anticipated Arrival***	***Location***
Weapons of Mass Destruction Civil Support Team	To support civil authorities and the local incident commander at a domestic CBRNE incident site, assess suspected WMD events, advise civilian responders on appropriate actions, and assist and expedite the arrival of follow-on emergency or military forces as needed. Response missions involve sampling unknown or suspected hazards; performing standby missions in support of designated National Security Special Events, VIP visits, and major community events; and providing technical advice and coordinating exercises	There are 36 certified teams in 49 states. Once DOD certifies all 55 teams, each state and territory will have one WMD-CST, with California possessing two.	Each team includes 22 highly skilled, full-time members of the Army and Air National Guard. Teams are designed to deploy rapidly, assist first responders in determining the nature of an attack, provide medical and technical advice, and facilitate the identification and arrival of military response assets. WMD-CSTs are designed to support local first responders with assessment, unknown sample characterization and identification, hazard prediction modeling, situational awareness, liaison, medical, communication, and logistics issues. Each team has one physician's assistant, a nuclear medical science officer, a medical operations officer, and a medic. The team's command and control, operations, and	WMD-CSTs are on call 24/7. To provide national coverage, the National Guard Bureau identifies a selected number of WMD-CSTs to maintain three response levels to support different emergencies: Tier 1, three-hour alert to deployment; Tier 2, 24 hours to deploy; and Tier 3, 72 hours to deploy. Unit arrival on site is based on travel distance.	

with first responders, emergency managers, and citizens about WMD-CST capabilities.

logistics personnel will work with the local incident commander to determine and fill unmet needs. A WMD-CST is neither designed nor intended to replace functions carried out under the Incident Command System, nor to replace those functions normally performed by the emergency first responder in the community.

Teams are equipped with state-of-the-art high-end detection, analytical, and protective equipment. Each WMD-CST has an analytical laboratory system designed to provide characterization and identification of unknown samples and then transmit digital sample information to federal agencies and the laboratory response network. The units use satellite, secure, and cellular telephone communications to connect civil and military forces within the operational conditions. They are equipped to detect a range of substances, including toxic industrial chemicals, organic substances, and chemical and biological warfare agents.

WMD-CSTs are part of a state-level emergency management response force. A CST will be deployed in a consequence management support role at the direction of the governor or to support another state's response under a supported governor. The

(continued)

WMD-CST *(continued)*

Name	Response Role	Size/Strength	General Capabilities	Anticipated Arrival	Location
			units are federally resourced, trained, and equipped, and they are trained to military, NFPA, and OSHA standards. The state National Guard provides the personnel, stationing, and command support.		

HHS

Name	Response Role	Size/Strength	General Capabilities	Anticipated Arrival	Location
U.S. Department of Health and Human Services	Led by the secretary of HHS and works with states to build the strongest possible network of protection and response capability for Americans faced with health and medical problems in emergencies.		The Office of Public Health Emergency Preparedness closely coordinates the medical response from the Centers for Disease Control and Prevention, Food and Drug Administration, National Institutes of Health, Health Resources and Services Administration, and other HHS components in the event of a public health emergency. The assistant secretary directs the office's activities and oversees the HHS command center operations. Outfitted with computers, satellite videoconferencing capabilities, telephones, and extensive computer mapping tools, officials can monitor disease outbreaks, track the movement of medical supplies and emergency personnel, and provide situation reports 24 hours per day. Under the Surgeon General's Office of Force Readiness and Deployment, the U.S. Public Health Service Commissioned Corps trains and prepares for rapid response to help provide health and medical support in the event of an emergency. The corps provides on-site health and medical services and technical public health expertise.		

CDC

Name	***Response Role***	***Size/Strength***	***General Capabilities***	***Anticipated Arrival***	***Location***
Centers for Disease Control and Prevention	Helps assess incident effects and develops strategies for public health aspects of an emergency. Two key roles are disease surveillance and disease investigation.		The CDC's emergency preparedness and response activities are organized by the Office of Terrorism Preparedness and Emergency Response, which oversees the response of CDC's emergency operations center to requests for emergency and recovery assistance.	Based on request and needs, a preliminary assessment team could be available by phone within minutes. Depending on the situation, the team members might include toxicologists, chemists, physicians, environmental health scientists, or health physicists. Within eight hours of the initial call, an emergency response coordinator and team can be on site with the appropriate equipment if necessary. Requests made to CDC for assistance are coordinated with local and state health departments.	

NLRN

Name	***Response Role***	***Size/Strength***	***General Capabilities***	***Anticipated Arrival***	***Location***
National Laboratory Response Network	CDC's multilevel response to the covert release of biological agent.		Developed as a partnership among the CDC and state and local public health laboratories, the FBI, FDA, USAMRIID, and several other military sites, the program was begun to ensure local preparedness in the event of bioterrorism or other mass-casualty biological event. The network links public health agencies to state-of-the-art facilities that can analyze and identify biological agents.		

DOJ

Name	*Response Role*	*Size/Strength*	*General Capabilities*	*Anticipated Arrival*	*Location*
U.S. Department of Justice	Led by the United States Attorney General, the DOJ oversees the federal law enforcement community's efforts relating to terrorism.		DOJ was formerly the lead agency through the FBI, for the coordination of all aspects of federal response to a WMD incident. This authority now has been passed to the Department of Homeland Security. DOJ retains the primary responsibility for preventing and investigating terrorism incidents, and continues to oversee a number of important assets, including the Critical Incident Response Group and the Evidence Response Team.		

CIRG

Name	*Response Role*	*Size/Strength*	*General Capabilities*	*Anticipated Arrival*	*Location*
Critical Incident Response Group	Tactical and crisis management operations.		CIRG facilities the FBI's rapid response to, and the management of, crisis incidents. CIRG was established in 1994 to integrate tactical and investigate resources and expertise for critical incidents. Each of the three major areas of CIRG—Operations Support Branch, Tactical Support Branch, and the National Center for the Analysis of Violent Crime—furnishes distinctive operational assistance and training to FBI field officers as well as state and local law enforcement agencies.		

ERT

Name	*Response Role*	*Size/Strength*	*General Capabilities*	*Anticipated Arrival*	*Location*
Evidence Response Team	Crime scene documentation and evidence collection.	Varies from eight to 50 members. For example, the New York ERT has three 12-member teams	The FBI's ERTs include special agents and support personnel who specialize in organizing and collecting evidence using a variety of		

		that rotate on-duty status every two weeks. There are currently more than 1,100 FBI members in the field.	technical evidence recovery techniques. Participation in the ERT program is a collateral duty.		

FBI

Name	***Response Role***	***Size/Strength***	***General Capabilities***	***Anticipated Arrival***	***Location***
Federal Bureau of Investigation	Led by the FBI director, the FBI coordinates federal crisis management efforts. It is the lead federal investigative agency for WMD incidents through the WMD Operations Unit.		There are 56 FBI field offices, approximately 400 satellite offices known as resident agencies, four specialized field installations, and more than 50 Legal Attachés Offices around the world. Representatives from 20 separate federal agencies are assigned to the National Counterterrorism Center in the Washington, D.C., area. At the field operational level, the FBI has more than 100 Joint Terrorism Task Forces based out of its field offices. The task forces are designed to maximize interagency cooperation and coordination among federal, state, and local law enforcement. More than 800 full- and part-time federal, state, and local law enforcement personnel participate on these task forces, in addition to the FBI personnel. The FBI has five Rapid Development Teams that can respond to multiple incidents stationed in Washington, D.C.; New York City; Miami; and Los Angeles. The FBI's Hazardous Materials Response Unit, housed within the FBI Laboratory in Quantico, Virginia, is designed to enhance evidence-gathering capability and provides technical and scientific response to investigations involving hazardous materials, including weapons of mass destruction.		

DOE/NNSA

Name	*Response Role*	*Size/Strength*	*General Capabilities*	*Anticipated Arrival*	*Location*
U.S. Department of Energy/National Nuclear Security Administration, Office of Emergency Response	Serves as the nation's premier technical leader in responding to and successfully resolving nuclear and radiological threats and consequences worldwide.		DOE/NNSA operates its Emergency Management System and manages seven radiological emergency response assets to protect the public, environment, and emergency responders from both terrorist and nonterrorist events. This is done by providing a response, flexible, efficient, and effective radiological emergency response framework and capability for the nation by applying NNSA's unique technical expertise residing within the DOE/NNSA Nuclear Weapons Complex.		

AMS

Name	*Response Role*	*Size/Strength*	*General Capabilities*	*Anticipated Arrival*	*Location*
Aerial Measuring System	Aerial surveys and searches.		The AMS provides helicopters and fixed-wing aircraft to respond to radiological emergencies. Personnel and equipment aboard these aircraft provide aerial radiation surveys and search, real-time radiological aerial sampling, plume sampling and tracking, aerial photographic surveys, and aerial multispectral scanning surveys to determine the extent of contamination or to search for potential radiological sources.		Las Vegas, Nevada

ARG

Name	*Response Role*	*Size/Strength*	*General Capabilities*	*Anticipated Arrival*	*Location*
Accident Response Group	Technical response group for nuclear emergencies, particularly accidents or incidents involving nuclear weapons that are in EOR or DOD custody.		ARG is deployed to manage or support the successful resolution of a U.S. nuclear weapon incident anywhere in the world. The ARG provides the technical response to U.S. nuclear weapons accidents and assistance in assessing weapons damage and risk, and in identifying procedures for safe weapons recovery, packaging, transportation, and disposal of damaged weapons. ARG also provides expert technical advice to the local U.S. military Explosive Ordnance Disposal operators to determine weapon damage and conduct an initial risk assessment.		

IMAAC

Name	*Response Role*	*Size/Strength*	*General Capabilities*	*Anticipated Arrival*	*Location*
Interagency Modeling and Atmospheric Assessment Center	Established April 15, 2004, IMAAC is the single source of federal hazard predictions leading up to and during incidents of national significance involving atmospheric releases of nuclear, chemical, and biological materials. DHS assigned the NARAC facility at Lawrence Livermore National Laboratory the primary interim provider of IMAAC services.		Under the National Response Plan, IMAAC leverages existing federal dispersion modeling and analysis assets to provide the best scientific tools and assessments available for major atmospheric hazardous releases; distributes atmospheric predictions to federal, state, and local response agencies; and supports the Homeland Security Operations Centers and FEMA's National Response Coordination Center. IMAAC's membership includes DHS, DOD, EPA, NRC, DOC/NOAA, DOE, and NASA. IMAAC provides first responders and state and federal emergency operations with a common operational picture of the consequences		The current interim IMAAC reach-back center is located at the National Atmospheric Release Advisory Center at Lawrence Livermore National Laboratory. Access is coordinated by the Homeland Security Operations Center in Washington, D.C.

(continued)

***IMAAC** (continued)*

Name	Response Role	Size/Strength	General Capabilities	Anticipated Arrival	Location
			of major hazardous atmospheric releases such as spills, fires, and explosions. Users are given reach-back access to modeling chemical/biological events. As an emergency situation evolves into an incident of national significance, information previously supplied to local emergency responders is sent to the IMAAC directly or via the Homeland Secretary Operations Center (HSOC), which becomes the single point for the coordination and dissemination of dispersion modeling and hazard predictions. Federal, state, local, and tribal government emergency response organizations can request IMAAC support for major events through the HSOC.		

FRMAC

Name	Response Role	Size/Strength	General Capabilities	Anticipated Arrival	Location
Federal Radiological Monitoring and Assessment Center	FRMAC is the multi-federal agency radiological response organization created for responding to radiological consequence management emergencies impacting the United States. Under the National Response Plan, DOE is the coordinating agency.		FRMAC is an interagency group that includes representatives from federal, state, and local radiological response organizations. The group supports states and the coordinating agency, and it has the assets and the capability to provide additional logistical and communications support for the interagency organizations responding to a radiological incident. Specific responsibilities include: coordinating federal off-site radiological environmental monitoring and assessment activities; maintaining technical liaison with state		Las Vegas, Nevada

Name	Response Role	Size/Strength	General Capabilities	Anticipated Arrival	Location
			and local agencies responsible for monitoring and assessment activities; maintaining technical liaison with state and local agencies responsible for monitoring and assessment; developing a common set of all off-site radiological monitoring data; providing monitoring area and technical support from other federal agencies; arranging consultation and support services through appropriate federal agencies to all other entities (e.g., private contractors) with radiological mentoring functions and capabilities; and providing technical and medical advice on handling radiological contamination and population monitoring.		

NARAC

Name	Response Role	Size/Strength	General Capabilities	Anticipated Arrival	Location
National Atmospheric Release Agency Center	NARAC provides real-time computer predictions of the atmospheric transport of radioactivity from a nuclear accident of incident. These predictions use topographical and real-time meteorological data in a 3-D radiological dispersion model. Maps are produced that contain accumulated integrated doses, airborne concentrations, and contamination distributions. These predictions are produced for local and federal		A laboratory based program, NARAC is maintained at Lawrence Livermore National Laboratory, Livermore, California. The High Consequence Assessment and Technology group from Sandia National Laboratory location provides additional NARAC tools and expertise for nuclear and conventional explosive source characteristics/effects and dose/risk assessment. A suite of NARAC software tools includes a simple stand-alone, local-scale plume modeling tool for end-user computer, and Web-based software to access advanced 3-D modeling tools and expert analyses.		

(continued)

NARAC (continued)

Name	*Response Role*	*Size/Strength*	*General Capabilities*	*Anticipated Arrival*	*Location*
	leaders to determine the protective actions necessary to ensure the health and safety of people in affected areas. Through the Interagency Modeling Atmospheric and Assessment Center, NARAC supports federal responses to incidents of national significance involving hazardous material release to the atmosphere.				

NEST

Name	*Response Role*	*Size/Strength*	*General Capabilities*	*Anticipated Arrival*	*Location*
Nuclear Emergency Support Team	Provides specialized technical expertise to resolve and advice on nuclear/radiological terrorist incidents. NEST is an umbrella of capabilities maintained by DOE/NNSA that provide technical advice, radiological/nuclear search and identification, and resources for the safe rendering and disposition of WMDs.		NEST offers specialized response teams that work with other federal agencies for the preparedness, planning, response, and recovery activities associated with combating or responding to radiological/nuclear terrorist incidents. The teams are able to search for, locate, and identify devices or materials; move, render safe, or disable devices; and mitigate damage from device detonation or disablement action. NEST provides on-scene scientific and technical advice; coordination for DOE support; deployable search assets; advanced technical capabilities to move or neutralize nuclear WMDs; and resources to handle the disposition of a damaged or recovered nuclear weapon, improvised nuclear device, or radiological dispersion device.		

RAP

Name	*Response Role*	*Size/Strength*	*General Capabilities*	*Anticipated Arrival*	*Location*
Radiological Assistance Program	Assess radiological emergencies and advise decision makers on how to minimize hazards.		Since the late 1950s, RAP has provided first-response radiological assistance to protect the health and safety of the general public and the environment. RAP assists federal, state, tribal, local, and other government agencies in detection, identification, analysis, and response to radiological/nuclear events. Deployed RAP teams provide traditional field monitoring and assessment support as well as a search capability, which includes the maritime environment both in port and at anchor. To provide a timely response capability, RAP is implemented on a regional basis. This regional coordination is intended to foster a working relationship between DOE radiological response elements and those of state, local, and other federal agencies.	RAP ensures a 24-hour response capability that can be deployed within two hours of the request for assistance. The response team(s) will be on the site of a radiological emergency within six hours of a request for assistance. The RAP response capability is self-sustained for the initial 24 hours of an emergency or until more permanent support is deployed to the emergency site.	Regional headquarters in Upton, New York; Oak Ridge, Tennessee; Aiken, South Carolina; Albuquerque, New Mexico; Chicago, Illinois; Idaho Falls, Idaho; Oakland, California; and Richland, Washington

REAC/TS

Name	*Response Role*	*Size/Strength*	*General Capabilities*	*Anticipated Arrival*	*Location*
Radiation Emergency Assistance Center/Training Site	Offers 24-hour emergency response to support the medical management of radiation accidents.		Since its formation in 1976, REAC/TS has provided support to DOE, the World Health Organization, and the International Atomic Energy Agency in the medical management of radiation accidents. REAC/TS is a 24-hour emergency response program that trains, consults, or assists in the response to all types of radiation accidents or incidents. The center's specially trained team of physicians, nurses, health physicists, radiobiologists, and emergency coordinators is prepared to provide around-the-clock		Oak Ridge, Tennessee

(continued)

REAC/TS *(continued)*

Name	Response Role	Size/Strength	General Capabilities	Anticipated Arrival	Location
			assistance at the local, national, international level. REAC/TS is recognized around the world for its expertise and often is called on to assist the global community in providing medical care to radiation accident victims, either directly or indirectly as consultants.		

DHS

Name	Response Role	Size/Strength	General Capabilities	Anticipated Arrival	Location
U.S. Department of Homeland Security	Created in 2003, the 180,000-person department is the nation's lead agency in dealing with terrorism and weapons of mass destruction. Headed by the secretary of the Department of Homeland Security, the department's goals are to make the nation less vulnerable to terrorism, prevent attacks, minimize the damage from attacks, and oversee recovery efforts. DHS works closely with other federal agencies and state, tribal, territorial, and local governments to coordinate disaster planning and relief efforts.		DHS operates under the National Response Plan, which provides a comprehensive framework for the management of domestic incidents. The NRP provides the structure and mechanisms for the coordination of federal support to state, tribal, territorial, and local incident managers, and for exercising federal responsibilities. The Homeland Security Operations Center (HOSC) collects, consolidates, and analyzes information from a variety of sources and oversees domestic incident management. HSOC represents more than 35 agencies, ranging from state and local law enforcement to federal intelligence agencies. The Office of Intelligence and Analysis analyzes intelligence from other agencies, including the CIA, FBI, DIA, and NSA, that involve threats to homeland security. It also evaluates vulnerabilities in the nation's infrastructure.		

DHS is responsible for border and transportation security. Major components in this area include the Transportation Security Administration, U.S. Customs and Border Protection, Immigration and Custom Enforcement, and Citizenship and Immigration Services. The Secret Service and the Coast Guard also are located in DHS, remaining intact and reporting directly to the DHS secretary.

The Directorate for Preparedness works with government agencies and the private sector to identify potential vulnerabilities and to focus resources on high-target areas. The directorate provides grants and training to local and national first responders.

The Science and Technology Directorate coordinates research and development efforts, including preparing for and responding to attacks involving weapons of mass destruction. The directorate includes the Homeland Security Advanced Research Projects Agency, the Office of Research and Development, and the Office of Systems Engineering and Development.

The Domestic Nuclear Detection Office works to improve the nation's capability to detect and report unauthorized attempts to import, possess, store, develop, or transport nuclear or radiological material. DNDO works closely with government agencies at all levels to improve detection and response capabilities. The office also conducts research and development programs on new technologies.

DHS-FEMA

Name	Response Role	Size/Strength	General Capabilities	Anticipated Arrival	Location
Federal Emergency Management Agency	Response Division leads, manages, and coordinates the national response to acts of terrorism, natural disasters, or other emergencies.		FEMA retains the responsibility as the lead agency for all-hazards consequence management when and if a presidential disaster or emergency is declared under the Stafford Act. FEMA coordinates federal response and recovery activities in support of state and local governments in responding to disasters and emergencies declared by the president, and it promotes the effective response of federal agencies. In 2003, FEMA assumed responsibility for the National Disaster Medical System and the Metropolitan Medical Response System from the Department of Health and Human Services (see DMAT and MMRS). The agency also has operational control over the Department of Energy's Nuclear Emergency Support Team (see DOE: NEST) and management control over the multi-agency Domestic Emergency Support Team (see DIST). FEMA also coordinates 28 urban search and rescue teams (see US&R). FEMA oversees domestic disaster preparedness training and coordinates government disaster response to both terrorist attacks and natural disasters. Major components include: • National Disaster Medical System • Metropolitan Medical Response System • Domestic Emergency Support Teams • National Incident Management Teams • Mobile Emergency Response Support • National Urban Search and Rescue Response System		

SNS

Name	*Response Role*	*Size/Strength*	*General Capabilities*	*Anticipated Arrival*	*Location*
Strategic National Stockpile	Managed by the Centers for Disease Control and Prevention, SNS purchases and maintains a national repository for pharmaceuticals and medical supplies. The supplies, including antidotes, are used to support state and local agencies in the event of a biological or chemical terrorism incident.		In a biological or chemical terrorism event, state, local, and private stocks of medical material will deplete quickly. State and local first responders and health officials can request the stockpile supplies to bolster their response to a biological or chemical terrorism attack, thereby increasing their capacity to more rapidly mitigate the results of this type of terrorism. SNS is organized into several packages. There are eight identical 12-hour push packages stored in cargo containers to facilitate rapid deployment. Each push pack is composed of pharmaceuticals, intravenous and airway supplies, emergency medications, and bandages and dressings.	The first push package, along with a technical adviser response unit, will arrive within 12 hours of a state-requested or federal decision to deploy. Additional supplies can be shipped within 24 to 36 hours, as needed. Supplies can be tailored to meet the requirements of a specific emergency.	

USCG

Name	*Response Role*	*Size/Strength*	*General Capabilities*	*Anticipated Arrival*	*Location*
U.S. Coast Guard	Multimission military service and the nation's maritime first responder.		The Coast Guard has a highly trained cadre of military professionals who maintain and rapidly deploy with specialized equipment to support state and local officials in preparing for and responding to chemical incidents. The Coast Guard also is responsible for maritime law enforcement and search and rescue. It is a lead federal agency in providing security for the nation's harbors, ports, and coastlines.		

USCG NSF

Name	Response Role	Size/Strength	General Capabilities	Anticipated Arrival	Location
U.S. Coast Guard National Strike Force	ICS implementation and operations oversight; damage assessment; site safety; and contractor oversight for maritime oil spills, chemical releases, and WMD events.	The NSF is capable of responding to two major oil and two chemical incidents at one time. This response capability includes 20 personnel per oil spill and 10 people per chemical response. There are three strike teams, one each responsible for the Atlantic, the Gulf of Mexico, and the Pacific.	When responding to incidents, strike team members join local emergency response forces to eliminate the source of a discharge, collect and store spilled material, prevent damage to sensitive environmental areas, and mitigate shoreline effects.	Will dispatch two people immediately, four people within two hours, 12 people within six hours, and requested equipment within four hours.	

DEST

Name	Response Role	Size/Strength	General Capabilities	Anticipated Arrival	Location
Domestic Emergency Support Team			The DEST is a specialized interagency team designed to expeditiously provide expert advice, guidance, and support to the Principal Federal Official (PFO) and the FBI On-Scene Commander (OSC) during a WMD incident or credible threat. The DEST is made up of crisis and consequence management components. It augments the FBI's Joint Operations Center with tailored expertise and assessment and analysis capabilities. The DEST also provides the PFO and OSC with expert advice and guidance in the following areas: interagency crisis management assistance; information management support; enhanced communications capability; contingency planning for consequence management support; explosive devices and their components; chemical, biological, and/or nuclear weapons/		

Name	Response Role	Size/Strength	General Capabilities	Anticipated Arrival	Location
			devices and their components; radiological dispersal devices; technical expertise and equipment to operate in a contaminated environment for various sampling activities; and follow-up on response assets and capabilities.		

IMT

Name	Response Role	Size/Strength	General Capabilities	Anticipated Arrival	Location
National Incident Management Teams	These interagency teams are traditionally deployed for wildland fires and other large-scale natural disasters, but they also provide a variety of support functions under Emergency Support Functions 4 and 7 in the National Response Plan.	There are 16 Type 1 teams that typically manage large, complex incidents with more than 500 people assigned, addressing complicated logistical, fiscal, planning, operational, and safety issues. There are 35 Type 2 teams, which manage less complex incidents. Both types of teams are strategically located, with a majority in the 11 wildfire-prone western states.			

NDMS

Name	Response Role	Size/Strength	General Capabilities	Anticipated Arrival	Location
National Disaster Medical System	NDMS is a nationwide partnership embracing communities with world-class medical and emotional care in the wake of a natural or human-made disaster.	More than 8,000 federal, state, local, and private sector medical and support personnel organized into more than 100 response teams, including at least 57 Disaster Medical Assistance Teams (DMAT). A typical deployable DMAT unit consists of 35 individuals. DMATs and all other NDMS teams	DMATs are deployed to provide immediate medical attention to the sick and injured during disasters when local emergency response systems become overwhelmed. DMATs deploy to disaster sites with adequate supplies and equipment to support themselves for 72 hours while providing medical services at a fixed or temporary medical site. DMATs are principally a community resource to support local, regional, and state requirements. They can be		

(continued)

NDMS *(continued)*

Name	Response Role	Size/Strength	General Capabilities	Anticipated Arrival	Location
		can be tailored to fit the scenario. Teams can be as large as three times their usual size or as small as three.	federalized to provide interstate aid. In addition to the standard DMATs, there are other highly specialized response teams that deal with specific disaster-related conditions, such as crush injury, burn and mental health emergencies, mortuary affairs, and veterinarian services. During an emergency response, Disaster Mortuary Operational Response Teams, or DMORTs, and other teams may work under the guidance of state and local authorities. They provide a spectrum of services, including the recovery, identification, and processing of deceased victims; the care of pets; and the prevention of animal-borne disease. All teams are directed by NDMS headquarters in concert with an area coordinator in each of the 10 geographic federal regions.		

MERS

Name	Response Role	Size/Strength	General Capabilities	Anticipated Arrival	Location
Mobile Emergency Response Support	MERS support FEMA's effective responses to disasters with robust teams that provide communications, life support, and logistical and operational support to rapidly create effective event field offices.		There are five MERS detachments equipped with fixed and mobile facilities and equipment and 240 people. The detachments are located in Bothell, Washington; Maynard, Massachusetts; Denton, Texas; Denver, Colorado; and Thomasville, Georgia. The detachments may support any event. Each detachment operates a MERS Operations Center with 24/7, real-time support to the regions. The Thomasville MERS Operations Center is the alternate FEMA Operations Center.		

MMRS

Name	*Response Role*	*Size/Strength*	*General Capabilities*	*Anticipated Arrival*	*Location*
Metropolitan Medical Response System	Provide initial on-site response and safe patient transport to emergency rooms in the event of a terrorist attack. Also provides medical and mental health care to victims of such attacks, and can move victims to other regions should local resources be overwhelmed.	Since 1999, MMRS has expanded from 27 to 125 jurisdictions nationwide. The federal government has been funding enhancements to the system's management and response capabilities.			

US&R

Name	*Response Role*	*Size/Strength*	*General Capabilities*	*Anticipated Arrival*	*Location*
National Urban Search and Rescue Response System	Responds to natural and human-made disasters to locate and rescue victims of structural collapse.		Established under the authority of FEMA in 1989, US&R is a framework for structuring local emergency personnel into integrated disaster response task forces. Task force members include structural engineers and specialists in the areas of hazmat; heavy rigging; search, including highly trained search dogs; logistics; rescue; and medicine. By design, there are two task force members assigned to each position for the rotation and relief of personnel, allowing for around-the-clock operations. After a request for federal assistance is received, task forces may be activated or placed on alert when a major disaster threatens or strikes a community. There are 28 task forces in the National US&R System. Each task force,	Each task force is charged with having all its personnel and equipment at the embarkation point within four to six hours of activation. The task force may then deploy immediately by ground for distances up to 1,000 miles or by air for longer distances, usually arriving at its destination in six to 14 hours.	

(continued)

US&R *(continued)*

Name	Response Role	Size/Strength	General Capabilities	Anticipated Arrival	Location
			composed of a minimum of 62 personnel, has an extensive cache of tools, supplies, and equipment and also is capable of addressing defensive WMD and hazmat operations. To responde effectively to an emergency, a typical task force has more than 130 members at the ready. FEMA deployed all 28 US&R task forces after Hurricane Katrina struck the Gulf Coast. The teams aided recovery efforts in New Orleans and throughout the region.		

EPA

Name	Response Role	Size/Strength	General Capabilities	Anticipated Arrival	Location
U.S. Environmental Protection Agency	EPA's emergency response program provides a safety net to state, tribal, and local first responders for oil spills and hazmat incidents. EPA's overall mission is to provide this support in response to releases or threats of releases of hazardous substances and oil that endanger public health, welfare, or the environment. EPA evaluates the need for federal response for all notifications of hazardous		The EPA emergency response program has On-Scene Coordinators in each of EPA's 10 regional offices. The OSCs perform functions in four main areas: notification and evaluation, monitoring and assessment, protective action guidance, and advising/implementing on-site response actions. Typical support includes monitoring and assessing chemical, biological, and radiological threats to humans, infrastructure, and environmental assets. Oil and hazardous substances releases or		

Name	Response Role	Size/Strength	General Capabilities	Anticipated Arrival	Location
	substances releases and oil spills. While state, tribal, local, and private first responders address the majority of domestic environmental responses, EPA will respond to support, monitor, lead, or direct an emergency response when others' technical, financial, and/or jurisdictional authorities are exceeded.		threats should be reported to the National Response Center. EPA has OSCs in 32 locations nationwide, which are accessible through the National Response Center (NRC).		

ERT

Name	Response Role	Size/Strength	General Capabilities	Anticipated Arrival	Location
Environmental Response Team	Chemical agent detection and multimedia environmental characterization, including subsurface, groundwater, and aquatic; air modeling and monitoring; and alternative treatment technology application.		ERT supports EPA's OSCs. The response teams have portable chemical agent instrumentation to detect and identify agents in low and sub parts-per-million quantities. The teams also can measure alpha, beta, and gamma radiation. The ERTs offer 24-hour emergency response capabilities, including access to chemical decontamination equipment. ERT members have the ability to enter contaminated areas using all levels of personnel protective gear. The team has specialized hazmat response expertise in health and safety, alternative treatment techniques, environmental engineering, risk and ecological assessment, and scientific support coordination.		

NDT

Name	*Response Role*	*Size/Strength*	*General Capabilities*	*Anticipated Arrival*	*Location*
National Decontamination Team	Emergency response decontamination support.		NDT supports emergency responders, normally through the EPA's OSCs. The NDT works within the Incident Command System to guide the decontamination of buildings, open spaces, transportation systems, water systems, and private infrastructure. The team has specialized WMD expertise in health and safety, remote sensing and analysis, decontamination methods, engineering, risk assessment and toxicology, waste treatment and disposal, and scientific support coordination.		

NHSRC

Name	*Response Role*	*Size/Strength*	*General Capabilities*	*Anticipated Arrival*	*Location*
National Homeland Security Research Center	Provides technical information and guidance on WMD-CBRNE response, including human exposure risk, prevention measures, sampling and detection, and decontamination and waste disposal.		The NHSRC provides rapid technical information and guidance to assess risk and support prevention, detection, containment, treatment, and the decontamination of water systems and buildings subject to chemical and biological acts of terror.		

RERT

Name	*Response Role*	*Size/Strength*	*General Capabilities*	*Anticipated Arrival*	*Location*
Radiological Emergency Response Team	Radiological environmental characterization, monitoring, and risk assessment.	There are approximately 75 team members stationed at EPA's two national radiation laboratories and EPA headquarters. EPA can send a few specialists or all of the team members to the site emergency. Headquarters RERT members support field operations activities from the EPA Emergency Operations Center.		Field teams can begin deployment six hours after notification. The team is on standby alert at all times and, if needed, can drive to any site within two to four days.	

APPENDIX C Assessing Your Local Capability to Respond to a Disaster

No. 1. Laws and Authorities: Federal, state, and local issuances and any implementing regulations that establish authority for development and maintenance of the emergency management program and organization, and define the emergency powers, authorities, and responsibilities of the chief executive officer and emergency preparedness manager.

Capabilities Rating		*Fully Prepared*	*Very Prepared*	*Generally Prepared*	*Marginally Prepared*	*Not Prepared*
1.1	A legal basis for the emergency management program exists.					
1.1.1	The emergency management program/responsibility is established in local response.					
1.1.2	The capability for the declaration of a local proclamation of emergency or disaster exists.					
1.1.3	The jurisdiction has adopted an executive order (EO) or other mechanisms for coordination among local agencies.					
1.1.4	Development of interagency mutual-aid agreements, including specific provisions (e.g., liabilities, responsibilities, participants, and review process), is supported by local law.					
1.1.5	Legal authority for evacuations (e.g., hurricane and hazardous materials [hazmat]) is defined.					
1.1.6	Legal authority for in-place sheltering/quarantine is defined.					
1.1.7	Local law exists that enables adoption and enforcement of building/fire codes and land use ordinances at the local government level.					
1.2	Legal authorities exist in local instructions for Continuity of Government (COG) activities.					

Adapted from United States Department of Homeland Security Scorecard for the Nationwide Plan Review, Washington, D.C., 2006.

1.2.1	The chief elected/appointed official's emergency powers are outlined in local law.					
1.2.2	Legal authorities exist for a line of succession for the chief elected/other elected officials/key appointed officials.					
1.2.3	Legal authorities, proclamations, and/or emergency orders exist for a line of succession for the heads of local departments and agencies.					
1.2.4	Legal authorities exist for successors to have pre-delegates authorities to take emergency actions during an emergency.					
1.2.5	Legal authorities exist for the preservation of vital local records.					

No. 2. Hazard Identification and Risk Assessment: The process of identifying situations or conditions that have the potential of causing injury to people, damage to property, or damage to the environment and the assessment of the likelihood, vulnerability, and magnitude of incidents that could result from exposure to hazards.

	Capabilities Rating	***Fully Prepared***	***Very Prepared***	***Generally Prepared***	***Marginally Prepared***	***Not Prepared***
2.1	A methodology exists to identify and evaluate natural, technological, and human-caused hazards within the area of responsibility.					
2.1.1	The jurisdiction identifies all hazards and the likelihood of their occurrence (hazards to be considered shall include, but not be limited to, natural, technological, and human events).					
2.1.2	Hazards identified by local, state, federal, and private agencies or organizations are included in the hazard analysis process.					
2.1.3	Hazards are assimilated into a common format, such as a digital geographic information system (GIS), for comparative evaluation.					
2.2	A vulnerability assessment of people and property to identified hazards has been performed.					
2.2.1	The jurisdiction uses a scientifically sound risk assessment methodology.					
2.2.2	Structural inventory data (e.g., critical facilities, residential and commercial structures, lifelines, transportation, and commercial industrial) is collected from available sources and assessed.					
2.2.3	Demographic data (e.g., daily population patterns, traffic patterns, seasonal population changes, and special needs populations) is collected from public and private sources and assessed.					
2.2.4	The risk assessment includes historic information for all disasters.					
2.2.5	The risk assessment is used as the basis for both mitigation and emergency operations planning.					
2.2.6	A life-cycle plan has been developed and implemented to ensure that demographic and structural inventory data is regularly updated.					

No. 3. Hazard Management: Systematic management approach to eliminate hazards that constitute a significant threat within the area of responsibility or to reduce the effects of hazards that cannot be eliminated through a program of hazard mitigation.

	Capabilities Rating	***Fully Prepared***	***Very Prepared***	***Generally Prepared***	***Marginally Prepared***	***Not Prepared***
3.1	A pre-disaster hazard mitigation program exists.					
3.1.1	The jurisdiction participates in all state mitigation programs for which it is eligible.					
3.1.2	The jurisdiction participates in the National Flood Insurance Program.					
3.1.3	The jurisdiction has the capability to track areas of repetitive loss in both declared and non-declared events.					
3.1.4	The jurisdiction develops a mitigation strategy based on the results of the hazard identification and risk assessment, program assessment, and operational experience to eliminate or mitigate the effects of hazards.					
3.1.5	The jurisdiction provides incentives that encourage mitigation activities sponsored by public and private sector partnerships.					
3.1.6	The mitigation strategy considers, but is not limited to, building construction standards; hazard avoidance through land use practices; relocation, retrofitting, or removal of structures at risk; and removal or elimination of the hazard.					
3.1.7	The jurisdiction implements mitigation projects and/or initiatives with or without state grant assistance, according to a plan that sets priorities based on the highest potential for damage.					
3.1.8	The jurisdiction provides policy leadership and coordination promoting hazard mitigation programs and initiatives.					
3.1.9	The jurisdiction provides technical assistance to business and industry for developing and implementing mitigation strategies.					
3.2	The jurisdiction manages the post-disaster mitigation activities.					
3.2.1	Procedures exist for handling the surge in building permit requests after an event.					
3.2.2	Procedures exist for the post-disaster inspection of damaged and potentially damaged structures.					
3.2.3	Procedures for post-disaster mitigation activities include environmental aspects.					
3.2.4	The jurisdiction identifies mitigation opportunities during the development of the Disaster Survey Report.					
3.2.5	The jurisdiction follows established state procedures for including local post-disaster mitigation projects into the state Hazard Mitigation Grant Program.					
3.2.6	Can the jurisdiction modify the mitigation strategy based on disaster experience?					

No. 4. Resource Management: Systematic development of methodologies for the prompt and effective identification, acquisition, distribution, accounting, and use of medical personnel, equipment, and supplies for essential emergency functions.

Capabilities Rating		***Fully Prepared***	***Very Prepared***	***Generally Prepared***	***Marginally Prepared***	***Not Prepared***
4.1	The local emergency management organization has the human resources required to carry out assigned day-to-day responsibilities.					
4.1.1	A director/coordinator has been appointed in writing.					
4.1.2	The local emergency management program has adequate staffing.					
4.1.3	Local jurisdiction staff is provided adequate training opportunities for professional development to enhance their emergency management qualifications.					
4.1.4	Jurisdiction staffs are adequately trained in the fiscal management aspects of the emergency management program.					
4.1.5	Emergency management personnel have the skills to obtain, distribute, and manage state emergency management grant programs.					
4.1.6	Jurisdiction staffs are adequately trained to efficiently acquire, manage, and upgrade equipment, technology, support services, and property assets.					
4.2	The local emergency management organization possesses or has access to the human resources required to carry out assigned emergency responsibilities.					
4.2.1	Local staff has specific emergency assignments as part of their written job descriptions in the event of an emergency or disaster.					
4.2.2	Local staff are trained and qualified to fulfill local responsibilities under local, state, and federal declarations.					
4.2.3	The local emergency management agency has the capability to obtained trained personnel from the jurisdiction's other agencies for augmentation.					
4.2.4	Mutual aid agreements/Memoranda of Understanding are in place with other jurisdictions to obtain trained personnel for augmentation.					
4.3	The jurisdiction has a system to manage solicited and convergent volunteers.					
4.3.1	The jurisdiction has identified the legal liabilities associated with the use of solicited and convergent volunteers during disasters.					
4.3.2	The jurisdiction has developed policies and procedures to lessen the liability associated with the use of solicited and convergent volunteers.					
4.4	Local departments or agencies are assigned responsibility for coordinating resource management issues in the related emergency operations plan.					
4.4.1	Responsibilities of local departments or agencies are identified in the EOP.					
4.4.2	Responsibilities of the private sector are identified in the EOP.					
4.4.3	Responsibilities of voluntary agencies and organizations are identified in the EOP.					
4.4.5	The local jurisdiction has identified the resource requirements to maintain operations at critical infrastructure facilities during disasters (e.g., backup generators).					
4.5	Resource inventories are developed and kept current.					
4.5.1	Local agencies update resource inventories and maintain a list of main resource supply points.					

(continued)

***Capabilities Rating** (continued)*

4.5.2	The local jurisdiction has developed procedures to request and coordinate local government and private sector personnel and equipment.					
4.5.3	The local emergency management program maintains liaison with local agencies that have primary responsibility for resources in short supply and helps coordinate requests for assistance.					
4.6	Mutual-aid agreements are addressed in the EOP.					
4.6.1	The local jurisdiction participates in a statewide emergency management mutual-aid agreement (emergency management assistance compact).					
4.6.2	Mutual-aid agreements/compacts are developed with neighboring jurisdictions.					
4.6.3	Mutual-aid agreements/compacts are developed with tribal nations.					
4.6.4	Mutual-aid agreements/compacts are developed with private sector businesses and industries.					
4.6.5	Guidance and assistance are provided to local government agencies to develop mutual-aid agreements/compacts.					
4.7	The jurisdiction has a knowledge of state resources that can be expected as a result of a request for assistance.					
4.7.1	Personnel are trained on the provisions of the state response plan.					
4.7.2	The jurisdiction complies with established procedures for requesting state resources.					
4.7.3	The jurisdiction provides guidance to departments/agencies on procedures to receive resources during a disaster.					
4.8	Staging areas have been pre-identified to receive resources in the event of an emergency.					
4.8.1	Locations throughout the jurisdiction are identified and arrangements made for use as staging areas for receipt and distribution of critical resources.					
4.8.2	Personnel are assigned to staff staging areas.					
4.8.3	Standard operations procedures are developed to manage the receipt and distribution of resources at staging areas.					
4.8.4	The local EOP identifies the locations of state mobilization centers and staging areas.					
4.9	The jurisdiction has acquired appropriate and sufficient equipment for response to weapons of mass destruction (WMD), terrorism, and/or chemical, biological, radiological, nuclear, or high-yield explosive events.					
4.9.1	Hazmat response resources are sufficient for a WMD terrorism incident requiring mass decontamination.					
4.9.4	Available hazmat response resources are adequately equipped and able to identify a wide variety of chemical agents, including nerve, blister, blood, and choking agents.					
4.9.5	Available response resources are adequately equipped and able to detect and make preliminary identification of biological agents, including but not limited to anthrax, smallpox, tularemia, Q fever, and Venezuelan Equine Encephalitis (VEE).					

4.9.6	The jurisdiction has identified other WMD capable response teams and WMD-related equipment sources, such as private/corporate, intrastate, and military.					
4.9.7	The jurisdiction has the capability to obtain adequate pharmaceuticals in the event of a WMD incident.					
4.9.8	The jurisdiction has identified the resource requirements (personnel and equipment) for site remediation.					

No. 5. Planning: The collection, analysis, and use of information, and the development, promulgation, and maintenance of the comprehensive emergency management, action, and mitigation plans.

	Capabilities Rating	***Fully Prepared***	***Very Prepared***	***Generally Prepared***	***Marginally Prepared***	***Not Prepared***
5.1	A comprehensive mitigation plan has been developed.					
5.1.1	The plan identifies hazards by associating them with a geographic location, a geological feature (e.g., river and earthquake fault), and a political jurisdiction.					
5.1.2	A scale is used to identify the likelihood of each hazard-related event.					
5.1.3	Based upon location, potential magnitude, and likelihood of an event, the jurisdiction has estimated the risk for each hazard (e.g., the number of people, the number of buildings, the number/type of critical facilities, and the type of infrastructure.)					
5.1.4	The plan contains a description and analysis of the local hazard management policies, programs, and capabilities to mitigate the hazards of the area.					
5.1.5	The plan contains hazard mitigation goals and objectives and proposed strategies, programs, and actions to reduce or avoid long-term vulnerability to hazards.					
5.1.6	The plan contains statements of goals and objectives related to the local mitigation-planning program.					
5.1.7	The plan contains a method for annual evaluation and updating.					
5.1.8	The plan contains long-term recovery strategies, including sustainability issues and corresponding resources to address them.					
5.1.9	The plan contains strategies, programs, and actions to reduce or avoid long-term vulnerability to those hazards identified.					
5.1.10	The plan prioritizes projects and/or initiatives based on the greatest opportunity for loss reduction.					
5.1.11	The plan documents how specific mitigation actions can contribute to the overall risk reduction.					
5.1.12	The plan addresses an education and outreach strategy.					
5.1.13	The plan is modified based on pre- and post-disaster mitigation requirements.					
5.2	The jurisdiction has developed an EOP that prescribes roles and responsibilities during disaster operations.					

(continued)

Capabilities Rating *(continued)*

5.2.1	The EOP has functional annexes; hazard-specific annexes where appropriate and has been promulgated by the chief elected/appointed official.					
5.2.2	A plan maintenance program has been established, including a schedule for updating annexes and SOPs.					
5.2.3	The EOP interfaces with all appropriate state/local plans.					
5.2.4	Voluntary organizations' emergency plans interface with the EOP.					
5.2.5	The EOP addresses the response to all hazards identified in the Hazard Identification and Risk Assessment.					
5.3	Direction, control, and coordination are addressed in the EOP.					
5.3.1	Roles and responsibilities of organizations and individuals are described.					
5.3.2	Coordination links among state, local, and private sector organizations that are part of the overall response organization are identified.					
5.3.3	The primary and alternate emergency operations centers (EOCs) are identified.					
5.3.4	Mobile or fixed EOCs are established with appropriate coordination links to the local EOC.					
5.3.5	Alternate methods of communication have been identified for direction, control, and coordination.					
5.4	Altering and notification are addressed in the EOP.					
5.4.1	Primary and alternate alerting and notification systems are established.					
5.4.2	Roles and responsibilities for alerting and notification are addressed in the plan (e.g., primary/secondary agency, levels of alert status based on situation).					
5.4.3	Alerting and notification SOPs are reviewed and updated at least annually.					
5.4.4	Primary and alternate systems are tested on a regular basis.					
5.5	Warning is addressed in the EOP.					
5.5.1	Primary and alternate warning systems are established.					
5.5.2	Roles and responsibilities assigned to local individuals and organizations are described.					
5.5.3	The EOP addresses the emergency alert system (EAS) and backup warning systems.					
5.5.4	A regular schedule for testing and maintenance of the warning system(s) and training of personnel is addressed in the EOP.					
5.5.5	The EOP describes procedures for warning special locations (e.g., schools, hospitals, nursing homes, and business/industry).					
5.5.6	The EOP describes procedures for warning special needs populations (e.g., hearing impaired).					
5.5.7	Warning SOPs and checklists are developed and updated at least annually.					
5.6	Communications is addressed in the EOP.					
5.6.1	The roles and responsibilities for communications are addressed in the EOP.					
5.6.2	A current inventory of local communications resources/capabilities is maintained.					

5.6.3	Communications SOPs and checklists are developed and updated at least annually.					
5.6.4	The EOP includes a schedule for communications equipment testing and maintenance.					
5.6.5	The EOP addresses the integration of private sector/voluntary agency communications capabilities (e.g., radio amateur, civil emergency services, and amateur radio emergency services)					
5.7	Emergency Public Information (EPI) is addressed in the EOP.					
5.7.1	Roles and responsibilities for EIP personnel, including the emergency public information officer (PIO), are addressed.					
5.7.2	Pre-scripted information bulletins are developed for release to the media in the event of an emergency.					
5.7.3	PIO SOPs and checklists are developed, reviewed, and updated at least annually.					
5.7.4	The EOP addresses the concept of a joint information system, including the joint information center (JIC), as a location for the coordination of local, state, and federal agency information releases.					
5.8	Resource management is addressed in the EOP.					
5.8.1	The roles and responsibilities for resource management are addressed.					
5.8.2	A concept of operations to identify, acquire, and then direct and control the flow of critical resources in an emergency is developed.					
5.8.3	The EOP addresses the prioritization strategy for resource allocation (e.g., life safety and property protection).					
5.9	Evacuation is addressed in the EOP.					
5.9.1	The roles and responsibilities for evacuation are addressed.					
5.9.2	The EOP identifies the authorities required to order evacuations.					
5.9.3	The EOP describes procedures for evacuating special needs populations (e.g., schoolchildren and nursing home residents.					
5.9.4	The EOP includes procedures to keep evacuees and the general public informed of evacuation activities and specific actions that should be taken.					
5.9.5	The EOP addresses evacuation options and primary and alternate routes.					
5.9.6	The EOP addresses the transportation resources required for evacuation.					
5.9.7	The EOP identifies assembly areas for people without transportation.					
5.9.8	The EOP is coordinated with the mass care plan.					
5.9.9	The EOP addresses provisions to control access to evacuated areas.					
5.9.10	The EOP addresses provisions to provide security in areas evacuated.					
5.9.11	The EOP addresses reentry procedures.					
5.9.12	The EOP addresses host community status of the jurisdiction for evacuees from other jurisdictions.					
5.10	Mass care is addressed in the EOP.					
5.10.1	The roles and responsibilities for the local agencies charged with mass care operations are addressed.					

(continued)

Capabilities Rating *(continued)*

5.10.2	Coordination among local and private sector, nonprofit, and public services organizations for mass care operations is addressed.					
5.10.3	The EOP addresses local resources that may be used to support mass care operations (e.g., mass care facilities, communications, food, water, health/medical, registration, a system to reunite families, and disaster housing).					
5.10.4	The EOP addresses the mass care requirements for special needs populations.					
5.10.5	Mass care planning is coordinated with animal control planning to address sheltering/mass care.					
5.10.6	The plan addresses the requirements of both short-term and long-term sheltering/mass care.					
5.10.7	SOPs/checklists for mass care operations are developed and updated at least annually.					
5.11	In-place sheltering is addressed in the EOP.					
5.11.1	The EOP identifies the person/position having authority for initiating in-place sheltering.					
5.11.2	The EOP addresses local government capabilities for in-place sheltering (e.g., public information, shelter management, special needs populations, and registration).					
5.11.3	The EOP outlines the decision-making criteria used for initiating in-place sheltering.					
5.12	Needs assessment is addressed in the EOP or other documents.					
5.12.1	The document addresses the roles and responsibilities of the agency(ies) charged with conducting needs assessments.					
5.12.2	The EOP provides criteria to identify life support requirements, critical infrastructure needs, human needs, etc.					
5.12.3	The document is coordinated with allied plans such as resource management and damage assessment.					
5.12.4	The document addresses the differences between the needs assessments at various intervals (e.g., rapid [24 hours], intermediate [72 hours], and unmet [+72 hours]).					
5.12.5	The document addresses the responsibility of voluntary agencies to assess the personal needs of citizens (e.g., clothing, food, and medications) destroyed during the disaster event.					
5.13	Damage assessment is addressed in the EOP.					
5.13.1	The roles and responsibilities for damage assessment are identified.					
5.13.2	Local damage assessment teams are identified, trained, and equipped with the necessary forms, transportation, and communications.					
5.13.3	The EOP addresses assessing both public infrastructure/facilities and private property.					
5.13.4	The EOP addresses the coordination between damage assessment and needs assessment in communicating identified needs.					
5.13.5	Provisions for integrating state damage assessment teams in the event of a major disaster are addressed.					

5.13.6	Provisions for integrating federal rapid needs assessment teams in the event of a major disaster are addressed.					
5.14	The EOP addresses the uses of Department of Defense (DOD) (active military units) resources during a disaster event.					
5.14.1	Staff is aware of state procedures to request military support for both active duty units and national guard.					
5.14.2	The EOP addresses the limitation of military support for both active duty units and the national guard.					
5.15	Donations management is addressed in the EOP.					
5.15.1	The roles and responsibilities for the donations manager are addressed.					
5.15.2	The EOP has provisions for soliciting, receipt, storage, and distribution of donated goods.					
5.15.3	Donations management planning is coordinated with allied components of the EOP, such as resource management.					
5.16	The role of voluntary organizations during disasters is addressed in the EOP.					
5.16.1	The EOP addresses the roles and responsibilities of voluntary organizations.					
5.16.2	Policies and procedures are developed for the utilization of local voluntary organizations (e.g., ARC, Salvation Army, religious organizations, and service organizations) during the response phase of a disaster.					
5.16.3	The role of voluntary organizations active in disaster is addressed in the EOP.					
5.16.4	SOPs/checklists are developed and updated at least annually.					
5.17	Response to WMD terrorism is addressed in the EOP.					
5.17.1	The roles and responsibilities of local agencies for WMD terrorism are addressed in the EOP.					
5.17.2	The jurisdiction has established an interagency working group to identify targets, develop resource SOPs, integrate mutual-aid resources, identify resource shortfalls, and gather and analyze intelligence data for threat assessments.					
5.17.3	Pre-incident planning/surveys are developed for identified target sites.					
5.17.4	WMD terrorism planning addresses the staging, transportation, and security issues for the movement of response assets (e.g., personnel and equipment).					
5.17.5	The EOP addresses the use of a unified command structure for multi-agency, intergovernmental response to WMD incidents.					
5.17.6	Local personnel are identified to support Federal Bureau of Investigation–led command and coordination structures (e.g., crisis and consequences management groups in the joint operations center).					
5.17.7	The EOP addresses alerting and notifying the cognizant FBI field office and state EMO of potential terrorist incidents.					
5.17.8	Protocols exist for local consequence management agencies to obtain information on credible threats from local, state, federal law enforcement agencies, as necessary.					
5.17.9	The EOP addresses the use of a joint information system for multi-agency, intergovernmental release of information.					

(continued)

***Capabilities Rating** (continued)*

5.17.10	The EOP addresses site remediation (e.g., "How clean is clean?") for the different types of WMD agents.					
5.17.11	Fact sheets and "canned" messages are developed regarding chemical and biological (C/B) agents and radiation dispersal.					
5.17.12	SOPs/checklists/field operations guides are developed and updated at least annually.					
5.18	Law enforcement is addressed in the EOP.					
5.18.1	The roles and responsibilities for law enforcement agencies are identified.					
5.18.2	The coordination of law enforcement personnel during an emergency is addressed.					
5.18.3	SOPs/checklists are developed and updated at least annually.					
5.19	Fire protection is addressed in the EOP.					
5.19.1	The roles and responsibilities for fire protection are identified.					
5.19.2	An inventory of local and mutual-aid fire protection resources has been developed and maintained.					
5.19.3	SOPs/checklists are developed and updated at least annually.					
5.20	Search and Rescue (SAR) is addressed in the EOP.					
5.20.1	The primary agency(ies) responsible for SAR are identified.					
5.20.2	Roles and responsibilities for SAR are identified.					
5.20.3	The EOP addresses the response to various types of SAR missions (e.g., wilderness, water, air, urban, and below ground).					
5.20.4	An inventory of local and mutual-aid resources has been developed and maintained.					
5.20.5	The EOP addresses the process for requesting SAR resources.					
5.20.6	SOPs/checklists are developed and updated at least annually.					
5.21	Public health is addressed in the EOP.					
5.21.1	The roles and responsibilities for public health services are identified.					
5.21.2	An inventory of public health resources (e.g., laboratory facilities and pharmaceutical/vaccine caches) is developed and maintained.					
5.21.3	"Boilerplate" public health advisories are developed.					
5.21.4	The roles and responsibilities for allied professions (e.g., medical and emergency management) in public health emergencies are identified.					
5.21.5	Provisions for conducting analysis to determine cause, origin, and scope of epidemics are addressed.					
5.22	Emergency medical services (EMS) is addressed in the EOP.					
5.22.1	The roles and responsibilities of the medical community (hospitals, urgent care facilities, and EMS) are identified.					
5.22.2	An inventory of local medical resources is developed and maintained.					
5.22.3	A plan for mass casualty/mass fatality operations is developed and coordinated among appropriate agencies and private sector organizations.					
5.22.4	The EOP addresses the mental health needs of individuals affected by a disaster.					

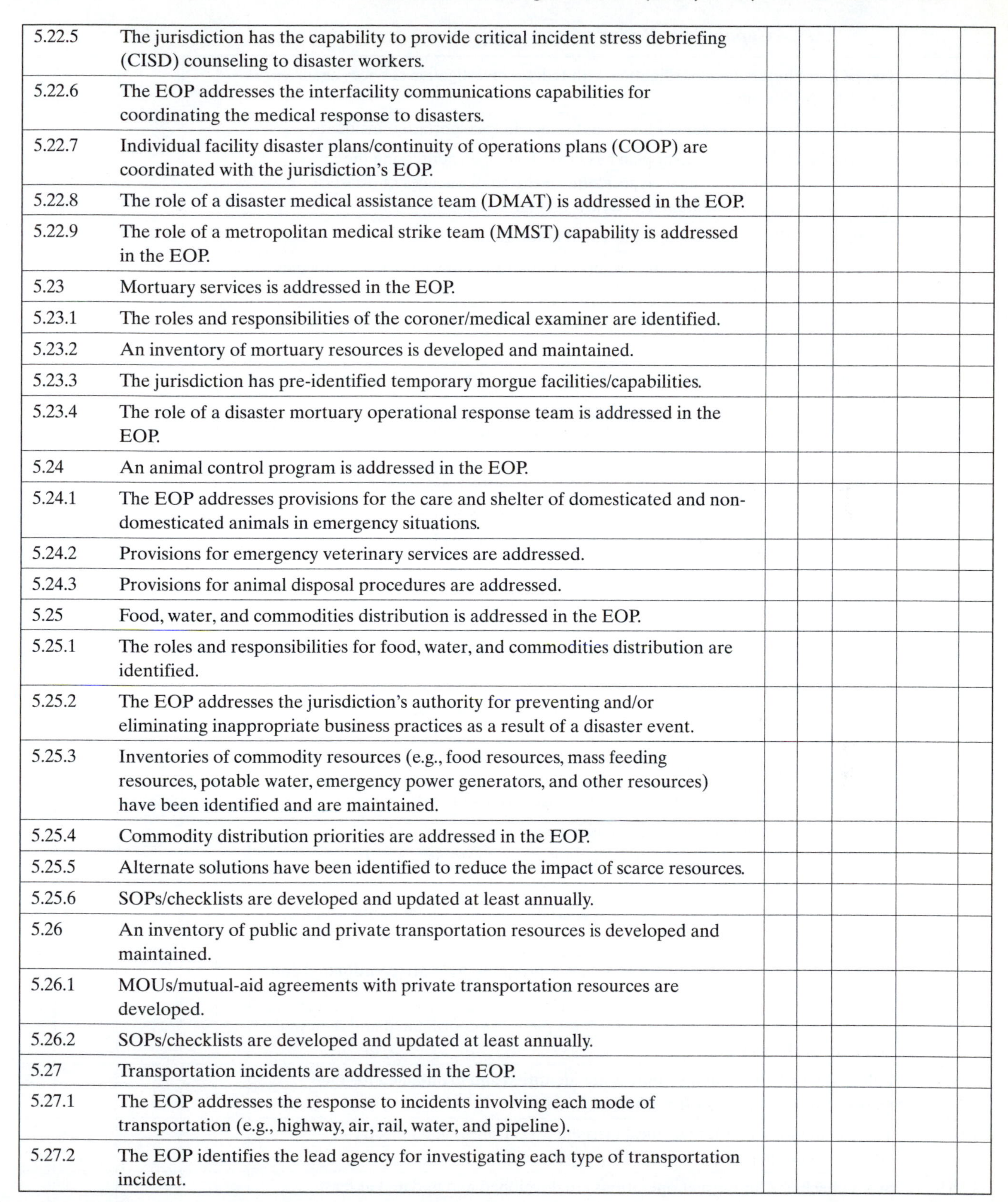

5.22.5	The jurisdiction has the capability to provide critical incident stress debriefing (CISD) counseling to disaster workers.					
5.22.6	The EOP addresses the interfacility communications capabilities for coordinating the medical response to disasters.					
5.22.7	Individual facility disaster plans/continuity of operations plans (COOP) are coordinated with the jurisdiction's EOP.					
5.22.8	The role of a disaster medical assistance team (DMAT) is addressed in the EOP.					
5.22.9	The role of a metropolitan medical strike team (MMST) capability is addressed in the EOP.					
5.23	Mortuary services is addressed in the EOP.					
5.23.1	The roles and responsibilities of the coroner/medical examiner are identified.					
5.23.2	An inventory of mortuary resources is developed and maintained.					
5.23.3	The jurisdiction has pre-identified temporary morgue facilities/capabilities.					
5.23.4	The role of a disaster mortuary operational response team is addressed in the EOP.					
5.24	An animal control program is addressed in the EOP.					
5.24.1	The EOP addresses provisions for the care and shelter of domesticated and non-domesticated animals in emergency situations.					
5.24.2	Provisions for emergency veterinary services are addressed.					
5.24.3	Provisions for animal disposal procedures are addressed.					
5.25	Food, water, and commodities distribution is addressed in the EOP.					
5.25.1	The roles and responsibilities for food, water, and commodities distribution are identified.					
5.25.2	The EOP addresses the jurisdiction's authority for preventing and/or eliminating inappropriate business practices as a result of a disaster event.					
5.25.3	Inventories of commodity resources (e.g., food resources, mass feeding resources, potable water, emergency power generators, and other resources) have been identified and are maintained.					
5.25.4	Commodity distribution priorities are addressed in the EOP.					
5.25.5	Alternate solutions have been identified to reduce the impact of scarce resources.					
5.25.6	SOPs/checklists are developed and updated at least annually.					
5.26	An inventory of public and private transportation resources is developed and maintained.					
5.26.1	MOUs/mutual-aid agreements with private transportation resources are developed.					
5.26.2	SOPs/checklists are developed and updated at least annually.					
5.27	Transportation incidents are addressed in the EOP.					
5.27.1	The EOP addresses the response to incidents involving each mode of transportation (e.g., highway, air, rail, water, and pipeline).					
5.27.2	The EOP identifies the lead agency for investigating each type of transportation incident.					

(continued)

Capabilities Rating *(continued)*

5.27.3	The EOP addresses the coordination with federal investigative services agencies (e.g., FBI, NTSB, etc.).					
5.27.4	The EOP addresses the role of family assistance in transportation incidents.					
5.27.5	A traffic incident management plan has been developed addressing the coordinated response to highway incidents or the consequences of disasters (e.g., evacuations) to maintain traffic flow on highways.					
5.28	Energy and utilities services are addressed in the EOP.					
5.28.1	The roles and responsibilities of local energy and utility departments/companies during disaster events are addressed.					
5.28.2	Each utility has identified the consequences to their respective systems based on the jurisdiction's hazards.					
5.28.3	The jurisdiction has assessed the risk to critical infrastructure systems by each utility system.					
5.28.4	The jurisdiction and the utility departments/companies have developed a critical infrastructure restoration plan.					
5.28.5	An inventory of energy and utilities resources is identified and maintained by utility departments/companies and is available to emergency management officials.					
5.28.6	SOPs/checklists are developed and updated at least annually.					
5.29	Public works and engineering services are addressed in the EOP.					
5.29.1	Roles and responsibilities of public works and engineering services are addressed.					
5.29.2	Public and private resources are inventoried and maintained.					
5.29.3	Procedures for the removal and disposal of disaster-related debris and other damage from public and private lands or waters are developed.					
5.29.4	SOPs/checklists are developed and updated at least annually.					
5.30	Hazmat planning is addressed in the EOP.					
5.30.1	Roles and responsibilities for hazmat responses are identified.					
5.30.2	Tier 1 and Tier 2 hazmat facilities are identified.					
5.30.3	Pre-incident response plans are developed for each of the Tier 1 and Tier 2 facilities.					
5.30.4	A local emergency planning committee is established to coordinate the mandates of state and federal hazmat laws.					
5.30.5	Chemical stockpile emergency preparedness program plans are coordinated and reviewed annually by the state.					
5.30.6	Local radiological emergency preparedness plans are coordinated and reviewed annually by the state.					
5.30.7	Public and private hazmat resources are identified and maintained for the response to, and recovery from, hazmat incidents.					
5.30.8	The jurisdiction provides technical assistance to business and industry in developing hazmat response and recovery plans.					
5.30.9	SOPs/checklists for hazmat operations are developed and updated at least annually.					

5.31	The EOP addresses emergency action plans developed by dam owners for the response required in the event of a dam-related emergency.					
5.31.1	The jurisdiction provides technical assistance to the dam owners in the development of EAPs.					
5.31.2	The jurisdiction develops response plans based on the consequences of a dam failure.					
5.31.3	Response resources are identified and updated annually.					
5.31.4	The jurisdiction maintains a copy of each EAP in accordance with the state dam safety program.					
5.32	The jurisdiction has developed plans for continuity of operations in the event of disaster/disruption.					
5.32.1	Each department/agency has conducted an internal hazard identification and risk assessment related to their facilities.					
5.32.2	Each department/agency has identified its mission critical systems and developed contingency plans in the event of a disaster/failure.					
5.32.3	Each department/agency has identified the key staff required to maintain continuity of operations in the event of a disaster at its facility.					
5.32.4	Each department agency has developed a plan for safeguarding vital records.					
5.32.5	Each department/agency has developed its individual requirements for a relocation site to maintain operations (e.g., space, utilities, and personnel).					
5.32.6	SOPs/checklists are developed for each department/agency.					
5.32.7	The emergency management agency provides technical assistance to other departments/agencies in the development of continuity of operations plans.					
5.32.8	The emergency management agency provides assistance to business and industry in the development of continuity of operations plans.					

No. 6. Direction, Control, and Coordination: Development of the capability for the senior HSS officer to direct, control, and coordinate medical response and recovery operations.

	Capabilities Rating	***Fully Prepared***	***Very Prepared***	***Generally Prepared***	***Marginally Prepared***	***Not Prepared***
6.1	The emergency operations center (EOC) operating procedures are developed and tested at least annually.					
6.1.1	An EOC operations manual is developed and maintained.					
6.1.2	An incident command system (ICS) is the foundation for direction, control, and coordination.					
6.1.3	EOC staff procedures, action guides, and SOPs are developed and updated at least annually.					

(continued)

***Capabilities Rating** (continued)*

6.1.4	Activation and deactivation of the EOC for emergency operations are defined.					
6.1.5	The duties of all EOC staff groups, including other local agency representatives, have been developed and are available to the EOC personnel.					
6.1.6	Written procedures exist for information handling.					
6.1.7	EOC equipment (e.g., phones, faxes, and computers) are tested and maintained and/or updated regularly.					
6.2	Intrajurisdiction coordination is established within the local operational area and other entities.					
6.2.1	Procedures for coordination are established with local elected/appointed officials and departments and agency heads.					
6.2.2	Procedures for coordination are established with the local media.					
6.2.3	Procedures for coordination are established with the local private sector business, industry, and professional organizations.					
6.2.4	Procedures for coordination are established with the elected/appointed officials of subjurisdictions (e.g., county to city).					
6.3	Interjurisdictional coordination is established between the local operational area agencies/organizations.					
6.3.1	Procedures for coordination are established with neighboring jurisdictions.					
6.3.2	Procedures for coordination are established with state-level voluntary organizations.					
6.4	The jurisdiction establishes coordination with the state response agencies and other state supporting agencies.					
6.4.1	The jurisdiction participates in state emergency management organization meetings, conferences, and working groups.					
6.4.2	The jurisdiction has procedures for coordination with the state emergency management organization in accordance with the state emergency operations/response plan.					

No. 7. Communications and Warning: Development and maintenance of a reliable communications capability to alert officials and emergency response personnel, warn the public, and effectively manage response to an actual or impending emergency.

Capabilities Rating		***Fully Prepared***	***Very Prepared***	***Generally Prepared***	***Marginally Prepared***	***Not Prepared***
7.1	Communications system capabilities are established.					
7.1.1	Procedures exist to coordinate available public and private communications systems and equipment.					
7.1.2	Primary and backup communications systems are developed, tested, and maintained.					

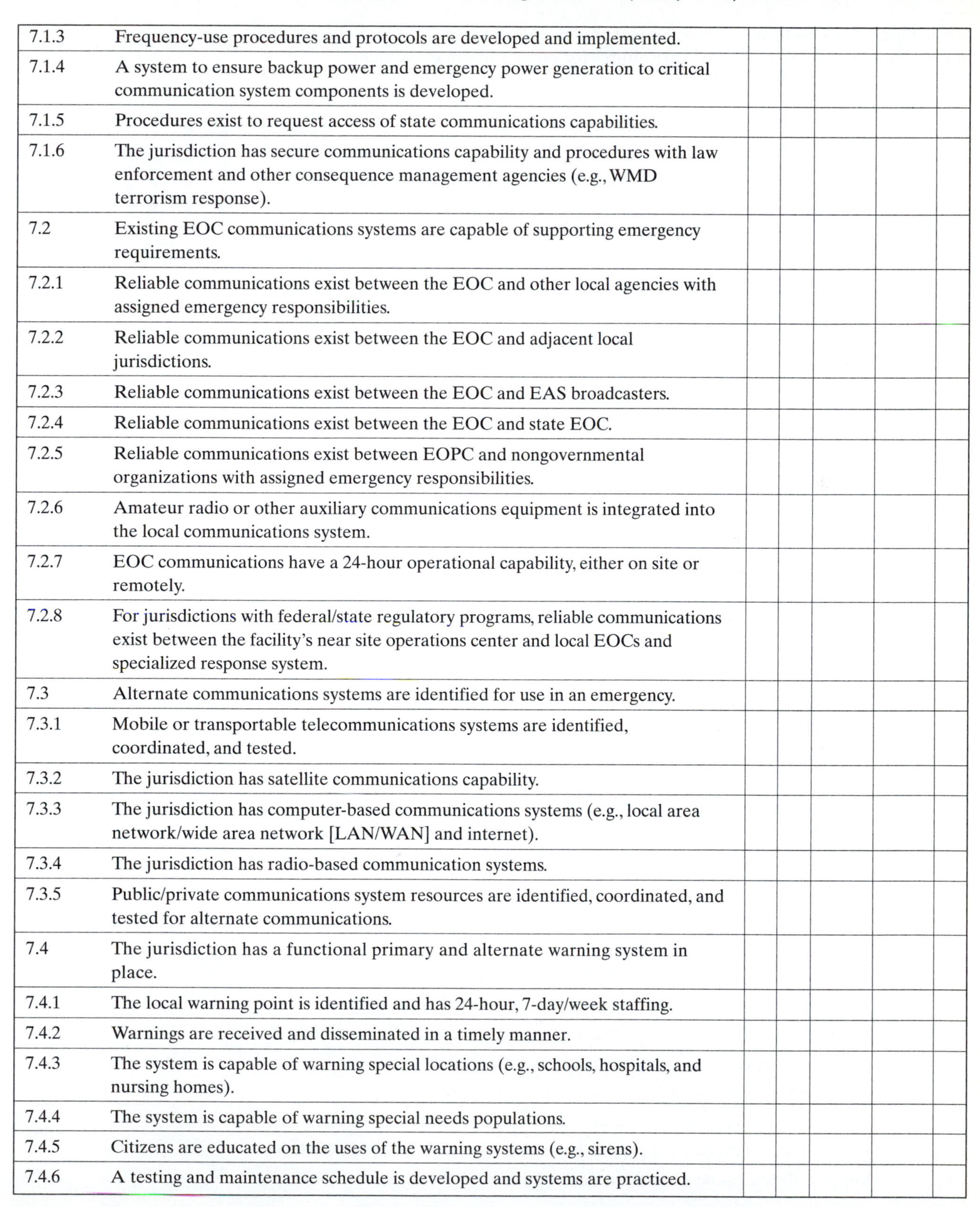

7.1.3	Frequency-use procedures and protocols are developed and implemented.					
7.1.4	A system to ensure backup power and emergency power generation to critical communication system components is developed.					
7.1.5	Procedures exist to request access of state communications capabilities.					
7.1.6	The jurisdiction has secure communications capability and procedures with law enforcement and other consequence management agencies (e.g., WMD terrorism response).					
7.2	Existing EOC communications systems are capable of supporting emergency requirements.					
7.2.1	Reliable communications exist between the EOC and other local agencies with assigned emergency responsibilities.					
7.2.2	Reliable communications exist between the EOC and adjacent local jurisdictions.					
7.2.3	Reliable communications exist between the EOC and EAS broadcasters.					
7.2.4	Reliable communications exist between the EOC and state EOC.					
7.2.5	Reliable communications exist between EOPC and nongovernmental organizations with assigned emergency responsibilities.					
7.2.6	Amateur radio or other auxiliary communications equipment is integrated into the local communications system.					
7.2.7	EOC communications have a 24-hour operational capability, either on site or remotely.					
7.2.8	For jurisdictions with federal/state regulatory programs, reliable communications exist between the facility's near site operations center and local EOCs and specialized response system.					
7.3	Alternate communications systems are identified for use in an emergency.					
7.3.1	Mobile or transportable telecommunications systems are identified, coordinated, and tested.					
7.3.2	The jurisdiction has satellite communications capability.					
7.3.3	The jurisdiction has computer-based communications systems (e.g., local area network/wide area network [LAN/WAN] and internet).					
7.3.4	The jurisdiction has radio-based communication systems.					
7.3.5	Public/private communications system resources are identified, coordinated, and tested for alternate communications.					
7.4	The jurisdiction has a functional primary and alternate warning system in place.					
7.4.1	The local warning point is identified and has 24-hour, 7-day/week staffing.					
7.4.2	Warnings are received and disseminated in a timely manner.					
7.4.3	The system is capable of warning special locations (e.g., schools, hospitals, and nursing homes).					
7.4.4	The system is capable of warning special needs populations.					
7.4.5	Citizens are educated on the uses of the warning systems (e.g., sirens).					
7.4.6	A testing and maintenance schedule is developed and systems are practiced.					

No. 8. Operations and Procedures: Development, coordination, and implementation of operational policies, plans, and procedures for emergency management.

	Capabilities Rating	***Fully Prepared***	***Very Prepared***	***Generally Prepared***	***Marginally Prepared***	***Not Prepared***
8.1	The jurisdiction has developed procedures for conducting needs assessments.					
8.1.1	The jurisdiction conducts a needs assessment in a disaster to identify life support requirements, critical infrastructure needs, and other key issues.					
8.1.2	Procedures are established to analyze and prioritize the needs, determine local availability of resources, or submit requests to the state.					
8.2	The jurisdiction has developed procedures for conducting damage assessments.					
8.2.1	The jurisdiction has developed procedures to activate and deploy damage assessment teams to collect damage information.					
8.2.2	The jurisdiction has the ability to obtain pre-disaster maps, photographs, and other documents.					
8.2.3	The jurisdiction has developed procedures to compile local data and submit preliminary damage assessment reports.					
8.2.4	The jurisdiction has developed procedures to identify unsafe structures.					
8.2.5	Personnel have received training in damage assessment procedures consistent with state requirements.					
8.3	The jurisdiction has developed procedures for requesting disaster assistance.					
8.3.1	Staff is familiar with the provisions for requesting disaster assistance according to the state response plan.					
8.3.2	Staff is familiar with the local requirements associated with state and federal disaster assistance.					
8.4	The jurisdiction has developed administrative procedures supporting pre-trans, and post-disaster response and recovery operations.					
8.4.1	Emergency response procedures incorporate ICS.					
8.4.2	Staff action guides/SOPs have been developed.					
8.4.3	Procedures are developed to augment existing human resources during disaster operations.					
8.4.4	Procedures are developed for information gathering and assessment.					
8.4.5	Procedures are developed to generate post-emergency disaster reports.					
8.4.6	Procedures are developed to conduct formal post-emergency disaster critiques.					
8.4.7	Procedures are developed for a corrective action program.					
8.4.8	Procedures are developed to support regulatory programs.					
8.5	The jurisdiction has developed procedures for military support operations.					
8.5.1	Procedures are developed to provide for coordination with DOD military units.					
8.6	The jurisdiction has developed procedures for law enforcement operations.					

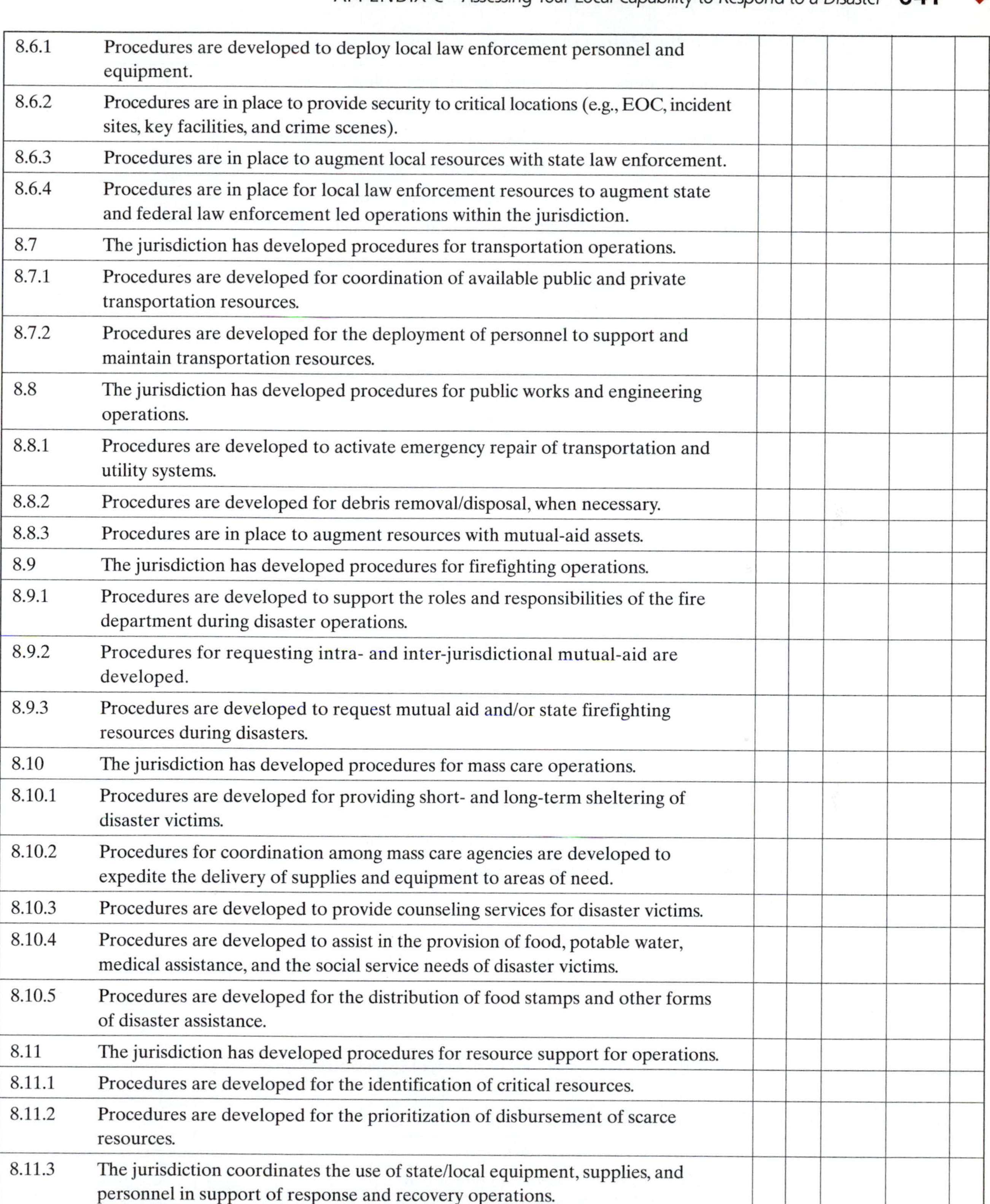

8.6.1	Procedures are developed to deploy local law enforcement personnel and equipment.					
8.6.2	Procedures are in place to provide security to critical locations (e.g., EOC, incident sites, key facilities, and crime scenes).					
8.6.3	Procedures are in place to augment local resources with state law enforcement.					
8.6.4	Procedures are in place for local law enforcement resources to augment state and federal law enforcement led operations within the jurisdiction.					
8.7	The jurisdiction has developed procedures for transportation operations.					
8.7.1	Procedures are developed for coordination of available public and private transportation resources.					
8.7.2	Procedures are developed for the deployment of personnel to support and maintain transportation resources.					
8.8	The jurisdiction has developed procedures for public works and engineering operations.					
8.8.1	Procedures are developed to activate emergency repair of transportation and utility systems.					
8.8.2	Procedures are developed for debris removal/disposal, when necessary.					
8.8.3	Procedures are in place to augment resources with mutual-aid assets.					
8.9	The jurisdiction has developed procedures for firefighting operations.					
8.9.1	Procedures are developed to support the roles and responsibilities of the fire department during disaster operations.					
8.9.2	Procedures for requesting intra- and inter-jurisdictional mutual-aid are developed.					
8.9.3	Procedures are developed to request mutual aid and/or state firefighting resources during disasters.					
8.10	The jurisdiction has developed procedures for mass care operations.					
8.10.1	Procedures are developed for providing short- and long-term sheltering of disaster victims.					
8.10.2	Procedures for coordination among mass care agencies are developed to expedite the delivery of supplies and equipment to areas of need.					
8.10.3	Procedures are developed to provide counseling services for disaster victims.					
8.10.4	Procedures are developed to assist in the provision of food, potable water, medical assistance, and the social service needs of disaster victims.					
8.10.5	Procedures are developed for the distribution of food stamps and other forms of disaster assistance.					
8.11	The jurisdiction has developed procedures for resource support for operations.					
8.11.1	Procedures are developed for the identification of critical resources.					
8.11.2	Procedures are developed for the prioritization of disbursement of scarce resources.					
8.11.3	The jurisdiction coordinates the use of state/local equipment, supplies, and personnel in support of response and recovery operations.					

(continued)

***Capabilities Rating** (continued)*

8.11.4	Procedures are developed to identify and resolve unmet disaster infrastructure needs.					
8.11.5	Procedures are developed to identify and resolve unmet disaster human needs.					
8.11.6	The jurisdiction maintains an inventory control and delivery system.					
8.12	The jurisdiction has developed procedures for public health operations.					
8.12.1	Procedures are developed for a public health response to disasters including disease, vector, waste, food, and water control.					
8.12.2	Procedures are developed for the coordination with state/federal health officials in the event of a public health emergency.					
8.12.3	Procedures for the coordination of information between pubic health officials and allied health professionals are developed in the event of a public health emergency.					
8.12.4	Procedures for the direction, control, and coordination of the jurisdiction's response to a public health emergency are developed.					
8.12.5	Procedures are developed to inform the public of major environmental and health hazards.					
8.13	The jurisdiction developed procedures for emergency medical services operations.					
8.13.1	Procedures are developed for coordinated medical response to disasters.					
8.13.2	Procedures are developed for interhospital communications/information sharing of the signs and symptoms of potential public health emergencies.					
8.13.3	Facility disaster plans are tested at least annually.					
8.13.4	A system for alerting doctors' offices/urgent care centers/other medical facilities of possible epidemics is developed and tested.					
8.13.5	The medical community has developed procedures for the decontamination of patients prior to entry into the facility.					
8.13.6	Procedures have been developed for the use of mutual-aid EMS resources in the event of a disaster.					
8.13.7	Procedures have been developed and personnel have been trained in the use of appropriate PPE by medical personnel, depending on the nature of the incident.					
8.13.8	Procedures have been developed for mental health/crisis counseling in a disaster situation.					
8.13.9	Procedures are developed for mass casualty operations.					
8.13.10	Procedures are developed for the integration of a DMAT into disaster operations.					
8.13.11	Procedures are developed for the integration of an MMST into disaster operations.					
8.13.12	Procedures are developed for the care and transportation of special needs populations.					
8.14	The jurisdiction has developed procedures for mortuary operations.					
8.14.1	Procedures for mass fatality operations are developed.					
8.14.2	Procedures for family assistance requirements are developed.					
8.14.3	MOUs are developed with local funeral homes in the event of a mass fatality incident.					

8.14.4	Security procedures are developed for mass fatality operations.						
8.14.5	The jurisdiction has exercised its mass fatality plan to include the activation and setup of temporary morgues.						
8.15	The jurisdiction has developed procedures for SAR operations.						
8.15.1	The jurisdiction maintains a current inventory of equipment and trained personnel for SAR operations, as appropriate.						
8.15.2	Procedures are developed to activate SAR personnel, based on the type of SAR mission.						
8.15.3	Procedures are developed to integrate state/federal SAR assets into the jurisdiction's incident command system.						
8.15.4	The jurisdiction's SAR personnel are provided training to support the different requirements of a SAR incident.						
8.16	The jurisdiction has developed procedures for HAZMAT operations.						
8.16.1	Procedures are developed in accordance with local/state/federal regulations.						
8.16.2	Procedures are developed to activate and deploy a HAZMAT Response Team, if needed.						
8.16.3	Procedures are developed to request intra- and inter-jurisdictional HAZMAT mutual aid.						
8.16.4	Procedures are developed to coordinate the response and recovery issues with public/private HAZMAT resources.						
8.16.5	HAZMAT Response Team members train and participate in disaster exercises annually, as appropriate.						
8.17	The jurisdiction has developed procedures for the management of donated goods.						
8.17.1	Procedures are developed to facilitate the distribution of donations.						
8.17.2	An inventory and database of donated goods are available.						
8.17.3	Procedures are developed for the coordination of the infrastructure facility requirements for donated goods.						
8.17.4	Procedures are developed for the implementation and management of a donations hotline.						
8.18	The jurisdiction has developed procedures for the coordination of volunteer services.						
8.18.1	An inventory and database of volunteer services are developed.						
8.18.2	Procedures are developed to facilitate the deployment of volunteers in a disaster.						
8.18.3	Procedures are developed to reduce the jurisdiction's liability in the use of convergent volunteers.						
8.18.4	Procedures are developed to manage convergent volunteers.						
8.19	The jurisdiction has developed procedures for the coordination of energy and utilities services during disaster operations.						
8.19.1	Procedures are developed for prioritizing the repair and restoration of critical infrastructures.						

(continued)

***Capabilities Rating** (continued)*

8.19.2	Critical energy facilities are pre-identified.					
8.20	The jurisdiction has developed procedures to provide for animal populations in disaster response.					
8.20.1	Procedures are developed to provide for the health and safety of farm/work animals in disaster operations.					
8.20.2	Procedures are developed to provide for the health and safety of family pets in disaster operations.					
8.20.3	Procedures are developed for establishing shelter facilities for animals.					
8.21	The jurisdiction has developed procedures for WMD terrorism operations.					
8.21.1	The jurisdiction has developed deployment procedures for WMD incidents.					
8.21.2	Procedures include the awareness of, and activities associated with, the potential use of secondary devices.					
8.21.3	The jurisdiction has identified local, state, and federal resources capable of responding to WMD terrorism incidents.					
8.21.4	Response procedures incorporate the use of specialized teams (e.g., EOD, MMST, DMAT).					
8.21.5	Procedures address both crisis management and consequence management aspects of a WMD terrorism event, including activities during a threat period.					
8.21.6	Procedures are developed for the coordinated response to nuclear, biological, and chemical emergencies.					
8.21.7	Procedures are developed for the mass decontamination of victims of a WMD event, both on site and self-referring victims to medical facilities.					
8.22	The jurisdiction has developed procedures for the protection and storage of vital records.					
8.22.1	The jurisdiction has identified the types of vital records that it produces as a result of conducting business.					
8.22.2	The jurisdiction has identified the types of vital records that are required by statutory authority to be maintained and the length of time they are required to be maintained.					
8.22.3	The jurisdiction has developed procedures for the on-site protection of these vital records.					
8.22.4	The jurisdiction has developed procedures for the backing up of vital records in the event of damage to the original record.					
8.22.5	The jurisdiction has developed procedures for the storage of the original and backup copies of these records.					
8.22.6	The jurisdiction addresses the procedures for the recovery of damaged vital records.					

No. 9. Logistics and Facilities: Identification, location, acquisition, distribution, and accounting for services, resources, materials, and facilities to support emergency HSS management. Logistics actions fall into one of four major categories: material management, property management, facility management, and transportation management.

Capabilities Rating		*Fully Prepared*	*Very Prepared*	*Generally Prepared*	*Marginally Prepared*	*Not Prepared*
9.1	The primary and alternate EOCs have the capability to sustain emergency operations for the duration of the emergency.					
9.1.1	The design/construction of the EOC is based on the hazards of the jurisdiction (e.g., flood plain, tornadoes).					
9.1.2	The EOC has adequate space for the emergency operations staff and equipment.					
9.1.3	The EOC complies with requirements of the Americans with Disabilities Act (ADA).					
9.1.4	The EOC has adequate furnishings, office equipment, supplies, and replacement parts.					
9.1.5	The jurisdiction has incorporated adequate information technology in the design of EOC processes and equipment.					
9.1.6	The EOC has adequate sanitary facilities.					
9.1.7	The EOC has an emergency generator with an adequate fuel supply.					
9.1.8	EOC equipment is tested and maintained on a regular basis.					
9.1.9	The EOC has built-in fire protection and other safety devices (e.g., sprinklers).					
9.1.10	The EOC has adequate facilities for food storage/preparation.					
9.1.11	The EOC has security procedures in place to prevent entrance of unauthorized personnel.					
9.1.12	All government facilities have an emergency generator with adequate fuel supply.					
9.2	The jurisdiction has developed logistics management and operations plans.					
9.2.1	SOPs are developed for logistics management.					
9.2.2	A resource inventory process is developed supporting the logistics management plan (e.g., conducting inventories, rotating supplies with shelf life, and establishing critical stocking levels).					
9.2.3	Warehouse sites are identified to store critical resources.					
9.2.4	Procedures are in place to obtain critical commodities or services from contractors.					
9.2.5	Processes are in place to track the movement of ordered commodities.					
9.2.6	Procedures are developed for receiving commodities and reporting discrepancies.					
9.2.7	Specifications are developed and suppliers are pre-identified for commonly needed commodities.					
9.2.8	Procedures are developed for disposing of items that are damaged, expendable, or destroyed.					
9.2.9	A program for scheduled maintenance of physical equipment is performed.					
9.2.10	Procedures are developed for retrieving and rehabilitating (for later reuse) appropriate equipment and supplies.					

No. 10. Training: Assessments, development, and implementation of a training/education program for officials and emergency HSS response personnel.

	Capabilities Rating	***Fully Prepared***	***Very Prepared***	***Generally Prepared***	***Marginally Prepared***	***Not Prepared***
10.1	The jurisdiction conducts an annual training needs assessment.					
10.1.1	On an annual basis, the jurisdiction systematically identifies performance problems that can be solved through training and determines those existing courses that can solve or mitigate the performance shortfalls.					
10.1.2	For identified performance problems and where no training activity exists, the jurisdiction analyzes identified performance tasks and incorporates those findings into future course designs.					
10.1.3	On an annual basis, the jurisdiction develops and publishes a schedule of training activities that meet identified needs.					
10.1.4	The jurisdiction maintains records of training (e.g., personnel trained, courses completed, certifications, and expiration dates).					
10.2	The jurisdiction's emergency management training program incorporates courses conducted by various federal, state, local, private, and voluntary agencies.					
10.2.1	The training program includes locally developed emergency management courses.					
10.2.2	The training program includes state emergency management courses.					
10.2.3	The training program includes federal emergency management courses.					
10.2.4	The training program includes voluntary agency courses (e.g., American Red Cross).					
10.2.5	The training program includes private sector emergency management courses.					
10.2.6	The training program includes non-emergency management-specific courses (e.g., computer, leadership/management, and customer service).					
10.3	The jurisdiction provides/offers training to all personnel with assigned emergency management responsibilities.					
10.3.1	Training is provided/offered to the elected/public officials.					
10.3.2	Training is provided/offered to department/agency heads.					
10.3.3	Training is provided/offered to the emergency response organizations (e.g., fire department, law enforcement, and mass care).					
10.3.4	Training is provided/offered to EOC staff.					
10.3.5	Training is provided/offered to nongovernment employees with emergency assignments.					
10.3.6	Training is provided/offered to voluntary agencies.					
10.3.7	Training is provided/offered to the identified response teams (e.g., damage assessment and needs assessment).					
10.4	Jurisdiction-specific training courses have been developed for staff with assigned emergency management responsibilities.					
10.4.1	Jurisdiction-specific training courses provide an overview of the EOP.					

10.4.2	Jurisdiction-specific training courses provide students with the established emergency management roles and responsibilities specific to their organization/agency.					
10.4.3	Jurisdiction-specific training courses provide students with the policies and procedures for performing their assigned duties.					
10.4.4	Students are provided with checklists, job aids, and/or field guides.					
10.5	The jurisdiction promotes professionalism among the collective emergency management personnel.					
10.5.1	Staff attends local, state, regional, and nationally sponsored seminars and conferences.					
10.5.2	Staff participates as instructors/facilitators of courses at the local and/or state levels.					
10.5.3	Staff belongs/participates in local, state, and/or national emergency management associations.					

No. 11. Exercises, Evaluations, and Corrective Actions: Assessment and evaluation of emergency response plans and capabilities through a program of regularly scheduled tests and exercises.

Capabilities Rating		***Fully Prepared***	***Very Prepared***	***Generally Prepared***	***Marginally Prepared***	***Not Prepared***
11.1	The jurisdiction has established an emergency management exercise program.					
11.1.1	A staff position is identified to coordinate the development and implementation of emergency management exercises.					
11.1.2	The local exercise coordinator has adequate access to policy- and decision-making officials and to budget and support staff resources to conduct the exercises.					
11.1.3	The jurisdiction utilized technical assistance from various levels of government and private entities during exercise planning.					
11.1.4	Actual disaster/emergency operations experience is factored into exercise planning.					
11.1.5	Results of the hazard identification and risk assessment are factored into the exercise planning.					
11.1.6	Data from the corrective action program or lessons learned are factored into exercise planning.					
11.2	The jurisdiction exercises the EOP on an annual basis.					
11.2.1	The jurisdiction sponsors and conducts a functional, full-scale, or tabletop exercise annually.					
11.2.2	The jurisdiction participates in one or more functional, full-scale, or tabletop exercises sponsored by a federal, state, or local government department/agency annually.					

(continued)

***Capabilities Rating** (continued)*

11.3	A multi-year exercise schedule is published and maintained.					
11.3.1	The schedule incorporates regulatory required, intra-jurisdictional, multi-jurisdictional, state, and federally sponsored programs.					
11.3.2	The schedule is updated and published at least semiannually.					
11.3.3	The schedule is developed in consultation with local and state departments and agencies.					
11.4	Exercises for hazard-specific programs comply with necessary regulatory requirements.					
11.4.1	REP exercises are conducted.					
11.4.2	CSEPP exercises are conducted.					
11.4.3	Dam EAP exercises are conducted as required by the Federal Energy Regulatory Commission.					
11.4.4	Superfund Amendments and Reauthorization Act Title III exercises are conducted.					
11.5	The jurisdiction has exercised the EOP and implementing documents using a WMD terrorism response scenarios in the past two years.					
11.5.1	The jurisdiction has conducted or participated in a tabletop or functional exercise using a WMD terrorism scenario.					
11.5.2	The jurisdiction has conducted or participated in a full-scale exercise using a WMD terrorism scenario.					
11.5.3	The jurisdiction has participated in a state or federal tabletop or functional exercise using a WMD terrorism scenario.					
11.5.4	The jurisdiction has participated in a state or federal full-scale exercise using a WMD terrorism scenario within the past two years.					
11.6	The jurisdiction's emergency management exercise program contains an evaluation component.					
11.6.1	The jurisdiction's exercise evaluation methodology is based on clearly delineated evaluation principles.					
11.6.2	The evaluation principles are formally documented, designed for easy use and implementation, and reviewed to ensure their ongoing validity.					
11.7	The jurisdiction utilizes lessons learned to strength their emergency management program.					
11.7.1	The jurisdiction has developed corrective action guidance documents.					
11.7.2	Corrective action guidance is applicable to local agencies with emergency management responsibility.					
11.7.3	Lessons learned from exercises and actual disasters are used to modify the EOP and associated SOPs, checklists, field guides, and training.					

No. 12. Crisis Communications, Public Education, and Information: Procedures to disseminate and respond to requests for pre-disaster, disaster, and post-disaster

information involving employees, responders, the public, and the media. Also, an effective public education program regarding hazards affecting the area of responsibility.

Capabilities Rating		***Fully Prepared***	***Very Prepared***	***Generally Prepared***	***Marginally Prepared***	***Not Prepared***
12.1	An emergency preparedness public education program is established.					
12.1.1	A program of public awareness to inform citizens about hazards and risk reduction is established using means such as public education materials (e.g., brochures, articles published in newspapers, and public service announcements [PSAs]).					
12.1.2	Seasonal hazard information supplements are published in newspapers and aired on radio and television.					
12.1.3	Emergency preparedness program information is inserted in telephone directories.					
12.1.4	Annual school programs are developed to enhance knowledge of emergency preparedness.					
12.1.5	Family and neighborhood disaster-planning programs are established.					
12.1.6	Business and industry programs are developed to enhance knowledge of emergency preparedness.					
12.1.7	Programs are developed for key government employees to enhance their knowledge of emergency preparedness.					
12.1.8	Outreach to professional associations, community organizations, and special event planners (e.g., fairs) is established.					
12.1.9	The jurisdiction participates in a neighborhood emergency response program such as the Community Emergency Response Team (CERT) program.					
12.2	Procedures are established for disseminating and managing emergency public information in a disaster.					
12.2.1	Procedures for recording and releasing casualty figures are established.					
12.2.2	PSA scripts are prepared and updated.					
12.2.3	Agreements are in place with local radio, TV, cable TV, newspapers, and other media.					
12.2.4	Alternate methods for contacting media are established.					
12.2.5	The dissemination of information on disaster assistance programs is coordinated with the local PIO(s) Community Relations staff.					
12.2.6	Procedures to minimize family separation are established (e.g., information on known dead, missing persons, patients in hospitals, and evacuees).					
12.2.7	Procedures for rumor control are established (e.g., hotline).					
12.2.8	Procedures are developed for the dissemination of individual and human needs information.					
12.3	Procedures are developed to establish and operate a joint incident center.					
12.3.1	The jurisdiction has pre-identified potential joint incident center locations.					
12.3.2	The jurisdiction pre-identifies fixed and mobile equipment (e.g., computers, phone lines, two-way radios, faxes, and copiers).					
12.3.3	SOPs for the joint incident center are established and maintained.					

(continued)

Capabilities Rating *(continued)*

12.3.4	A protocol for handling media inquires has been established.					
12.3.5	MOUs with other local agency PIOs are in place.					
12.3.6	A PIO contact list is established for communications with federal/state agencies and other local governments.					
12.3.7	JIC SOPs include training and exercise requirements.					
12.3.8	Media lists are established and updated annually.					
12.3.9	"Boilerplate" news releases, flyers, and public service announcements are established.					

No. 13. Finance and Administration: Development of finance and administrative procedures to support emergency HSS measures before, during, and after disaster events, and to preserve vital records.

Capabilities Rating		*Fully Prepared*	*Very Prepared*	*Generally Prepared*	*Marginally Prepared*	*Not Prepared*
13.1	The jurisdiction has established an administrative system for day-to-day operations.					
13.1.1	Employee job descriptions are developed.					
13.1.2	The jurisdictions comply with Equal Opportunity Commission (EEOC) requirements.					
13.1.3	The jurisdiction has developed personnel policies and procedures.					
13.1.4	The jurisdiction has an established auditing process.					
13.2	A local emergency management program administration system has been established.					
13.2.1	A strategic plan is reviewed and updated based on an annual review of the program.					
13.2.2	Requirements associated with state emergency management performance grants are part of the local strategic plan.					
13.3	An emergency administrative program is established for emergency operations.					
13.3.1	The jurisdiction has developed administration plans (e.g., emergency hiring and purchasing).					
13.3.2	Emergency job descriptions are developed.					
13.3.3	Procedures are developed for performing essential administrative activities during emergency operations.					
13.3.4	The jurisdiction has a purchasing/procurement unit within its finance and administrative section to administer all contractual matters during emergency operations.					
13.3.5	The jurisdiction maintains "standby" contracts for resources in the event of disasters.					

13.3.6	The jurisdiction has the capability to suspend competitive procurement procedures during disaster events.					
13.3.7	The jurisdiction has procedures for handling all compensations and claims issues.					
13.3.8	Procedures are developed for the management of temporary emergency personnel to meet disaster response needs.					
13.3.9	Departments/agencies are responsible for keeping vital records (e.g., personnel and equipment).					
13.4	An emergency fiscal program is established. Has funding been requested to procure equipment and supplies identified as being deficient?					
13.4.1	A budget and accounting system is established to track and document costs during emergency operations.					
13.4.2	Procedures are established to ensure the safety of cash, checks, and accounts receivable, and assist in the protection of other valuable documents/records, as well as the issue of necessary checks.					
13.4.3	An emergency payroll system is established.					
13.4.4	Departments/agencies have provisions for emergency fiscal record keeping in their emergency procedures.					
13.4.5	Departments/agencies have identified personnel to be responsible for documentation of disaster costs.					
13.4.6	Procedures are developed for the coordination and acquisition of supplies, equipment, and services in support of emergency response efforts.					
13.4.7	Procedures are developed for the governing body to appropriate or allocate funds to meet disaster expenditure needs.					
13.4.8	Copies of current regulations, applications, forms, and program guidance concerning federal and state emergency response and recovery reimbursement programs are maintained at the local government level.					

Glossary

acceptable level of risk The question of whether a risk is acceptable must be gauged against a benchmark or standard that has been deemed adequate by a particular AHJ at a specific point in time. Acceptable level of risk for both life safety and property is set through adoption of public policy through law, regulation, or level of service. A level of risk can be set by default; for example, when voters reject a bond issue or the increase to purchase apparatus or provide fire stations or increase staffing.

accredit Give official authorization to or approval of; to provide with credentials; to recognize or vouch for as conforming to a standard or an established set of nationally recognized guidelines.

accreditation A process by which an association or agency evaluates and recognizes a program of study or an institution as meeting certain predetermined standards or qualifications. It applies only to institutions and their programs of study or their services.

ACE American Council on Education.

achievement test A test designed to determine how much an individual has learned or the level of proficiency attained in a specific subject matter, usually conducted to assess the effects of instruction.

act phase The fourth phase of the plan-do-check-act (PDCA) cycle. Decisions are made whether to adopt the changes that were tested, propose new changes, or run through the cycle once more.

action plan A basic project management tool defining clearly the actions to be taken, the responsibilities, the goals, the date to start, and the date for completion.

adequacy The quality or state of being adequate; sufficient for a purpose; equal to; proportionate to; or fully sufficient for a specified or implied requirement.

adequate Providing what is needed to meet a given objective without being in excess.

advanced life support A sophisticated level of pre-hospital care that builds upon basic life support procedures and includes the use of invasive techniques such as advanced airway management, cardiac monitoring and defibrillation, intravenous therapy, and the administration of specified medications to save a patient's life.

adverse impact The receipt (collectively) of lower scores or ratings by members of a group or groups that are legally protected against unfair discrimination in employment.

AFGP Assistance to Firefighters Grant Program.

agency A division of government with a specific function offering a particular kind of assistance. In ICS, agencies are defined either as jurisdictional (having statutory responsibility for incident management) or as assisting or cooperating (providing resources or other assistance).

agency representative A person assigned by a primary, assisting, or cooperating federal, state, local, or tribal government agency or private entity that has been delegated authority to make decisions affecting that agency's or organization's participation in incident management activities following appropriate consultation with the leadership of that agency.

AHJ Authority having jurisdiction.

aircraft crash rigs Specialized vehicles capable of immediate fire suppression using foam or other extinguishing agents.

alarm processing time The elapsed time from the receipt of an alarm by the dispatch center and the notification of specific fire companies that are to respond. May also be defined as dispatch processing time.

alternative Different means of achieving the same objective; one of two or more ways of conducting an activity or offering different approaches to the same conclusion.

analysis Examination of a system, its elements, and their relationships and interaction.

APHIS Animal and Plant Health Inspection Service.

apparatus Fire suppression equipment such as engine companies, aerial trucks, crash fire rescue vehicles, and command vehicles.

area command (unified area command) An organization established (1) to oversee the management of multiple incidents that are each being handled by an ICS organization or (2) to oversee the management of large or multiple incidents to which several incident management teams have been assigned. Area command has the responsibility to set overall strategy and priorities, allocate critical resources according to priorities, ensure that incidents are properly managed, and ensure that objectives are met and strategies followed. Area command becomes unified area command when incidents are multijurisdictional. Area command may be established at an EOC facility or at some location other than an ICP.

arson The willful or malicious burning of property with criminal or fraudulent intent.

assessment A systematic method of determining the state or condition of something that involves the collection, analysis, and interpretation of data.

assumption A situation or condition that must be considered as existing if the organization is forced to operate in a specific manner and over which the organization does not exercise any control.

ATF U.S. Bureau of Alcohol, Tobacco, and Firearms.

average on-scene to patient time The average time it takes to actually reach the patient in need of care after arriving on the scene of a call.

baseline data Internal information about a process that is collected repeatedly over a period of time prior to introducing a change.

baseline system The current method of doing things in an organizational context. The activities that are currently in place to achieve the organization's goals and objectives.

basic life support A primary level of prehospital care that includes the recognition of life-threatening conditions and the application of simple emergency lifesaving procedures, including the use of adjunctive equipment, aimed at supporting life.

benchmark A benchmark is defined as a standard from which something can be judged. A benchmark is the best practice found in a specific service and/or product and will help define what superior performance is of a product, service, or process.

benchmarking Searching for the best internal or external practices or competitive practices that will help define superior performance of a product, service, or support process. Benchmarking is the concept of comparing processes and performance against another business to determine where improvements can be made.

BLM Bureau of Land Management.

bond Promise to repay the principal along with interest on a specified date, that is, when the bond reaches maturity; form of investment in which an individual purchases the bond with the intention of selling it upon maturity.

boundary Something that fixes a limit or extent. Boundaries may be spatial, temporal, or organizational, or they may be established by agreement or by definition. In the practice of quality improvement, it is necessary to establish process boundaries in order to establish process identity.

brainstorming A group process for generating creative and diverse ideas.

British Standards (BS) Institute British Standards is the new name of the British Standards Institute and is part of BSI Group, which also includes a testing organization. British Standards has a royal charter to act as the standards organization for the United Kingdom.

budget A financial plan for the coordination of resources and programs of the organization that articulates quantitative allocations of resources for specific activities and lists both the proposed expenditures and the expected revenue resources.

budget system Model or format to which a budget process conforms.

budget type How costs or revenues are divided between capital and operating purchases. Also see capital budget; line-item budget; operating budget; performance budget; planning, programming, and budgeting system; program budget; and zero-based budget.

buildings in service area by occupancy classifications The total number of buildings within the service area as identified by the building and fire code adopted by the agency.

bureaucratic hierarchy A form of organization and management characterized by specialization of functions, adherence to fixed rules, and a hierarchy of authority.

capital budget Budget that includes funds for projected major purchases—items that cost more than a certain specified amount of money and are expected to last more than one year, usually three or more years.

catastrophic incident Any natural or human-made incident, including terrorism, that results in extraordinary levels of mass casualties, damage, or disruption severely affecting the population, infrastructure, environment, economy, national morale, and/or government functions. A catastrophic event could result in sustained national impacts over a prolonged period of time; almost immediately exceeds resources normally available to state, local, tribal, and private sector authorities in the impacted area; and significantly interrupts governmental operations and emergency services to such an extent that national security could be threatened. All catastrophic events are incidents of national significance.

category The divisions used to separate subject areas within the Commission on Fire Accreditation International.

causal system The combination of influences or sources of variation that determines the nature of an output characteristic at a point in time.

CBO Community-based organization.

CDC U.S. Centers for Disease Control and Prevention.

CDRG Catastrophic disaster response group.

CERCLA Comprehensive Environmental Response, Compensation, and Liability Act.

CERT Community Emergency Response Team.

certification A process whereby an individual is tested and evaluated in order to determine his or her mastery of a specific body of knowledge or some portion of a body of knowledge.

CFAI Commission on Fire Accreditation International.

CFO Chief fire officer.

CFSI Congressional Fire Service Institute.

chain of command A series of command, control, executive, or management positions in hierarchical order of authority.

characteristic An attribute or a feature of something.

check phase The third phase of the plan-do-check-act cycle. The effects of having made a change are studied and assessed during this phase.

Chemical Stockpile Emergency Preparedness Program (CSEPP), 50 U.S.C. 1521(c)(4) and (5) Pursuant to the CSEPP program, FEMA works with the Defense Department in the course of the department's efforts to destroy the United States' stockpile of chemical weapons. FEMA's role in the implementation of this program is to provide assistance to ensure state and local governments located in the vicinity of the chemical weapons that are being destroyed have adequate emergency preparedness and response plans in place.

chief executive officer The person in a community who is charged with carrying out the policies established by the legally established authority having jurisdiction. Examples would vary according to the level of government involved, that is, city, fire district, county, regional, state, or federal government.

CI/KR Critical infrastructure/key resources.

CIP Critical infrastructure protection.

CMC Crisis management coordinator.

CMO Chief Medical Officer.

CNMI Commonwealth of the Northern Mariana Islands.

cognitive styles The cognitive styles describe how the individual acquires knowledge (cognition) and how an individual processes information (conceptualization). The cognitive styles are related to mental behaviors, habitually applied by an individual to problem solving and generally to the way that information is obtained, sorted, and utilized. Cognitive styles are usually described as personality dimensions that influence attitudes, values, and social interaction.

command staff In an incident management organization, the command staff consists of the incident command and the special staff positions of public information officer, safety officer, liaison officer, and other positions as required, who report directly to the incident commander. They may have an assistant or assistants, as needed.

commitment The trait of sincere and steadfast fixity of purpose; the act of binding yourself (intellectually or emotionally); an engagement by contract involving financial obligation; a message that makes a pledge; a pledge or promise, usually of a specified time period, given by an individual or an organization to complete a task or meet the obligations of an agreement.

community recovery In the context of the NRP and its annexes, the process of assessing the effects of an incident of national significance, defining resources, and developing and implementing a course of action to restore and revitalize the socioeconomic and physical structure of a community.

community risk assessment The evaluation of fire and other risks in a community that take into account all the pertinent facts that increase or decrease these risks. Such evaluation is used to help define standards of coverage (distrubution and concentration of resources) and to help frame other related policy issues.

CONPLAN U.S. Government Interagency Domestic Terrorism Concept of Operations Plan.

consequence management Predominantly an emergency management function that includes measures to protect public health and safety, restore essential government services, and provide emergency relief to governments, businesses, and individuals affected by the consequences of terrorism. The requirements of consequence management and crisis management are combined in the NRP.

content validity The extent to which an examination or selection procedure reflects a representative proportion of the content and response factors that are important in actual job performance and analysis.

contingency plan Alternative plan implemented in the event of uncontrollable circumstances.

continuous improvement Perhaps the central tenet of the quality revolution is that improvement must be continuous. The goal of quality improvement is zero defects. Continuous improvement

(CI) is the cornerstone of a lasting competitive advantage.

control systems process Process of establishing and implementing mechanisms to ensure that objectives are attained.

COO Chief operating officer.

coping styles The concept of coping styles attempts to capture the variety of ways in which individuals try to cope with the demands of the environment. A common distinction is between "problem" and "emotion" focused strategies. There are many competing frameworks for describing and understanding coping styles. One important question is whether they genuinely capture the diversity of the stress experience.

core values Attitudes and beliefs thought to pattern a culture uniquely.

correlation A statistical term that refers to the degree to which two sets of scores or facts are associated with each other. A correlation coefficient is one index of association. The range goes from −1.0 to +1.0. The closer an index is to either of these numbers, the stronger the relationship. A −1.0 correlation means a perfect inverse relationship. A +1.0 correlation means a perfect positive relationship. A 0.0 coefficient means no relationship.

cost-benefit A term used to express the value of a component of a system. It is expressed usually as a ratio of a cost (which is an expenditure) to a benefit (which is an outcome of some type).

courage A quality of spirit that enables a person to face danger or pain without showing fear.

CPC Commission on Professional Credentialing.

CPR Cardiopulmonary resuscitation.

CPSC U.S. Consumer Product Safety Commission.

criterion/criteria A measure or index (or set of measures or indexes) on which a judgment or decision may be based to evaluate performance or define a major area within each category. *Criterion* is singular; the plural is *criteria*.

critical incidents A method of evaluation based on specific examples of above or below average performance.

critical infrastructures Systems and assets, whether physical or virtual, so vital to the United States that the incapacity or destruction of such systems and assets would have a debilitating impact on security, national economic security, national public health or safety, or any combination of those matters.

cross-functional teams A cross-functional team focuses on a work process rather than on the structural hierarchy of members. All members of the team, regardless of their position within the hierarchy (i.e., rank and file, supervisory, management, executive) work to analyze and improve a specific work process.

CSB U.S. Chemical Safety and Hazard Investigation Board.

CSG Counterterrorism Security Group.

current Occurring in the present time frame, within the last calendar or budget year.

customer The person or group that establishes the requirements or expectations of a process and receives or uses the output of that process.

customer service Providing a product or service that meets the needs of a group of people, called customers, who use and consume the produce or service.

customer–supplier relationship That relationship between an individual or group that establishes, receives, uses, and judges the output of a process, and the individual or group providing output that serves as the user's input.

customer survey A questionnaire or interview designed to elicit information on customer satisfaction, unmet customer needs, and customers' expectations. It is used to obtain information that can be used in product, service, or process improvement.

cycle time The moment from the time a customer first requests service to the time that all aspects of the service experience are completed.

data-based decision making A decision-making process based on facts and other objective information as opposed to intuition or hunches.

DCE Defense coordinating element.

DCO Defense coordinating officer.

defect A defect is anything that fails to meet customer expectations. This includes all aspects not only of physical or tangible products but also all of the intangibles of the service experience.

Defense Production Act of 1950, 50 U.S.C. App. 2061 et seq. This statute is one of the nation's primary authorities for ensuring the availability of resources needed for military requirements and civil emergency preparedness and response. Executive Order 12919 delegates to the director of FEMA authority to use the Defense Production Act (DPA) for emergency preparedness, response, mitigation, and recovery activities.

deficiency The condition of not being able to achieve the desired results.

delegation The act of empowering someone to act for another.

demographic Pertaining to the study of human population characteristics including size, growth rates, density, distribution, migration, birth rates, and mortality rates.

deployment The strategic assignment and placement of fire agency resources such as fire companies, fire stations, and specific staffing levels for those companies.

DEST Domestic emergency support team.

detection method of quality control The traditional method for quality control in which quality is achieved through inspection after production.

DFO Disaster field office.

DHS Department of Homeland Security.

diagnosis The process of studying symptoms, taking and analyzing data, conducting experiments to test theories, and establishing relationships between causes and effects.

disaster Any occurrence that inflicts widespread destruction or distress.

disaster recovery center (DRC) A facility established in a centralized location within or near the disaster area at which disaster victims (individuals, families, or businesses) apply for disaster aid.

dispatch processing time The portion of a fire department's response time that begins when the dispatcher receives an alarm and ends when the dispatcher assigns the proper companies to respond to the emergency. Also referred to as *alarm processing time*.

diversity The presence of a wide range of variation in the qualities or attributes under discussion. When used to describe people and population groups, diversity encompasses such factors as age, gender, race, ethnicity, ability, and religion, as well as education, professional background, and marital and parental status.

DMAT Disaster Medical Assistance Team.

DMORT Disaster Mortuary Operational Response Team.

do phase The second phase of the plan-do-check-act (PDCA) cycle. Changes that are expected to improve processes are tried or made during this phase.

DOC Department of Commerce.

DOD Department of Defense.

DOE Department of Energy.

DOI Department of the Interior.

DOJ Department of Justice.

DOL Department of Labor.

DOS Department of State.

DOT Department of Transportation.

DPA Defense Production Act.

drawdown The resource level an agency will not go below when asked for mutual aid.

DRC Disaster recovery center.

DRM Disaster recovery manager.

DSCA Defense support of civil authorities.

DTRIM Domestic threat reduction and incident management.

Earthquake Hazards Reduction Act, 422 U.S.C. 7701 et seq. This statute authorizes FEMA, in coordination with the United States Geological Survey, the National Science Foundation, and the National Institute of Standards and Technology, to administer the National Earthquake Hazard Reduction Program (NEHRP). The act designates FEMA as the lead agency in the

NEHRP program, which Congress created to promote the implementation of earthquake hazard reduction measures by the federal government, as well as state and local governments.

EAS Emergency assistance personnel or emergency alert system.

education The activities of educating or instructing; activities that impart knowledge or skill; knowledge acquired by learning and instruction; the profession of teaching (especially at a school or college or university.

effective Doing the right thing(s) to meet a given objective(s), producing or capable of producing an intended result, exerting force or influence.

effective response force The minimum amount of staffing and equipment that must reach a specific emergency zone location within a maximum prescribed travel or driving time and is capable of initial fire suppression, EMS, and/or mitigation.

effectiveness In a measure of achieving the desired result.

efficiency Capacity to produce desired results with a minimum expenditure of time, energy, money, or materials.

EFOP Executive Fire Officer Program.

EIN Employer identification number.

emergency As defined by the Stafford Act, an emergency is "any occasion or instance for which, in the determination of the President, Federal assistance is needed to supplement State and local efforts and capabilities to save lives and to protect property and public health and safety, or to lessen or avert the threat of a catastrophe in any part of the United States."

emergency operations center (EOC) The physical location at which the coordination of information and resources to support domestic incident management activities normally takes place. An EOC may be a temporary facility or may be located in a more central or permanently established facility, perhaps at a higher level of organization within a jurisdiction. EOCs may be organized by major functional disciplines (e.g., fire, law enforcement, and medical services), by jurisdiction (e.g., federal, state, regional, county, city, tribal), or by some combination thereof. A central location where those in authority congregate to allow for exchange of information and conduct face-to-face coordination in the making of decisions. The center, often referred to as the EOC, provides for centralized emergency management in major natural disasters and other emergencies.

emergency operations plan (EOP) The comprehensive plan maintained by various jurisdictional levels for managing a response to a catastrophic event.

emergency public information Information disseminated primarily in anticipation of an emergency or during an emergency. In addition to providing situational information to the public, it also frequently provides directive actions required to be taken by the general public.

emergency response provider Includes federal, state, local, and tribal emergency public safety, law enforcement, emergency response, emergency medical (including hospital emergency facilities), and related personnel, agencies, and authorities. (See Section 2[6], Homeland Security Act of 2002, Public Law 107-296, 116 Stat. 2135 [2002].) Also known as "emergency responder."

emergency support function (ESF) A grouping of government and certain private sector capabilities into an organizational structure to provide the support, resources, program implementation, and services that are most likely to be needed to save lives, protect property and the environment, restore essential services and critical infrastructure, and help victims and communities return to normal, when feasible, following domestic incidents. The ESFs serve as the primary operational-level mechanism to provide assistance to state, local, and tribal governments or to federal departments and agencies conducting missions of primary federal responsibility.

EMI Emergency Management Institute/FEMA.

empathy More than feeling compassion or sympathy "for" another person, empathy puts you in that person's shoes to feel "with" him or her or "as one" with him or her. First used in English in the early Twentieth century to translate the German psychoanalytic term *Einfühlung*, meaning "to feel as one with," though in practice more closely translating the German *Mitgefühl* "to feel with" someone.

empirical Originating in or based upon actual observation or experience verifiable in the literal sense that it can be determined to be truthful and accurate, as opposed to theory, generalities, or intuition.

empowerment The process of increasing the capacity of individuals or groups to make choices and to transform those choices into desired actions and outcomes. Central to this process are actions that both build individual and collective assets, and improve the efficiency and fairness of the organizational and institutional context, that govern the use of these assets.

EMR-ISAC Emergency Management and Response—Information Sharing and Analysis Center.

EMS Emergency medical services.

enterprise fund Fund established to finance and account for acquisition, operation, and maintenance of government facilities and services that are entirely or predominantly self-supporting by user fees.

environment Natural and cultural resources and historic properties as those terms are defined in this glossary and in relevant laws.

environmental response team Established by EPA, the environmental response team includes expertise in biology, chemistry, hydrology, geology, and engineering. The environmental response team provides technical advice and assistance to the on-scene coordinator for both planning and response to discharges and releases of oil and hazardous substances into the environment.

EOC Emergency operations center.

EPA Environmental Protection Agency.

EPCRA Emergency Planning and Community Right-to-Know Act.

EPLO Emergency preparedness liaison officer.

EPR Emergency preparedness and response.

equal employment opportunity A personnel management responsibility to be sensitive to the social, economic, and political needs of a jurisdiction and/or labor market.

Equal Employment Opportunity Commission (EEOC) To administer its responsibilities, the EEOC accepts written charges filed against an employer alleging that it has engaged in unlawful employment practice in violation of Title VII or other federal civil rights laws. It has the power to bring suits; subpoena witnesses; issue guidelines, that have the force of law; render decisions; and provide legal assistance to complainants in regard to fair employment.

ERL Environmental Research Laboratories.

ERT Emergency response to terrorism; also environmental response team (EPA).

ERT-A Emergency response team—advance element.

ERT-N National emergency response team.

ESF Emergency support function.

ESFLG Emergency support function leaders group.

esprit de corps The spirit of a group that makes the members want the group to succeed.

EST Emergency support team.

ethics (ethos) Analysis of the principles of human conduct in order to determine between right and wrong; philosophical principle used to determine correct and proper behavior by members of a society; sometimes called moral philosophy.

evacuation Organized, phased, and supervised withdrawal, dispersal, or removal of civilians from dangerous or potentially dangerous areas, and their reception and care in safe areas.

evaluating Establishing the worth or value of each part or the sum total of the whole based on desired outcome.

Evaluation determines how effective or efficient an item, a program, or a process is compared to a benchmark or established set of criteria.

evaluation Analysis and comparison of actual performance versus prior plan and stated goals and objectives. The systematic and thoughtful collection of information and decision making. Evaluation consists of having a criteria, collecting evidence, and making judgments.

evaluation methodology Use of statistics and other methods to determine the efficiency, effectiveness, quality, and coverage of a program, policy, or activity.

exhibit(s) Evidence that a particular activity has occurred. Demonstration that documentation or hard evidence exists for the proof of achieving a performance indicator.

experimental A tentative procedure or policy; an operation carried out under controlled circumstances to test or establish a hypothesis; the process of testing before adoption into practice.

exploratory methodology A system of evaluation that reviews possible policies, practices, and procedures for expansion of given capabilities without expansion of staffing or physical resources.

external customer An individual or group outside the boundaries of the producing organization who receives or uses the output of a process.

facility management Facility selection and acquisition, building services, information systems, communications, safety and health, and physical security.

FAS Freely associated states.

FBI U.S. Federal Bureau of Investigation.

FCC U.S. Federal Communications Commission.

FCO Federal coordinating officer.

federal Of or pertaining to the federal government of the United States of America.

federal coordinating officer (FCO) The federal officer who is appointed to manage federal resource support activities related to Stafford Act disasters and emergencies. The FCO is responsible for coordinating the timely delivery of federal disaster assistance resources and programs to the affected state and local governments, individual victims, and the private sector.

federal emergency communications coordinator (FECC) The person, assigned by GSA, who functions as the principal federal manager for emergency telecommunications requirements in major disasters, emergencies, and extraordinary situations, when requested by the FCO or FRC.

Federal Fire Prevention and Control Action of 1974, as amended, 15 U.S.C. 2201 et seq. This statute created the United States Fire Administration (USFA) within FEMA. The USFA is headed by an administrator, who is authorized to provide training and grant assistance to state and local fire service organizations. In addition to providing training to a wide variety of emergency response personnel, the USFA administers a program to provide grants on a competitive basis to fire departments for the purpose of protecting the health and safety of both the public and firefighting personnel against fire and fire-related hazards.

federal on-scene coordinator (FOSC or OSC) The federal official predesignated by the EPA or the USCG to coordinate responses under subpart D of the NCP, or the government official designated to coordinate and direct removal actions under subpart E of the NCP.

federal resource coordinator (FRC) The federal official appointed to manage federal resource support activities related to non–Stafford Act incidents. The FRC is responsible for coordinating support from other federal departments and agencies using interagency agreements and MOUs.

feedback system (customer) A system for obtaining information from customers about relevant quality characteristics of products and services. In a passive feedback system, feedback is voluntarily supplied by the customer or user. There is no requirement to provide feedback. In an active feedback system, the supplier actively seeks and acts upon information from the customer on a continuing basis.

FEMA Federal Emergency Management Agency.

FESHE Fire and Emergency Services Higher Education.

FESSAM *Fire and Emergency Service Self-Assessment Manual.*

fibrillation Erratic electrical activity of the heart. Atrial fibrillation, or erratic electrical activity of the upper chambers, causes irregular contraction of the lower chamber and palpitations, characterized by the feeling of galloping horses. Although palpitations can be disconcerting, atrial fibrillation is generally not a serious condition. Ventricular fibrillation results in inefficient pumping of the lower chamber and can cause death if it is not treated immediately.

FICEMS Federal Interagency Committee on Emergency Medical Services.

Fire Act The president is authorized to provide assistance under the Robert T. Stafford Disaster Relief and Emergency Assistance Act, Public Law 93-288, as amended, 42 U.S.C. § 5121, et seq. (hereinafter referred to as the Stafford Act). Federal assistance is coordinated through the Department of Homeland Security's Federal Emergency Management Agency (FEMA). Under the Fire Management Assistance Grant Program (FMAGP), FEMA provides assistance, including grants, equipment, supplies, and personnel, to any state or local government for the mitigation, management, and control of any fire on private forest land or grassland that threatens such destruction as would constitute a major disaster.

fire confined to structure Responses to fire calls in which the fire is contained to the structure or structures that were involved when the responding unit first arrived at the scene of the fire.

fire flow available The amount of water available for firefighting on a continuous basis. The highest demand upon the water distribution system.

fire flow delivered The amount of water that can be delivered at the scene of an emergency. It is a combination of three factors: pump capacity that is available, hose and nozzle configurations, and staffing levels.

fire flow required (or estimated) The quantity of water that should be available for a period of two to three hours at a minimum pressure of 20 psi in a water distribution system. The estimated amount of water is calculated for individual structural conditions based upon a structure's size, construction type, fire loading, compartmentation, installation of built-in fire protection devices, and relationship to other exposed structures. May be compared to the available fire flow to determine relative ability to control specific problems.

fire management area A geographic area of a jurisdiction that is classified according to one or more risk categories. The size and classification of a fire analysis area is usually based upon either a specific area or a building.

fire out on arrival Responses to fire calls in which the fire that initiated the call is completely extinguished when the responding unit arrives at the scene of the fire.

fire pre plan Plan developed by the fire department to identify hazardous building information and owner information, used during emergency incidents to determine the best course of mitigating an emergency.

fire protection duties Fire protection duties may include, but are not limited to, any or all of the following: fire suppression (including structural, wildland, transportation, and/or all other types of fires); fire prevention activities (including code enforcement, inspections, public education and fire investigation); emergency medical services (including basic and advanced life support and ambulance transport services); managed health care services; hazardous materials response and preparedness; technical rescues (such as extrication, swift water, high angle, or confined space); urban search and rescue (involving compromised structural rescues); disaster management and preparedness; community service activities; public safety calls (including animal rescues, lockouts and standbys); response to civil disturbances and terrorism incidents; nonemergency functions (such as training, preplanning, housekeeping, maintenance, and physical conditioning), and other related emergency and nonemergency service tasks as may be assigned or required.

fire protection environment The conditions, circumstances, and influences under which a fire protection system must operate. It includes the population, the geographical area, land use, occupancy factors, weather conditions, structural and nonstructural physical situations, and financial, political, legislative and regulatory criteria.

fire protection system The regular interaction of dependent and independent sources of fire protection services, includes both public and private organizations, apparatus, equipment, fixed and mobile, facilities, methods, human resources, and policies by the authority having jurisdiction.

fire scenario(s) A tabulation of data from previous incidents or community hazard, risk, response, and analysis that describes factors that lead up to losses of life or property.

FIRST Federal incident response support team.

first-due area The portion of a jurisdiction that each response company has been assigned to be the first unit to arrive at the scene of an emergency. Usually the first-due company is responsible for most activities in that area.

first responder A term used for the person who is trained and/or certified to be the first to arrive at a scene of a specific type of emergency, that is, EMS or hazardous materials. Local and nongovernmental police, fire, and emergency personnel who, in the early stages of an incident, are responsible for the protection and preservation of life, property, evidence, and the environment, including emergency response providers as defined in section 2 of the Homeland Security Act of 2002 (6 U.S.C. 101); as well as emergency management, public health, clinical care, public works, and other skilled support personnel (such as equipment operators) who provide immediate support services during prevention, response, and recovery operations. First responders may include personnel from federal, state, local, tribal, or nongovernmental organizations.

flowchart A schematic diagram that depicts the nature and flow of the steps in a process.

FMC Federal Mobilization Center.

FNS Food and Nutrition Service.

FOC FEMA Operations Center.

FOG Field operations guide.

FOIA Freedom of Information Act.

Food and Nutrition Service (FNS) disaster task force The Food Security Act of 1985 (Public Law 99-198) requires the secretary of agriculture to establish a disaster task force to assist states in implementing and operating various disaster food programs. The FNS disaster task force coordinates the overall FNS response to disasters and emergencies. It operates under the general direction of the administrator of FNS.

FRC Federal resource coordinator.

FRERP Federal Radiological Emergency Response Plan.

FRP Federal Response Plan.

FY Fiscal year.

Gantt Chart Graph presenting the different steps of a project in a structured and chronological way.

GAR Governor's authorized representative.

general operating guidelines Written guidelines that suggest courses of action, usually provided in a manual format.

GIS Geographical information system.

goal The term that designates the general end toward which an effort is directed. In the context of a fire protection effort, goals must be directed toward the resolution of life and property safety, based on the stated mission of the organization, and be future oriented. A major organizational outcome or end toward which efforts, expressed as objectives, are directed.

goal achievement The means of verifying through indicators, either quantitative or qualitative, that the end result is accomplished. Goal measurement does not imply that the goal is totally resolved.

government infrastructure Entity that has the responsibility of providing the citizens or customers within the jurisdiction with an agreed-upon level of service.

grants Source of funds provided by government agencies and nongovernment organizations. Grants are intended to provide equipment that an organization may not have the funds to purchase.

GSA General Services Administration.

guiding principles A set of statements about the values and philosophy of an organization that guides the behavior of its members toward one another, toward customers, and toward suppliers and governs the way the organization approaches its work. Sometimes presented as an organization's "philosophy of operations," or "credo."

hard data Anytime an element can clearly be reconstructed from electronic or mechanical devices that make an automatic indication of time once activated.

Hawthorne experiments The Hawthorne experiments conducted at Western Electric's Hawthorne plant outside Chicago, starting in 1927 and running through 1932, were intended to bring about a greater understanding of the effects of working conditions on worker productivity. The results of the experiments were contrary to the management theory of the time (see *scientific management*) and were key in bringing about an understanding of motivation factors in employment.

hazard Something that is potentially dangerous or harmful, often the root cause of an unwanted outcome.

hazard mitigation Any cost-effective measure that will reduce the potential for damage to a facility from a disaster event.

hazardous material Hazardous material is a substance or material, including a hazardous substance, that has been determined by the secretary of transportation to be capable of posing an unreasonable risk to health, safety, and property when transported in commerce, and which has been so designated (see 49 CFR 171.8).

hazardous substance Any substance designated pursuant to section 311(b)(2)(A) of the Clean Water Act; any element, compound, mixture, solution, or substance designated pursuant to section 102 of the Comprehensive Environmental Response, Compensation, and Liability Act (CERCLA); any hazardous waste having the characteristics identified under or listed pursuant to section 3001 of the Solid Waste Disposal Act (but not including any waste the regulation of which under the Solid Waste Disposal Act [42 U.S.C. § 6901 et seq.] has been suspended by act of Congress); any toxic pollutant listed under section 307(a) of the Clean Water Act; any hazardous air pollutant listed under section 112 of the Clean Air Act (42 U.S.C. § 7521 et seq.); and any imminently hazardous chemical substance or mixture with respect to which the EPA administrator has taken action pursuant to section 7 of the Toxic Substances Control Act (15 U.S.C. § 2601 et seq.).

heavy extrication All rescues of persons trapped in road, rail, air, and water vehicles that require specialized tools and training.

HHS Department of Health and Human Services.

high hazard risks Built-up areas of substantial size with a concentration of property presenting a substantial risk of life loss, a severe financial impact on the community, or unusual potential damage to property in the event of fire.

historic property Any prehistoric or historic district, site, building, structure, or object included in or eligible for inclusion in the National Register of Historic Places, including artifacts, records, and remains that are related to such district, site, building, structure, or object (16 U.S.C. § 470[w][5]).

honor Comprises the reputation, self-perception, or moral identity of an individual or group; a primary principle and prime virtue, is the feeling and the expression of admiration, respect, or esteem accorded to another as a right or as due. Honor refers to virtue, purity, a keen sense of ethical conduct, and integrity.

HQ Headquarters.

HSAS Homeland Security Advisory System.

HSC Homeland Security Council.

HSOC Homeland Security Operations Center.

HSPD Homeland Security Presidential Directive.

IAFC International Association of Fire Chiefs.

IAFF International Association of Fire Fighters.

IAIP Information analysis and infrastructure protection.

IC Incident command.

ICP Incident command post.

ICS Incident command system.

IFSTA International Fire Service Training Association.

IIMG Interagency incident management group.

IMT Incident management team.

incident An occurrence or event, natural or human caused, that requires an emergency response to protect life or property. Incidents can, for example, include major disasters, emergencies, terrorist attacks, terrorist threats, wildland and urban fires, floods, hazardous materials spills, nuclear accidents, aircraft accidents, earthquakes, hurricanes, tornadoes, tropical storms, war-related disasters, public health and medical emergencies, and other occurrences requiring an emergency response.

incident action plan (IAP) An oral or written plan containing general objectives reflecting the overall strategy for managing an incident. It may include the identification of operational resources and assignments. It may also include attachments that provide direction and important information for management of the incident during one or more operational periods.

incident command post (ICP) The field location at which the primary tactical-level, on-scene incident command functions are performed. The ICP may be collocated with the incident base or other incident facilities and is normally identified by a green rotating or flashing light.

incident command system (ICS) A standardized on-scene emergency management construct specifically designed to provide for the adoption of an integrated organizational structure that reflects the complexity and demands of single or multiple incidents, without being hindered by jurisdictional boundaries. ICS is the combination of facilities, equipment, personnel, procedures, and communications operating with a common organizational structure, designed to aid in the management of resources during incidents.

incident commander (IC) The individual responsible for all incident activities, including the development of strategies and tactics and the ordering and release of resources. The IC has overall authority and responsibility for conducting incident operations and is responsible for the management of all incident operations at the incident site.

incident management team (IMT) The incident commander and appropriate command and general staff personnel assigned to an incident.

incident mitigation Actions taken during an incident designed to minimize impacts or contain the damages to property or the environment.

incident of national significance Based on criteria established in HSPD-5 (paragraph 4), an actual or potential high-impact event that requires a coordinated and effective response by an appropriate combination of federal, state, local, tribal, nongovernmental, and/or private sector entities in order to save lives and minimize damage, and provide the basis for long-term community recovery and mitigation activities.

Industrial Revolution Historical period, lasting throughout most of the 1800s, when the economic base of the United States and many European nations began to shift from an agricultural to a manufacturing base.

information officer See *public information officer.*

infrastructure The human-made physical systems, assets, projects, and structures, publicly and/or privately owned, that are used by or provide benefit to the public. Examples of infrastructure include utilities, bridges, levees, drinking water systems, electrical systems, communications systems, dams, sewage systems, and roads.

innovation The application of knowledge that leads to the development of new or improved processes, products, or services.

INRP Initial National Response Plan.

Insurance Services Office (ISO) A national organization that evaluates public fire protection and provides rating information to insurance companies. Insurers use this rating to set basic premiums for fire insurance.

integrity Integrity composes the personal inner sense of "wholeness" deriving from honesty and consistent uprightness of character. The etymology of the word relates it to the Latin adjective *integer* (whole, complete). Evaluators, of course, usually assess integrity from some point of view, such as that of a given ethical tradition or in the context of an ethical relationship. Moral soundness; "he expects to find in us the common honesty and integrity of men of business"; "they admired his scrupulous professional integrity."

internal customer An individual or group inside the boundaries of the producing organization that receives or uses the output of a process within the organization.

IOF Interim operating facility.

ISAO Information-sharing and analysis organization.

ISFSI International Society of Fire Service Instructors.

JFO Joint field office.

JIC Joint information center.

JIS Joint information system.

JOC Joint operations center.

Johari Window A metaphorical tool created by Joseph Luft and Harry Ingham—thus the Joe-Harry/Johari window—in 1955 in the United States; used to help people better understand their interpersonal communication and relationships.

joint field office (JFO) A temporary federal facility established locally to provide a central point for federal, state, local, and tribal executives with responsibility for incident oversight, direction, and/or assistance to effectively coordinate protection, prevention, preparedness, response, and recovery actions. The JFO will combine the traditional functions of the JOC, the FEMA DFO, and the JIC within a single federal facility.

joint information center (JIC) A facility established to coordinate all incident-related public information activities. It is the central point of contact for all news media at the scene of the incident. Public information officials from all participating agencies should collocate at the JIC.

joint information system (JIS) Integrates incident information and public affairs into a cohesive organization designed to provide consistent, coordinated, timely information during a crisis or incident operations. The mission of the JIS is to provide a structure and system for developing and delivering coordinated interagency messages; developing, recommending, and executing public information plans and strategies on behalf of the IC; advising the IC concerning public affairs issues that could affect a response effort; and controlling rumors and inaccurate information that could undermine public confidence in the emergency response effort.

joint operations center (JOC) The JOC is the focal point for all federal investigative law enforcement activities during a terrorist or potential terrorist incident or any other significant criminal incident, and is managed by the senior law enforcement official. The JOC becomes a component of the JFO when the NRP is activated.

JTF Joint task force.

JTTF Joint terrorism task force.

Jungian Type Inventory Model based on the types and preferences of Carl Gustav Jung, who wrote *Psychological Types* in 1921.

Junto In 1727 Benjamin Franklin organized a group of friends to provide a structured forum for discussion. The group, initially composed of 12 members, called itself the Junto. The members of the Junto were drawn from diverse occupations and backgrounds, but they shared a spirit of inquiry and a desire to improve themselves and their community, and to help others. Among the original members were printers, surveyors, a cabinetmaker, a cobbler, a clerk, and a merchant.

jurisdiction A range or sphere of authority. Public agencies have jurisdiction at an incident related to their legal responsibilities and authorities. Jurisdictional authority at an incident can be political or geographical (e.g., city, county, tribal, state, or federal boundary lines) or functional (e.g., law enforcement, public health). A population area wherein there is clearly defined responsibility, based on statutory authority, to provide fire and/or emergency medical services. Also called authority having jurisdiction, or AHJ.

Keirsey Temperament Sorter A personality test that attempts to identify which of four temperaments, and which of 16 types, a person prefers. Hippocrates, a Greek philosopher who lived circa 460–377 B.C., proposed four temperaments, which are related to the four humors. These were sanguine, choleric, phlegmatic, and melancholic.

KSA Knowledge, skills, and abilities. A document that describes the knowledge, special skills, and abilities of a person, commonly used in the personnel file.

ladder truck Vehicle that carries a variety of equipment such as ladders, forcible entry tools, and rescue equipment.

layering The process of creating hierarchical levels in an organization.

leadership The process of directing and inspiring workers to perform the task-related activities of the group; art of influencing and directing people to accomplish a mission or task.

leadership functions The group maintenance and task-related activities that must be performed by the leader, or someone else, for a group to perform effectively.

leadership styles The various patterns of behavior favored by leaders during the process of directing and influencing workers.

leading Act of controlling, directing, conducting, guiding, and administering employees through the use of personal behavioral traits or personality characteristics that motivate them to complete the organization's goals successfully.

level of service The resources needed to meet the stated service-level objective(s). Level of service is defined only in terms of what resources are provided and not in terms of effectiveness or quality.

liability A broad, comprehensive term that describes a person's or an organization's responsibility under the law. This responsibility implies that if damage occurs, a person or an organization may have to respond to legal allegations and pay damages to the person or organization that has been damaged.

liaison officer A member of the command staff responsible for coordinating with representatives from cooperating and assisting agencies.

line-item budget Lists of revenue sources and proposed expenditures for a budget cycle; most common budget system used in North America.

local government A county, municipality, city, town, township, local public authority, school district, special district, intrastate district, council of governments (regardless of whether the council of governments is incorporated as a nonprofit corporation under state law), regional or interstate government entity, or agency or instrumentality of a local government; an Indian tribe or authorized tribal organization or, in Alaska, a Native Village or Alaska Regional Native Corporation; or a rural community, unincorporated town or village, or other public entity (as defined in section 2[10] of the Homeland Security Act of 2002, Public Law 107-296, 116 Stat. 2135, et seq. [2002]).

low risk Small commercial structures that are remote from other buildings; detached residential garages and outbuildings.

MAC entity Multi-agency coordinating entity.

MACC Multi-agency command center.

major disaster As defined by the Stafford Act, any natural catastrophe (including any hurricane, tornado, storm, high water, wind-driven water, tidal wave, tsunami, earthquake, volcanic eruption, landslide, mudslide, snowstorm, or drought), or, regardless of cause, any fire, flood, or explosion, in any part of the United States, which in the determination of the president causes damage of sufficient severity and magnitude to warrant major disaster assistance under this act to supplement the efforts and available resources of states, local governments, and disaster relief organizations in alleviating the damage, loss, hardship, or suffering caused thereby.

Malcolm Baldrige National Quality Award An annual national quality award used to promote quality awareness, to recognize quality achievements of U.S. companies, and to publicize successful quality strategies. The award formally recognizes companies that attain preeminent quality leadership.

management by objectives (MBO) A formal set of procedures that establishes and reviews progress toward common goals for managers and subordinates. Management by objectives specifies that superiors and subordinates will jointly set goals for a specified period of time and then meet again to evaluate the subordinate's performance in terms of the previously established goals.

management information system (MIS) A formal, usually computerized, structure for providing management with information, often through an MIS department.

management science (MS) Mathematical techniques for modeling, analysis, and solution of management problems.

management science school A group of management scholars trained in quantitative methods who develop mathematical techniques for analyzing and solving organizational problems.

managerial grid model A behavioral leadership model developed by Robert Blake and Jane Mouton. This model identifies five different leadership styles based on the *concern for people* and the *concern for production*.

managers Individuals who plan, organize, lead, and control other individuals in the process of pursuing organizational goals.

Maslow's hierarchy A theory in psychology that Abraham Maslow proposed in his 1943 paper, *A Theory of Human Motivation*, which he subsequently extended. His theory contends that as humans meet their basic needs, they seek to satisfy successively higher needs that occupy a set hierarchy.

material management Requisitioning and sourcing (requirements processing); acquisition, asset visibility (resource tracking), receipt, storage, and handling; security and accountability; inventory, deployment, issue, and distribution; and recovery, reuse, and disposition.

maximum prescribed travel time The maximum travel time to a specific fire management area that is deemed necessary for the responding fire company to be considered as an effective response force in controlling emergencies.

maximum/worst risk Occupancies classified as maximum risk will be of substantial size and contain a concentration of properties that present a very high risk of life loss, loss of economic value to the community, or large loss or damage to property in the event of fire. These risks frequently impact the need for the fire department to have multiple alarm capability and have an adequate assessment of their ability to concentrate resources.

measurable terms A word, expression, or number that has a precise meaning that can be quantified.

median age of population The median age of the population as reported in the most recent census.

median household income The median household income as reported by the U.S. Bureau of Labor Statistics for the most recent period reported.

mentors Individuals who pass on the benefits of their knowledge to other individuals who are usually younger and less experienced.

MERS Mobile emergency response support.

method(s) The utilization of a systematic body of procedures and techniques specifically designed to meet goals and objectives.

minimum staffing per unit The minimum number of personnel assigned to staff each type of equipment or vehicle.

mission An organization's mission is an enduring statement of purpose that describes what the organization does, how it does it, and for whom it does it. Defines the fundamental, unique purpose that sets an organization apart from other forms of its type and identifies the scope of operations in product and market terms. It provides the foundation for priorities, strategies, plans, and work assignments.

mitigation Activities designed to reduce or eliminate risks to persons or property or to lessen the actual or potential effects or consequences of an incident. Mitigation measures may be implemented prior to, during, or after an incident. Mitigation measures are often developed in accordance with lessons learned from prior incidents. Mitigation involves ongoing actions to reduce exposure to, probability of, or potential loss from hazards. Measures may include zoning and building codes, floodplain buyouts, and analysis of hazard-related data to determine where it is safe to build or locate temporary facilities. Mitigation can include efforts to educate governments, businesses, and the public on measures they can take to reduce loss and injury.

MOA Memorandum of agreement.

mobilization The process and procedures used by all organizations—federal, state, local, and tribal—for activating, assembling, and transporting all resources that have been requested to respond to or support an incident.

mobilization center An off-site temporary facility at which response personnel and equipment are received from the point of arrival and are prepositioned for deployment to an incident logistics base, to a local staging area, or directly to an incident site, as required. A mobilization center also provides temporary support services, such as food and billeting, for response personnel prior to their assignment, release, or reassignment and serves as a place to out-process following demobilization while awaiting transportation.

moderate risk Built-up areas of average size, where the risk of life loss or damage to the property in the event of a fire in a single occupancy is usually limited to that occupancy. In certain areas, such as small apartment complexes, the risk of death or injury may be relatively high. The moderate/typical risks are often the greatest factor in determining fire station locations and staffing due to the frequency of emergencies in this category. To assure an equitable response and to provide adequate initial attack/rescue capability to the majority of incidents, the typical risk is often used in determining needed resources.

motivation The factors that cause, channel, and sustain an individual's behavior.

MOU Memorandum of understanding.

Myers-Briggs Type Inventory Katherine Briggs and Isobel Briggs Myers, a mother-and-daughter team, built the modern system that is probably the most popular type of system in the world today based upon psychologic differences according to the theory of Cart Jung. In particular, they devised a written test (the Myers-Briggs Type Inventory, or MBTI®) to identify the person's type. Other variants have been evolved that are also based on the Jung typology. The most well known of these is David Keirsey's Temperament Sorter.

NAHERC National Animal Health Emergency Response Corps.

NASA National Aeronautics and Space Administration.

National Disaster Medical System (NDMS) A coordinated partnership between DHS, HHS, DOD, and the Department of Veterans Affairs established for the purpose of responding to the needs of victims of a public health emergency. NDMS provides medical response assets and the movement of patients to health care facilities where definitive medical care is received when required.

National Flood Insurance Act of 1968, as amended, 42 U.S.C. 4001 et seq. This statute authorizes FEMA to administer the National Flood Insurance Program (NFIP). Under the NFIP, FEMA is authorized to provide flood insurance for commercial and residential structures that are built in communities that agree to adopt standards for the construction of buildings located within flood-prone areas of the communities.

National Incident Management System (NIMS) A system mandated by HSPD-5 that provides a consistent, nationwide approach for federal, state, local, and tribal governments; the private sector; and NGOs to work effectively and efficiently together to prepare for, respond to, and recover from domestic incidents, regardless of cause, size, or complexity. To provide for interoperability and compatibility among federal, state, local, and tribal capabilities, the NIMS includes a core set of concepts, principles, and terminology. Provides for the systematic establishment of a complete, functional command organization for multi-agency incidents that involve the U.S. government as well as state and local agencies.

National Interagency Coordination Center (NICC) The organization responsible for coordinating allocation of resources to one or more coordination centers or major fires within the nation and is located in Boise, Idaho.

National Interagency Fire Center (NIFC) A facility located in Boise, Idaho, that is jointly operated by several federal agencies and is dedicated to coordination, logistical support, and improved weather services in support of fire management operations throughout the United States.

National Preparedness Program Pursuant to Presidential Decision Directives No. 39 and No. 62, FEMA has been assigned responsibilities relating to weapons of mass destruction (WMD) terrorism response preparedness. Under these directives, FEMA is responsible for coordinating the federal response to the consequences of terrorist incidents and for ensuring the adequacy of state plans for responding to terrorist events.

National Response Center A national communications center for activities related to oil and hazardous substance response actions. The National Response Center, located at DHS/USCG Headquarters in Washington, DC, receives and relays notices of oil and hazardous substances releases to the appropriate federal OSC.

National Response System The mechanism for coordinating response actions by all levels of government (40 CFR § 300.21) for oil and hazardous substances spills and releases.

National Response Team (NRT) The NRT, comprised of the 16 federal agencies with major environmental and public health responsibilities, is the primary vehicle for coordinating federal agency activities under the NCP. The NRT carries out national planning and response coordination and is the head of a highly organized federal oil and hazardous substance emergency response network. EPA serves as the

NRT chair, and DHS/USCG serves as vice chair.

natural resources Natural resources include land, fish, wildlife, domesticated animals, plants, biota, and water. Water means salt and fresh water, surface and ground water, including water used for drinking, irrigation, aquaculture, and recreational purposes, as well as in its capacity as fish and wildlife habitat, including coral reef ecosystems as defined in 16 U.S.C. 64501. Land means soil, surface and subsurface minerals, and other terrestrial features.

NAWAS National Warning System.

NCP National Oil and Hazardous Substances Pollution Contingency Plan.

NCR National Capital Region.

NCS National Communications System.

NCTC National Counterterrorism Center.

NDMS National Disaster Medical System.

NEP National Exercise Program.

NETC National Emergency Training Center.

NFA National Fire Academy/USFA.

NFDC National Fire Data Center/USFA.

NFIC National Fire Information Council.

NFIRS National Fire Incident Reporting System.

NFP National Fire Programs/USFA.

NFPA National Fire Protection Association. Standards publications adopted by the NFPA through the consensus process setting a level of standard for fire service–related dimensions or equipment specification.

NFPA 1710 Standard for the Organization and Deployment of Fire Suppression Operations, Emergency Medical Operations, and Special Operations to the Public by Career Fire Departments.

NFPA 1720 Standard for the Organization and Deployment of Fire Suppression Operations, Emergency Medical Operations, and Special Operations to the Public by Volunteer Fire Departments.

NFPCA National Fire Protection and Control Administration.

NGO Nongovernmental organization.

NHTSA U.S. National Highway Traffic Safety Administration.

NICC National Infrastructure Coordinating Center or National Interagency Coordination Center.

NIFC U.S. National Interagency Fire Center.

NIJ U.S. National Institute of Justice.

NIMS National Incident Management System.

NIOSH U.S. National Institute for Occupational Safety and Health.

NIPP National Infrastructure Protection Plan.

NIRT Nuclear Incident Response Team.

NIST U.S. National Institute of Standards and Technology.

NJTTF National Joint Terrorism Task Force.

NMRT National Medical Response Team.

no lost injury time Fire-related firefighter injuries resulting in no time lost from work.

NOAA National Oceanic and Atmospheric Administration.

nongovernmental organization (NGO) A nonprofit entity that is based on the interests of its members, individuals, or institutions and that is not created by a government, but may work cooperatively with government. Such organizations serve a public purpose, not a private benefit. Examples of NGOs include faith-based charity organizations and the American Red Cross.

nontransport Responses in which no individuals are transported to a medical facility.

NRC Nuclear Regulatory Commission.

NRCC National Response Coordination Center.

NRCS Natural Resources Conservation Service.

NRP National Response Plan.

NRT National Response Team.

NSC National Security Council.

NSP National Search and Rescue Plan.

NSSE National special security event.

NTIS U.S. National Technical and Information Service.

NTSB U.S. National Transportation Safety Board.

Nuclear Incident Response Team (NIRT) Created by the Homeland Security Act to provide DHS with a nuclear/radiological response capability. When activated, the NIRT consists of specialized federal response teams drawn from DOE and/or EPA. These teams may become DHS operational assets providing technical expertise and equipment when activated during a crisis or in response to a nuclear/radiological incident as part of the DHS federal response.

number of population by age The number of persons in each category within the service area as reported in the most recent census.

NVFC National Volunteer Fire Council.

NVOAD National Voluntary Organizations Active in Disaster.

NWCG National Wildfire Coordinating Group.

objective, fair, and impartial Use of facts, undistorted by personal feelings, bias, or prejudice. Items scored or observed in accordance with strict instructions and scoring techniques that would be the same for all raters.

objectively verifiable indicators What is being measured is compared against how it is being accomplished. For example, fire suppression is a function that has two different dimensions, level of service versus loss experience in the community. Both can be set at goals: that is, response time and per capita fire loss. In the case of response time, one measures it by the clock and determines whether the firefighting resources arrive on the scene when they are supposed to. In the case of fire loss, one can measure the total fire loss for the year, divide it by the population, and define the per capita fire loss.

objectives The specific end toward which effort is being directed. In the context of fire protection, objectives are considered independent of the means in which they are achieved. Objectives must be achievable within a certain planning period and be able to be measured in some quantifiable way. Specific, measurable midterm and short-term performance targets necessary for achieving long-term strategic goals. See *service-level objectives*.

occupancy The classification given to a building in accordance with a specific building code.

occupancy risk assessment An assessment of the potential severity of a specific structure in relation to the fire agency's ability to handle the types and severity of emergencies within that structure. Occupancy risk assessment often includes classifying these risks into categories.

ODP U.S. Office of Domestic Programs/DHS.

OIA Office of the Assistant Secretary for Information Analysis.

on-scene coordinator (OSC) See *federal on-scene coordinator.*

operating budget Type of budget that includes funds for recurring expenses of day-to-day operation of the fire and emergency services organization; expenses include personnel salaries, facility utilities, office supplies, apparatus and vehicle fuel, janitorial supplies, and other items needed on a daily basis.

optimization A process of orchestrating the efforts of all components toward achievement of the stated goal or objective.

organization chart A diagram of an organization's structure, showing the functions, departments, or positions of the organization and how they are related.

organizational conflict Disagreement between individuals or groups within the organization stemming from the need to share scarce resources or engage in interdependent work activities, or from differing statuses, goals, or cultures.

organizational design The determination of the organizational structure that is most appropriate for the strategy, people, technology, and tasks of the organization.

organizational goals The purpose, mission, and objectives that are the reason for an organization's existence and that form the basis of its strategy.

organizational structure Framework that permits an organization to accomplish its mission, strategy, goals, and objectives; defines the functions of each portion of the organization and establishes the communication and control lines associated with the operation of the organization; visualized in an organization chart.

organizational transformation The result of making fundamental changes in the strategies, design, and management of an organization.

organizing The process of arranging an organization's structure and coordinating its managerial practices and use of resources to achieve its goals.

OSC On-scene coordinator.

OSHA Occupational Safety and Health Administration.

OSHA 29 CFR 1910.120 (q) (3) The citation for the Federal Occupational Safety and Health Administration program.

OSLGCP Office of State and Local Government Coordination and Preparedness.

outcome The way the customer responds to the product or service. A description of the intended result, effect, or consequence that will occur from carrying out a program or activity. A long-term, ultimate measure of success or strategic effectiveness. A measure of the extent to which a service has achieved its goals or objectives and, as defined, met the needs of its beneficiaries. The result of the organization's taking inputs and transforming them into products or services.

paradigm A set of rules based on an explicit set of assumptions that explains how things work or ought to work.

paradigm shift A change in the way one perceives the way things work or ought to work. Requires a reexamination of basic assumptions. From the Greek word *paradhma* (paradigma), the term *paradigm* was introduced into science and philosophy by Thomas Kuhn in his landmark book, *The Structure of Scientific Revolutions* (1962). A paradigm is a set of rules and regulations (written or unwritten) that does two things: (1) establishes or defines boundaries; and (2) tells you how to behave inside the boundaries in order to be successful. A shared set of assumptions. The paradigm is the way we perceive the world: water to the fish. The paradigm explains the world to us and helps us to predict its behavior.

Pareto chart A special form of vertical bar that displays information in such a way that priorities for process improvement can be established. It shows the relative importance of all the data and is used to direct efforts to the largest improvement opportunity by highlighting the "vital few" in contrast to the "many others."

participative management A management technique that allows nonmanagement personnel an opportunity to participate in decision making, provide input to management decisions, and/or execute management actions.

PCC Policy coordination committee.

PDA Preliminary damage assessment.

PDCA cycle See *plan-do-check-act (PDCA) cycle.*

PDD Presidential decision directive.

peer evaluation review The process in the accreditation program whereby a peer group reviews the self-assessment information and the criteria to determine whether the agency is in compliance with its proposed courses of action.

percentage below poverty level The percentage of the total population reported in the most recent census.

performance The quantity and quality of work accomplished by an individual, group, or organization.

performance appraisal The process of evaluating an individual's performance by comparing it to existing standards or objectives.

performance budget Type of budget categorized by function or activity; each activity is funded based on projected performance; similar to program budgets or outcome-based budgets.

performance goals An individual or organizational output or outcome that results from performance, which is measurable and desired.

performance indicator The desired level of achievement toward a given objective and the ability to demonstrate doing a particular task as specified in the accreditation process.

persuasion A process by which individuals or groups attempt to influence others to accept their fixed point of view.

PFO Principal federal official.

PIO Public information officer.

plan phase The first phase of the plan-do-check-act (PDCA) cycle. A plan identifying what needs to be improved, how it is to be implemented, and how the results are to be evaluated is developed by the organization.

plan-do-check-act (PDCA) cycle The systematic approach used to guide managers to quality. Also known as the Shewhart or Deming cycle. It is a scientific method useful for organizational decision making. The initials refer to specific phases that occur during this process: plan, do, check, act.

planning The process of establishing objectives and suitable courses of action before taking action. Managerial function that determines in advance what organizations, subunits, or individuals should do and how they will do it.

planning assumption An expectation about how future events, both internal and external to the organization, are likely to affect the achievement of desired results. These assumptions are taken into account during the planning process and may affect the actual goals, objectives, and strategies adopted by the organization.

planning, programming, and budgeting system (PPBS) Budget system that provides a framework consistent with the organization's goals, objectives, policies, and strategies for making decisions about current and future programs using three interrelated phases: planning, programming, and budgeting; developed in the 1970s to coordinate planning, program development, and budget processes; used currently in industry and by the U.S. Department of Defense.

POC Point of contact.

policy A standing plan that establishes general guidelines for decision making.

pollutant or contaminant As defined in the NCP, includes, but is not limited to, any element, substance, compound, or mixture, including disease-causing agents, which after release into the environment and upon exposure, ingestion, inhalation, or assimilation into any organism, either directly from the environment or indirectly by ingestion through food chains, will or may reasonably be anticipated to cause death, disease, behavioral abnormalities, cancer, genetic mutation, physiological malfunctions, or physical deformations in such organisms or their offspring.

population served The total number of residents counted as living within the service area as reported in the most recent census.

position power The power that is inherent in the formal position someone holds. This power may be great or small, depending upon the specific position.

power The ability to exert influence, that is, the ability to change the attitudes or behavior of individuals or groups.

PPE Personal protective equipment.

preparedness The range of deliberate, critical tasks and activities necessary to build, sustain, and improve the operational capability to prevent, protect against, respond to, and recover from domestic incidents. Preparedness is a continuous process involving efforts at all levels of government and between government and private sector and nongovernmental organizations to identify threats, determine vulnerabilities, and identify required resources.

prevention Actions taken to avoid an incident or to intervene to stop an incident from occurring. Prevention involves actions taken to protect lives and property.

principal federal official (PFO) The federal official designated by the secretary of Homeland Security to act as his or her representative locally to oversee, coordinate, and execute the secretary's incident management responsibilities under HSPD-5 for incidents of national significance.

private sector Organizations and entities that are not part of any governmental structure. Includes for-profit and not-for-profit organizations, formal and informal structures, commerce and industry, private emergency response organizations, and private voluntary organizations.

procedure A standing plan of detailed guidelines for handling organizational actions that occur regularly.

process The combination of people, machines and equipment, raw materials, methods, and the environment that produces a given product or service. A set of causes and conditions that repeatedly come together to transform inputs into output. The inputs may include people, methods, material, equipment, environment, and information. There can be several stages to the process, or each stage could be viewed as a process. The output is a product or service.

process flowchart map A diagram that shows the sequential steps of a process or a workflow around a product or service.

process improvement The continuous endeavor to learn about the cause system in a process and to use this knowledge to change the process to reduce variation and complexity and to improve customer satisfaction.

process management Actions taken every day to ensure that the right tasks are identified and performed in the way they were intended and improved at every opportunity to meet customer expectations.

process mapping This is a process used to identify, map, and analyze all significant steps in a work process. Cross-functional process mapping takes it a step further by including all functions within a process in a single map. The result: non-value-added steps are eliminated, cycle time is reduced, and quality is improved.

product Any tangible or measurable commodity. It does not necessarily need to be a physical item. Service providers generate products as well. For instance, a visit to a restaurant has several tangible aspects: how clean the restaurant is, how quickly the food was brought, whether the food tasted good and was warm enough, and so forth. All these things can be directly compared to the same service provided at another restaurant; that makes them part of the product. A product is the output of any process and may be classified as (1) (a) physical things such as automobiles, television sets, or rotor blades; (b) information as in conversations, annual reports, plans, or advice; or (2) (c) services such as work performed for someone else, for example, recruiting, transportation, or plant maintenance.

production The transformation of organizational resources into finished goods and services.

productivity A measure of the performance of a worker or an operation system relative to resource utilization: output divided by input.

productivity measurement Indexes that reflect an organization's efficiency. These indexes are usually reported as an organization's ability to provide various outputs per unit of time or cost.

professional development A personally initiated obligation to build discipline expertise, to enhance personal growth, to improve abilities, and to contribute to the industry of the fire service.

program A single-use plan that covers a relatively large set of organizational activities and specifies major steps, their order and timing, and unit responsible for each step.

program budget Budget system used to categorize funds by program or activity.

programmatic change Organizational change focused on improving one specific area of performance, having specific measurable results, usually under the direction of a program manager, and generally not integrated with other improvement efforts or the general conduct of organizational work.

protective clothing Personal items of clothing and equipment issued to individual firefighters for protection against heat, flame, abrasion, puncture, or other traumatic injury during combat operations. Includes, but is not limited to, coats, trousers, boots, gloves, helmets, personal alarm devices, fire shelters, and any other special equipment issued for evaluating exposure such as dosimeters, communicable disease shields, and so on.

PSA Public service announcement.

PSAP Public safety answering point.

public assistance program The program administered by FEMA that provides supplemental federal disaster grant assistance for debris removal and disposal; emergency protective measures; and the repair, replacement, or restoration of disaster-damaged, publicly owned facilities and the facilities of certain private nonprofit organizations.

public health Protection, safety, improvement, and interconnections of health and disease prevention among people, domestic animals, and wildlife.

public information officer (PIO) A member of the command staff or designate responsible for interfacing with the public and media or with other agencies with incident-related information requirements.

public works Work, construction, physical facilities, and services provided by governments for the benefit and use of the public.

purpose That which is expected to be achieved if the organization is successful in completing its mission. It can be expressed in either qualitative or quantitative terms, within the parameters to be able to verify them objectively. That which we hope to create, accomplish, or change with a view toward influencing the solution to a problem.

quality A characteristic or the value of a product or service from the perspective of the user. The extent to which a product or service meets or exceeds customer requirements and expectations. Good quality does not necessarily mean high quality. It means a predictable degree of uniformity and dependability at low cost, with a quality suited to the market.

quality control Measures output and identifies defects as they occur, with the intention of preventing them from reaching the customer. Although it is important to catch defects before they escape to customers, it is much more important to prevent them from happening in the first place.

quality improvement The actions taken to increase the value to the customer by improving the effectiveness and efficiency of processes and activities throughout the organizational structure.

quality improvement teams Any team that has been formed by management to improve quality, usually through the improvement of an organization's processes.

R&D Research and development.

RA Reimbursable agreement.

Radiological Emergency Preparedness (REP) Program Pursuant to a memorandum of understanding between FEMA and the Nuclear Regulatory Commission (NRC), as well as Executive Order 12657, FEMA works with state and local jurisdictions, in cooperation with the operators of licensed commercial nuclear power plants, to ensure they have adequate radiological emergency preparedness plans in place to satisfy the NRC's licensing requirements and to ensure the safety of the public in the vicinity of the plants in the event of an accident at any licensed plant.

radiological emergency response teams (RERTs) Teams provided by EPA's Office of Radiation and Indoor Air to support and respond to incidents or sites containing radiological hazards. These teams provide expertise in radiation monitoring, radionuclide analyses, radiation health physics, and risk assessment. RERTs can provide both mobile and fixed laboratory support during a response.

RAMP Remedial action management program.

ranking The ordering of quantitative scores or qualitative ratings from the highest to the lowest.

RCP Regional contingency Plan.

RCRA Resource Conservation and Recovery Act.

reciprocity The principle that characterizes an exchange, a mutual action, a give and take. In the fire service, *reciprocity* is a term used in respect to certifications, academics, and pensions.

recovery The development, coordination, and execution of service- and site-restoration plans for impacted communities and the reconstitution of government operations and services through individual, private sector, nongovernmental, and public assistance programs that identify needs and define resources; provide housing and promote restoration; address long-term care and treatment of affected persons; implement additional measures for community restoration; incorporate mitigation measures and techniques, as feasible; evaluate the incident to identify lessons learned; and develop initiatives to mitigate the effects of future incidents.

recruitment The development of a pool of job candidates in accordance with a human resource plan.

reliability The degree to which a test or other examination is free from chance errors of measurement. The extent to which scores are stable, dependable, and similar upon repeated measurements; consistent scores in successive ratings even with different raters. Also see *response reliability*.

remote and isolated rural risks Areas may be classified as remote/isolated rural risks if they are isolated from any centers of population and contain few buildings, for example, rural land with no occupied structures or recreational areas.

REPLO Regional emergency preparedness liaison officer.

residential single-family dwelling One- and two-family units.

response Activities that address the short-term, direct effects of an incident. Response includes immediate actions to save lives, protect property, and meet basic human needs. Response also includes the execution of emergency operations plans and of incident mitigation activities designed to limit the loss of life, personal injury, property damage, and other unfavorable outcomes.

response reliability The probability that the required amount of staffing and apparatus that is regularly assigned will be available when a fire or emergency call is received, that is, the percentage of time that all response units are available for a dispatch. This is a function of the average amount of time that a fire unit is unavailable for a dispatch because it is already committed to another response. When a response unit is unavailable, the response time to an emergency in its first-due area will be longer because a more distant unit will have to respond to the call. Response reliability is a statement of the probability that an effective response force may not be provided when a call is received.

response time The total amount of time that elapses from the time that a communications center receives an alarm until the responding unit is on the scene of the emergency and prepared to control the situation. Response time is composed of several elements.

result(s) The actual achievement of predetermined goals and objectives with measurable outcomes. A consequence, effect, issue, or conclusion that is achieved, obtained, or brought about by calculation, activity, investigation, or any direct action.

RFI Request for information.

RFO Request for qualifications.

RFP Request for proposal.

RFSI Residential Fire Safety Institute.

RISC Regional interagency steering committee.

risk Exposure to a hazard based on the probability of an outcome when combined with a given situation with a specific vulnerability. The level of risk can be described as the probability of a specified loss over a given period of time. All structures, for example, are subject to destruction by fire; however, individual structures vary considerably as to the possibility of loss as a result of their construction, contents, and built-in protection.

risk category A rank or category assigned to a fire management area that reflects the degree of risk to life and property, and hence a demand upon services by a responding agency. Risk categories are defined by the authority having jurisdiction.

ROC Regional Operations Center.

RRCC Regional Response Coordination Center.

RRT Regional response team.

rules Standing plans that detail specific actions to be taken in a given situation.

SAC Special agent-in-charge.

SAFER Staffing for adequate fire and emergency response grant program.

safety equipment Tools and equipment used by individual firefighters to perform firefighting, hazardous entry, or rescue work upon which the individual must rely for personal safety. This equipment is normally not assigned to the individual, but rather carried on the apparatus. Includes, but is not limited to, respiratory equipment, hazardous materials entry suits, carabineers, and lifelines. Does not include nozzles, hoses, ladders, and the like.

SAR Search and rescue.

SCC Secretary's command center (HHS).

schema In psychology and cognitive science, a mental structure that represents some aspect of the world. People use schemata to organize current knowledge and provide a framework for future understanding. Examples of schemata include stereotypes, social roles, scripts, worldviews, and archetypes.

scientific management A management approach, formulated by Frederick Taylor and others between 1890 and 1930, that sought to determine scientifically the best methods for performing any task and for selecting, training, and motivating workers.

SCO State coordinating officer.

senior federal official (SFO) An individual representing a federal department or agency with primary statutory responsibility for incident management. SFOs utilize existing authorities, expertise, and capabilities to aid in management of the incident working in coordination with other members of the JFO Coordination Group.

service-level objectives Statements of performance unique to a given jurisdiction. These statements should be developed by the agency based upon nationally recognized standards and practices for fire and ancillary services. The service-level objectives should be written based upon a community's specific profile, which includes both existing and future risk levels. The community risk profile should examine the makeup of occupancies, types of uses, what the probability/consequences are of anticipated incidents, and the historical response trends and patterns.

SFLEO Senior federal law enforcement official.

SFO Senior federal official.

SIOC Strategic Information and Operations Center.

situation assessment The evaluation and interpretation of information gathered from a variety of sources (including weather information and forecasts, computerized models, GIS data mapping, remote sensing sources, ground surveys, etc.) that, when communicated to emergency managers and decision makers, can provide a basis for incident management decision making.

situational leadership theory An approach to leadership developed by Paul Hersey and Kenneth H. Blanchard that describes how leaders should adjust their leadership style in response to their subordinates' evolving desire for achievement, experience, ability, and willingness to accept responsibility.

Six Sigma This is the statistical measurement for Motorola's quality initiative. It represents no more than 3.4 errors per million opportunities.

soft data Time elements that are estimated or added to documents by reference to wristwatches, estimates, or any form of data collection that is done retrospective to the incident.

SOG Standard operating guideline.

SOP Standard operating procedure.

span of control The spatial, temporal, or resource limits accorded to a member of management.

span of management The number of subordinates who report to a given manager.

special risks These are areas, whether composing a single building or complexes, that require a first due response over and above that appropriate to the risk that predominates the surrounding area. These premises or small areas should be treated as special risks and given an appropriate predetermined response.

square miles served The total number of square miles contained within the boundaries of the service area.

staffing The level of personnel assigned to perform the anticipated emergency tasks of a specific fire company for the risk identified in a given district or community. The number of personnel required to perform multiple emergency operations functions such as fire suppression versus EMS or hazardous materials operations.

Stafford Act This statute authorizes the president to provide assistance to state and local governments, as well as some nonprofit entities and individual disaster victims, in the aftermath of presidential-declared emergencies and major disasters. Most of the Stafford Act authorities have been delegated to the director of FEMA pursuant to Executive Order 12148, as amended. Title II of the Stafford Act provides authority for a variety of federal disaster preparedness activities. Title III is comprised of the act's administrative provisions, while Titles IV and V of the act authorize programs of responding to major disasters and emergencies, respectively. Title VI contains authorities formerly in the Federal Civil Defense Act of 1950 for emergency preparedness and cross references the Defense Production Act to include "emergency preparedness" in the definition of "national defense."

stakeholders The groups and individuals inside or outside the organization who affect and are affected by the achievement of the organization's mission, goals, and strategies.

standard operating procedures A term used to describe written direction provided to personnel in a manual format. Similar to the general operating guideline, but may be more specific, requiring specific actions.

standardization A process by which a product or service is assessed against some standard, performance, or quality.

standards of response coverage A written statement that combines service-level objectives with staffing levels to define how and when a fire agency's resources will respond to call for service.

START Scientific and technical advisory and response team.

state Any state of the United States, the District of Columbia, the Commonwealth of Puerto Rico, the U.S. Virgin Islands, Guam, American Samoa, the Commonwealth of the Northern Mariana Islands, and any possession of the United States (as defined in section 2[14] of the Homeland Security Act of 2002, Public Law 107-296, 116 Stat. 2135, et seq. [2002]).

statistics The use of mathematical tools (e.g., averages, spread, and shapes of distributions) to either (1) describe characteristics of a set of data or (2) make inferences to the population from which the data were drawn.

Steward B. McKinney Homeless Assistance Act, as amended Title III of this statute, 42 U.S.C. 11331–11352, created the Federal Emergency Management Agency Food and Shelter Program. This authority enables FEMA, in coordination with the Emergency Food and Shelter (EFS) National Board, to provide great assistance to local governments for the use of private nonprofit organizations or local public entities within such local governments for emergency food and shelter purposes.

stewardship In general, stewardship is responsibility for taking good care of resources entrusted to one. A function of a government responsible for the welfare of the population, and concerned with the trust and legitimacy with which its activities are viewed by the citizenry. It requires vision, intelligence, and influence, which must oversee and guide the working and development of the nation's health actions on the government's behalf.

strategic framework The combination of an organization's mission, vision, and guiding principles that serves as a context for practicing strategic management.

strategic goal A long-range performance target consistent with an organization's mission, usually requiring a substantial commitment of resources and achievement of midterm and short-term supported plans. Achievement of strategic goals moves an organization closer to realizing its vision.

strategic intent An active management process that focuses attention on future threats and opportunities as part of an enduring quest for global leadership. It is a driving force impelling management toward its vision.

strategic leadership Actions focused on setting a long-term direction and vision for the future, communicating that vision to those who have the knowledge, commitment, and power to help achieve the vision and inspiring them to keep moving in the right direction.

strategic management A process that links strategic planning and strategic intent with day-to-day operational management into a single management process.

strategic plan A document by which the guiding members of an organization envision its future and develop the necessary procedures and operations to achieve that future. Strategic plans are comprehensive plans designed to define and achieve the long-term objectives of the organization.

strategic planning The active formulation by top management of an organization's objectives and the definition of strategies for achieving them.

strategy The broad program for defining and achieving an organization's objectives; the organization's response to its environment over time. A plan or other means for achieving a long-range strategic goal.

stratification The process of classifying data into subgroups based on characteristics or categories.

stress The tension and pressure that result when an individual views a situation as presenting a demand that threatens to exceed his or her capabilities and resources.

structure The arrangement and interrelationships of the components of an organization.

supervising Act of directing, overseeing, or controlling the activities and behavior of employees who are assigned to a particular supervisor.

supervision Processes of directing, overseeing, and controlling the activities of other individuals.

swift water The rescue of persons trapped in rivers or flood control waterways.

synergy The Greek root of this word means "working together." The concept of synergy is that every individual working on a team achieves more than he or she would as an individual. The situation in which the whole is greater than its parts. In organizational terms, the fact that departments that interact cooperatively can be more productive than if they operate in isolation.

system Parts that interact with one another to function as a whole. A series of functions or activities within an organization that work together for the aim of the organization.

systems approach View of the organization as a unified, directed system of interrelated parts.

target An indicator with a magnitude that can be realized at a specific time or location or date; an explicit and objectively measured output. For example, if a goal is to arrive at the scene of the emergency with an engine company within five minutes, 90 percent of the time, this target might be achieved under "normal conditions" only. To note that on rainy days the response time may be slower by 25 percent or that on Sunday nights at 2100 hours the response time may be 25 percent faster is to identify that there are other targets to be measured. These targets can be verified without eroding the integrity of the overall goal or objective for the system; used in relation to both benchmarks and baselines.

team A group convened to improve processes, products, or services to serve the needs of the organization and its customers. Optimally, a team has five to seven members. It often includes line workers along with supervisors and managers who get input from people actually working on the process each day.

team building A method of improving organizational effectiveness at the team level by diagnosing barriers to team performance and improving interteam relationships and task accomplishment.

team performance model Allan Drexler and David Sibbet developed a comprehensive model of team performance that shows the predictable stages involved in both creating and sustaining teams. The *Drexler/Sibbet Team Performance™ Model (TPM)* illustrates team development as seven stages, four to create the team and three to describe levels of performance.

terrorism Any activity that (1) involves an act that (a) is dangerous to human life or potentially destructive of critical infrastructure or key resources; and (b) is a violation of the criminal laws of the United States or of any state or other subdivision of the United States; and (2) appears to be intended (a) to intimidate or coerce a civilian population; (b) to influence the policy of a government by intimidation or coercion; or (c) to affect the conduct of a government by mass destruction, assassination, or kidnapping.

Theory X Developed by Douglas McGregor in the 1960s. The assumptions that the average employee dislikes work, is lazy, has little ambition, and must be directed, coerced, or threatened with punishment to perform adequately.

Theory Y Developed by Douglas McGregor in the 1960s. Assumes employees are ambitious, self-motivated, and anxious to accept greater responsibility, and exercise self-control and self-direction. It is believed that employees enjoy their mental and physical work activities, and have the desire to be imaginative and creative in their jobs if they are given a chance.

Theory Z According to William Ouchi, the management belief that the key to productivity and quality is the development and participation of all employees.

threat An indication of possible violence, harm, or danger.

total operational expenditures Personnel costs (salaries and fringe benefits) and other actual direct expenses for most recent year completed, such as personnel, equipment, administration, materials and supplies, utilities, and professional and consulting services. Does not include capital expenditures or debt ser-vice payments. Adjusted by regional cost of living factor.

total quality leadership (TQL) The application of quantitative methods and people to assess and improve materials and services supplied to the organization and all significant processes within the organization and to meet the needs of the end user now and in the future.

total response time The total elapsed time from the point of notification to a responding fire company and the arrival of that unit at the scene. Total response time equals notification, plus alarm processing/dispatch time plus turnout time plus travel time.

TRADE Training resources and data exchange.

transformation A shift from one way of being to a new way of being.

transformational process The operations required to change inputs into a product or service. This process consists of the steps necessary to change inputs into outputs.

travel time The portion of response time that is utilized by responding companies to drive to the scene of the emergency. Travel time begins when assigned fire companies actually begin to drive to the emergency and ends when they arrive.

tribe Any Indian tribe, band, nation, or other organized group or community, including any Alaskan Native Village as defined in or established pursuant to the Alaskan Native Claims Settlement Act (85 Stat. 688 [43 U.S.C.A. and 1601 et seq.]), that is recognized as eligible for the special programs and services provided by the United States to Indians because of their status as Indians.

TSA Transportation Security Administration.

TSC Terrorist Screening Center.

Turnout Clothing A synonym for protective clothing, also called "bunker gear."

Turnout Time The portion of response time when fire companies are donning personal protective clothing and boarding their apparatus. The time begins once the companies have been notified to respond and ends when they begin travel time.(vehicle is moving)

UL Underwriters Laboratories.

unaffiliated volunteer An individual who is not formally associated with a recognized voluntary disaster relief organization; also known as a "spontaneous" or "emergent" volunteer.

uncertainty The degree to which a particular examination or test cannot be reproduced under similar conditions from one time to the next.

unified command An application of ICS used when there is more than one agency with incident jurisdiction or when incidents cross political jurisdictions. Agencies work together through the designated members of the unified command to establish their designated incident commanders at a single ICP and to establish a common set of objectives and strategies and a single incident action plan. Incident management system based on a team of command personnel from different jurisdictions and agencies who work together to fulfill goals and objectives defined in the incident action plan.

unit dispatched to arrival See *travel time* and *total response time.*

United States The term *United States,* when used in a geographic sense, means any state of the United States, the District of Columbia, the Commonwealth of Puerto Rico, the U.S. Virgin Islands, Guam, American Samoa, the Commonwealth of the Northern Mariana Islands, any possession of the United States, and any waters within the jurisdiction of the United States (as defined in section 2[16] of the Homeland Security Act of 2002, Public Law 107-296, 116 Stat. 2135, et seq. [2002]).

unsolicited goods Donated items offered by and/or sent to the incident area by the public, the private sector, or other source that have not been requested by government or nonprofit disaster relief coordinators.

urban search and rescue Operational activities that include locating, extricating, and providing on-site medical treatment to victims trapped in collapsed structures.

USACE U.S. Army Corps of Engineers.

USAR Urban search and rescue.

USCG U.S. Coast Guard.

USDA U.S. Department of Agriculture.

USFA U.S. Fire Administration.

USSS U.S. Secret Service.

Utstein criteria A recommended guideline for uniform reporting of data from out-of-hospital cardiac arrest.

validity The extent to which an examination or test measures or predicts what it is supposed to measure or predict. Validity is not an absolute quality of any test, but is relevant only in the context of the specific use to which a test has been put. Validity implies the existence of evidence that it is legitimate to make inferences from the rating or score of a particular assessment.

value-added If an organization measures itself by a satisfaction index, that means that it should measure itself by the three component indices: customer, employee, and financial satisfaction. Making your products and services value-added means that they fulfill customer requirements and exceed their expectations. Empowering employees to improve customer satisfaction, in turn, increases employee satisfaction. The result is better financial performance. In other words, adding value is not a bonus for the customer only—everyone benefits.

values Enduring beliefs and assumptions about specific modes of conduct or states of existence that are preferable to opposite or converse modes of conduct or states of existence. Values are the general guiding principles that govern our actions.

vertical communication Any communication that moves up or down the chain of command.

vision statement A written document describing an idealized view of where or what an organization would like to be in the future.

VMAT Veterinarian Medical Assistance Team.

volunteer Any individual accepted to perform services by an agency that has authority to accept volunteer services when the individual performs services without promise, expectation, or receipt of compensation for services performed. (See, for example, 16 U.S.C. §742f[c] and 29 CFR § 553.101.)

volunteer and donations coordination center Facility from which the volunteer and donations coordination team operates. It is best situated in or close by the state EOC for coordination purposes. Requirements may include space for a phone bank, meeting space, and space for a team of specialists to review and process offers.

volunteer firefighter A person that volunteers his or her time to perform fire and rescue services to a community. A volunteer is not compensated for his or her response.

vulnerability A measure of adverse consequence that might occur to a structure as a result of exposure to an uncontrolled fire. It is usually expressed as an indication of the difference between a level of risk and a level of service. For example, if a building has a calculated fire flow of 5,000 gpm and the level of service can deliver only 3,000 gpm, the structure is vulnerable to total loss unless the fire is controlled at the compartment level. Vulnerability is increased as the size and complexity of a risk exceeds the resources available to contain a fire to a limited level.

WAWAS Washington Area Warning System.

weapon of mass destruction (WMD) As defined in Title 18, U.S.C. § 2332a: (1) any explosive, incendiary, or poison gas, bomb, grenade, rocket having a propellant charge of more than 4 ounces, or missile having an explosive or incendiary charge of more than one-quarter ounce, or mine or similar device; (2) any weapon that is designed or intended to cause death or serious bodily injury through the release, dissemination, or impact of toxic or poisonous chemicals or their precursors; (3) any weapon involving a disease organism; or (4) any weapon that is designed to release radiation or radioactivity at a level dangerous to human life.

WMD Weapon of mass destruction.

work ethic A set of values based on the moral virtues of hard work and diligence. Qualities of character believed to be positive and promoted by work.

work uniform Items of clothing worn to perform routine day-to-day operations of maintenance and noncombat work. Includes shirt, pants, hats, underclothing, and other items. Work uniforms are not considered protective clothing.

zero-based budget (ZBB) Budget type whereby all expenditures must be justified at the beginning of each new budget cycle, as opposed to simply explaining the amounts requested that are in excess of the previous cycle's funding. During ZBB planning, it is assumed that there is zero money available to operate the organization or program; then the contribution that the organization or program makes to the jurisdiction must be justified.

zero defects A situation that exists when all quality characteristics are produced within design specifications. This concept is reflected in the attitude that defects can be prevented, especially if more attention is given to the task at hand. It is the theme that most embodies the concept of "do it right the first time."

Index

Note: Italicized *page numbers* represent glossary entries. Page numbers followed by *f* and *t* represent figures and tables respectively:

A

A Book of Five Rings (Musashi), 205
A Force for Change (Kotter), 92
A Theory of Human Motivation (Maslow), 290
A Tipping Point: Little Things Can Make A Big Difference (Gladwell), 547
Above-the-line accountability, 206, 206*f*
 vs. victim mentality, 207
Abrashoff, Captain D. Michael, 186
Aburdene, Patricia, 200
Acceptable level of risk, *652*
Acceptance, as individual need, 295
Accident Response Group (ARG), 607
Accountability:
 above-the-line, 206, 206*f*
 in authority, 342–43
 in team empowerment, 321–22
Accredit, *652*
Accreditation, 63, *652*
 certification vs., 82
 CFAI program model, 423, 424. *See also* Commission on Fire Accreditation International (CFAI)
 by NBFSPQ, 83
Accretion, in schema modification, 155
ACE (American Council on Education), 55
Achievement, motivation and, 117, 118*t*, 119
Achievement test, *652*
Achievers, characteristics of, 298
The Achieving Society (McClelland), 119
Act phase, *652*
 TQM and, 11, 12
Action(s):
 corrective, in disaster response, 647–48
 disciplinary. *See* Disciplinary action
 of organizations, NBES findings relating to, 244
 work-to rule, 362, 364
Action plan, *652*
ad valorem tax, 354
ADA (Americans with Disabilities Act), 17, 356, 363
Adams, Samuel, 28
The Adaptive Corporation (Toffler), 218
ADEA (Age Discrimination in Employment Act of 1967), 356
Adequacy/Adequate, *652*
Administration, for disaster response, 650–51
Administration Industrielle et General (Fayol), 103
Administrative Behavior: A Study of Decision-making Processes in Administration Organization (Simon), 108
Administrative fire officer rank:
 experiential elements for, 75*t*
 self-development elements for, 75*t*
 training elements for, 73*t*
Administrative management:
 Barnard's insights on, 103–4
 Fayol's principles of, 102–3
Advanced life support (ALS):
 defined, 439, *652*
 NFPA standards and, 439, 439*f*, 440
Advantage, personal, ethics law principles pertaining to, 274
Adverse impact, *652*
AED. *See* Automatic external defibrillator (AED)
Aerial Measuring System (AMS), 606
Affiliation, motivation and, 117, 118*t*
Affilitative leader, 145, 147*t*
AFRAT (Air Force Radiation Assessment Team), 594–95
African Americans, citizenship for, 9
Age Discrimination Act (1975), 17–18
Age Discrimination in Employment Act (1967), 356
Agency(ies):
 coordinating entity for, 521
 involvement in emergencies, 505–6, 505*f*
 response roles, 594–623, *652*
 sector-specific, 535
Agency representative, *652*
AHJ (authority having jurisdiction), *653*
AICPA (American Institute of Certified Public Accountants), 352–53
AIDS/HIV, civil rights protection for employees with, 17
Air Force Radiation Assessment Team (AFRAT), 594–95

Aircraft crash rigs, *653*
Aircraft rescue firefighting, 554–55, 554*f*
  NFPA 1710 standards for, 440
Alarm processing time, 406, *653*
Alarm time line, in NFPA 1720, 444*t*
Albemarle Paper v. Moody, 359
Alcohol abuse, disciplinary action and, 375–76
Alfred P. Murrah Federal Building, Oklahoma City, attack on, 31–33, 33*f,* 34*f*
All I Really Needed to Know I Learned in Kindergarten (Fulghum), 284
Allbaugh, Joe M., 504
ALS. *See* Advanced life support (ALS)
Alternative, *653*
America at Risk, America Burning Recommissioned (2002), 20–21
America Burning (1973), 20, 564
American Council on Education (ACE), 55
American Institute of Certified Public Accountants (AICPA), 352–53
American Red Cross (ARC), 524
American Society of Mechanical Engineers (ASME), 98
Americans with Disabilities Act (1990), 17, 356, 363
AMS (Aerial Measuring System), 606
Analysis, *653*
Animal and Plant Health Inspection Service (APHIS), *653*
Annual Symposium in the Sun, 60
Antidiscrimination laws, 17
APHIS (Animal and Plant Health Inspection Service), *653*
Apparatus, *653*
  NFPA needs assessment report on, 191
  quint, 437
Appeal process, disciplinary action and, 379–80
Appointed officials:
  conflict of interest checklist for, 275
  ethics codes for, 271–73
  ethics law principles for, 273–74
Appointments, interfering with, 271
Appraisal, of performance. *See* Performance appraisal
Appraisal judgments, 368
Appreciation, employee morale and, 317–18
Aquarius rank, in *Vigiles,* 3
Arbitration, in labor disputes, 364
ARC (American Red Cross), 524
Area command, 519, 520, *653*
ARG (Accident Response Group), 607
Aristotle, 237, 239
Arson, *653*
The Art of War (Tzu), 194
ASME (American Society of Mechanical Engineers), 98
Aspirational ethics, 249
Assembly line, 101
Assessment, *653*
  occupancy risk, 429
  preliminary damage, 510
  in promotional process, 372–73
  of risk. *See* Community risk assessment; Risk assessment
  situation, *678*
  of skills, 53
Associate's degree programs, 49–51
Assumption, *653*
ATF (U.S. Bureau of Alcohol, Tobacco and Firearms), *653*
Attendance, disciplinary action and, 374
Attention, employee morale and, 317
Attorney General, NRP role and responsibility of, 518*t*–519*t*
Audits, ethics, 273
Augustus Caesar, 2, 3
Authority:
  centralized vs. decentralized, 341
  delegation of, 341–42
  distribution of, 340–41
  forms of, 341
  and power, 105
  responsibility and accountability in, 342–43
Authority having jurisdiction (AHJ), *653*
Authority-compliance management, 142, 143*t*
Autocratic leadership style, 144
Automatic external defibrillator (AED), 438
  first responder with, 439, 439*f*
Automobile production, 101
Average on-scene to patient time, *653*

B

Babbage, Charles (1791-1871), management theory of, 94
Babel, tower of, 4
Baby boomers:
  characteristics of, 13, 304, 305*t*–306*t*
  recruiting, managing, and retaining, 310–11
  strengths of, 309
  views of, 311–12
Babylon, temple tower in, 4
Baccalaureate degree programs, 51
Backdraft (movie), 28
Badge(s):
  chief officer, 36*f*
  symbolism of, 22–23, 26–27, 26*f,* 31, 32*f,* 34
Bakke, E. Wright (1932-1971), 115–16
Ballista catapults, 2
Baltimore conflagration (1904), 413–14
Bank Wiring Observation Room, in Hawthorne experiments, 108
Bargaining tactics, 365. *See also* Negotiating, in labor disputes
Barker, Joel, 185
Barnard, Chester (1886-1961), 103–4
BARS (Behaviorally Anchored Rating Scales), 368
Base line system, *653*
Baseline data:
  in benchmarking, 476
  defined, 470, 476, *653*
  response time benchmarks and, 430–33, 431*t*–432*t*

Baseline data (*continued*)
types of, 471*t*
use of, 471, 472*f*
Basic life support (BLS):
defined, 438, *653*
NFPA standards and, 439, 440
Bed key, 7
Bedaux, Charles E. (1897-1944), 101–2
60/80 scale of, 102
Behavioral movement, 104–6
and Hawthorne effect, 106–8
Behaviorally Anchored Rating Scales (BARS), 368
Beliefs, 239
Benchmark(s), 470–71
defined, 430, *653*
for response time criteria, 430–33, 431*t*–432*t*
types of, 471*t*
use of, 471, 472*f*
Benchmarking:
benefits of, 475
for competitiveness, 474–75, 475*f*
defined, 474, *653*
as learning tool, 476
measurements for, 471, 473
opportunities for, 476
overview of, 472*f*
process flowchart, 474, 474*f*
steps in, 476, 477*f*
for top performance, 470–77
Benchmarking survey, how to complete, 589–91
Benevolent dictatorship leadership style, 144
Bennis, Warren, 128, 197, 198
Bentham, Jeremy, 268
BFRL (Building of Fire Research Lab), 567
Binding arbitration, 364
Bioterrorism defense, in homeland security strategic plan, 541
Black, meaning of, 27
Black Americans. *See* African Americans
Blake, Robert R., 114, 142
Blanchard, Kenneth H., *678*
Blind spot, 153, 160
BLM (Bureau of Land Management), *653*
BLS. *See* Basic life support (BLS)
Blue, meaning of, 27
Blumenfeld, Samuel, 238
Board of Selectmen, 336–37
"Boiled frog syndrome," 187–88, 193
Bond, *653*
Border, on badges, symbolism of, 26–27
Border and Transportation Security Directorate, of DHA, 227
Border security, in homeland security strategic plan, 541
Boston, Massachusetts:
building construction in, 5
Great Fire of, 8–9
Boundary, *654*
Boundaryless organization, 339
Boyatzis, Richard, 145
Boycotts, 364
Brainstorming, *654*
Brass trumpets, symbolism of, 26, 26*f*
Bridges, William, 223
Briggs, Katherine, 167
British Standards (BS) Institute, 102, *654*
Brito v. Zia Company, 370
Brown, Mike, 504
Bruegman, Randy R., biographical and career details for, 161–65
BS (British Standards) Institute, 102, *654*
Bucket Brigade, 29*f*
Buckets, leather, 389, 389*f*
Budget(s):
defined, 343, *654*
line-item, 344, 345*f*
NFPA needs assessment report on, 189
operating, *671*
performance, 346, 347*f*, *673*
program, 346, *675*
type of, *654*
zero-based, 348–49, *683*
Budget reform:
federal efforts in, 349
local efforts in, 350
performance budgeting and, 349
stages of, 350, 351*t*
state efforts in, 350
in twenty-first century, 350
Budget system, 346, 348, *654*
Budgeting:
fiscal management and, 350–54
origins and historical context of, 343–44
reform efforts in, 349–50
Building High Performance Teams (Ross and Isgar), 298
Building of Fire Research Lab (BFRL), 567
Buildings:
mitigation measures and, 512
in service area, by occupancy classifications, *654*
wildfire risk and, 395
"Bunker gear," *681*
Burchard, Pete, 163
Bureau of Land Management (BLM), *653*
Bureaucracy, 95–96, 95*f*
activities of, 340
characteristics of, 96
dysfunctional aspects of, 96–97
Bureaucratic hierarchy, *654*
Bureaucratic organizations, 337
Burns, Tom, 337
Burr, Aaron, 28
Bush, George H. W., 356
Bush, George W., 225, 396, 504
Business ethics, 238, 240
timeline in development of, 241*t*
2003 NBES and, 242–45

C

CAD (computer-aided dispatch) systems, 561
Camp, Robert, 476
Capes, use of, 24–25
Capital budget, *654*
Cardiac arrest resuscitation, 411, 411*f*

Career firefighters, 548
  workplace issues and, 11
Career goals, generational views on, 311
Carnegie, Andrew, 342
Carter, Jimmy, 503
Cartwright, Dorwin, 176*t*
Cascade of events, in response to emergency call, 405–6, 405*f*
Catapults, 2
Catastrophic incident, *654*
Catastrophic threats, in homeland security strategic plan, 541
Category, *654*
Causal system, *654*
CBIRF (Chemical Biological Incident Response Force), 595–96
CBRNE (Chemical, Biological, Radiological, Nuclear, and High Yield Explosive):
  fire cover standards and, 391
  WMD response role, 596–98
CDC (Centers for Disease Control and Prevention), 603
CEM (comprehensive emergency management), 504–5
Center for Public Safety Excellence (CPSE), 19, 121, 391
  organizational chart, 422*f*
Center for Public Service Excellence, 63
Centers for Disease Control and Prevention (CDC), 603
Centralization:
  of authority, 341
  as management principle, 103
CEO (chief executive officer), *655*
CERT (Community Emergency Response Team) training, 525–26, 527
Certification, 55, 62–63, *654*
  accreditation vs., 82
  by NBFSPQ, 83–84
CFAI. *See* Commission on Fire Accreditation International (CFAI)
CFO. *See* Chief fire officer (CFO)
Chain of command, 341, *654*
Challenger tragedy, 33
Change:
  evaluating, 222–23
  psychology of, 223–25
Change continuum, 223–24, 223*f*
Character: America's Search for Leadership (Sheehy), 239
Characteristic, *654*
Charismatic leadership, 146
Charlestown, Massachusetts, 7
Cheaper by the Dozen (Gilbreth and Careyone), 100
Cheating, 237
Check phase, *654*
  TQM and, 11, 12
Chemical Biological Incident Response Force (CBIRF), 595–96
Chemical Stockpile Emergency Preparedness Program (CSEPP), *655*
Chertoff, Michael, 504
Chiaramonte, Michael, xxi
  advice to new firefighters, 570
  on career development, 85
  on disaster management and response, 541–42
  on ethics, 277
  on future directions, 382
  on future of fire service, 570
  on generational issues, 325
  on leadership and management, 135
  on leadership style, 177
  on organizational changes, 229
  on quality improvement, 491
  on risk management, 445
  on tradition, 36
Chicago, 1871 fire in, 7–8, 7*f*
Chicago Tribune, 7–8
Chief administrative officer (CAO), in mayor-council government model, 334
Chief executive officer (CEO), *655*
Chief fire officer (CFO):
  badge of, 36*f*
  code of professional conduct for, 258–60
  designation, 65, 79
    portfolio for, 79*f*
    revocation of, cause for, 258
  path toward becoming, 56–57, 57*f*
  personal success of. *See* Personal success strategies
  political process and, 257–58
  political survival of. *See* Political survival
  role of, mayoral view, 256, 256*f*
  and volunteer firefighters, relationship between, 313
Chief Fire Officer Program, 258
Chief medical officer (CMO), 79
Chimney sweeps, 5, 6
Chimneys, regulations governing, 5, 6
Churchill, Winston, 142, 238, 268
CI. *See* Continuous improvement (CI)
CI/KR. *See* Critical infrastructure/key resources (CI/KR) protection
CIRG (Critical Incidence Response Group), 604
Citizen Corps Guide for Local Officials, 525
Citizen involvement, in emergency management, 525–27
Citizenship:
  for African Americans, 9
  of Rome, 2
City commission government model, 334–35
City manager, 334
Civil rights:
  for disabled workers, 17
  of volunteer firefighters, 14–15
Civil Rights Act (1957), 10
Civil Rights Act (1964), 10, 14, 357
  Title VII, 355–56, 358
Civil Service Reform Act (1978), 361, 363
Civil War, 9
Clackamas Fire District No.1, code of conduct of, 264–65
Clark, Bett, xxi
  advice to new firefighters, 570
  on career development, 85

Clark, Bett (*continued*)
on disaster management and response, 542
on ethics, 277
on future directions, 382
on future of fire service, 570
on generational issues, 326
on leadership and management, 135
on leadership style, 177
on organizational changes, 229
on quality improvement, 492
on risk management, 445–46
on tradition, 37
Cleveland Board of Education v. Laudermill, 380
Clinton, Bill, 504
Clothing. *See also* Uniform(s)
protective, *675*
turnout, *681*
CMO (chief medical officer), 79
COA (Commission on Accreditation), 83
Coaching leader, 145, 147*t*
Cochran, Kelvin, xxi–xxii
advice to new firefighters, 570
on career development, 85–86
on disaster management and response, 542
on ethics, 277
on future directions, 382–83
on future of fire service, 570
on generational issues, 326
on leadership and management, 135
on leadership style, 177–78
on organizational changes, 229–30
on quality improvement, 492
on risk management, 446
on tradition, 37
Code administration, performance measures for, 487
Code development:
natural disasters and, 395
performance measures for, 487
Code enforcement:
future vision for, 556, 556*f,* 557*t*–559*t*
NFPA needs assessment report on, 190–91
Code of conduct:
of Clackamas FD No.1, 264–65
personal, 266–67
professional, 258–60
Code of ethics:
applicability of, 273
for appointed officials and employees, 271–73
in decision-making process, 251
developing, 270
IAFC, 260–62
identifying, 240, 242–45
as living document, 275
mandatory, 248–49
values-based vs. rule-based, 270
Code of Professional Conduct, 258
Cognitive styles, 296–97, *655*. *See also* Keirsey Temperament Sorter; Myers-Briggs Type Inventory (MBTI)
Collective bargaining, 363
Collins, Jim, 151, 196
Colors, on uniform, symbolism of, 27
Columbia tragedy, 33
Combustion products, time vs., 408, 409*f*
Command:
chain of, *654*
in emergency management, 513, 514*f*
in ICS, 530, 531*f*
in NIMS, 529
unified, *681*
Command staff, *655*
Commanding leader, 146, 147*t*
Commission on Fire Accreditation International (CFAI), 53, 421–27
accreditation program model
benefits of, assessing, 426–27
criteria and performance indicators in, 423–24
effects on service levels, 425–26, 425*f,* 426*f*
objectives of, 424, 424*t*
benchmarking survey and, 484–87
cycle time example and, 473, 473*f*
mission of, 421–22
organizational chart for, 422*f*
response time benchmarks and performance baselines, 430–33, 431*t*–432*t*
risk categories and, 430
self-assessment process from, 423
Commission on Professional Credentialing (CPC), 63, 121
CFO code of professional conduct and, 258–60
CMO designation and, 79
Commitment, *655*
to continuous quality, 460, 461*f*
Committee on Accreditation (COA), 83
Communication(s):
with customers, measurement criteria, 576
in disaster response, 539, 539*f*–540*f*
organizational, 638–39
with public, 648–50
employee morale and, 318
with employees, 275–76, 277
leadership dimension of, 168–69, 170*t*
with media, 509
NFPA needs assessment report on, 191–92
NIMS and, 529–30
within organization, 187, 187*f*
team building and, 286–87
Community(ies):
changing demographics of, 12
diversity in, 12
profile of, in risk assessment, 429
risk assessment in. *See* Community risk assessment
risk factors in, 430, 431*f*
risk management in, 392–96, 402–4
StormReady, 512–13
Community education, performance measures for, 485
Community Emergency Response Team (CERT) training, 525–26, 527
Community ratings schedule, 390
Community ratings system (CRS), 512
Community recovery, *655*

Community risk assessment, *655*
life-safety threats, 392–93
property loss, 395
responder risk, 393–94
risk and planning, 395–96
Company(ies):
FSRS criteria for
distribution, 418
personnel, 418
individual property rating schedule and, 419
Competition/competitiveness:
benchmarking for, 474–75, 475*f*
organizational culture and, 208–9
Complacency:
in organizations, 187–88, 193
post-World War II, 453
performance and, 193
Complaints, tracking form for, 584
Comprehensive emergency management (CEM), 504–5
Computer-aided dispatch (CAD) systems, 561
Comradeship, 28
Concentration, in risk assessment, 429
"Conditional reflex," 150
Confidential information, disclosure of, 272
Conflict:
organizational, *672*
of values, 13
Conflict of interest:
in CFO code of professional conduct, 259
checklist for public officials, 275
in code of ethics for public officials, 272–73
Conflict resolution, 105
Congressional Act (1803), 503
Consciousness-raising, 319
Consequence:
in probability and consequence matrix, 401*t*, 403*t*
in risk assessment, 429
Consequence management, *655*
Consolidation, within and between organizations, 227–28
Consultation, about perceived wrongdoing, 252
Consultative leadership style, 144
Contaminant (pollutant), *673*
Content validity, *655*
Contingency approach, 120–21
Contingency organization, 340
Contingency plan, *655*
Continuous improvement (CI), 455, *655–56*
checklist for, 578
in CI/KR protection, 538*f*
commitment to, 460, 461*f*
customer-centered, steps in, 489–90
customer-driven, 453, 454*f*, 461*f*
elements of, 460, 461*f*, 462–63
fire service challenges in, 455–60
measurement criteria, 575
sample surveys evaluating, 578–84
tenets of, 489
Continuous quality improvement. *See* Continuous improvement (CI)
Control:
in disaster response, 637–38
span of, *678*
Control systems process, *656*
Coordination, in disaster response, 538
assessing local capability for, 637–38
Coping styles, 295–96, *656*
Core value(s), *656*
absent, problems caused by, 217
defined, 199
ethics as, 275–77
generational differences in, 304, 305*t*–306*t*
of service, 198–204
Corporate values, implementing, 276–77
Corps of *Vigiles*, 2, 3
Corrective actions, in disaster response, 647–48
Correlation, *656*
Cost cutting/cost containment, 456
Cost-benefit, *656*
Council-manager government model, 332, 334
Counter-dependent coping style, 296
defined, 296
Counterterrorism, in homeland security strategic plan, 541
Country club management, 142, 143*t*
Courage, *656*
John F. Kennedy on, 21–22
"Court tests":
job-related validation strategy and, 359
in performance appraisal, 370
CPC. *See* Commission on Professional Credentialing (CPC)
CPSE (Center for Public Safety Excellence), 19, 121, 391
Crassus (Roman emperor), 2–3
Creative conflict resolution, 105
Credibility, political dynamic and, 262
Criminal records, 359
Crisis communications, in disaster response, 648–50
Criterion/criteria:
CFAI example, 423
defined, 423, *656*
Critical Incidence Response Group (CIRG), 604
Critical incidents, *656*
Critical infrastructure:
defined, 396, 398–99, *656*
functions of, 399
in homeland security strategic plan, 541
in risk assessment, 393
sector categories for, 398
Critical infrastructure/key resources (CI/KR) protection, 535–38
challenges in, 399, 400*t*
continuous improvement in, 538*f*
homeland security and, 396, 397–98
long-term resilience and, 400–405, 400*t*
physical protection, 396–97
significance of, 398–99
spectrum of, 400*t*
Crosby, Philip, 491
Cross-functional teaming, 460

Cross-functional teams, *656*
CRS (community ratings system), 512
Crusades, fire as weapon during, 24–25
CSEPP (Chemical Stockpile Emergency Preparedness Program), *655*
Culture:
  coping styles and, 296
  defined, 10–11
  in United States, 9–10
Culture of discipline, 196
Current, *656*
Customer, *656*
  external, *660*
  feedback system, *661*
  internal, *665*
  needs of, 488, 488*f*
  quality improvement benefiting, 469–70, 469*f*
Customer alignment, measurement criteria, 577
Customer expectations, 453, 454*f*
  identification of, 460, 461*f*
  performance measures and, 483
Customer problems, eliminating, measurement criteria, 576
Customer satisfaction:
  defined, 466, 467*f*
  measuring, 487–91, 575–77
  philosophy of, 466
  requirements for, 467
  sample surveys evaluating, 579–83
  total, 450
Customer service:
  and benchmarking, 476
  defined, 476, *656*
  measurement criteria, 576–77
  sample surveys evaluating, 578–84
Customer survey(s), *656*
  of satisfaction
    with emergency responses, 579–84
    with public education programs, 584–88
Customer-supplier relationship, *656*
Cycle time, 473, *656*
  and benchmarking, 476
  CFAI example of, 473, 473*f*
  in quality improvement, 473–76

D

dantotsu (striving to be the best), 476–77
Data:
  hard, *663*
  soft, *678*
Data-based decision making, *656*
Deaths:
  civilian, sprinkler systems reducing, 556, 557*t*–559*t*
  firefighter (1977-2004), 563, 563*t*
Decentralization, of authority, 341
Decision making:
  data-based, *656*
  in emergency management, 513, 514*f*
  ethical. *See* Ethical decision-making
  40/70 formula for, 222
  information for, 221–22
  and solution commitment, 222
  team building and, 287
Dedicated service, by public officials, 271
Defect(s), *657*
  zero, *683*
Defense Production Act (1950), *657*
Deficiency, *657*
de-Gaulle, Charles, 263
Degree programs, 49–51, 55
Degrees at a Distance Program (DDP), 51, 80
  frequently asked questions about, 81
  standards "crosswalks" and, 48
Delegation:
  of authority, 341–42
  defined, 342, *657*
Deming, W. Edwards (1900-1993), 134, 452, 490
  management principles of, 124–25
    implementing, 126*f*
    obstacles to, 126–27, 126*f*
Democratic leader, 145, 147*t*
Democratic leadership style, 144
Demographic, *657*
Demographic information, in benchmarking survey, 589
Demographic shifts, impact on fire service, 12
Department of Defense (DOD):
  fire protection and, 555
  WMD response role, 594
Department of Energy/National Nuclear Security Administration (DOE/NNSA), 606
Department of Health and Human Services (HHS), 602
Department of Homeland Security (DHS), 504
  NFPA needs assessment with, 188–93
  organizational changes within, case study, 225–28
  strategic goals of, 228–29
  vision and mission statements of, 228
  WMD response role, 612–13
Department of Justice (DOJ), 604
Dependent coping style, 297
  defined, 296
Deployment, *657*
Deployment policies, in community risk assessment, 395–96
Design, organizational, *672*
DEST (Domestic Emergency Support Team), 616–17
Detection method of quality control, *657*
DHS. *See* Department of Homeland Security (DHS)
DHS-FEMA. *See* Federal Emergency Management Agency (DHS-FEMA)
Diagnosis, *657*
Dimensions, in promotional process, 372–73
Direction, in disaster response, 637–38
Directorates, within DHS, 227

Disaster(s). *See also* Emergency(ies)
community-based planning for, 501–2, 501*f*
defined, 498, *657*
emergency vs., 498
major, *667*
natural. *See* Natural disasters
preparation for, 499
risk for, 499–500
Disaster Medical Assistance Teams (DMATs), 617–18
Disaster recovery center (DRC), *657*
Disaster Relief Act (1974), 503
Disaster response. *See also* Emergency management
assessing local capability for, 624–51
government's role in, 501–2
in homeland security strategic plan, 541
performance measures for, 487
plan for, 500
Disciplinary action, 374, 375
and appeal process, 379–80
due process and, 378–79
general guidelines for, 378
grievance procedures and, 380–81
just cause and, 380
Disciplinary actions, guidelines for, 376
Disciplinary interview, 377–78
Discipline:
application of, 376
culture of, 196
discussions on, 374
documentation of employee behavior and, 376–77
problems with, 373–74
progressive, 374, 375
purpose of, 375–76
and volunteers, 381
Discrimination:
defined, 14, 355
illegal. *See* Illegal discrimination
Disney, Walt, 453
Dispatch circuits, Fire Suppression Rating Scale criteria for, 417
Dispatch processing time, 405, *657*
Disputes, labor, 364–65
Distance learning programs, 81–82. *See also* Degrees at a Distance Program (DDP)
Distribution:
of authority, in organization, 340–41
of fire companies, FSRS criteria for, 418
in risk assessment, 429
"stair stepping," technology and, 463
Diversity, *657*
acceptance of, 18, 19
defined, 10
in fire service, 18–19
in United States, 9–10
Divisional organization, 338–39
DMATs (Disaster Medical Assistance Teams), 617–18
Do phase, *657*
TQM and, 11
Dobbin, Frank, 227
Documentation, of employee behavior, 376–77
DOD. *See* Department of Defense (DOD)
DOE/NNSA (Department of Energy/National Nuclear Security Administration), 606
DOJ (Department of Justice), 604
Domestic counterterrorism, in homeland security strategic plan, 541
Domestic Emergency Support Team (DEST), 616–17
DPP. *See* Degrees at a Distance Program (DDP)
Drawdown, *657*
DRC (disaster recovery center), *657*
Drexler, Allan, 290
Drucker, Peter F. (1909-2005), 127, 132–34
Drug abuse, disciplinary action and, 375–76
Due process, 378–79

E

Eagle, symbolism of, 27
EAP (employee assistance program), 381–82
Early adopters, in organizational change, 210, 210*f*
Earthquake Hazards Reduction Act, *657–58*
Earthquake-resistant construction, as mitigation measure, 512
Earthquakes, 503
Ebbs, Susan L., 269
Economics, empowerment approach in, 319
Edison, Thomas, 106
Education, *658*. *See also* Training and education
for fire officer ranks
administrative, 73*t*–74*t*
executive, 76*t*–77*t*
managing, 70*t*–71*t*
supervising, 67*t*–68*t*
fire prevention through, 556–57, 559*f*
IAFC benchmarks in, 63–64
impact on fire service, 13–14
as misconduct preventive measure, 255–56
safety and, 20
Education programs, for public, 525–26
EEOC. *See* Equal Employment Opportunity Commission (EEOC)
Effective leader/leadership:
delegation in, 342
team building and, 284
Effective response force, 409, *658*
Effective/effectiveness, *658*. *See also* Quality
as performance measure, 482–83
Efficiency, *658*
as performance measure, 482
EFOP (Executive Fire Officer Program), 80
8-Minute Initial Attack, 438*f*
80/20 rule, 214–15, 215*f*
Einstein, Albert, 182
Eisenhower, Dwight D., 500
Elected officials:
and CFO political survival, 262–65
conflict of interest checklist for, 275
duty of, 269–70
ethics codes for, 271–73
ethics law principles for, 273–74
trust in, 268–69

Ellison v. Brady, 16
Elwood, Charles A., 105
Emblems, symbolism and significance of, 22–27, 25*f*, 26*f*
Emergency(ies). *See also* Disaster(s)
  defined, 498, *658*
  disaster vs., 498
  handling process for, 193–94
  managing. *See* Emergency management
  NFPA needs assessment report on, 192
  responses to. *See* Emergency responses
  WMD, agency response roles in, 594–623
Emergency calls, cascade of events in response to, 405–6, 405*f*
Emergency management:
  command/decision making in, 513, 514*f*
  communication in, 539, 539*f*–540*f*
  components of, 504–5
  comprehensive, 504–5
  concept of, 504
  coordination in, 538
  funding for, 540–41
  and hazard assessment, 514–15
  incident command system in. *See* Incident command system (ICS)
  integrated system of, 513–14
  NGOs and. *See* Nongovernmental organization (NGO)
  NIPP and. *See* National Infrastructure Protection Plan (NIPP)
  NRP and. *See* National Response Plan (NRP)
  phases of, 505
    mitigation, 511–12
    preparedness, 506–8
    recovery, 509–11, 510*f*
    response, 508–9
  planning for, 505–6, 505*f*
Emergency Management Institute (EMI), 526
Emergency medical services (EMS):
  community risk assessment and, 392–93
  evaluating capabilities of, 410–12, 411*f*
  future vision for, 549–51, 550*f*
  as integral part of fire service, 209
  NFPA needs assessment report on, 190, 192
  NFPA standards for, 434–35, 437–40, 439*f*
  performance measures for, 485, 487
  sample surveys evaluating, 580–83
Emergency operations center (EOC), 506*f*
  defined, 521, *658*
  Homeland Security, 521
  local and state, 521
  in NRP structure, 527
Emergency operations plan (EOP), 500, *658*
Emergency Preparedness and Response Directorate, 504
Emergency Preparedness Division, of DHA, 227
Emergency public information, *658*
Emergency response provider, *658*
Emergency response team–advance element (ERT–A), 522
Emergency responses:
  cascade of events in, 405–6, 405*f*
  in homeland security strategic plan, 541
  performance measures for, 486–87
  surveys evaluating, 579–84
  to WMD events, 594–623
Emergency support function (ESF), 519, *658–59*
  scope of, 520*t*–521*t*
EMI (Emergency Management Institute), 526
Empathy, *659*
Empirical, *659*
Employee(s):
  application and interview guidelines for, 357
  changing role of, 339–40
  communicating with, 275–76, 277
  disabled, civil rights protection for, 17
  empowerment of. *See* Employee empowerment
  ethics and, NBES findings, 243
  and labor relations, 360–67
  morale of. *See* Employee morale
  problems with. *See* Discipline
  public, code of ethics for, 271–73
  quality improvement benefiting, 468–69
Employee assistance program (EAP), 381–82
Employee behavior, documentation of, 376–77
Employee empowerment, 11
  delegation and, 342
  measurement criteria, 577
  model of, 324*f*
  paradox of, 324–25
  PDCA cycle and, 11–12
  quality service promoted by, 322–24
Employee morale, 313–14
  improving, 314–15
    guidelines for, 315–17
  leadership focus and, 317–19
Employee-employer contracts, rights proscribed in, 363
Employment:
  incompatible with public duty, 272
  regulations relating to
    for equal opportunity implementation, 358–59
    in equal opportunity laws, 355–57
    for firefighters, 359–60
    during selection and testing, 357–58
Employment application, guidelines for, 357
Empowerment, *659*
  and access to information, 320–21
  changing leadership processes and, 130
  defined, 11
  elements of, 319–20, 320*f*
  of employees. *See* Employee empowerment
  of leaders and managers, 464
  leaders influence on, 322
  paradox of, 324–25
  team building and, 319–25
Empowerment model, 324*f*
EMS. *See* Emergency medical services (EMS)
EMS Agenda for the Future: Implementation Guide, 549–51
Enforcement, safety and, 20
Engagement, employee morale and, 318

Engine companies, FSRS criteria for, 417
Engineering, safety and, 20
Enterprise fund, *659*
Environment, *659*
Environment Safety Occupational Health Service Center (ESOH), 598
Environmental Protection Agency (EPA):
  industrial fire protection and, 555
  quality improvement example and, 451
  WMD response role, 620–21
Environmental Response Team (ERT), 621, *659*
EOC. *See* Emergency operations center (EOC)
EOD (Explosive Ordnance Disposal), 598
EOP (emergency operations plan), 500, *658*
EPA (U. S. Environmental Protection Agency). *See* Environmental Protection Agency (EPA)
Equal employment opportunity, 14–15, 355, *659*
  laws and regulations governing, 355–57
Equal Employment Opportunity Commission (EEOC), 14–15, *659*
  employment selection and testing and, 357
  performance appraisal and, 371
  sexual harassment and, 15–16
  volunteer firefighter civil rights and, 14–15
Equal Pay Act (1963), 355
Equipment:
  of early fire companies, 389
  for modern firefighting, issues surrounding, 463
  NFPA needs assessment report on, 191
  safety, *677*
ERT (Environmental Response Team), 621, *659*
ERT (Evidence Response Team), 604–5
ERT–A (Emergency response team–advance element), 522
ESF. *See* Emergency support function (ESF)
ESOH (Environment Safety Occupational Health Service Center), 598
Esprit de corps, 22, *659*
  as management principle, 103
  organizational change and, 185
Etemenanki, 4
Ethic codes:
  identifying, 240, 242–45
  mandatory, 248–49
Ethical behavior, modeling, 245–46
Ethical conduct, building, 276
Ethical decision-making, 249–50
  analytical process in, 250–56
  outcomes of, 267
  testing questions for, 267
Ethical dilemma, 249
  example of, 249–50
  nature of, in decision-making process, 251
  resolving, analytical steps in, 250–56
Ethical leader/leadership, 235, 237–40
  characteristics of, 268
  components of, 246–47
  modeling ethical behavior and, 245–46
  modes of, 247–48
  NBES findings and, 244–45
Ethical organization:
  characteristics of, 268
  commitment to, 275
  safeguards in, 275
Ethics (ethos), *659–60*
  aspirational, 249
  in business. *See* Business ethics
  as core value, 275–77
  decline in, 235
  defined, 236
  employee perceptions of, NBES findings, 243
  impact of, 236–40
  mandatory, 248–49
  in organizations, 236–37
  origins of, 236
  personal orientation and, 249
  political process and, 257–58
  talking about, 245, 246
  vs. the law, 251–52
Ethics audits, 273
Ethics laws, principles for public officials, 273–75
Ethics programs, NBES findings on, 243–44
Ethics-related problems, NBES findings on, 243
Ethnic groups, U.S. census statistics for, 10
Evacuation, *660*
Evaluating, *660*
Evaluation(s):
  in disaster response, 647–48
  methodology, *660*
  of performance. *See* performance appraisal
"Everyone Goes Home" firefighter-safety initiative, 564, 564*f*
  objectives of, 564–65
Evidence Response Team (ERT), 604–5
Excellence, as core value, 202–3
Executive, functions of, 104
Executive Fire Officer Program (EFOP), 80
Executive fire officer rank:
  experiential elements for, 78*t*
  program for, 80
  self-development elements for, 78*t*
  training elements for, 76*t*
Exercises, in disaster response, 647–48
Exhibit, *660*
Existence, in ERG model of needs, 111, 111*f*
Expectations:
  changing nature of, 128–29
  communicating to employees, 276
  customer. *See* Customer expectations
Expenditures:
  of fire departments, 353*f*
  of fire districts, 354*f*
  total operational, *680–81*
Experience:
  for fire office ranks
    administrative, 75*t*
    executive, 78*t*
    managing, 68*t*
    supervising, 71*t*
  IAFC benchmarks in, 64
Experimental, *660*
Expertise, leader's power base and, 176, 176*t*

Exploitive authoritative leadership style, 144
Exploratory methodology, *660*
Explorers, in organizational change, 210, 210*f*
Explosive Ordnance Disposal (EOD), 598
External customer, *660*
Extroversion, vs. introversion, 168–69, 170*t*

F

FAA (Federal Aviation Administration), 554, 555
Facilitators, characteristics of, 298
Facilities:
 in disaster response, 645
 NFPA needs assessment report on, 191
Facility management, *660*
Fact-finding arbitration, 364
Fair Labor Standards Act, 363
Fairness:
 ethics law principles pertaining to, 274
 public officials and, 271
familia publica, 2
Family, changing demographics of, 12
Family Preparedness Program, 506, 507
FASB (Financial Accounting Standards Board), 352–53
Favors:
 conflict of interest and, 272
 ethics law principles pertaining to, 274
Fayol, Henri (1841-1925), 102–3
 management principles of, 103
FBI. *See* Federal Bureau of Investigation (FBI)
FCC (Federal Communications Commission), 539
FCO (federal coordinating officer), 523, *660*
FECC (federal emergency communications coordinator), *660*
Federal Aviation Administration (FAA), 554, 555
Federal Bureau of Investigation (FBI):
 SIOC of, 522
 WMD response role, 605
Federal Communications Commission (FCC), 539
Federal coordinating officer (FCO), 523, *660*
Federal emergency communications coordinator (FECC), *660*
Federal Emergency Management Agency (DHS-FEMA):
 in disaster response, 501
 history of, 503–6, 505*f*
 Project Impact of, 512
 suggestions for personal and family safety, 507
 training institute of, 526
 WMD response role, 614
Federal Fire Prevention and Control Action (1974), *660*
Federal fire protection, 555
Federal government:
 budget reform and, 349
 in emergency management, 505–6, 505*f*
  mitigation phase, 511–12
  preparedness phase, 506–7
  response phase, 508–9
 role in disaster response, 501–2
Federal Labor Relations Authority (FLRA), 361, 363
Federal on-scene coordinator (FOSC), *660*
Federal Radiological Monitoring and Assessment Center (FRMAC), 608–9
Federal resource coordinator (FRC), *660–61*
Federal Response Plan (FRP), 501, 509–11
Federal Uniform Guidelines on Employee Selection, 19
Feedback:
 employee morale and, 318–19
 generational views on, 311
 from peer group, 381
 in performance appraisal, 368
Feedback system, customer, *661*
Feeling, thinking vs., 171–72, 173*t*
Felony convictions, 359
FEMA. *See* Federal Emergency Management Agency (DHS-FEMA)
FESHE. *See* Fire and Emergency Services Higher Education (FESHE)
FESHE National Fire Service Curriculum Committee, 48
FESSAM (Fire and Emergency Service Self-Assessment Manual), 392
Fibrillation, 412, *661*
Finance/Administration section, in ICS, 534, 534*f*
Financial Accounting Standards Board (FASB), 352–53
Financial interests, personal:
 ethics law principles pertaining to, 274
 public officials and, 272
Financing:
 challenges for public sector agencies, 455
 for disaster response, assessing local capability for, 650–51
Fire:
 growth stages of, 407–8
 as weapon, 24–25
Fire Act, 567, *661*
Fire and Emergency Service Self-Assessment Manual (FESSAM), 392
Fire and emergency services:
 higher education for. *See* Fire and Emergency Services Higher Education (FESHE)
 professional status and, 44–47
 specializations within, 46
Fire and Emergency Services Higher Education (FESHE), 48
 conferences of, 49
 distance learning program and, 80
 future directions, 55–56
 professional development model, 49–51, 50*t*
 program mark, 49*f*
Fire chief(s). *See also* Chief fire officer (CFO)
 badge of, 36*f*
 Florian von Cetum as, 3
 path toward becoming, 56–57, 57*f*
 responsibilities of, 30

Fire codes, in colonial America, 5–9
Fire confined to structure, *661*
Fire Corps, 526
Fire cover, national standards of, 390–92
Fire department:
  nonunionized, labor relations in, 362–63
  organizational chart for, 333*f*
  volunteer, labor relations in, 362–63
Fire departments, 419. *See also* Company(ies)
  applying motivational theories to, 111
  changes and challenges in, 456, 456*f*
  class 9, Fire Suppression Rating Scale criteria for, 419
  class 8B, Fire Suppression Rating Scale criteria for, 418–19
  evaluating
    early history of, 413–14
    permanent grading system for, 414–15
    property insurance ratings and, 412
  expenditures pie chart for, 353*f*
  fiscal management by, 353
  Fire Suppression Rating Scale criteria for, 416–19
  image of, maintaining and improving, 463
  revenue pie chart for, 353*f*
Fire districts:
  expenditures pie chart for, 354*f*
  fiscal management by, 354
  revenue pie chart for, 354*f*
Fire flow:
  available, *661*
  delivered, *661*
  required (or estimated), 408, *661*
  in risk assessment, 428
  tasks in, 408–9, 409*f*, 410*t*
Fire insurance, 7
  ISO Grading Schedule for, 9
Fire management area:
  defined, 429, *661*
  in risk assessment, 429
Fire marks, 7
Fire out on arrival, *661*
Fire pre plan, *661*
Fire prevention:
  code enforcement and, 556, 556*f*, 557*t*–559*t*
  NFPA needs assessment report on, 190–91
  performance measures for, 485–86
  sample surveys evaluating, 579–80
  through education, 556–57, 559*f*
Fire protection:
  duties, *661–62*
  environment, *662*
  federal and military, 555
  grading schedule for, 414–15, 415*t*
    modern, 416–19
    municipal, 416
    updating, 415–16
  industrial, 555–56
  performance measures and, 483
  system, *662*
  wildfires and, 395
Fire Rescue International (FRI), 58
Fire scenario, *662*
Fire service:
  challenges to, 307–8, 308*f*
    quality improvement and, 455–60
  changes within, 183–85
  core values of, 198
  creating national doctrine for, 568–69
  demographics of, 548
  evaluation of, 578–84
  evolution of, 21–23, 389–90
  hallmarks of, 455
  human relations in, 18–19
  impacts and challenges to, 18–21
  mission after 9/11 terrorist attacks, 183
  NFPA needs assessment of, 188–93
  opportunities and challenges to, 30–31
  primary purpose of, CFAI program and, 424, 424*t*
  research into, 566–68
  responsiveness of, 209
  societal impact on, 9–14
  television shows and movies depicting, 28
  Truman's perspective on, 20
  workforce quality of, 464
Fire Service Leadership Partnership workshops, 313
Fire stations, growth and placement of, 389
Fire suppression:
  effective response force in, 409–10
  evaluating capabilities for, 406–7
  fire ground tasks in, 408–9, 409*f*, 410*t*
  fire growth stages and, 407–8, 408*t*
  NFPA 1710 and, 435–37
  in risk management, 402–3, 403*t*
Fire suppression activities:
  evolution of, 389–90
  future of, 548, 549*f*
Fire Suppression Rating Schedule (FSRS), 416–4190
Fire wardens, volunteer, in New Amsterdam, 5
Firefighters:
  career. *See* Career firefighters
  demands upon, 462–63
  employment of, laws and regulations governing, 359–60
  at Ground Zero, 183*f*
  opportunity for, 569
  patron saint of, 3
  sacrifice of, 28, 34
  safety of, future vision for, 563–65, 563*t*, 564*f*
  tests for hiring and promoting, 359
    in performance appraisal, 370
  training and education continuum for, 42–43, 43*t*
  volunteer. *See* Volunteer firefighters
Firefighting:
  in colonial America, 5–9
  early history of, 1–2
  as profession
    evolution of, 42–43, 43*t*
    opportunity in, 569
  salvage operations and, 7
  in special situations, NFPA 1710 standards for, 440–41
Fire-related higher education, national system for, 51–54, 52*f*, 52*t*

Fire-resistant construction, as mitigation measure, 512
First responder:
with AED, 439, 439*f*
defined, 440, *662*
First-due area, *662*
Fiscal management:
at local level, 353–54, 353*f*–354*f*
oversight agencies for, 350–53
Flame stage, in fire growth, 407
"Flashover" stage, in fire growth, 407–8, 408*t*
significance of, 408, 409*f*
Flawless Execution Model^{sm}, 193–94
Flexibility, NIMS and, 528
Flexner, Abraham, 46
Flood Control Act, 503
Flood mitigation:
measures for, 512
NFPA needs assessment report on, 192
Florian von Cetum (St. Florian), 3–4
Flowchart, *662*
FLRA (Federal Labor Relations Authority), 361, 363
FNS (Food and Nutrition Service), *662*
Follett, Mary Parker (1868-1933), 105
Follower, as leader, 129
Food and Nutrition Service (FNS), *662*
Forces, in total quality leadership, 464, 464*f*
Ford, Henry, 99, 101
Ford, Henry, Sr., 238
Forest fire, destroys Peshtigo, Wisconsin (1871), 8
FOSC (federal on-scene coordinator), *660*
Founding Fathers, as volunteer firefighters, 28
"4 Minutes to Excellence" theme, 165, 166*f*, 219
40/70 formula, for decision-making, 222
Frame of reference, 161
author's, 161–65
Franklin, Benjamin, 6–7, 6*f*, 28, 389
FRC (federal resource coordinator), *660–61*
Free-burning stage, in fire growth, 407
"Freeman," 9
Freeman, R. Edward, 238
French, John R. P., Jr., 176*t*
Fresno Fire Department:
core values of, 203–4, 204*f*
"4 Minutes to Excellence" theme, 165, 166*f*, 219
FRI (Fire Rescue International), 58
FRMAC (Federal Radiological Monitoring and Assessment Center), 608–9
FRP (Federal Response Plan), 501, 509–11
FSRS (Fire Suppression Rating Schedule), 416–4190
Fulghum, Robert, 284
Functional organization, 338
The Functions of the Executive (Barnard), 103–4
Funding, for emergency management, 540–41
Fusion process, in personnel motivation, 115–16

G

GAAP (Generally Accepted Accounting Principles), 351, 352–53
Gaebler, Ted, 349, 475
Galvin, Bob, 451
Gandhi, Mahatma, 238
Gantt, Henry (1861-1919), 101
Gantt Chart, 101, *663*
Gardner, John, 239
Garvey, Marcus, 9
GASB (Governmental Accounting Standards Board), 351–52
General and Industrial Management (Fayol), 103
General operating guidelines, *663*
Generally Accepted Accounting Principles (GAAP), 351, 352–53
Generation X:
characteristics of, 304, 305*t*–306*t*
recruiting, managing, and retaining, 310–11
strengths of, 309
views of, 311–12
Generation Y (millennials):
characteristics of, 304, 305*t*–306*t*
recruiting, managing, and retaining, 310–11
strengths of, 309–10
views of, 311–12
Generations:
conflict between, 309–10
defining events for, 306–7
profiles of, 304, 305*t*–306*t*
Geographical designations:
in benchmarking survey, 589–90
CFAI response time benchmarks and performance baselines for, 431–33, 431*t*–432*t*
Geographical information systems (GIS):
fire suppression and, 548
integrated risk management and, 561
standards of response coverage and, 427
Gerard (medieval monk), 23
Gifts:
conflict of interest and, 272
ethics law principles pertaining to, 274
Gilbreth, Frank (1868-1924), 100–101
Gilbreth, Lillian (1878-1972), 100–101
GIS. *See* Geographical information systems (GIS)
Gladwell, Malcolm, 547
Glass bombs, naphtha in, 24
Goal(s), *663*
communicating to employees, 276
future, PSCA cycle and, 468
organizational. *See* Organizational goals
performance, *673*
strategic, *679*
Goal achievement, *663*
Gold (color), meaning of, 27
Golden Rule, 284
Goleman, Daniel, 145
Good to Great (Collins), 151, 196
Goodwill, leader's power base and, 176, 176*t*
Governance, local models of, 332, 334–37
Government. *See also* Federal government; Local government; State government
infrastructure of, *663*
leadership in, 269–75

Government Performance and Results Act (1993), 349, 350
Governmental Accounting Standards Board (GASB), 351–52
Governmental Finance Officers Association, 350
Governor, NRP role and responsibility of, 517*t*
Grading schedule, for fire protection, 415, 415*t*
  modern, 416–19
  municipal, 416
  updating, 415–16
Grants, *663*
"Great" fires:
  Boston, 8–9
  Chicago (1871), 7–8, 7*f*
  Peshtigo, Wisconsin (1871), 8
  Rome, Italy, 3
Greenleaf, Robert, 129–30, 152
Grievance procedures, 380–81
Group dynamics:
  leading change and, 197–98
  phases of, 194, 195*f*
Group process:
  defined, 297
  in teamwork, 297–98
Groups:
  peer, 381
  as teams. *See* High-performance teams; Team *entries*
Growth, in ERG model of needs, 111, 111*f*
Guiding principles, *663*

H

Hamilton, Alexander, 28
Hammurabi code, 4, 4*t*
Hancock, John, 28
Hand in Hand Insurance Company, 7
Hard data, *663*
Harlem Renaissance, 9
Harvard Business Review, 127
Hawthorne effect, 106–8
Hawthorne experiments/studies, 105, 106–8, *663*
Hazard(s):
  defined, 429, *663*
  management of, in disaster response, 626
  mitigation of, *663*
  in risk assessment, 430
    disaster risk, 514–15
  single-family residence and, 393
Hazard identification, 515
  in disaster response, 625–26
  performance measures for, 485
Hazardous material/substance, *663*
  response to. *See* Hazmat incident
Hazmat incident:
  NFPA needs assessment report on, 192
  performance measures for, 487
  response to, 551–52, 551*f*
Heavy/extrication rescue, *663–64*
Height/weight restrictions, 358
Hersey, Paul, *678*
Herzberg, Frederick (1923-2000), 116–17
HHS (Department of Health and Human Services), 602
Hierarchy, bureaucratic, *654*
High hazard risks, 428, *664*
Higher education:
  FESHE model, 49–51, 50*t*. *See also* Fire and Emergency Services Higher Education (FESHE)
  fire-related, national system for, 51–54, 52*f*, 52*t*
High-performance programming model, 293, 293*f*
High-performance teams:
  characteristics of, 294–97, 303–4
  components of, 303–4, 303*f*
  employee morale and. *See* Employee morale
  group hierarchy of needs and, 292, 292*t*
  leadership of, 294, 294*f*
    measurement exercise in, 298–302
  Maslow's hierarchy of needs and, 290–92, 291*t*
  programming model for, 293, 293*f*
  research on, 302
Hill, Anita, 15
Hippocrates (460-377 B.C.), 168*t*
Historic property, *664*
History of the British Fire Service (Blackstone), 2
Hitler, Adolf, 238
Holmes, Oliver Wendell, 221
Homans, George, 106
Homeland security:
  and infrastructure protection, 397–98
  strategic plan for, 396–97, 541
Homeland Security Act (2002), 225, 228
Homeland Security Operations Center (HSOC), 521
Homeland Security Presidential Directive (HSPD), 515–16. *See also* National Incident Management System (NIMS)
Honesty, team building and, 285
Honor, *664*
Hopkins, Esek, 201
"Hospitaller" knights. *See* Knights of Saint John of Jerusalem
Hostile environment, in sexual harassment, 15, 358
HSOC (Homeland Security Operations Center), 521
HSPD (Homeland Security Presidential Directive), 515–16. *See also* National Incident Management System (NIMS)
Hudson Institute of Indianapolis Researchers, 12–13
Human behavior, future fire service challenges and, 560
The Human Group (Homans), 106–7
Human relations, in fire service, 18–19
Human relations management theories, 108–19
Human resources, regulations relating to, 308
Human resources management theory, 119
The Human Side of Enterprise (McGregor), 112, 113
Hurricane Hugo, 509

Hurricane Katrina, 499*f*, 566
  Federal Response Plan and, 510
  FEMA and, 504
  risk assessment and, 393
  risk mitigation and, 394
Hurricane Rita, risk assessment and, 393
Hurricanes, 503
Hussein, Sadam, 238
Hutchinson, Peter, 226
Hydrants, FSRS criteria for, 418
Hygiene factors, job attitudes and, 116–17, 116*t*

I

IAFC Professional Development Guide. *See* International Association of Fire Chiefs (IAFC)
IAFF (International Association of Fire Fighters), 19, 313, 361
IAP (incident action plan), 532–33, *664*
IC (incident commander), 530, *664*
ICMA. *See* International City/County Management Association (ICMA)
ICP (incident command post), 519, 520, *664*
ICS. *See* Incident command system (ICS)
Identity, as individual need, 294
IEMS (Integrated Emergency Management System), 513–14, 514*f*
IFSAC (International Fire Service Accreditation Congress), 46, 63, 82–83
Ignition stage, in fire growth, 407
IIMG (Interagency Incident Management Group), 521
Illegal discrimination, 14
  detection of, 16–17
  legislation against, 17–18
IMAAC (Interagency Modeling and Atmospheric Assessment Center), 607–8
Image, maintaining and improving, 463
Immigration, effects of, 10
Immigration Reform and Control Act (1986), 357
Impartiality:
  ethics law principles pertaining to, 274
  public officials and, 271
Impoverished management, 142, 143*t*
Imprinting, of values, 13
IMT (incident management team), *664–65*
IMT (National Incident Management Teams), 617
In Search of Excellence (Peters and Waterman), 195
Incident(s), *664*. *See also* Emergency(ies)
  catastrophic, *654*
  critical, *656*
  hazardous materials. *See* Hazmat incident
  mitigation of, *665*
  of national significance, 516, *665*
Incident action plan (IAP), 532–33, *664*
Incident command post (ICP), 519, 520, *664*
Incident command system (ICS), 500, *664*
  organizational structure of, 530–34, 531*f*–534*f*, 531*t*
Incident commander (IC), 530, *664*
Incident management team (IMT), *664–65*
Inclusion:
  as individual need, 294
  team empowerment and, 321
Indicator, of performance. *See* Performance indicator(s)
Individual(s):
  assessing potential harm to, 392–93
  in high-performance teams
    characteristics of, 303
    cognitive styles of, 296–97
    coping styles of, 295–96
    needs of, 294–95
  level of preparedness of, 507, 507*f*
  NRP role and responsibility of, 519*t*
  responsibility for safety of, 565
  service and, 198
Individual property, fire suppression rating schedule for, 419
Industrial fire protection, 555–56
Industrial Revolution, *665*
Influence, as individual need, 294
Information:
  about disaster response, assessing local capability for, 648–50
  access to, empowerment and, 320–21
  in benchmarking survey, completing, 589–91
  confidential, disclosure of, 272
  for decision-making, 221–22
  emergency, for public, *658*
  for problem understanding, 220–21, 221*f*
  web-based, risk management and, 561
Information Analysis and Infrastructure Protection Directorate, of DHA, 227
Information management, NIMS and, 529–30
Infrastructure, *665*
  critical. *See* Critical infrastructure
  government, *663*
  national protection plan for, 535–38
Ingham, Harry, 152
Initial rapid intervention team (IRIT), 437
Ink blot (Rorschach) test, 359
Innovation, *665*
Innovators:
  in organizational change, 210, 210*f*
  as team member, characteristics of, 298
Input, as performance measure, 479
Insignia, symbolism of, 22–23, 26–27, 26*f*
  Maltese Cross, 22, 23–25, 25*f*
Institutionalization:
  leadership and, 158
  organizational culture and, 217
Insurance Services Office (ISO), 395, 512, 560
  defined, 412, *665*
  fire department evaluation and, 412
  grading schedule of, 415, 415*t*
    modern, 416–19
    for municipalities, 416
    updating, 415–16
  logo of, 413*f*
  PPC ratings and, 420–21
Integrated Emergency Management System (IEMS), 513–14, 514*f*

Integrated risk management, 560–61
CFAI response time benchmarks and performance baselines, 430–33, 431*t*–432*t*
measurement standards and, 428–29
modeling, 426, 427–28
plan presentation, 433
and standards of response coverage, 429–30. *See also* Standards of response coverage
Integrated Risk Management Model (2004), 394
Integrated Risk Management Planning (IRMP), 390–91
Integrity, *665*
defined, 202
organizational, 200
certifying, 244
political dynamic and, 262
Intelligence, in homeland security strategic plan, 541
Intent, strategic, *679*
Intention, employee morale and, 318
Interagency Incident Management Group (IIMG), 521
Interagency Modeling and Atmospheric Assessment Center (IMAAC), 607–8
Internal customer, *665*
Internal investigation, of alleged misconduct, 253–54
International Association of Fire Chiefs (IAFC), 19, 57–58, 58*f*
code of ethics, 260–62
labor-management workshops, 313
logo of, 58*f*, 261*f*
mapping definition, 62
national fire cover standards and, 391–92
national mutual-aid system and, 565–66
Officer Development Handbook, 56, 57*f*, 58–59, 63–64
post 9/11 vision for, 212–13, 212*f*–215*f*
Professional Development Committee of, 53, 55–56
International Association of Fire Fighters (IAFF), 19, 313, 361
International City/County Management Association (ICMA), 391, 477
due process considerations and, 378–79
International Fire Service Accreditation Congress (IFSAC), 46, 63, 82–83
Interoperability, of radio systems, 539, 539*f*–540*f*
Intervention, with potential wrongdoer, 252–53
Interview(s):
disciplinary, 377–78
for employee selection, guidelines for, 357
in firefighter selection, 359
in Hawthorne experiments, 107
in promotional process, 373
Interviewing stage, in Hawthorne experiments, 107
Intrinsic rewards, 112
Introversion, extroversion vs., 168–69, 170*t*
Introversion/extroversion scale, in Myers-Briggs inventory, 167
Intuiting, sensing vs., 169–71, 172*t*
Investigations, performance measures for, 487
IRIT (initial rapid intervention team), 437
IRMP (Integrated Risk Management Planning), 390–91
Isgar, Tom, 298
ISO. *See* Insurance Services Office (ISO)
ISO Fire Insurance Grading Schedule, 9
It's Your Ship (Abrashoff), 186

J

JAMA (Journal of the American Medical Association), 411
James, William, 239
Jamestown, Virginia, 5
Jay, John, 28
Jefferson, Thomas, 28
JFO. *See* Joint Field Office (JFO)
JIC (joint information center), *665*
JIS (joint information system), *665–66*
Job attitudes, factors affecting, 116–17, 116*t*
Job changing, generational views on, 311
Job evaluation process, 371
Job performance:
disciplinary action and, 374, 375
employee assistance program and, 381–82
Job relatedness:
evaluation factors and, 371
validation strategy and, 359
JOC (Joint Operations Center), 522–23, *666*
Johari Window model, 152–53, 153*t*, 160, *665*
Johnson, Admiral Harvey, 228
Joint Field Office (JFO), 522, *665*
Coordination Group members and, 523
Joint information center (JIC), *665*
Joint information system (JIS), *665–66*
Joint Operations Center (JOC), 522–23, *666*
Joint Task Force Civil Support (JTF-CS), 599
Joint Terrorism Task Forces (JTTFs), 522
Jones, Cliff, xxii
advice to new firefighters, 571
on career development, 86
on disaster management and response, 542–43
on ethics, 277–78
on future directions, 383
on future of fire service, 571
on generational issues, 326
on leadership and management, 136
on leadership style, 178
on organizational changes, 230
on quality improvement, 492
on risk management, 446
on tradition, 37
Jones, John Paul, 201
Journal of the American Medical Association (JAMA), 411
JTF-CS (Joint Task Force Civil Support), 599
JTTFs (Joint Terrorism Task Forces), 522
Judging, vs. perceiving, 172–74, 174*t*
Judging/perceptive scale, in Myers-Briggs inventory, 168
Judicial decisions, on sexual harassment, 16
Jung, Carl Gustav, 167
Jungian Type Inventory, 167–68, 167*t*, *666*

Junto, 6–7, 7*f*, *666*
Juran, Joseph, 490–91
Jurisdiction, *666*
Just cause, 380

K

Kaye, Allan, 382
Keirsey Temperament Sorter, 168–74, 170*t*, 172*t*–174*t*, *666*
Kennedy, John F., 33, 270, 361
  on courage, 21–22
Key assets:
  importance of, 399
  protecting, challenges of, 399. *See also* Critical infrastructure/key resources (CI/KR) protection
Killen, Bill, 566, xxii
  advice to new firefighters, 571
  on career development, 86
  on disaster management and response, 543
  on ethics, 278
  on future directions, 383
  on future of fire service, 571
  on generational issues, 326–27
  on leadership and management, 136
  on leadership style, 178
  on organizational changes, 230
  on quality improvement, 492–93
  on risk management, 446
  on tradition, 37
Killing the messenger, 287
King, Kevin, xxii–xxiii
  advice to new firefighters, 571
  on career development, 86
  on disaster management and response, 543
  on ethics, 278
  on future directions, 383
  on future of fire service, 571
  on generational issues, 327
  on leadership and management, 136
  on leadership style, 178
  on organizational changes, 230
  on quality improvement, 493
  on risk management, 446–47
  on tradition, 38
King, Martin Luther, Jr., 33
KITA factors, 117
Knights of Saint John of Jerusalem, 22, 24*f*, 25*f*
  colors used by, symbolism of, 27
  origins of, 23
Knights Templar, 27
Knowledge-worker productivity, factors for, 134
Koresh, David, 238
KSA (Knowledge-Skills-Abilities), *666*

L

Labor, and management, relationship between, 312–13
Labor disputes, settling, 364–65
Labor Management Relations Act (1947). *See* Taft-Hartley Act (1947)
Labor relations:
  collective bargaining and, 363
  dispute settlement and, methods for, 364–67
  employees and, 360–67
  societal trends in, 361–62
  in volunteer and nonunionized departments, 362–63
Labor strikes, alternatives to, 362
Labor unions. *See* Organized labor
Ladder 49 (movie), 28
Ladder and service companies, FSRS criteria for, 417
Ladder truck, *666*
Laissez-faire leadership style, 144
Landrum-Griffin Act (1959), 361, 363
Late adopters, in organizational change, 210, 210*f*
Law enforcement involvement, in alleged misconduct, 254–55
Law of the Situation, 105
Laws and Authorities, for disaster response, assessing local capability for, 624–25
Layering, *666*
Lead ducks, in organizational change, 210, 210*f*
Leaders:
  changed expectations and, 128–29
  effective, team building and, 284
  empowering influence of, 322
  followers as, 129
  legacy of, 158
  managers and, 92, 159
    differences between, 198
  power base of, 174, 176–77, 176*t*
  qualities of, 196–98
  resonance created by, 146
  styles of. *See* Leadership style(s)
Leadership, *666*
  art of, 284–85
  assessment and measurement of, 239
  core value characteristics of, 205
  and core values of service, 198–204
  80/20 rule in, 214–15, 215*f*
  empowerment of, 464
  ethical. *See* Ethical leader/leadership
  functions of, *666*
  government, 269–75
  in managerial grid model, 114–16, 115*f*
  moral, 238
  needs and, 142–44
  for organizational change, 184–85
    principles for, 222–23
    qualities required, 196–98
  organizational stagnation and, 186–87
  politics and ethics and, 256–58
  reflections on, 30–36
  servant, 151–52
  strategic, *679*
  strategies for personal success. *See* Personal success strategies
  styles of. *See* Leadership style(s)
  task of, 105
  twenty-first century changes in, 128–32
  values-based, 235, 237–40. *See also* Ethics

Leadership style(s), *666*
author's, events framing, 165, 166*f*
charismatic, 146
determining one's own, 165, 167–74
emotional, 145–46, 147*t*
failure to understand one's own, 159–60
importance of, 174, 176–77
Lewin's, 144–45
Likert's, 144
in managerial grid model, 142–44, 143*f*, 143*t*
measurement exercise, 298–301
scoring key for, 302–3
participative, 144, 148, 148*t*
personal framework shaping, 156–57
quiet, 151
results affected by, 157–61
servant, 129–30
situational, 148–49
transactional, 149–50
transformational. *See* Transformational leadership
Leading, *666*
Leather buckets, 389, 389*f*
Legal issues, 307
Legal profession, and fire service compared, 45–46
Legislation:
affecting employment, 355–60
disclosure of interest in, 272–73
Legislative mandates, 14–18
diversity acceptance stimulated by, 19
Levee construction, as mitigation measure, 512
Level of service:
defined, 428, *666*
principles of, 428–29
Lewin, Kurt (1890-1947), 144–45
Liability issues, 307, *666–67*
Liaison officer, *667*
Life safety, 392–93
Life support:
advanced, *652*
basic, *653*
Light, Mark, xxiii
advice to new firefighters, 571
on career development, 86–87
on disaster management and response, 543
on ethics, 278
on future directions, 383–84
on future of fire service, 571
on generational issues, 327
on leadership and management, 136–37
on leadership style, 178
on organizational changes, 230–31
on quality improvement, 493
on risk management, 447
on tradition, 38
Likert, Rensis (1903-1981), 144
Lincoln, Abraham, 46, 238, 239
Line authority, 341
Line-item budget, 344, 345*f*
defined, 344, *667*
Listening, employee morale and, 317–18
Local community:
disaster planning in, 501–2, 501*f*
FEMA and, 503–6, 505*f*
hazard assessment in, 515
preparedness and, 502
preparedness variations in, 534–35
recovery decisions and, 510
Local government, *667*
budget reform and, 350
disaster planning by, 501–2, 501*f*
in emergency management, 505–6, 505*f*
mitigation phase, 512
preparedness phase, 507
recovery phase, 511
response phase, 508–9
governance models in, 332, 334–37
NRP role and responsibility of, 517*t*
Logical thinker:
as coping style, 296
defined, 295
Logistics, in disaster response, 645
Logistics section, in ICS, 533, 553*f*
Loma Prieta Earthquake, 504, 509
Lombardi, Vince, 200
Low risk, 427, *667*
Loyalty, in public officials, 271
Luft, Joseph, 152

M

Machiavelli, Niccolò, 228
Macy, John, 503
Mail-order business, 101
Major disaster, *667*
Making the Corps (Ricks), 204–5
Malcolm Baldridge National Quality Award, *667*
winners of, 452, 475
Maltese Cross, 22, 23–25, 25*f*
Management:
administrative, 102–4. *See also* Administrative management
author's style of, events framing, 165, 166*f*
behavioral approach to, 104–8
bureaucratic, 95–97, 95*f*
consequence, *655*
Deming's theory of, 124–25
implementing, 126*f*
obstacles to, 126–27, 126*f*
in dispute settlement, 364–65
80/20 rule in, 214–15, 215*f*
emerging trends in, 121–28
empowerment approach in, 319
generational differences and, 310
human relations theories of, 108–19
human resources theory of, 119
integrating theories of, 119–21
and labor, relationship between, 312–13
material, *668*
new assumptions of, 133–34
old assumptions for discipline and practice of, 133
participative, 322–23

Management (*continued*)
rights in employee-employer contracts, 363
scientific, 97–102. *See also* Scientific management
span of, *678*
strategic, *679*
twenty-first century challenges for, 132–35
Management by objectives (MBO), 121, 127–28, *667*
defined, 127
in performance appraisal, 368
Management Challenges for the 21st Century (Drucker), 132–33
Management information system (MIS), *667*
Management results, leadership style affecting, 157–61
Management science (MS), *667*
Management Science School, *667*
Management theory and thought:
classical school of, 95–104
criticisms of, 108
early development of, 93–94
historical perspectives on, 92–93
integrating, 119–21
Management trends:
Deming and, 121–22
Drucker and, 127–28
Managerial grid model, 114–15, 115*f*, *667*
leadership style and, 142–44, 143*f*, 143*t*
Managers, *667*
empowerment of, 464
leaders and, 92, 159
differences between, 198
Managing by walking around (MBWA), 275–76
Managing fire officer rank:
experiential elements for, 68*t*
NFPA standards pertaining to, 69*t*
self-development elements for, 72*t*
training elements for, 70*t*
Managing Transitions, Making the Most of Change (Bridges), 223
Mandates:
regulatory, discipline and, 381–82
unfunded, impact on fire service, 307
Manning, Bill, 564
Mapping, IAFC definition of, 62
March, James, 109
Marine firefighting, NFPA 1710 standards for, 440–41
MARS (Military Affiliate Radio System), 599
Marshall, Peter, 211
Martyrdom, of Florian von Cetum, 3
Maslow, Abraham H., 109, 290
Maslow's hierarchy of needs, 109, 110–12, 110*f*, 291*t*, *667*
and ERG model, 111, 111*f*
high-performance team building and, 290–93
Mass casualties, risk assessment and, 393
Massey, Morris, 12, 13
Massey-Shaw, Sir Eyre, 44
Material management, *667*
Maximum prescribed travel time, *668*
Maximum/worst risk, *668*
Mayo, Elton, 106
Mayor-council government model, 334
MBO. *See* Management by objectives (MBO)
MBTI. *See* Myers-Briggs Type Inventory (MBTI)
MBWA (managing by walking around), 275–76
McClelland, David C. (1917-1998), 117, 119
McGregor, Douglas, 109
McKee, Annie, 145
Measurable terms, *668*
Mechanistic structure (organizational), 337
Media communication, 509
Media images, of fire service:
movies depicting, 28
in terrorist attacks, 31–33, 33*f*, 34*f*
TV shows depicting, 28
Media involvement, in alleged misconduct, 255–56
Median age of population, *668*
Median household income, *668*
Mediation, in labor disputes, 364
Medical calls, risk assessment and, 393
Medical profession, and fire service compared, 45–46
Medical Research Institute of Infectious Diseases, U.S. Army (USAMRIID), 600
Medical Reserve Corps (MRC), 526
Mediocrity. *See* Complacency
Megatrends 2000 (Aburdene), 200
Megatrends 2010 (Aburdene), 200
Memorandum of understanding, between ICMA and IAFC, 421
Memorial services, 34, 35*f*
Mentoring/mentors, 61, *668*
Meritor Savings Bank v. Vinson, 16
MERS (Mobile Emergency Response Support), 618
Merton, Robert K. (1910-2003), 96–97
Method, *668*
Metropolitan area, CFAI response time benchmarks for, 431, 431*t*
Metropolitan Medical Response System (MMRS), 619
Michigan Leadership Studies, 143–44
Middle of the road management, 143, 143*t*
Military Affiliate Radio System (MARS), 599
Military fire protection, 555
Mill, John Stuart, 268
Millennials. *See* Generation Y (millennials)
Minimum staffing per unit, *668*
Minnesota Multiphasic Personality Inventory (MMPI), 359
MIS (management information system), *667*
Misconduct in workplace:
disciplinary action and, 374
example of, 249–50
NBES findings on, 243, 244
preventing recurrence, 255–56
reporting, ethical analysis in, 250–56
unreported, consequences of, 252
Mission, *668*
organizational, team-building and, 287–88
Mission areas, of DHS, 396
Mission statement, of DHS, 228

Mitigation:
defined, 394, *668*
in emergency management, 511–12
in risk assessment, 394
Mitigation center, 511*f*
MMPI (Minnesota Multiphasic Personality Inventory), 359
MMRS (Metropolitan Medical Response System), 619
Mobile Emergency Response Support (MERS), 618
Mobilization, *668*
Mobilization center, *668*
Modeling:
ethical behavior, 245–46
integrated risk management, 426, 427–28
Moderate risk, 427, *668*
Moore-Merrell, Lori, xxiii–xxiv
advice to new firefighters, 571–72
on career development, 87
on disaster management and response, 543–44
on ethics, 278–79
on future directions, 384
on future of fire service, 571–72
on generational issues, 327
on leadership and management, 137
on leadership style, 179
on organizational changes, 231
on quality improvement, 493–94
on risk management, 447
on tradition, 38
Moral compass. *See* Integrity
Moral leadership, 238
Mother Teresa, 238
Motion study, therbligs in, 100
Motion Study (Gilbreth), 100
Motivation(s), 109–10, *669*
in ethical decision-making process
for media involvement, 255
of potential wrongdoer, 252–53
of reporter of incident, 250–51
framework for, 110*f*
fusion process in, 115–16
in managerial/leadership grid, 114–16, 115*f*
Maslow's hierarchy of needs and, 110–12, 110*f*
motivation-hygiene model, 116–17, 116*t*
patterns of, 117, 118*t*, 119
Theory X and, 112, 114*t*
Theory Y and, 112, 113, 114*t*
Theory Z and, 113–14, 114*t*
Motivation-hygiene model, 116–17, 116*t*
implications for management, 117
Motivators, job attitudes and, 116–17, 116*t*
Motorola Corporation:
benchmarking and, 475
Six Sigma quality and, 450, 451–52, 451*f*, 457, 458*f*–459*f*, *678*
MOU (memorandum of understanding), between ICMA and IAFC, 421
Mouton, Jane S., 114, 142
Movies, fire service depicted in, 28
MRC (Medical Reserve Corps), 526
MS (management science), *667*
Multifamily dwellings, safety of, 395
Multiple casualties, risk assessment and, 393
Municipalities:
ISO grading schedule for, 416
types of, 335–36
Musashi, Miyamoto, 205
Mutual gains bargaining, 365
Mutual-aid system, national, 565–66
principles of, 566
My American Dream (Powell), 222
Myers, Isobel Briggs, 167
Myers-Briggs Type Inventory (MBTI), *669*
characteristic types, 168*t*, 175*t*
scales, 167–68, 296–97

N

NAACP (National Association for the Advancement of Colored People), 9, 10
Nair, Keshavan, 198, 199
Naphtha, in glass bombs, 24
NARAC (National Atmospheric Release Agency Center), 609–10
Narcissistic perspective, in leadership ethics, 240
"Narcissistic Process and Corporate Decay" (Schwartz), 240
National Association for the Advancement of Colored People (NAACP), 9, 10
National Atmospheric Release Agency Center (NARAC), 609–10
National Board of Fire Underwriters (NBFU), 390, 391
fire department evaluations and, 413–14
fire protection grading system and, 414–15
National Board on Fire Service Professional Qualification (NBFSPQ), 46
accreditation by, 83
benefits of, 84
standards "crosswalks" and, 48
certification by, 83–84
organization of, 83
National Business Ethics Survey (NBES), 242–45
of 2003, 242–45
National Commission on Fire Prevention and Control:
1973 report, 20
2002 report, 20–21
National Conference of State Legislatures (NCSL), 350
National Construction Safety Team Act (NCST), 567
National Decontamination Team (NDT), 622
National Disaster Medical System (NDMS), 617–18, *669*
National Earthquake Hazard Reduction Program (NEHRP), *657*
National Fallen Firefighters Memorial, 35*f*
National Fire Academy (NFA), 47, 47*f*
baccalaureate degree program of, 51
memorial services at, 34, 35*f*
standards "crosswalks" and, 55

National Fire Protection Association (NFPA), *670*
  fire service needs assessment by, 188–93
  operational standards, 394, 395, 560, *670*. *See also* NFPA 1710; NFPA 1720
  professional standards. *See* NFPA professional qualification standards
National Flood Insurance Act (1968), 503, *669*
  provisions of, 503
National Governors' Association, 503
  budget reform and, 350
National Homeland Security Research Center (NHSRC), 622
National Incident Management System (NIMS), 516, 528–30, *669*
National Incident Management Teams (NMT), 617
National Infrastructure Protection Plan (NIPP), 535–38
  goal of, 535–36, 536*f*
  roles and responsibilities in, 536
  sector partnership model in, 537–38, 537*f*
  value of, 536–37
National Institute of Standards and Technology (NIST), 567
National Interagency Fire Center (NIFC), *669*
National Joint Terrorism Task Force (NJTTF), 522
National Labor Relations Act (1935), 357, 360, 363
National Labor Relations Board (NLRB), 360, 363
  disciplinary appeal and, 380
National Laboratory Response Network (NLRN), 603
National mutual-aid system, 565–66
  principles of, 566
National Organization for Women (NOW), 10
National Preparedness Program, *669*
National Response Center, *669–70*
National Response Coordination Center (NRCC), 521
National Response Plan (NRP), 515–16
  incidents of national significance and, 516
  organizational structure of, 527
  roles and responsibilities in, 516, 517*t*–519*t*
  supporting and structural provisions of, 519–23
National Response System, *670*
National Response Team (NRT), *670*
National security, homeland security vs., 397–98
National special security event (NSSE), 521
National Strategy for Homeland Security (2002), 396–97, 398
National Urban Search and Rescue Response System (US&R), 619–20
National Voluntary Organizations Active in Disaster (NVOAD), 524
National Volunteer Fire Council (NVFC), 59
National Weather Service (NWS), 512–13
Natural disasters, 393, 503
  code development and, 395
Natural resources, *670*
NBES (National Business Ethics Survey), 242–45
NBFSPQ. *See* National Board on Fire Service Professional Qualification (NBFSPQ)
NBFU. *See* National Board of Fire Underwriters (NBFU)
NCSL (National Conference of State Legislatures), 350
NCST (National Construction Safety Team Act), 567
NDMS (National Disaster Medical System), 617–18, *669*
NDT (National Decontamination Team), 622
Needs:
  adapting leadership style to, 142–44
  group hierarchy of, 292–93, 292*t*
  higher-order. *See* Theory Y
  of individuals, in high-performance teams, 294–95
  lower-order. *See* Theory X
  Maslow's hierarchy of. *See* Maslow's hierarchy of needs
  for organizational change, establishing, 219–20
Negotiating, in labor disputes, 364
  general process for, 365–67
  group development climates in, 366, 366*f*
  mutual gains bargaining and, 365
  power and, 365
  principled negotiation, 366, 367
NEHRP (National Earthquake Hazard Reduction Program), *657*
Neighborhood Watch Program (NWP), 526–27
NEST (Nuclear Emergency Support Team), 610
Neutral zone, 225
New Amsterdam (New York), fire regulations in, 5
NFA. *See* National Fire Academy (NFA)
NFPA. *See* National Fire Protection Association (NFPA)
NFPA 1710:
  defined, 433, *670*
  labor relations and, 360
  origin and development of, 433–41
NFPA 1720:
  alarm time line in, 444*t*
  components of, 442*t*–443*t*
  defined, 441, *670*
  labor relations and, 360
  origin and development of, 441, 442*t*–444*t*
  staffing and response time in, 444*t*
NFPA professional qualification standards, 64–65
  for administrative rank, 72*t*
  for executive rank, 76*t*
  for managing rank, 69*t*
  for supervising rank, 66*t*
NGO. *See* Nongovernmental organization (NGO)
NHSRC (National Homeland Security Research Center), 622
Nicomachean Ethics (Aristotle), 237
NIFC (National Interagency Fire Center), *669*

NIMS (National Incident Management System), 516, 528–30, *669*
9/11 terrorist attacks. *See* September 11 terrorist attacks
9-1-1 telephone call, response time and, 406
NIPP. *See* National Infrastructure Protection Plan (NIPP)
NIRT (Nuclear Incident Response Team), *671*
NIST (National Institute of Standards and Technology), 567
Nixon, Richard, 225, 239, 361
NJTTF (National Joint Terrorism Task Force), 522
NLRA (National Labor Relations Act of 1935), 357
NLRB. *See* National Labor Relations Board (NLRB)
NLRB v. Weingarten, 380
NLRN (National Laboratory Response Network), 603
"No smoking" ban, in colonial America, 5
Nongovernmental organization (NGO):
  defined, 524, *670–71*
  in emergency management, 523–24
  NRP role and responsibility of, 518*t*
Nontransport, *671*
Nonunionized fire departments, labor relations in, 362–63
Normalcy, return to, 510
Norris–La Guardia Act (1932), 360, 363
NOW (National Organization for Women), 10
NRCC (National Response Coordination Center), 521
NRP. *See* National Response Plan (NRP)
NRT (National Response Team), *670*
NSSE (national special security event), 521
Nuclear Emergency Support Team (NEST), 610
Nuclear Incident Response Team (NIRT), *671*
Number of population by age, *671*
NVFC (National Volunteer Fire Council), 59
NVOAD (National Voluntary Organizations Active in Disaster), 524
NWP (Neighborhood Watch Program), 526–27
NWS (National Weather Service), 512–13

O

Oath of service, 33, 34
Objective, fair, and impartial, *671*
Objectively verifiable indicators, *671*
Objectives, *671*
  service-level, *677*
Obligations, of public officials to citizens, 271
Occupancy, *671*
  classification of buildings in service area by, *654*
  risk assessment and, *671*
Occupancy risk assessment, 429
Occupational Safety and Health Administration (OSHA):
  citation for, *672*
  industrial fire protection and, 555
Office of Emergency Response. *See* Department of Energy/National Nuclear Security Administration (DOE/NNSA)
Officer Development Handbook, 56, 57*f,* 58–59
  objectives and benchmarks in, 63–64
Ohio State Leadership Studies, 144
Older Workers Benefit Protection Act (1990), 356
O'Leary, Mr. Mrs. Patrick, 8
Onieal, Dr. Denis, 43, 44, 44*f,* 56, 62
On-scene coordinator (OSC). *See* Federal on-scene coordinator (FOSC)
On-scene to patient time, average, *653*
Open government, ethics law principles pertaining to, 274
Operating budget, *671*
Operating guidelines, general, *663*
Operating procedures:
  developing for disaster response, assessing local capability for, 640–44
  standard, *678*
Operations section, in ICS, 532, 532*f*
Operators, FSRS criteria for, 417
Optimization, *671*
Oral presentation, in promotional process, 373
Order of Saint John. *See* Knights of Saint John of Jerusalem
Organic structure (organizational), 337
Organization(s):
  benchmarking comparisons between, 474–75, 474*f*
  bureaucratic, 337
  communication and, 286–87
  as communication systems, 104
  consolidation within and between, 227–28
  core values of, 198–204
  culture of, ethical decision-making and, 251–52
  80/20 time and energy rule in, 214–15, 215*f*
  employee as resource in, 342
  ethical, characteristics of, 268
  ethics in, 236–37
  evolutionary nature of, 131–32
  goals of. *See* Organizational goals
  honesty in, 285
  local government models, 332, 334–37
  management of. *See* Management
  mediocrity and complacency in, 187–88, 193
  mission of, team-building and, 287–88
  political dynamics in, understanding, 257–58
  power base of, 174, 176*t*
  problem solving vs. decision making in, 287
  "professional" disagreements in, 366
  quality improvement benefiting, 469
  quality planning process in, 122–24, 123*f*
  redefinition and reorientation of, helpful approaches to, 225
  size of, NBES findings relating to, 244
  socially responsive practice trends in, 200
  stagnation in, 186–87
  structure of. *See* Organizational structure
  SWOT analysis of, 285–86, 286*t*
  team building in, factors affecting, 285–88

Organization(s) (*continued*)
totalitarianism phenomenon of, 240
with transformational leadership, characteristics of, 132
trust in, 285, 287
volunteer. *See* Volunteer organizations
Organization chart:
for CFAI, 422*f*
for CPSE, 422*f*
defined, 332, *672*
for fire department, 333*f*
Organizational change, 183–85
agenda for, 207–23
case study, 225–29
challenges in, 207–8
culture and, 208–11
externally driven, 209, 210*f*
internally driven, 208–9
leadership for, 184–85
principles of, 222–23
qualities required, 196–98
neutral zone and, 225
paradigms for, 185–87
psychological aspects of, 223–25
requirements for, 211
responsiveness and, 209–11, 210*f*
strategies for accomplishing, 218–24
transition and, 223–24, 223*f*
vision and, 211–16
Organizational conflict, *672*
Organizational culture, 185
above-the-line accountability vs. victim mentality, 206–7, 206*f*
and agenda for change, 208–11, 210*f*
analogy of, 216–17, 216*f*
cultural norms affecting, 216–17, 218
lore in, capturing and repeating, 246
nature of, 208–11
Organizational design:
defined, 338, *672*
factors affecting, 338–39, 338*f*
Organizational goals, *672*
performance measurements and, 479
process for achieving, 193–94
Organizational structure:
authority in, 340–41
boundaryless, 339
classification of, 337
contingency, 340
defined, 337, *672*
design factors, 338–40, 338*f*
divisional, 338–39
functional, 338
informal, 338
work activity grouping methods and, 340
Organizational totalitarianism phenomenon, 240
Organizational transformation, *672*
Organized labor:
actions of, 362
in dispute settlement, 364
union membership and, 360–61
Organizers, characteristics of, 298
Organizing, *672*
Osborne, David, 226, 349, 475
OSC (on-scene coordinator). *See* Federal on-scene coordinator (FOSC)
OSHA. *See* Occupational Safety and Health Administration (OSHA)
Ouchi, William, 113–14
Outcome(s), *672*
as performance measure, 480–81
in total quality leadership, 463, 463*f*
Output, as performance measure, 479–80
OWBPA (Older Workers Benefit Protection Act of 1990), 356
Owen, Robert (1771-1858), management theory of, 94

P

Pacesetting leader, 145–46, 147*t*
Paradigm shift, *672*
Paradigms, for organizational change, 185, *672*
Paramilitary organizations:
fire service and, 240
origins of, 2
Pareto chart, *672*
Pareto's 80/20 rule, 214–15, 215*f*
Participation, team empowerment and, 321
Participative leadership style, 144, 148, 148*t*
Participative management, 322–23, *672*
PASS (personal alert system) devices, 192
Patriot Act, critical infrastructure defined in, 398–99
Patron saint of firefighters, 3, 4
Paulison, R. David, 504
Pavlov, Ivan Petrovich (1849-1936), 149–50
PDA (Pregnancy Discrimination Act of 1978), 356
PDA (preliminary damage assessment), 510
PDCA (plan-do-check-act) cycle, 11–12, *673*
Peer evaluation review, *672*
Peer groups, 381
Pennsylvania Gazette, 6
Pentagon:
terrorist attacks on. *See* September 11 terrorist attacks
unfolding of flag at, symbolism of, 33–34, 35*f*
Perceiving, judging vs., 172–74, 174*t*
Percentage below poverty level, *673*
Performance, *673*. *See also* Job performance
benchmarking for, 470–77, 472*f*
critical objectives of mission and, 193–94
group dynamics and, 194–98, 195*f*
perception vs., 193
Performance appraisal:
appraisal conference agenda, 369
appraisal judgments in, 368
conducting, 369–73
defined, 368, *673*
methods for, 368–69
objectives of, 369
Performance budget, 346, 347*f*
budget reform and, 349
defined, 346, *673*
Performance goals, *673*

Performance indicator(s):
CFAI example, 423–24
defined, 423, *673*
for response time criteria, 430–33, 431*t*–432*t*
Performance measurements, 477–79
dimensions in, 479–83
ethical conduct as, 246
NFPA 1720 standards for, 434
operational, 478
organizational goals and, 479
in organizational performance cycle, 370–71
pitfalls of, 483–84
political decisions based on, 478–79
uses for, 478
workload, 484–86
Perks:
conflict of interest and, 272
ethics law principles pertaining to, 274
Personal alert system (PASS) devices, 192
Personal development, empowerment approach to, 319
Personal frame of reference, 161
author's, 161–65
leadership style shaped by, 156–57
Personal orientation, ethics and, 249
Personal success strategies, 265–66
code of conduct, 266–67
guidelines and resources, 267–69
Personality tests, 359
Personnel:
behavior of, 112
honesty between, team building and, 285
motivation of. *See* Motivation
NFPA needs assessment report on, 189–90, 189*t*
problems with. *See* Discipline
Persuasion, *673*
Peshtigo, Wisconsin, forest fire destroys (1871), 8
Peters, Tom, 195, 238, 276
PFO (principal federal official), 523, *674*
Philadelphia Contributorship for the Insurance of Houses, 7
Phoenix, symbolism of, 27
Physical fitness standards, 565
Picketing, 362, 364
Picture, in team-building, 288
Pig Iron Experiment, 97–98
PIO (public information officer), *675*
Plan:
contingency, *655*
emergency operations, *658*
prior to fire, *661*
strategic, *679*
Plan phase, *673*
TQM and, 11
Plan-do-check-act (PDCA) cycle, 11–12, *673*
benchmarking for competitiveness and, 474–75, 475*f*
Deming and, 490
in problem-solving process, 468, 468*f*
in total quality improvement, 465–66, 466*f*
Planning:
community risk assessment and, 395–96
defined, 529, *673*
in disaster response, 629–37
for emergency management, 505–6, 505*f*
leadership dimension of, 169–74, 172*t*–174*t*
NIMS and, 529
for quality improvement, 465–66, 465*f*, 466*f*
strategic, *679*
in team-building, 288
Planning, programming, and budgeting system (PPBS), 346, 348, *673*
Planning assumption, *673*
Planning section, in ICS, 532–33, 533*f*
Plymouth, Massachusetts, 5
Police service, volunteers for, 527
Policy, declaration of, 271, *673*
Political ethics, 257
Political issues:
credibility and integrity and, 262
leadership in face of, 256–58
Political survival:
code of conduct as starting point for, 264–65
establishing positive climate for, 262–63
guidelines for, 265
ten commandments of, 263–65
Pollutant, *673*
Popcorn, Faith, 208
Population:
number by age, *671*
percentage below poverty level, *673*
Population served, *673*
Position power, *673*
Poverty level, percentage below, *673*
Powell, Colin, 222
Power, *673*
defined, 117
motivation and, 117, 118*t*
in negotiations, 365
position of, *673*
structures, changing composition of, 12
PPBS (planning, programming, and budgeting system), 346, 348, *673*
PPC. *See* Public Protection Classification (PPC)
prefectus vigilum (fire chief), 3
Preference scales, in Jungian inventory, 167, 167*t*
Pregnancy discrimination, 358
Pregnancy Discrimination Act (1978), 356
Preliminary damage assessment (PDA), 510
Preparedness:
defined, 506, *673–74*
in emergency management, 506–8, 506*f*, 507*f*
example of, 506
FEMA and, 503–6, 505*f*
in homeland security strategic plan, 541
at individual level, 507, 507*f*
at local level
community-based planning and, 502
variations in, 534–35
NIMS and, 529
Presidential Disaster Declaration process, 511–12
Prevention, *674*
of fire. *See* Fire prevention
of wrongdoing recurrence, 255–56
The Price of Government (Osborne and Hutchinson), 226

Primal Leadership (Goleman et al.), 145
Principal federal official (PFO), 523, *674*
Principled negotiation, 366, 367
The Principles of Scientific Management (Taylor), 98
Priorities, communicating to employees, 276
Private interests, public officials representing, 272
Private sector, *674*
  NRP role and responsibility of, 518*t*–519*t*
Pro Board. *See* National Board on Fire Service Professional Qualification (NBFSPQ)
Probability:
  and consequence matrix, 401*t*, 403*t*
  in risk assessment, 429
Probationary periods, 359–60
Problem solving:
  leadership dimension of, 168–69, 170*t*
  in PDCA cycle, 468, 468*f*
  in performance appraisal, 369
  principled negotiation and, 366, 367*t*
  in promotional process, 373
  team building and, 287
  and understanding, for organizational change, 220–21, 221*f*
Problems:
  with employees. *See* Discipline
  ethics-related, NBES findings on, 243
Procedure, *674*
  developing for disaster response, 640–44
Process:
  flowchart map, *674*
  improvement, *674*
  management, *674*
  mapping, *674*
  roles in teamwork, 297–98, *674*
Product, *674*
Production, *674*
Production assembly line, 101
Productivity, *674*
  manual- vs. knowledge-worker, 134
  measurement of, *674*
Professional development, *674–75*. *See also* Training and education
  career process in, 64–65, 65*f*
  elements and standards in, 66*t*–78*t*
  FESHE conferences and, 49
  fragmented system of, 48–49
  history of, 60–61
  model for, 49–51, 50*t*
  national model for, 62*f*
  portfolio setup for, 79, 79*f*
  volunteer advocates for, 59–60
Professional standards, establishing, 45–46. *See also* NFPA professional qualification standards
Professional status:
  and benefits of a common training system, 54
  defined, 45
  requirements for, 44–47
Professionalism, 22, 32, 84, 197
Professions:
  comparisons between, 45–46
  specializations within, 46
Profound knowledge, Deming's system of, 122, 122*f*
Program, *675*
Program budget, *675*
Programmatic change, *675*
Progressive discipline, 374, 375
Project Impact, 512
Promotion practices, 55
  assessment processes and, 372–73
  differences in, 84–85
Property loss, risk assessment and, 395
Protection:
  from fire hazard. *See* Fire protection
  from other hazards and risks, 400–405, 400*t*. *See also* Risk management
Protective clothing, *675*
Protopapas, George, 263
PSAP. *See* Public safety answering point (PSAP)
"Psychic secretion," 149
Psychological tests, 359
Psychological Types (Jung), 167
Psychology of recovery, 510
The Psychology of Work (Gilbreth), 100
PTI (Public Technology Institute), 391
Public education:
  about disaster response, assessing local capability for, 648–50
  evaluation of program for, 584–88
Public employees, code of ethics for, 271–73
Public health, *675*
Public information officer (PIO), *675*
Public officials. *See* Appointed officials; Elected officials
Public petitions, 364
Public property, use of, 271
Public Protection Classification (PPC), 412
  community reliance upon, 420
  community-wide, 419–20
  effect in insurance premiums, 420
  for individual properties, 419
Public safety, fire chief's role in, 256, 256*f*
Public safety answering point (PSAP), 406
  response time and, 435*f*
Public Technology Incorporated (PTI), 391
Public works, *675*
Pump capacity, FSRS criteria for, 417
Purpose:
  defined, 433, *675*
  in team-building, 288

Q

QIP. *See* Quality improvement process (QIP)
Quality. *See also* Continuous improvement (CI); Customer satisfaction; Effective/effectiveness; Total quality leadership (TQL); Total quality management (TQM)
  customer-driven, 469–70, 469*f*
  defined, 487–88, *675*
  rationale for, 450–55
  Six Sigma. *See* Six Sigma quality
Quality control, 490, *675*
  detection method of, *657*

Quality council, 491
Quality customer service, 488–89, 488*f*, 489*f*
Quality improvement, 451. *See also* Continuous improvement (CI)
cycle time in, 473–76
elements of, 460, 461*f*, 462–63, 462*f*, 462*t*
examples of, 457–58, 457*f*, 458*f*–459*f*
good vs. poor, impact of, 458, 459*f*
need for, 456–58
pioneers of, 490–91
value of, 458–60, 459*f*
Quality improvement process (QIP), 450
benefits of, 468–70, 469*f*
components of, 465–66, 466*f*
Quality improvement teams, *675*
Quality initiatives:
designing, factors in, 453–55
EPA guidelines and, 451
impact of, 452–53
Quality management, 453
Quality operations, in quality improvement process, 465. *See also* Plan-do-check-act (PDCA) cycle
Quality planning, in quality improvement process, 465–66, 465*f*, 466*f*
Quality service:
employee empowerment and, 322–24
surveys measuring
to emergency responses, 578–84
to public education programs, 584–88
Quality teams, in quality improvement process, 465
Quid pro quo sexual harassment, 15, 358
Meritor Savings Bank v. Vinson, 16
organizational liability for, 16
Quiet leader, 151
Quint apparatus, 437

R

Radiation Emergency Assistance Center/Training Site (REAC/TS), 611–12
Radio system interoperability, 539, 539*f*–540*f*
Radiological Assistance Program (RAP), 611
Radiological Emergency Preparedness (REP) Program, *675–76*
Radiological emergency response team (RERT), *676*
WMD response role of, 623
Rand Institute, 391
Ranking, *676*
RAP (Radiological Assistance Program), 611
Rapid intervention team (RIT), 437
Rasmussen, Jesse, 228
Rating, defined, 102
Rating scales, 102
Rationalization, in ethical decision-making process, 253
Rattle watch, in New Amsterdam, 5
Raven, Bertram H., 176*t*
Rawls, John, 237
REAC/TS (Radiation Emergency Assistance Center/Training Site), 611–12
Reality, and change continuum, 223–24, 223*f*
Reardon, James P. (Jay), xxiv–xxv
advice to new firefighters, 572
on career development, 87
on disaster management and response, 544
on ethics, 279
on future directions, 384
on future of fire service, 572
on generational issues, 328
on leadership and management, 137
on leadership style, 179
on organizational changes, 231
on quality improvement, 494
on risk management, 447
on tradition, 38–39
"Reasonable woman" rule, 16
Reciprocity, *676*
common system for, benefits of, 54
independent assessment of, 53–54
Reconstruction Finance Corporation, 503
Recovery:
defined, 509, *676*
in emergency management, 509–11, 510*f*
psychology of, 510
Recovery center, 510*f*
Recruitment, *676*
in fire service, 19
generational differences and, 310, 312
Recruitment, generational views on, 311
Red, meaning of, 27
References, in firefighter selection, 359
Refusers, in organizational change, 210*f*, 211
Regional Response Coordination Center (RRCC), 521
Reinventing Government (Osborne and Gaebler), 349, 475
Relatedness, in ERG model of needs, 111, 111*f*
Relationships, building, for organizational change, 220
Relay Assembly test, in Hawthorne experiments, 107
Reliability, *676*
Remote and isolated rural risks, *676*
REP (Radiological Emergency Preparedness) Program, *675–76*
RERT. *See* Radiological emergency response team (RERT)
Rescue, specialty. *See* Specialized responses
Rescue 9-1-1 (TV show), 28
Reserve ladder and service trucks, FSRS criteria for, 417
Reserve pumpers, FSRS criteria for, 417
Residential single-family dwelling, *676*
Resilience, of critical infrastructure, 400–405, 400*t*
Resisters, in organizational change, 210, 210*f*
Resonance, leaders and, 146
Resource(s):
employees as, 342
natural, *670*
Resource management:
in community risk assessment, 402–3, 403*t*
in disaster response, 626–29
NIMS and, 529

Responder(s):
first, *662*
risk to, 393–94
Response:
defined, 508, *676*
in emergency management, 508–9. *See also* Emergency responses
performance measures for, 486–87
reliability of, 410, *676*
time for. *See* Response time
Response reliability, 410, *676*
Response time:
CFAI benchmarks and performance baselines, 430
combustion products and, 408, 409*f*
defined, 434, *676*
NFPA 1710 and, 434–35, 435*f*
NFPA 1720 and, 444*t*
total, 406, *681*
turnout time and, 406, *681*
Responsibility(ies):
in authority, 342–43
in CFO code of professional conduct, 258–59
of public office, 271
Responsiveness, organizational change and, 209–11, 210*f*
Restructuring, in schema modification, 155
Result, *677*
Retention, generational differences and, 310
Revenue:
for fire departments, 353*f*
for fire districts, 354*f*
NFPA needs assessment report on, 189
Revere, Paul, 28
Rewards, intrinsic, 112
Ricks, Thomas E., 204–5
Ridge, Tom, 504
"Right to know" laws, 361
Rights, 239
"Right-to-work" laws, 363
"Right-to-work" states, 362
Risk(s):
acceptable level of, *652*
assessment of. *See* Risk assessment
in community, 395–96
defined, 430, *677*
of disaster, 499–500
high hazard, *664*
low, *667*
management of. *See* Risk management
moderate, *668*
to responder, 393–94
in risk assessment, 430
rural, remote and isolated, *676*
special, *678*
Risk assessment:
challenges of, 445
community, 392–93, *655*
in disaster response, 625–26
elements in, 428–29
fire suppression and, 402–3, 403*t*
model of, 404, 404*t*, 427–28
occupancy, *671*
purpose of, 430
Risk category, *677*
Risk management:
in community, 392–96
community response in, 402
fire suppression in, 402–3, 403*t*
firefighter safety and, 565
integrated, 560–61
network for, in NIPP, 538, 538*f*
plan development, 427, 428*f*
probability and consequences in, 401, 401*t*, 403*t*
response coverage standards in, 403–4
in training and education model, 50*t*
Risk-benefit analysis, in ethical decision-making, 249
RIT (rapid intervention team), 437
Robert T. Stafford Disaster Relief and Emergency Assistance Act. *See* Stafford Act
Robinson v. Jacksonville Shipyards, 16
Roethlisberger, Fritz, 106
Role play, in promotional process, 373
Roles, communicating to employees, 276
Rome, Italy:
citizenship of, 2
Great Fire of, 3
Roosevelt, Eleanor, 238
Roosevelt, Franklin D., 238
Rorschach test, 359
Ross, Richard, 298
RRCC (Regional Response Coordination Center), 521
Rules, *677*
Rural area, CFAI response time benchmarks for, 432, 432*t*

S

SAC (senior agent-in-charge), 523
Safety:
of firefighters, future vision for, 563–65, 563*t*, 564*f*
future fire service challenges and, 560
three E's of, 20
Safety equipment, *677*
Safety issues, 307
"Saltier," 27
Salvage bags, 7
San Francisco, California:
earthquake of 1906, 413–14
NBFU report on, 414
Sandbagging, 512
SCBA (self-contained breathing apparatus), 192
Schema, 160, *677*
creating and modifying, 154–56
defined, 154
types of, 154
Schmidt, Warren H., 149
Schwartz, Howard S., 240
Science and Technology Directorate, of DHA, 227
"Science of shoveling," 98–99

Scientific management, *677*
basic framework and results of, 99
Bedaux's work on, 101–2
drawbacks of, 99
Gantt's work on, 101
Gilbreths' work on, 100–101
principles of, 99
production assembly line and, 101
Taylor's work on, 97–99
SEA (service efforts and accomplishments), 349
SEC (Securities and Exchange Commission), 352–53
Secretary of Defense, NRP role and responsibility of, 518*t*–519*t*
Secretary of Homeland Security, NRP role and responsibility of, 518*t*–519*t*
Secretary of State, NRP role and responsibility of, 518*t*–519*t*
Sector partnership model, in NIPP, 537–38, 537*f*
Sector-specific agencies (SSAs), 535
Securities and Exchange Commission (SEC), 352–53
Seghini, Joann B., 256
Selectmen, Board of, 336–37
Self-actualization concept, 109, 110–11
Self-contained breathing apparatus (SCBA), 192
Self-delusion, in ethical decision-making process, 253
Self-development:
for fire officer ranks
administrative, 75*t*
executive, 78*t*
managing, 72*t*
supervising, 69*t*
IAFC benchmarks in, 64
roadmap for, 65*f*
Self-knowledge, leadership and, 152–53, 153*t*
Senior agent-in-charge (SAC), 523
Senior federal law enforcement official (SFLEO), 523
Senior federal official (SFO), 523, *677*
Sensing, vs. intuiting, 169–71, 172*t*
Sensing/intuition scale, in Myers-Briggs inventory, 168
September 11 terrorist attacks, 28, 29*f*
communication problems and, 539
fire service mission after, 183, 548
fire service research since, 567–68
firefighter deaths on, 564
media images of, 33–34, 35*f*, 36
as national tipping point, 547
responder risk and, 394
risk assessment models before and after, 397
volunteer response following, 525–27
Servant leadership, 129–30, 151–52
Servant Leadership (Greenleaf), 129–30
Service:
before/over self, as core value, 199, 202
core values of, 198–204
level of, *666*
square miles served, *678*
Service area, CFAI benchmarks for, 431–33, 431*t*–432*t*
Service efforts and accomplishments (SEA), 349
Service-level objectives, *677*
"Seven deadly diseases," in management, 126–27, 126*f*
Sex discrimination, 358
Sexual harassment:
court cases on, 16
detection of, 16–17
EEOC definitions of, 15
organizational liability and, 16
supervisor's responsibility regarding, 15–16
types of, 15, 358
underlying causes of, 15
SFLEO (senior federal law enforcement official), 523
SFO (senior federal official), 523, *677*
Shackleton, Ernest, 151–52
Sheehy, Gail, 239
Sherman Act (1890), 360, 363
Sibbet, David, 290
Sick-outs, 362, 364
Silver (color), meaning of, 27
Simon, Herbert (1916-2001), 108, 109
Single-family residence:
hazards, 393
safety of, 395
SIOC (Strategic Information and Operations Center), 522
Siphonarius rank, in *Vigiles,* 3
Situation assessment, *678*
Situational analysis, 285–86, 286*t*
Situational (contingency) approach, 120–21
Situational leadership, 148–49, *678*
"Six sense," 170
Six Sigma quality, 450, 451–52, 451*f*, *678*
examples of, 457–58, 457*f*, 458*f*–459*f*
Skelly v. State Personnel Board, 379–80
Skills, assessment of, 53
Slavery, 9
Slowdowns, 362, 364
Smith, Captain John, 5
Smith, Doug, 222
Smith, Gary W., 494, xxv
advice to new firefighters, 572–73
on career development, 87
on disaster management and response, 544
on ethics, 279
on future directions, 384
on future of fire service, 572–73
on generational issues, 328
on leadership and management, 137
on leadership style, 179
on organizational changes, 231
on quality improvement, 494
on risk management, 447
on tradition, 39
Smoldering stage, in fire growth, 407
SNS (Strategic National Stockpile), 615
Social motives, 117, 118*t*, 119
Socially responsive practice trends, in organizations, 200

Societal trends, in employee/labor relations, 361–62
Sociology, empowerment approach in, 319
Soft data, *678*
Soldiering analysis, 97–98
Solution to problems, commitment to, 222
Sovereign immunity, 434
Span of control, 341, *678*
Span of management, *678*
Special risks, *678*
Specialized responses, 551–59. *See also individual types of responses*
  NFPA 1710 standards for, 440
  performance measures for, 487
  in urban areas. *See* Urban search and rescue (USAR)
Spokesperson, purpose of, 509
Sprinkler systems, commercial and residential, 556, 557*t*–559*t*
Square miles served, *678*
SSAs (Sector-specific agencies), 535
"St. Andrew's Cross," 27
St. Florian (Florian von Cetum), 3, 4
Staff authority, 341
Staffing, 548, *678*. *See also* Personnel
  in NFPA 1720, 444*t*
Stafford Act, 498, *678*
  Federal Response Plan and, 509–10
  provisions of, 501
Stagnation, within organizations, 186–87
Stakeholders, *678*
Stalin, Joseph, 238
Stalker, G. M., 337
Standard operating procedures:
  defined, 441, *678*
  NFPA 1710 standards and, 441
Standardization:
  defined, 528, *678*
  NIMS and, 528–30
Standards, for professional services:
  establishing, 45–46
  of performance for fire service, 46
Standards "crosswalks":
  NBFSPQ and, 48
  in training and education model, 50*t*
Standards of response coverage:
  CFAI benchmarks and performance baselines for, 430–33, 431*t*–432*t*
  at community level, 403–4
    national, development of, 390–92
  defined, 427, *678*
  developing, 429–30
  measurement standards and, 428–29
  risk assessment model in, 427–28
Star Wars (movie), 206
State, *678–79*
State government:
  budget reform and, 350
  in emergency management, 505–6, 505*f*
    mitigation phase, 512
    preparedness phase, 507
    response phase, 508–9
  NRP role and responsibility of, 517*t*
Statements of National Significance to the Fire Problem in the United States, 60–61
Statistics, *679*
Steel industry, scientific management and, 97–99
Stein, David L., 207
Steward B. McKinney Homeless Assistance Act, *679*
Stewardship, *679*
  core values of, 198–204
  ethical leadership and, 248
Stone, W. Clement, 277
StormReady communities, 512–13
Strategic framework, *679*
Strategic goals, *679*
  of DHS, 228–29
Strategic Information and Operations Center (SIOC), 522
Strategic intent, *679*
Strategic leadership, *679*
Strategic management, *679*
Strategic National Stockpile (SNS), 615
Strategic plan, *679*
Strategic planning, 104, *679*
Strategy, *679*
Stratification, *679*
Strength/weakness opportunity threat (SWOT) analysis, 285–86, 286*t*
Stress, *679*
Strikes, alternatives to, 362
Structure, *679*
  fire confined to, *661*
  organizational, *672*
Stuyvesant, Peter (governor of New Amsterdam), 5
Suburban area, CFAI response time benchmarks for, 432, 432*t*
Sumurai philosophy, principles of, 205
Supervising fire officer rank:
  experiential elements for, 68*t*
  NFPA standards pertaining to, 66*t*
  self-development elements for, 69*t*
  training elements for, 66*t*
Supervising/supervision, *679*
Supervisor, responsibility regarding sexual harassment, 15–16
Surgical teams, as Six Sigma quality example, 457–58, 457*f*
Survey(s):
  benchmarking. *See* Benchmarking survey
  of customer satisfaction
    with emergency responses, 579–84
    with public education programs, 584–88
Sweitzer, Albert, 199
Swift water rescue, 552, 552*f*
  defined, 552, *680*
SWOT (strength/weakness opportunity threat) analysis, 285–86, 286*t*
Symbol, defined, 22
Synergy, *680*
  in systems theory, 120
Synthesizers, characteristics of, 13. *See also* Generation X; Generation Y (millennials)

System, *680*
Systems approach, *680*
Systems theory, 119–20

T

Tactical scenario, in promotional process, 373
Taft-Hartley Act (1947), 360–61, 362, 363
Taking Charge of Change (Smith), 222
Tannenbaum, Robert, 149
Tao Te Ching (Tzu), 151
Target, *680*
Task of leadership, 105
Task roles, in teamwork, 297, 298
Task system, 98
Task vs. person preference, 142–43, 143*t*
TAT (Thematic Apperception Test), 359
Taylor, Frederick Winslow (1856-1916), 97–99, 134
Taylor System, 98
Team(s):
  cross-functional, *656*
  management of, 143, 143*t*
  member categories in, 298
Team authority, 341
Team building, *680*
  employee morale and. *See* Employee morale
  factors affecting, 285–88
  four P's of, 288
  reorganization during, resistance toward, 288, 290
Team performance model, 288, 289*f*, 290, *680*
Teaming, cross-functional, 460
Teamwork, 283–84
  employee roles and, 339
  task/process roles in, 297–98
Technical response:
  future vision for, 552–53, 552*f*
  NFPA needs assessment report on, 192
Technology(ies):
  and changing role of employees, 339
  future visions for, 561–63, 562*f*
  impact on fire suppression, 548
  integrated risk management and, 561
  new and emerging, NFPA needs assessment report on, 193
  for radio system interoperability, 540*f*
  role of, 196
  "stair stepping" distribution and, 463
  supporting, NIMS and, 530
Telephone operators, FSRS criteria for, 417
Telephone service, FSRS criteria for, 417
Television shows, fire service depicted in, 28
Terrorism:
  April 19 (1995) attacks, 31–36, 33*f*–35*f*
  defined, 397, *680*
  September 11 (2001) attacks. *See* September 11 terrorist attacks
Tests, for firefighter hiring and promotion, 359
Teutonic Knights, 27
Thematic Apperception Test (TAT), 359
Theory X, 109, 112, 114*t*, *680*
Theory Y, 109, 112, 113, 114*t*, *680*
Theory Z, 113–14, 114*t*, *680*
Theory Z: How American Management Can Meet the Japanese Challenge (Ouchi), 113–14
Therbligs, in motion study, 100
Theurer, Jim, 483
Thinking, vs. feeling, 171–72, 173*t*
Thinking/feeling scale, in Myers-Briggs inventory, 168
Third Watch (TV show), 28
Thomas, Clarence, 15
Threat, *680*
Time. *See* Response time *and individual related entries*
Tipping point, concept of, 547
Title VII, 14
Toffler, Alvin, 218
Total credit and divergence, FSRS criteria for, 418
Total customer satisfaction, 450
Total operational expenditures, *680–81*
Total quality:
  elements of, 454*f*, 462*f*
  implementation obstacles, 455*f*
Total quality improvement, principles of, 465–66, 466*f*
Total quality leadership (TQL), 122–24, 123*f*, 463–70, *681*
  defined, 123
  forces of, 464, 464*f*
Total quality management (TQM), 11–12, 121
Total response time, 406, *681*
Tower of Babel, 4
Town government model, 335–37
Town meeting, 336
TQL. *See* Total quality leadership (TQL)
TQM (total quality management), 11–12, 121
TRADE (Training Resources and Data Exchange) network, 48
Traditionalists. *See* Veterans (traditionalists)
Tragedy, media images of, 31–33, 33*f*, 34*f*
Training and education. *See also* Professional development
  common system for, benefits of, 54
  complementary and supplementary systems in, 47–51
  continuum of, 42–43, 43*t*
  cost benefits of a common system for, 54
  in disaster response, 646–47
  in FESHE professional development model, 50*t*
  for fire officer ranks
    administrative, 73*t*
    executive, 76*t*
    managing, 70*t*
    supervising, 66*t*
  FSRS criteria for, 418
  future directions for, 54–56
  of generation X and Y, 312
  generational views on, 312
  IAFC benchmarks in, 63, 64
  independent systems of, 46
  integrating independent elements of, 47

Training Resources and Data Exchange (TRADE) network, 48
Transactional leadership, 149–50
Transformation, *681*
  organizational, *672*
Transformational leadership, 129–31, 150, 196
  organizational characteristics and, 132
Transformational process, *681*
Transportation, in homeland security strategic plan, 541
Transportation Security Administration (TSA), 554
Travel time, 406, *681*
  maximum prescribed, *668*
Tribe/tribal government, *681*
  NRP role and responsibility of, 517*t*
Truman, Harry S., 20
Trumpets:
  rank designated by, 31
  symbolism of, 26, 26*f*
Trust:
  business ethics and, 268
  in organizations, 200
  team building and, 285, 287
TSA (Transportation Security Administration), 554
Tuning, in schema modification, 155
Turnout clothing, *681*
Turnout time:
  defined, 406, *681*
  NFPA 1710 and, 434
Tutu, Desmond, 294
Tzu, Lao, 151

U

Unaffiliated volunteer, *681*
Uncertainty, *681*
Uncinarius rank, in *Vigiles,* 3
Underestimation, in ethical decision-making process, 253
Unified area command, 519, 520, *653*
Unified command, *681*
Uniform(s):
  badges and insignia on, symbolism of, 22–27, 25*f*, 26*f*
  of Florian von Cetum, 3
  work, *683*
Union Fire Company, 6, 28, 29*f*, 389
Unit dispatched to arrival. *See* Total response time; Travel time
United States, *681*. *See also* U.S. *entries*
  changing culture and diversity in, 9–11
    1770-1870, 9
    1870-1970, 9–10
    1970 onward, 10
    ethnic groups in, census statistics for, 10
    immigration and, 10
  as collection of diverse cultures, 19
  fire problems and needs in, 20
  governmental departments and agencies. *See individually named departments and agencies*
  local governance models in, 332, 334–37
  "right-to-work" states in, 362
University of Pennsylvania, 7
Unsolicited goods, *681*
Urban area, CFAI response time benchmarks for, 431, 431*t*
Urban search and rescue (USAR):
  defined, 552, *681–82*
  future vision for, 552, 552*f*
  teams for, 32
Urban Search and Rescue Response System, 552
U.S. Air Force, core values of, 202–3
U.S. Army:
  core values of, 203
  Medical Research Institute of Infectious Diseases of, WMD response role, 600
U.S. Army Corps of Engineers, 503
U.S. Army 20th Support Command. *See* CBRNE (Chemical, Biological, Radiological, Nuclear, and High Yield Explosive)
U.S. Bureau of Alcohol, Tobacco and Firearms (ATF), *653*
U.S. Capitol, 269*f*
U.S. Citizen Corps, 525
U.S. Coast Guard (USCG), 615
U.S. Coast Guard National Strike Force (USCG NSF), 616
U.S. Constitution, amendments to, 10
U.S. Environmental Protection Agency. *See* Environmental Protection Agency (EPA)
U.S. Fire Administration (USFA), 47
  and FESHE conferences, 49
  recommendations from, 20
U.S. Marine Corps:
  Chemical Biological Incident Response Force of, WMD response role, 595–96
  core values of, 201–2
    leadership and, 205
  and fire service compared, 568
U.S. Navy, core values of, 201
U.S. v. City of Chicago, 370
USAMRID (Medical Research Institute of Infectious Diseases, U.S. Army), 600
USAR. *See* Urban search and rescue (USAR)
USCG (United States Coast Guard), 615
USCG NSF (United States Coast Guard National Strike Force), 616
USFA. *See* U.S. Fire Administration (USFA)
USFA/NFA curricula, 47–48
  in common training system, benefits of, 54
  standards "crosswalks" and, 55
USMC (War Fighting), 568
US&R (National Urban Search and Rescue Response System), 619–20
USS Benfold (warship), 186
Utstein criteria, 411–12, 411*f*, *682*

V

Validation strategy, job-related, 359
Validity, *682*
  content, *655*
  test scores and, 359

Value(s):
core. *See* Core value(s)
corporate, implementing, 276–77
defined, 13, *682*
future fire service challenges and, 560
impact on fire service, 13
of NIPP, 536–37
team-building and, 287–88
Value-added practices, 200, *682*
Values-based leadership, 235, 237–40. *See also* Ethics
Van Tassel, Bob, 193
VCOS (Volunteer and Combination Officers Section), 59–60
Vertical communication, *682*
Veterans (traditionalists):
characteristics of, 304, 305*t*–306*t*
recruiting, managing, and retaining, 310–11
strengths of, 309
views of, 311–12
Veterans' Readjustment Assistance Act (1974), 357
Victim mentality, 206*f*, 207
Vigiles, 2, 3
VIPS (Volunteers in Police Service), 527
Virtual collaboration, 339
Vision, organizational:
measurement criteria, 575–76
in organizational change, 211–16, 212*f*
team-building and, 287–88
transforming into reality, 213–16
Vision statement, *682*
of DHS, 228
Visionary leader, 145, 147*t*
Volunteer(s), *682*. *See also* Volunteer firefighters
discipline and, 381
unaffiliated, *681*
Volunteer advocates, 59–60
Volunteer and Combination Officers Section (VCOS), 59–60
Volunteer and donations coordination center, *682*
Volunteer fire wardens, in New Amsterdam, 5
Volunteer firefighters, 548, *682*
civil rights of, 14–15
Founding Fathers as, 28
and local fire chiefs, relationship between, 313
workplace issues and, 11
Volunteer organizations:
in emergency management, 523–24
funding for, 354
labor relations in, 362–63
Volunteers in Police Service (VIPS), 527
Voting, women and, 10
Voting Rights Act (1965), 10
Vulnerability:
defined, 430, *682*
in risk assessment, 430, 431*f*

W

Wagner Act. *See* National Labor Relations Act (1935)
Wall Street Journal, 127
War Fighting (USMC), 568
Warning system:
in disaster response, 638–39
in homeland security strategic plan, 541
Washington, George, 28
Water supply:
FSRS criteria for, 418
individual property rating schedule for, 419
Watergate affair, 239–40
Waterman, Robert H., 195, 238
Watson, John B., 150
Weapon of Mass Destruction (WMD):
defined, 502, *682–83*
incident involving, agency response roles, 594–623
Stafford Act and, 501
Weapons of Mass Destruction Civil Support Team(WMD-CST), 600–602
Weather, severe, 512–13
Weber, Max (1864-1920), 95–96
Westermann, Steve, 494–95, xxv
advice to new firefighters, 573
on career development, 88
on disaster management and response, 544–45
on ethics, 279–80
on future directions, 385
on future of fire service, 573
on generational issues, 328
on leadership and management, 137
on leadership style, 179
on organizational changes, 231–32
on quality improvement, 494–95
on risk management, 448
on tradition, 39
White, meaning of, 27
Whitehead, T. N., 106
Why Leaders Can't Lead (Bennis), 128, 197
Wieczorek, Thomas, xxv–xxvi
advice to new firefighters, 573
on career development, 88
on disaster management and response, 545
on ethics, 280
on future directions, 385
on future of fire service, 573
on generational issues, 328–29
on leadership and management, 137
on leadership style, 180
on organizational changes, 232
on quality improvement, 495
on risk management, 448
on tradition, 39
Wilcox, John R., 269
Wildfires, community risk assessment and, 395
Wildland firefighting:
CFAI response time benchmarks and performance baselines for, 432, 432*t*
future vision for, 553–54, 553*f*
NFPA 1710 standards for, 441
Wildland/urban interface fire:
NFPA 1710 standards for, 441
NFPA needs assessment report on, 192
"Will to believe," 239

Wilmoth, Janet, xxvi
  advice to new firefighters, 573
  on career development, 88
  on disaster management and response, 545
  on ethics, 280
  on future directions, 385
  on future of fire service, 573
  on generational issues, 329
  on leadership and management, 137
  on leadership style, 180
  on organizational changes, 233
  on quality improvement, 495
  on tradition, 39
Windisch, Fred C., xxvi
  advice to new firefighters, 573
  on career development, 89
  on disaster management and response, 545
  on ethics, 280
  on future directions, 385
  on future of fire service, 573
  on generational issues, 329
  on leadership and management, 137
  on leadership style, 180
  on organizational changes, 233
  on quality improvement, 495
  on risk management, 448
  on tradition, 40
Wingspread Conference, statements from, 60–61
Win-lose theory, dispute settlement and, 365
Win-win theory, dispute settlement and, 365
Witt, James L., 504
WMD. *See* Weapon of Mass Destruction (WMD)
WMD-CST (Weapons of Mass Destruction Civil Support Team), 600–602
Women's suffrage, 10
Work activity grouping methods, 340
Work ethic, *682–83*
Work uniform, *683*
Workforce:
  future changes in, 307–8, 308*f*
  generational differences and, 304–7
  key employee issues and, 308
  quality in fire service, 464
  traditional nature of, 12, 307
  in twenty-first century, 12–13
Workforce 2000, 12–13
Workplace:
  consolidation effects on, 227–28
  hostile, 15, 358
  impact on fire service, 11
  misconduct in, NBES findings on, 243, 244
Work-to rule actions, 362, 364
World Trade Center attacks, 28, 29*f*. *See also* September 11 terrorist attacks
  firefighter deaths at, 564
Written presentation, in promotional process, 373
Wrongdoing. *See* Misconduct in workplace

X

Xerox Corporation, benchmarking at, 475, 476

Y

Yellow, meaning of, 27
Yukl, Gary, 148–49

Z

Zamlen v. City of Cleveland, 357–58
ZBB (zero-based budget), 348–49, *683*
Zero defects, 457, 491, *683*
Zero-based budget (ZBB), 348–49, *683*
Zero-sum theory, dispute settlement and, 365